PENGUIN REFERENCE BOOKS

THE PENGUIN DICTIONARY OF BIOLOGY

Michael Thain was born in Hampstead in 1946 and educated at University College School and at Oxford and London Universities. He holds degrees in Zoology, Philosophy and History of Technology and a Diploma in Human Biology and is currently collaborating on a Penguin *Dictionary of Zoology* and the *Larousse Encyclopedia*. An Assistant Master at Harrow School since 1969, where his practical interests include natural history and nature conservation, he lives in Harrow with his wife and two children.

Michael Hickman was born in Worcestershire in 1943. He gained a First Class Honours BSc in Biology (Botany) from the University of Western Ontario, London, Ontario, Canada in 1966, and a PhD in Phycology/Freshwater Ecology from the University of Bristol, England, in 1970. He has been Professor of Botany (now Biological Sciences) at the University of Alberta, Canada, since 1981, having been initially an Assistant and then Associate Professor in the same department since 1970. He has published numerous papers on his research activities into the ecology of benthic and planktonic algae in lakes and rivers of Alberta, and has been retained as a consultant by various organizations, both private and governmental. His current research interests lie in the field of Palaeoecology, specifically palaeolimnology, and how climatic changes affect lakes and their biota. He lives in Edmonton with his wife and two sons.

THE PENGUIN
DICTIONARY OF BIOLOGY

M. Thain
M. Hickman

Ninth Edition

being the Ninth Edition of the work
originally compiled by M. Abercrombie,
C. J. Hickman and M. L. Johnson

Diagrams by Raymond Turvey

PENGUIN BOOKS

PENGUIN BOOKS

Published by the Penguin Group
Penguin Books Ltd, 27 Wrights Lane, London W8 5TZ, England
Penguin Books USA Inc., 375 Hudson Street, New York, New York 10014, USA
Penguin Books Australia Ltd, Ringwood, Victoria, Australia
Penguin Books Canada Ltd, 10 Alcorn Avenue, Toronto, Ontario, Canada M4V 3B2
Penguin Books (NZ) Ltd, 182–190 Wairau Road, Auckland 10, New Zealand

Penguin Books Ltd, Registered Offices: Harmondsworth, Middlesex, England

First published as *The Penguin Dictionary of Biology*, 1951
Second edition 1954
Third edition 1957
Fourth edition 1961
Fifth edition 1966
Sixth edition 1973
Seventh edition 1980
Eighth edition, entitled *The New Penguin Dictionary of Biology*, 1990
Reprinted with amendments 1991
Reprinted with amendments 1992
Reprinted, with amendments, as *The Penguin Dictionary of Biology*, 1992
Ninth edition 1994
Reprinted with minor revisions 1995
3 5 7 9 10 8 6 4 2

The acknowledgements on pages ix–xi constitute an extension of this copyright page

Filmset by Datix International Limited, Bungay, Suffolk
Printed in England by Clays Ltd, St Ives plc
Set in 8.5/10.25 pt Monophoto Bembo

FOR KATEY, AVRIL AND HARRY, AND MARGARET

PREFACE TO NINTH EDITION

Some spectacular advances in biology have occurred in the few years since the eighth edition was written. One can illustrate this by mention of just two, very different, fields of research: cell biology and palaeontology. Several key molecules in the regulation of the cell cycle and various forms of cell signalling have been isolated and their genetic determinants clarified. The mechanics of RNA processing and protein targeting have become clearer as have some of the details of B and T cell interaction required for antibody production. Within a few years of their invention, the techniques of DNA sequencing and the polymerase chain reaction have become routine. In palaeontology, the tetrapod and primate fossil records have been expanded remarkably in both range and depth, the hominid record especially so, while dating techniques have made enormous strides in parallel. Indeed, the two are inseparable. Different fields indeed. Yet in the debate surrounding the usefulness of the molecular clock approach in working out phylogenies they are brought together. The modern biologist needs perhaps to keep abreast of a greater diversity of research and to read more widely than ever before.

We have tried to bring before the interested general public, as well as biologists, these and other developments. It has been a huge help for us to have had so many excellent biology journals and textbooks to hand, all brilliantly illustrated. Our unfailing admiration is directed towards the writers and editors of those journals, and the publishers who are brave enough to produce them. But above all we congratulate the scientists among them, to whom this dictionary is ultimately dedicated. Without their hard endeavour we should be ignorant of many of the things within the covers of this little book.

MT
MH
July 1994

ACKNOWLEDGEMENTS

For Figures

We gratefully acknowledge the following publishing houses for permission to reproduce material as indicated below. Figures may have been modified slightly to fit the contexts of entries, and the legends will generally have been rewritten. In several cases, authors' permissions were also received. While it is very much hoped that all necessary permission for work reproduced here will have been obtained at the time of going to press, the authors sincerely apologize for use of any uncredited material and will be happy to hear from any such interested parties.

Benjamin/Cummings Publishing Company: Watson, J. D., *Molecular Biology of the Gene* (3rd edn, 1976), for **Fig. 67**.

Blackwell Scientific Publications Ltd: May, R. M. (ed.), *Theoretical Ecology* (2nd edn, 1981), for **Table 6;** Lewin, Roger, *Human Evolution* (2nd edn, 1989), for **Fig. 58**.

The Blakiston Company (McGraw-Hill Book CO.): Kingley, J. S., *Outline of Comparative Anatomy of Vertebrates* (1928), with permission of McGraw-Hill, Inc., for **Fig. 45a**.

British Museum (Natural History): Charig, A. *A New Look at the Dinosaurs* (1979), for **Fig. 112**.

The British Ornithologists Union: Landsborough Thompson, A. (ed.), *A New Dictionary of Birds* (1964), for **Fig. 38**.

Butterworth & Company (Publishers): Cohen, J. *Reproduction* (1977), for **Fig. 111**.

Cambridge University Press: Willmer, P., *Invertebrate Relationships* (1990), for **Table 1 and Fig. 11**; Lowrie, Pauline and Wells, Sue, *Microorganisms, Biotechnology & Disease* (student's book) (1991), for **Figs. 7 and 10**; Austin, C. R., and Short, R. V., *Reproduction in Mammals, Book 2: Embryonic and Fetal Development* (2nd. edn, 1982), for **Figs. 35 and 94**; Slack, J. M. W., *From Egg to Embryo* (1983), for **Fig. 44**; van Emden, H. F., *Pest Control* (1992), for **Fig. 70**.

Chapman and Hall Ltd: Mather, K., *Genetical Structure of Populations* (1973), for **Fig. 43**.

The Company of Biologists Ltd, Cambridge: Grimstone, A. V., Harris, H. and Johnson, R. T., *Prospects in Cell Biology* (1986), for **Fig. 65**.

Current Science: **Fig. 1** is modified from Fig. 1, P. J. Goldschmidt-Clermont, et al., *Current Biology, 2*:669 (1992); **Fig. 15c** is from Fig. 1, Andrew W. Murray, *Current Biology, 3*:291 (1993); **Fig. 28** is from Fig. 1, G. I. Evan and T. D. Littlewood, *Current Opinion in Genetics & Development, 3*:46 (1993); **Fig. 64** is from Fig. 1, Jacopi Meldolesi, *Current Biology, 3*:911 (1993); **Fig. 99a** is from Fig. 1, T. Pawson and Schlessinger, *Current Biology, 3*:435 (1993); **Fig. 103** is from Fig. 1, F-U. Hartl and M. Weidmann, *Current Biology*, 3.86 (1993); **Fig. 108** is from Fig. 2, Grace Gill, *Current Biology, 2*:566 (1992); **Fig. 109** is from Fig. 1, Anthony L. DeFranco, *Current Biology 2*:478 (1992); **Fig. 110** is from Fig. 1, Brigitte T. Huber, *Current Biology 2*:495 (1992).

Elsevier Trends Journals, Cambridge, UK: **Fig. 5** is from Fig. 1, Robert Foley, *Trends in Ecology and Evolution, 8*:196 (1993); **Fig. 54** is modified from David Schubert, *Trends in Cell Biology, 2*:65 (1992); **Fig. 76** is modified from Fig. 1, Robert DeMars and Thomas Spies, *Trends in Cell Biology, 2*:82 (1992); **Fig. 93e** is from Fig. 1, Deborah Brown, *Trends in Cell Biology, 2*:338 (1992); **Fig. 99b** contains information originally published in Fig. 1, Stephen P. Jackson, *Trends in Cell Biology, 2*:105 (1992).

Garland Publishing Inc.: Alberts, B., et al., *Molecular Biology of the Cell* (1st edn, 1983), for **Figs. 18, 20a, 23, 32, 42, 75 and 106**; Alberts, B., et al., *Molecular Biology of the Cell* (2nd edn, 1989), for **Figs. 16, 52, 60, 81, 88, 97 and Fig. 12**, modified from a drawing by Jane Richardson; **Fig. 81**, which includes data supplied by Linda Amos; **Fig. 99c**, from N. H. Boke, *Am. J. Bot.* 36:535–547 (1949).

Gower Medical: Staines, N., Brostoff, J. and James, K., *Introducing Immunology* (1985), for **Fig. 74**.

Harcourt Brace Jovanovich: Frobisher, M, et al., *Fundamentals of Microbiology* (9th, edn, 1974), for **Figs. 14c and 53**; Hopkins, C. R., *Structure and Function of Cells* (1978), for **Figs. 17a and 17b**.

HarperCollins: Tortora, G. J. and Anagnostokos, N. P., *Principles of Anatomy and Physiology* (4th edn, 1984), for **Figs. 47a, 49, 56, 69, 80, 85, 90a, 100 and 107a**; Tortora, G. J. and Anagnostokos, N. P., *Principles of Anatomy and Physiology* (6th edn, 1990), for **Figs. 6, 14a and 89**.

Harvard University Press: Hughes, G. M., *Comparative Physiology of Vertebrate Respiration* (1963), for **Figs. 45b and 45c**.

Heinemann: Freeman, W. H., and Bracegirdle. B., *An Advanced Atlas of Histology* (1976), for **Fig. 46**.

Hodder & Stoughton: Chapman, R. F., *The Insects* (2nd edn, 1971), for **Fig. 84**.

CBS College Publishing (Holt-Saunders): Goodenough, U., *Genetics* (3rd edn, 1984), for **Fig. 66;** Romer, A. S., *The Vertebrate Body* (5th edn, 1977), for **Figs. 90b and 91**.

IRL Press: **Fig. 3** is from Fig. 1.2, *Immune Recognition*, M. J. Owen & J. R. Lamb (1988); **Fig. 20b** is from Fig. 3.4, *Gene Structure & Transcription*, T. Beebee & J. Burke (1988); **Fig. 26** is from Fig. 5.9, *Protein Targeting & Secretion*, B. M. Austen & O. M. R. Westwood (1991); **Fig. 41** is from Fig. 3.2, *Gene Structure & Transcription*, T. Beebee & J. Burke (1988); **Fig. 51a** is from Fig. 1.2, *Protein Targeting & Secretion*, B. M. Austen & O. M. R. Westwood (1991).

S. Karger AG, Basel: **Fig. 118** is from *Intervirology 12*:3–5 (1979).

Longman Group Ltd: Lewis K. R., and John, B., *The Matter of Mendelian Heredity* (2nd edn, 1972), for **Fig. 72(i)**.

McGraw-Hill Inc.: Katz, B., *Nerve, Muscle and Synapse* (1966), for **Fig. 62.**

Macmillan Magazines: **Fig. 33** is from Fig. 2, Regis B. Kelly, *Nature*, *364*:488 (1993); **Fig. 51b** is from Fig. 1a, Regis B. Kelly, *Nature*, *364*:487 (1993); **Fig. 59a** is from Box 4, Bernard Wood, *Nature*, *355*:789 (1992); **Fig. 59b** is from Fig. 4, Bernard Wood, *Nature*, *355*:789 (1992); **Fig. 96** is Fig 1 from Robert D. Martin, *Nature*, *363*:224 (1993); **Fig. 104** is derived from Sean E. Egan and Robert A. Weinberg, *Nature*, *365*:782 (1993); **Table 8** is modified from Table 1, Barry R. Bloom, *Nature*, *342*:119 (1989).

John Murray (Publishers) Ltd: Clegg, C. J., and Cox, G., *Anatomy and Activities of Plants* (1988), for **Fig. 114**.

Thomas Nelson & Sons, Ltd: Barrington E. J. W., *Invertebrate Structure and Function* (2nd edn, 1979), for **Fig. 36**; Roberts, M. B. V., *Biology: A Functional Approach* (3rd edn, 1982), for **Fig. 92**.

Philip Allen Publishers: **Fig. 57** is from *Biological Sciences Review*, *6*:13 (September 1993).

Sinauer Associates Inc.: Scott F. Gilbert., *Developmental Biology* (3rd edn, 1991) – **Fig. 8** is from Fig. 3, p. 653, and Fig. 16, p. 663, **Fig. 17c** is from Fig. 17 (A), p. 537, **Fig. 17d** is from Fig. 17 (B), p. 537, **Fig. 113** is from Fig. 7, p. 416; Jeffrey M. Camhi, *Neuroethology* (1984), for **Fig. 87**.

University of London: ULSEB (1990), Human Biology A Level Specimen Paper, for **Fig. 9**.

Viking Penguin Inc: Hartman, P. E., and Suskind, S. R.: *Gene Action* (1965), for **Fig. 24**.

Wadsworth Inc.: Salisbury, F. B., and Ross, C. W., *Plant Physiology* (4th edn, 1992), for **Fig. 14b**.

Worth Publishers Inc.: Lehninger, A. L., *Biochemistry* (2nd edn, 1975), for **Figs. 27, 47b, 48, 50, 86 and 116**.

Fig. 77 is reproduced here from Prof. R. D. Martin's forthcoming 2nd edition of *Primate Origins and Evolution*, Chapman Hall: London.

For Assistance

It is a great pleasure to record our thanks to the several people who have made the task of producing this dictionary easier. Our families, once again, have taken by far the largest share of the burden, and it is to them that we dedicate the work. Help on the contents of the book has come from several sources, but we would particularly like to thank Bob Martin, Robert Foley and Bill Richmond. Thanks also go to Brenda Metherell for doing a great job on software conversion in Edmonton. But above all, we wish to record the enormous encouragement from the staff at Penguin. Our editors, Ravi Mirchandani and Stefan McGrath have been most supportive. But the team of workers led by Richard Duguid, including the illustrator Raymond Turvey, has excelled in both patience and courteousness. It is doubtful whether there are many such teams around and we are extremely lucky to have had their expert guidance and friendship. It goes without saying that any errors which remain in these pages are entirely the authors' responsibility. We should be happy to have them pointed out to us.

UNITS AND SYMBOLS

$>$ greater than
$<$ less than
$\sim$ approximately

Prefixes

M = mega = 10^6
K = kilo = 10^3
d = deci = 10^{-1}
c = centi = 10^{-2}
m = milli = 10^{-3}
μ = micro = 10^{-6}
n = nano = 10^{-9}

Suffixes

°C = degrees Celsius
(0°C = 273° Kelvin)
J = Joule
m = metre
s = second
h = hour
g = gram
mole = 6.023×10^{23} particles
(ions, atoms, molecules)
dm^3 = 1000 cm^3 or litre (l)
ha = $10^4 m^2$
ppm = parts per million
yr = year
BP = before present
Ig = immunoglobulin
spp. = species (plural)
atm = atmosphere
N = Newton
bar = $10^5 N m^{-2}$

ABA See **abscisic acid**.

A-band See **striated muscle**.

Abaxial (Of a leaf surface) facing away from stem. Compare **adaxial**.

Abdomen (1) Vertebrate body region containing viscera (e.g. intestine, liver, kidneys) other than heart and lungs; bounded anteriorly in mammals but not other classes by a diaphragm. (2) Posterior arthropod trunk segments, exhibiting **tagmosis** in insects, but not in crustaceans.

Abducens nerve Sixth vertebrate **cranial nerve**. Mixed, but mainly motor, supplying external rectus eye-muscle.

Abductor Muscle moving a bone away from the midline. Contrast **adductor**.

Aberrant chromosome behaviour Departures from normal mitotic and meiotic chromosome behaviour, often with a recognized genetic basis. Includes (1) *achiasmate meiosis*, where chiasmata fail to form (e.g. in *Drosophila* spermatogenesis; see **suppressor mutation**); (2) *amitosis*, where a dumb-bell-like constriction separates into two the apparently 'interphase-like', but often highly polyploid, ciliate macronucleus prior to fission of the cell; (3) *chromosome extrusion* or loss, as with X-chromosomes in egg maturation of some parthenogenetic aphids (see **sex determination**); and in *Drosophila* where gynandromorphs may result; but notably in some midges (e.g. *Miastor*, *Heteropeza*) where paedogenetic larvae produce embryos whose somatic cells contain far fewer chromosomes than **germ line** cells, owing to selective elimination during cleavage (see **Weismann**). In some scale insects, males and females develop from fertilized eggs, but males are haploid because the entire paternal chromosome set is discarded at cleavage (see **heterochromatin**, **parasexuality**, **gynogenesis**). In the haplodiploid parasitic wasp *Nasonia vitripennis*, cytoplasmic microorganisms in different strains cause failure of condensation and eventual loss of the paternal chromosome set in fertilized (diploid) eggs from incompatible strains, giving all-male offspring. Such a mechanism may promote reproductive isolation and rapid speciation; (4) *meiotic drive*, where segregation of chromosomes after meiosis is non-random, one member of a particular pair being inherited more frequently than the other owing to differential effects on gamete development. If sex chromosomes are involved, offspring of a cross may be all, or nearly all, of the same sex. Two loci are usually involved, one gene (the driving gene) encoding information to kill gametes (e.g. the segregation distorter, *Sd*, in *Drosophila*), the other determining sensitivity of the cell to the killer gene product. Most cases of meiotic drive involve either non-homologous regions of sex chromosomes or inverted repeat sequences on autosomes (high copy number of the sequence brings insensitivity to the driving gene) when linkage disequilibrium between driving and insensitivity genes allows the trait to spread (crossing-over would provide an antidote to such spread, and may have contributed to its selective advantage). The *t* haplotype in mice and the Spore killer system in the mould *Neurospora* are functionally analogous to *Sd*, but each apparently operates through a different molecular mechanism. See **selfish DNA/genes, sex**. (5) *premeiotic chromosome doubling* (see **automixis**); (6) **endomitosis**, where chromosomes replicate and separate but the nucleus and cell do not divide; (7) **polyteny**, where DNA replication occurs but the strands remain together to form thick, giant chromosomes.

Abiotic Environmental features, such as climatic and **edaphic** factors, that do not derive directly from the presence of other organisms. See **biotic**.

Abomassum The 'true' stomach of **ruminants**.

Abscisic acid (ABA, abscisin, dormin) Inhibitory growth sub**stance**. A sesquiterpene present in a variety of plant organs: leaves, buds, seeds, fruits, roots and tubers. Promotes senescence and abscission in leaves; that produced in the core cells of the root cap has been implicated in the response of roots to gravity; induces dormancy in buds and seeds; induces stomata to close, helping reduce water loss from the plant. Antagonizes the influences of growth-promoting substances. Believed to inhibit protein and nucleic acid synthesis while apparently acting as a calcium agonist, causing influxes of the second messenger calcium (Ca^{2+}) into a wide variety of cells, from cyanobacteria to plants and animals, suggesting genetic conservation of ABA receptors and calcium transport mechanisms during evolution. Inhibits fecundity and egg-viability in several insects and has been located in mammalian brain tissue.

Abscission layer Layer at base of leaf stalk in woody dicotyledons and gymnosperms, in which the parenchyma cells become separated from one another through dissolution of the middle lamella before leaf-fall.

Absorption spectrum Graph of light absorption versus wavelength of incident light. Shows how much light (measured as quanta) is absorbed by a pigment (e.g. plant pigments) at each wavelength. Compare **action spectrum.**

Abyssal Inhabiting deep water, roughly below 1000 metres.

Acanthodii Class of primitive, usually minnow-sized, fossil fish abundant in early Devonian freshwater deposits. Earliest known gnathostomes. Bony skeletal tissue. Fins supported by very stout spine; several accessory pairs of fins common. Row of spines between pectoral and pelvic fins. Heterocercal tail. Relationships with osteichthyan fishes uncertain, but probably not directly ancestral. See **Placodermi**.

Acanthopterygii Spiny-rayed fish. Largest superorder of (teleost) fishes. Spiny rays in their fins consist of solid pieces of bone (and not numerous small bony pieces); are unbranched and pointed at their tips. Radial bones of each ray are sutured or fused, preventing relative lateral movement. Often have short, deep bodies, and relatively large fins, making these fish very manoeuvrable. See **Teleostei**.

Acari (Acarina) Order of **Arachnida** including mites and ticks. External segmentation much reduced or absent. Larvae usually with three pairs of legs, nymphs and adults with four pairs. Of considerable economic and social importance as many are ectoparasites and vectors of pathogens.

Accessory bud A bud generally situated above or on either side of main axillary bud.

Accessory chromosome See **supernumerary chromosome**.

Accessory molecules Membrane glycoproteins of the Ig superfamily on certain **T cell** surfaces, additional to the **T cell receptor** itself (see Fig. 110), regulating **adhesion** between T cell and **antigen-presenting cells** (APCs). These include many 'cluster of differentiation' (CD) antigens, first identified using monoclonal antibodies. The CD4 (or T4) gene product assists helper T cells (T_H) bind MHC Class II-bound antigens on the APC surface and binds a p56 tyrosine kinase on the cytoplasmic side, while the heterodimer CD8 (or T8) formed from CD8α and CD8β helps cytotoxic T cells (T_C) bind MHC Class I-bound antigens externally and binds a p56 tyrosine kinase on the cytoplasmic side. CD8 binds the $\alpha3$ domain of MHC Class I molecules; CD2 glycoprotein binds leucocyte functional antigen-3, promoting adhesion of target cell prior to TCR engagement. CD4 is a receptor for HIV-1 (AIDS virus).

Accessory nerve Eleventh cranial nerve of tetrapod vertebrates, unusual in originating from both brain stem and spinal cord. A mixed nerve, whose major motor output is to muscles of throat, neck and viscera.

Accessory pigment Pigment that captures light energy and transfers it to chlorophyll *a*, e.g. chlorophyll *b*, carotenoids, phycobiliproteins.

Accommodation Changing the focus of the eye. In man and a few other mammals occurs by changing the curvature of the lens; at rest lens is focused for distant objects and is focused for near objects by becoming more convex with contraction of the ciliary muscles in the **ciliary body**. See **eye, oculomotor nerve**.

Acellular Term sometimes applied to organisms or their parts in which no nucleus has sole charge of a specialized part of the cytoplasm, as in unicellular organisms. Applicable to coenocytic organisms (e.g. many fungi), and to tissues forming a **syncytium**. Sometimes preferred to 'unicellular'. See **multicellularity**.

Acentric (Of chromosomes) chromatids or their fragments lacking any **centromeres**.

Acetabulum Cup-like hollow on each side of hip girdle into which head of femur (thigh bone) fits, forming hip joint in tetrapod vertebrates. See **pelvic girdle**.

Acetification See **fermentation**.

Acetylcholine (ACh) **Neurotransmitter** of many interneural, neuromuscular and other *cholinergic* effector synapses. Relays electrical signal in chemical form, with transduction back to electrical signal at the postsynaptic membrane.

Initiates depolarization of postganglionic membranes to which it binds, but hyperpolarizes vertebrate cardiac muscle membranes (slowing heart rate). Different ACh receptors mediate these effects: the cardiac receptor is a non-channel-linked **receptor** (*muscarinic*, activated by the fungal toxin muscarine); channel-linked ACh receptors are *nicotinic* (activated by nicotine). Nicotine receptors (ganglia, neuromuscular junctions and maybe some brain and spinal cord regions) are blocked by curare, muscarinic receptors (peripheral autonomic interneural synapses) by atropine. An anaesthetist can relax skeletal muscles by administering curare, leaving heart muscle unaffected. ACh relaxes smooth muscle of blood vessels by causing endothelial release of nitric oxide (NO), a relaxing factor, which in turn activates a secondary messenger **cascade** involving cGMP. Stored in **synaptic vesicles** inside axon terminals and released there in quantal fashion in response to calcium ion uptake on arrival of an **action potential** (the greater the Ca^{2+} entry, the greater the ACh release). It diffuses across the synaptic cleft and binds to receptor sites on the postsynaptic membrane, whereupon these ion channels open and allow appropriately sized positive ions to enter cell, initiating membrane depolarization. Hydrolysis to choline and acetate by cholinesterase in the synaptic cleft ensures that the chemical signal is appropriately brief (see **summation**). Vertebrate ACh postsynaptic receptors are distinguished as nicotinic or muscarinic on the results of alkaloid administration. Nicotinic receptors (ganglia, neuromuscular junctions and possibly some brain and spinal cord regions) are blocked by curare, muscarinic (peripheral autonomic interneural synapses) by atropine. ACh is found in some protozoans. Compare **adrenergic**.

Acetylcholinesterase See **cholinesterase**.

Acetyl coenzyme A See **coenzyme A**.

Achene Simple, dry, one-seeded fruit formed from a single carpel, without any special method of opening to liberate seed; seed coat is not adherent to the pericarp; may be smooth-walled (e.g. buttercup), feathery (e.g. traveller's joy), spiny (e.g. corn buttercup), or winged (when termed a *samara*) as in sycamore and maple.

Achiasmate Of meioses lacking chiasmata. One form of **aberrant chromosome behaviour**. See **suppressor mutation**.

Achlamydeous (Of flowers) lacking petals and sepals; e.g. willow.

Acid dyes Dyes consisting of an acidic organic compound (anion) which is the actively staining part, combined with an inorganic cation, e.g. eosin. Stain particularly cytoplasm and collagen. See **basic dyes**.

Acid hydrolase Any hydrolytic enzyme whose optimum pH of activity is in the acidic range. Many different examples occur in **lysosomes**. Pepsin is an acid protease.

Acid phosphatase One of several acid hydrolases located in **lysosomes** and concentrated in the trans-most cisternae of the **Golgi apparatus**.

Acid rain Phrase introduced in 1872

by the first British air pollution inspector, Robert Angus Smith, when he discovered that rain falling around Manchester contained sulphuric acid. Today, 'acid rain' is used more generally to describe the total deposition of atmospheric pollutants on vegetation, soils and aquatic ecosystems. This deposition can take three possible forms: (1) *Wet deposition* – rain, hail or snow; (2) *dry deposition* – where pollutants as gases or particles are absorbed on to plant and soil surfaces or settle on water surfaces; and (3) *occult deposition* – where pollution is incorporated into cloud droplets which are deposited on vegetation when enveloped in the cloud. Generally, rainwater is considered acid if the pH value falls below 5.6. Rain dissolves carbon dioxide, forming carbonic acid, giving it a normal pH of 5.6. This arbitrary pH value does not allow for natural sources of sulphur and nitrogen. Gaseous sulphur compounds of terrestrial and marine origin (e.g. dimethyl sulphide, carbonyl sulphide, hydrogen sulphide) can be oxidized to sulphur dioxide. Similarly, the release of nitric acid by natural processes of **denitrification** can contribute nitric acid to the precipitation. So, the pH of unpolluted rainwater varies between 4.6 and 5.6, with a mean value of 5.0. Pollutants, attributable to human activities, can further reduce the pH of rainwater, while occult precipitation can act as a further concentrating mechanism, especially in regions where orographic cloud sits for considerable periods over upland regions. Polluted air rising into clouds can leave the pollutants dissolved in water droplets which then deposit a concentrated solution on surfaces with which they come into contact. Similarly, rainfall as drizzle possesses a greater total drop surface area than the equivalent amount of heavy rain; pH values between 2.5 and 3.0 can result from such light precipitation.

Consequences of acid precipitation include leaching of cations, particularly calcium, from soils, causing knock-on Ca^{2+} depletion in lakes and rivers – a natural process which has been accelerated this century by pollution from strong acids by industry, and by car exhausts. The result (especially in Scandinavia, northern Scotland and eastern North America) has been the progressive transfer of sulphuric acid and acidity into surface waters, so that many lakes cannot support fish populations. High acidity alone interferes with fish reproduction by inhibiting softening of the hard egg case, preventing hatching. Fish are also killed by a secondary consequence of acid run-off from the soil by soluble aluminium, which upsets gill operations by injuring epithelial cells; gills become clogged with mucus and the oxygen content of the blood falls. All toxicity is greatest when Ca^{2+} concentrations are low. Damage to forest trees leading to forest decline has also occurred. Owing to leaching of Mg^{2+} ions from the soil, leaves become chlorotic and growth is reduced. Sensitivity of trees to this pigment loss may depend upon the overall nutritional status of the soil.

Acinar cells See **acini**.

Acinar gland (Zool.) Multicellular gland (e.g. seminal vesicle) with flask-like secretory portions.

Acini (Zool.) Cells lining tubules of pancreas and secreting digestive juices. Their secretory vesicles (*zymogen granules*) concentrate the enzymes and fuse with the apical portion of the plasmalemma under the stimulus of **acetylcholine** or **cholecystokinin**, releasing their contents into the lumen of the duct. Much used in the study of secretion.

Acoelomate Having no **coelom**. Refers to some lower animal phyla, e.g. coelenterates, platyhelminths, nemerteans and nematodes.

Acoustic nerve See **vestibulocochlear nerve**.

Acoustico-lateralis system See **lateral line system**.

Acquired characteristics, inheritance of See **Lamarckism**.

Acquired immune response Secondary antibody response to presence of antigen and differing from the initial response (which may precede it by a matter of years) in that it appears more quickly, achieves a higher antibody titre (concentration) in the blood and in that the principal **immunoglobulin** species present is IgG rather than IgM. See **antibody, immunity**.

Acrania See **Cephalochordata**.

Acrasiomycota Cellular slime moulds (Kingdom **Protista**). Slime moulds which may exist as separate amoebae (*myxamoebae*), and which retain their original identities within the pseudoplasmodium (the 'slug') formed by swarming. Probably more closely related to the amoebae (Phylum **Rhizopoda**) than to any other group. Compare **Myxomycota**.

Acrocentric Of chromosomes and chromatids in which the **centromere** is close to one end.

Acromion Point of attachment of clavicle to scapula in mammals and mammal-like reptiles. A bone process.

Acropetal (Bot.) Development of organs in succession towards apex, the oldest at base, youngest at tip (e.g. leaves on a shoot). Also used in reference to direction of transport of substances within a plant, i.e. towards the apex. Compare **basipetal**.

Acrosome Specialized penetrating vesicular organelle, formed from **Golgi apparatus** and part of the nuclear envelope at the tip of a spermatozoon. It contains **hyaluronidase**, several lytic enzymes and acid hydrolases released when the sperm cell membrane fuses at several points with the acrosome during the *acrosome reaction*, dissolving the jelly-like zona pellucida so that the sperm head can fuse with the ovum membrane (see **cell fusion**). Some sperm discharge an *acrosomal process* composed of rapidly polymerizing **actin** which punctures the egg membranes prior to fusion with ovum (e.g. in some echinoderms).

ACTH (adrenocorticotropic hormone, corticotropin) A polypeptide of thirty-nine amino acids secreted by anterior lobe of the pituitary, involved in the growth and secretory activity of adrenal cortex. Has a minor positive effect on aldosterone secretion, but an important role in glucocorticoid secretion. Both stress and low blood glucocorticoid levels cause release from the hypothalamus of *corticotropin releasing factor* (*CRF*) which initiates ACTH release. See **adrenal gland, cortisol**.

Actin Diagnostic eukaryotic protein, absent from prokaryotes. Filamentous actin (*F-actin*) is composed of 43Kd globular protein monomers (*G-actin* molecules) polymerized to form long fibrous molecules, two of which coil round one other in the thin actin filaments of muscle and other eukaryotic cells, where they are termed *microfilaments*. Each *G-actin* molecule binds one calcium ion and one ATP or ADP molecule, when it polymerizes to form *F-actin* with ATP hydrolysis. Like **microtubules**, the opposite ends of actin filaments grow and depolymerize at different rates and play a vital role in **cytoskeleton** structure. Stress fibres are bundles of actin filaments and other proteins at the lower surfaces of cells in culture dishes and will contract if exposed to ATP *in vitro*. Microfilaments are involved in the building of **filopodia**, microspikes and **microvilli** where, as in stress fibres, they form paracrystalline bundles. Filaments of actin and **myosin** are capable of contracting together as **actomyosin** in both muscle and non-muscle cells, e.g. in the contractile ring of dividing cells, in belt **desmosomes** and in **cytoplasmic streaming**. Numerous actin-related proteins (ARPs), with at least 40–50% sequence identity to actin, have been discovered.

Actinomorphic (Of flowers) regular; capable of bisection vertically in two (or more) planes into similar halves, e.g. buttercup. Such flowers are also said to exhibit **radial symmetry**.

Actinomycete Member of an order (Actinomycotales) of Gram-positive bacteria with cells arranged in hypha-like filaments. Mostly saprotrophs, some parasites. Source of streptomycin.

Actinomycin D Antibiotic derived from species of the bacterial genus *Streptomyces*. Binds to DNA between two G–C base pairs and prevents movement of RNA polymerase, so preventing transcription in both prokaryotes and eukaryotes. Penetrates into intact cells. See **antibiotic**.

Actinopterygii Ray-finned fishes. Generally regarded as subclass of Osteichthyes, and includes all common fish except sharks, skates and rays. Earliest forms (chondrosteans) represented in the Devonian by the palaeoniscoids and today by e.g. *Polypterus*; later forms (holosteans) were predominantly Mesozoic fishes but represented today by e.g. *Lepisosteus* (gar pikes); teleosts are the dominant fish of the modern world and represent the subclass in almost every part of the globe accessible to fish. Internal nostrils absent; **scales** ganoid in primitive forms, but reduced or even absent in teleosts. The paired fins are webs of skin braced by horny rays (like ribs of a fan), each a row of slender scales, there being no fleshy fin lobes except in the most primitive forms. A swim bladder is present and the skeleton is bony. Internal groupings given here probably represent **grades** rather than **clades**. See **Teleostei**, **Acanthopterygii**.

Actinozoa (Anthozoa) Sea anemones, corals, sea pens, etc. A class of Coelenterata (Subphylum Cnidaria). The body is a polyp, there being no medusoid stage in the life cycle. Polyp more complexly organized than that of other coelenterates; coelenteron divided by vertical mesenteries. May have external calcareous skeleton as in well-known corals, but some forms have internal skeleton of spicules in mesogloea.

Action potential Localized reversal and then restoration of electrical potential between the inside and outside of a nerve or muscle cell (or fibre) which marks the position of an impulse as it travels along the cell. See **impulse, resting potential, activation**.

Action spectrum Plot of the quanta of different wavelengths required for a photochemical response against the wavelength of light used. Its reciprocal indicates photochemical efficiency.

Activation (Of eggs). When the membrane of the sperm **acrosome** fuses with the egg plasma membrane, an *activation reaction* passes over the egg surface involving an **action potential** of longer duration than in nerve or muscle. It signifies the onset of embryological development and may be achieved merely by pricking of some eggs (e.g. frog).

Active site Part of an enzyme molecule in its natural hydrated state which, by its three-dimensional conformation and charge distribution, confers upon the enzyme its substrate specificity. It binds to a substrate molecule, forming a transient enzyme–substrate complex. Enzymes may have more than one active site and so catalyse more than one reaction. Competitive inhibitors of an enzyme reaction bind reversibly to the active site and reduce its availability for normal substrate. Active sites may only take on their appropriate conformation after the enzyme has combined at some other site with an appropriate modulator molecule. Some active sites require metal ions as prosthetic groups (e.g. human carboxypeptidase requires a zinc atom). See **enzyme**.

Active transport The energy-dependent carriage of a substance across a cell membrane, accumulating it on the other side in opposition to chemical or electrochemical gradients (i.e. 'uphill'). The process involves 'pumps' composed of protein molecules in the membrane (often traversing it) which carry out the transport. Requires an appropriate energy supply, commonly ATP, or a gradient of protons across the membrane itself usually generated by redox, photochemical or ATP-hydrolysing reactions. Collapse of this gradient drives proton-linked symports or antiports (see **transport proteins**). Alternatively, a membrane potential arising from ion asymmetry across the membrane may drive specific ions through special transport systems. Probably all cells engage in active transport. See **sodium pump, electron transport system, facilitated diffusion**.

Active zone A specialized region of a presynaptic membrane to which synaptic vesicles fuse prior to release of transmitter.

Activin (TGF-β) See **growth factors**.

Actomyosin Complex formed when the pure proteins **actin** and **myosin** are mixed, resulting in increased viscosity of the solution. Actomyosin undergoes dissociation in the presence of ATP and magnesium ions (Mg^{2+}), when ATP hydrolysis occurs. Completion of this hydrolysis results in reaggregation of the two proteins. Live muscle cells have an absolute requirement for calcium ions (Ca^{2+}) before myosin and actin filaments will interact, and when Ca^{2+} is removed the actin and myosin dissociate. Such interactions form the basis of many biological force-generating events, notably during **muscle contraction**, **cytoplasmic streaming**, **cell locomotion** and blood clot contraction.

Adaptation (1) Evolutionary. Some property of an organism is normally regarded as an adaptation (i.e. fits the organism in its environment) if (a) it occurs commonly in the population, and (b) the cause of its commonness was **natural selection** in its favour. Adaptations are not, therefore, fortuitous benefits, the implication being that they have a genetic basis, since selection operates only upon genetic differences between individuals. Alternatively, we often in practice identify an adaptation by its effects rather than its causes. Learned abilities which improve an individual's **fitness** or inclusive fitness, but without clear genetic causation, are cases in point. See **teleology**. (2) Physiological. A change in an organism, resulting from exposure to certain environmental conditions, allowing it to respond more effectively to them. (3) Sensory. A change in excitability of a sense organ through continuous stimulation, increasingly intense stimuli being required to produce the same response.

Adaptive enzyme Inducible enzyme. See **enzyme**.

Adaptive immune response See **immunity**, **programmed cell death**.

Adaptive radiation Evolutionary diversification from a single ancestral (prototype) population of descendant populations into more and more numerous **adaptive zones** and ecological **niches**. May involve both **anagenesis** and **cladogenesis**. See **character displacement**.

Adaptive zone A more or less distinctive set of ecological niches established and occupied by an evolutionary lineage with time.

Adaxial (Of a leaf surface) facing the stem. Compare **abaxial**.

Adductor Muscle moving a bone closer to the midline. Contrast **abductor**.

Adenine A purine base of DNA, RNA, some nucleotides and their derivatives.

Adenohypophysis See **pituitary gland**.

Adenosine diphosphate See **ADP**.

Adenosine monophosphate See **AMP**.

Adenosine triphosphate See **ATP**.

Adenovirus One kind of DNA tumour virus of animals. See **virus**.

Adenyl cyclase (adenylate cyclase) A plasma membrane-bound enzyme converting ATP to **cyclic AMP**. Many peptide hormones and local chemical signals operate through activation of this enzyme. See **receptor**.

ADH See **antidiuretic hormone**.

Adhesion Mechanisms by which cells are able, by forming focal contacts (focal adhesions, see Fig. 1) with one another and/or with an appropriate substratum, to link the **extracellular matrix** to the intracellular **actin** network. This GTP-dependent process is often made possible through the several affinities of their cell-surface receptors, often acting as components of **intercellular junctions**. Involved in **multicellularity**, **morphogenesis** and the functions of the immune system.

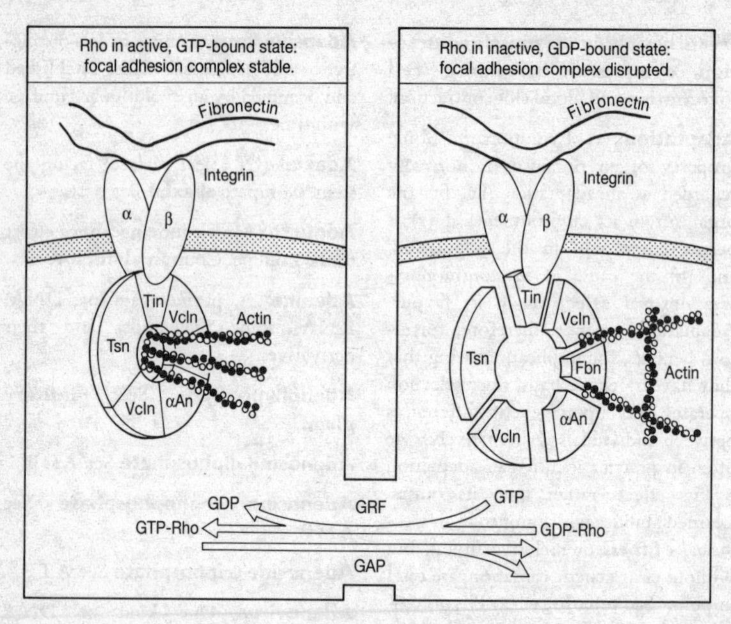

Fig. 1 *A model for focal adhesion regulation by the GTP-ase, Rho. The complex lipid and protein structure is stabilized by GTP-Rho. Some of the proteins involved are shown: the heterodimer integrin; Tin = talin; Tsn = tensin; Vcln = vinculin; Fbn = fimbrin; αAn = α-actinin; GRF = guanine nucleotide release factor; GAP = GTP-ase activity protein; Pi = inorganic phosphate. The shapes and stoichiometry of molecules are not to be taken literally.*

Several families of adhesion receptors have been recognized (see **accessory molecules**): the immunoglobulin superfamily (including the **T cell** and **B cell** receptors, MHC Class I and II receptors, and **CD2**, **CD4** and **CD8 proteins**); the **integrin** family, involved in regulation of cell adhesion and migration; **selectins**, prominent in interactions of lymphocytes and neutrophils with the vascular endothelium; and cadherins, Ca^{2+}-dependent homophilic molecules, expression of one of which (E-cadherin, possibly encoded by a **tumour suppressor gene**) is sufficient to cause cell–cell adhesion even in non-adherent cell lines, its under-expression promoting invasiveness (a feature of transformed **cancer cells**). Adhesive interactions between animal epithelial cells and the **basement membrane** or **extracellular matrix** determine cell shape through changes in cytoskeletal organization (in the former case in the absence of induced protein synthesis) through mediation of **cyclic nucleotides** and divalent ions. These cell shape changes may in turn modulate a cell's response to specific **growth factors**. Some adhesion molecules are exploited as **virus receptors**.

Adipose tissue A connective tissue. (1) Brown adipose tissue (brown fat) comprises cells whose granular cytoplasm is due to high concentration of cytochromes and whose function appears to be release of heat in the neonatal mammal. Distributed around neck and between scapulae in these and hibernating mammals but not otherwise extensively in adults. Richly innervated and vascularized. (2) White adipose tissue is distributed widely in animal bodies, comprising large cells (fat cells) each with single large fat droplet inside a thin rim of cytoplasm. This depot fat is composed largely of triglyceride. **Adrenaline**, **glucagon**, **growth hormone** and **ACTH** all stimulate release of fatty acids and glycerol via activation of intrinsic lipases, probably via **cyclic AMP**. It is less richly innervated than brown adipose tissue.

In humans, adipocyte precursors (fibroblast-like cells) divide through the first year of life, after which increase in tissue volume is by **hypertrophy** accentuated by giving high-fat diets. Cases of genetic obesity may be due to prolongation of the period for which adipocyte precursors can divide. See **hibernation**.

ADP (adenosine diphosphate) A nucleoside diphosphate found universally inside cells. Phosphorylated to **ATP** during energy-yielding catabolic reactions and produced in turn when ATP itself is hydrolysed.

Adrenal gland (suprarenal g., epinephric g.) Endocrine gland of most tetrapod vertebrates lying paired on either side of the mid-line, one atop each kidney. Each is a composite of an outer cortex derived from coelomic mesoderm, making up the bulk of the gland, and an inner medulla derived from neural crest cells of the ectoderm. Rarely found as a composite gland in fish. Cortex comprises three zones, the outermost secreting aldosterone which promotes water retention by kidneys by increasing renal potassium excretion and sodium retention; the middle zone secretes cortisol and other glucocorticoids under **ACTH** control and enhancing **gluconeogenesis**; the innermost zone secretes sex (mainly male) hormones. The medulla comprises *chromaffin cells*, effectively postganglionic sympathetic nerve cells that have lost their axons and secrete adrenaline (epinephrine) and a little noradrenaline (norepinephrine) into the large surrounding blood-filled sinuses. These mimic effects of the sympathetic nervous system (see **autonomic nervous system**), release being under hypothalamic control via the splanchnic nerve. They promote liver and muscle glycogenolysis via **cyclic AMP**, lipolysis in **adipose tissue**, vasodilation in skeletal and heart muscle and brain, and vasoconstriction in skin and gut. They relax bronchi and bronchioles and increase rate and power of heart beat, raising blood pressure. All adrenal hormones are known as 'stress' hormones, those of the cortex responding to internal physiological stress such as low blood temperature or volume, while medullary hormones are released in response to stress situations (often auditory or visual) outside the body. See **L-dopa**.

Adrenaline (In USA, **epinephrine**.) Hormone derivative of amino acid tyrosine secreted by chromaffin cells of **adrenal gland** and to a lesser extent by sympathetic nerve endings.

Adrenaline

Noradrenaline

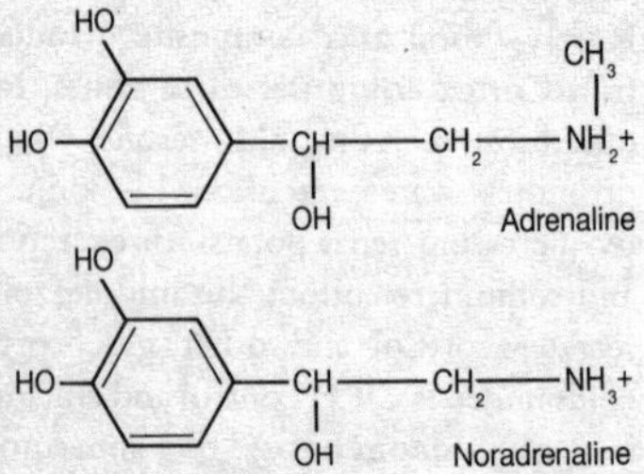

Adrenergic Of a motor nerve fibre secreting at its end noradrenaline (norepinephrine) or, less commonly, adrenaline. Characteristic of postganglionic sympathetic nerve endings. Compare **cholinergic**.

Adrenocorticotropic hormone (ACTH) See **ACTH**, **adrenal gland**.

Adventitious Arising in an abnormal position; of roots, developing from part of the plant other than roots (e.g. from stem or leaf cutting); of buds, developing from part of the plant other than a leaf axil (e.g. from a root).

Aecium A cup-shaped structure in which binucleate (dikaryotic) spores (aeciospores) of rust fungi (**Basidiomycota**, Order **Uredinales**) are produced. The aeciospores develop in chains and are dispersed by wind; they only infect wheat or related grasses. See **pycnium**, **uredium**, **Uredinales**, **telium**.

Aerenchyma Secondary spongy tissue of some aquatic plants, with intercellular air spaces formed by the activity of a **cork** cambium or phellogen. May develop in a lesser way from the lenticels of land plants such as willow (*Salix*), and poplar (*Populus*) if partially submerged. Seems to function mainly in a flotation capacity rather than as a respiratory aid.

Aerobic Requiring free (gaseous or dissolved) oxygen. In most cases the oxygen is utilized in aerobic respiration, but a few enzymes (oxygenases) insert oxygen atoms directly into organic substrates. See **respiration**.

Aerobic respiration See **aerobic**, **respiration**.

Aestivation (Bot.) Arrangement of parts in a flower-bud. (Zool.) **dormancy** during summer or dry season as e.g., in lungfish (dipnoans). See **hibernation**.

Aethalium See **Myxomycota**.

Afferent Leading towards, as of arteries leading to vertebrate gills or of nerve fibres (sensory) conducting an input towards the central nervous system. Opposite of **efferent**.

Affinity maturation Process involving somatic mutation in variable region of immunoglobulin genes (hypermutation) whereby average affinity of antibodies (esp. IgG) for antigen increases after T cell-dependent priming. See **antibody diversity**.

A-form helix (A DNA, DNA-A) Right-handed, double helical form of **DNA**. Less hydrated and consequently more compact and stable than the more common **B-form helix**. Unlike that form, the axis of the molecule does not pass through the hydrogen bonds holding the base pairs together, so the minor groove is narrower. See Fig. 29a.

After-ripening Dormancy exhibited by certain seeds (e.g. hawthorn, apple) which, although embryo is apparently fully developed, will not germinate immediately seed is formed. Even when removed from seed coat and provided with favourable conditions, the embryo has to undergo certain chemical and physical changes before it can grow. Possibly associated with delay in production of required growth substances, or with gradual breakdown of growth inhibitors. See **dormant**.

Agamospecies See **species**.

Agamospermy Any plant **apomixis** in which embryos and seeds are formed but without prior sexual fusion. Excludes vegetative reproduction (vegetative apomixis). Occurs widely in higher plants, both ferns and flowering plants. Unknown in gymnophytes. See **pseudogamy**.

Agamospory Asexual formation of an embryo and the subsequent development of a seed.

Agar A phycocolloid of red algal (**Rhodophyta**) origin; insoluble in hot water. A 1.5% solution is clear and forms a solid but elastic gel on cooling to 32–39°C, not dissolving again at a temperature below 85°C. Comprises two polysaccharides, agarose and agaropectin, and is commercially extracted from such red algal species as *Gelidium*, *Pterocladia*, *Acanthopeltis*, *Ahnfeltia* and *Gracilaria*. Greatest use of agar occurs in the food and pharmaceutical industries for gelling and thickening purposes (e.g. canning of fish and meat, processed cheese, mayonnaise, puddings, jellies), emulsions, ointments and lotions. It is also widely used as a solidifying agent for culture media in bacteriology, mycology and phycology.

Agarose Polysaccharide used as gel in column chromatography and in electrophoresis. See **northern blot technique**, **southern blot technique**.

Ageing (senescence) Progressive deterioration in function of cells, tissues, organs, etc., related to the period of time since that function commenced. By dividing indefinitely, bacteria and many protozoans avoid ageing; higher plants often seem capable of unlimited vegetative propagation. Regeneration and renewal in many simple invertebrates seem to permit escape from senescence. **Germ lines** of sexual metazoa are potentially immortal (see **Weismann**). Expressed as disintegration of somatic tissue, ageing may be due to gradual accumulation of somatic mutations or to late expression of genes not subject to strong selection. Some evidence suggests loss of **DNA methylation** may be involved. In the population context, it may be due to inbreeding or to some other factor reducing genetic variation. In many plants, particularly annuals and biennials, senescence and death of the whole organism follow flowering and fruiting (**cytokinins** may delay senescence until fruiting is over). New leaves often induce senescence in older leaves, which may be reversed if the former are removed. Several **growth substances**, including **ethylene**, **abscisic acid** and **auxins** are known to affect senescence in one or more plant organs. See **programmed cell death**.

Agglutination Sticking together or clumping; as of bacteria (an effect of antibodies), or through mismatch of **agglutinogens** of red blood cells and plasma **agglutinins** in blood transfusions. See **lectin**.

Agglutinins (isoantibodies) Plasma and cell-surface proteins that by interacting with **agglutinogens** (antigens) on foreign cells can cause cell clumping (**agglutination**). Commonly **lectins**.

Agglutinogen Proteins acting as cell-surface antigens of red blood cells and interacting with **agglutinins** to cause red cell clumping and possible blockage of blood vessels. Genetically determined, and the basis of **blood groups**.

Aggregate fruit Fruit which develops from several separate carpels of a single flower (e.g. magnolia, raspberry, strawberry).

Agnatha Class of Subphylum Vertebrata (if treated as a class, other vertebrates form Superclass Gnathostomata). Modern forms (cyclostomes) include lampreys (Subclass Monorhina) and hagfishes (Subclass Diplorhina), but fossil forms included anaspids, osteostracans and heterostracans. Jawless vertebrates. Buccal chamber acts as muscular pump sucking water in, serving for filter-feeding in lamprey larvae as well as ventilating gills – an advance over ciliary mechanisms. Paired appendages almost unknown. Earliest forms (heterostracans) appear in the late Cambrian.

Agonistic behaviour Intraspecific behaviour normally interpreted as attacking, threatening, submissive or fleeing. Actual physical injury tends to be rare in most apparently aggressive encounters.

Agrobacterium Bacterial genus noted for crown gall tumour-inducing ability. Oncogenic strains are host to a tumour-inducing (Ti) **plasmid** which can be transmitted between species. A segment (T) of the Ti plasmid is transmitted to the plant host cell and is the immediate agent of tumour induction. See **oncogene**, **protoplast**.

Ahnfeltan A complex phycocolloid substance occurring in the cell walls of some red algae (Rhodophyta).

AIDS Acquired immune deficiency syndrome. See **HIV**.

Air bladder See **gas bladder**.

Air sacs (1) Expanded bronchi in abdomen and thorax of birds, initially in five pairs but one or more pairs fusing to form thin-walled passive sacs with limited vascularization. Ramify throughout the body and within bones. Connected to lung by small tubes whose relative diameters are probably crucially important in establishing a unidirectional passage of air from lung to sacs and back to lung. The avian ventilation system lacks a tidal rhythm characteristic of mammals. (2) Expansions of insect tracheae into thin-walled diverticulae whose compression and expansion assist **ventilation**.

Akinete Vegetative cell which becomes transformed into a thick-walled, resistant spore. Formed by certain Cyanobacteria and some algae (e.g. some Chlorophyta).

Albinism Failure to develop pigment, particularly melanin, in skin, hair and iris. Resulting *albinos* light-skinned with white hair and 'pink' eyes due to reflection from choroid capillaries behind retina. In mammals, including humans, usually due to homozygous autosomal recessive gene resulting in failure to produce enzyme tyrosine 3-monooxygenase.

Albumen Egg-white of birds and some reptiles comprising mostly solution of **albumin** with other proteins and fibres of the glycoprotein *ovomucoid*. Contains the dense rope-like **chalaza** and with yolk supplies protein and vitamins to embryo, but is also major source of water and minerals.

Albumin Group of several small proteins produced by the liver, forming up to half of human plasma protein content, with major responsibility for transport of free fatty acids, for blood viscosity and **osmotic potential**. If present in low concentration oedema may result, as in kwashiorkor.

Albuminous cells Ray and parenchyma cells in gymnosperm phloem, closely associated morphologically and physiologically with sieve cells.

Alcohols Organic compounds with at least one (monohydric) or more (dihydric, trihydric, etc.) hydroxyl groups (-OH) attached directly to carbon atoms. Aliphatic dihydric alcohols are known as glycols; cyclic alcohols include phenols. Alcohols ending -ol are named after the corresponding paraffin hydrocarbon, e.g. ethanol for ethyl alcohol. Monohydric alcohols react with metals such as sodium, calcium and aluminium to give alkoxides and give esters on reacting with acids. On oxidation, primary alcohols (RCH_2OH) give aldehydes; secondary alcohols ($RR'CHOH$) give ketones, while tertiary alcohols ($RR'R''COH$) give a mixture of carboxylic acids with fewer carbon atoms than the original alcohol. Humans have probably produced ethanol by **fermentation** since prehistory, and first used it as a motor fuel in 1890. Fuel shortages in World War II encouraged this technology, and at present most industrial alcohol is produced as a by-product of the petrochemical industry, some of it mixed with petrol to form 'gasohol'. Brazil leads the field in the use of industrial fermenters to produce sufficient alcohol to replace its petrol needs, *Zygomonas mobilis* rather than *Saccharomyces* species being used as a more efficient converter of sugar to ethanol. When immobilized in an inert support column, *Zygomonas* cells are particularly good for continuous culture production (see **bioreactor**).

Alcyonaria Order of coelenterates within the Class Actinozoa. Sea pens, soft corals, etc. Have eight pinnate tentacles and eight mesenteries. Polyps colonial, with continuity of body wall and enteron. Skeleton, often of calcareous spicules, within mesogloea and occasionally externally.

Aldosterone Hormone of **adrenal** cortex. See **osmoregulation**.

Aleurone grains Membrane-bound granules of storage protein occurring in the outermost cell layer of the endosperm of wheat and other grains.

Aleurone layer Metabolically active cells of outer cereal endosperm (in contrast to metabolically inactive cells of most of the endosperm) containing *aleurone grains*, several hydrolytic enzymes and reserves of *phytin* (releasing inorganic phosphate and inositol on digestion by phytase). During germination, aleurone cells secrete α-amylase into the endosperm, initiating its digestion. Recent work suggests that the synthesis of enzymes by aleurone cells may not be as specifically induced by gibberellins from the embryo axis as was once thought, although these growth substances are certainly implicated in the control of endosperm digestion.

Aleuroplast Colourless plastid (leucoplast) storing protein; found in many seeds, e.g. brazil nuts.

Algae An informal grouping of many simple photosynthetic organisms within the Kingdom **Protista**, all possessing chlorophyll *a* as their primary photosynthetic pigment, and lacking a sterile covering of cells around reproductive cells. The body comprises a thallus (i.e. lacks roots, stems and leaves). Within this informal grouping are several forms, not necessarily closely related, but included with the algae by most phycologists (e.g. the monerans, **Cyanobacteria** and **Prochlorophyta**). In some classifications, these are included with the bacteria rather than the other algal groups, which belong to the Kingdom *Protista*. Algae are predominantly aquatic (marine or freshwater) or live in damp habitats (e.g. damp rock faces, tree trunks, moss hummocks or soils). A few live endolithically in deserts, relying upon night-time dew for their source of water; others grow on melting snow, attached to the undersurface of floating ice. Some algae, particularly species of red (**Rhodophyta**) and brown (**Phaeophyta**) algae, are harvested and eaten as a vegetable, or the mucilages are extracted for use as gelling and thickening agents. (See **agar**,**carrageenan**).

The formal eukaryotic divisional taxon 'Algae' has been abandoned, component groups being considered sufficiently distinct to merit divisional status on the basis of comparisons between pigments, assimilatory (storage) products, flagella, cell-wall chemistry and structure, and aspects of cell ultrastructure. Electronmicroscopy has revealed the presence and structures of flagella, their hairs, swellings and scales; chloroplasts; **chloroplast endoplasmic reticulum** (CER); thylakoids; phycobilisomes; external organic and inorganic scales; silica deposition vesicles; theca; nuclear structure and division; and cell division – all employed in algal systematics. Simultaneously, biochemists analysed the molecular details of pigments, assimilatory products and cell walls. Today, four quite distinct evolutionary groups are recognized as 'algae': (1) Prokaryotic algae, the first to evolve (e.g. Cyanobacteria and Prochlorophyta); (2) Eukaryotic algae with a chloroplast envelope of two membranes. Probably evolved via an evolutionary event involving capture of a prokaryotic cell in a food vesicle by a phagocytic, non-photosynthetic protozoan. Eventually, the plasmalemmas of the endosymbiont and food vesicle became the inner and outer chloroplast envelope membranes, respectively (e.g. **Glaucophyta**, representing an intermediate stage in this evolutionary process; Rhodophyta and Chlorophyta both represent completion of this evolutionary process resulting in chloroplasts); (3) Eukaryotic algae with chloroplasts surrounded by an additional membrane of CER. Probably evolved via an event involving the capture of a eukaryotic alga whose chloroplast entered a food vesicle of a phagocytic protozoan; eventually, the food vesicle membrane of the host became the single membrane of CER surrounding the chloroplast (e.g. **Euglenophyta**, **Dinophyta**); (4) Eukaryotic algae with two membranes

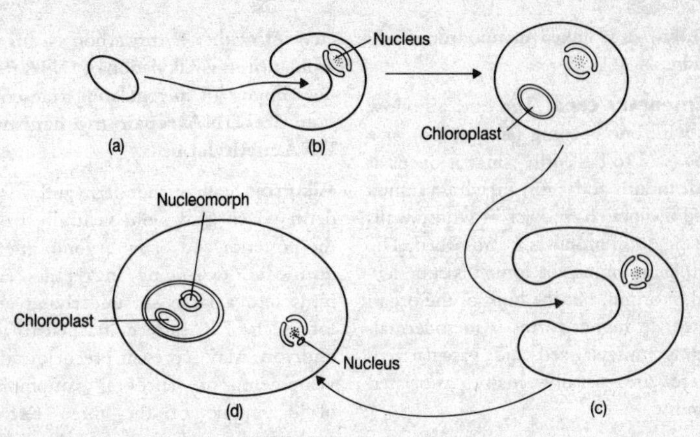

Fig. 2 *The possible steps involved in the origin of the cryptophyte algae such as* Cryptomonas. *Two endosymbiotic events are required: first, a photosynthetic prokaryote (a) is engulfed by eukaryote I (b) which is subsequently engulfed by a second eukaryote (c) to form the present* Cryptomonas *cell (d). The membranes surrounding the chloroplast in (d) form part of the* **chloroplast endoplasmic reticulum**, *in which the nucleomorph is trapped.* (From *Recombinant DNA 2/E* by Watson, Gilman, Witkowski, and Zoller. Copyright © James D. Watson, Michael Gilman, Jan Witkowski and Mark Zoller, 1992. Reprinted with permission of W. H. Freeman and Company.)

of CER. Evolutionary origin probably involved ingestion of a red alga into a protozoan's food vesicle where it remained as a symbiont, its nucleus being reduced to a nucleomorph. The protozoan, with its algal symbiont, was probably taken into a food vesicle of another phagocytic protozoan where it remained as an endosymbiont. The nucleus of the first protozoan took over the functioning of the cell, while that of the second protozoan was lost, along with the food vesicle and the outer nuclear membrane of the second and first protozoans, respectively. Thus, two membranes of CER resulted (e.g. **Cryptophyta**, **Chrysophyta**, **Bacillariophyta**, **Prymnesiophyta**, **Xanthophyta**, **Eustigmatophyta**, **Raphidophyta**, and **Phaeophyta**).

The oldest algal group known (**Cyanobacteria**) possesses a definite fossil **stromatolite** record dating back almost 3000 Myr. They evolved when the atmosphere contained no or very little oxygen; methane, ammonia and other reduced compounds dominated. Photosynthesis by these early blue-green algae eventually increased oxygen concentrations in the atmosphere to today's levels. Eukaryotic algae did not evolve until about 700–800 Myr BP. It is difficult to provide an accurate date because algae comprise soft tissues which do not fossilize well.

Algal bloom See **bloom**.

Alginic acid Complex phycocolloid occurring in cell walls and intercellular spaces of brown algae (**Phaeophyta**), commercially marketed. Polymer of mostly β-1,4, linked mannouronic acid residues.

Alimentary canal The gut; a hollow sac with one opening (an enteron) or a tube (said to be 'entire' since it opens at both mouth and anus) in whose lumen food is digested, and across whose walls the digestion products are absorbed. The epithelium lining the lumen is endodermal in origin, but the bulk of the organ system in higher forms is mesodermal, and is muscularized and vascularized. There are usually many associated glands.

Alkaline phosphatase Broad specificity enzyme, hydrolysing many phosphoric acid esters with an optimum activity in the basic pH range. Breaks down pyrophosphate in vertebrate blood plasma, enabling bone mineralization. In the yeast *Saccharomyces cerevisea*, it releases phosphates in the periplasmic space between plasma membrane and cell wall, in which (like invertase) it is somehow anchored.

Alkaloids Wide group of basic nitrogenous organic compounds, many of them lipids, produced by a few families of flowering plants (e.g. Solonaceae, Papaveraceae); often derivatives of pyridine, pyrimidine, quinoline or isoquinoline. Include atropine (anti-cholinergic; application may double human heart rate; generally blocks effect of **muscarine** on vagus nerve effectors); **caffeine** (stimulant of central nervous system); cocaine (local anaesthetic); morphine (analgesic); **nicotine** (general stimulant) and **quinine**. See **ergot**, **Cyanobacteria**, **glycoside**.

Alkylating agent A substance introducing alkyl groups (e.g. $-CH_3$, $-C_2H_5$, etc.) into either hydrocarbon chains or aromatic rings. Alkylation of DNA residues important in regulating transcription. See **DNA repair mechanisms**, **DNA methylation**.

Allantois Stalk of endoderm and mesoderm which grows out ventrally from the posterior end of embryonic gut in **amniotes**, expanding in reptiles and birds into a large sac underlying and for much of its surface attached to the **chorion**. May represent precocious development of ancestral amphibian bladder. One of the three **extra-embryonic membranes**. A richly vascularized organ of gaseous exchange within cleidoic eggs, also functioning as a bladder to store embryo's nitrogenous waste. In higher primates and rodents, persists into later life as the urinary bladder.

Alleles (allelomorphs) Different sequences of genetic material occupying the same gene **locus**, and therefore on different chromosomes; said to be alleles of (allelomorphic to) one another – a relational property. In classical genetics, alleles were ascribed to the same gene on the basis of two criteria: (i) failure to recombine with one another at meiosis, as if occupying the same locus, and (ii) failure, when mutant, to exhibit **complementation** when present together in a diploid. Alleles of the same gene differ by **mutation** at one or more nucleotide sites within the same length of DNA, and back-mutation from one to another may occur. There may be many alleles of a gene in a population, but normally only two in the same diploid cell. See **gene**, **multiple allelism**, **allelic complementation**.

Allelic complementation Interaction between individually defective mutant alleles of the same gene to give a phenotype more functional than either could produce by itself. Due to interaction (hybridization) of protein products. A source of confusion in the delineation of **cistrons**. See **complementation**.

Allelopathy Inhibition of one species of plant by chemicals produced by another plant (e.g. by *Salvia leucophylla* – purple sage).

Allen's rule States that the extremities (tail, ears, feet, bill) of **endothermic** animals tend to be relatively smaller in cooler regions of a species range. See **Bergmann's rule**.

Allergic reaction Pointless immune response to some non-threatening foreign protein, initiated when IgE-bearing **B cells** are activated by antigen to secrete IgE antibody. These bind **mast cells** and **basophils** resulting in degranulation and release of **histamine** and other potent mediators causing allergic symptoms such as long-term inflammatory effects.

Alloantibody Antibody introduced into an individual but produced in a different member of the same species.

Alloantigen (isoantigen) Antigen stimulating antibody response in genetically different members of the same species.

Allochronic Of species or species populations that are either sympatric at different times of the year or otherwise have non-overlapping breeding seasons (e.g. different flowering seasons in anthophytes). See **allopatry, sympatric**.

Allochthonous Originating somewhere other than where found.

Allogamy (Bot.) Cross-fertilization.

Allogeneic (allogenic) With different genetic constitutions. Often refers to intraspecific genetic variations. See **infraspecific variation**.

Allograft (homograft) Graft between individuals of the same species but of different genotypes (allogeneic). See **autograft, isograft, xenograft**.

Allometry Study of relationships between size and shape. Organisms do not grow isometrically; rather proportions change as size changes. Thus juvenile mammals have relatively large heads, while limb proportions of arthropods alter in successive moults. Summarized by the exponential equation $y = bx^a$, where y = size of structure at some stage, b = a constant for the structure, x = body size at the stage considered and a = allometric constant (unity for isometric growth). The analysis is open to multivariate generalization. See **heterochrony**.

Allopatric Geographical distribution of different species, or subspecies or populations within a species, in which they do not occur together but have mutually exclusive distributions. Populations occupying different vertical zones in the same geographical area may still be fully allopatric. See **allochronic, sympatric**.

Allopolyploid Typically, a **tetraploid** organism derived by chromosome doubling from a hybrid between diploid species whose chromosomes have diverged so much that little or no synapsis occurs between them at meiosis, so that only bivalents are formed (e.g. New

World cottons, *Gossypium* spp.). This clearly distinguishes the term from **autopolyploid**, but some polyploids do not fall readily into either category. Allopolyploids may back-cross with one or other diploid parent stock; hence allotetraploids, which are generally themselves fully fertile (since they form bivalents at meiosis), behave in effect as new reproductively isolated species. However, if the original diploid progenitors were closely related species, or even ecotypes of the same species, then **multivalents** may arise in meioses, which then resemble meioses in typical autopolyploids. Nevertheless, as a result of their greater fertility classical allopolyploids have been more significant in evolution than have classical autopolyploids. Many new plant species have arisen this way. Cultivated wheat (*Triticum aestivum*) is an allohexaploid, combining doubling in a triploid hybrid between an allotetraploid and a diploid.

All-or-none response Ability of certain excitable tissues, under standardized conditions, to respond to stimuli of whatever intensity in just two ways: (a) no response (stimulus sub-threshold), or (b) a full-size response (stimulus at or above threshold). **Action potentials** of nerve and muscle membranes are characterized by all-or-none behaviour. Where thresholds of different units in a response differ, as in the many motor fibres of the sciatic nerve, or the various **motor units** of an entire muscle, an increase in stimulus intensity may bring progressively more units to respond. In muscle, this constitutes spatial **summation**. Nerve signals cannot use such amplitude variations.

Allosteric Of those molecules (typically proteins) whose three-dimensional conformations alter in response to their environmental situation, normally registered by a change in molecule function. Often the key to regulation of critical biochemical pathways, serving as a feedback monitoring device in cybernetic circuits both inside and outside cells (see **regulatory enzymes**). At least as significant is allosteric control of **gene expression** by regulatory proteins. Among non-enzyme proteins, the haemoglobin molecule is allosteric under different blood pH values, with marked effects upon its oxygen saturation curve (see **Bohr effect**). For allosteric inhibition and induced fit of enzymes, see **enzyme**.

Allotetraploid An **allopolyploid** derived by doubling the set of chromosomes resulting from fusion between haploid gametes from more or less distantly related parental species. In classical cases, there is no meiotic **synapsis** between the chromosomes of different origin, and more or less complete fertility is achieved. Far more common in plants than animals, probably through comparative rarity of vegetative habit and/or parthenogenesis in the latter, in which it is difficult to rule out autopolyploidy as the source. See **polyploidy**.

Allotopic Of closely related sympatric populations, whose distributions are such that both occupy the same geographical range, but each occurs in a different habitat within that range.

Allotype Genetic variant within a **locus** of a given species population, such as allelic forms within a **blood group system** or variants of heavy chain constant regions of **antibody** molecules. See **idiotype**, **isotype**.

Allozymes Forms of an enzyme encoded by different allelic genes.

Alpha-actinin (α-actinin) An accessory protein of muscle, anchoring actin filaments at the Z-disc and cross-linking adjacent sarcomeres; also cross-links actin in many other cells to contribute to the **cytoskeleton**.

Alpha blocker Drug blocking **adrenergic** alpha receptors, preventing activity of the sympathetic neurotransmitter **noradrenaline**.

Alpha helix (Of proteins) a common secondary structure, in which the chain of amino acids is coiled around its long axis. Not all proteins adopt this conformation, it depending upon the molecule's primary structure. When adopted there are about 3.6 amino acids per turn (corresponding to 0.54 nm along the axis), amino acid R-groups pointing outwards. Hydrogen bonds between successive turns stabilize the helix. The α-helix may alternate with other secondary structures of the molecule such as β-sheets or 'random' sections. See **protein**.

Alpha receptor **Adrenergic** membrane receptor site binding **noradrenaline** in preference to **adrenaline**. May be excitatory or inhibitory, depending on the tissue. As with beta receptors, effects are mediated through an adenylate cyclase molecule adjacent in the membrane. The commonest receptors on postsynaptic membranes of postganglionic cells of sympathetic system. See **cholinergic**, **autonomic nervous system**, **alpha blocker**.

Alpha-richness Number of species present in a small, local, homogeneous area. See **diversity**.

Alternation of generations Either (1) *metagenesis*, a life cycle alternating between a generation reproducing sexually and another reproducing asexually, the two often differing morphologically; or (2) the alternation within a life cycle of two distinct cytological generations, one being haploid and the other diploid. See **life cycle**.

Metagenesis occurs in a few animals, e.g. **Cnidaria** and parasitic flatworms, where both generations are normally diploid. The alternation of distinct cytological generations is clearest in some algae and ferns, where the two generations (gametophyte and sporophyte) are independent and either identical in appearance (alternation of *isomorphic* generations) or quite dissimilar (alternation of *heteromorphic* generations). In mosses and liverworts the dominant (vegetative) plant is the gametophyte while the sporophyte (the capsule) is more or less nutritionally dependent on the gametophyte. In flowering plants, the male (micro-) and female (macro-) gametophytes are reduced to microscopic proportions, the male gametophyte being shed as the pollen grain and the female gametophyte (embryo sac) being retained on the sporophyte in the ovule. A clearcut alternation of physically distinct plants is avoided here, although alternating cytological phases are still discernible. In vascular plants generally, the sporophyte generation is the vegetative plant itself, be it a fern, herb, shrub or tree.

Alternative splicing See **RNA processing**.

Altricial Animals born naked, blind and immobile (e.g. rat and mouse pups, many young birds). See **nidicolous**.

Altruism Behaviour benefiting another individual at the expense of the agent. Widespread and apparently at odds with Darwinian theory, which predicts that any genetic component of such behaviour should be selected against. Theories of altruism in biology tend to be concerned with cost-benefit analysis, as dictated by the logic of natural selection. One component of Darwinian **fitness** may be the care a parent bestows upon its offspring, although this is not usually considered altruism. **Hamilton's rule** indicates the scope for evolutionary spread of genetic determinants of altruistic character traits, compatibly with Darwinian theory, and explains the evolution of parental care, while showing that *reciprocal altruism* can evolve even in the absence of relatedness between participants (e.g. members of different species). **Multicellularity** may afford opportunities for sacrifice of somatic cells (e.g. leucocytes) for a genetically related germ line harbouring the potentially immortal **units of selection**. See **arms race**.

Alveolus (1) Minute air-filled sac, grouped together as *alveolar sacs* to form the termini of bronchioles in vertebrate lungs. Their thin walls are composed of squamous epithelial and surfactant-producing cells. A rich capillary network attached to the alveoli supplies blood for gaseous exchange across the huge total alveolar surface. A surfactant (lecithin) layer reduces surface tension, keeping alveoli open from birth onwards, and provides an aqueous medium to dissolve gases. Macrophages in the alveolar walls remove dust and debris. (2) Expanded sac of secretory epithelium forming internal termini of ducts of many glands, e.g. mammary glands. (3) Bony sockets into which teeth fit in mandibles and maxillae of jawed vertebrates, lying in the *alveolar processes* of the jaws. (4) An elongated chamber on the cell wall of some diatoms (Bacillariophyta) from the central axis to the margin, and opening to the inside of the cell wall.

Amacrine cell One of three classes of neurone in mid-layer of vertebrate retina. Conducts signals laterally without firing action potentials.

Amastigomycota In older fungal classifications, that division of fungi lacking a motile stage. See **Zygomycota, Basidiomycota, Ascomycota, Deuteromycota**.

Amber mutation One of three mRNA **codons** not recognized by transfer RNAs commonly present in cells, and bringing about normal polypeptide chain termination. Its triplet base sequence is UAG. *Missense* or *stop mutation*. See **ochre** and **opal mutations**, **genetic code**.

Amensalism Interaction in which one animal is harmed and the other unaffected. See **symbiosis**.

Ames test Test assessing mutagenic potential of chemicals. Strains of the bacterium *Salmonella typhimurium* having qualities such as permeability to chemicals, inability to repair DNA damage, or ability to convert DNA damage into heritable mutations, are made **auxotrophic** for histidine. After mixing with potential mutagen prior to plating, increase in normal (**prototrophic**) colonies indicates mutagenicity.

Ametabola Primitively wingless insects (**Apterygota**).

Amino acid Amphoteric organic compounds of general structural formula

$$\begin{matrix} & & H & & \\ H & \searrow & | & & \nearrow O \\ & N- & C & -C & \\ H & \nearrow & | & & \searrow OH \\ & & R & & \end{matrix}$$

(where R may be one of twenty atomic groupings)

occurring freely within organisms, and polymerized to form dipeptides, oligopeptides and polypeptides. Amino acids differ in their R-groups (seven of which are readily ionizable) and the amino acid sequence in a polypeptide determines not only its charge sequence but also its conformation in solution (the distribution of acidic, basic and non-polar R-groups depends upon R-group composition, the pH and the microenvironment). Relative molecular masses of the common forms vary from 75 (glycine) to 204 (tryptophan). Only three commonly contain sulphurous R-groups: methionine, cysteine and cystine (formed from two oxidized cysteines, providing 'sulphur bridges'). During **protein synthesis** the carboxy- and amino-terminal ends of adjacent amino acids condense to form peptide bonds, leaving only the N-terminal and C-terminal ends of the protein and appropriate R-groups ionizable. About 20 amino acid radicals occur commonly in proteins, encoded by the **genetic code**. Their modification after attachment to a transfer RNA molecule may result in rare non-encoded amino acids occurring in proteins. Some amino acids (e.g. ornithine) never occur in proteins. Most naturally occurring amino acids have a free carboxyl group and (except proline) are alpha amino acids, bearing a free amino group on the α carbon atom (that adjacent to the free carboxyl group). Aspartic acid is required for pyrimidine synthesis, as glutamine is for purine synthesis. *Essential amino acids* are required by an organism from its environment, due to inability to synthesize them from precursors (see **vitamins**, which they are not); in humans, these are lysine, phenylalanine, leucine, threonine, methionine, isoleucine, tryptophan, histidine and valine.

Amitosis See **aberrant chromosome behaviour** (2).

Ammocoete Filter-feeding larva of lamprey, capable of attaining lengths of over 10 cm if conditions for metamorphosis do not prevail.

Ammonification Decomposition of amino acids and other nitrogenous organic compounds; results in production of ammonia (NH_3) and ammonium ions (NH_4^+). Bacteria involved are *ammonifying bacteria*. See **nitrogen cycle**.

Ammonites Group of extinct cephalopod molluscs (Subclass Ammonoidea, Order Ammonitida) dominating the Mesozoic cephalopod fauna. Had coiled shells, with protoconch (calcareous chamber) at origin of the shell spiral. Of great stratigraphic value.

Ammonotelic (Of animals) whose principal nitrogenous excretory material is ammonia. Characterizes aquatic, especially freshwater, forms. See **ureotelic**, **uricotelic**.

Amniocentesis See **amnion**.

Amnion Fluid-filled sac in which **amniote** embryo develops. An **extra-**

embryonic membrane (Fig. 35) formed in reptiles, birds and some mammals by extraembryonic ectoderm and mesoderm growing up and over embryo, the (amniotic) folds overarching and fusing to form the amnion surrounding the embryo, and the **chorion** surrounding the amnion, **allantois** and **yolk sac**. The amnion usually expands to meet the chorion. In humans and many other mammals the amnion originates by rolling up of some of the cells of the **inner cell mass** during **gastrulation**. Amniotic fluid (amounting to about one dm^3 at birth in humans) is circulated in placental mammals by foetal swallowing, enabling wastes to pass to the placenta for removal. Provides a buffering cushion against mechanical damage, helps stabilize temperature and dilate the cervix during birth. In *amniocentesis*, amniotic fluid containing cells from the human foetus is withdrawn surgically at around the sixteenth week of pregnancy for signs of congenital disorder, such as **Down's syndrome**.

Amniote Reptile, bird or mammal. Distinguished from anamniotes by presence of **extraembryonic membranes** in development.

Amniotic egg Egg type characteristic of reptiles, birds and **Prototheria** (much modified in placental mammals). Shell leathery or calcified; **albumen** and yolk typically present. **Extraembryonic membranes** occur within it during development. See **cleidoic egg**.

Amoeba Genus of sarcodine protozoans. Protists of irregular and protean shape, moving and feeding by use of **pseudopodia**. Some slime mould cells are also loosely termed 'amoebae', while any **cell locomotion** resembling an amoeba's is termed 'amoeboid'.

Amoebocyte Cell (haemocyte) capable of active amoeboid locomotion found in blood and other body fluids of invertebrates; in sponges, an amoeboid cell type implicated in mobilization of food from the feeding **choanocytes** and its conveyancing to non-feeding cells in absence of true vascular system.

Amoeboid Describing cells resembling those of the genus ***Amoeba***.

AMP Adenosine monophosphate. Nucleotide component of DNA and RNA (in deoxyribosyl and ribosyl forms respectively), and hydrolytic product of ADP and ATP. Converted to **cyclic AMP** by **adenylate cyclase**, intracellular concentrations of cAMP rising rapidly in response to extracellular (esp. hormonal) signals and falling rapidly due to activity of intracellular phosphodiesterase. Its level dictates rates of many biochemical pathways, depending upon cell type. See **cascade**, **second messenger**, **G-protein**, **GTP**, **receptor**.

Amphibia Class of tetrapod vertebrate, its first fossil representatives being Devonian ichthyostegids and its probable ancestors rhipidistian crossopterygian fishes. A **polyphyletic** origin has not been ruled out. Many early forms had scaly skins, almost entirely lost in the one modern Subclass (Lissamphibia) of three orders: Apoda, legless caecilians; Urodela, salamanders and newts; Anura, toads and frogs. Compared with their mainly aquatic ancestors, the more terrestrialized amphibians have: vertebrae with larger, more articulating neural arches and larger intercentra (see **vertebral column**); greater freedom of the

pectoral girdle from the skull, allowing some lateral head movement; **pelvic girdle** composed of three paired bones (pubis, ischium and ilium) with some fusion to form the rigid **pubic symphysis**; eardrums (homology with part of the spiracular gill pouch of fish) and a single middle ear ossicle, the columella, homologous with the hyomandibular bone of fish. Fertilization is internal or external (but intromittant organs are lacking). Most return to water to lay anamniote eggs, although some are viviparous. The skin is glandular for gaseous exchange. Modern forms specialized and not representative of the Carboniferous amphibian radiation.

Amphicribral (Bot.) Type of vascular arrangement where phloem surrounds the xylem. Compare **amphiphloic**.

Amphidiploid See **allotetraploid**.

Amphimixis Normal sexual reproduction, involving meiosis and fusion of haploid nuclei, usually borne by gametes. See **automixis**, **apomixis**, **parthenogenesis**.

Amphineura Class of **Mollusca**, including the chitons. Marine, mostly on rock surfaces; head reduced and lacking eyes and tentacles; mantle all round head and foot; commonly eight calcareous shell plates over visceral hump; nervous system primitive, lacking definite ganglia.

Amphioxus Lancelets (Subphylum **Cephalochordata**). Widely distributed marine filter-feeding burrowers up to 5 cm long. Two genera (*Branchiostoma*, *Asymmetron*). Giant larva resulting from prolonged pelagic life once given separate genus (*Amphioxides*) and develops premature gonads, providing support for the evolutionary origin of vertebrates by **progenesis**.

Amphipathic Of protein molecules with one surface containing hydrophilic and the other hydrophobic amino acid residues. See **LDL**.

Amphiphloic (Bot.) Type of vascular arrangement where phloem is on both sides of the xylem. Compare **amphicribral**.

Amphipoda Order of Crustacea (Subclass Malacostraca). Lack carapace; body laterally flattened. Marine and freshwater forms; about 3600 species. Very important detritus feeders and scavengers. Includes gammarids.

Amphistylic Method of upper jaw suspension in a few sharks, in which there is support for the jaw both from the hyomandibular and the brain case. See **autostylic**, **hyostylic**.

Ampulla (Of inner ear) see **vestibular apparatus**.

Amygdala (amygdaloid bodies or **nuclei)** Basal ganglia of the subcortical region of the most ancient part of the vertebrate **cerebral hemispheres**, gathering olfactory and visceral information. They appear to be involved in the generation of emotions. Removal in humans increases sexual activity.

Amylases (diastases) Group of enzymes hydrolysing starches or glycogen variously to dextrins, maltose and/or glucose; α-amylase (in saliva and pancreatic juice) yields maltose and glucose; β-amylase (in malt) yields maltose alone. Present in germinating cereal seeds (see **aleurone layer**), where only α-amylase can digest intact starch grains, and produced by some microorganisms.

Amylopectin Highly branched polysaccharide component of the plant storage carbohydrate **starch**. Consists of homopolymer of α[1,4]-linked glucose units, with α[1,6]-linked branches every 30 or so glucose radicals. Like **glycogen** it gives a red-violet colour with iodine/KI solutions. See **amylases**.

Amyloplast Colourless plastid (leucoplast) storing **starch**; e.g. found in cotyledons, endosperm and storage organs such as potato tubers.

Amylose Straight-chain polysaccharide component of **starch**. Comprises α[1,4]-linked glucose units. Forms hydrated micelles in water, giving the impression of solubility. Gives a blue colour with iodine/KI solutions. Hydrolysed by **amylases** to maltose and/or glucose.

Anabolism Enzymatic synthesis (build-up) of more complex molecules from more simple ones. Anabolic processes include multi-stage photosynthesis, nucleic acid, protein and polysaccharide syntheses. ATP or an equivalent needs to be available and utilized for the reaction(s) to proceed. See **catabolism**, **growth hormone**, **metabolism**.

Anadromous Animals (e.g. lampreys, salmon) which must ascend rivers and streams from the sea in order to breed. See **osmoregulation**.

Anaerobic (Of organisms) ability to live *anoxically* i.e. in the absence of free (gaseous or dissolved) oxygen. (Of processes) occurring in the absence of such oxygen. *Anaerobic respiration* is the enzyme-mediated process by which cells (or organisms) liberate energy by oxidation of substances but without involving molecular oxygen. This involves less complete oxidation of substrates, with less energy released per g of substrate used, enabling *anaerobes* to exploit environments unavailable to obligate aerobes. *Facultative anaerobes* can switch metabolism from aerobic to anaerobic under anoxic conditions, as required of many internal parasites of animals, some yeasts and other microorganisms. **Glycolysis** is anaerobic but may require aerobic removal of its products to proceed. Relatively anoxic environments include animal intestines, rumens, gaps between teeth, sewage treatment plants, polluted water, pond mud, some estuarine sediments and infected wounds. See **oxygen debt**, **respiration**.

Anagenesis (1) Process by which characters change during evolution within species, by **natural selection** or **genetic drift**. (2) Any non-branching speciation in which species originate along a single line of descent yet only one species represents the lineage after any speciation event (contrast **cladogenesis**). Gradual anagenetic speciation is not possible within the *biological species concept*, for reproductive isolation is never completed between ancestral and descendant species. **Cladistics** excludes anagenetic speciation by definition, but the term is retained in the context of characters. See **species**.

Analogous A structure present in one evolutionary lineage is said to be analogous to a structure, often performing a similar function, within the same or another evolutionary lineage if their phyletic and/or developmental origins were independent of one another; i.e. if there is **homoplasy**. Tendrils of peas and vines and eyes of squids and vertebrates are pairs of analogous structures. See **convergence**, **homology**, **parallel evolution.**

Anamniote (Of vertebrates) more primitive than the **amniote** grade. Includes agnathans, all fish, and amphibians.

Anandrous (Of flowers) lacking stamens.

Anaphase Stage of mitosis and meiosis during which either bivalents (meiosis I) or sister chromatids (mitosis, meiosis II) separate and move to opposite poles of the cell. See **spindle**.

Anaphylaxis A type of hypersensitivity to antigen (allergen) in which IgE antibodies attach to mast cells and basophils. May result in circulatory shock and asphyxia. See **allergic reaction**.

Anatropous (Of ovule) inverted through 180°, micropyle pointing towards placenta. Compare **campylotropous**.

Androdioecious Having male and hermaphrodite flowers on separate plants. Compare **andromonoecious.**

Androecium A collective term referring to the stamens of a flower. Compare **gynoecium**.

Androgen Term denoting any substance with male sex hormone activity in vertebrates, but typically steroids produced by vertebrate testis and to a much lesser extent by adrenal cortex. See **testosterone**.

Androgenesis Process by which an embryo (an *androgenone*) is produced solely from sperm-derived chromosomes. Occurs when the female pronucleus is absent (or removed) from an egg and the sperm chromosomes duplicate on entry, restoring the diploid number (similar to production of hydatiform mole in mammals). *Gynogenones* are equivalent, but totally female-derived, embryos. Experimental work in this field, including pronuclear transplantation, suggests that sperm-derived genes are required for normal chorion development in mammals, while egg-derived genes are required for normal development of the embryo itself. See **chromosomal imprinting**.

Andromonoecious Having male and hermaphrodite flowers on the same plant. Compare **androdioecious**.

Anemophily Pollination of flowers by the wind. Compare **entomophily**.

Anergic (Of **T cells**) failure to respond to a normally stimulatory MHC-peptide complex on an **antigen-presenting cell**.

Aneuploid (heteroploid) Of nuclei, cells or organisms having more or less than an integral multiple of the typical haploid chromosome number. See **euploid**, **monosomy**, **trisomy**, **nullisomy**.

Angiosperm Literally, a seed borne in a vessel (carpel); thus one of a group of plants (the flowering plants) whose seeds are borne within a mature ovary (fruit). See **Anthophyta**, which replaces Angiospermae.

Angiotensins *Angiotensin I* is a decapeptide produced by action of kidney enzyme, renin, on the plasma protein angiotensinogen when blood pressure drops. It is in turn converted by a plasma enzyme in the lung to the octapeptide *angiotensin II*, an extremely powerful vasoconstrictor which raises blood pressure and also results in sodium retention and potassium excretion by kidney. See **osmoregulation**.

Ångström unit (Å) Unit of length, 10^{-10} metres (0.1 nm); 10^{-4} microns. Not an SI unit.

Animalia Animals. Kingdom containing those heterotrophic eukaryotes lacking cell wall material and having a **blastula** stage in their development. One proposal for the defining character of the taxon (its autapomorphy, or **zootype**) is the presence of ***Hox* gene** clusters and their characteristic mode of expression during development.

Animal pole Point on surface of an animal egg nearest to nucleus, or extended to include adjacent region of cell. Often marks one end of a graded distribution of cytoplasmic substances. See **polarity**.

Anisogamy Condition in which gametes which fuse differ in size and/or motility. In **oogamy**, gametes differ in both properties. Significantly, the sperm often contributes the sole centriole for the resulting zygote. See **fertilization**, **isogamy**, **parthenogenesis**.

Annelida (Annulata) Soft-bodied, metamerically segmented coelomate worms with, typically, a closed blood system; excretion by **nephridia**; a central nervous system of paired (joined) nerve cords ventral to the gut, and a brain comprising paired ganglia above the oesophagus, linked by commissures to a pair below it. Cuticle collagenous, not chitinous. Chitin present in **chaetae**, which may be quite long, bristle-like and associated laterally with fleshy parapodia (e.g. ragworms, Class Polychaeta) or shorter and not housed in parapodia (e.g. earthworms, Class Oligochaeta). Leeches (Class **Hirudinea**) have 34 segments, confused by surface annulations. **Clitellum** present in both oligochaetes and leeches. Septa between segments often locally or entirely lost. The coelom acts as a hydrostatic skeleton against which longitudinal and circular muscle syncytia (and diagonal muscles in leeches) contract. Cephalization most pronounced in polychaetes (largely marine); eyes and mandibles often well developed but oligochaetes lack specialized head structures. Gametes leave via **coelomoducts**. Oligochaetes and leeches are typically hermaphrodite, polychaetes frequently dioecious.

Annual Plant completing its life cycle, from seed germination to seed production followed by death, within a single season, regardless of the number of times reproduction occurs (see **semelparous**). Compare **biennial**, **ephemeral**, **perennial**. See **desert, *r*-selection**.

Annual ring Annual increment of secondary wood (xylem) in stems and roots of woody plants of temperate climates. Because of sharp contrast in size between small wood elements formed in late summer and larger elements formed in spring the limits of successive annual rings appear in a cross-section of stem as a series of concentric lines.

Annular thickening In protoxylem, internal thickening of a xylem vessel or tracheid wall, in rings at intervals along its length. Provides mechanical support, permitting longitudinal stretching as neighbouring cells grow.

Annulus (1) Ring of tissue surrounding the stalk (stipe) of fruit bodies of certain Basidiomycota (e.g. mushrooms); (2) line of specialized cells involved in opening moss capsules and fern sporangia to liberate spores.

Anoestrus Period between breeding seasons in mammals, when **oestrous cycles** are absent. See **oestrus**.

Anoplura See **Siphunculata**.

Anoxia Deficiency or absence of free (gaseous or dissolved) oxygen.

Antenna Paired, preoral, tactile and olfactory sense organs developing from second or third embryonic somites of all arthropod classes other than Onychophora and Arachnida. Usually much jointed and mobile. In some crustaceans locomotory or for attachment, a pair of **antennules** (often regarded as antennae) typically occurring on the segment anterior to that with antennae. **Onychophora** have pair of cylindrical preantennae on first somite. See **tentacles**.

Antenna complex Clusters of several hundred chlorophyll molecules fixed to the thylakoid membranes of chloroplasts by proteins in such a way as to harvest light energy falling on them, and relaying it to a special chlorophyll molecule in an associated **photosystem**. See **photosynthesis** and Fig. 32b.

Antennapedia complex (Ant-C) Complex of **homeotic** and segmentation loci in *Drosophila* which, when homozygously mutant, may result in conversion of antennal parts into leg structures. Intensely studied in contexts of **morphogenesis**, and **positional information**. Some loci in the complex appear to be expressed only in specific embryonic **compartments**. The *Drosophila* Antennapedia and Bithorax genes form one (split) cluster of ***Hox* genes** (four in each), while vertebrate antennapedia-like **homeogenes** are homologues.

Antennule Paired and most anterior head appendages of crustaceans; uniramous, whereas antennae like most appendages in the class are biramous.

Anther Terminal portion of a **stamen**, containing pollen in *pollen sacs*.

Antheridiophore In some liverworts, a stalk that bears the antheridia.

Antheridium 'Male' sex organ (gametangium) of fungi, and of plants other than seed plants (e.g. algae, bryophytes, lycophytes, sphenophytes and pterophytes).

Antherozoid Synonym of **spermatozoid**.

Anthocerotopsida Hornworts. Class of **Bryophyta**. Small, widely distributed group, especially in tropical and warm temperate regions, growing in moist, shaded habitats. Plant a thin, lobed, dorsiventral **thallus**, anchored by rhizoids. Each cell usually has a single large chloroplast rather than the many small discoid ones found in cells of other bryophytes and vascular plants; and each chloroplast possesses a **pyrenoid**, all features suggesting algal affinities. Some (e.g. *Anthoceros*) contain Cyanobacteria (e.g. *Nostoc* spp.), supplying fixed nitrogen to their host plants.

Anthocyanins Group of water-soluble,

flavonoid pigments (glycosides) occurring in solution in vacuoles in flowers, fruits, stems and leaves. Change colour, depending on acidity of solution. Responsible for most red, purple and blue colours of plants, especially in flowers; contribute to autumn (fall) colouring of leaves and tint of young shoots and buds in spring. Colours may be modified by other pigments, e.g. yellow flavonoids.

Anthophyta Flowering plants (formerly Angiospermae). Division of plant kingdom. Seed plants whose ovules are enclosed in a carpel, and with seeds borne within fruits. Vegetatively diverse; characterized by **flowers**; pollination basically by insects, but other modes (e.g. **anemophily**) have evolved in a number of lines. Gametophytes much reduced; male gametophyte, initiated by pollen grain (microspore), comprising two non-motile gamete nuclei and a tube cell nucleus each associated with a little cytoplasm in the pollen tube; female gametophyte developing entirely within wall of megaspore which at maturity is a large cell containing eight nuclei, the **embryo sac**. Characteristic **double fertilization**.

Two classes: *Monocotyledonae* (monocots, about 65,000 spp.), with flower parts usually in threes, leaf venation usually parallel, primary vascular bundles in the stem scattered, true secondary growth absent, and just a single cotyledon present; *Dicotyledonae* (dicots, about 170,000 spp.), with flower parts usually in fours or fives, leaf venation usually net-like, primary vascular bundles in the stem forming a ring, often with true secondary growth and vascular cambium, and two cotyledons present.

Anthozoa See **Actinozoa**.

Anthropoid apes Members of what used used to be the primate Family Pongidae (see **Ponginae**): the so-called 'great apes'. Include orangutan, chimpanzees and gorillas. Common ancestor of pongids and hominids ('men') probably Miocene in age. Gibbons (Family Hylobatidae) are in same suborder (Anthropoidea) as 'great apes' (pongids) and occasionally included in the term 'anthropoid ape'. Much ape anatomy may be regarded as adaptation to brachiation. Fundamentally quadrupedal, with limited (but famous) tendency to bipedalism. Markedly prognathous, with diastemas. All are Old World forms.

Anthropoidea (Simii) Suborder of **Primates**. Three living superfamilies: Ceboidea (New World monkeys); Cercopithecoidea (Old World monkeys); Hominoidea (gibbons, great apes and man). Eyes large and towards front of face; brain expansion associated with relative expansions of frontal, parietal and occipital bones of skull; great manual dexterity. The earliest fossils bearing clear anthropoid (simian) characters now seem to be from the late Eocene of Algeria; but the most numerous are from the Fayum (now dated in the Lower Oligocene) of Egypt – once forest, now desert. The skull and postcranial skeleton of the monkey-sized *Aegyptopithecus* from Fayum suggest a proto-hominoid of **dryopithecine** affinities, whereas the squirrel-sized *Apidium* and Parapithecus appear closer to prosimians and New World monkeys. There may have been an early simian radiation before the end of the Eocene, for *Algeripithecus minutus* – a putative simian from the Glib Zegdou deposits of Algeria – may be of middle or even early Eocene age. A date of 55 Myr BP rather than 35 Myr BP for simian origins now seems likely.

Antiauxins Chemicals which can prevent the action of **auxins** in plants, e.g. 2,6-dichlorophenoxyacetic acid; 2,3,5-triiodobenzoic acid.

Antibiotic Diverse group of generally low molecular mass organic compounds (see **secondary metabolite**). Produced by spore-forming soil microorganisms (esp. moulds such as *Penicillium*, filamentous bacteria, e.g. *Streptomyces*, and other bacteria of the genus *Bacillus*) during or just after sporulation (noticeable in culture). Although they are often regarded as growth inhibitors of potentially competing microorganisms (bacteristatic when reversibly so; bactericidal when irreversibly), this is not totally convincing and may obscure their original role. They may be relics from the prebiotic era (prior to the evolution of cells) possibly interacting with **ribozyme**-like RNA molecules ('RNA-life'), as some do within cells today. Certainly many antibiotics serve mankind well by interfering with pathogen protein synthesis by adhering to ribosomal RNA, a property for which they have been artificially selected; but these effects may be due to overproduction of substances normally present in much smaller quantities within the cell. *Streptomycins* affect DNA, RNA and protein synthesis; *penicillins* (produced by *Penicillium chrysogenum*) prevent cross-linking of the glycan chains of peptidoglycan molecules in bacterial cell walls (cell growth producing wall-deficient or wall-less cells, which burst); **actinomycin** prevents RNA transcription; *puromycins* prevent translation; *anthracyclins* block DNA replication and RNA transcription. These effects generally involve the antibiotic complexing with, or inserting into, a nucleic acid. Antibiotics have been widely used as clinical drugs against bacterial pathogens, but are ineffective against viruses. Their use has put new selection pressures on the target microorganisms, resulting in the troublesome spread of **antibiotic resistance elements** in what resembles an **arms race**. See **Deuteromycota** (for a little history), **plasmids**.

Antibiotic resistance element Genetic element, composed of DNA and often borne on a **transposon**, conferring bacterial resistance to an antibiotic. Often with **insertion sequences** at either end, when capable of moving between **plasmid**, viral and bacterial DNA and selecting insertion sites, sometimes turning off expression of genes it inserts into or next to. Able to spread rapidly across species and other taxonomic boundaries, making design of new antibiotic drugs even more urgent. Many common pathogenic bacterial strains are now resistant to some of the best-known drugs. Non-homologous recombination between plasmids can give rise to multiple-resistance plasmids, bacterial plasmid R1 conferring resistance to chloramphenicol, kanamycin, streptomycin, sulphonamide and ampicillin. See **plasmid**.

Antibody Member of the **immunoglobulin superfamily** of glycoproteins secreted by mature vertebrate **B cells**, binding selectively to epitopes of antigens and clumping them (agglutination) prior to phagocytic engulfment.

Five major classes differ principally in their type of heavy protein chain, and

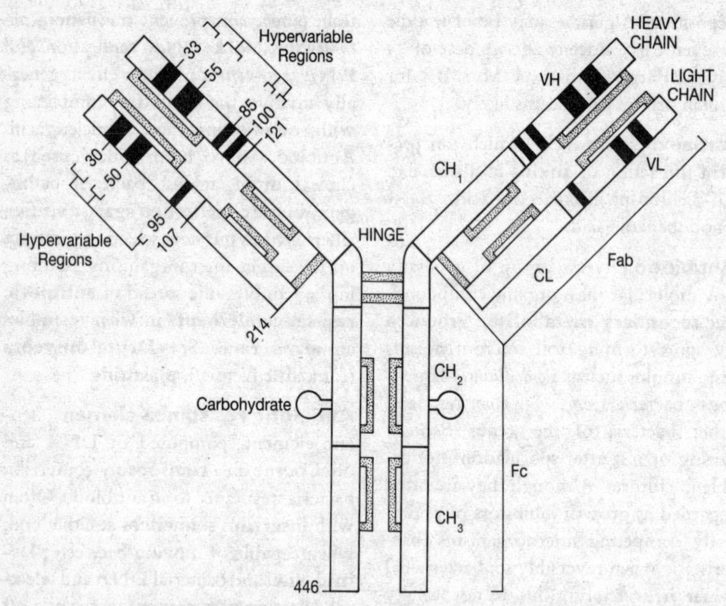

Fig. 3 *Diagram of the basic IgG structure. The hypervariable regions (amino acid residues numbered) are the antigen-binding sites. Disulphide bridges are indicated within and between light and heavy by stippling. Fab regions are those released by papain digestion, along with the complement-fixing Fc portions, which also bind Fc receptors. See* **antibody**.

the degree to which the molecule is a polymer of immunoglobulin 'monomers'. Each immunoglobulin unit comprises two identical H- (heavy) and two identical L- (light) polypeptide chains forming mirror images of each other and joined by a flexible hinge region involving disulphide bridges. They bind to antigen at specific antigen-binding regions provided uniquely by the combination of H- and L-chain amino-terminal portions (see Fig. 3), which are extremely variable in their amino acid sequences between different antibodies, in contrast to constant regions at their carboxy-terminal portions. Only about 20–30 amino acids of the variable regions of H- and L-chains contribute to the antigen-binding site, these being located in three short hypervariable regions of each variable region. These lie themselves within relatively invariant 'framework regions' of the variable regions. The other biological properties of the molecule are determined by the constant domains of the heavy chains.

Digestion of antibody with papain produces two identical Fab (antigen-binding) fragments and one Fc (crystallizing) fragment. The latter region in the intact Ig (immunoglobulin) molecule is responsible for determining which component of the immune system the antibody will bind to (for Fc receptor, see **receptor**

(2)). The Fc region of IgG may bind phagocytes and the first component of **complement**. Only the IgG antibody can cross the mammalian placenta. IgM is the major Ig type secreted in a primary immune response, but IgG dominates in secondary immune responses (see **B cell**).

Transformation of B cells into differentiated antibody-producing plasma cells generally requires both antigen-presenting cells and a signal from a helper T cell (see **T cell**). Because B cells have only a a few days' life in culture they are not suitable for commercial antibody production: however, if an antibody-producing B cell from an appropriately immunized mouse is fused to an appropriate mutant tumour B cell, the hybrid cell formed may continue dividing and producing the particular antibody required. The resulting **hybridoma** can be sub-cloned indefinitely, giving large amounts of antibody. Initial isolation of the appropriate B cell follows discovery of the required antibody in the growing medium. The purity of the resulting *monoclonal antibody* (mAb) and its production in response to what is possibly a minor component of an impure antigen mixture are both desirable features of the technique. All the **accessory molecules** known to participate in T cell recognition of their targets were first identified by monoclonal antibodies raised against these cell-surface markers. See **antibody diversity**, **antigen–antibody reaction**, **IgA–IgM**.

Antibody diversity (a. variation) Production of different **antibody** molecules by different **B cells** (see Figs. 3 and 4 for symbols). Light and heavy chains are encoded by different gene clusters. In humans, light chain genes lie on chromosomes 2 and 22, heavy chain genes on chromosome 14, and the light chains of a particular immunoglobulin molecule are encoded either by chromosome 2 or chromosome 22, not both. Any particular B cell assembles in a line all the heavy chain genes needed to make its own unique antibody type, joining first the genes for variable (V), hypervariable (HV) and joining (J) regions of the molecule, then linking this combination to the genes for the constant (C) regions of the molecule, with different constant regions for different immunoglobulin classes (see **immunoglobulin** references). The enzymes bringing together genes from different parts of a chromosome are performing a form of **recombination**. Diversity arises from the randomness with which particular genes from heavy and light chain clusters are brought together. In addition, extra short nucleotide sections (N segments) get inserted, probably in some rule-following way, into the DNA encoding the antigen-binding regions of the molecule, and this together with variation in **RNA processing** of the hnRNA transcript increases still further the total antibody diversity, often classified as follows: (1) Allotypic: variation in the C_H1, C_H3 and C_L antibody regions caused by allelic differences between individuals at one or more loci for a subclass of immunoglobulin chains; (2) Idiotypic: variation in the V_L and V_H regions (especially in the hypervariable regions) that are generally characteristic of a particular antibody clone, and therefore not present in all members of a population; (3) Isotypic: variation in the C_L, and in the C_H1–3

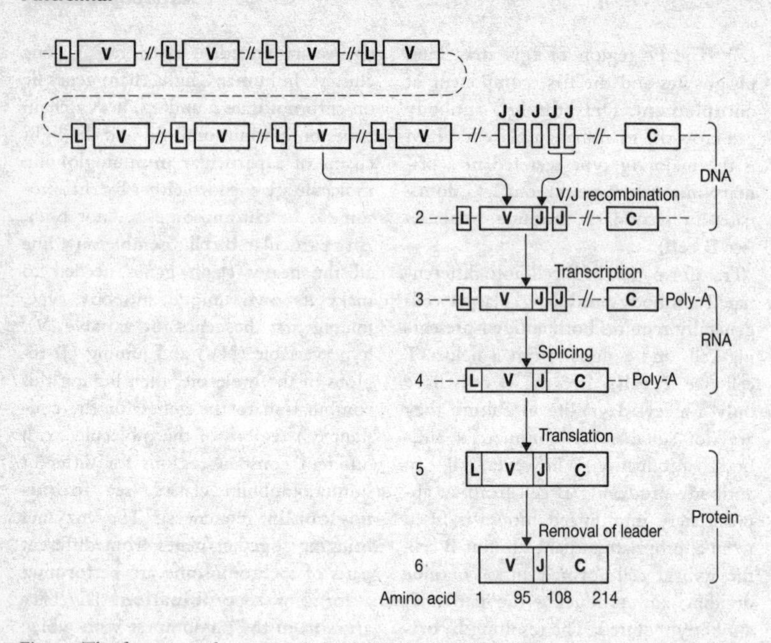

Fig. 4 *The formation of an active light chain of an antibody. The DNA is organized into families of related elements: one copy of the gene for the constant (C) region, five repeats of a joining (J) sequence, and up to 150 related but different variable (V) sequences, each separated by a leader (L) sequence. Through recombination, one of the V elements with its L sequence is inserted into a J segment. The entire VJC complex is transcribed into RNA, and excess J and linking sequences to C are excised as introns. The mature mRNA is translated, and the amino acids coded by the L portion, which are necessary to secrete the polypeptide, are removed from the protein. Active heavy chains arise by similar rearrangements. Poly-A indicates polyadenosine tail. See* **antibody diversity**. (From *Introduction to Genetic Analysis 5/e* by Griffiths, Miller, Suzuki, Lewontin and Gelbert. Copyright © W. H. Freeman and Company, 1993. Reprinted with permission.)

antibody regions, determined by loci whose representative alleles are shared by all healthy members of a population. See **affinity maturation**.

Anticlinal (Bot.) Alignment of the plane of cell division approximately at right angles to the outer surface of the plant part. Compare **periclinal**. See **cell division** for diagram.

Anticoagulant Any substance preventing blood clotting. Blood naturally contains such substances: fibrin and antithrombin III absorb much of the thrombin formed in the clotting process and **heparin** inhibits conversion of prothrombin to thrombin. Blood-sucking animals (leeches, insects, bats, etc.) frequently produce anticoagulants in their saliva. Artificial anticoagulants (e.g. dicumarol) are either helpful to patients, or prevent blood samples from clotting in blood banks (e.g. EDTA). The rat poison warfarin is an anticoagulant. See **blood clotting**.

Anticodon The triplet sequence of tRNA nucleotides capable of base-pairing with a codon triplet of an mRNA molecule. See **protein synthesis**.

Antidiuretic hormone (ADH, vasopressin) Ring-structure octapeptide hormone produced by hypothalamic neurosecretory cells and released into posterior pituitary circulation if blood water potential drops below the homeostatic norm. Has marked vasoconstrictor effects on arterioles, raising blood pressure, and increases water permeability of collecting ducts and distal convoluted tubules to the 10–20% of the initial glomerular filtrate still remaining (see **kidney**), resulting in water retention. See **nicotine**, **osmoregulation**, **oxytocin**.

Antigen Any substance (often protein or glycoprotein) that can be recognized by an already induced immune response and initiate production of further specific **antibody**, to which the latter binds at a specific conformational domain of the antigen molecule called the antigenic determinant, or epitope. Not every antigen is immunogenic, i.e. capable of initial induction of an immune response. See antigen references and cross-references below.

Antigen–antibody reaction Non-covalent bonding between antigenic determinant of **antigen** and antigen-combining site on an immunoglobulin molecule (see **antibody**). Several such bonds form simultaneously. The reactions show high specificity but cross-reactivity may result if some determinants of one antigen are shared by another. Antibodies seem to recognize the three-dimensional configuration and charge distribution of an antigen rather than its chemical make-up as such. Such reactions form the basis of humoral and of many cell-mediated immune responses. See **agglutinin**, **complement**, **precipitin**, **immunity**.

Antigenic variation Ability of some pathogens, notably viruses (see **haemagglutinin**), bacteria and protozoa, to change their coat antigens during infection. Trypanosomes and some stages in the malarial life-cycle achieve it, making the search for vaccines to some devastating human diseases very difficult. Can result from **gene conversion**.

Antigen-presenting cell (APC) Few antigens bind directly to antigen-sensitive **T cells** or **B cells** but are generally 'presented' to these lymphocytes on the surfaces of other cells, the antigen-presenting cells. Dendritic cells with a large resultant surface area for antigen-attachment are widely distributed in the human body and trap antigens. These cells, notably in spleen and lymph nodes, along with macrophages and B cells, can trap lymph- and blood-borne antigens and, after internalization and degradation, present antigenic peptide fragments, bound to cell-surface molecules of the **major histocompatibility complex**, to T cells. APCs may then activate T cells to clonal expansion, and these daughter cells may in turn activate B cells with the same MHC-bound antigen to clonal expansion and specific antibody production. **Interleukins** are involved in the activation processes. See **immunity**.

Antigen receptors Membrane glycoproteins on **T cell** and **B cell** surfaces. Those on B cells are membrane-bound antibodies; those on T cells are antibody-like heterodimers (see **T cell receptor**).

Antigibberellins Organic compounds of varied structure causing plants to grow with short, thick stems or with appearance opposite to that obtained with **gibberellin**, which can reverse the action of most of these compounds. Of agricultural importance, they include phosphon and maleic hydrazide (retarding growth of grass, reducing frequency of cutting).

Antipodals Three (sometimes more) cells of the mature **embryo sac**, located at the end opposite the micropyle.

Antiport See **transport proteins**.

Antiseptic Substance used on a living surface (e.g. skin) to destroy microorganisms and sterilize it. Ethyl and isopropyl alcohol, diluted 70% with sterile water, destroy vegetative bacteria and some viruses, but not spores of bacteria or fungi. Iodine (dissolved with potassium iodide in 90% ethanol) is rapidly bactericidal, killing both vegetative cells and spores. It does, however, stain. See **disinfectant**, **autoclave**.

Antiserum Serum containing antibodies with affinity for a specific antigenic determinant (see **antigen**) to which they bind. May result in cross-reactivity (see **antigen–antibody reaction**) within recipient.

Antler Bony projection from skull of deer. Unlike **horn** (which is matted hair) they are often branched, are shed annually, and are confined to males (except in reindeer).

Anura (Salientia) Frogs and toads. An order of the Class **Amphibia**. Hind legs modified for jumping and swimming; no tail; often vocal.

Anus The opening of the alimentary canal to the exterior through which egested material, some excretory material and water may exit. When present, the gut is said to be entire. Absent from coelenterates and platyhelminths. See **proctodaeum**.

Aorta Term applied to some major vertebrate arteries. See **aorta**, **dorsal**; **aorta**, **ventral**; **aortic arch**.

Aorta, dorsal Major vertebrate (and cephalochordate) artery through which blood passes to much of body, supplying arteries to most major organs. In sharks a single dorsal aorta collects oxygenated blood from the gills, but in bony fish paired dorsal aortae on either side in the head region perform this task before uniting as a single median vessel. Oxygenated blood then passes backwards to the body; but in fish too blood flowing up through the third **aortic arch** tends to pass anteriorly through the aorta(e) rather than posteriorly (see **carotid artery**). In adult tetrapods, those parts of the single or paired dorsal aortae between the third and fourth aortic arches tend to disappear, blood from the fourth (systemic) arch(es) passing back within two uniting dorsal aortae (terrestrial salamanders, lizards) or within a single dorsal aorta (most reptiles, birds and mammals) derived from the right arch (reptiles and birds) and from the left arch (mammals). Protected throughout in vertebrates by proximity of bone above (typically vertebrae).

Aorta, ventral Large median artery of fish and embryonic amniotes leading anteriorly from ventricle of heart, either giving off branches to gills or running uninterrupted as **aortic arches** to dorsal aorta(e). In lungfish, branches differ in this respect. In living amphibians it has disappeared, while in other tetrapods it serves merely as a channel supplying blood to aortic arches III, IV and VI.

Aortic arches Paired arteries (usually six, but up to fifteen in hagfishes) of vertebrate embryos connecting ventral aorta with dorsal aorta(e) by running up between gill slits or gill pouches on each side, one in each **visceral arch**. The study of their comparative anatomy in embryos and, where they persist, in adults provides striking support for macroevolutionary change. Each is given a Roman numeral, beginning anteriorly. Arches I and II do not persist in post-embryonic tetrapods, but arch II at least is present in sharks, some bony fish and lungfish. Arch III usually serves (with parts of the dorsal aortae) as the tetrapod carotid arteries, but in fish is usually interrupted by gills; arch IV is separated from the anterior arches in most tetrapods and becomes the systemic arch (see **aorta, dorsal**); arch V is absent from adult tetrapods other than urodeles, but serves as the **ductus arteriosus** in development prior to lung function; arch VI then shifts to supply the lungs.

Ape General term for **hominoid** primates of families Hylobatidae (gibbons, siamangs) and Pongidae ('great apes'). See **anthropoid ape**.

Apetalous Lacking petals, e.g. flower of wood anemone.

Aphaniptera See **Siphonaptera** (fleas).

Aphid Green fly or black fly. Homopteran insect (Superfamily Aphidoidea) notorious for sucking plant juices, for transmitting plant viral diseases, and for phenomenal powers of increase by viviparous **parthenogenesis**.

Aphyllous Leafless.

Apical dominance (Bot.) Influence exerted by a terminal bud in suppressing growth of lateral buds. See **auxins**.

Apical meristem Growing point (zone of cell division) at tip of root and stem in vascular plants, having its origin in a single cell (*initial*), e.g. Pterophyta, or in a group of cells (initials), e.g. Anthophyta. In the latter, the growing point apex (*promeristem*) consists of actively dividing cells. Behind this, division continues and differentiation begins, becoming progressively greater towards mature tissues. One (older) concept of growing point organization in flowering plants recognizes differentiation into three regions (*histogens*): *dermatogen*, a superficial cell layer giving rise to the epidermis; *plerome*, a central core of tissue giving rise to the vascular cylinder and pith; and *periblem*, tissue lying between dermatogen and plerome, that gives rise to cortex. It is now evident that respective roles assigned to these histogens are by no means universal; nor can periblem and plerome always be distinguished, especially in the shoot apex. Becoming widely accepted is the *tunica-corpus* concept, an interpretation of the shoot apex recognizing two tissue zones in the promeristem: *tunica*, consisting of one or more peripheral layers, in which the planes of cell division are predominantly anticlinal, enclosing the *corpus* or central tissue of irregularly arranged cells in which the planes of cell division vary. No relation is implied between cells of these two regions and differentiated tissue behind the apex as in the histogen concept. Although epidermis arises from the outermost tunica layer, underlying tissue may originate in tunica or in corpus, or in both, in different plant species.

Besides providing for growth in length of main axis, apical meristem of stem is the site of origin of leaf and bud primordia. In roots, two types of apical meristem occur, one in which vascular cylinder, cortex and root cap can be traced to distinct layers of cells in the promeristem, and a second type in which all tissues have a common origin in one group of promeristem cells. In contrast to those of stems, apical meristems of roots provide only for growth in length, lateral roots originating some distance from apex and, endogenously, from pericycle.

Apocarpous (Of the gynoecium of flowering plants) having separate carpels, e.g. buttercup. See **flower**.

Apocrine gland Type of gland in which only the apical part of the cell from which the secretion is released breaks down during secretion, e.g. mammary gland. Compare **holocrine gland**, **merocrine gland**.

Apoda (Gymnophiona) Caecilians. Order of limbless burrowing amphibians with small eyes and, sometimes, a few scales buried in the dermis of the skin, and a pair of tentacle-like structures in grooves above the maxillae.

Apoenzyme The protein component of a holoenzyme (enzyme-cofactor complex) when the **cofactor** is removed. It is catalytically inactive by itself.

Apogamy See **apomixis**.

Apogeotropic Growth of roots away from the earth and from the force of gravity (i.e. into the air).

Apomict Plant produced by **apomixis**.

Apomixis Most common in botanical contexts. (1) **Agamospermy**, reproduction which has the superficial appearance of ordinary sexual reproduction (amphimixis) but occurs without fertilization and/or meiosis. Affords the advantages of the seed habit (dispersal, and survival through unfavourable conditions) without risks in achieving pollination. Often genetically equivalent to asexual reproduction. See **parthenogenesis**. (2) Vegetative apomixis; **asexual** methods of propagation such as by rhizomes, stolons, runners and bulbils.

Apomorphous In evolution, of a character derived as a novelty from pre-existing (plesiomorphous) character. The two form a homologous pair of characters, termed an *evolutionary transformation series* in **cladistics**. See **synapomorphy**.

Apoplast The cell wall continuum of a plant or organ; movement of substances in the cell walls is termed apoplastic movement or transport. Compare **symplast**.

Apoptosis The process by which non-pathological animal cells die and are phagocytosed. See **programmed cell death**.

Aposematic Colour, sound, behaviour or other quality advertising noxious or otherwise potentially harmful qualities of an animal. See **mimicry**.

Apospory See **apomixis, parthenogenesis**.

Apostatic selection Selection by visual predators, held to increase the divergence between morphs in a polymorphic population by favouring any variation departing from the predator's **search image**.

Apothecium Cup or saucer-shaped fruit body (**ascocarp**) of certain **Ascomycota** and **lichens**); lined with a hymenium of asci and paraphyses. They are sessile, often brightly coloured, varying from a few mm to more than 40 cm across.

Appendage A functional projection from an animal surface; termed *paired appendages* if bilaterally symmetrical. Two such pairs (e.g. limbs, fins) generally occur in gnathostome vertebrates. Primitively one pair per segment in arthropods (walking legs, mouthparts, antennae) and polychaetes (parapodia).

Appendicular skeleton See **endoskeleton**.

Appendix, vermiform Small diverticulum of human caecum, of many other primates, and of rodents, containing lymphoid tissue. Not a vestigial structure, contrary to common belief.

Appetitive behaviour Behaviour (e.g. locomotory activity) variable with circumstances, increasing the chances of an animal satisfying some need (e.g. for food, nesting material) usually through a more stereotyped **consummatory act**, such as eating. To this extent it is *goal-oriented*.

Apposition (Bot.) Growth in thickness of cell walls by successive deposition of material, layer upon layer. Compare **intussusception**.

Apterous Wingless; either of insects which are polymorphic for winged and wingless forms, e.g. aphids, many social insects; or of insects which have discarded wings, as do some ants and termites; or of primitively wingless (apterygotan) insects.

Apterygota (Ametabola) Subclass of primitively wingless insects. Comprises orders Thysanura (bristletails, silverfish), Collembola (springtails), Protura and Diplura. Probably a polyphyletic assemblage. Some abdominal segments in members of all four orders have small paired lateral appendages, another primitive characteristic. Metamorphosis slight or absent. See **Pterygota**.

Aqueous humour Fluid filling the space between cornea and **vitreous humour** of vertebrate **eye**. The iris and lens lie in it. Much like cerebrospinal fluid in composition. Continuously secreted by ciliary body, and drained by canal of Schlemm into blood. Much less viscous than vitreous humour. Links circulatory system to lens and cornea, neither having blood vessels for optical reasons; also maintains intraocular pressure.

Arabidopsis thaliana Common wall cress, a cruciferous weed (Tribe Sisymbriae) of no economic value but of increasing importance as an experimental plant for molecular and developmental biology on account of its (1) very small genome size (7×10^7 base pairs; about five times that of the yeast *Saccharomyces cerevisiae*) in five chromosome pairs, over 60% of it encoding protein; (2) rapid life cycle (6–8 weeks); (3) small size (15–30 cm height). Because of the small amount of repetitive DNA, cloning procedures involving chromosome walking and gene tagging are applicable.

Arachnida Class of chelicerate arthropods. Most living forms terrestrial, using lung books (scorpions), lung books and tracheae (spiders), tracheae alone (e.g. pseudoscorpions, larger mites), or just the body surface (smaller mites) for gaseous exchange. Usually there is **tagmosis** into a *prosoma* of eight adult segments anteriorly and an *opisthosoma* of 13 segments posteriorly. No head/thorax distinction. Prosoma lacks antennae and mandibles; first pair of appendages clawed and prehensile chelicerae; second pair (pedipalps) may be prehensile and sensory, copulatory or stridulatory devices. Remaining four pairs of prosomal appendages are legs. Bases (gnathobases) of second and subsequent pairs of appendages are often modified for crushing and 'chewing' (in absence of true jaws). Includes orders: Acari (mites and ticks), Araneae (spiders), Scorpiones (scorpions), Pseudoscorpiones (false scorpions), Palpigrada (palpigrades), Solifugae (solfugids) and Opiliones (harvestmen). Xiphosura (king crabs), and the predatory and extinct Eurypterida are usually placed as subclasses of the **Merostomata**.

Arachnoid membrane One of the **meninges** around vertebrate spinal cord and brain.

Araneae (Araneida) Order of **Arachnida**. Spiders. Abdomen (opisthosoma) almost always without any trace of segmentation and joined to prosoma (cephalothorax) by 'waist'; silk produced from two to four spinning glands (spinnerets); pedipalps in male modified as intromittant organs for copulation; ends of chelicerae modified as poisonous fangs.

Archaea In taxonomic systems that regard the prokaryote/eukaryote distinction as unnatural and failing to represent an inner phylogenetic division within prokaryotes, an alternative to the more traditional **Archaebacteria**. On the basis of protein sequencing, these are then regarded as having eukaryotes as their most recent common ancestor.

Archaean (Archaeozoic) Geological division preceding **Proterozoic**; earlier than about 2600 Myr BP.

Archaebacteria Ancient lineage of bacteria distinct from other bacteria (eubacteria) and from eukaryotes (but see **Archaea**). Many live in hot acidic conditions (i.e. they are thermophilic and acidophilic), growing best at temperatures approaching 100°C. Formerly in two groups, either aerobic (Sulfolobales) or anaerobic (Thermoproteales), facultative anaerobic forms are now known. Many unusual biochemical characteristics including possession of a novel 16S-like ribosomal RNA component in the small ribosome subunit, which with their peculiar membrane composition indicates that there may be a deep divide among prokaryotes between archaebacteria and eubacteria. Halophiles, methanogens and sulphur-dependent thermophiles occur.

Archaeopteryx Most ancient recognized fossil bird (late Jurassic, 150–145 Myr BP). Exhibits mixture of reptilian and bird-like characters, having feathered wings and tail (impressions clear in limestone) and furcula (fused clavicles and interclavicles); but with teeth, bony tail, and claws on three digits of forelimbs. Only known representative of Subclass Archaeornithes of Class **Aves**.

Archegoniophore In some liverworts, a stalk bearing archegonia.

Archegonium 'Female' sex organ of liverworts, mosses, ferns and related plants, and of most gymnosperms. Multicellular, with neck composed of one or more tiers of cells, and swollen base (venter) containing egg-cell.

Archenteron Cavity within early embryo (at gastrula stage) of many animals, communicating with exterior by **blastopore**. Formed by invagination of mesoderm and endoderm cells at gastrulation; becomes the gut cavity.

Archesporium Cells or cell from which spores are ultimately derived, e.g. in developing pollen sac, fern sporangium.

Archonta Mammalian superorder employed by some authors, who generally regard it as monophyletic, containing primates, dermopterans, tree shrews (Order Scandentia) and bats (see **Chiroptera**). The interpretation of morphological similarities between these groups is questionable and no consistent picture has emerged from molecular comparisons.

Archosaurs 'Ruling reptiles'; the Subclass Archosauria. Originating with thecodonts in the Triassic, it includes the bipedal carnivorous dinosaurs (saurischians) and the bird-like dinosaurs (ornithiscians). Crocodiles and alligators are living representatives. Birds are descendants. See **dinosaur**.

Aril Accessory seed covering, often formed from an outgrowth at the base of the **ovule** (e.g. yew); often brightly coloured, aiding dispersal by attracting animals that eat it and carry seed away from the parent plant.

Arista See **awn**.

Arms race Term sometimes used to express the dialectical changes in selection pressure that occur when regular, often unavoidable, conflicts of interest between two or more 'ways of life' favour an adaptation for one party which creates a fresh 'challenge' for the other to respond by adapting to. Such conflicts are common: predator/prey; parasite/host; parent/offspring and male/female. It has been argued that selection will be the stronger where one party has more to lose by 'not evolving' and minimizing the probability of losing the conflict. Consequence to a prey organism of losing a predator/prey conflict is probably more serious than to a predator on any occasion. Much depends on how likely such conflict encounters are as to whether selection will favour whatever 'costs' may be involved in evolving a ploy to avoid or win the conflict. Conflicts are best generalized as conflicts of 'ways of life', or strategies, rather than as conflicts between individuals *per se*. Some conflicts of interest may resolve in favour of one of the parties through inability of the genetic system to 'represent' the other in the arms race. See **altruism**, **coevolution**, **LD_{50}**.

Arousal General causal term (and factor) invoked to account for the fact that animals are variably alert and responsive to potential stimuli. There may be a general 'sleeping/waking' difference; but it is less clear that there is a continuum of levels of awareness or responsiveness during either of these states. The phenomenology of arousal may be correlated with neural activity in the **reticular formation** of the medulla, hypothalamus and cortex of the vertebrate brain. Physiological processes which facilitate certain behaviours include hormone release and endogenous rhythms. Both may then be said to be arousal mechanisms, or to affect motivation.

Arrhenotoky See **male haploidy**.

Artery Any relatively large blood vessel carrying blood (not necessarily oxygenated) from the heart towards the tissues. Vertebrate arteries have thick elastic walls of smooth muscle and connective tissue (larger ones have capillaries in them), damping blood pressure changes. Their innermost layer is endothelium, as with all vertebrate blood vessels. They divide repeatedly to form arterioles. See **sclerosis**.

Arthropoda The largest phylum in the animal kingdom in terms of both number of taxa and biomass. Bilaterally symmetrical and metamerically segmented coelomates, with appendages on some or all segments (somites). A chitinous cuticle provides the exoskeleton, flexible to provide joints. Haemocoele is the main body cavity (coelom reduced). They lack true nephridia and cilia (onychophorans have the latter); with an annelid-like central nervous system and one or more pairs of coelomoducts acting as gonoducts or excretory ducts. Taxonomy varies. Thirteen classes are widely recognized, including: Onychophora (peripatids), Myriapoda (centipedes and millipedes), Insecta (insects), Trilobita (trilobites, extinct), Merostomata (king, or horseshoe crabs and extinct eurypterids), Arachnida (scorpions, spiders, harvestmen, solfugids, mites and ticks), Crustacea (crabs, prawns and shrimps, water-fleas). However, onychophorans are sometimes excluded from extant arthropods, while merostomatans and arachnids are often grouped into the **Chelicerata** (see Table 1). Some take the view that Crustacea, Insecta, Myriapoda and Chelicerata are phyla in their own rights within an athropod superphylum. See **biramous appendage** for diagram of limbs. The extent and patterning of **tagmosis** reflects the locomotory method, while appendages have proved marvellously adaptable and account in large measure for the success of the group. There appear to be three major evolutionary lineages: the *Onychophora-Myriapoda-Insecta* group, the *Merostomata-Arachnida-Trilobita* group, and the *Crustacea*. The phylum may be regarded as a **grade**, a polyphyletic origin not yet discounted.

Articular cartilage Cartilage providing the articulating surfaces of vertebrate joints.

Artificial insemination Artificial injection of semen into female reproductive tract. Much used in animal breeding.

Artificial key Any **identification key** not based upon evolutionary relationships but rather upon any convenient distinguishing characters. See **classification**.

Artificial selection Directional selec-

	Appendages			*Tagmosis*	*Other features*
	Antennae	Legs	Mouthparts		
Crustacea	2 pairs	*n* pairs, biramous	mandibles maxillae 1 maxillae 2	variable	nauplius larva
Insecta	1 pair	3 pairs, appear uniramous	mandibles maxillae 1 maxillae 2	3: head thorax abdomen	2 pairs wings (usually)
Myriapoda	1 pair	*n* pairs, appear uniramous	mandibles maxillae 1 (+ variable)	2: head trunk	
Chelicerata	0	4 pairs, appear uniramous	chelicerae (+ pedipalps)	2: prosoma opisthosoma (= cephalothorax, abdomen)	

Table 1. *A simple diagnosis of major extant arthropod groups.*

tion imposed by humans, deliberately or otherwise, upon wild or domesticated organisms. Crop plants originated in many cases from such deliberate crosses, sometimes involving one or more polyploid stocks. Procedures employed in harvesting these crops commonly involve unintentional (but still artificial) selection upon plants growing with the crops, favouring weed properties (see **weeds**). The phenomenon was well known to **Darwin** and examples of conscious human selection provided an analogy through which his readers could grasp the theory of **natural selection**. See **antibiotic**.

Artiodactyla Order of eutherian mammals. Ungulates (hoofed), with even number of toes upon which they walk. Includes pigs and hippopotamuses (four toes) and camels, deer, antelopes, etc. (two toes – the cloven-hoof). See **Perissodactyla**, **ruminant**.

Aschelminthes A phylum of pseudocoelomate animals, probably representing a grade rather than a clade. The body cavity is a persistent blastocoele. Many are worm-like, and most either aquatic, soil-dwelling or parasitic. They are bilaterally symmetrical, unsegmented, and frequently minute, composed of remarkably few cells in view of their anatomical complexity. The classes (often regarded as phyla) include: **Nematoda**, **Rotifera**, **Kinorhyncha**, **Gastrotricha** and **Nematomorpha**. See **pseudocoelom**.

Ascidian Sea squirt, or tunicate; member of Class Ascidiacea of Subphylum **Urochordata**.

Ascocarp Fruiting body of **Ascomy-**

cota; consists of tightly interwoven hyphae. Many are macroscopic and may be open and cup- or saucer-shaped (**apothecium**); closed and spherical (**cleistothecium**); or flask-shaped with a small pore through which the **ascospores** escape (**perithecium**). Asci are usually borne on the inner surface of the ascocarp.

Ascogenous hyphae Hyphae containing haploid 'male' and 'female' nuclei, which develop from ascogonia (the oogonia or female gametangia of the **Ascomycota**); eventually give rise to asci.

Ascogonium Oogonium or 'female' gametangium of the **Ascomycota**.

Ascomycota Formerly Ascomycotina (Ascomycetes). Division of the Kingdom **Fungi**. Terrestrial and aquatic fungi possessing septate hyphae whose septa are perforated; complete septa delineate reproductive structures (e.g. sporangia, gametangia). Cell walls comprise primarily chitin. Sexual reproduction involves the formation of a characteristic cell, the **ascus**; meiosis and spore formation take place in the ascus. Ascus formation takes place within a complex structure composed of tightly interwoven hyphae, the **ascocarp**. Asexual reproduction takes place through production of non-motile spores (conidia). Ascomycota comprise about 30,000 described species. Most of the blue-green, red and brown moulds that cause food spoilage are members of the Ascomycota, including *Neurospora*, which has played a central role in modern genetics. Many **yeasts** are also members of the Ascomycota, as are the delectable morels and truffles.

Ascorbic acid (vitamin C) A water-soluble sugar acid; a **vitamin** for a few arthropods and vertebrates (including humans), readily undergoing oxidation without loss of vitamin activity. Especially abundant in citrus fruits, potatoes and tomatoes. Hydrolysis of its lactone ring destroys its vitamin activity, as often occurs in cooking. Mild dietary deficiency affects connective tissue as it appears to be required for collagen formation; its complete dietary absence in humans results in symptoms of scurvy, with poor wound-healing.

Ascospore Fungal spore of **Ascomycota**.

Ascus In some **Fungi** (i.e. **Ascomycota**) a spherical, cylindrical or club-shaped cell in which fusion of haploid nuclei (**karyogamy**) occurs during sexual reproduction, followed by meiosis and formation of, usually, eight ascopores.

Aseptate Those algal filaments and fungal hyphae devoid of cross-walls (septa).

Asexual In non-zoological contexts, indicating reproduction involving spore production, but in which meiosis, gamete production, fertilization (leading to genome or nuclear union), transfer of genetic material between individuals and **parthenogenesis** do not occur. In this sense, therefore, not synonymous with vegetative reproduction (see **life cycle**). In zoological contexts, the stipulation that spores be produced does not apply; but the other criteria included above do. Often employed, with vegetative reproduction and parthenogenesis, as a means of rapidly increasing progeny output during a favourable period (these having practically uniform genotype);

hence common in internal parasites (see **polyembryony**). The basis of natural cloning (artificially imposed in the propagation of plants by cuttings). May alternate with sexual phase in **life cycle** (see **alternation of generations**). Some organisms (e.g. *Amoeba*, trypanosomes) are obligately asexual, and this raises questions about the evolutionary and ecological significance of **sex**.

A-site (Of ribosome) binding site on ribosome for charged (amino-acyl, hence A for acyl) tRNA molecule in **protein synthesis**. See **P-site**.

Assemblage (Of plants) collection of populations; merely a collective term and is best used when no attempt is made to define dominance. See **association**.

Assimilation Absorption of simple substances by an organism (i.e. across cell membranes) and their conversion into more complicated molecules which then become its constituents.

Association (Of plants) climax plant community dominated by a particular species and named according to them (e.g. oak–beech association of a deciduous forest). It can be defined as a segment of the **biocoenosis**. The association is an assemblage of species that recurs under comparable ecological conditions in different places; keystone of ecological studies, since it forms a recognizable entity which can be described and within which species are in intimate interaction.

Assortative mating (a. breeding) Non-random mating, involving selection of breeding partner, usually based on some aspect of its phenotype. This 'choice' (consciousness not implied) may be performed by either sex, and may be positively assortative (choice like self in some respect) or negatively assortative (disassortative, choice unlike self in some respect). Likely to have consequences for degree of inbreeding and maintenance of **polymorphism**. Assortative mating by size may be due to mate choice (large individuals choosing large mates is common), to mate availability or to physical mating constraints. Natural selection and sexual selection may be involved. Some **incompatibility** mechanisms in plants are analogies. See **sexual selection**.

Aster (1) (Zool.) A star-like perinuclear structure containing an **MTOC** and appearing as a system of cytoplasmic striations radiating from the centriole, comprising **microtubules**. Often conspicuous during cleavage of egg, or during fusion of nuclei at fertilization. Also probably present in many other animal cells during division. Absent from higher plants. (2) (Bot.) The common name applied to a family of flowering plants, the Asteraceae (Compositae), possessing evolutionarily specialized flowers. Individual flowers are subordinated to the overall display of the head, which functions as a single large flower in attracting pollinators. Individual flowers are epigynous, rather small, and possess an inferior ovary comprising two fused **carpels** within a single **ovule** in a locule. Stamens are reduced to five; usually fused to one another and to the **corolla**. Petals, also five, are fused to one another and to the ovary. Sepals absent or reduced to a series of bristles or scales (the pappus). Pappus often aids in seed dispersal by wind (e.g.

dandelion, *Taraxacum*); may be barbed so becoming attached to passing animals. In many members each flower head comprises two flower types: (a) *disk flowers* – make up the central portion of the head; (b) *ray flowers* – arranged on the outer periphery; often carpellate but sometimes completely sterile. Flower head matures over several days with individual flowers opening serially in a centripetal spiral. Very successful family with about 22,000 species; probably the largest family of flowering plants.

Asteroidea Class of **Echinodermata**. Starfishes. Star-shaped; arms, containing projections of gut, not sharply marked off from central part of body; mouth downwards; suckered **tube feet**; spines and pedicillariae. Carnivorous (some notoriously on oysters or corals).

Astrocyte One type of **glial cell** of central nervous system. Star-shaped, with numerous processes, they provide mechanical support for transmitting cells by twining round them and attaching them to their blood vessels. Different types are found in white and grey matter of the CNS.

Atherosclerosis See **sclerosis**.

Atlas First **vertebra**, modified for articulation with skull. Modified further in amniotes, which have freer head movement, than in amphibians. Consists of simple bony ring, while a peg (odontoid process) of the next vertebra (the **axis**) projects forward into the ring (through which the spinal cord also runs). This peg represents part of the atlas (its centrum) which has become detached and fused to the axis. Nodding the head takes place at the skull–atlas joint; rotation of head at atlas–axis joint.

ATP Adenosine triphosphate. Adenyl nucleotide diphosphate. The common 'energy currency' of all cells, whose hydrolysis accompanies and powers most cellular activity, be it mechanical, osmotic or chemical. Its two terminal phosphate groups have a more negative **standard free energy** of hydrolysis than phosphate compounds below it on the thermodynamic scale (e.g. sugar phosphates), and less negative than those higher (e.g. phosphocreatine, phosphoenolpyruvate), but this varies with intracellular concentrations of ATP, ADP and free phosphate as well as pH. A **high-energy phosphate** (see Fig. 57), it tends to lose its terminal phosphate to substances lower on the scale, provided an appropriate enzyme is present, and its mid-position on the scale enables it to serve as a common intermediate in the bulk of enzyme-mediated phosphate-group transfers in cells. Its relationship with ADP and AMP may be summarized:

$$\mathrm{ATP} + \mathrm{H_2O} \rightleftharpoons \mathrm{AMP} + \mathrm{PP_i} - 10\ \mathrm{kcal\ mol^{-1}}$$

$$\mathrm{ATP} + \mathrm{AMP} \rightleftharpoons \mathrm{ADP} + \mathrm{ADP}$$

$$\mathrm{ATP} \rightleftharpoons \mathrm{ADP} + \mathrm{P_i} - 7.3\ \mathrm{kcal\ mol^{-1}}$$

The energy values are ***standard free energy*** *changes at pH 7, standard temperature and pressure, at 25°C. 1 kcal = 4.184 kJ.*

Cells normally contain about ten times as much ATP as ADP and AMP, but when metabolically active the drop in the ATP/(ADP + AMP) ratio results in acceleration of **glycolysis** and aerobic respiration (see **respiration**), the signal being detected by **allosteric** enzymes in these pathways whose modulators (see **enzyme**) are ATP, ADP or AMP. ATP is not a reservoir of chemical energy in the cell but rather a transmitter or carrier of it. The bulk of ATP in eukaryotic cells is provided by mitochondria, where these are present. Some extra-mitochondrial ATP is produced anaerobically in the cytosol, and chloroplasts produce it but do not export it. ATP hydrolysis is used to transfer energy when work is done in cells.

ATPase activity is found in **myosin** (e.g. **muscle contraction**) and **dynein** (e.g. ciliary/flagellar beating). Membrane ion pumps (e.g. sodium and calcium pumps) and macromolecular syntheses of all kinds involve ATPase activity. Ultimately the energy source for ATP formation in the biosphere is solar energy trapped by autotrophs in photosynthesis – plus some lithotrophy. All heterotrophs depend upon respiratory oxidation of these organic compounds to power their own ATP synthesis. ATP is, like the other common nucleoside triphosphates in cells (CTP, GTP, TTP, UTP), a substrate in nucleic acid synthesis, and its hydrolysis provides the energy needed to build the resulting AMP monomer into the growing polynucleotide chain. These other triphosphates may participate in some other energy transfers; but ATP has by far the major role. See **AMP**, **ADP**, **phosphagen**, **bacteriorhodopsin**.

ATPase Any member of several families of enzymes bringing about either (i) orthophosphate (P_i) cleavage of ATP yielding ADP and inorganic phosphate, or (ii) pyrophosphate (PP_i) cleavage of ATP to yield AMP and pyrophosphate. The latter provides a greater decrease in free energy and is involved where a 'boost' is needed for an enzyme reaction. ATPase activity is found in cell motor molecules such as kinesins and dyneins (both microtubule-associated), myosins (actin-associated), and in choloroplast thylakoids and inner mitochondrial membranes (as ATP synthetases). See **mitochondrion**, **chloroplast**, **bacteriorhodopsin**.

Atrium (1) Chamber, closed except for a small pore, surrounding gill slits of Amphioxus and urochordates. (2) A type of heart chamber of vertebrate chordates synonymous with 'auricle'; receives blood from major vein and passes it to ventricle. Walls not as muscular as those of ventricle. Fishes have single atrium, but tetrapods, breathing mainly or entirely by lungs, have two: one (the left) receives oxygenated blood from lungs, the other (the right) receives deoxygenated blood from the body. Much of the blood flow though the atria is passive (see **heart cycle**). Non-chordates may have an atrial component of the heart, e.g. some polychaete worms and most molluscs, in which the term 'auricle' is sometimes preferred. (3) A space or cavity in some invertebrates (e.g. platyhelminths, some molluscs) known as the genital atrium, which houses the penis and/or opening of the vagina, and into which these may open.

Atrophy Diminution in size of a structure, or in the amount of tissue of part of the body. Generally involves destruction of cells, and may be under genetic and hormonal control, as is frequently the case in metamorphosis. May also result from starvation. Compare **hypertrophy**.

Atropine See **acetylcholine**, **alkaloids**.

Attenuation (Of pathogenic microorganisms) loss of virulence. May be achieved by heat treatment. See **vaccine**.

Auditory (otic) capsule Part of vertebrate skull, enclosing auditory organ.

Auditory nerve See **vestibulocochlear nerve**.

Auditory organ Sense organ detecting pressure waves in air ('sound'), in vertebrates represented by the cochlea of the inner ear, but the term often intended to include **vestibular apparatus** detecting positional and vectorial information as well. See **lateral line system**.

Auerbach's plexus That part of the autonomic nervous system in vertebrates (mostly from the vagus nerve) lying between the two main muscle layers of the gut and controlling its peristaltic movements.

Auricle (Zool.) (1) Often used synonymously with atrial heart chamber (see **atrium**). (2) External ear of vertebrates. (Bot.) Small ear- or claw-like appendage occurring one on each side at the bases of leaf-blades in certain plants.

Australian region Zoogeographical region consisting mainly of Australia, New Guinea and the Celebes; demarcated from south-east Asia by **Wallace's Line**.

Australopithecine Member of genus *Australopithecus*, now extinct; of the primate family Hominidae (see **hominid**). It appears to have been a long extant genus (4.4–1 Myr BP), with perhaps as many as six African species. *A. ramidus* (Awash, Ethiopia, at 4.4 Myr BP) is the oldest to date. *A. afarensis* (4.0–3.0 Myr BP) appears to have been a conservative species near the common ancestry of later forms; *A. aethiopicus* and *A. africanus* were later contemporaries (3–2.1 Myr BP) and possibly sister species with *A. afarensis* as common ancestor; *A. robustus* and *A. boisei* were later still (approx. 2–1.2 Myr BP) and shared several derived features (synapomorphies). *A. aethiopicus, A. robustus* and *A. boisei* were all 'robust' (megadontic) forms, with heavy skulls and facial features; but megadonty may have evolved in *A. aethiopicus* earlier than, and independently of, its appearance in the other two species. *A. africanus* had more 'gracile' features and may have been ancestral to *Homo*, although some cladistic analysis suggests it has strong links with neither *Homo* nor the robust australopithecines and was possibly too specialized to have been either ancestral to one of these clades or, as was once thought, the common ancestor of all later hominines (see ***Homo***, **Hominini**). One possible phylogeny, showing the hypothetical independent evolution of megadontic characters (darker shading), is illustrated below (Fig. 5). Cranial capacities of typical australopithecines were 400–600 cm^3 (modern humans average 1400 cm^3); **Broca's area** was seemingly absent. They had bipedal posture (and, from the footprints at Laetoli 3.6 Myr BP, were capable of bipedal gait). They are not generally thought to have been responsible for the stone choppers of the Oldowan tool culture of *c.* 1.75

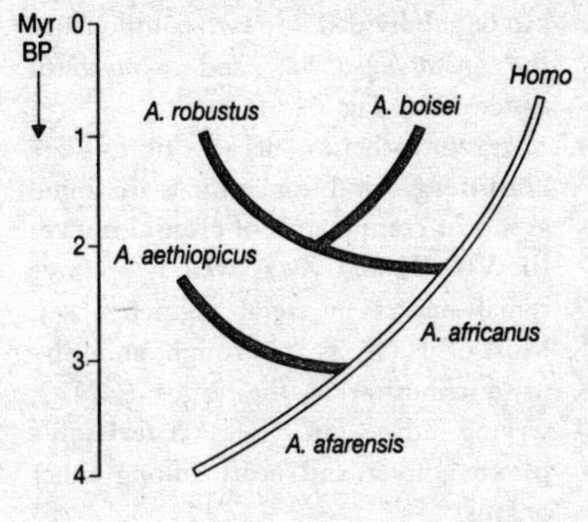

Fig. 5 *A possible phylogeny of early African hominines as implied by some cladistic analysts. The darker shading refers to megadontic lineages, the paler shading to more gracile lineages as indicated in the* **australopithecine** *entry*.

Myr BP (*Homo habilis* preferred). Physiologically, bipedalism would have reduced thermal stress and this in turn would have reduced water requirement; it may not have been an adaptation to locomotion in a savannah environment since modern chimpanzees are usually only bipedal when static and feeding under trees. *A. afarensis* may have been such a postural biped; hominid locomotory bipedalism may not have appeared until *Homo erectus* or *H. ergaster* (see ***Homo***). Further revisions of phylogenetic status are to be expected as new fossils and improved dating techniques occur. All fossil material currently comes from Africa.

Autapomorphy In **cladistics**, one of the character states uniquely defining a taxon.

Autecology Ecology of individual species, as opposed to communities (synecology).

Autoantigen Molecular component of organism, normally regarded as 'self' by its immune system, but here recognized as 'non-self' and eliciting an **autoimmune reaction**.

Autocatalyst Any molecule catalysing its own production. The more produced, the more catalyst there is for further production. Most likely, some such process was involved in the origin of pre-biological systems which, once enclosed in a membrane, might be called 'living'. The current process whereby nucleic acid codes for enzymes that decode and replicate it is, in a strong sense, autocatalytic. See **antibiotic**, **origin of life**.

Autochthonous (1) Of soil micro-organisms whose metabolism is relatively unaffected by increase in organic content of soil. See **zymogenous**. (2) The earliest inhabitant or product of a region (aboriginal). In this sense contrasted with *allochthonous* (not native to a region).

Autoclave Widely-used equipment for heat-sterilization. Commonly air is either pumped out prior to introduction of steam, or simply replaced by steam as the apparatus is heated under pressure. Material being sterilized is usually heated at 121°C and 138–172 kNm pressure for 12 minutes, which destroys vegetative bacteria, all bacterial spores and viruses; but these figures will vary with the size and nature of material.

Autoecious Of rust **Fungi** (**Basidiomycota**) having different spore forms during the life cycle, all produced upon one host species, as in mint rust. Compare **heteroecious**.

Autogamy Fusion of nuclei derived from the same zygote but from different meioses. Includes all forms of self-fertilization. See **automixis**.

Autograft Tissue grafted back onto the original donor. See **allograft**, **isograft**, **xenograft**.

Autoimmune reaction Response by an organism's immune system to molecules normally regarded as 'self' but which act as antigens. Quite often the thyroid gland, adrenal cortex or joints become damaged by the action there of antibodies or sensitized lymphocytes. Autoimmunity seems to arise from autoantibody production by defective (mutant) **B cells** rather than from any defect in helper **T cells**.

Autolysis (1) Self-dissolution that tissues undergo after death of their cells, or during metamorphosis or atrophy. Involves **lysosome** activity within cell. (2) Prokaryotic self-digestion, dependent upon enzymes of cell envelope.

Automixis Fusion of nuclei derived both from the same zygote and from the same meiosis. See **autogamy**, **parthenogenesis**.

Autonomic nervous system (ANS) Term sometimes referring to the entire vertebrate visceral nervous system, but more often restricted to the efferent (motor) part of it (the visceral motor system), supplying smooth muscles and glands. Neither anatomically nor physiologically autonomous from the central nervous system. Sometimes termed the 'involuntary' nervous system; but here again its effects (in humans) can largely be brought under conscious control with training. For convenience the ANS can be subdivided into two components: the *parasympathetic* and *sympathetic* systems. See Fig. 6.

Parasympathetic nerve fibres are **cholinergic** and in mammals are found as motor components of **cranial nerves** III, VII, IX and X, as well as of three spinal nerves in sacral segments 2–4. Most of its effects are brought about by its distribution in the vagus (CNX), serving the gut (see **Auerbach's plexus**), liver and heart among other organs.

Fibres of the sympathetic system originate within the spinal cord of the thoracic and lumbar segments, but beyond the vertebrae each departs from the cord and turns ventrally in a short white ramus (rami communicantes) to enter a sympathetic ganglion, a chain of which lies on either side of the mid-line. In the sympathetic ganglia many of the preganglionic fibres relay with postganglionic fibres that innervate target organs (e.g. mesenteries and gut); others pass straight through as splanchnic nerves and meet in plexi (collectively termed the *solar plexus*) beneath the lumbar vertebrae. From here postganglionic fibres innervate much of the gut, liver, kidneys and adrenals. Postganglionic fibres usually liberate catecholamines, particularly noradrenaline. Preganglionic parasympathetic fibres are relatively long and usually synapse in a ganglion on or near the effector, postganglionic fibres being short. In general ANS preganglionic fibres are myelinated, postganglionic unmyelinated and usually (there are exceptions) where the sympathetic stimulates, the parasympathetic inhibits, and vice versa; but organs are not always innervated by both. Both are, however, under central control, notably via the

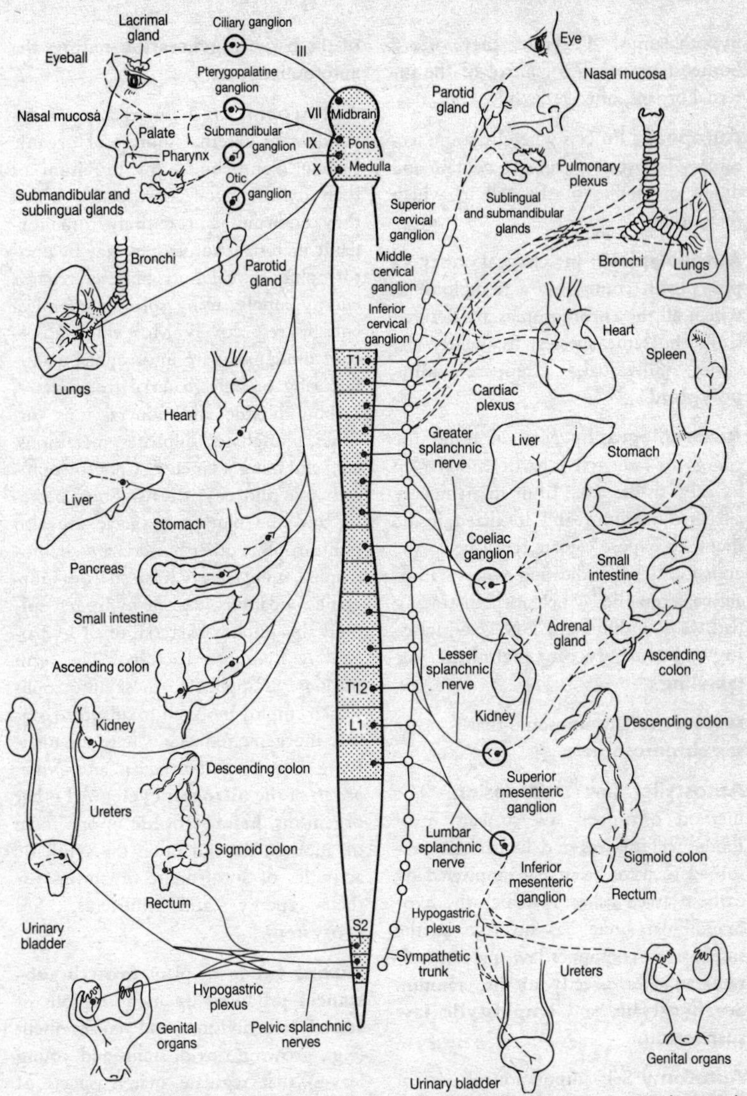

Fig. 6 *The human* **autonomic nervous system**. *Parasympathetic components are shown only on the left side, sympathetic components only on the right (in the body, both occur on both sides). Cranial nerve components have roman numerals. Thoracic and lumbar segments are indicated by T1–T12, L1, etc., and sacral segments by S2, etc. Preganglionic fibres = solid lines, postganglionic fibres = broken lines.*

hypothalamus. Together they afford homeostatic nervous control of the internal organs, often reflexly.

Autophagy Process whereby some secondary **lysosomes** come to contain and digest organelles of the cell in which they occur.

Autopolyploid In classical cases, a **polyploid** (commonly a tetraploid) in which all the chromosomes are derived from the same species, frequently the same individual. Compare **allopolyploid**.

Autoradiography Method using the energy of radioactive particles taken up by cells, tissues, etc., from an artificially enriched medium and localized inside them, to expose a plate sensitive to the emissions, thus indicating where radioactive atoms lie. Much used in tracing pathways within cells, **DNA sequencing** and **southern blot technique**. See **labelling**.

Autosome Chromosome that is not a **sex chromosome**.

Autostylic jaw suspension The method of upper jaw suspension of modern chimaeras and lungfishes; presumed to have been that employed by earliest jawed fishes, in which the hyomandibular bone has no role in the suspension. The upper jaw (palatoquadrate) attaches directly to the cranium. See **hyostylic**, and **amphistylic jaw suspension.**

Autotomy Self-amputation of part of the body. Some lizards can break off part of the tail when seized by a predator, muscular action snapping a vertebra. Both here and in many polychaete worms which can shed damaged parts of the body, **regeneration** restores the autotomized part.

Autotrophic Of those organisms independent of external sources of organic carbon (compounds) for provision of their own organic constituents, which they can manufacture entirely from inorganic material. Autotrophs may be phototrophic or chemotrophic as regards energy supply, using solar or chemical energies respectively. Most chlorophyll-containing plants are autotrophic, manufacturing organic material from water, carbon dioxide and mineral ions (nitrates, phosphates, sulphates, metal ions, etc.) and using solar energy phototrophically (see **photosynthesis**). Some bacteria and the blue-green algae are also phototrophic; other bacteria are chemotrophic, using energy released from inorganic oxidations (e.g. of hydrogen sulphide by sulphur bacteria, or of hydrogen by hydrogen bacteria). The term *lithotroph* is applied to those autotrophs which employ inorganic oxidants to oxidize inorganic reductants (e.g. the nitrifying bacteria *Nitrosomonas* and *Nitrobacter*, of the **nitrogen cycle**). All other organisms, **heterotrophic** in one form or another, depend upon the synthetic activities of autotrophic organisms for their energy and nutrients. See **ecosystem**.

Auxins Group of plant **growth substances** produced by many regions of active cell division and enlargement (e.g., growing tips of stems and young leaves) that regulate many aspects of plant growth. Promote growth by increasing the plasticity of the cell wall. When the cell wall softens, the cell enlarges because of turgor pressure. As the turgor pressure is reduced by cell expansion, the cell takes up more water, and the cell continues to enlarge until it encounters sufficient resistance from the wall. Cell wall softening results from changes in the cell's metabolism that causes a rapid pumping of protons across the plasmalemma, pumping H^+ ions out of the cell. The resulting acidification of the cell wall leads to hydrolysis of restraining bonds within the wall, and consequently, to cell elongation driven by the cell's turgor pressure. This is termed the *acid growth hypothesis*. But it cannot account for the continued effect of auxin treatment upon the plant. It appears that auxin has two different effects upon cell elongation, a rapid short-term effect caused by acid growth, and a second continued effect due to the regulation of gene expression. Auxin causes expression of at least ten specific genes, all presumably involved in the growth effect, the step affected by auxin being transcription: auxin's effect on gene expression in plants parallels that of several animal hormones. Auxins are transported basipetally in shoots at a rate of about 1 cm/hr^{-1}, act synergistically with **gibberellin** in stem elongation and with **cytokinin** in control of buds behind the apical bud (apical dominance). Effects of auxin on cell growth include curvature responses such as **geotropism** and **phototropism**. They may also have mitogenic effects, as in the initiation of cambial activity in association with cytokinins (and the differentiation of vascular tissue in association with sucrose), and adventitious root formation. Interactions between auxins and gibberellins determine rates of production of secondary phloem and secondary xylem. Auxins are implicated in flower initiation, sex determination, fruit growth (via production of auxin, or a factor promoting its synthesis, by pollen tube), delays in leaf fall and fruit drop.

Naturally occurring auxins include indole-3-acetic acid (IAA) and indole-3-acetonitrile (IAN). IAA has been isolated from such diverse sources as corn, endosperm, fungi, bacteria, human saliva and, the richest source, human urine. In addition to naturally occurring auxins, many substances with plant growth regulatory activity (*synthetic auxins*) have been developed. Some are used on a very large scale for regulating growth of agriculturally and horticulturally important plants, as in inhibition of sprouting in potato tubers, prevention of fruit drop in orchards, achievement of synchronous flowering (hence fruiting) in pineapple, in parthenocarpic fruit production (as in tomatoes, so avoiding risk of poor pollination). At increased, yet still relatively low, concentrations, auxins inhibit growth, sometimes resulting in death. Some synthetic auxins have differential toxicity in different plants: toxicity of 2,4-dichlorophenoxyacetic acid (2,4-D) to dicotyledonous and non-toxicity to monocotyledonous plants is perhaps the best known example, being exploited successfully in control of weeds in cereal crops and lawns. Naphthalenacetic acid, another synthetic auxin, is commonly employed to induce formation of adventitious roots in cuttings, and to reduce fruit drop in commercial crops.

Auxospore See **Bacillariophyta**.

Auxotroph Mutant strain of bacterium, fungus or alga requiring nutritional supplement to the **minimal medium**

upon which the wild-type strain can grow. See **prototroph**, **replica plating**.

Aves Birds. A class of vertebrates. Feathered **archosaurs** whose earliest known fossil, *Archaeopteryx*, was of upper Jurassic date and the sole known representative of the Subclass Archaeornithes. All other known birds (including fossils) belong to the Subclass Neornithes. The two superorders with living representatives are the Palaeognathae (ratites) and Neognathae (20 major orders; about half the 2900 living species, including songbirds, belonging to the Order Passeriformes, or 'perching' birds). Distinctive features include: **feathers**; furcula (**wishbone**); forelimbs developed as wings. Bipedal and homeothermic, laying cleidoic eggs and (excluding *Archaeopteryx* and two other fossil genera) lacking teeth, but with the skin of the jaw margins cornified to form a beak (bill), whose diversity of form is in large part responsible for the Cretaceous, and subsequent, adaptive radiation of the group.

Awn (arista) Stiff, bristle-like appendage occurring frequently on the flowering glumes of grasses and cereals.

Axenic (Of cultures of organisms) a pure culture.

Axial skeleton See **endoskeleton**.

Axial system In secondary xylem and secondary phloem, collective term for those cells originating from fusiform cambial initials. Long axes of these cells are orientated parallel with the main axis of the root or stem.

Axil (Of a leaf) the angle between its upper side and the stem on which it is borne; the normal position for lateral (axillary) buds.

Axillary Term used to describe buds, etc., occurring in the **axil** of a leaf.

Axis (1) Embryonic axis of animals. There are generally three such: antero-posterior, dorso-ventral and medio-lateral, established very early in development, and sometimes by the **polarity** of the egg. The genetics of early development is under active study. See **homeotic**, **compartment**. (2) Second amniote **vertebra**, modified for supporting the head. See **atlas**.

Axon The long process which grows out from the cell bodies of some neurones towards a specific target with which it connects and carries impulses away from the cell body. See **neurone**, **nerve fibre**.

Axoneme Complex microtubular core of **cilium** and flagellum. Contains actin in *Chlamydomonas*.

Axopod **Pseudopodium** of some sarcomastigophoran protozoans in which there is a thin skeletal rod of siliceous material upon which streaming of the cytoplasm occurs. They may bend to enclose prey items.

Bacillariophyta (diatoms) Microscopic protistans (*c.* 10,000 spp.), accounting for *c.* 20% of all primary production. Unicellular, eukaryotic algae, occurring singly or grouped in colonies; pigmented and photosynthetic, containing in their chloroplasts chlorophylls *a*, c_1 and c_2 plus the major carotenoid fucoxanthin, which gives the cells their characteristic brown colour. Plastid structure similar to that of the **Phaeophyta**, **Chrysophyta**, **Raphidophyta**, and **Xanthophyta**, but often larger and more elaborate. Oil droplets, reserve polysaccharides (chyrsolaminarin) and other inclusions occur within the cytoplasm, often visibly. Some species can live heterotrophically in the dark, if there is a source of organic carbon. Less than ten species are obligate heterotrophs, all colourless (apochlorotic), and in two genera (*Nitzschia*, *Hantzschia*).

The diatom cell is surrounded by a characteristic and highly differentiated cell wall, almost always heavily impregnated with silica ($SiO_2.nH_2O$), whose uptake and deposition require less energy than formation of an equivalent organic wall. The diatom cell wall is multipartite, comprising two large intricately sculptured halves (*valves*) together with several thinner linking structures (*girdle elements or cincture*) and mostly comprises quartzite or hydrated silica. The valves lie at the end of the cell, and the girdle elements surround the region in between. Wall components (the *frustule*) fit together closely, so movement of materials occurs mainly via pores or slits in it. Probably all diatoms secrete polysaccharides, some of which may diffuse into the surrounding medium while some remain as a gelled capsule around the cell or as threads, pads, stalks or tubes.

Each frustule possesses one valve formed just after the last cell division, and an older valve which may have existed for several generations since its own formation.

Some diatoms form thick ornamented *resting spores*, usually in response to stress; others form *resting cells*, identical morphologically to vegetative cells but possessing no protective layer. Auxospores represent another way for re-establishing the original size of a cell; they result from sexual reproduction, which can involve uniflagellate male and non-motile female gametes, or the fusion of amoeboid gametes.

Many diatoms, particularly those possessing a *raphe* (longitudinal slit in the valve of some diatoms) or labiate process (appendage at a valve's periphery through which mucilage is secreted) can move over a substratum. Movement is through secretion of material from the raphe or labiate process; the motile force being generated by interaction between actin filaments and transmembrane structures, which are free to move within the cell and raphe but fixed to the substratum at their distal ends. The transmembrane structure probably includes ATPase and a protein able to make translational movements with the plasmalemma. The transmembrane structure is itself connected to filaments of acid mucopolysaccharide which can become attached to the substratum at their distal ends. Thus, transmembrane structures are moved along the raphe by their interactions with actin filament bundles beneath, and the cell moves relative to the substratum. Epipelic and epipsammic diatoms exhibit endogenous vertical migration rhythms.

Diatoms are abundant in both the benthos and plankton of both fresh and marine waters. Their fossil record extends back to the mid-Cretaceous. Past deposits of countless numbers of diatom frustules have formed siliceous or diatomaceous earths; industrial uses of such earths are diverse (e.g. as a mild abrasive in toothpaste and metal polishes, as an absorbent for nitroglycerin in making dynamite, and in the filtration of liquids, especially in sugar refineries).

Diatoms are important and ecologically sensitive microfossils, being used extensively in palaeolimnology, enabling both qualitative and quantitative reconstructions of past lake histories, as well as palaeoclimates, pH values, nutrient loading and palaeosalinity; very important microfossils in determining the responses of lakes to acid deposition, and anthropogenic effects upon lakes.

Bacillus General term for any rod-shaped bacterium. Also a genus of bacteria: *Bacillus*.

Backbone See **vertebral column**.

Backcross Cross (mating) between a parent and one of its offspring. Employed in **chromosome mapping**, when the parent is homozygous and recessive for at least two character traits, and to ascertain genotype of offspring (i.e. whether homozygous or heterozygous for a character), the parent used being the homozygous recessive. Where the organism of known genotype in the cross is not a parent of the other, the term *testcross* is often used.

Bacteria Unicellular, filamentous and mycelial **prokaryotes**, of the Kingdom Monera. Among the simplest of all known organisms (see **cell** for diagram). Opinions differ on whether to include the blue-green algae within bacteria (see **Cyanobacteria**). Work on **Archaebacteria** suggests that these form a distinctive side-branch. It is not known how many bacterial species exist, not even the correct order of magnitude, since bacteria cannot be differentiated under the microscope; but a phylogenetic system based on 5S rRNA sequences is being developed and is a useful tool in estimating bacterial species richness. The description which follows relates to 'true' bacteria, or **Eubacteriales**, which vary greatly in shape, being rod-like (bacilli), spherical (cocci), more or less spiral (spirilli), filamentous and occasionally mycelial. Multiplication is by simple fission; other forms of asexual reproduction, e.g. production of aerially dispersed spores or flagellated swarmers, occur in some bacteria. As prokaryotes, they have no meiosis or syngamy, but genetic recombination occurs in many of them (see **F factor**, **pili**, **plasmid**, **recombination**).

Bacteria are ubiquitous, occurring in a large variety of habitats. 1 g of soil may contain from a few thousand to several million; 1 cm^3 of sour milk, many millions. Most are saprotrophs or parasites;

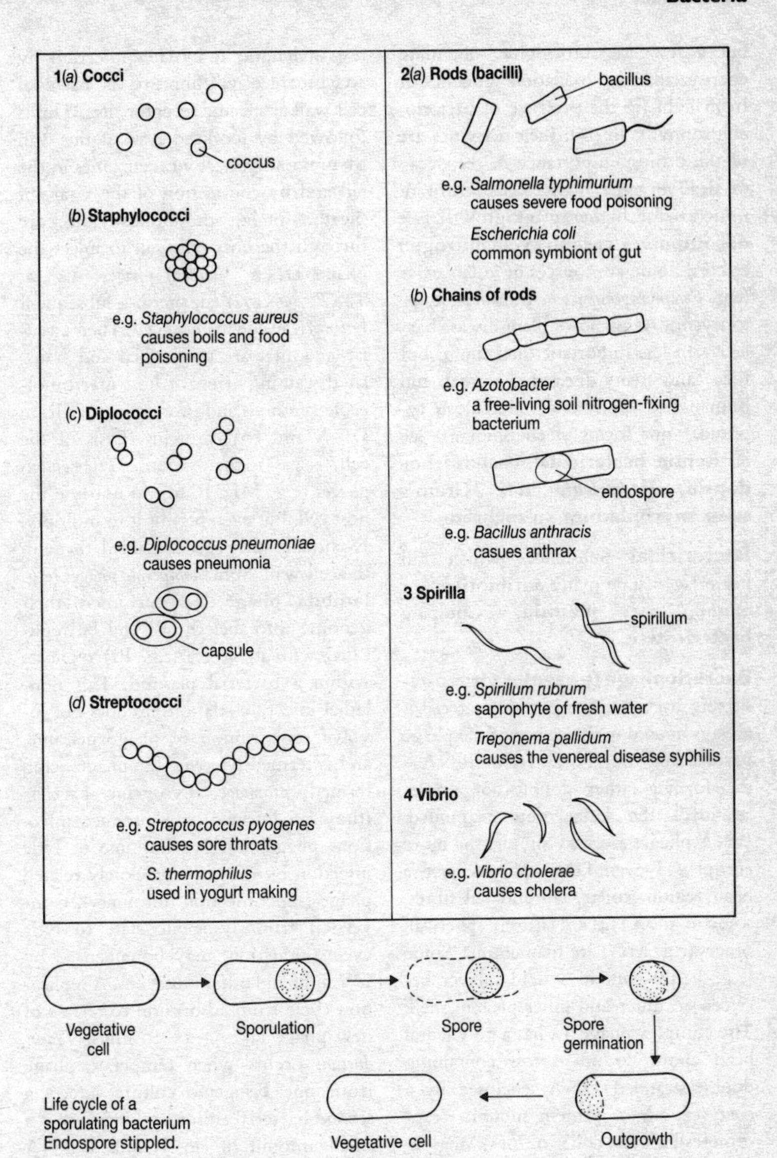

Fig. 7 *Some bacterial forms and diseases caused by particular species in humans.*

but a few are autotrophic, obtaining energy either by oxidation processes or from light (in the presence of bacteriochlorophyll). In soil, their activities are of the utmost importance in the decay of dead organic matter and return of minerals for higher plant growth (see **decomposers**, **carbon cycle**, **nitrogen cycle**). Some are sources of **antibiotics** (e.g. *Streptomyces griseus* produces streptomycin). As agents of plant disease bacteria are less important than fungi; but they cause many diseases in animals and humans (e.g. diphtheria, tuberculosis, typhoid, some forms of pneumonia). See **antiseptic**, **bactericidal**, **bacteriorhodopsin**, ***Escherichia coli***, **Gram's stain**, **mycoplasmas**, **spirochaetes**.

Bactericidal Substances which kill bacteria. Include many **antibiotics**. For *colicins*, see **plasmid**. Compare **bacteriostatic**.

Bacteriophage (phage) A **virus** parasitizing bacteria. The genetic material is always housed in the centre of the phage particle and may be DNA or RNA – the former either double- or single-stranded, the latter double-stranded. RNA phages are very simple; the more complex T-even DNA phages have a head region, collar, tail and tail fibres; some (e.g. φX174) are entirely spherical; others (e.g. M13) are filamentous. Some (e.g. PM2) contain a lipid bilayer between an outer and inner protein shell. The complex phage T4 has a polyhedral head about 70 nm across containing double-stranded DNA, enclosed by a coat (capsid) of protein subunits (capsomeres); a short collar or 'neck' region; a cylindrical tail (or tail sheath) region and six tail fibres. The particle (virion) of T4 phage is quite large – about 300 nm in length. It initiates infection by attachment of tail fibres to the bacterial cell wall at specific receptor sites. This is followed by localized lysis of the wall by previously inert viral enzymes in the tail and by contraction of the viral tail sheath, forcing the hollow tail core through the host cell wall to inject the phage DNA. *Virulent phages* (such as T4, T7, φX174) engage in a subsequent *lytic cycle* inside the host cell. Their circular genomes are transcribed and translated, causing arrest of host macromolecule synthesis and production of virion DNA and coat proteins. Lysis of the cell releases the virions. *Filamentous phages* (e.g. MI3, fl, fd) do not lyse the host cell, but even permit it to multiply. Eventually they get extruded through the cell wall. Some *temperate phages* (e.g. **lambda phage**, λ) can insert their genome into their host's and be replicated with it; others (e.g. P1) replicate within a bacterial plasmid. This non-lethal infective relationship does not involve transcription of phage genome, and is termed *lysogeny*, the phage being termed *prophage*. Lysogenic bacteria (those so infected) can produce infectious phage, but are immune to lytic infection by the same or closely related phage (superinfection immunity). Conversion from lysogenic state to lytic cycle (induction) may be enhanced by UV light and other mutagens. A replication cycle from adsorption to release of new phage takes $\simeq$ 15–20 mins. *Transduction* occurs when temperate phage from one lysogenic culture infects a second bacterial culture, taking with it a small amount of closely-linked DNA which remains as a stable feature of the recipient cell. Antibiotic resistance can be transferred this way (see **antibiotic**

resistance element). See **phage conversion**, **phage restriction**.

Bacteriorhodopsin Conjugated **receptor** protein of the 'purple membrane' of the photosynthetic bacterium *Halobacterium halobium*, forming a proton channel composed of seven α-helices, each spanning the membrane once, and whose prosthetic group (retinal) is light-absorbing and identical to that of vertebrate rod cells. Allosteric change on illumination of the pigment results in proton ejection from the cell, the resulting proton gradient being used to power ATP synthesis. Its integration into **liposomes** along with mitochondrial ATP synthetase showed that the latter is also a proton channel, driven by a proton gradient. Bacteriorhodopsin absorbs strongly at about 570 nm (in the green region of the spectrum), and if chlorophyll-based photosynthesis evolved after a rhodopsin-based variety then absence of green-absorption in chlorophyll's action spectrum may be due to selection in favour of a pigment avoiding competition with abundant rhodopsin-based forms. See **cell membranes**, **chloroplast**, **electron transport system**, **mitochondrion**, **rod cell**.

Bacteriostatic Inhibiting growth of bacteria, but not killing them. See **bactericidal**.

Balanced lethal system Genetic system operating when the two homozygotes at a locus represented by two alternative alleles each produce a lethal phenotype, yet the two alleles persist in the population through survival of heterozygotes, which thus effectively breed true. Compare **heterozygous advantage**.

Balance of nature Phrase glossing observations that in natural ecosystems, communities and the biosphere at large, herbivores do not generally overgraze, predators do not generally over-predate nor parasites decimate host populations; that populations of all appear to be held roughly in equilibrium, and that drastic (sometimes trivial) disturbance of this harmony between organisms and the physical environment will have inevitable and generally unfavourable consequences for mankind. Causal processes involved in the complex systems with which ecologists deal are increasingly amenable to computer simulation. See **chaos**, **density-dependence**, **homeostasis**.

Balbiani ring See **puff**.

Baldwin effect Reinforcement or replacement of individually acquired adaptive responses to altered environmental conditions, through selection (artificial or natural) for genetically determined characters with similar functions. Not regarded as evidence of Lamarckism. See **genetic assimilation**, **Lamarck**.

Barbs See **feather**.

Barbules See **feather**.

Bark Protective corky tissue of dead cells, present on the outside of older stems and roots of woody plants, e.g. tree trunks. Produced by activity of cork cambium. Bark may consist of cork only or, when other layers of cork are formed at successively deeper levels, of alternating layers of cork and dead cortex or phloem tissue (when it is known as *rhytidome*). Popularly regarded as everything outside the wood.

Baroreceptor (pressoreceptor) Receptor for hydrostatic pressure of blood. In man and most tetrapods, located in carotid sinuses, aortic arch and wall of the right atrium. Basically a kind of stretch receptor. When stimulated, those in the first two locations activate the cardio-inhibitory centre and inhibit the **cardio-acceleratory centre**, while those in the atrium stimulate the cardio-acceleratory centre, helping to regulate blood pressure.

Barr body Heterochromatic X-chromosome occurring in female placental mammals. Paternally- or maternally-derived X-chromosomes may behave in this way, often a different X-chromosome in different cell lineages. See **heterochromatin**.

Basal body Structure indistinguishable from centriole, acting as an organizing centre (nucleation site) for eukaryotic cilia and flagella, unlike which its 'axoneme' is a ring of nine triplet microtubules, each comprising one complete microtubule fused to two incomplete ones. It is a permanent feature at the base of each such flagellum or **cilium**. See **centriole** for details.

Basal lamina Thin layer of several proteins, notably collagen and the glycoprotein laminin, about 50–80 nm thick, varying in composition from tissue to tissue. Underlying and secreted by sheets of animal epithelial cells and tubes they may form (e.g. many glands, and endothelial linings of blood vessels). In kidney glomerulus and lung alveolus, may be an important selective filter of molecules between cell sheets. See **basement membrane**.

Basal metabolic rate (BMR) The respiratory rate of a resting animal, normally measured by oxygen demand. The 'background' respiration rate, as required for unavoidable muscle contractions (e.g. heart), growth, repair, temperature maintenance, etc. See **thyroid hormones**.

Basal placentation (Bot.) Condition in which ovules are attached to the bottom of the locule in the ovary.

Base Either a substance releasing hydroxyl ions (OH^-) upon dissociation, with a pH in solution greater than 7, or a substance capable of acting as a proton acceptor. In this latter sense, the nitrogenous cyclic or heterocyclic groups combined with ribose to form nucleosides are termed bases. See **purine**, **pyrimidine**.

Basement membrane Combination of **basal lamina** with underlying reticular fibres and additional glycoproteins, situated between many animal epithelia and connective tissue.

Base pairing Hydrogen bonding between appropriate purine and pyrimidine bases of (antiparallel) nucleic acid sequences, as during DNA synthesis, mRNA transcription and during translation (see **protein synthesis**). If two DNA strands align then adenine in one strand pairs with a thymine and a guanine with a cytosine (i.e. A:T, G:C); but if a DNA strand aligns with an RNA strand, then adenines in the DNA pair align with uracils in the RNA (i.e. A:U, G:C). Without these 'rules' there could be no **genetic code** or heredity as we know it. See **DNA hybridization**.

Base ratio The (A + T):(G + C)

ratio in double-stranded (duplex) DNA. It varies widely between different sources (i.e. from different species). The amount of adenine equals the amount of thymine; the amount of guanine equals the amount of cytosine. See **base pairing**.

Basic dyes Dyes consisting of a basic organic grouping of atoms (cation) which is the actively staining part, usually combined with an inorganic acid. Nucleic acids, hence nuclei, are stained by them and are consequently referred to as **basophilic**.

Basidiomycota Formerly Basidiomycotina (Basidiomycetes). Fungal division comprising a diverse variety of forms (e.g. jelly fungi, bracket fungi, mushrooms, toadstools, stinkhorns, puffballs, smuts and rusts). Most such common names refer to the visible part of the fungus, the conspicuous reproductive or fruiting body (basidiocarp) supported nutritionally by an extensive assimilative mycelium that penetrates the plant or soil and derives nutrients. Basidiomycota are characterized by production of **basidiospores** borne outside on a club-shaped spore-producing structure, the **basidium**. Mycelium is always septate; when a spore germinates, a primary mycelium is formed, which is initially multinucleate but septa soon form such that the mycelium becomes divided into monokaryotic (uninucleate) cells. A secondary mycelium is produced by fusion of primary hyphae from the same (homokaryotic) or different mating types (heterokaryotic) resulting in the formation of a **dikaryotic** (binucleate) mycelium. Apical cells of secondary mycelium usually divide by formation of highly characteristic **clamp connections**. Basidiomycota are components of most ectomycorrhizal associations.

Some 25,000 species of Basidiomycota have been described, the best known being *Agaricus campestris* (field mushroom) and the commercially important *Agaricus bisporus*, which together with the oriental shiitake mushroom, *Lentinus edodes*, comprise about 80% of the global crop. Other mushrooms are highly poisonous (e.g. *Amanita*), while others contain hallucinogenic compounds similar to LSD. Rusts and smuts are unlike other Basidiomycota since they do not form basidiocarps; they are of tremendous economic importance, because they cause damage to many crop plants throughout the world (e.g. *Puccinia graminis*, black stem rust of wheat).

Basidiospore Characteristic spore produced by the **Basidiomycota**; produced within a **basidium** by meiosis and borne on its outside.

Basidium Specialized reproductive cell of **Basidiomycota**; often club-shaped, cylindrical or divided into four cells. Nuclear fusion and meiosis occur within the basidium, resulting in the formation of four **basidiospores**, borne externally on minute stalks or sterigmata.

Basipetal (Bot.) (Of organs.) Developing in succession towards the base, oldest at the apex, youngest at the base. Also used of the direction of transport of substances within a plant: away from apex. Compare **acropetal**.

Basophil Blood **polymorph**. Very similar to **mast cell** in structure and probably function, but with peroxidase rather than acid and alkaline phosphatase activity in its granules.

Basophilic Staining strongly with basic dye. Especially characteristic of nucleic acids, and hence of nucleus (during division of which the condensed chromosomes are strongly basophilic), and of cytoplasm when actively synthesizing proteins (due to rRNA and mRNA).

B cell (B lymphocyte) A **leucocyte**, derived from a **lymphoid tissue** stem cell which has migrated from foetal liver to bone marrow and has not entered the thymus but has settled either in a lymph node or in the spleen. B cells express a specific immunoglobulin (Ig, or **antibody**) on their plasma membranes. This can bind appropriate antigen, when the cell becomes activated to divide repeatedly by mitosis and produce a clone of specific antibody-coated cells. This primary immune response (see **immunity**) is also characterized by their secretion of specific IgM antibody into the blood, the combined effect being to remove antigen. Some B cells do not greatly participate in Ig production, but circulate in the body and may persist for years as memory cells, capable of clonal expansion and rapid Ig secretion (a secondary immune response) if activated by the initial antigen. This provides immunological memory. Memory B cell differentiation involves somatic **hypermutation** in antibody variable region (*V*) genes (see **antibody diversity**) and selection of those cells expressing high-affinity variants of this antigen receptor. Still other B cells mature into plasma cells (the major Ig-producer in a secondary response) after multiplication, and although these progeny cells may have Igs of more than one class, they all have the same antigen specificity. A fully mature plasma cell will have little surface Ig but will be secreting an Ig of one class and of one antigen specificity. See **IgA-IgM, myeloid tissue**, **T cell**.

B-chromosome See **supernumerary chromosome**.

Belt desmosome See **desmosome**.

Benedict's test A modification of Fehling's test for sugar using just one solution. Contains sodium citrate, sodium carbonate and copper sulphate dissolved in water in the ratio 1.7: 1.0:0.17 g:30 cm^3 water. Five drops of test solution are added to 2 cm^3 Benedict's solution. If a **reducing sugar** is present then a rust-brown cuprous oxide precipitate forms on boiling.

Bennettitales See **Cycadophyta**.

Benthos General term originally introduced to describe those animals found living on the sea bottom; now used to describe organisms of aquatic ecosystems that live associated with a substratum (crawling or burrowing there, or attached to rocks as with seaweeds and sessile animals). Many schemes have been proposed for subdividing the benthos. With respect to higher aquatic plants and algae, the benthos can be subdivided into four categories which have characteristic communities. (1) **rhizobenthos**, vegetation rooted via roots, rhizomes or rhizoids into the submerged sediments, e.g. emergent and submersed higher plants such as *Typha, Scirpus, Potamogeton, Nuphar*, and the algae *Chara* and *Nitella*; (2) Haptobenthos, communities of algae attached to a solid substratum, e.g. epilithon, attached to rocks and stones, **epiphyton**, attached to plant surfaces, **epi-**

zoon, attached to animals, and **epipsammon**, attached to sand grains; (3) Herpobenthos, algae living on and or moving through the very top few mm of submerged sediment, mostly motile but also some non-motile form, e.g. **epipelon**; (4) endobenthos, living within, and often boring into, a substratum, e.g. endolithon, within pores of rocks, **endopelon**, within sediments, **endozoon**, within animals, and **endophyton**, within plants. Organisms feeding primarily upon the benthos are termed benthophagous. See **pelagic**, **nekton**, **plankton**.

Bergmann's rule States that in geographically variable species of **homeothermic** animals, body size tends to be larger in cooler regions of a species range. See **Allen's rule**.

Berry Many-seeded succulent fruit, in which the wall (*pericarp*) consists of an outer skin (*epicarp*) enclosing a thick fleshy *mesocarp* and inner membranous *endocarp*, as in gooseberry, currant, tomato. Compare **drupe**.

Beta blocker Substance, such as the drug propanolol, selectively blocking **beta receptors**. Clinical use is to slow heart rate and lower blood pressure. See **opiates**.

Beta cells See **pancreas**.

Beta-globulins A class of vertebrate plasma proteins including certain lipoproteins, **transferrin** and plasminogen (precursor of fibrinolysin, see **fibrinolysis**).

Beta-oxidation (**β-oxidation**) An iterative series of reactions shortening a fatty acid chain by two carbon atoms at a time. Has been shown to occur in **peroxisomes**. See **fatty acid oxidation**.

Beta receptor **Adrenergic** receptor sites (associated with adenylate cyclase in appropriate postsynaptic membranes, but not identical with it) binding preferentially to adrenaline rather than noradrenaline. Heart muscle has beta receptors predominantly which result in increased heart rate and blood pressure when stimulated; other beta effects include dilation of arterioles supplying skeletal muscle, bronchial relaxation and relaxation of the uterus. All these effects are mimicked by the synthetic drug isoprenaline. See **alpha receptor**, **beta blocker**, **autonomic nervous system**.

Beta-sheets (beta-conformation) One type of protein secondary structure. See **protein**, **amino acid**.

B-form helix (B DNA, DNA-B) The (paracrystalline) form of **DNA** believed to be adopted most frequently in living cells, as in aqueous media generally. Compared with the less common **A-form helix**, the molecule is more hydrated (hence less stable and compact) and the axis of the molecule passes through the hydrogen bonds of the base pairs, producing a wider minor groove. See **Z-form helix**.

***bicoid* gene** (*bcd*). A pattern-specifying **maternal gene** in *Drosophila*, whose mRNA transcript passes from maternal ovary nurse cells to anterior poles of developing oocytes where it becomes localized, apparently trapped by the oocyte cytoskeleton. Its eventual protein product is a transcription activator (with a DNA-binding homeodomain), is located in nuclei of the syncytial blastoderm, and is translocated half-way along

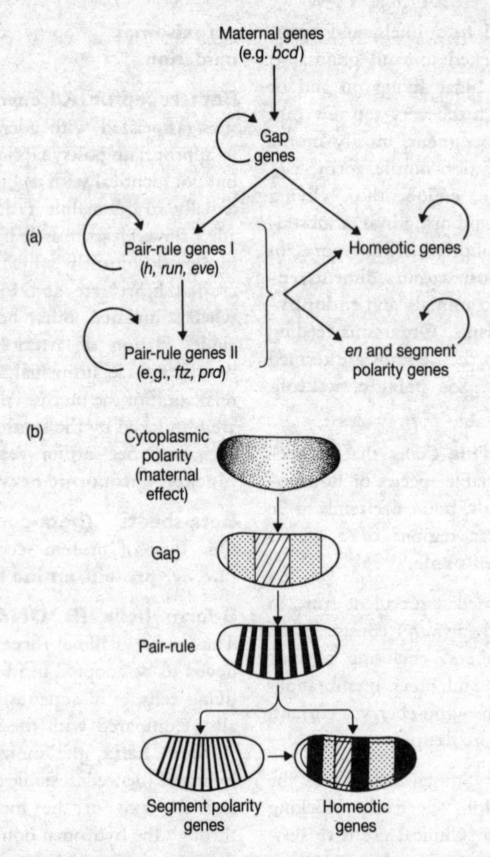

Fig. 8 (a) *Proposed cascade of segmentation gene activity in* Drosophila *and* (b) *the pattern formation in the fly resulting from this. The pattern is established by maternal effect genes which set up morphogenic gradients of proteins as described in the entry for* **gap genes**. *Italicized acronyms refer to the relevant genes. See also* **pair-rule genes**.

the embryo from the anterior pole. With a similar translocation of nanos gene product (see ***oskar* gene**) in the opposite direction, two opposing gradients provide quantitative **positional information** which the embryo genome converts into qualitative phenotypic differences. Genes involved in this conversion process, i.e. *Krüppel* (*Kr*), *hunchback* (*hb*) and *knirps* (*kni*), belong to the **gap gene** class, and all three encode proteins with DNA-binding **finger domains**. Bicoid product is responsible for production of a normal head and thorax, structures absent in embryos from females lacking functional *bcd*, and both it and the oskar product are examples of **morphogens**.

By repressing transcription of *Kr* in anterior and posterior embryo regions, bicoid and oskar morphogens only permit *Kr* expression in mid-embryo. High levels of *bcd* product activate *hb* transcription, while low levels permit *Kr* transcription. However, their abilities to elicit such zygotic (embryo) gene expression depend upon the affinities of their respective DNA-binding sites for them, and these are known to vary. The *hb* and *Kr* products are mutual repressors of each others' transcription, so that their 'domains' of product expression become stable and restricted. These *gap gene domains* delimit the domains of **homeotic gene** expression but also enable position-specific regulation of the **pair-rule genes**, whose expression leads to metameric segmentation of the embryo.

Bicollateral bundle See **vascular bundle**.

Biennial Plant requiring two years to complete life cycle, from seed germination to seed production and death. In the first season, biennials store up food which is used in the second season to produce flowers and seed. Examples include carrot and cabbage. Compare **annual**, **ephemeral**.

Bilateral symmetry Property of most metazoans, having just one plane in which they can be separated into two halves, approximately mirror images of one another. This is usually the antero-posterior and dorso-ventral plane, separating left and right halves. Major metazoan phyla excluded are coelenterates and echinoderms (having radial symmetry as adults). In flowers, the condition is termed zygomorphy.

Bile Fluid produced by vertebrate liver cells (hepatocytes), containing both secretory and excretory products, and passed through bile duct to duodenum. Contains (i) **bile salts** (e.g. sodium taurocholate and glycocholate) which emulsify fat, increasing its surface area for lipolytic activity. Also form micelles for transport of sterols and unsaturated fatty acids towards intestinal villi (see **chylomicron**); (ii) bile pigments, breakdown products of haemoglobin such as bilirubin, which are true excretory wastes and colour faeces; (iii) **lecithin** and **cholesterol** as excretory products. Bile is aqueous and alkaline due to $NaHCO_3$, providing a suitable pH for pancreatic and subsequent enzymes. Stored in gall bladder. See **micelle**.

Bile duct Duct from liver to duodenum of vertebrates, conveying **bile**. See **gall bladder**.

Bile salts Conjugated compounds of bile acids (derivatives of **cholesterol** and taurine and glycine), forming up to two thirds of dry mass of hepatic **bile**.

Bilharzia Schistosomiasis. See ***Schistosoma***.

Binary fission Vegetative reproduction occurring when a single cell divides into two equal parts. Compare **multiple fission**.

Binocular vision Type of vision occurring in primates and many other active, predatory vertebrates; eyeballs can be so directed that an image of an object falls on both retinas. Extent to which eyes converge to bring images on to the foveas of each retina gives proprioceptive information needed in judging distance of object. Stereoscopic vision (perception of shape in depth) depends on two slightly different images of an object being received in binocular vision, the eyes viewing from different angles.

Binomial nomenclature The existing method of naming organisms scientifically and the lasting contribution to taxonomy of **Linnaeus**. Each newly described organism (usually in a paper published in a recognized scientific journal) is placed in a genus, which gives it the first of its two (italicized) Latin names – its generic name – and is always given a capital first letter. Thus the genus *Canis* would probably be given to any new placental wolf discovered. Within this genus there may or may not be other species already described; in any case, the new species will receive a second (specific) name to follow its generic name, this time with a lower case first letter. The wolf found in parts of Europe, Asia and North America belongs to the species *Canis lupus*; jackals, coyotes and true dogs (including the domesticated dog) also belong to the genus *Canis*, but each species is further identified by a different specific name: the domestic dog, in all its varieties, is *C. familiaris* (often only the first letter of the genus is given, if it is contextually clear what the genus is). If a previously described species is subsequently moved into another genus, it takes the new generic name, but carries the old specific name. The author who published the original description of a species often receives credit in the form of an abbreviation of their name after the initial mention of the species in a paper or article. Thus one might see *Canis lupus* Linn., since Linnaeus first described this species. The specimen upon which the initial published description was based is termed the type specimen, or 'type', and is probably housed in a museum for comparison with other specimens. If sufficient variation exists within a species range for **subspecies** or varieties to be recognized, then a trinomial is employed to identify the subspecies or variety. The British wren rejoices under the trinomial *Troglodytes troglodytes troglodytes*.

Bioassay Quantitative estimation of biologically active substances by measurement of their activity in standardized conditions on living organisms or their parts. A 'standard curve' is first produced, relating the response of the tissue or organism to known quantities of the substance. From this the amount giving a particular experimental response can be read. See **biosensor**.

Biochemical oxygen demand (BOD) Amount of dissolved oxygen (in mg/dm^3 water) which disappears from a water sample in a given time at a certain temperature, through decomposition of organic matter by micro-organisms. Used as an index of organic pollution, especially sewage.

Biochemistry The study of the chemical changes within, and produced by, living organisms. Includes molecular biology, or molecular genetics.

Biocoenosis Totality of organisms living in a biotope, itself defined as any segment of the environment which has convenient but arbitrary upper and lower limits, characterized by dominant or characteristic species. Virtually synonymous with **community**.

Biodiversity The level of global or local biological diversity, often quantified crudely as numbers of species or of higher taxa; increasingly taken as indicative of genetic diversity level (see **genetic variation, genome**). Total species estimates for Earth vary from 3×10^6 to 30×10^6, but these are almost entirely terrestrial estimates. Terrestrial insects account for more than half of all named and recorded species, and it has been argued that marine invertebrates alone may provide a further 10×10^6 species. In terms of diversity of basic body plan (phyla), more than 80% of all phyla are found only in the sea, and most of these in sediments.

There is a growing consensus that species numbers of predators and parasites within a community are not simply summative: weighted (ordinal) measures of diversity contrast with summative (cardinal) ones. Weightings often reflect either the functional importance of taxa within communities or their degree of taxonomic variability, species number within a community (its richness) being a poor indicator of its structure. Much interest surrounds **tropical rain forests** where, e.g., an individual tree may harbour as many species and genera of ants as are found in the entire British Isles. Few of the species present in a community are usually abundant (i.e. have a large population size, large biomass, productivity, or some other measure of importance). The large number of rare species (see **rarity**) mainly determines the species diversity, and the ratios between the number of species and their 'importance values' are termed *species diversity indices*. The particular index employed depends on the type of community studied. Over geological time, global species number will be governed by rates of **speciation** and **extinction**.

Biogenesis The theory that all living organisms arise from pre-existing life forms. The works of Redi (1688) for macroorganisms, and of Spallanzani (1765) and **Pasteur** (1860) in particular for microorganisms, stand as landmarks in the overthrow of the theory of **spontaneous generation**. Since their time it has become generally accepted that every individual organism has a genetic ancestry involving prior organisms. The first appearance of living systems on the Earth (see **origin of life**) is still problematic, but there is no reason in principle to dispense with faith in natural and terrestrial causation, and every reason to pursue that line of inquiry.

Biogenetic Law Notorious view propounded by Ernst Haeckel in about 1860 (a more explicit formulation of his mentor Muller's view) that during an animal's development it passes through ancestral adult stages ('ontogenesis is a brief and rapid recapitulation of phylogenesis'). Much of the evidence for this derived from the work of embryologist Karl von Baer. It is now accepted that embryos often pass through stages resembling related *embryonic*, rather than adult, forms. Such comparative embryology provides important evidence for **evolution**. See **paedomorphosis**.

Biogeography A major branch of geography, concerned with the study and interpretation of geographical distributions of animals (**zoogeography**), plants (**plant geography**), and man (cultural biogeography), both extant and extinct. One approach (*dispersal biogeography*) stresses the role of dispersal of animals and plants from a point of origin or across pre-existing barriers. Another approach (*vicariance biogeography*) takes barriers which occur within a pre-existing continuous distribution to be more important. See **Gondwanaland, Laurasia**.

Biological clock In its widest sense, any form of biological timekeeping, such as heart beat or ventilatory movements; more often used in context of **photoperiodism**, **dormancy** and **diapause**, but mostly associated with physiological, behavioural, etc., rhythms relating to environmental cycles, notably **circadian**, tidal, lunar-monthly and annual. Mechanisms counting the number of cell divisions would also be included. In plants some recent evidence suggests that cell membranes may be the key to the biological clock by regulating ion movements into and out of cells and subcellular compartments, and through regulation of energy metabolism via energy-transducing membranes of chloroplasts and mitochondria. Enables plant or animal to respond to changing seasons and other environmental factors.

Biological control Artificial control of **pests** and parasites by use of organisms or their products; e.g. of mosquitoes by fish and aquatic insectivorous plants which feed on their larvae; or of the prickly pear (*Opuntia*) in Australia by the moth *Cactoblastis cactorum*. There is increasing use of **pheromones** in attracting pest insects, which may then be killed or occasionally sterilized and released. Sometimes attempts are made to encourage spread of genetically harmful factors in the pest population's gene pool, although this needs great care. Success depends on a thorough grasp of relevant ecology.

Biology Term coined by **Lamarck** in 1802. The branch of science dealing with properties and interactions of physico-chemical systems of sufficient complexity for the term 'living' (or 'dead') to be applied. These are usually cellular or acellular in organization; but since viruses share some of the same polymers (nucleic acid and protein) as cells, and moreover are parasitic, they are regarded as biological systems but not usually as organisms. See **life**.

Bioluminescence Emission of light by 'photogenic' organisms, including bacteria, fungi, dinoflagellates, jellyfish, brittle stars, worms, fireflies, molluscs and fish. Bacteria often form symbiotic associations with animals such as fish, which thereby become luminous (see **photophore**). Some animals are self-luminous, having photogenic organs containing photocysts. In the coelenterate *Obelia* these are scattered through the endoderm. Two types of light emission occur in living organisms: (1) bioluminescence (chemiluminescence), in which energy from an exergonic chemical reaction is transformed into light energy; and (2) photoluminescence, which is dependent on the prior absorption of light.

Dinoflagellates are the main contributors to marine bioluminescence, emitting a bluish-green (maximum wavelength 474 nm) flash of light of 0.1 s duration when the cells are stimulated. The compound responsible for the bioluminescence is *luciferin*, which is oxidized with the aid of the enzyme luciferase. The chemistry of luciferin varies with the organism, e.g., in bacteria, insects, and dinoflagellates it is a flavin, a (benzo)thiazole nucleus and a tetrapyrrole respectively. Luciferases share the feature of being oxygenases (enzymes that add oxygen to compounds). In the dinoflagellate *Gonyaulax polyedra* bioluminescence is rhythmic and circadian.

Flashing that occurs in luminous organs of many animals (often serving in mate or prey attraction) is often under nervous control, the organs having a rich supply of nerve terminals. Bioluminescent flashing in dinoflagellates has been suggested to serve as an antipredation device. Feeding upon such dinoflagellates by copepods produces a bright spot of light making the copepod conspicuous to its predators. Hence such copepods feeding upon bioluminescent dinoflagellates would be at a selective disadvantage.

Biomass The total quantity of matter (the non-aqueous component frequently being expressed as dry mass) in organisms, commonly of those forming a trophic level or population, or inhabiting a given region. See **standing crop**, **pyramid of biomass**.

Biome Major terrestrial, climatically controlled, regional ecological complex or set of ecosystems possessing characteristic vegetation, and among which there is exchange of water, nutrients and biotic components (e.g. **tropical rain forest**, coral reefs, grasslands). Biomes occupy large areas of the land surface, and typically occur on more than one continent. The actual distribution of biomes results from (1) geological factors, e.g., proximity of mountain ranges, their height and orientation; (2) direction in which the prevailing moisture-laden winds are blowing, hence global climate patterns; and (3) seasonal changes in insolation arising from the sun.

Biometry Branch of biology dealing specifically with application of mathematical techniques to the quantitative study of varying characteristics of organisms, populations, etc.

Biophysics Fields of biological inquiry in which physical properties of biological systems are of overriding interest. Biophysics departments tend to work in such areas as crystal structures of macromolecules and neuromuscular physiology.

Biopoiesis The generation of living from non-living material. See **origin of life**.

Bioreactor A chamber within which the activity of a biocatalyst can be optimally controlled, often for industrial production. They include fermenters such as the **chemostat** (Fig. 9). In *batch fermentation*, the nutrients and microorganism are put into a closed reactor, nothing is added while fermentation occurs and nothing except waste gases leaves the chamber, and most environmental conditions except temperature are normally allowed to vary unchecked. The product is separated from the microorganism when nutrients have been utilized; the exponential growth phase usually lasts only a short time. In *continuous fermentation*, open reactors (chemostats) are used and nutrients are added continuously, at a rate which balances their removal rate. Microorganisms are thus kept in their exponential growth phase, for which monitoring and constancy of environmental variables (pH, temperature, O_2, substrate, product and waste levels) are

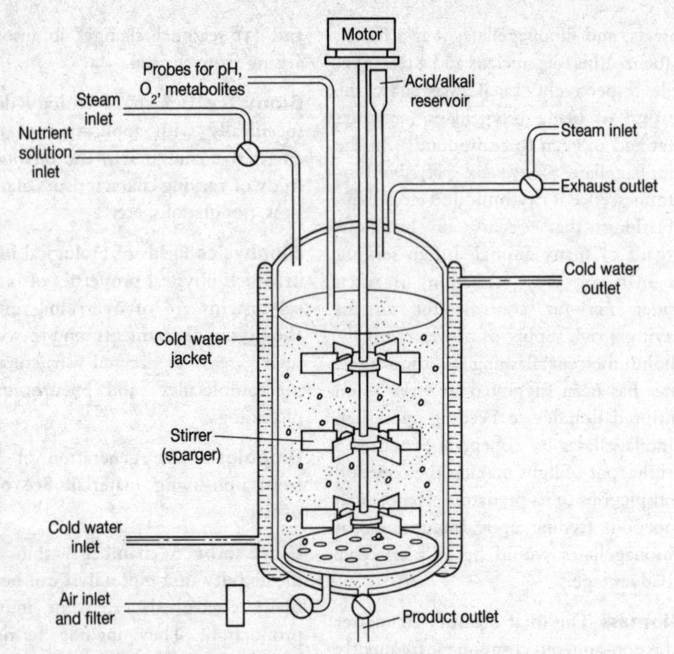

Fig. 9 *An industrial fermenter suitable for continuous cultures.*

essential. The difficulty of achieving such steady-state conditions could be offset by the greater cost-effectiveness achievable. See **single cell protein**.

Biosensor Device in which a bioactive layer lies in intimate contact with a transducer whose responses to changes in the bioactive layer generate electronic signals for subsequent interpretation. The bioactive layer may consist of membrane-bound enzymes, antibodies or receptors. The membrane itself may be synthetic (e.g. acetyl cellulose) and coat an inorganic chip. The potential for this fusion of electronics and biotechnology includes the direct assay of clinically important substrates (e.g. blood glucose) and of substances too unstable for storage or whose concentrations fluctuate rapidly.

Biosphere That part of Earth and its atmosphere inhabited by living organisms.

Biotechnology Application of discoveries in biology to large-scale production of useful organisms and their products. Centres on development of enzyme technology in industry and medicine, and of **gene manipulation**, often in the service of plant and animal breeding (e.g. see **protoplast**). Together, these constitute biomolecular engineering. Branches include fermentation technology, waste technology (e.g. recovery of metals from mining waste) and renewable resources technology, such as the

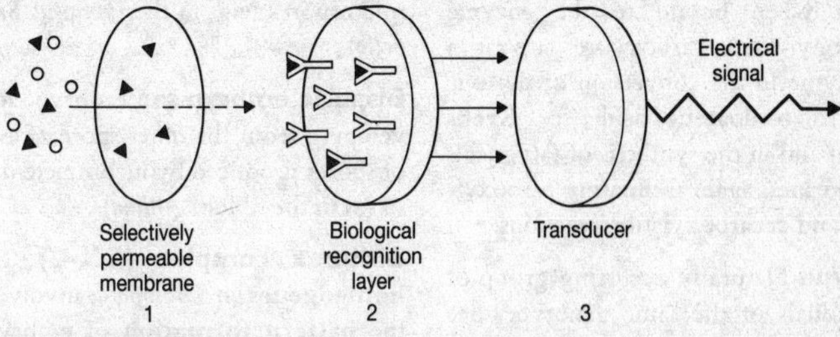

(a) *The parts of a biosensor*

1 The substance to be detected passes through the membrane
2 The substance binds to the recognition layer which may contain antibodies or enzymes; a reaction takes place.
3 The product of the reaction passes to a transducer and this produces an electrical signal

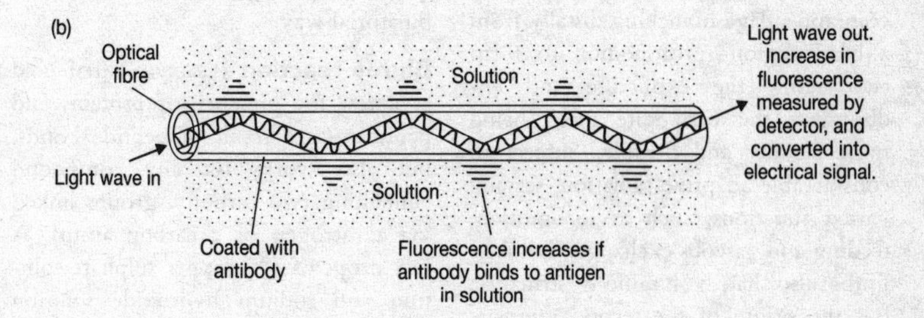

Fig. 10 (a) *The principal components of a biosensor, illustrated in* (b) *by an optical fibre on which antibodies to a particular antigen have been tagged with a fluorescent molecule. Light travelling along the fibre is reflected back off the parts where antigen has bound, so increasing light transmitted along the fibre.*

use of **lignocellulose** to generate more usable energy sources. Organisms involved tend to be microorganisms, and their traditional involvement in the brewing, baking and cheese/yoghurt industries is also affected by the new technology.

Biotic factors Those features of the environments of organisms arising from the activities of other living organisms; as distinct from such abiotic factors as climatic and edaphic influences.

Biotin Vitamin of the B-complex made by intestinal bacteria, so difficult to prevent its uptake and assimilation. However avidin, a component of raw egg white, binds tightly to biotin and prevents its uptake from the gut lumen, resulting in biotin deficiency in those eating raw eggs too avidly. A modified biotin when bound to the enzyme propionyl-CoA carboxylase acts as a coenzyme in the conversion of pyruvic acid to oxaloacetic acid (see **Krebs cycle**) and in the synthesis of fatty acids and purines, again facilitating carboxylation and decarboxylation reactions.

Biotype Naturally occurring group of individuals of the same genotype. See **infraspecific variation**.

Bipedalism For hominid bipedalism – postural and locomotory – see **Australopithecines**.

Bipolar neurone See **neurone**.

Biramous appendage Paired crustacean appendage branching distally from a basal region (protopodite) to form two rami, the exopodite and endopodite, the exopodite often being more slender and flexible. Subject to considerable adaptive radiation, serving varied functions, such as locomotion, feeding and gaseous exchange. Trilobite limbs also had a biramous structure, but the origin of the second ramus is different. *Stenopodia* has been the term used for crustacean appendages in which one or more processes, epipodites, lie on the outer sides of protopodites; *phyllopodia* are broader and flatter than most stenopodia and may be unjointed, bearing lobes known as endites and exites. See **uniramous appendage**, and Fig. 11.

Bisaccate Pollen grains possessing two air sacs or bladders; mainly coniferous pollen, e.g. pine pollen.

Bisexual See **hermaphrodite**.

Bisporangiate cone (Bot.) Cone containing both megaspores and microspores (e.g. in the lycopsid *Selaginella*).

Bisporic embryo sac Embryo sac developing from the inner spore of a diad of spores produced by incomplete meiosis (as in the onion, *Allium*).

Bithorax complex (BX-C) Three **homeogenes** in *Drosophila* involved in the **pattern formation** of embryonic **parasegments** 5–14. The proteins encoded by all three genes (*Ultrabithorax*, *Ubx*; *abdominal-A*, *abd-A*; *Abdominal-B*, *Abd-B*) vary through alternative splicing during **RNA processing**. Each works in a particular parasegmental domain within the embryo, sometimes in a combinatorial way.

Biuret reaction Reaction often used as a test for presence of proteins and peptides (as a result of peptide bonds; but also works for any compound containing two carbonyl groups linked via a nitrogen or a carbon atom). A few drops of 1% copper sulphate solution and sodium hydroxide solution are added to the test sample, when a purple Cu^{2+}-complex is produced if positive; solution stays blue otherwise.

Bivalent Pair of homologous chromosomes during pairing (synapsis) at the first meiotic prophase and metaphase. See **meiosis**.

Bivalve Broadly speaking, any animal with a shell in two parts hinged together, e.g. bivalve molluscs (with which the term is often equated) and **Brachiopoda**. Occasionally ostracod crustaceans are said to have a bivalve carapace, although this is not a shell. See **Bivalvia**.

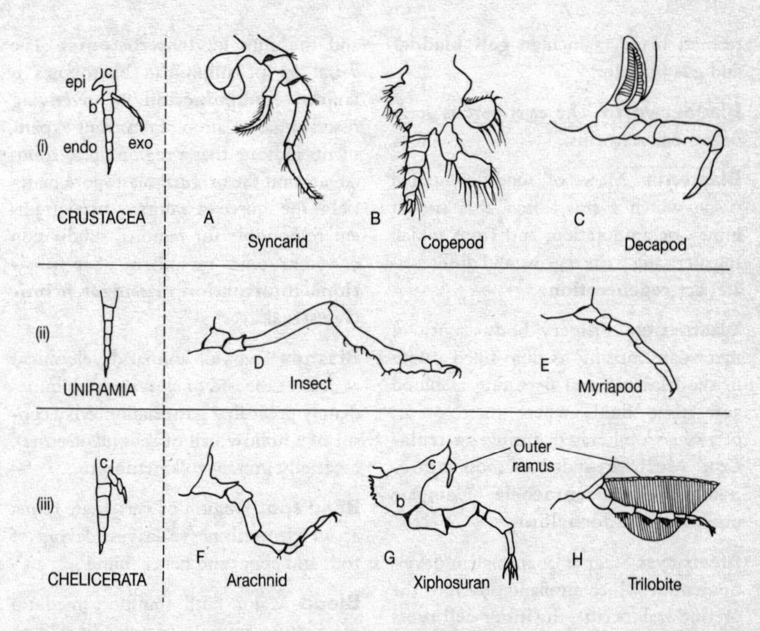

Fig. 11 *A diagram indicating the main features of adult limb structures in arthropods. (i) biramous, (ii) uniramous, (iii) primitively biramous, with the exite/outer ramus more proximally placed than in crustaceans. epi = epipodite; endo = endopodite; exo = exopodite; b = basipodite; c = coxopodite.*

Bivalvia (Lamellibranchia, Pelecypoda) Class of **Mollusca**, with freshwater, brackish and marine forms with greatly reduced head (no eyes, tentacles or radula), and body laterally compressed and typically bilaterally symmetrical. Shell composed of two hinged valves (both lateral) under which lie two large gills (*ctenidia*) used for gaseous exchange and generally for filter-feeding. Sexes nearly always separate; trochosphere and veliger larvae, but occasionally a parasitic glochidium. Considerable radiation, including fixed forms (e.g. mussels, clams) and burrowers (e.g. shipworm, razorshell). See **Brachiopoda**.

Bladder (1) Urinary bladder. Part of the anterior **allantois** of embryonic amniotes, which persists into adult life and receives the openings of the ureters either directly (mammals) or via a short part of the cloaca (lower tetrapods). In fish the urinary ducts themselves may enlarge as 'bladders', or fuse for part of their length forming a single sac, receiving also the oviducts in females, as a urogenital sinus. In all cases the bladder serves for urine retention, and in tetrapods is distensible with a thick smooth-muscle wall. In lower tetrapods it opens into the cloaca, but in mammals its contents leave via the urethra. (2) Other sac-like fluid-filled structures in animals termed bladders include **gall bladder** and **gas bladder**.

Bladderworm The **cysticercus** stage of some tapeworms.

Blastema Mass of undifferentiated tissue which forms, often at a site of injury or amputation, and from which regenerating parts regrow and differentiate. See **regeneration**.

Blastocoele Primary body cavity of metazoans, arising as fluid-filled cavity in the **blastula** and persisting as blood and tissue fluids where these occur; otherwise obliterated during **gastrulation**. Much expanded in most arthropods as the **haemocoele**. Compare **coelom, pseudocoelom**.

Blastocyst Stage of mammalian development at which implantation into the uterine wall occurs, the **inner cell mass** spreading inside the blastocoele as a flat disc.

Blastoderm Sheet of cells, usually one or just a few cells thick, surrounding the blastocoele in non-yolky eggs, or covering it in yolky ones. Consists usually of small, tightly packed cells.

Blastomere Cell produced by cleavage of an animal egg, up to the late blastula stage. In yolky eggs especially this results in smaller *micromeres* at the animal pole, with larger yolky *macromeres* at the vegetal pole where the rate of division is slower. See **cleavage**.

Blastopore Transitory opening on surface of gastrula by which the internal cavity (archenteron) communicates with the exterior; produced by invagination of superficial cells during **gastrulation**. It becomes the mouth in **protostomes** and the anus in **deuterostomes**. The *dorsal lip* of amphibian blastopores is famous for **induction** of overlying tissue in gastrulation. Transplant experiments indicate that a region up to about 60° around the original blastopore material is the source of a dorsoventral gradient responsible for regional subdivision of embryonic mesoderm. See **positional information, organizer, primitive streak**.

Blastula Stage of animal development at or near the end of cleavage and immediately preceding gastrulation. May consist of a hollow ball of cells (*blastomeres*), especially in non-yolky embryos.

Blind spot Region of vertebrate retina at which optic nerve leaves; devoid of rods and cones and hence 'blind'.

Blood Major fluid transport medium of many animal groups, including nemerteans, annelids, arthropods, molluscs, brachiopods, phoronids and chordates. Derived from the **blastocoele**. Comprises an aqueous mixture of substances in solution (e.g. nutrients, wastes, hormones, gases and osmotically active compounds) in which are suspended cells (haemocytes) functioning either in defence (e.g. phagocytes) or oxygen transport (e.g. **red blood cells**). Blood is moved by muscle contraction in some of the vessels it passes through. Hearts are such specialized vessels, the hydrostatic pressure generated being employed in filtration processes (in capillaries generally, and kidney glomeruli in particular), in locomotion (e.g. many bivalve molluscs) and other activities besides solute translocation. Arthropod blood (haemolymph) is hardly confined to vessels (open circulation), insects and onychophorans having least vascularization, the haemocoele being much expanded. When blood is confined to vessels, the circulation is said to be closed, as it is throughout annelid and vertebrate bodies, although expanded blood sinuses are a feature of lower vertebrates. Respiratory pigment (absent in most insects) is in simple solution in invertebrate blood, but confined to red blood cells (erythrocytes) in vertebrates. See **tissue fluid**.

Blood-brain barrier Phrase indicating the relative resistance to diffusion of molecules across (i) the unfenestrated capillaries of the brain, whose cells have tight junctions, and (ii) the fatty glial cell (astrocyte) sheath surrounding the capillaries. The latter obstructs polar solutes more than it does such lipid-soluble solutes as psychoactive drugs.

Blood clotting (b. coagulation) Adaptive response to haemorrhage involving local conversion of liquid blood to a gel, which plugs a wound. In vertebrates, blood does not clot in normal passage through vessels since the smooth endothelial lining prevents damage to platelets. The altered surface of damaged vessel walls plus turbulence of blood flow results in ADP release from platelets, and the exposed collagen in vessel walls causes adherence of platelets, which release Ca^{2+} and cause more aggregation. Vasoconstriction, stemming blood flow, results from **serotonin** released when platelets fragment. A temporary plug of platelets is made permanent by a **cascade** of enzymic reactions caused by release of phospholipid and the protein thromboplastin (thrombokinase) by damaged cells and platelets. Ten clotting factors in addition to the above have been isolated; but the main sequence may be summarized as: thromboplastin (Factor III) converts prothrombin (Factor II) to thrombin, which converts fibrinogen (Factor I) to an insoluble fibrin meshwork, trapping erythrocytes, platelets and plasma to form clot, stabilized by Factor XIII in presence of Ca^{2+} (Factor IV). Fibrin absorbs 90% of the thrombin formed and prevents the clot from spreading away from the damaged area. Several genetic disorders cause poor clotting. Factor VIII, one of the substances required for thromboplastin formation, is absent in classical X-linked haemophilia. See **anticoagulant**, **fibrinolysis, vitamin K**, and Fig. 12.

Blood group Either a group of people bearing the same antigen(s) on their red blood cells within a particular **blood group system**; or the blood characteristic used to distinguish groups of individuals within a blood group system. Main clinical significance lies in blood transfusion and Rhesus incompatibility between mother and foetus. Without due matching of donor and recipient blood, death or severe illness of a recipient may result from a single transfusion (in ABO system) or after repeated transfusion (in Rhesus system). Causes of death include haemolysis (red cell rupture), agglutination (red cell clumping) with blockage of capillaries etc., and tissue damage. Danger in transfusion comes when donor antigens meet antibodies to them in recipient plasma and elsewhere resulting in **innate immune response**. Antibodies to antigens of the ABO system occur naturally in plasma; those to Rhesus antigens occur in plasma only as a result of immunization,

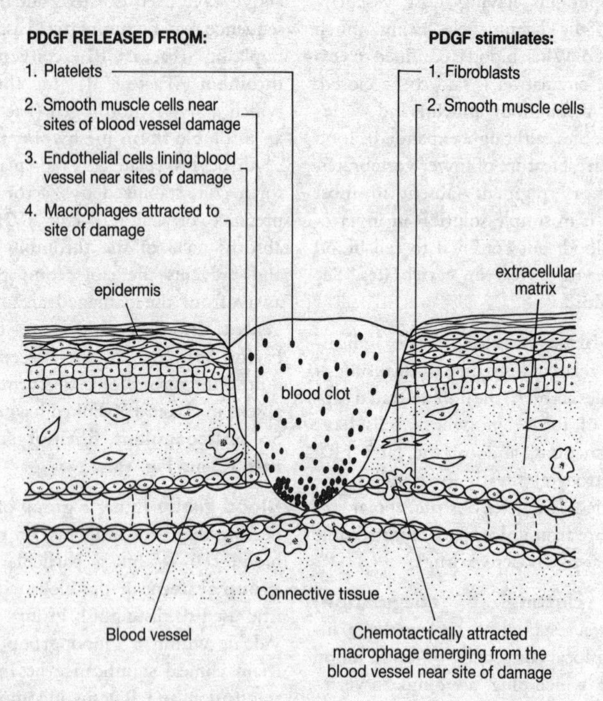

Fig. 12 *Probable role of PGDF (platelet-derived* **growth factor***) in wound healing. Its release stimulates division of fibroblasts and smooth muscle cells, the former producing extracellular matrix material, while PGDF attracts macrophages.*

during pregnancy or transfusion. People with an antigen of the ABO system on their red cells automatically lack the antibody to it in their plasma, whereas lack of an ABO antigen on the red cells is coupled to presence of antibody to it in plasma (Table 2a).

Success or failure of transfusions is determined by the data outlined in Table 2b. People with O-type are universal donors; those with AB-type blood are universal recipients. The Rhesus antibody may be found in plasma of Rhesus negative (Rh −) women who have been pregnant with Rhesus positive (Rh +) babies, who have Rhesus antigen on their red cells. This is because mothers may become sensitized to the Rhesus antigen if foetal blood leaks across the placenta during birth of a Rh + baby. Later Rh + foetuses are at risk if anti-Rh antibody crosses the placenta from the mother, who had produced it as an immune response to the earlier leaked foetal blood. Foetal haemolysis can be averted by injecting anti-Rh antibody (anti-D) to a Rh − mother prior to delivery of any Rh + offspring; immedi-

	Blood Group A	B	AB	O
Antigen(s) on red cells	A	B	A & B	—
Antibody (ies) in plasma	anti-B	anti-A	—	anti-A anti-B

Table 2a. *Relationship of antigen and antibody distributions in the ABO blood group system.*

		Donor Blood Group A	B	AB	O
Recipient Blood Group	A	✓	×	×	✓
	B	×	✓	×	✓
	AB	✓	✓	✓	✓
	O	×	×	×	✓

Table 2b. *Success [✓] or failure [×] of blood transfusions from specific donor blood groups to specific recipient blood groups.*

Blood Group [phenotype]	*Possible Genotypes*
A	I^AI^A or I^AI^O
B	I^BI^B or I^BI^O
AB	I^AI^B
O	I^OI^O

Table 3a *Genetics of ABO system.*

P_1 phenotypes	*A* Case 1	*A* Case 2	×	O
P_1 genotypes	I^AI^A	I^AI^O	×	I^OI^O
Gametes	I^A	I^A & I^O	×	I^O
F_1 genotypes	I^AI^O	I^AI^O: I^OI^O		
F_1 phenotypes	all A	A : O		

Table 3b *Example showing expected offspring from marriage between group A and group O people.*

ate blood transfusion of babies ('blue babies') born with haemolytic anaemia is an alternative. Blood groups are a classic instance of **polymorphism** in man. The genetics involved is often complex (there are at least three loci responsible for the Rh system), but the ABO system is a simple multiple allelism, as shown in Table 3a, where blood groups A and B are codominant and O recessive. The antigens of the ABO system are found on, and appear to have evolved earlier than, cell types (mainly epithelial cells) other than red blood cells and should therefore be termed histo-blood group antigens. The I^o allele contains a frame-shift mutation causing loss of the transferase activity which modifies the cell surface H antigen into either A or B antigens. Adenocarcinomas of most organs of O-type individuals are less frequent than in A-type individuals.

The ratios of different blood groups within a blood group system differ geographically and indicate either **natural selection** or **genetic drift** among different populations. Their genetic basis may be used in cases of contested parentage. See **Mendelian heredity**.

Blood group system A person's blood groups are genetically determined, each person belongs to several, and most are determined by loci situated on autosomes. The alleles at these loci determine antigens on a person's red blood cell membranes. A blood group system refers to the range of red cell antigens determined by the alleles at one such locus, or sometimes a group of closely linked loci. Some non-human primates share certain human blood group systems. In humans there are at least fourteen such systems (ABO, Rhesus, MNS, P, Kell, Lewis, Lutheran, Duffy, Kidd, Diego, Yt, I, Dombrock and Xg). Only the Xg locus is sex-linked. No linkage between any of these systems is apparent, other than between Lewis and Lutheran.

Blood plasma Clear yellowish fluid of vertebrates, clotting as easily as whole blood and obtained from it by separating out suspended cells by centrifugation. An aqueous mixture of substances, including plasma proteins. See **blood**.

Blood platelets. See **platelets**, **blood clotting**.

Blood pressure Usually refers to pressure in main arteries; in humans, normally between 120 mm Hg at **systole** and 80 mm Hg at **diastole** (i.e. 120/80). Pressure drops most rapidly in arterioles, falling further in capillaries and more slowly in venules and veins. In the venae cavae it is 2 mm Hg, and 0 mm Hg in the right atrium. Homeostatically controlled (see **baroreceptor**).

Blood serum Fluid expressed from clotted blood or from clotted blood plasma. Roughly, plasma deprived of fibrinogen and other clotting proteins.

Blood sugar Glucose dissolved in blood. Homeostatically regulated in humans at about 90 mg glucose/100 cm^3 blood. Hormones affecting level include: **insulin**, **glucagon**, **adrenaline**, **growth hormone** and **thyroxine**.

Blood system (blood vascular system) The system of blood vessels (in sequence: arteries, arterioles, metarterioles, capillaries, venules and veins) and/or spaces (often sinuses) through which blood flows in most animal bodies. See **blood** for other details.

Bloom A term used to describe the dense growth of planktonic algae which imparts a distinct colour to the water. Commonly, blue-green algae (Cyanobacteria) form such blooms in **eutrophic** lakes. See **eutrophication**.

Blue-green algae See **Cyanobacteria**.

BOD See **biochemical oxygen demand**.

Body cavity (1) Primary body cavity; see **blastocoele**. (2) Secondary body cavity; see **coelom**.

Bohr effect The effect of dissolved carbon dioxide on the oxygen equilibria curves or respiratory pigments (see **haemoglobin**), whereby increased CO_2 concentration (pCO_2) shifts the curve to the right, i.e. decreases the percentage O_2-saturation of the pigment for a given pO_2, increasing the rate of oxygen unloading in regions of high respiratory activity (the tissues) and loading of oxygen in regions of low pCO_2 (e.g. lungs, gills). Brought about by **allosteric** effect of pH on haemoglobin molecule.

Bone Vertebrate connective tissue laid

down by specialized mesodermal cells (osteocytes) lying in **lacunae** within a calcified matrix which they secrete, containing about 65% by weight inorganic salts (mainly hydroxyapatite, providing hardness); remainder largely organic and comprising mostly **collagen** fibres providing tensile strength. *Compact bone* forms outer cylinder of shafts of long bones of limbs, and is typified by **Haversian systems**; *spongy bone* forms vertebrae, most 'flat' bones (e.g. skull), and ends of long bones (epiphyses), typified by presence of **trabeculae**. Bone is a living tissue, supplied with nerves and blood vessels. Its constitution changes under hormone influence (see **calcitonin**, **parathyroid hormone**). Besides its skeletal role in providing levers for movement and support for soft parts of the body, it protects many delicate tissues and organs. See **ossification**, **haemopoiesis**, **cartilage**, and Fig. 89.

Bony fish See **Osteichthyes**.

Botany Branch of biological science concerned with plants. Since animal life depends for its existence on the process of **photosynthesis**, its importance is self-evident. Its roots lie deep in antiquity, for our interest in plants as sources of food and medicine goes back to the very origins of human society; but the scientific study of plants (a requirement of which is the ability to distinguish and record plants in question) did not begin until the sixth century BC in Asia Minor and the Greek-speaking cities of Ionia. Modern botany commenced in the mid to late nineteenth century when some splendid textbooks were published and research schools became established at universities.

Bouton Knob-like terminal of nerve axon, containing synaptic vesicles. See **synapse**.

Bowman's capsule Cup-like receptacle composed of two epithelial layers, formed by invagination of a single layer, surrounding a glomerulus and forming part of a vertebrate **kidney** nephron.

Brachial Relating to the arm; e.g. brachial plexus, the complex of interconnections of spinal nerves V–IX supplying the tetrapod forelimb.

Brachiation The arm-over-arm locomotion adopted by many arboreal primates.

Brachiopoda Lamp shells. A small phylum of marine coelomate bivalve invertebrates, valves being dorsal and ventral (see **Bivalvia**); with teeth and sockets along the hinge in articulate, but absent from inarticulate forms. Ciliated **lophophore** serves for feeding and gaseous exchange. Of enormous value in dating rock strata, appearing in Lower Cambrian, with major extinctions in Devonian and Permian and expansions in the Ordovician, Carboniferous and Jurassic. One of the few surviving genera, *Lingula* (inarticulate) is almost unchanged since the Ordovician.

Bract Small leaf with relatively undeveloped blade in axil of which arises a flower or a branch or an inflorescence.

Bracteole Small **bract**.

Bradycardia Slowing of heart (hence pulse) rate. Compare **tachycardia**.

Bradykinin Nonapeptide hormone of submaxillary salivary gland, inducing vasodilation and increased capillary permeability of tissues. Released by parasympathetic stimulation. Implicated in **allergic reactions**.

Brain Enlargement of the central nervous system of most bilaterally symmetrical animals with an antero-posterior axis. Its development anteriorly is a major component of the process of **cephalization**.

(1) In vertebrates, an anterior enlargement of the hollow neural tube, sharing features with the spinal cord, such as relative positioning of white matter (myelinated axons) externally and grey matter (unmyelinated neurones) internally – although in higher forms there is migration of grey matter cell bodies above the white matter to form a third layer, reaching its zenith in the human cerebrum. Three dilations of the neural tube give rise to forebrain, midbrain and hindbrain.

The *forebrain* comprises the diencephalon anteriorly and the telencephalon posteriorly. Primitively, this is olfactory but in higher vertebrates the diencephalon roof is expanded to form the paired cerebral hemispheres and dominates the rest of the brain in its sensory and motor function. The telencephalon retains its association with smell, forming a pair of olfactory lobes (see Fig. 13).

The *midbrain* is primitively an optic centre, the pair of optic nerves entering after **decussation** at the optic chiasma. Although terminations of these fibres enter the visual cortex of the cerebrum in higher vertebrates, the midbrain still retains integrative functions (see **brainstem**) and is the origin of some **cranial nerves**.

The *hindbrain* generally has its anterior roof enlarged to form a pair of cerebellar hemispheres associated with proprioceptive coordination of muscle activity in posture and locomotion. Its floor is thickened to form the pons anteriorly and the medulla oblongata posteriorly (see **brainstem**). The central canal of the spinal cord expands to form the brain ventricles, and the whole is surrounded by **meninges**.

(2) In invertebrates, there is enlargement of paired anterior **ganglia** associated with cephalization, those above the gut often fusing to form complex integrative centres; but in segmented forms, segmental ganglia and local reflexes are just as important in nervous integration. In molluscs 'brain functions' may be divided between several pairs of distinct ganglia or, as in cephalopods (octopus, squid, etc.), as many as 30 integrated brain lobes, enabling a complexity of behaviour on a par with the lower vertebrates.

Brainstem The vertebrate midbrain, pons and medulla oblongata. It links the cerebrum to the pons and spinal cord, has reflex centres for control of eyeballs, head and trunk in response to auditory and other stimuli, and houses part of the **reticular formation**. The respiratory, **vasomotor** and **cardiac centres** are here, associated with appropriate cranial nerves. See **ventilation**.

Branchial Relating to gills. The aortic arches serving the gills in fish (the third arch onwards) are termed *branchial arches*.

Branchiopoda Crustacea with at least four fairly uniform pairs of phyllopodial trunk limbs (see **biramous appendage**) used in gaseous exchange, and a variable

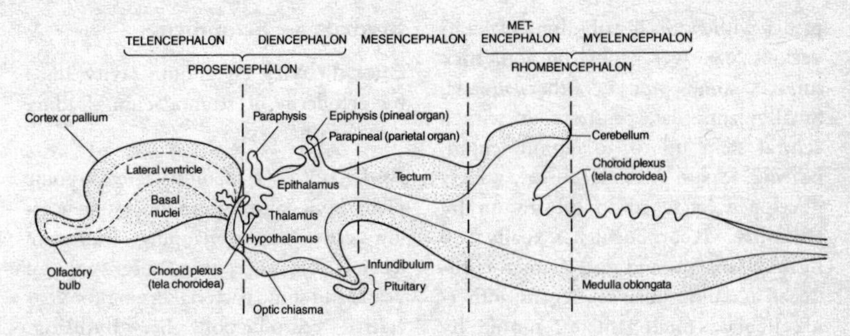

Fig. 13 *Diagram showing median section of typical mature vertebrate brain, illustrating the principal divisions and structures.*

but usually large number of somites. Primitive. Includes water fleas (e.g. *Daphnia*) and brine shrimps (e.g. *Artemia*).

Breathing See **ventilation**.

Breeding system All factors, apart from mutation, affecting the degree to which gametes which fuse at fertilization are genetically alike. Includes population size, and levels of **inbreeding** and **outbreeding** controlled by variable selfing, incompatibility mechanisms, assortative mating, heterostyly, dichogamy, arrangement and distribution of sex organs (e.g. size of plant and inflorescence number are factors), mechanism of sex-determination, etc. See **genetic variation**, **rarity**.

Brefeldin A Fungal product causing disintegration of the **Golgi apparatus**.

Bristletails See **Apterygota**.

Brittle stars See **Echinodermata**.

Broca's area A language-associated centre, visible as a bump on the left side of the frontal lobe of the brain in modern humans; apparently also present in all members of the genus *Homo* (from endocranial casts), but absent from *Australopithecus*.

Bronchiole Small air-conducting tube (less than 1 mm in diameter) of tetrapod lung, arising as a branch of a bronchus and terminating in alveoli. Smooth muscle abundant in walls, controlling lumen diameter. Lacks cartilage and mucous glands of bronchi.

Bronchus Large air tube of tetrapod lung. One per lung, connecting it to trachea; divided into smaller and smaller bronchi, and finally into **bronchioles**. Each bronchus has cartilaginous plates and smooth muscle in its walls, and mucus from glands traps bacteria and dust, the whole being beaten by cilia up to the pharynx for swallowing.

Brown fat See **adipose tissue**.

Brush border Animal cell surface (often apical), often a whole epithelial surface, covered in **microvilli** and serving for absorption and/or **digestion**.

Bryophyta Division of plant kingdom comprising **Hepaticopsida** (liverworts), **Bryopsida** (mosses) and **Anthocerotopsida** (hornworts). Small group of plants with wide distribution. Habitats various, e.g. wet banks, on soil, rock surfaces; some epiphytic, others aquatic. Small plants, flat, prostrate or with a central stem up to 30 cm in length, bearing leaves. Vascular tissue poorly developed; attached to substratum by **rhizoids**. Reproducing sexually by fusion of macro- and microgametes produced in multicellular sex organs; antherida liberate microgametes, motile by flagella; archegonia contain a single macrogamete (egg cell). Sexual reproduction is followed by development of capsule containing spores giving rise via **protonemata** to new plants. A well-marked **alternation of generations**: the leafy vegetative plant is the gametophyte generation; the capsule (and seta) comprise the sporophyte generation, partially dependent nutritionally upon the gametophyte. Origins linked to **Charophyceae**.

Bryopsida (Musci) Mosses. Class of **Bryophyta**, having fairly conspicuous **protonemata** and multicellular **rhizoids**. The sporophyte develops from an apical cell, and the capsule has a complex opening mechanism. **Elaters** are absent. Cosmopolitan, occurring in damp habitats (e.g. moist humus, peat, wet cliff faces, dry boulders, etc.) and as epiphytes on branches and tree trunks. Sex organs (antheridia and archegonia) borne either terminally or in lateral bud-like structures (perigonia and perichaetia), either on separate or the same plants. Fertilization is followed by development of the sporophyte, the capsule elevated on a seta and covered when young by a gametophytic **calyptra**. A filamentous or thalloid protonema gives rise to new moss plants from lateral buds.

Bryozoa See **Ectoprocta**.

Buccal cavity The mouth cavity, lined by ectoderm of stomodaeum, leading into the **pharynx**.

Bud (Bot.) (1) Compact embryonic plant shoot comprising a short stem bearing crowded overlapping immature leaves. (2) Vegetative outgrowth of yeasts and some bacteria serving for vegetative reproduction. See **budding**. (Zool.) Outgrowth of organism that may detach, as in **budding**; or a morphogenic feature of a growing region, as in vertebrate limb buds (see **regeneration**).

Budding (Bot.) (1) Grafting in which grafted part is a bud. (2) Asexual reproduction in which a new cell is formed as an outgrowth of a parent cell, e.g. in yeast. Compare **binary fission**. (Zool.) **Asexual** method of reproduction common in many invertebrates such as sponges, coelenterates (e.g. *Hydra*), ectoproctans and urochordates, but also a feature of the **hydatid cysts** of tapeworms. Rarely known as gemmation. See **polyembryony**.

Buffer A solution resisting pH change on addition of acid (i.e. H^+) or alkali (OH^-), absorbing protons from acids and releasing them on addition of alkali. Usually consists of a mixture of a weak acid and its conjugate base, or vice versa. Intracellular and extracellular buffers may differ; thus the commonest intracellular buffer is the acid-base pair $H_2PO_4^-$–HPO_4^{2-} and such organic phosphates as ATP, but the bicarbonate buffer system (H_2CO_3– HCO_3^-) is common extracellularly, as in vertebrate blood plasma, where plasma proteins are also a major buffer system. **Haemo-**

globin acts as a buffer during the **chloride shift**. If pH varies, **protein** shape and function may be affected. See **kidney**.

Buffon, G.L.L. de French naturalist (1707–1788), with comparable influence to **Linnaeus** on contemporaries. His great work is *Histoire Naturelle* (1749–1788), a natural history in 44 volumes. Espoused **nominalism** with regard to species and other taxa, but had noted that species seemed to 'breed true'. He held that the environment had important effects on animal characteristics, amounting in time to a sort of degeneration from original types rather than being in any way creative. He favoured the theory of the **great chain of being**, indicating his distance from the later **Darwinism**.

Bug See **Hemiptera**.

Bulb Organ of vegetative reproduction; modified shoot consisting of very much shortened stem enclosed by fleshy scale-like leaves (e.g. tulip) or leaf-bases (e.g. onion). See **bulbil**. Compare **corm**.

Bulbil Dwarf shoot occurring in place of a flower (e.g. *Saxifraga*, *Festuca*, *Allium*), borne either in lower part of the inflorescence or in axils of leaves (lesser celandine) and serving as an organ of vegetative reproduction. See **apomixis**.

Bulk flow Overall movement of water or some other liquid induced by gravity or pressure or the interplay of both.

Bulla, auditory Bony protection of middle ear cavity in most placental, but not most marsupial, mammals. Absent also in earliest mammals.

Bundle scar The scar from a vascular bundle remaining on a leaf scar after **abscission**.

Bundle sheath Layer of cells which surrounds a vascular bundle, comprising cells of **parenchyma**, **sclerenchyma**, or both. See **Kranz anatomy**.

Bursa of Fabricius Thymus-like **lymphoid** organ found in birds, but not in mammals, developing dorsally from **cloaca**. Like the mammalian **thymus**, it appears to cause differentiation of circulating stem cells into immunoglobulin-producing cells; hence bursa-derived cells (**B cells**). Contains **antigen-presenting cells**. See **T cell**.

Buttress root An **adventitious** root on a stem, functioning in support; found mainly in monocotyledonous plants.

c- Genes which are cellular counterparts of **oncogenes** are prefixed by c-, as in c-*fos* – the cellular counterpart of the oncogene v-*fos*. Such genes are also termed c-oncs (cellular oncogenes).

Cadherins See **adhesion**.

Caducous (Bot.) Not persistent. Of sepals, falling off as flower opens (e.g. poppy); of stipules, falling off as leaves unfold (e.g. lime).

Caecum Blind-ending diverticulum, commonly of gut. One or two may be present at junction of vertebrate ileum and colon, housing cellulose-digesting bacteria. Thin-walled, sometimes with spiral valve for increased surface area, terminating in vermiform **appendix**. Mesenteric (midgut) caeca in some annelids (e.g. leeches) and many arthropods are secretory and absorptive, and may be generally 'liver-like' en masse.

Caenorhabditis elegans A small (1 mm) transparent, hermaphroditic, free-living soil nematode of relatively small genome (about 3000 genes) and few cell types, whose 16-hour embryogenesis can be achieved in a petri dish, making it highly suitable for the study of developmental and behavioural genetics. See **sex determination**.

Caffeine A powerful **alkaloid** stimulant of the central nervous system inducing increased mental alertness, clearer flow of thought, wakefulness and restlessness; in consequence, a psychoactive drug. Present in coffee, tea, cocoa, cola drinks and chocolate. An average cup of coffee contains 100 mg caffeine; a cup of tea contains about 50 mg. About 80% of adult Americans consume 3–5 cups of coffee per day on average. It seems to operate by blocking the action of adenosine receptors located on the membranes of cells in the CNS and peripheral nervous system; as a result, it prevents the sedating effects of adenosine but also reduces asthma and migraine. It is only a slight heart stimulant. High doses cause *caffeinism*: nervousness, irritability, muscle hyperactivity and twitching, raised body temperature and ventilation, heart palpitations and arrhythmias. Persons prone to panic attacks may be particularly sensitive to even moderate doses (4–5 cups of coffee). Total caffeine intake does not seem to increase risk of coronary artery disease or increase cancer; however, it appears to induce physiological dependence, with withdrawal symptoms including headache, irritability, fatigue, muscle pain and stiffness. It does not appear to be a **mutagen**, but should be regarded as possibly hazardous to the human foetus and neonate since it crosses the placenta and is concentrated in breast milk. See **nicotine**.

CAK (cdk-activating kinase) See **CDKs**.

Calciferol See **vitamin D**.

Calcitonin (CT) Polypeptide hormone

of parafollicular cells (*C-cells*) of thyroid gland. Lowers plasma calcium and phosphate levels by inhibiting bone degradation and stimulating their uptake by bone. May have evolved alongside conquest of land by vertebrates, given its role in regulating plasma ion levels. Antagonized by **parathyroid hormone**. See **ossification**.

Calcium pump An ATP-driven **transport protein**. Calcium ions (Ca^{2+}) act as **second messengers** in the cell cytosol and their changing concentration there, particularly in eukaryotes, is significant. Although total cell Ca^{2+} concentration approximates to that of the environment, it is unevenly distributed and Ca^{2+} pumps in the plasma membrane expel Ca^{2+} when it enters. Much is accumulated by pumps in **mitochondria** and other organelles, causing a thousandfold drop in Ca^{2+} concentration across the plasma membrane, a gradient down which the ion moves. The calcium pump in the sarcoplasmic reticulum of striated muscle accumulates Ca^{2+} from the cytosol, enabling **muscle contraction**. In plants, calcium pumps exist at the plasmalemma (see **guard cells**), tonoplast and endoplasmic reticulum, keeping Ca^{2+} levels low until an appropriate signal. See **calmodulins**, and Fig. 64.

Callose A complex branched polysaccharide associated with the **sieve areas** of sieve elements. May form in reaction to injury of these and parenchyma cells and be deposited so that their activity is impaired or finished, permanently or seasonally.

Callus (Bot.) Superficial tissue developing in woody plants, usually through cambial activity, in response to wounding, protecting the injured surface. Often used in tissue culture, when the effects of hormones upon cell differentiation can be studied. (Zool.) Fibrocartilage produced at bone fracture, developing into bone as blood vessels grow into it and pressure and tension are applied.

Calmodulins Small multiply-allosteric proteins required for Ca^{2+}-dependent activities of many cellular (esp. membrane-bound) enzymes. Said to be *activated* when bound to Ca^{2+}. Ubiquitous cellular component related to troponin C (see **muscle contraction**) which, once activated can in turn bind to several cell proteins (e.g. adenylate cyclase, some ATPases and membrane pumps) and regulate their activities and is a component of muscle **phosphorylase kinase**, accounting for its Ca^{2+}-dependence. Present in all plant species so far tested and implicated in geotropic response of roots. Its presence in plant cell walls probably affects the level there of free Ca^{2+}, which in turn seems to influence both growth and development (see **inositol 1,4,5-triphosphate**). Calmodulin gene expression in tobacco seedlings is greatly increased by touch. See **signal transduction**.

Calvin cycle Series of enzymic photosynthetic reactions in which carbon dioxide is reduced to 3-phosphoglyceraldehyde, while the carbon dioxide acceptor ribulose-1,5, bisphosphate is regenerated. For every six molecules of carbon dioxide that enter the cycle, a net gain of two molecules of glyceraldehyde-3,phosphate results. See **photosynthesis**.

Calyptra Hood-like covering of moss

and liverwort capsules, developing from the archegonial wall.

Calyptrogen Layer of actively dividing cells formed over apex of growing part of roots in many plants, giving rise to **root cap**.

Calyx Outermost part of a flower, consisting usually of green, leaf-like members (sepals) that in the bud stage enclose and protect other flower parts. See **flower**.

CAM See **crassulacean acid metabolism**.

Cambium (Bot.) A meristem giving rise to parallel rows of cells. See **vascular cambium**, **cork cambium**.

Cambrian The earliest period of the **Palaeozoic** era, and hence the start of the phanerozoic age (evident animals). Although Precambrian fossils are known, the earliest structural fossils are found in the Cambrian. Its fauna included trilobites, crustaceans, king crabs, eurypterids (see **Arthropoda**), annelids and brachiopods. The flora included bacteria, blue-green algae, large coenocytic green algae and red algae. Extended from 600–500 Myr BP.

Campylotropous (Of ovule) curved over so that funicle appears to be attached to the side, between chalaza and micropyle. Compare **anatropous**, **orthopterous**.

Cancer cells Cells which have escaped normal controls regulating growth and division, producing clones of dividing daughter cells which invade adjacent tissues and may interfere with their activities. Despite a normal oxygen uptake, cancer cells tend to use several times the normal glucose requirement. In vertebrates they produce lactic acid under aerobic conditions (termed *aerobic glycolysis*). This places a burden on the liver, which must use ATP to get rid of lactate. Cancer cells that proliferate but stay together form benign tumours; those that not only proliferate but also shed cells, e.g. via the blood or lymphatic system (metastasis) to form colonies elsewhere form malignant tumours, and cancer generally refers to a disease resulting from either. Among these, *carcinomas* are malignant tumours of epithelial cells. *Teratocarcinomas* are carcinomas that can be cultured *in vitro* and serially grafted to other hosts (see **teratoma**); *sarcomas* are cancers of connective tissue; *myelomas* are malignant tumours of bone marrow. Several types of DNA viruses can initiate tumours in animals, each encoding one or more **oncogenes** whose products promote growth and division of the host cells, some by binding to (sometimes promoting degradation of) negative cell growth regulators such as the proteins Rb (retinoblastoma gene product) and **tumour necrosis factor** p53. These may initiate cellular DNA replication, regulating entry into S phase of the **cell cycle**, or perhaps regulate transcription of those genes which do so (see also ***ras* genes**, **tumour suppressor genes**).

Canine tooth Dog- or eye-tooth of mammals; usually conical and pointed, one on each side of upper and lower jaws between incisors and premolars. Missing or reduced in many rodents and ungulates, they are used for puncturing flesh, threat, etc. Sometimes enlarged as tusks (e.g. in wild boar).

Capacitation Final stage in maturation

of, at least, bird and mammal sperm, without which fertilization is impossible. Generally occurs in female tract (sometimes *in vitro*) where substances, perhaps secreted by the ovary or by the uterine lining, must be encountered for the sperm to undergo the acrosome reaction (see **acrosome**). Sperm do 'wait' at specific points on the uterus wall, and may be capacitated then.

Capillary (1) (Of blood system) an endothelial tube, one cell thick and 5–20 μm in internal diameter, on a **basal lamina**, and linking a narrow metarteriole to a venule. Permits exchange of water and solutes between blood plasma and tissue fluid (hence called exchange vessels). Their walls lack smooth muscle and connective tissue, and their permeability depends on the junctions between the endothelial cells. Three main types: (i) *continuous capillaries* (e.g. in muscle), where just one endothelial cell with overlapping ends forms the whole tightly sealed structure; (ii) *fenestrated capillaries* (as in intestine, endocrine glands), where pores through the cell are closed by just a cell membrane diaphragm, offering little resistance to small solute molecules; or (e.g. in glomeruli) pores occur between the adjacent cells, the basal lamina alone restricting solute passage; (iii) *sinusoids*, discontinuous capillaries (as in liver and spleen), where complete gaps occur between fenestrated endothelial cells. These incomplete capillaries have the highest permeability and largest diameters of the three, proteins passing through, although few blood cells do. *Precapillary sphincters* at the junctions of capillaries and metarterioles can slow or shut off blood flow in response to pH, carbon dioxide, oxygen, temperature, dilator and constrictor agents (e.g. **adrenaline**, **noradrenaline**). Blood pressure squeezes water and solutes from plasma across capillaries, forming the tissue fluid bathing body cells. Blood cells and plasma proteins are retained on grounds of size, the latter causing the relatively low **water potential** of plasma at the venule end of a capillary, returning water to the blood. Solutes diffusing across the endothelium include oxygen, glucose, amino acids and salts (all outwards), carbon dioxide and metabolic wastes (e.g. urea in liver) inwards. Capillaries are absent from animals with open blood systems, but most vertebrate body cells are no further than 50 μm from a capillary. (2) (Of the **lymphatic system**) structurally similar to blood capillaries, but blind-ending and with non-return valves, draining off surplus water from the tissue fluid. See **blood**, **inflammation**.

Capillitium (Bot.) (1) Tubular protoplasmic threads in fruiting bodies of Myxomycota (slime moulds), assisting discharge of spores in some species by their movements in response to changes in humidity. See **elaters**. (2) Sterile hyphae in the fruiting bodies of certain fungi, e.g. puff-balls.

Capitulum (Bot.) (1) In flowering plants, inflorescence composed of dense aggregation of sessile flowers. (2) In the Sphagnidae (**Bryopsida**), a dense tuft of branches at the apex of the gametophyte.

cAPK Cyclic-AMP-dependent protein kinase.

Capping (1) See **RNA capping**. (2) *Cell capping*. Process by which antibodies or other membrane components are attached by cross-linking ligands (see **lectin**) to cell-surface antigens and then swept along the surface to one end (cap) of a motile cell (commonly the rear) where they may be ingested by endocytosis. Unlinked membrane components diffuse fast enough in the membrane to avoid being swept back.

Capsid Coat of virus particle, composed of one or a few protein species whose molecules (capsomeres) are arranged in a highly ordered fashion. See **virus**, **bacteriophage**.

Capsomere See **capsid**.

Capsule (Bot.) (1) In flowering plants, dry indehiscent fruit developed from a compound ovary; opening to liberate seeds in various ways, e.g. by longitudinal splitting from apex to base, separated parts being known as *valves* (e.g. iris); by formation of pores near top of fruit (e.g. snapdragon) or in the *pyxidium*, by detachment of a lid following equatorial dehiscence (e.g. scarlet pimpernel). (2) In liverworts and mosses, organ within which spores are formed. (3) In some kinds of bacteria, a gelatinous envelope surrounding the cell wall. (Zool.) Connective tissue coat of an organ, providing mechanical support.

Carapace (1) Bony plates, often fused, beneath the horny scutes of the chelonian dorsal skin (turtles, tortoises). See **plastron**. (2) Dorsal skin fold of many crustaceans arising from posterior border of head and reaching to varying extent over trunk somites. May enclose whole body (ostracods), the thorax (malacostracans), or be absent altogether (e.g. copepods). May enclose chamber in which gills are housed, embryos protected, etc.

Carbohydrate The class of organic compounds with the approximate empirical formula $C_x(H_2O)_y$, (i.e. literally 'hydrated carbon'), where $y = x$ (monosaccharides) or $y = x - [n - 1]$ (di-oligo- and polysaccharides) where n is the number of monomer units in the molecule. Of enormous biological importance both structurally and as energy stores. Sometimes atoms of nitrogen and other elements are also present (e.g. acetylglucosamine). They include **cellulose**, **chitin**, **glucose**, **glycogen**, **ribose**, **starch** and **sucrose**, but also occur as components of **glycolipids** and **glycoproteins**.

Carbon cycle The constant recycling of carbon atoms between inorganic (carbon dioxide, carbonates, bicarbonates) and organic sources. Both abiotic factors (e.g. rock-weathering) and biotic factors are involved. The major carbon-fixing process is **photosynthesis** by plants, phytoplankton, marine and freshwater algae and **Cyanobacteria**, during which they incorporate CO_2 from the atmosphere into organic carbon-containing compounds and release oxygen into the atmosphere. During **respiration** by all organisms, these compounds are broken down and CO_2 is again released. These processes result in the cycling of carbon. About 75 billion metric tons of carbon a year are bound into carbon compounds by photosynthesis. Some carbohydrates are used by the autotrophs themselves; plants release CO_2 from their roots, stems and leaves, and algae release carbon dioxide into the water where it maintains an equilibrium with that in the atmosphere. The seas and atmosphere have carbon sinks of some 500 and 700 billion metric tons respectively. Heterotrophs (herbivores, secondary consumers, decay organisms) use the organic products of photosynthesis for their own metabolic processes. CO_2 is released into the reservoir of the air and the oceans by almost all organisms through respiration for re-fixing. Another even larger carbon reservoir exists below the earth's surface as coal and oil, deposited there some millions of years ago. Another occurs in limestone rocks and as peat in the world's vast tracts of wetlands.

Natural processes of photosynthesis and respiration balance one another out. But since 1850 atmospheric CO_2 concentrations have been increasing dramatically, due in part to the burning of fossil fuels (coal, oil), the ploughing of the soil and the destruction of forests, particularly in the tropics, where, however, in contrast to the northern hemisphere, they are being replanted. See **ecosystem**, **food web**, **greenhouse effect**.

Carbonic anhydrase Enzyme of vertebrate red blood cells, brush borders of kidney proximal convoluted tubule, and other body cells. Essential in catalysing the reaction: $CO_2 + H_2O \leftrightarrow H^+ + HCO_3^-$ (reversibly under different blood pH conditions in lungs and tissues), speeding CO_2 transport. Also involved in blood pH regulation by kidney. See **red blood cell**.

Carboniferous A **palaeozoic** period, lasting from 345–280 Myr BP, notable for its coal measures, with lycopods (*Lepidodendron*) dominating along with sphenophytes (*Calamites*). During it, the lycopods influenced the evolution of many other groups of organisms. The multilayered forest ecosystems, dominated by lycopods, cordaites and calamites, produced a shaded environment which probably influenced the selection of plants tolerant of such conditions. This environment would have provided shelter for many organisms intolerant of more exposed conditions present at that time. Also, thick limestone deposits formed, rich in brachiopods. The present-day continents were in greater contact than today, but **Pangaea** had yet to form. Amphibians radiated during it, and reptiles appeared in its lowest deposits.

Carboxylase Enzyme fixing carbon dioxide or transferring COO^- groups. Important carboxylases occur in both respiration and photosynthesis.

Carboxypeptidase (carboxypolypeptidase) Pancreatic enzyme, digesting peptides to amino acids.

Carboxysome Structure in some bacteria (e.g. chemoautotrophic *Thiobacillus*), housing the CO_2-fixing enzyme ribulose-1,5, bisphosphate carboxylase.

Carcinogen Any factor resulting in transformation of a normal cell into a **cancer cell**. The **Ames test** assesses carcinogenicity of a substance. Gut bacteria and other fermenting organisms often produce carcinogens as by-products, many of them glycosides (sugar-containing). When bacterial glycosidase cleaves the sugar group, they become mutagenic and potentially carcinogenic. Red wine and tea appear more carcinogenic than white wine and coffee. Ultraviolet and X-radiation and mustard gas are classic carcinogens, their effects usually being attributable to mutation. See **mutagen.**

Carcinoma See **cancer cells**.

Cardiac Of the **heart**; hence cardiac cycle (see **heart cycle**), **cardiac muscle**, cardiac sphincter (at the junction of oesophagus and stomach, near the heart).

Cardiac centres See **cardio-acceleratory centre**.

Cardiac muscle One of three vertebrate **muscle** types; restricted to the heart walls. Striated, and normally involuntary. Myogenic (see **pacemaker**). Most obvious structural distinctions from skeletal muscle are its anastomosing (branching and rejoining) fibres, and the periodic irregularly thickened sarcolemma, forming *intercalated discs* which appear dark in most stained light microscope preparations. Unlike skeletal muscle, cardiac muscle tissue is not a multinucleate syncytium; each fibre is uninucleate and limited by its sarcolemma. Cardiac muscle tissue has a longer **refractory period** than skeletal muscle and consequently does not fatigue (see **muscle contraction**). Both pacemaker and accompanying Purkinje fibres are modified cardiac tissue, but with neurone-like properties.

Cardinal veins Paired veins dorsal to the gut of fish and tetrapod embryos, taking blood towards the heart from the head/front limb region (anterior cardinals) to join posterior cardinals (from trunk), forming a common cardinal (Cuvierian) duct which enters the sinus venosus. Replaced by venae cavae in adult tetrapods.

Cardio-acceleratory and cardio-inhibitory centres (cardiac centres) Association centres in medulla oblongata (see **brainstem**) of homeothermic vertebrates, with reciprocal effects on heart rate. The former employs sympathetic nerves, the latter parasympathetic (vagus). Regulated by hypothalamus and cerebrum. Adjustments of heart rate involve **baroreceptors**. See **carotid sinus**, **pacemaker**, **vasomotor centre**.

Carinate Of those birds (the majority apart from **ratites**) with a keel (*carina*) on the sternum. The group so formed is not now regarded as more than a **grade**.

Carnassial teeth Modified last premolars in each upper half-jaw and corresponding first lower molars of carnivorous mammals. Between them they shear and slice, e.g. tendons and bones, when jaw closes.

Carnivora Order including all living carnivorous mammals. Fossil **creodonts**, also carnivorous, form a separate order. Two suborders: **Fissipedia** (dogs, cats, weasels, civets), and **Pinnipedia** (seals, walrus). Canine and carnassial teeth and retractile claws usually present.

Carnivore Any meat-eater. Sometimes indicates a member of the **Carnivora**.

Carotenoids Group of yellow, orange and red lipid-soluble pigments found in all chloroplasts, **Cyanobacteria** and some bacteria and fungi, and chromoplasts of higher plants. Chromoplasts lack chlorophyll but synthesize and retain carotenoids, which are responsible for the colour of many fruits and roots (e.g. carrots). Like chlorophylls, carotenoids are embedded in thylakoid membranes. Carotenoids are not essential to photosynthesis but are involved as accessory pigments in the capture of light energy, absorbing photons of different energy which is then transferred to chlorophyll *a*; they cannot substitute for chlorophyll *a* in photosynthesis. Chemically, carotenoid pigments are long-chain compounds (tetraterpenes) and include carotenes (oxygen-free hydrocarbons) and **xanthophylls** (oxygenated derivatives of carotenes). The most widespread carotene is β-carotene, a principal source of vitamin A required by humans and other animals. A large number of xanthophylls exist, some unique to particular groups of **algae** and important in their systematics (e.g. *myxoxanthophyll*, *aphanizophyll* and *oscillaxanthin* are unique to the **Cyanobacteria**). In several photosynthetic and non-photosynthetic bacteria, carotenoids protect cells from photochemical damage.

Carotid artery Major paired vertebrate artery, one on each side of neck, supplying oxygenated blood to head from heart. Derived from third **aortic arch**. Each common carotid branches into internal and external carotids; their origins from the aorta vary.

Carotid body Small neurovascular structure near branch of internal and external carotids (near carotid sinus); supplied by vagus and glossopharyngeal nerves (see **cranial nerves**). Sensitive to oxygen content of blood, assisting in homeostatic reflex control of ventilation. May also assist carotid sinus reflexes.

Carotid sinus Small swelling in internal carotid artery (therefore paired) in whose walls lie **baroreceptors** innervated by the glossopharyngeal nerve (cranial nerve IX). Increase in arterial pressure and sinus stimulation causes reflex homeostatic drop in heart rate and vasodilation, involving the cardio-inhibitory and vasomotor centres of the **brainstem**.

Carpal bones Bones of proximal part of hand (roughly the wrist) of vertebrates. Compact group of primitively 10–12 bones, reduced to 8 in man. Articulate with radius and ulna on proximal side, and with metacarpals on distal side. See **pentadactyl limb**.

Carpel Female reproductive organ (megasporophyll) of flowering plants. Consists of ovary containing one or more **ovules** (which become seeds after fertilization), and a **stigma**, a receptive surface for pollen grains. Often borne at apex of a stalk, the **style**. See **flower**.

Carpellate See **pistillate**.

Carpogonium Female sex organ of red algae (**Rhodophyta**). Consists of swollen basal portion containing the egg, and an elongated terminal projection (trichogyne) receiving the microgamete.

Carpospore In red algae (**Rhodophyta**), the single diploid protoplast found within a containing cell (the carposporangium). Formed after fertilization and borne at the end of an outgrowth of the mature carpogonium.

Carpus Region of vertebrate fore-limb containing carpal bones. Approximates to wrist in man.

Carrageenan Complex phycocolloid found in the cell walls of various red algae (**Rhodophyta**, e.g. *Chondrus crispus, Gigartina stellata*). Similar to **agar**, except carrageenan has a higher ash content and requires a higher concentration to become a gel. Carrageenan is used extensively for many of the same purposes as agar; however, it provides a lower gel strength than agar. Used for stabilization of emulsions in cosmetics, paints, pharmaceutical preparations, stiffening of ice cream, instant puddings, sauces and creams.

Carrier (1) An individual **heterozygous** for a recessive character and who does not therefore express it, but half of whose gametes would normally contain the allele for the character (sex linkage excepted). (2) An individual infected with a transmissible pathogen and who may or may not suffer from the disease.

Carrier molecule See **ionophore**, **permease**.

Cartilage With **bone**, the most important vertebrate skeletal connective tissue. Cells (chondroblasts) derive from mesenchyme and become *chondrocytes* when surrounded within lacunae by the ground substance they secrete. This amorphous matrix (chondrin) contains glycoproteins, basophilic chondroitin and fine collagen fibres, varying proportions of which determine whether it is hyaline (gristle), elastic or fibrocartilage. The surface of cartilage is surrounded by irregular connective tissue forming the perichondrium. Growth may be interstitial (endogenous) resulting from chondrocyte division and matrix deposition within existing cartilage; or appositional (exogenous) resulting from activity of deeper cells of the perichondrium. Lacks blood vessels or nerves. Cartilage is more compressible than bone and in the form of intercostal cartilage absorbs stresses generated throughout the vertebral column during locomotion, lifting, etc.; costal cartilage caps the articulating bone surfaces of **joints**. The trachea is kept open by rings of hyaline cartilage; the pinnae of ears and auditory tubes contain elastic cartilage. In some kinds of **ossification** cartilage is destroyed and replaced by bone. The **Chondrichthyes** have entirely cartilaginous skeletons.

Cartilage bone See **ossification**.

Caruncle Warty outgrowth on seeds of a few flowering plants, e.g. castor oil; obscures **micropyle**.

Caryopsis A simple, dry, single-seeded indehiscent fruit. An **achene** with ovary wall (pericarp) firmly united with seed coat (testa). Characteristic of grasses (Fam. Poaceae).

Cascade Biological process by which progressive amplification of a signal via a sequence of biochemical/physiological events results in a very localized response. Such a sequence might involve a hormone or other ligand binding to a membrane receptor site, activation of membrane adenylate cyclase producing many cAMP molecules, each activating many kinase molecules, which in turn activate many enzyme molecules, each producing quantities of product. Activation of **complement**, **blood clotting**, **fibrinolysis**, **rhodopsin** activity and embryonic acquisition of **positional information** all result from cascades (see ***bicoid*** **gene** and **receptor** for Figs.)

Casein Conjugated milk protein. A phosphate ester of serine residues.

Rennin and calcium precipitate it to produce curd; also a major component of cheese.

Cassettes See **expression signals**, **mating type**, **mutagen**.

Caste In **eusocial** insects, a structurally and functionally specialized individual: a **morph**. Caste determination may depend upon the state of ploidy (e.g. haploid bees, ants and wasps are male), or a combination of ploidy (the number of haploid chromosome sets) and environmental factors (e.g. worker and queen bees are diploid and female, but only queens are fed royal jelly as larvae); in lower termites at least it appears to be non-genetic, pheromones produced by king and queen controlling differentiation of caste. Hymenopteran castes are: queen, worker (some ant species having soldier and non-soldier subcastes of worker) and drone; termite castes include: primary reproductives (king, queen), supplementary reproductives, workers and soldiers. See **polymorphism**.

Catabolism The sum of enzymatic breakdown processes, such as digestion and respiration in an organism. Opposite of **anabolism**.

Catabolite Metabolite broken down enzymatically.

Catabolite repression Suppression by a fuel molecule, or one of its breakdown products, of synthesis of inducible enzymes which would make use of alternative fuel molecules in the cell. Glucose represses production of galactosidase and some respiratory enzymes (the glucose effect) in bacteria. This involves gene repression, a glucose breakdown product combining with the cell's cyclic AMP (cAMP), reducing the amount available for transcription of the operon. See **gene regulation**, **Pasteur effect**.

Catalase Haem enzyme of **peroxisomes** of many eukaryotic cells. Converts hydrogen peroxide, produced by certain dehydrogenases and oxidases, to water and oxygen. Used commercially in converting latex to foam rubber and in removing hydrogen peroxide from food.

Catalyst Substance speeding up a reversible chemical reaction without altering its equilibrium point. Biological catalysts are **enzymes** and **ribozymes**.

Cataphyll Small scale-like leaf in flowering plants, often serving for protection.

Catarrhine Old World monkeys and apes, and all humans; i.e. all cercopithecoids and hominoids (the anthropoid Infra-order Catarrhini). Characterized by narrow nasal septum, thirty-two teeth (two premolars in each jaw quadrant) and by menstrual cycle. No prehensile tail. See **Anthropoidea**, **platyrrhines**.

Catecholamines Monoamine derivatives of amino acids having a catechol ring. Examples include the hormones/neurotransmitters **adrenaline**, **noradrenaline** and **dopamine**.

Caterpillar Larval stage of Lepidoptera, Mecoptera and some Hymenoptera, bearing abdominal prolegs in addition to thoracic legs. Generally poorly sclerotized and inactive, living close to food.

Catharanthus Genus of flowering plant; the rosy periwinkle (*C. roseus*), native to Madagascar, is the natural source of two highly effective anti-cancer drugs: *vinblastine* (used to treat Hodgkin's disease) and *vincristine* (used in cases of acute leukemia).

Cathepsins A group of proteolytic enzymes occurring in **lysosomes**.

Catheter Tube, often plastic, inserted into gut, blood vessels, etc. for withdrawal/introduction of material. Balloon catheters have an inflatable tip and may be used to dilate blocked vessels (e.g. the coronary artery).

Cation A positively charged ion.

Caudal (Of the tail) caudal vertebrae are tail vertebrae, the caudal fin of a fish is its tail fin.

Cauline Belonging to the stem, or arising from it.

Cauline bundle A **vascular bundle** forming part of the stem tissue.

Caveolae See **potocytosis**.

Cavitation Occurrence of air pockets and/or bubbles in xylem vessels when tension exerted on the water column exceeds that enabling cohesion. It may occur during water stress. An alternative route for the **transpiration** stream would be needed, bypassing the blockage; however, air pockets so formed may be squeezed out again by **root pressure**.

CD2, CD4, CD8, etc. (**cluster of differentiation**) **proteins**. See **accessory molecules**.

cdc genes Cell-division-cycle genes of yeasts, conserved in all eukaryotes so far examined, forming a regulatory network involved in the timing of mitosis. Their products are sometimes invovled in **cascades**. See **cell cycle**.

cdk (CDK) Cyclin-dependent **kinase**.

cDNA (complementary DNA) DNA complementary to RNA and produced by **reverse transcriptase** activity. Initially single-stranded, can be converted to double-stranded cDNA by **DNA polymerase** activity. cDNA complementary to mRNA lacks **intron** sequences, useful when cloning functional DNA.

Cell Mass of protoplasm made discrete by an enveloping plasma membrane (plasmalemma). Any cell wall material is, strictly speaking, extracellular (e.g. in most plants and fungi); but distinctions between intracellular and extracellular may be arbitrary (see **glycocalyx**).

The two basic types of cell architecture are those of **prokaryotes** and **eukaryotes**. See Fig. 14. In the former, cells consist entirely of cytoplasm (lacking nuclei); in the latter, cells have (or had) in addition one or more nuclei. Eukaryotic cells have greater variety of organelles, many enclosed in one or more membranes (see **cell membranes**). They are further distinguished by the presence of distinctive proteins, particularly **actin**, **myosin**, **tubulin** and **histone**, that have very significant uses and are entirely absent from prokaryotic cells. Actin is paramount in the structure of the eukaryotic **cytoskeleton**; tubulin is fundamental in cilium and flagellum structure, and in mitotic and meiotic spindles – none of these being found in prokaryotes, whose flagella are rigid and of a completely different structure. These and other features indicate how similar even such apparently dissimilar cells as those of plants and animals are

Low Speed (1000 G, 10 min)	*Medium Speed* (20,000 G, 20 min)	*High Speed* (80,000 G, 1 h)	*Very High Speed* (150,000 G, 3 h)
nuclei	mitochondria	microsomes	ribosomes
whole cells	lysosomes	rough & smooth ER	viruses
cytoskeletons	peroxisomes	small vesicles	large macro-molecules

Table 4. *Centrifugation of cell components.*

when compared with those of prokaryotes (bacteria and blue-green algae). Basic eukaryotic cell architecture is elaborated upon in many ways, notably by fungi, where true cells are commonly absent in much of the vegetative body, organization being **coenocytic**. A similar multinucleate situation, without intervening cell membranes, arises where eukaryotic cells fuse to form a **syncytium**. Both may be termed **acellular**. The plasmodesmata uniting many plant cells may be regarded as producing an intermediate condition. See **multicellularity**, **origin of life**.

Cell body (perikaryon) Region of a neurone containing the nucleus and its surrounding cytoplasm. Generally swollen compared with rest of cell. Some ganglia consist of aggregations of cell bodies.

Cell centre Alternative term for **centrosome**.

Cell cycle The period during which events involved in successful eukaryotic nuclear and cell reproduction are completed (see Fig. 15 for some of the details). In proliferative cells this includes all the events taking place between the completion of one round of mitosis and cytokinesis and the next. Cells which commence upon their differentiation pathway have generally left the cell cycle for good. The molecular details of the cell cycle have been the subject of intense research in recent decades, not least because of their implications for our understanding of the origins of many cancers (see **cancer cells**).

The onset of DNA replication at S-phase (within interphase) and of mitosis (M-phase), are the cycle's two key events (some embryonic cleavage divisions dispense with G1 and/or G2). The controls of entry to both may be common to all eukaryotic cells, involving successive waves of **cyclin**-dependent kinase (**cdk**) activity leading to phosphorylation of certain key proteins (e.g. cyclins, H1 histone, **lamins, elongation factors** and RNA polymerase II). Members of the **protein kinase** p34 family (also variously known as $p34^{cdc2}$, $p34^{cdc2/CD28}$, phosphoprotein pp34, or cdk1) encoded by the *cdc2* gene (see ***cdc* genes**) are key players in these events. Activity of $p34^{cdc2}$ kinase is itself regulated by phosphorylation at three or

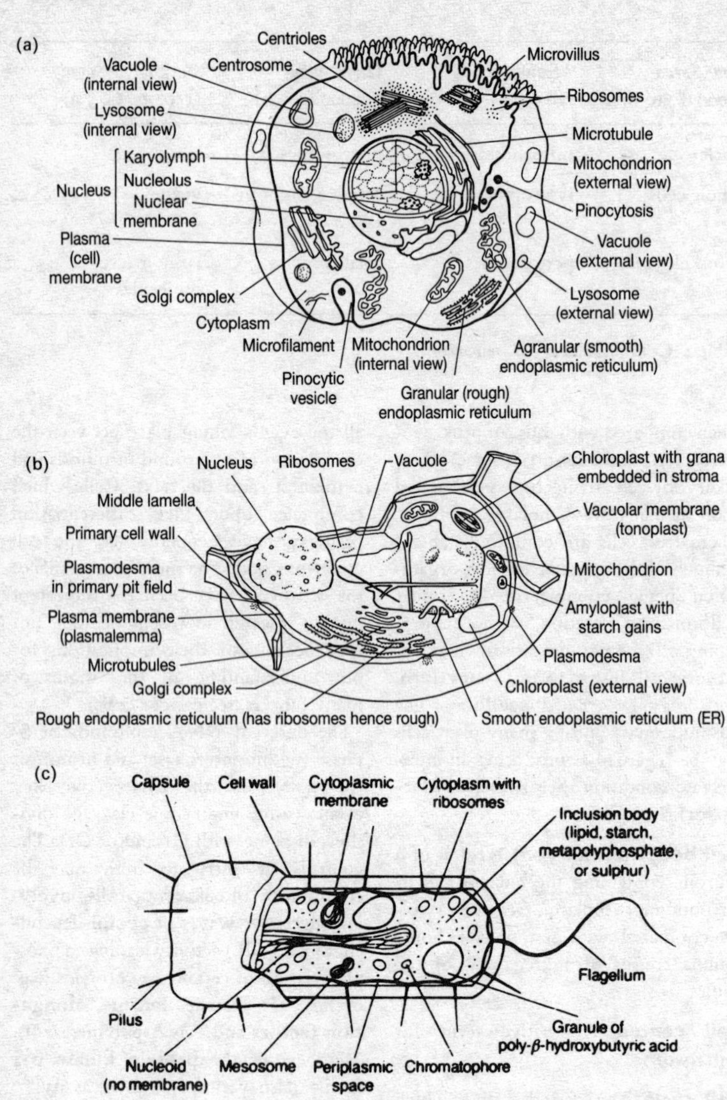

Fig. 14 (a) *Generalized animal cell structure based on electron microscope observations (approx. diameter 50 μm).* (b) *Generalized photosynthetic plant cell (approx. diameter 100 μm).* (c) *Generalized prokaryotic (bacterial) cell structure (approx. length 1–2 μm).*

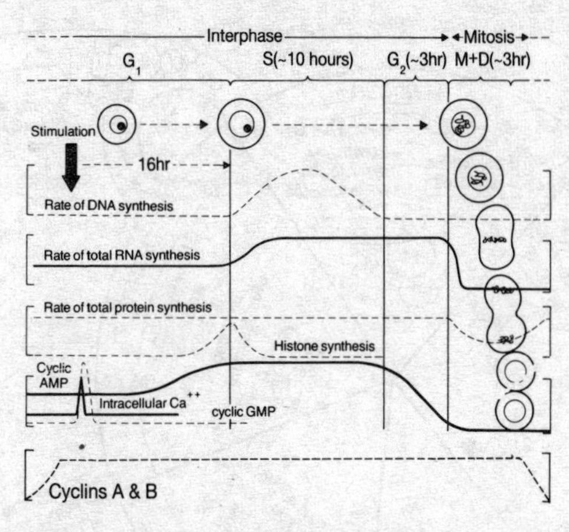

Fig. 15a *Breakdown of the phases of the eukaryotic cell cycle. Times are only approximate and vary for different systems.*

more sites, by phosphatases and by cyclins, with which it must complex prior to activation. 'Mitotic cyclin' degradation may potentiate the $p34^{cdc2}$ kinase inactivation required for exit from M-phase. Other phases of the cell cycle are mediated by other key regulatory protein kinases (e.g. the Abl cytoplasmic tyrosine kinase family) and their inhibitors. The c-Abl protein binds both actin and specific nuclear DNA sequences and is regulated by a cdc2-mediated phosphorylation (c-Abl has three phosphorylated serine/threonine sites in interphase cells, but seven more in mitotic cells). See **CKIs**, **centrosome**, **maturation promoting factor**, **tumour necrosis factor**.

Cell division Process by which a cell divides into two. (a) *Prokaryotic*: one event achieves separation of both DNA and cytoplasm into daughter cells. Since the two sister chromosomes are attached separately to the cell membrane they can become separated by the cleavage furrow formed between them as the cell membrane invaginates. Fission occurs as membrane intuckings fuse. No microtubules occur in prokaryotes, so there is no mitosis or meiosis; (b) *eukaryotic*: nuclear and cytoplasmic divisions are achieved by separate mechanisms. In higher eukaryotes the nuclear membrane breaks down and chromosomes attach to microtubules of the spindle by their **kinetochores**; cytoplasmic division (*cytokinesis*) usually starts in mitotic anaphase and proceeds by a furrowing of the plasmalemma in the plane of the metaphase plate, achieved by a contractile ring of **actin** filaments. Fusion of the invaginating plasmalemma then occurs. In plant cells with walls the new wall is built upon a **cell plate**. The plane of plant cell division may be anticlinal, periclinal transverse, as illustrated in Fig. 16. Golgi vesicles travelling towards it on microtubules deposit their wall precursor molecules, extending the plate to

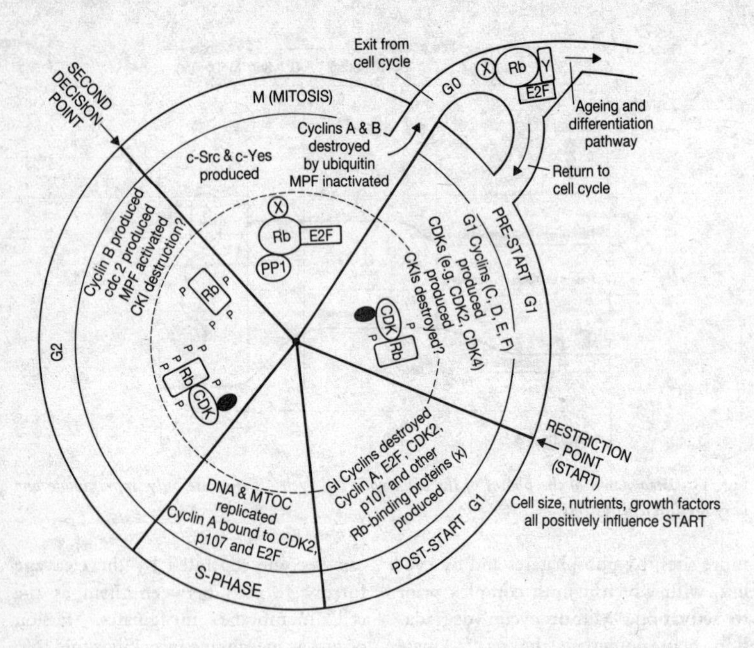

Fig. 15b *The major mammalian cell-cycle stages, with some of their prominent molecular influences and accompaniments. CDK = cyclin-dependent kinase; CKI = cyclin-dependent inhibitors; Rb = retinoblastoma protein, its phosphorylation states (p) indicated by figures; E2F = transcription factor E2F; c-Src and c-Yes are non-receptor protein kinases (cellular oncogenes). The table below indicates some of the complexes required for passage into, or through, critical cell-cycle stages in mammalian cells. Cyclin equivalents (Cln, Clb) in yeasts are shown in parenthesis.*

Stage in cell cycle	*Progress requires these complexes*
	cdk4/cyclin D (Cln3)
G1→S	cdk2/cyclin E (Cln1/2)
S	cdk2/cyclin A (Clb5/6)
G2→M	cdc2/cyclin A (Clb5/6)
M	cdc2/cyclin B (Clb1/2)

the cell membranes and pinching off the cell into two. Cells often need to coordinate cell division and cell mass. Cells in early embryos divide repeatedly with little increase in mass, periodicity of cell division being controlled by an internal timer, or oscillator. In somatic cells, M-phase is coupled to completion of S-phase (see **cell cycle**). In others, rate of mass increase dictates periodicity of cell division. See **mitosis** and **meiosis** for nuclear division.

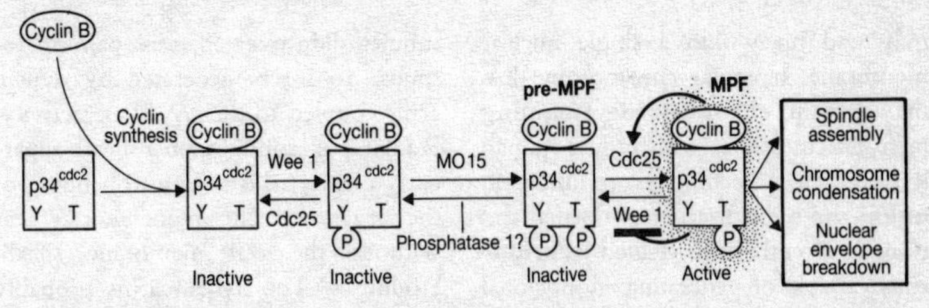

Fig. 15c *Figure to show mechanism of* **maturation promoting factor** *(MPF) activation. Y = tyrosine 15; T = threonine 161 (amino acid residues of* $p34^{cdc2}$*). MO15 is a kinase. Proteins Cdc25 and Wee1 modulate* $p34^{cdc2}$ *in opposite directions. Molecules phosphorylated on both Y and T residues accumulate in G2 and are converted to active MPF by Cdc25, in turn accompanied by Cdc25 activation and Wee1 inactivation.*

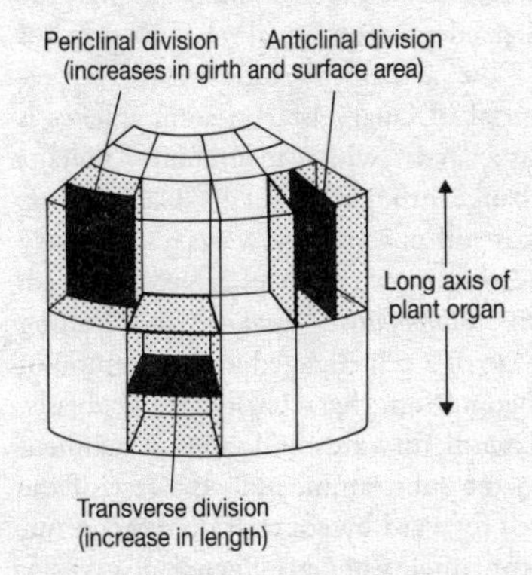

Fig. 16 *The three main planes in which plant cells can divide.*

Cell-division-cycle genes See ***cdc* genes, cell cycle**.

Cell fractionation Process whereby cells are first appropriately buffered (often in sucrose solution) and then disrupted (by osmotic shock, sonic vibration, maceration or grinding with fine glass, sand, etc.); the cell fragments are then spun in a refrigerated centrifuge. Different cell components descend to the bottom of the centrifuge tube at different speeds, and these can be increased progressively. Forces generated may be 500,000 times that of gravity (G). The G-forces and times required to spin down different cell constituents are shown in Table 4.

Cell fusion Process involving fusion of plasma membranes of two cells to form one resultant cell. All such membrane fusion events require the release of calcium ions (Ca^{2+}) from intracellular stores and are often initiated by receptor/ligand- and phospholipase C-mediated release of **inositol 1,4,5-triphosphate** (see Fig. 64). Naturally occurring cell fusion may or may not result in hybridization (unity of genomes). Fusion of **myoblasts** in skeletal muscle development, and other syncytial organizations, does not normally involve hybridization. The processes of **plasmogamy** and **karyogamy** are temporally separated in those fungal life cycles where a **dikaryon** occurs at some stage. In fertilization, separation of plasmogamy and karyogamy is usually brief. Artificial cell fusion is often achieved by treatment with inactivated viruses, or a glycol. The heterokaryon, with its separate nuclei intact, may then divide, in which case all chromosomes may end up within a single nuclear membrane. Irregular chromosome loss may permit **chromosome mapping** in tissue culture, as with mouse-human hybrid cells. Techniques resulting in fusion and hybridization of normal and tumour **B cells** have yielded **hybridomas** capable of generating monoclonal antibodies on a commercial or clinical scale. Protoplasts resulting from enzymic digestion of plant cell walls can be encouraged to fuse, and may generate heterokaryons or even fusion hybrids. Appropriate horticulture can generate somatic hybrid plants between species that would not normally hybridize. As with mouse-human somatic hybrids, chromosome loss often prevents a genetically stable product.

Cell hybridization See under **cell fusion**.

Cell locomotion There are various methods by which cells move, those of prokaryotes having apparently little in common with those of eukaryotes. For the latter, most mechanisms seem to involve protein tubules or filaments sliding past one another and generating force. The details of how force is transmitted to the substratum remain largely unknown.

(1) *Bacterial:* H^+ gradients across the inner cell membrane provide the motive force for rotation of the **flagellum**, whose fixed helix of protein subunits permits clockwise and counter-clockwise rotations, like a corkscrew. This involves an extraordinary 'wheel-like' rotor in the inner membrane, and cylindrical fixed bearing in the outer membrane. Reversal of flagellar rotation alters the behaviour of the cell.

(2) *Eukaryotic.* (*a*) *Ciliary/flagellar:* see **cilium** for structure. Paired outer microtubules slide over adjacent pairs in response to forces generated by dynein arms coupled to their ATPase activity. Radial spokes and the inner sheath apparently convert this sliding to bending of the organelle. The axoneme can beat without the cell membrane sheath around it. The dynein arms probably act in an equivalent fashion to myosin heads during **muscle contraction** and make contact with adjacent microtubule pairs during their power stroke. Control of ciliary/flagellar beat appears to be independent of Ca^{2+} flux, but may be dependent upon signal relay via proteins of the actual structure. However, reversal of ciliary beat in some ciliates is associated with membrane voltage change brought about by Ca^{2+} influx. It is still uncertain how waves of ciliary beating in cell surfaces are coordinated.

(*b*) '*Fibroblastic*' *crawling*: the leading edge of a cell engaged in this method of locomotion, characteristic of fibroblasts, extends forwards and, after attachment to the substratum, pulls the rest of the cell forward by contraction of actin microfilaments under influence of myosin. Typical features associated with this method are lamellipodia and microspikes (see **cell membranes**, **filopodium**), which both pass backwards in waves along the upper cell surface ('ruffling'), typically when the anterior of the cell has failed to attach to the substratum. Molecular mechanisms involved are not clear, but it seems that random endocytosis of plasmalemma and its subsequent restricted exocytosis (incorporating the membrane pieces) at the anterior of the cell generates a circulation of membrane akin to movement of tank caterpillar tracks. The protein **fibronectin** is involved in fibroblast crawling.

(*c*) '*Amoeboid*' (*pseudopodial*): the cell's outermost layer is gel-like (plasmagel) while the core is a fluid sol (plasmasol). It is possible that contraction of the thick cortical plasmagel squeezes the plasmasol and generates pseudopodial extensions of the cell, at the tips of which sol-to-gel transformation occurs. Gel-to-sol changes accompany this elsewhere in the cell, e.g. as a pseudopod retracts. Just how these **cytoplasmic streaming** events are coupled to locomotion is not clear, but motive force must act against regions where the cell adheres to its substratum (see **fibronectin**). Apparently, surfaces of large amoebae are relatively permanent, undergoing folding and unfolding to accommodate pseudopod extension and retraction. **Actin** is implicated in the process. Characteristic of amoebae, macrophages. Cell migration plays a crucial part in development (see **gastrulation**, **positional information**). See **capping**, **desmid**.

Cell markers See **genetic marker**.

Cell-mediated immunity See **immunity**.

Cell membrane (plasma membrane, plasmalemma) The membrane surrounding any cell. See **cell membranes**.

Cell membranes Cells may have a wide variety of membranes (often called 'unit' membranes) varying from 5–10 nm in thickness; but all have a plasma membrane (plasmalemma), the outer limit of the cell proper, which is generally quite distinct from any cell wall material present (which is extracellular). See **self-assembly**.

Major membrane functions include: restriction and control of movements of molecules (e.g. holding the cell together) enabling scarce metabolites to reach local concentrations sufficient to enhance enzyme-substrate interactions (see **potocytosis**); to act as platforms for the spatial organization of enzymes and their cofactors, holding otherwise scattered molecules in functional contact; and separation and localization of incompatible reactions. Many eukaryotic organelles have one or two membranes around them, chloroplasts having yet a third system within. The currently accepted structure of most cell membranes is that proposed in the *fluid mosaic model*, the evidence coming from X-ray crystallography, freeze-fracture and freeze-etching electron microscopy (see **microscope**), radiolabelling, electron spin resonance spectroscopy and fluorescence depolarization. The last two involve insertion of *molecular probes* with particular spectroscopic features adding peaks or troughs to the lipid spectrum.

In this model an outer and an inner phospholipid monolayer (major components phosphatidyl ethanolamine and lecithin) lie with their polar phosphate heads in the direction of the water which the bilayer thus separates. Specific, and different, proteins lie in one or other layer or traverse the bilayer, making the membrane *asymmetric*. Proteins which span the membrane (transmembrane proteins) have their topologies established in the endoplasmic reticulum. Their hydrophobic portions (esp. the outer surfaces of α-helices) often span the phospholipid bilayer several times (e.g. **bacteriorhodopsin, rhodopsin, receptor tyrosine kinases** and the acetylcholine receptor), such multi-spanning proteins often acting as **ion channels** or ion gates – the hydrophilic inner portions of their α-helices allowing water through. Such integrated membrane proteins commonly require detergent for their release, but proteins linked by ester bonds to the membrane fatty acids are removed by 1 M hydroxylamine or high pH. Proteins for incorporation into the plasmalemma will have appropriate signal regions targeting them there (see **protein targeting**).

The whole structure has fluid properties resulting from rapid lateral movement of most of its molecular components through thermal agitation (1 $\mu m.s^{-1}$ for lipids, 10 $\mu m.min^{-1}$ for proteins). Thus fused mouse and human cells, each with differently labelled membrane proteins, exhibit rapid mixing of labels over the entire cell surface. The ionophore gramicidin functions only when the two halves of the molecule, one in each half of the bilayer, come together – which they do in a quantized way, indicating membrane fluidity. Endocytosis, exocytosis and other processes involving membrane fusion (e.g. fertilization) are made possible by this fluidity.

The plasmalemmas of animal cells typically have the oligosaccharide chains of their **glycolipids** and **glycoproteins** exposed freely on their outer surfaces, playing important roles in immunological responses, in cell–cell **adhesion** and identification, and in cell surface changes. The plasmalemma of the bacterium *Halobacterium halobium*, unique among biological membranes, has terpenoids and not fatty acids in its phospholipid molecules. Most plasmalemmas comprise about 40–50% lipid and 50–60% protein by weight. The phospholipid bilayer has a non-polar hydrophobic interior, preventing passage of most polar and all charged molecules. Small non-polar molecules readily dissolve in it, and uncharged polar molecules (e.g. H_2O) can also diffuse rapidly across it, possibly assisted by the polar phospholipid heads, or by such ionospheres as gramicidin. Lipid bilayers are impermeable to carbohydrates and ions at the diffusion rates needed by cells; but membranes contain various **transport proteins** which speed transfer of metabolites across them so that small and otherwise inaccessible ions and molecules may be carried across cell membranes by **gated channels**, **facilitated diffusion**, **ionophores** or **active transport**. One by-product of this activity may be to generate ionic imbalances across the membrane which may be used to power ATP synthesis, or drive symports and antiports.

Large molecules or even solid particles gain access to the cell's geographic interior by pinocytosis and phagocytosis, and may be jettisoned by **exocytosis**. All these involve enclosure of transported molecules within membranous vesicles which fuse only with appropriate cell membranes. This recognition ability probably resides in the specificity of proteins exposed at a membrane surface.

Not all membrane phospholipids are identical, and this prevents their crystallization at low temperatures (**cholesterol** has an important role here in animal plasma membranes) as well as permitting local loss of fluidity as at synapses and **desmosomes**.

The carbohydrate content of plasma membrane glycolipids and glycoproteins may be such as to create a cell coat

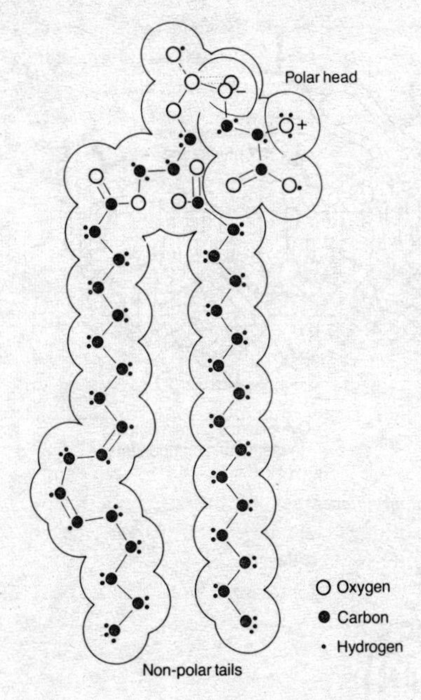

Fig. 17a *A phospholipid molecule (phosphatidylserine).*

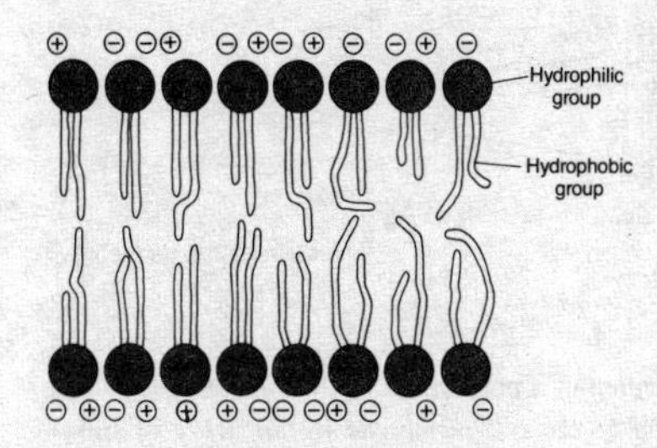

Fig. 17b *A phospholipid bilayer.*

or **glycocalyx**. Other membrane proteins act as *receptor sites* binding specific ligands (e.g. see **cascade**, **gated channels**). The eukaryotic plasma membrane is involved in the structures of **cilia** and **flagella**, **microvilli**, **lamellipodia**, **microspikes** and several sorts of **intercellular junction**. See appropriate organelles for further membranes. For membrane movement through the cell, see **Golgi apparatus**, **lysosome**.

Cell plate (Bot.) 'Plate' of differentially staining material which appears at telophase in the **phragmoplast** across the equatorial plane of the spindle. Believed to be forerunner of **middle lamella**. See **cell wall**.

Cell shape For some of the factors involved, see **adhesion**, **cytoskeleton**.

Cell theory The theory, first proposed by Schwann in 1839, that organic structure originates through formation and differentiation of units, the cells, by whose divisions and associations the complex bodies of organisms are formed. Much of the original theory is now untenable. Schleiden's name is also associated with the theory. See **Virchow**.

Cellular immunity and cellular response See **immunity**.

Cellular memory See **chromosomal imprinting**.

Cellulose The most abundant organic polymer. A polysaccharide, occurring as the major structural cell wall material in the plant kingdom. Some fungi have it as a component of their hyphal walls, and it may occur in animal cell coats (see **glycocalyx**). A long-chain polysaccharide of repeating *cellobiose* units, it may also be considered as a long chain of β[1, 4]-linked glucose units. Hydrogen bonding both within each molecule as well as between parallel molecules (producing crystalline *microfibrils*) gives

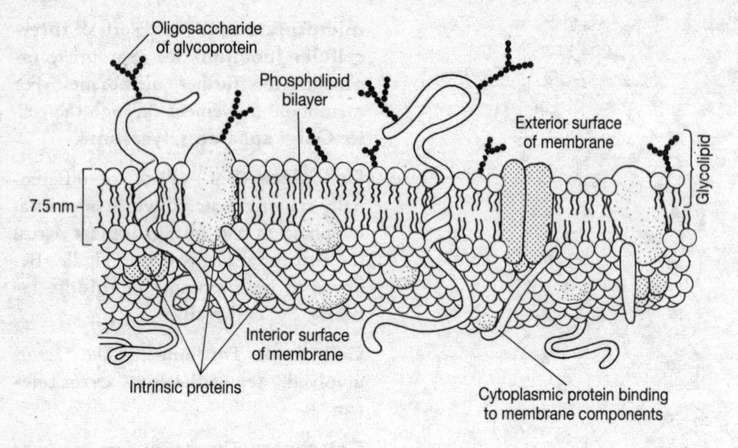

Fig. 17c *Depiction of the fluid mosaic model of cell membranes.*

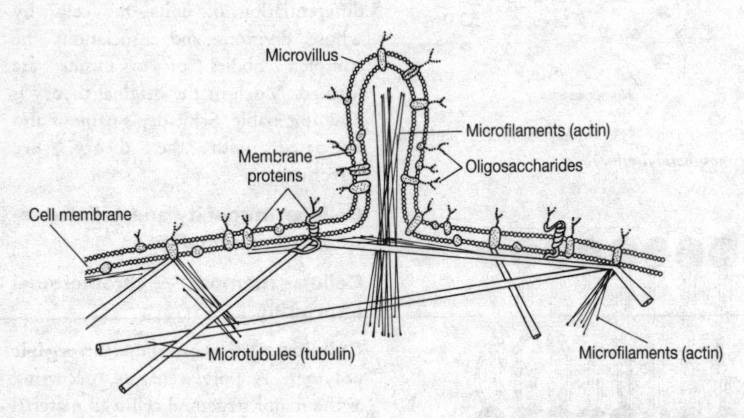

Fig. 17d *Components of the cytoskeleton in the region of a microvillus. Actin microfilaments and tubulin microtubules are linked to each other and to the cell membrane by an array of linker proteins not shown.*

cellulose its great tensile strength; but microfibrils can be loosened by lowered pH (an effect of **auxins** on the cell) allowing for wall extension in cell growth, when more cellulose may be laid down between existing microfibrils. With **lignin**, it forms *lignocellulose*. The fibrous texture of cellulose is responsible for its use in textile industries (cotton, linen, artificial silk). See **cell wall** for cellulose distribution.

Cell wall Extracellular coat of cells of bacteria, **Cyanobacteria** (blue-green algae), **Prochlorophyta**, **Plantae**, **Fungi** and many **Protista**; secreted by the protoplasm, and closely investing it. The bacterial wall is a component of its envelope and contains either a thick layer of **peptidoglycan**, or rather little (see **Gram's stain**). Mucins may also be present. The cell wall structure of the blue-green algae and prochlorophytes is very similar to the Gram-negative bacteria, but more complex, comprising several layers, while outside is a mucilage layer (sheath or capsule) which is fibrillar. Protruding from the wall of some blue-green algae are fimbriae or pili, which are possibly involved in prokaryotic–eukaryotic interactions (e.g. symbiosis). Comparatively rigid, these and the chitinous walls of fungal cells and hyphae provide mechanical support. Protistan algae may also have cell walls comprising fibrillar and amorphous components – again, the most common fibrillar component is cellulose; but in some siphonaceous green algae and red algae, mannan (polymer of 1,4-linked β-D-mannose) and xylans (of different polymers) replace cellulose. The amorphous mucilaginous components occur in greatest amounts in the brown and red algae, and are commercially exploited (see **alginic acid**, **fucoidin**, **agar**, **carrageenan**). Other protista may have incomplete cellulose walls (e.g. **lorica** or theca of *Dinobyron*, *Trachelomonas*); others have exquisitely ornate siliceous walls (e.g. **Bacillariophyta**–diatoms). In *Chlamydomonas*, the glycoprotein *extensin* forms the entire cell wall. See **Chrysophyta**, **pellicle**, **scale**.

Walls of newly formed plant cells (**Plantae**) are at first very thin and permeable, thickening as cells mature. At plant **cell division** (see Fig. 16), **pectic compounds** are laid down in the **cell plate** across the equatorial plane of the division's spindle forming the **middle lamella**, intercellular material cementing adjacent cells together. Each new cell lays down a *primary wall* consisting of **cellulose** (fibrillar component; polymer of 1,4-linked β-D-glucose), **hemicelluloses**, and negatively charged pectins (see **cutin**). Hydrogen bonds bind hemicellulose molecules to cellulose microfibrils, cross-linking them. Pectin molecules, being negatively charged, bind cations such as calcium (Ca^{2+}), and in doing so form a gel-like matrix (amorphous component) filling the interstices between the cellulose microfibrils, holding them together. Glycoprotein molecules probably attach to pectins (see Fig. 18a). At maturity, a cell may remain with just its primary wall (e.g. in some forms of **parenchyma**); in others, after cell growth has ceased, a *secondary wall* may develop inside the primary wall (see Fig. 18b). During deposition of these layers, certain small areas remain largely unthickened, forming **pits**. Pits of adjacent cells usually coincide, so that in these areas protoplasts are separated by the **pit membrane** on each side. Through the pit membrane pass the majority of the plasmodesmata, fine protoplasmic connections which are elements of the **symplast**. Some walls undergo further modifications, waxy cuticles developing on epidermal cells; others undergo suberization (e.g. cork cells which become impermeable to water). Lignification of fibre vessels and tracheids gives them more strength and rigidity.

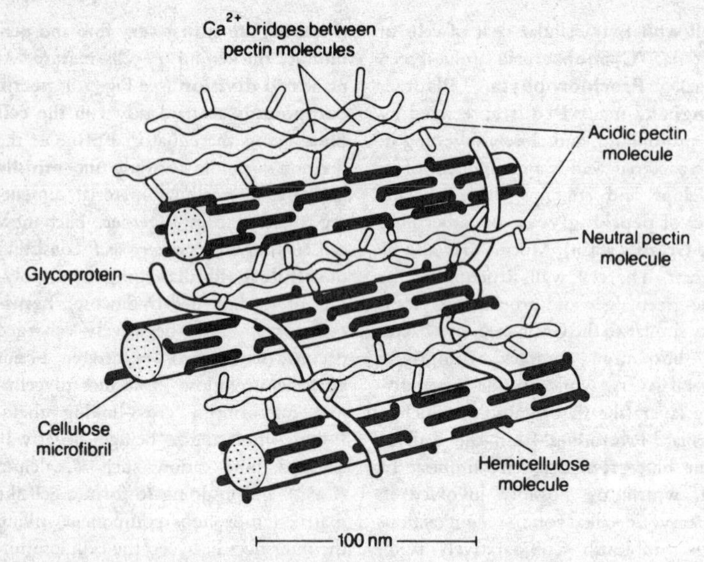

Fig. 18a *The relative arrangements of molecule types in a primary cell wall.*

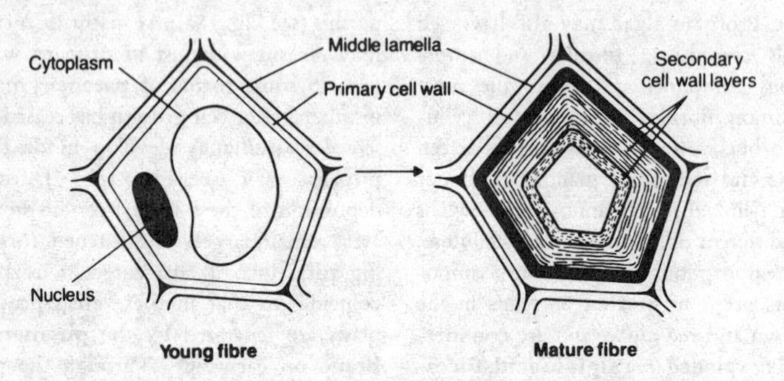

Fig. 18b *Secondary cell wall deposition by a phloem fibre cell, to show different wall layers.*

The plant cell wall limits cell growth (see **auxins**), is a barrier to digestion (especially when toughened by aromatic polymers; see **tannins**), glues adjacent cells together and plays an important role in plant morphogenesis. Its stretch-resistance is a major contributory factor to a plant cell's **water potential**. Cell walls can contain enzymes which incompletely digest its polysaccharides, releasing oligosaccharides that can act like growth substances and serve in cell-to-cell signalling (see **cellulose, chitin**). They also contain **calmodulin**, regulating the concentration of free Ca^{2+} in the wall, which in turn may affect cell growth and development.

Cement (cementum) Modified bone surrounding roots of vertebrate teeth (i.e. below gum), binding them to periodontal ligament by which tooth is attached to jaw. In some herbivorous mammals, occurs between folds of the tooth, forming part of the grinding surface.

Cenozoic (Cainozoic) The present geological era; extends from about 65 Myr BP to the present. The 'age of mammals'. Its two periods, the Tertiary and the Quaternary, are sometimes regarded as eras in their own rights.

Centimorgan (cM) Unit of relative distance between genes on a chromosome, 1 centimorgan corresponding to a cross-over value (COV) of 1%; in the human genome, this corresponds to about 10^6 base pairs. See **human genome project**.

Central dogma Proposal by F. H. C. Crick in 1958 that the flow of molecular information in biological systems is from DNA to RNA and then protein. RNA viruses (e.g. RNA tumour viruses) have since been shown to transcribe single-stranded DNA from RNA templates by means of the enzyme reverse transcriptase, providing exceptions to the generalization. See **cDNA**.

Central mother cells Relatively large vacuolated cells in a subsurface position in the apical meristem of a plant shoot.

Central nervous system (CNS) A body of nervous tissue integrating animal sensory and motor functions and providing through-conduction pathways to transmit impulses rapidly, usually medially, along the body. In vertebrates it comprises the **brain** and **spinal cord**; in annelids and arthropods a pair of solid ventral nerve chains, each with segmental ganglia, and a pair of dorsal ganglia anteriorly serving as a 'brain', united to the nerve chains by commisures. Impulses travel to and from the CNS via peripheral nerves (vertebrate spinal nerves), while local reflex arcs (vertebrate spinal reflexes) produce adaptive responses to stimuli independently of higher centres (the brain), although these centres initiate and coordinate actions and store memory. See **nervous system**, **spinal cord**.

Centric diatom Common term for a diatom (**Bacillariophyta**) which is radially symmetrical when viewed in valve view.

Centriole Organelle (probably of endosymbiotic origin) found in cells of those eukaryotic organisms which have cilia or flagella at some stage in their life cycle; hence absent from higher plants. Each comprises a hollow cylinder composed of nine sets of triplet microtubules held together by accessory proteins. Each is 300–500 nm long and 150 nm in diameter. Often functionally interconvertible with **basal body**. They occur at right angles to each other near the nucleus, separating at cell division amd organizing the spindle microtubules (which arise from material surrounding the centriole, but possibly in turn organized by it). Centrioles generally arise at right angles to existing centrioles. Normally an animal obtains its centrioles from the sperm cell at fertilization;

rarely, an egg may form its own (see **parthenogenesis**). Centrioles possess their own DNA and appear to be self-replicating, and there may be a link between centriole replication and nuclear DNA replication. Similar or identical structures (**basal bodies**), possibly functionally interconvertible, occur at the bases of cilia or flagella in cells which have these. See **centrosome**.

Centrolecithal Of eggs (typically insect) where yolk occupies centre of egg as a yolky core. See **telolecithal**.

Centromere (spindle attachment) A chromosome region holding sister chromatids together until mitotic or second meiotic anaphase. The position of a centromere defines the ratio between the lengths of the two chromosome arms. Centromeres may be associated with **repetitive DNA** sequences (not in the yeast *Saccharomyces cerevisae*) and centromeric DNA may be late-replicating. They either include or correspond to **kinetochores**, which attach to the spindle fibres and by replicating at late metaphase allow the forces pulling sister chromatids apart to operate only if chromosomes are properly aligned. Normally one per chromosome; but chromosomes with 'diffuse' centromeres (e.g. those of many lepidopterans) permit spindle fibre attachment along the whole chromosome length. See **acentric**, **acrocentric**, **metacentric**, **telocentric**.

Centrosome (cell centre) Amorphous, electron-dense material (in most animal cells surrounding and including the **centriole** pair) situated close to the nuclear envelope, serving as a microtubule-organizing centre (see **MTOC**). Despite a lack of clear physical boundaries, centrosomes have their own duplication cycle lasting through all phases of the **cell cycle**. Removal of the centriole itself seems not to prevent expression of a new MTOC but prevents centrosome partitioning and formation of a normal bipolar mitotic spindle. They must be accurately duplicated once per cell cycle, and activation of the $p34^{cdc2}$ kinase controlling M-phase of the cell cycle seems to depend on this. Functional details are expected to emerge as the biochemistry of this enigmatic organelle is better understood; e.g., the actin-related protein centractin is associated both with centrosomes and cytoplasmic dynein.

Centrum Bulky part of a vertebra, lying ventral to spinal cord. In function, as in development, replaces the notochord. Each is firm but flexible, attached to adjacent centra by collagen fibres.

Cephalaspida (Osteostraci) Extinct group of monorhine vertebrates. See **Agnatha**.

Cephalic index Measure of skull shape introduced by anatomist Retzius. It relates breadth as a percentage of length ((B/L) × 100).

Cephalization The tendency, during evolution of animals with an antero-posterior axis, for sense organs, feeding apparatus and nerve tissue to proliferate and enlarge at the anterior end, forming a head.

Cephalochordata (Acrania) Subphylum of marine chordates characterized by persistence of notochord in adult, extending (unlike in vertebrates) to the tip of the snout. Metameric segmentation, dorsal hollow nerve cord, gill slits and post-anal tail also present. Amphioxus (*Branchiostoma*) is typical. Compare **Urochordata**.

Cephalopoda Most advanced class of the phylum **Mollusca**. All are aquatic, and most marine, possessing a well-developed head surrounded by a ring of prehensile tentacles; and a muscular siphon derived from the foot through which water is forced from the mantle cavity during locomotion. Primitively (e.g. *Nautilus* and extinct **ammonites**) the animal inhabits the last chamber of an external spiral shell which also serves for buoyancy; in the cuttlefish *Sepia* the shell is internal, while in squids it is much reduced, and absent altogether in *Octopus*. The complexity of cephalopod eyes rivals that of vertebrates (and provides an example of convergent evolution), while the large brain enables powers of learning and shape recognition on a par with simple vertebrates. Much has to be learnt about cephalopod communication; some believe that cuttlefish employ their phenomenal powers of colour and pattern change to this effect.

Cephalothorax Term indicating either fusion of, or indistinctness between, head and some or all anterior trunk (thoracic) segments in crustacean and arachnid arthropods.

Cercaria The last larval stage of flukes (Order Digenea); produced asexually by **polyembryony** within preceding redia larva inside secondary host, often a snail, from which it emerges and swims with its tail to penetrate skin of primary host (e.g. man in *Schistosoma* causing bilharzia) or to encyst as a metacercaria awaiting ingestion by primary host.

Cerci A pair of appendages, often sensory, at the end of the abdomen of some insects. Long in mayflies, short in cockroaches and earwigs (where they are curved).

Cercopithecoidea Old World monkeys. See **Anthropoidea**.

Cereal Flowering plant of the family Gramineae, whose seeds are used as human food, e.g. wheat, oats, barley, rye, maize, rice, sorghum.

Cerebellum Enlargement of the hindbrain of vertebrates, anterior to the medulla oblongata. Coordinates posture (balance) during rest and activity through reflexes initiated by inputs mainly from the **vestibular apparatus** fed via acoustic regions of the medulla (the lower vertebrate **acoustico-lateralis system**), and from the **proprioceptors** in muscles and tendons. In mammals, covered in a cortex of grey matter. See Fig. 13 (**brain**).

Cerebral cortex (pallium) Layer of **grey matter** rich in synapses lying atop white matter, covering cerebral hemispheres of amniote and some anamniote vertebrates. In advanced reptiles and all mammals a new association centre, the neopallium, appears in the cortex receiving sensory inputs from the brainstem and initiating actions via motor bundles of the pyramidal tract. Its evolving dominance in mammalian brain involves its reception of increasingly wide ranges of sensory information via the thalamus and the emergence of higher neural (i.e. mental) activities based upon these data. Folding of the cortical surface in mammals provides a large surface area for synaptic association.

Cerebral hemispheres (cerebrum) Paired outpushings of vertebrate forebrain, originally olfactory in function, whose evolution has involved progressive movement of grey matter to its surface and an increasing role as an association and motor control centre. The **cerebral cortex** dominates the mammalian brain both physically and functionally.

Cerebroside **Sphingolipids** of the myelin sheaths of nerves, the commonest being *galactocerebrosides* with a polar head group containing D-galactose. Other tissues contain small amounts of glucose-containing cerebrosides.

Cerebrospinal fluid (CSF) Fluid filling the hollow neural tube and subarachnoid space of vertebrates. Secreted continuously into ventricles of the brain by the choroid plexuses and reabsorbed by veins. Clear and colourless fluid, with some white blood cells, supplying nutrients. Serves as shock absorber for the central nervous system. About 125 cm^3 present in humans. See **meninges**.

Cerebrum See **cerebral hemispheres**.

Cervical (Adj.) Of the neck; or **cervix**. Cervical vertebrae have reduced or absent ribs; almost all mammals (including giraffe) have seven.

Cervix Cylindrical neck of mammalian uterus, leading into vagina. Glands secrete mucus into vagina.

Cestoda Tapeworms. Endoparasites (Class **Platyhelminthes**) lacking gut and absorbing digested food from host gut lumen across microtriches, minute folds of the surface epithelial cell membranes similar to microvilli. Tapeworms are unsegmented, but body sections (proglottides) budded off from head region (scolex) give segmented appearance. Sequentially hermaphrodite, young proglottides male but become female with age. Self-fertile. Life cycle involves primary and secondary hosts. Larva a six-hooked onchosphere egested in proglottis with faeces of primary host. Sense organs reduced.

Cetacea Whales. Order of placental (eutherian) mammals. Entirely aquatic. Digit symmetry in the vestigial hindlimbs of the Eocene fossil whale *Basilosaurus* supports a common ancestry of cetaceans and artiodactyls with mesonychid **Condylarthra**. Morphology convergent with ichthyosaurs, with a dorsal fin, forelimbs developed as flippers, and tail a powerful fluked swimming organ. Traces only of pelvic girdle. Subcutaneous fat (blubber) for thermal insulation. Dorsal blowhole connects with lungs. Includes Odontoceti (toothed whales, including porpoises and dolphins) and Mysticeti (whalebone whales). Earliest fossils from Eocene.

CFCs (chlorofluorocarbons) Halocarbons containing carbon, chlorine and fluorine, all human-derived. Global mean tropospheric concentrations of the two most important CFCs in global warming through the **greenhouse effect**, i.e. CFC-11 ($CFCL_3$) and CFC-12 (CF_2CL_2), increased by about 6% between 1977 and 1986. Their main uses are in aerosols, air conditioning and refrigerators, while their major natural sink is photolytic breakdown in the stratosphere at 20–40 km, releasing free chlorine atoms which combine with **ozone**, decreasing the latter's concentration and UV-absorbing effect.

Their radiative effect per molecule relative to CO_2 varies from 11,000–14,000 times.

Chaeta Chitinous bristle characteristic of oligochaete (where few) and polychaete (where many) annelid worms. In polychaetes they are borne on parapodia. Assist in contact with substratum during locomotion. See **seta**.

Chaetognatha Arrow-worms. Small phylum of marine coelomate invertebrates, abundant in plankton. Hermaphrodite.

Chain elongation Growth of polypeptide on a ribosome during **protein synthesis**. See **elongation factors**.

Chalaza (Bot.) Basal region of ovule, where the stalk (funiculus) unites with the **integuments** and the **nucellus**. (Zool.) Of a bird's egg, the twisted strand of fibrous **albumen**; two are attached to the vitelline membrane, one each at opposite poles of the yolk, lying in the long axis of the egg. They stabilize the position of the yolk and early embryo in the albumen.

Chalone Term once used for substances now regarded as growth-inhibiting **growth factors**.

Chamaerophytes Class of **Raunkiaer's life forms**.

Channel proteins See **ion channels**.

Chaos Biological systems can often be studied at more than one level of organization or complexity and each may exhibit dynamic properties. Symptoms of what has come to be termed 'chaotic behaviour' arise when attempts to model these deterministic systems mathematically fail to predict their future states in all but the very short term, and when the system's characteristics show unexpected sensitivity to slight differences in initial conditions. Importantly, equilibrium points (points of stability) within the system do not recur despite stability of the system as a whole, which is either dominated by **positive feedback** or displays such over-compensating negative feedback that instability arises in just those regions where positive feedback occurs. In ecology, **density-dependence** is central to the notion of chaos, for populations can be affected by direct density-dependence (rapid negative feedback), delayed density-dependence (delayed negative feedback) and inverse density-dependence (positive feedback). Although the jury is still out on whether natural ecosystems exhibit chaotic dynamics, there is concern that they may be made chaotic by human actions. See **balance of nature**.

Chaperones Proteins helping non-covalently in the assembly and folding of a protein during translation without forming part of its final assembly. Preventing non-productive protein–protein interactions and premature attainment of mature protein conformation, they can enable a protein to pass through a membrane (pore) in open conformation. One class of chaperones are the *heat-shock proteins* (prokaryotic and eukaryotic types occur) whose synthesis and/or activity rate is increased when the cell is subjected to higher-than-optimal temperatures (the appropriate **enhancer** is sometimes itself heat-affected, but heat-induced phosphorylation may be a factor). They are also implicated in ATP-driven protein transport across (e.g. mitochondrial) membranes, which involves temporary loss of tertiary structure. Chaperones provide hydrophobic surfaces which stabilize other proteins and either prevent them from denaturing or promote refolding after heat-induced modification. Examples of the other major chaperone class, *chaperonins*, have been isolated from bacteria, chloroplasts and mitochondria. **Signal recognition** particles also have chaperoning functions.

Chaperonins A class of molecular **chaperone**.

Character displacement Evolutionary phenomenon whereby, it is believed, interspecific competition causes two closely related species to become more different in regions where their ranges overlap than in regions where they do not. Such differences are often anatomical, but may involve any aspect of phenotype. Few rigorously documented examples exist where such differences have been shown to be due to competition; but studies on fish in post-glacial lakes and lizards on Caribbean islands indicate multiple speciation events accompanied by similar patterns of ecological and morphological divergence involving, apparently, character divergence – implicit as this is in current thinking on **adaptive radiation**.

Charophyceae Stoneworts (an old term; from characteristic incrustations with calcium carbonate of some of the macroscopic members). Protists, formerly Division Charophyta, now considered a class of green algae (**Chlorophyta**). A diverse group of predominantly freshwater algae. Characterized, when present, by asymmetrical motile cells which are commonly scaly, biflagellate (with flagella inserted laterally), and usually lacking **eyespots**. The flagellar root comprises a broad band of microtubules and a second smaller microtubular root. A **multilayered structure** may be present, but no rhizoplast. An interzonal mitotic spindle persists through telophase. Many become heavily calcified, the female reproductive structure (the *nucule*) being an important macrofossil (called a *gyrogonite*, Class Charales), the earliest of which occur in the uppermost **Silurian**, while all extant Charales date back to the Upper **Carboniferous**. It is thought land plants evolved from this line of algal evolution. The motile cells of the advanced members of the Charophyceae are similar to the flagellated male gametes of bryophytes and vascular cryptogams.

Chela The last joint of an arthropod limb, if it can be opposed to the joint preceding it so that the appendage is adapted for grasping, as in pincers of lobster and some **chelicerae**. Such a limb is termed *chelate*.

Chelicerae Paired, prehensile first appendages of **Chelicerata**, contrasting with antennae of other groups. Often form **chelae** (when said to be *chelate*).

Chelicerata Probably natural assemblage containing those arthropods with chelicerae. Includes **Merostomata** and **Arachnida**. No true head, but an anterior tagma termed the **prosoma**. Mandibles absent. Probably closely related to trilobites. See **Arthropoda**, **biramous appendage** (for limb diagram), **mouthparts**.

Chelonia (Testudines) Tortoises and turtles. Anapsid reptile order, with bony plates enclosing body or covered by epidermal horny plates. Shoulder and pelvic girdles uniquely within rib cage. Teeth absent.

Chemiosmotic theory (chemiosmotic-coupling hypothesis) Hypothesis of P. Mitchell, now generally accepted, that chloroplasts and mitochondria require their appropriate membrane to be intact so that a proton gradient created across it by integral membrane pumps can be coupled to ATP synthesis as protons return across the membrane down their electrochemical gradient. See **chloroplast**, **electron transport system**, **mitochondrion**.

Chemoautotrophic (chemosynthetic) See **chemotrophic**.

Chemoheterotrophic See **chemotrophic**.

Chemokines Chemoattractants of inflammatory cells (see **inflammation**), sharing structural motifs. Include **interleukin-8** which attracts neutrophils, inflammatory proteins α and β, and RANTES (Regulated upon Activation Normal T cell Expressed presumed Secreted), which attract macrophages and lymphocytes.

Chemoreceptor (1) In bacteria, molecules (also termed chemosensors) located either in the periplasmic space or spanning the plasma membrane (when signal transducers) which, by binding attractor or repellant molecules from the environment, activate a motor system controlling **flagellum** behaviour and hence cell direction (chemotaxis). (2) Receptor cell responding to chemical aspects of internal or external environment (aqueous solubility of signal molecule being a prerequisite). Taste and olfaction are chemosenses. See **signal transduction** and **receptor** for some components of cellular pathways. See **carotid body**.

Chemostat A **bioreactor** (see Fig. 9) in which maintenance of microbial culture density is achieved by exhaustion of some limiting substrate (e.g. nutrient). As such, chemostatic control differs from control by optical density of culture (turbidostatic maintenance).

Chemosynthetic See **chemoautotrophic**.

Chemotaxis **Taxis** along a chemical gradient, involving a **chemoreceptor**.

Chemotrophic Of organisms obtaining energy by chemical reactions independent of light. Reductants obtained from the environment may be inorganic (**chemoautotrophic**), or organic (chemoheterotrophic). See **autotrophic**, **heterotrophic**, **phototrophic**.

Chemotropism (Bot.) **Tropism** in which stimulus is a gradient of chemical concentration, e.g. downward growth of pollen tubes into stigma due to presence of sugars. (Zool.) Rarely used as a synonym of **chemotaxis**.

Chiasma (pl. chiasmata) (1) The visible effects of the process of genetic **crossing-over** between chromosomes which have paired up (i.e. between bivalents) in appropriately stained meiotic cells, and hence indicators of homologous (non-random) **recombination**. Each chiasma may involve either of the two chromatids of each chromosome. Appreciation that chiasmata result from breakage and reciprocal refusion between chromatids during the first meiotic prophase was a major achievement of classical cytogenetics and is due largely to Jannsens and Darlington (their *chiasmatype theory*). The molecular mechanism involved may incorporate enzymes that were formerly part of a **DNA repair mechanism**. Several chiasmata may occur per bivalent, longer bivalents having more on average. Their frequency and distribution are not entirely random and are sometimes under genetic control. See **suppressor mutation**. (2) See **optic chiasma**.

Chilopoda Centipedes and their allies. Class (or Subclass) of **Arthropoda**. See **Myriapoda**.

Chimaera (1) Usually applied to organisms which (*contra* **mosaics**) comprise cells of two or more distinct genomes, resulting from experimental manipulation (e.g. grafting or aggregation), often early in development. Sometimes used in the context of cells and organisms (e.g. **Cryptophyta**) whose evolutionary history has involved **endosymbiosis**, where parental genomes remain more or less intact. (2) A genus of holocephalan fish (*Chimaera*).

Chiroptera Bats. Order of eutherian (placental) mammals; characterized by membranous wing spread between arms, legs, and sometimes tail, generally supported by greatly elongated fingers. Use of echolocation for avoidance of objects and food capture during commonly nocturnal insectivorous feeding. Some are plant pollinators. Diurnal fruit-eating bats (suborder Megachiroptera) share more derived (apomorphous) anatomical characters with euprimates and dermopterans than do the smaller-eyed echolocating bats (suborder Microchiroptera); but immunological evidence suggests that all bats form a cohesive phylogenetic unit. See **Archonta**.

Chi-squared (χ^2) test Statistical test for assessing the significance of departures of sets of whole numbers (those observed) from those expected by hypothesis, as when scoring phenotypic classes obtained from a genetic cross. The formula used is

$$\chi^2 = \sum \frac{(n_{obs} - n_{exp})^2}{n_{exp}}$$

The value obtained has to be assessed in relation to the number of *degrees of freedom*, which is the number of classes minus 1, and a χ^2 table will then give the probability (P) of finding as poor a fit with the expected results owing to random sampling error. If, for instance, $P < 0.05$, the data are said to be significantly different from expectation at the 5 per cent level. The χ^2 test becomes seriously inaccurate if any of the expected numbers is less than 5. See **null hypothesis**.

Chitin Nitrogenous polysaccharide found in many arthropod exoskeletons, hyphal walls of many fungi and cell walls of the protistan **Chytridiomycota**. Comprises repeated N-acetylglucosamine units (β[1,4]-linked). Strictly a **proteoglycan**, owing to peptide chains attached to its acetamido groups. Of considerable mechanical strength, hydrogen bonding between adjacent molecules stacked together forming fibres giving structural rigidity; also resistant to chemicals. With lignocellulose, among the most abundant of biological products.

Chlamydospore Thick-walled fungal spore capable of surviving conditions unfavourable to growth of the fungus as a whole; asexually produced from a cell or portion of a hypha.

Chloramphenicol Antibiotic, formed originally by *Streptomyces* bacteria, inhibiting translation of mRNA on prokaryotic ribosomes, eukaryotic translation being unaffected. Its use can thus distinguish proteins synthesized by mitochondrial/chloroplast ribosomes from those manufactured in the rest of eukaryotic cell. See **cycloheximide**.

Chlorenchyma Parenchymatous tissue containing chloroplasts.

Chloride shift Entry/exit of chloride ions across red blood cell membranes to balance respective exit/entry of hydrogen carbonate ions resulting from **carbonic anhydrase** activity. See **Bohr effect**.

Chlorocruorin Respiratory pigment (green, fluorescing red) dissolved in plasma of certain polychaete worms. Conjugated iron-porphyrin protein resembling haemoglobin.

Chlorophyll Green pigment found in all algae and higher plants except a few saprotrophs and parasites. Responsible for light capture in **photosynthesis**. Located in **chloroplasts**, except in **Cyanobacteria** (blue-green algae) where borne on numerous photosynthetic membranes (thylakoids) dispersed in the cytoplasm at the periphery of the cell. Each molecule comprises a magnesium-containing *porphyrin* group, related to the prosthetic groups of haemoglobin and the cytochromes, ester-linked to a long *phytol* side chain. Several chlorophylls exist (*a*, *b*, *c*, *d* and *e*), with minor differences in chemical structure. Chlorophyll *a* is the only one common to all plants (and the only one found in blue-green algae). In photosynthetic bacteria, other kinds of chlorophyll (*bacteriochlorophylls*) occur. Can be extracted from plants with alcohol or acetone and separated and purified by chromatography. See **accessory pigments**, **antenna complex**.

Chlorophyta Green algae. Protists possessing chlorophylls *a* and *b* which, unlike other eukaryotic algae, form starch within the chloroplast, usually in association with a **pyrenoid**. No **chloroplast endoplasmic reticulum** surrounds the chloroplast. Primarily a freshwater group; only 10% are marine. If sufficient moisture is present, some species grow aerially (e.g. on tree bark, attached to mosses or rocks). One genus, *Cephaleuros*, is a plant parasite responsible for the disease red rust, an economically important disease of tea plants in some regions of the world; it also causes economic losses in other crops (e.g. citrus and peppers). Morphologically diverse, including motile flagellated and non-motile non-flagellated unicells and coenobia, unbranched and branched filaments, parenchymatous and siphonaceous thalli. Asexual reproduction occurs by cell division, fragmentation, aplanospores, autospores or zoospores. Sexual reproduction may be isogamous or anisogamous, involving either flagellate or non-flagellate (amoeboid) gametes (e.g. Zygnematales), or oogamous.

Advances in electron microscopy and biochemistry have greatly influenced present-day concepts of green algal evolution and systematics. The current classification is based upon a series of complex cytological and biochemical features. Five classes may be recognized (Ulvophyceae, Chlorophyceae, Charophyceae, Micromonadophyceae and Pleurastrophyceae). Further changes will undoubtedly occur as additional evidence emerges.

Chloroplast Chlorophyll-containing plastid of eukaryotic organisms; the organelle within which both light and dark reactions of photosynthesis occur. Present in all **Plantae** (but not usually in all their cells), and most **algae** (see further details). The similarity of chloroplasts in diverse autotrophs suggests a common origin. Based upon close cytological and biochemical similarities between certain bacteria and chloroplasts, the most accepted theory is that chloroplasts evolved through a series of independent endosymbioses, involving different groups of photosynthetic bacteria. Chloroplasts contain certain DNA (see **cpDNA**) and protein-synthesizing machinery, including **ribosomes**, of a prokaryotic type. Hence they are semi-autonomous organelles. Isolated chloroplasts can synthesize RNA but only under the direction of chromosomal DNA. The ability to form chloroplasts and associated pigment complexes is largely controlled by chromosomal DNA, interacting with chloroplast DNA.

Where present, chloroplasts may be numerous per cell or single. In higher plants, they are usually discoid (disc-shaped), about 4–6 μm in diameter and arranged in a single layer in the cytoplasm but capable of changing both shape and position in relation to light intensity (see **cyclosis**). In algae, they are variously shaped (cup-shaped, spiral, stellate, reticulate, lobed and discoid). Often accompanied by **pyrenoids**, with which storage products are frequently associated. Those of green algae (**Chlorophyta**) and plants often contain starch grains and small lipid droplets.

Mature chloroplasts are typically bounded by two outer membranes enclosing a homogeneous *stroma* (where the dark reactions occur). Traversing the stroma are membranes in the form of flattened sacs (*thylakoids*), comprising two membranes. Stacks of thylakoids form the grana. Thylakoid membranes house the photosynthetic pigments and electron transport system involved in the light-dependent reactions of photosynthesis. Thylakoids of the grana are connected to each other by stromal thylakoids (*intergranal thylakoids*). In algae, the chloroplast may or may not be surrounded by one or two membranes of **chloroplast endoplasmic reticulum**, and the thylakoid groupings vary: bands of three with a girdle or peripheral band running parallel to the chloroplast envelope (e.g. **Euglenophyta**, **Chrysophyta**, **Xanthophyta**, **Raphidophyta**, **Bacillariophyta** and **Phaeophyta**); or bands of two to six, with thylakoids running from one band to the next or free from one another (e.g. **Rhodophyta**). Photosynthetic prokaryotes lack chloroplasts, the numerous thylakoids lying free in the cytoplasm and varying in arrangement and shape between species.

Chloroplast endoplasmic reticulum The outer one or two membranes, respectively, where a chloroplast is bounded by either three membranes (**Dinophyta**) or, in the **Cryptophyta, Bacillariophyta, Chrysophyta, Eumastigophyta, Phaeophyta, Prymnesiophyta** and **Xanthophyta**, by four. When composed of two membranes, the outer one is often continuous with the outer membrane of the nuclear envelope. See **nucleomorph**.

Chlorosis Disease of green plants characterized by yellow (chlorotic) condition of parts that are normally green; caused by conditions preventing chlorophyll formation e.g. lack of light or of appropriate soil nutrients.

Choanae (internal nares) Paired connections between nasal and oral cavities of typical crossopterygian (lobe-finned) fish, some teleosts, lungfish and higher vertebrates; probably evolved independently in different fish groups. Not used for respiratory purposes in any living jawed fish, but providing a passage for ventilation of lungs in tetrapods. Situated near front of roof of mouth, unless false palate (see **palate**) present, when they are at the back. See **Choanichthyes**, **nares**.

Choanichthyes (Sarcopterygii) A probably natural vertebrate clade, containing **Crossopterygii** (coelacanths), **Dipnoi** (lungfishes) and **Rhipidistia** (including porolepids, osteolepids and tetrapods). New fossils have been shown to link lungfishes and tetrapods to separate extinct rhipidistian groups.

Choanocyte (collar cell) Cell with single flagellum generating currents by which sponges (**Porifera**) draw water through their ostia and catch food particles which stick to the outside of cylindrical protoplasmic collar around base of flagellum. Affinities of sponges with the protozoan choanoflagellates problematical.

Cholecystokinin (pancreozymin) Hormone of mucosa of small intestine, released in response to presence of **chyme**. Causes pancreas to release enzymatic juice and gall bladder to eject bile. Promotes intestinal secretion but inhibits gastric secretion. Acts as a **neurotransmitter** in the brain, involved in satiety after eating and experience of fear. See **secretin**.

Cholesterol Sterol lipid derived from squalene, forming a major component of animal **cell membranes** where it affects membrane fluidity. Absent from higher plants and most bacteria. Precursor of several potent steroid hormones (e.g. corticosteroids, sex hormones) which are in turn converted back to it in liver. Synthesis by liver suppressed by dietary cholesterol. Most plasma cholesterol is transported esterified to long-chain fatty acids within a micellar lipoprotein complex. These structures, **low-density lipoproteins** (LDL), are about 22 nm in diameter and adhere to plasma membrane receptor sites produced and found on **coated pits** when a cell needs to make more membranes using the cholesterol in the LDL. Cholesterol ester transfer protein (see **phospholipid transfer proteins**) transfers a portion of HDL2 to triglyceride-rich lipoproteins while triglyceride is transferred in the opposite direction to modify HDL2. Cholesterol is excreted in **bile**, both in native form (as micelles) and conjugated with taurine or glycine as bile salts. See **lipoprotein**, **chylomicron**.

Choline An organic base (formula OH-$C_2H_4.N[CH_3]_3OH$), a vitamin for some animals, and a component of some lipids (e.g. lecithin), and **acetylcholine**.

Cholinergic Of nerve fibres which secrete **acetylcholine**. In vertebrates, motor fibres to striated muscle, parasympathetic fibres to smooth muscle, and fibres connecting CNS to sympathetic ganglia are cholinergic, as are some invertebrate neurones.

Cholinesterase Hydrolytic enzyme anchored to **basal lamina** between synapsing membranes of most (especially vertebrate) neuromuscular junctions and of cholinergic synapses. Degrades **acetylcholine** to choline and acetate.

Chondrichthyes Vertebrate class containing cartilaginous fish, first appearing in the Devonian. Includes **Holocephali** (e.g. ratfish, *Chimaera*) and **Elasmobranchii** (sharks, skates and rays). Cartilaginous skeleton; **placoid scales** (denticles), modified to form replaceable teeth; intromittant organs (claspers) formed from male pelvic fins. No **gas bladder**. See **Osteichthyes**.

Chondroblast, chondrocyte See **cartilage**.

Chondrocranium Part of the skull first formed in vertebrate embryos as cartilaginous protection of brain and inner ear. Usually ossified during development to form membrane bones. See **ossification**.

Chondroitin Sulphated **glycosaminoglycan** composed largely of D-glucuronic acid and N-acetylgalactosamine. Found in cartilage, cornea, bone, skin and arteries.

Chondrostei Group (often considered a superorder) of the **Actinopterygii**. Includes the primitive Palaeozoic palaeoniscoids represented today by the bichirs (*Polypterus*), paddlefishes and sturgeons. Ganoid scales of bichirs are lost altogether in paddlefish, sturgeons having rows of bony plates lacking ganoine. Ancestral bony internal skeleton largely substituted by cartilage. Primitive heterocercal tail present in sturgeons and paddlefish. Bichirs have lungs, sturgeons a gas bladder.

Chordata Animal phylum, characterized by presence at some stage in development of a **notochord**, by the dorsal hollow nerve cord, pharyngeal gill slits and a post-anal tail. Includes the invertebrate subphyla **Urochordata** and **Cephalochordata**, and vertebrates (Subphylum **Vertebrata**). See Fig. 19.

Chorion (1) One of three **extra-embryonic membranes** of amniotes. Comprises the **trophoblast** with an inner lining of mesoderm, coming to enclose almost the entire complement of embryonic structures. In reptiles and birds it forms with the **allantois** a surface for gaseous exchange within the egg. In most mammals it combines with the allantois to form the **placenta**. Chorionic villus sampling (CVS) involves removal by catheter, or ultrasound needle probe, of c. 30 mg villus tissue to test for chromosomal disorders between 6–10 weeks of pregnancy (earlier than in amniocentesis). See **androgenesis.** (2) Egg shell of insects, secreted by follicle cells of ovary, and often sculptured externally.

Choroid Mesodermal layer of vertebrate eyeball between outer sclera and retina within. Soft and richly vascularized (supplying nutrition for retina); generally pigmented to prevent internal reflection of light, but reflecting crystals of **tapetum**, part of the choroid, increase retinal stimulation in many nocturnal/deep-water vertebrates. Becomes the **ciliary body** anteriorly.

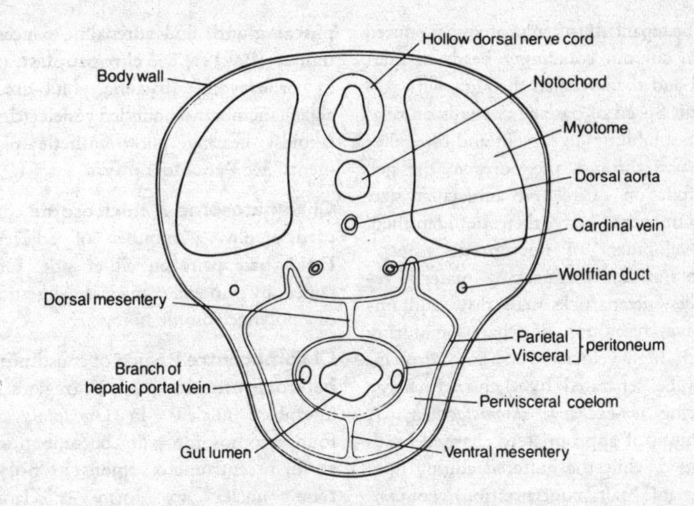

Fig. 19 *Transverse section through embryonic vertebrate, indicating the layout of the trunk region prior to origin of the skeleton.*

Choroid plexuses Numerous projections of non-nervous epithelium into ventricles of brain, secreting **cerebrospinal fluid** from capillary networks. One plexus occurs in the roof of each of the four ventricles in man.

Chromaffin cell, c. tissue Cells derived from **neural crest** tissue, which having migrated along visceral nerves during development come to lie in clumps in various parts of the vertebrate body (e.g. the adrenal medulla). They are really postganglionic neurones of the sympathetic nervous system, which have 'lost' their axons and secrete the catecholamines **adrenaline** and **noradrenaline** into the blood, the former more abundantly. Stain readily with some biometric salts (hence name).

Chromatid One of the two strands of **chromatin**, together forming one **chromosome**, which are held together after DNA replication during the cell cycle by one or more **centromeres** prior to separation at either mitotic anaphase or second meiotic anaphase. In mitosis the strands are genetically identical (barring mutation), but in meiosis crossing-over increases the likelihood of dissimilarity.

Chromatin (nucleohistone) The material of which eukaryotic **chromosomes** are composed. Consists of DNA and proteins, the bulk of them **histones**, organized into nucleosomes. **chromosomal imprinting**, See **euchromatin, heterochromatin**.

Chromatography Techniques involving separation of components of a mixture in solution through their differential solubilities in a moving solvent (*mobile phase*) and absorptions on, or solubilities in, a *stationary phase* (often gels, e.g. polyacrylamide or agarose; or special paper). In *gel filtration*, mixture to be separated (often proteins) is poured into column containing beads of inert gel and then washed through with solvent. Speed of passage depends on relative solubilities in solvent and on ability to pass through the pores in the gel, a function of relative molecular size. Components may then be identified. Development of microparticles (600–800 nm) for the packing material reduces intraparticle mass flow and improves resolution of separation during even high-velocity perfusion. Proteins can be separated by their net charge during *ion-exchange chromatography*. A column of appropriately charged beads is used while the buffered eluting fluid (variable salt concentration) contains metal ions which compete with positively charged groups on the protein for binding to the beads so that proteins with low positive charge tend to emerge first. See **electrophoresis**.

Chromatophore (Zool.) Animal cell lying superficially (e.g. in skin), with permanent radiating processes containing pigment that can be concentrated or dispersed within the cell under nervous and/or hormonal stimulation, effecting colour changes. When dispersed, the pigment of groups of such cells is noticeable; when condensed in centre of cells the region may appear pale. Three common types occur in vertebrates: *melanophores*, containing the dark brown pigment melanin; *lipophores*, with red to yellow carotenoid pigments; *guanophores*, containing guanine crystals whose light reflection may lighten the region when other chromatophores have their pigments condensed. **Melanocyte-stimulating hormone** disperses melanin, while melatonin (see **pineal gland**) and adrenaline concentrate it. (Bot.) (1) See **chromoplast**. (2) In prokaryotes (bacteria, blue-green algae), membrane-bounded vesicles (thylakoids) bearing photosynthetic pigments. See **Prochlorophyta**.

Chromatosome A **nucleosome** core particle plus a number of adjacent DNA base pairs on either side. Obtained by moderate nuclease digestion of a polynucleosome fibre.

Chromocentre Region of constitutive **heterochromatin** which aggregates in interphase nucleus. In *Drosophila* all four chromosome pairs become fused at their centromere regions in **polytene** nuclei to form a large chromocentre.

Chromomeres Darkly staining (heterochromatic) bands visible at intervals along chromosomes in a pattern characteristic for each chromosome. Especially visible in mitotic and meiotic prophases, and at bases of loops of **lampbrush chromosomes**. Probably reflects tight clustering of groups of chromosome loops (see **chromosome**). Dark bands of polytene chromosomes are probably due to multiple parallel chromomeres.

Chromonema Term usually used for chromosome thread while extended and dispersed throughout nucleus during interphase.

Chromoplast (chromatophore) Pigmented plant cell **plastid** of variable shape, lacking chlorophyll but synthesizing and retaining carotenoid pigments; often responsible for the yellow to orange or red colours of many flowers, old leaves, some fruits and roots. May arise from a chloroplast whose internal membrane structure disappears, when masses of carotenoids accumulate. Precise function is not well understood. Often used synonymously with **chloroplast**; in older literature a chloroplast that has a colour other than green is often called a chromoplast. See **leucoplast**.

Chromosomal imprinting Heritable change in gene expressibility possibly brought about by genes becoming heritably non-expressible in a cell line by incorporation into a **heterochromatin** cluster on a chromosome. **Homeogene** activity/inactivity may depend upon such events. See **androgenesis**, **genomic imprinting**, **position effect**.

Chromosome Literally, a coloured (i.e. stainable) body; originally observed as threads within eukaryote nuclei during mitosis and meiosis. Composed of nucleic acid, most commonly DNA, usually in conjunction with various attendant proteins, in which form the genetic material of all cells is organized. Chromosomes are linear sequences of **genes**, plus additional non-genetic (i.e. apparently non-functional) nucleic acid sequences. Gene sequence is probably never random, being the result of selection for particular **linkage** groups (but see **transposable element**). Prokaryotes and eukaryotes differ in the amount of genetic material which needs to be packaged, and in resulting complexities of their chromosomes. Thus the absence to date from prokaryotic chromosomes of the DNA-binding proteins, histones, has some taxonomic value (see **chromatin**). Non-histone proteins (e.g. *protamines*) form part of the structure of all chromosomes, however, and their roles, for example as activators of transcription, are being increasingly elucidated. The DNA of a normal individual chromosome or chromatid is probably just one highly folded molecule.

The *prokaryotic chromosome* (usually one main chromosome per cell) is just over 1 mm in length, contains about 4×10^6 base pairs of DNA, is circular and is attached to the cell membrane, at least during DNA replication. It lacks the nucleosome infrastructure of eukaryotic chromatin. Additionally, there may be one or more **plasmids**, some of which (*megaplasmids*) may constitute more than 2% of the cell's DNA. There is no nucleus to contain the chromosome, but the term 'nucleoid' may be used to indicate this region of the cell. The DNA appears to be packaged in a series of loops (see later). *Eukaryotic chromosomes* are made of chromatin, containing DNA and five different histone species roughly equal in total weight to the DNA; plus various attendant proteins. The fundamental organizational unit is the **nucleosome**, a polynucleosome giving rise in turn during nuclease digestion to mononucleosomes (200 DNA base pairs), chromatosomes (165 DNA base pairs) and nucleosome core particles (145 DNA base pairs). See Fig. 20.

The polynucleosome filament has a diameter of about 10 nm, but adopts a tight 30 nm helix under physiological ion concentrations. This reduces the DNA length 50-fold and may be the normal interphase state of chromatin. Further looping along a single axis forms a fibre 0.3 μm in width which may in turn form a helix of radially arranged loops about 0.7 μm in diameter, possibly the metaphase chromatin condition. Bands seen in stained mitotic chromosomes probably reflect tight clustering of groups of loops, which stain more densely. Polytene chromosome bands (see **polyteny**) would result from lateral amplification of these tightly clustered loops. The higher orders of chromatin packing are features of **heterochromatin** such as chromocentres, centromeres and pericentric regions (see **chromosomal imprinting**). Chromosomes can be stained in various ways to reveal different banding patterns. Routinely, G (Giemsa) banding is employed and generates 300–400 alternating light and dark bands in the human karyotype, reflecting differing levels of chromosome condensation. The convention for eukaryotic chromosomes is to call the two arms of a chromosome on either side of the centromere p (short arm) and q (long arm). Prominent bands then subdivide the arms further, each region being further subdivided by the next most prominent bands, and so on. Thus, band 5p15.2 is found in the short arm (p) of chromosome 5, in region 1, band 5, sub-band 2 (see Fig. 20c and **chromosome mapping**).

Chromosome crawling Alternative for **chromosome walking.**

Chromosome diminution Phenomenon in nematode (e.g. *Parascaris*) eggs whereby after an equatorial first cleavage division the upper (animal) blastomere's chromosomes fragment at their ends during the next division with only a portion of the chromosomes surviving. This contrasts with the vegetal blastomere, whose chromosomes remain normal but whose cell line is such that all but the eventual germ cell undergo chromosome diminution and differentiate into somatic cells. Cytoplasmic determinants in the egg are responsible. Similar events occur in some dipteran insects (e.g. *Wachtiella*), although in others (e.g. *Drosophila*) the **germ plasm** does not cause chromosome diminution. See **aberrant chromosome behaviour(3)**, **maternal effect**.

Chromosome inactivation See **Barr body, dosage compensation.**

Chromosome jumping Chromosome mapping (and DNA sequencing) technique in which quite large DNA fragments (80–150 kb long) are first cut out and isolated by gel separation from very high molecular mass DNA, then circularized and cloned inside a phage or cosmid (see **vector**) before plating out on bacteria designed to allow only phages or cosmids with the inserted fragments to form plaques. After isolating the inserts, an already mapped DNA sequence is then used as a probe, when those fragments containing the probe sequence will have DNA from a distantly linked site at their other end; intervening sequences can be filled in by **chromosome walking**. The combined methods have enabled mapping and sequencing of the human cystic fibrosis gene. See Fig. 21.

Chromosome map Linear map (circular in bacteria, plasmids, etc.) of the sequence of genes (cistrons or loci) on a chromosome as defined by **chromosome mapping** techniques. The **map distance** between two genes does not accurately reflect their physical separa-

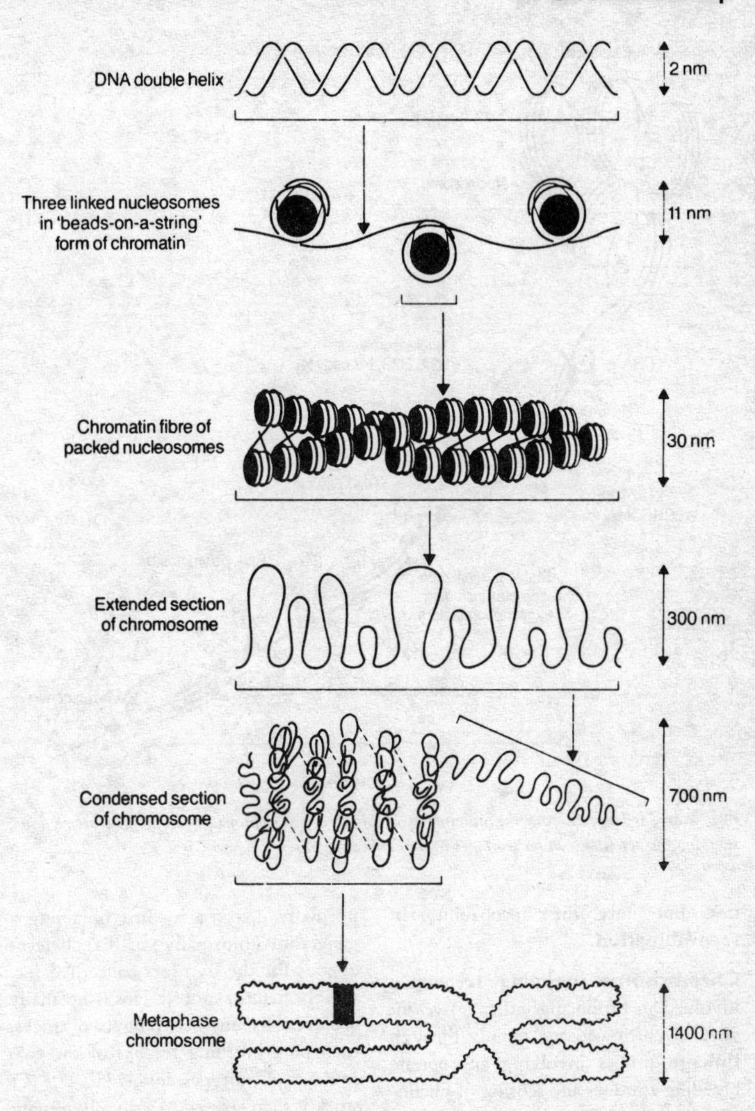

Fig. 20a *Possible progressive packing arrangement of a DNA duplex with histones to form nucleosomes and then subsequent packing of these, ultimately to form the chromosomes visible in light microscopy.*

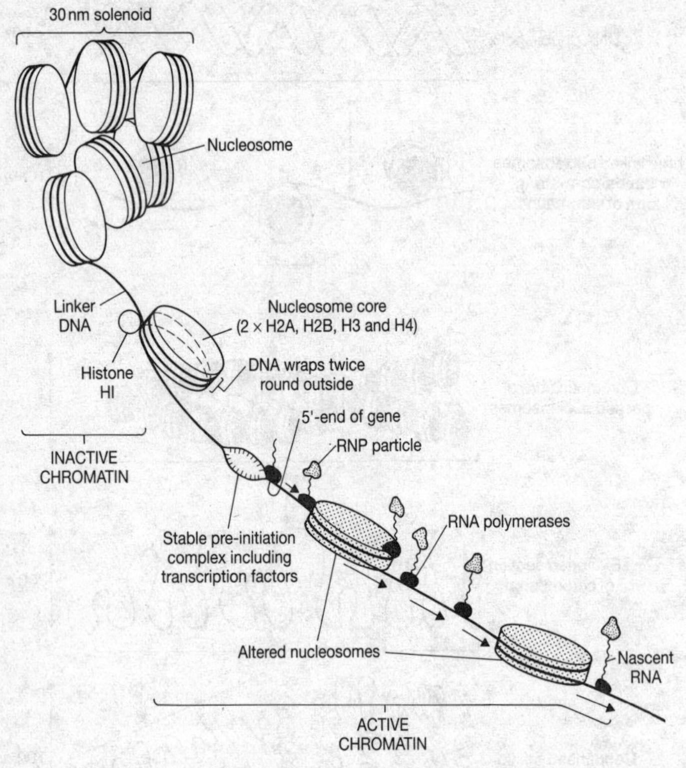

Fig. 20b *Model indicating the unwinding of inactive heterochromatin around an active gene, showing the persistence of nucleosomes in transcribed regions.*

tion but only their probability of **recombination**.

Chromosome mapping Techniques involved in producing either (a) *genetic maps* of chromosomes, mainly through **linkage** studies involving appropriate breeding routines and scoring of phenotypic ratios; or (b) *physical maps* of chromosomes through analysis of its entire DNA sequence. For most eukaryotes, linkage between two or more loci is normally detected by first obtaining a generation (normally an F1) heterozygous for the two loci concerned (i.e. doubly heterozygous). This is normally achieved by first crossing two stocks, each pure-breeding for *one* of the two mutant phenotypes involved. The F1 stock is then crossed to a doubly mutant stock and the resulting offspring scored for phenotypes. If all four possible phenotypes (assuming complete dominance of wild-type over the

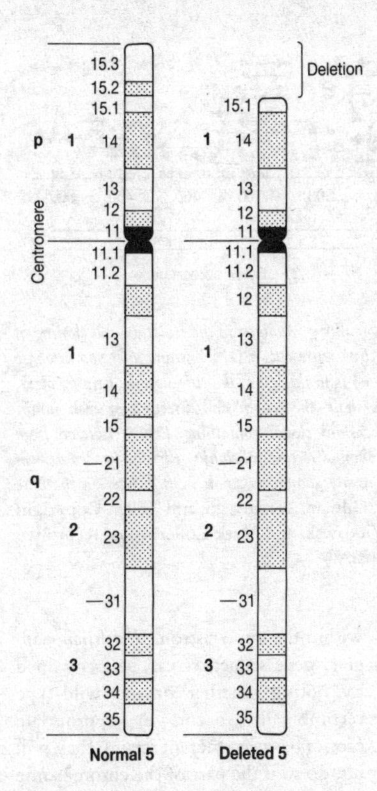

Fig. 20c *The cause of the* cri du chat *syndrome of abnormalities in humans is loss of the tip of the short arm of one of the homologues of chromosome 5. The figure indicates chromosome banding in human chromosome 5, both normal and with the deletion. See* **chromosome** *text for banding notation.* (from *Introduction to Genetic Analysis 5*E* by Griffiths, Miller, Suzuki, Lewontin and Gelbert. Copyright © W.H. Freeman and Company, 1993. Reprinted with permission.)

mutant phenotype) are present in equal ratio, linkage is not probable; but if there is a departure in the ratios from those expected on the null hypothesis of no linkage, then this departure can be tested for its significance (using **chi-squared test**). Where the ratio is obviously non-Mendelian (i.e. departs obviously from 1:1:1:1), with the parental classes outnumbering the recombinants, then a **cross-over value** can be determined giving a map distance between the two loci.

When we wish to know whether the loci bearing the alleles for black body and vestigial wing (both recessive characters) in *Drosophila* are linked, then using the symbols

b = black body
$+$ = wild-type body
v = vestigial wing
$+$ = wild-type wing

first pure-breeding black body/wild-type wing flies ($bb++$) are mated with pure-breeding wild-type body/vestigial wing flies ($++\ vv$). F1 offspring are then mated with a double recessive stock (i.e. pure-breeding black body/vestigial wing, $bbvv$) as a **test cross**. If all four resulting offspring phenotypes ($++$, $+v$, $b+$, bv) occur in equal ratio then, given adequate sample size, linkage is unlikely. If two phenotypic classes (the parental classes, $b+$, $+v$) outnumber the other two (the two recombinant classes, bv, $++$) then linkage is likely and a provisional map distance can be calculated, equal to the frequency of the recombinant offspring as a percentage of the total number of offspring. (The example is actually more complex, for only when male flies are used as the double recessive in the backcross do four phenotypic classes appear in the F2 generation. This is because in male *Drosophila* there is no crossing-over during meiosis (see **suppressor mutation**) so the males cited only produce two

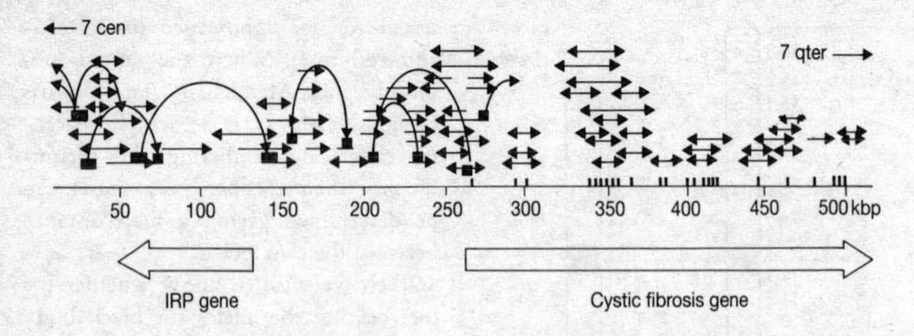

Fig. 21 *The chromosome jumping and subsequent walking involved in the isolation and cloning of the cystic fibrosis (CF) gene. The calibrated base line represents a DNA strand over 500 kilobase pairs long. The centromere on chromosome 7 (7 cen) is to the left of the chromosome's tip (7 qter), off to the right of the diagram. Curved arrows show the length and direction of each jump. Horizontal arrows are overlapping phage and cosmid clones containing DNA isolated from regions at the ends of each jump; arrow directions show directions of cloning (double-headed arrows show clones that went both ways). The CF exons are shown by small vertical bars on the base line.* (From *Recombinant DNA 2/E* by Watson, Gilman, Witkowski, and Zoller. Copyright © James D. Watson, Michael Gilman, Jan Witkowski and Mark Zoller, 1992. Reprinted with permission of W. H. Freeman and Company.)

gamete types, giving only two F2 phenotypes.) Sex-linked loci would give a different result, suitably modified to take account of the chromosome arrangement of the heterogametic sex. When testing for linkage between mutations for dominant characters, the recessive characters in the method employed above would be wild-type characters.

Chromosome mapping in bacteria can employ transformation, **transduction** or interrupted mating. In the latter, progress of the donor bacterial chromosome into the recipient cell during conjugation is interrupted, as by shaking (see **F factor**, for *Hfr* strain). The map of cistrons on the incoming chromosome will be a function of the time allowed for conjugation before interruption, and is deduced from recipient cell phenotypes. The ***cis*-trans test** may be used to determine whether two mutations lie within the same cistron. In *deletion mapping*, gene sequences can be ascertained by noting whether or not wild-type recombinants occur in appropriate crosses between mutant strains: they will not do so if the part of the chromosome needed for recombination is missing, so that fine mapping of such recombinants can indicate the limits of a deletion and the genes involved in it. Plasmid and viral chromosome maps may be constructed using *restriction fragment mapping* techniques in which different **restriction endonucleases** digest the chromosome, and electrophoretic patterns of resulting fragments are used to reconstruct the complete nucleotide sequences of the chromosomes. New electrophoretic techniques with infrequently cutting restriction endonucleases now permit restriction fragment mapping of even entire mammalian chromosomes

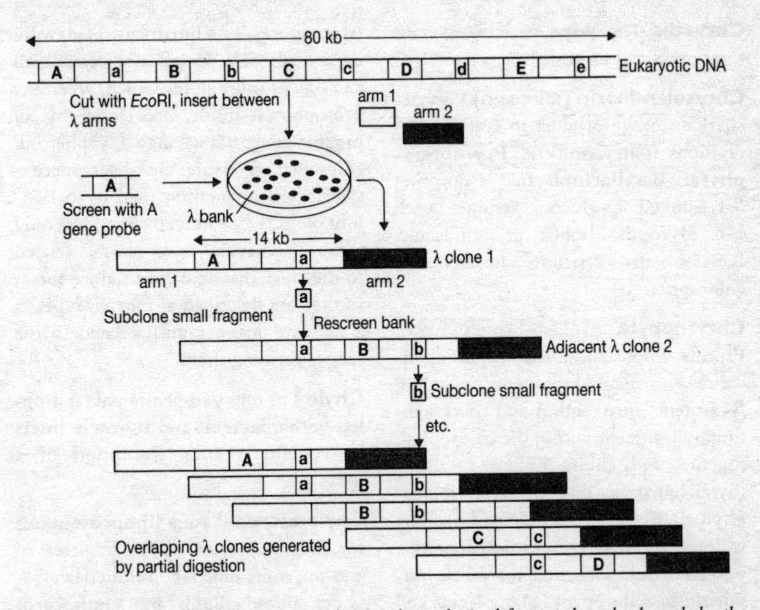

Fig. 22 *Chromosome walking. One recombinant phage obtained from a phage bank made by the partial* EcoRI *digest of a eukaryotic genome can be used to isolate another recombinant phage containing a neighbouring segment of eukaryotic DNA. In this case the recombinant phage contains the A gene, as detected by the A gene probe. It is cleaved into smaller fragments using different restriction enzymes, and these fragments are subcloned into plasmid* **vectors**. *One such subclone (a in the example) from the end of the λ phage clone 1 serves as a probe to detect a clone in the original* **gene library** *that also contains the sequence defined by probe a, which will have some sequences in common with clone 1 but will also contain new sequences further along the chromosome. This produces a new probe b, and this can be used in the same way as probe a, and so on iteratively.* (From *Introduction to Genetic Analysis 5/e* by Griffiths, Miller, Suzuki, Lewonton and Gelbert. Copyright © W. H. Freeman and Company, 1993. Reprinted with permission.)

and render chromosome mapping an extension of **DNA sequencing** in general. See **cell fusion**, **RFLP**, **vector**.

Chromosome puff See **puff**.

Chromosome walking Chromosome mapping (and DNA sequencing) technique in which a small DNA sequence from one end of a DNA clone is used as a probe to isolate other DNA sequences containing this and the next adjacent sequence, which is used in turn as a probe, and so on iteratively until an already known sequence is discovered. All intervening sequences will by then have been cloned. Often used in conjunction with **chromosome jumping** (see Fig. 21). See **cosmid**.

Chrysalis The **pupa** of lepidopterans (butterflies and moths).

Chrysolaminarin (leucosin) Polysaccharide storage product in certain algal divisions (**Chrysophyta**, **Prymnesiophyta**, **Bacillariophyta**). Comprises β-1,3-linked D-glucose residues with 1–6 glycosidic bonds per molecule. Resides in vesicles outside the chloroplast.

Chrysophyta Golden-brown algae. Protists, whose colour is due to the abundance of carotenoid pigments, including β-carotene, fucoxanthin and other xanthophylls present within the chloroplast, together with chlorophylls *a*, c_1, and c_2 (Synurophyceae do not have chlorophyll c_2). Reserve assimilatory product is stored as oils and **chrysolaminarin**, a polysaccharide deposited in a vesicle outside of the chloroplast. Many lack a cell wall; when present, the cell wall is composed of cellulose. **Loricas**, silicified, and organic scales occur in some species. Siliceous scales are radially or bilaterally symmetrical, possess species-specific morphology, and are formed by species which are very sensitive to environmental change. The scales preserve in lake sediments and provide yet another microfossil for the palaeolimnologist. Most chrysophytes are freshwater algae occurring in soft water (low in calcium); many freshwater species are planktonic and flagellate (both unicells and colonies); coccoid and filamentous species are mostly found in cold water springs and streams. Marine species also occur. A diverse group, which includes the **silicoflagellates**, having links to other protists (e.g. protozoa, dinoflagellates and brown algae) and fungi.

These algae form characteristic cysts (**statospores**, or stomatocysts) asexually or sexually and possess a siliceous wall and one or more pores which, when the protoplast is inside, are closed by an organic plug. Cysts may be spherical, ellipsoidal or ovate and bear species-specific ornamentation used by palaeolimnologists (see **heterokont**); but only some 5% of the cysts can be related to the algae that actually produce them, so they are described as morphotypes in a standard manner until related to the species forming them.

Chyle The milky suspension of fat droplets within **lacteals** and **thoracic ducts** of vertebrates after absorption of a meal.

Chylomicron Plasma **lipoprotein** (see for diagram) with mean diameter of 500 nm, containing reconstituted triglycerides, phospholipids and **cholesterol** produced by the epithelial cells of intestinal villi after long-chain fatty acids and monoglycerides have diffused across the microvilli. Also act as transport vehicles for dietary lipids within the **lacteals**, **lymphatic system** and **blood plasma**, being absorbed ultimately by the liver. See **fat**, **lipoprotein**.

Chyme Partially digested food as it leaves the vertebrate stomach. See **cholecystokinin**, **secretin**.

Chymotrypsin Proteolytic enzyme secreted as inactive chymotrypsinogen by vertebrate pancreas. An exopeptidase, it converts proteins to peptides and is activated by the enzyme enterokinase.

Chytridiomycota (chytrids) Predominantly aquatic protists, extremely varied in their sexual interactions and life histories. They produce motile cells (zoospores and gametes) possessing a single posterior smooth (whiplash) flagellum. Vary from simple unicellular organisms that do not form a mycelium, and which are **holocarpic**, to those possessing slender rhizoids that penetrate the substratum. Cell walls are composed of chitin; but other polymers may be present. Storage product is glycogen. Different species are parasites of algae, aquatic **Oomycota** and vascular plants; others are saprotrophs. Some are quite complex both structurally and reproductively, with some displaying alternation of isomorphic generations (e.g. *Allomyces*).

Ciliary body Anterior part of the fused **retina** and **choroid** of the eyes of vertebrates and cephalopod molluscs; containing ciliary processes secreting the aqueous humour, and ciliary muscles (circular smooth muscle) which may permit **accommodation** of the eye either by altering the focal length of the lens (amniotes), or by moving the lens to and fro (cephalopods, sharks and amphibians). In mammals the lens is suspended from it by ligaments, and the **iris** arises from the same region.

Ciliary feeding Variety of feeding mechanism (**microphagy**) by which many soft-bodied aquatic invertebrates draw minute water-borne food particles through e.g. gills or the pharyngeal region of the gut, when the particles are frequently trapped in mucus and moved either towards the gut (often by further cilia) or further along it (by peristalsis).

Ciliata Class of Protozoa (Subphylum Ciliophora) containing the most complex cells in the phylum. Covered typically in cilia, with meganucleus, micronucleus, and a cytostome (at the end of a depression, or 'mouth') at which food vacuoles form. Includes familiar *Paramecium* and *Vorticella*, and voracious predatory suctorians. **Conjugation** and **binary fission** both occur, as may **autogamy** (see **parthenogenesis**).

Ciliated epithelium Layer of columnar cells with apices covered in cilia whose coordinated beating enables **ciliary feeding**, movement of mucus in the respiratory tract, etc.

Ciliophora Protozoan Subphylum containing the solitary Class **Ciliata**.

Cilium Organelle of some eukaryotic cells. See Fig. 23. Tubular extension of the cell membrane, within which a characteristic 9 + 2 apparatus of **microtubules** and associated proteins occurs (nine paired outer tubules and a lone central pair). Used either for **cell locomotion** (see for details of ciliary action) or for movement of material past a **ciliated epithelium**; but frequently sensory, especially elongated cilia known as **flagella**. Cilia may beat in an organized **metachronal rhythm**, for which **kinetodesmata** are probably responsible. Such rows of beating cilia may fuse to form *undulating membranes*; or several cilia may mat together to beat as one, as in the conical *cirri* of some ciliates used for 'walking'. Factors known to increase ciliary beat include **serotonin** (mussel gills), nervous stimulation and mucus (mammalian trachea); a cyclic-AMP-dependent signal transduction pathway has been implicated, with phosphorylation of outer arm dynein a key component. For *stereocilium*, see **hair cell**.

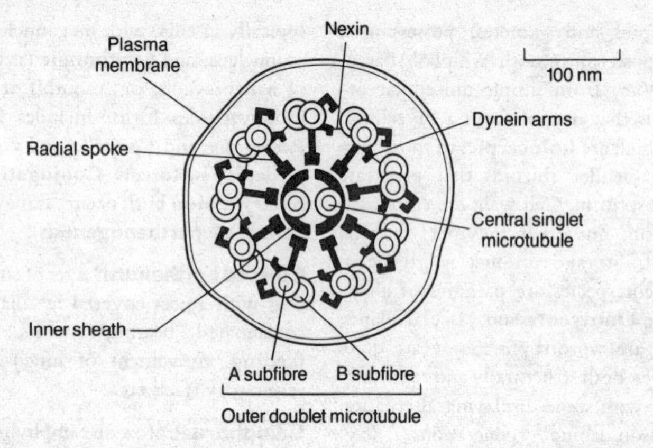

Fig. 23 *Diagram of a cilium or flagellum in cross-section, as viewed by light microscopy. The microtubule apparatus is termed the* **axoneme.**

Circadian rhythm (diurnal rhythm) Endogenous (intrinsic) rhythmic changes occurring in an organism with a periodicity of approximately 24 h; even persisting for some days in the experimental absence of the daily rhythm of environmental cycles (e.g. light/dark) to which circadian rhythm is usually entrained. Widely distributed, including leaf movements, growth movements, sleep rhythms and running activity. In animals, rhythms of hormone secretion have been implicated in some circadian rhythms, these in turn requiring explanation. Their existence indicates a **biological clock**, but the detailed chemistry is usually unknown.

Circinate vernation Coiled arrangement of leaves and leaflets in the bud; gradually uncoils as leaf develops further, as in ferns. See **phyllotaxy**.

Circulatory system System of vessels and/or spaces through which blood and/or lymph flows in an animal. See **blood system**, **lymphatic system**.

Circumnutation See **nutation**.

Cirripedia Barnacles and their relatives. Subclass of **Crustacea**. Typically marine, sessile and hermaphrodite. Unlike most of the Class in appearance, with a carapace comprising calcareous plates enclosing the trunk region. Usually a cypris larva, which becomes attached to the substratum by its 'head', remaining fixed throughout its adult life and filter-feeding using **biramous appendages** on its thorax. Several parasitic forms occur.

***cis*-acting control elements** Regulatory genetic elements (e.g. promoters, enhancers), mutations which affect synthesis of an mRNA molecule downstream on the same chromosome. Contrast ***trans*-acting control elements**.

Cisternae Flattened sac-like vesicles of **endoplasmic reticulum** and **Golgi apparatus** intimately involved in transport of materials via vesicles which either bud from or fuse with their membranous surfaces.

Cis-trans test (Complementation test) Genetic test to discover whether or not two mutations which have arisen on separate but usually homologous chromosomes are located within same **cistron**. See Fig 24. The two chromosomes, e.g. of phage or prokaryote origin, are artificially brought together in a single bacterial cell (e.g. by **transduction**) or in a diploid eukaryote by a sexual cross. If their co-presence in the cell rectifies their individual mutant expression, then **complementation** is reckoned to have occurred between the functional gene products of two distinct cistrons. However, if their resultant expression is still mutant, no such complementation has occurred and the two mutations are reckoned to lie within the same cistronic region. Two mutations lie in the *trans* condition if on separate chromosomes, but in the *cis* condition when on the same. *Trans*-complementation only occurs when two mutations lie in different cistrons, and by careful mapping of mutations the boundary between two cistrons can be located from the results of the cis-trans test.

Cistron A region of DNA within which mutations affect the same functions by the criterion of the **cis-trans test**. In molecular terms, the length of DNA (or RNA in some viruses) encoding a specific and functional product, usually a protein, in which case the cistron is 'read' via messenger RNA; but both ribosomal RNA and transfer RNA molecules have their own encoding cistrons. In modern terminology, 'cistron' is equivalent to 'gene', except that not all putative genes have been fully validated by complementation analysis.

Citric acid cycle See **Krebs cycle**.

CKIs (cyclin-dependent kinase inhibitors) Control elements of the eukaryotic **cell cycle** which bind to and inactivate cyclin-dependent kinases. Some of them are probably constitutive of the cell cycle, others are thought to be induced by such conditions as DNA damage, starvation or **growth factors**. One 'hydraulic' model of the cell cycle proposes that the peaks of cdks observed at G1-S and G2-M are caused by their build-up behind a 'CKI dam' until the excess cyclins somehow trigger CKI destruction or inactivation so that the cell becomes irreversibly committed to S-phase and M-phase respectively, after the latter of which cyclins are destroyed.

Clade Phylogenetic lineage of related taxa originating from a common ancestral taxon. See **cladistics**, **grade**.

Cladist Proponent of, or worker in, **cladistics**.

Cladistics Method of **classification** employing genealogies alone in inferring phylogenetic relationships among organisms – disregarding their phenetic adaptations, which may not be due to **homology**. The resulting diagram of relationships is termed a **cladogram**: see Fig. 25. A line of descent is characterized by the occurrence of one or more evolutionary novelties (*apomorphies*). Any character found in two or more taxa is Homologous in them if their most recent common ancestor also had it. Such a shared homologue may be *symplesiomorphous* (or *-morphic*) in these

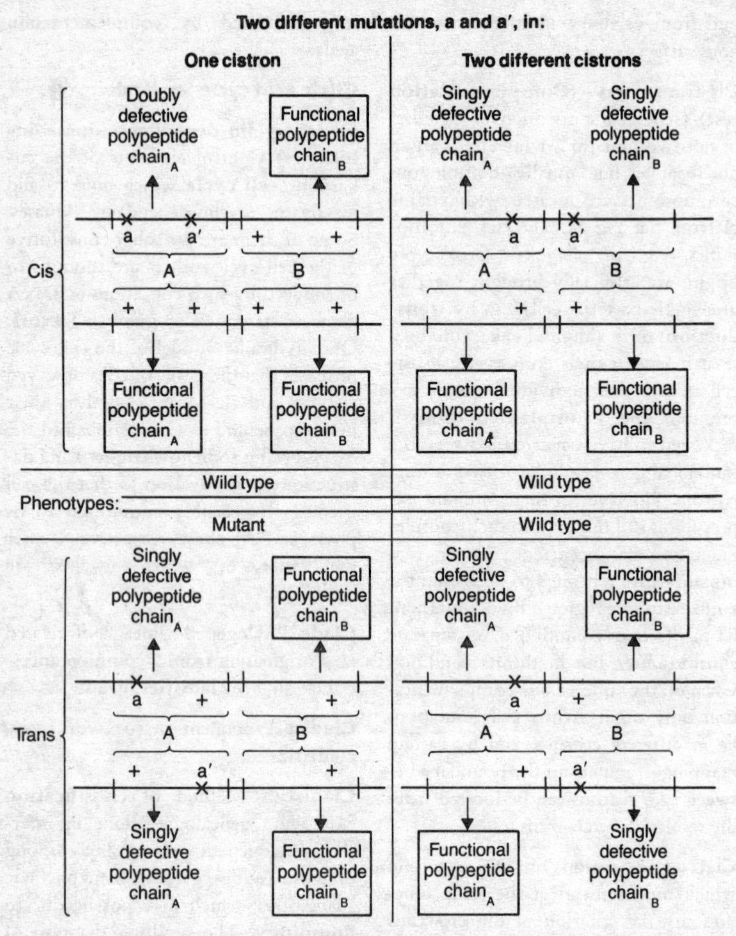

Fig. 24 *Theoretical basis of the* **cis-trans test**. *Where two mutations (—×—) occur within the same chromosome (*cis *configuration), complementation occurs through functional polypeptide production by the complementary cistron or cistrons of the other chromosome. Mutant phenotype only occurs where the two mutations occur within the same cistron but on different chromosomes. This effect enables precise mapping of the physical limits of cistrons within chromosomes.*

taxa if it is believed to have originated as a novelty in a common ancestor earlier than the most recent common ancestor, but *synapomorphous* (*-morphic*) if not. **Homoplasy** occurs between characters that share structural aspects but are thought to have arisen independently, either by **parallel evolution** or

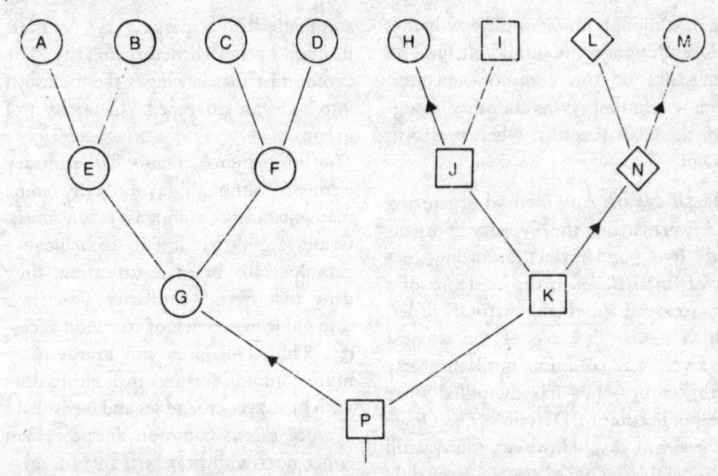

Fig. 25 *A cladogram illustrating terminology employed. Character states are represented by shapes of figures, and a change in these by an arrow. Character state in circles (○) is symplesiomorphous for taxa A & B and C & D but is synapomorphous for taxa E, F and G. It is homoplasous between taxa A – G (being an example of parallel development) and homoplasous between taxa A – H and M, but here it is also an example of convergent development.*

convergence: the cladistic method does not distinguish between these, because it does not need to; nor does it permit what it terms **paraphyletic** taxa. See **phenetics, phylogenetics**.

Cladocera Order of **Branchiopoda**. 'Water fleas', including *Daphnia*. Carapace encloses trunk limbs, used for feeding. Antennae used for swimming.

Cladode (phylloclade) Modified stem, having appearance and function of a leaf, e.g. butcher's broom.

Cladogenesis Branching **speciation**, in which an evolutionary lineage splits to yield two or more lineages. See **cladistics**.

Cladogram See **cladistics**.

Cladophyll Branch that resembles a foliage leaf.

Clamp connection Lateral connection between adjacent cells of a dikaryotic hypha, found in some of the **Basidiomy-cota**. Ensures that each cell of the hypha contains two genetically dissimilar nuclei. Compare **crozier formation**.

Class A taxonomic category in **classification**. Of higher rank (more inclusive) than **order** but of lower rank (less inclusive) than **phylum** (or **division**). Thus there may be one or more classes in a phylum or division and one or more orders in a class. In employing any taxonomic category one aim is to ensure that all its members share a common ancestor which is also a member of the taxon, although opinions differ whether the resulting group should include all *descendants* of the common ancestor. Thus evolutionary taxonomists recognize the Class Reptilia, whereas cladists do not.

Classification Any method organizing and systematizing the diversity of organisms, living and extinct, according to a set of rules. Belief in the existence of a pre-arranged (divine?) natural order, which it was the role of the scientist discover, was common until the early 19th century, but has dwindled since the publication of Darwin's *The Origin of Species* in 1859. However, there is still considerable dialogue between **essentialist** and **nominalist** accounts of biological classification as to whether the groups which different classifications recognize are 'natural' (real) or merely human constructs (artificial). Characters are not randomly distributed among organisms but tend to cluster together with high predictability, suggesting that all the taxonomist has to do is discover the various nested sets of characters to attain a 'natural' classification. However, even those taxa which seem to have most to recommend their objective realities in nature (i.e. species) lack some of the features of **natural kinds**.

The charge of arbitrariness (artificiality) over the rule of classification has led to the search for an objective methodology. The solution of *numerical taxonomists* has been to select not just one or a few characters which are given added weight 'often apparently arbitrarily) in comparisons between organisms, but to give all phenetic characters equal weight, in the expectation that natural (as opposed to artificial) groups will automatically emerge as clusters through overall phenetic similarity. The taxonomist then arranges these clusters into a rule-governed hierarchy of groups.

But this approach also has arbitrary elements, although favoured by some mathematically minded taxonomists. Critics argue that it fails to achieve a genealogically based classification: they deny that overall similarity alone is a sure guide to recency of common ancestry. Thus, crocodiles and lizards share more common features than either does with birds; yet crocodiles and birds share a more recent common ancestor than either does with lizards. If there is anything like an objectifying principle available to taxonomists it must surely be genealogy. Two principal schools of taxonomy which endeavour to objectify their methods by acknowledging the process of evolution in this way are *cladism* and *evolutionary taxonomy*. The principal difference between them (see **cladistics**) is that cladists include *all* descendant species along with the ancestral species within taxonomic groups; evolutionary taxonomists hold that different rates of adaptation within the descendant groups of a single ancestor should be reflected in the classification. This may require that those that have diverged more from the ancestral stock are given special (often higher) taxonomic ranking compared with those that have diverged less, and that consequently *not all* descendant species will be included in the taxon of the ancestral species. Much depends on the accepted definition of **monophyletic**. All biological classifictions are hierarchical: the higher the taxonomic category, the more inclusive. In descending order of inclusiveness, and omitting intermediate (sub-, infrap-, super-, etc.) taxa, the sequence is: kingdom, phylum (division in botany), class, order, family, genus, species. There are however enormous difficulties in establishing accurate genealogical classifications, not least with taxa which are extinct. See **identification keys**.

Clathrin One of the main proteins covering some **coated vesicles**, interconnecting molecules (triskeletons) forming varying numbers of lattice-like pentagonal/hexagonal facets on the cytosolic surfaces of coated pits, causing invagination to form the vesicle. The clathrin coat is shed after the vesicle is formed and internalized. See Fig. 26.

Clavicle Membrane bone of ventral side of **pectoral** (shoulder) **girdle** of many vertebrates. Collar-bone of man.

Clearing Process used to prepare many histological slides in light microscopy. The object is to remove any alcohol used in the dehydration of the material; the preparation is soaked in two or three changes of *clearing agent* (e.g. benzene, xylene, or oil of cloves). Clearing makes the material transparent and permits embedding in paraffin wax (insoluble in alcohol) prior to sectioning.

Cleavage (segmentation) Repeated subdivision of egg or zygote cytoplasm associated with, but not always accompanied by, mitoses. In animals it often produces a mass (the blastula) of small cells (blastomeres) which subsequently enlarge. Bilateral (radial) cleavage, in which **animal pole** blastomeres tend to lie directly on top of vegetal blastomeres, occurs in echinoderms and chordates. Spiral cleavage, in which the first four animal blastomeres lie over the junctions of the first four vegetal blastomeres, is characteristic of other animal phyla. Cleavage may be *deterministic* or *indeterministic*, depending respectively upon whether the fates of blastomeres are already fixed or are plastic. Cleavage is complete (*holoblastic*) in eggs with little yolk; partial (*meroblastic*) in yolky eggs where only the non-yolky portion engages in cell division; *superficial* in centrolecithal eggs, where nuclear division produces many nuclei towards the centre of the cell and which then migrate to the cytoplasmic periphery to become partitioned by cell membranes (see **mosaic development**). Mammalian development is unique in involving compaction, whereby cells of the 8-cell stage maximize their contact with each other to form a compact ball stabilized by tight junctions sealing off the sphere's interior prior to division to form the 16-cell morula.

Cleidoic egg Egg of terrestrial animal (e.g. bird or insect) enclosed within protective shell, largely isolating it from its surroundings and permitting gaseous exchange and minor water loss or gain. Contrasts with most aquatic eggs, in which exchange of water, salts, ammonia, etc., occurs fairly freely. See **uricotelic**.

Cleistocarp (cleistothecium) Completely closed spherical fruit body (ascocarp) of some of the **Ascomycota**, e.g. powdery mildews, from which spores are eventually liberated through decay or rupture of its wall.

Cleistogamy Fertilization within an unopened flower; e.g. in violet.

Climacteric (Bot.) Occurrence of a large increase in cellular respiration in fruit-ripening (e.g. tomatoes, avocadoes, apples, etc.). Such fruits are called *climacteric fruits* (those that display a steady decline or gradual ripening are called *non-climacteric fruits*). Exposure of the fruits to low temperatures stops the climacteric permanently. Fruits can be stored for long periods in a vacuum, and since the available oxygen is minimal, cellular respiration and ethylene are suppressed. After the climacteric, fruits senesce and become susceptible to fungal and bacterial attack. See **ethylene**.

Climax community Community of organisms, with composition more or less stable and in equilibrium with existing natural environmental conditions (e.g. oak forest in lowland Britain; spruce and aspen forests of the boreal forest region). Climax community may reflect either regional climate (*climatic climax community*) or may be on atypical parent material or in a poorly drained depression such that the normal community climax does not develop (*edaphic community climax*). Each region is usually characterized by a mosaic of regional (climate) and edaphic climaxes; can be described as a polyclimax landscape.

Cline Continuous gradation of phenotype or genotype in a species population, usually correlated with a gradually changing ecological variable. See **infraspecific variation**, **ecotype**.

Clisere Succession of **climax communities** in an area as a result of climatic changes.

Clitellum Saddle-like region of some annelid worms (Oligochaeta, Hirudinea), prominent in sexually mature animals. Contains mucus glands secreting a sheath around copulating worms binding them together; the resultant cocoon houses the fertilized eggs during their development.

Clitoris Small erectile organ of female amniotes, homologous to the male's penis; anterior to vagina and urethra.

Cloaca Terminal region of the gut of most vertebrates into which kidney and reproductive ducts open. There is only one posterior opening to the body, the cloacal aperture, instead of separate anus and urogenital openings (e.g. placental mammals). Also terminal part of intestine of some invertebrates, e.g. sea cucumbers.

Clonal selection theory The theory, originated by N. Jerne and M. Burnet, that during their development both **T cells** and **B cells** acquire specific antigen receptors, being activated (selected) to proliferate into clones of appropriate effector cells only on binding this antigen. See **programmed cell death**.

Clone (1) A group of organisms of identical genotype, produced by some kind of **asexual** reproduction and some sexual processes, such as haploid selfing, or inbreeding a completely homozygous line. Nuclear transplantation techniques introducing genetically identical nuclei into enucleated eggs can also produce clones in some animals, even *in utero*. (2) A group of cells descended from the same single parent cell. Often used of sub-populations of multicellular organisms (e.g. see **clonal selection theory**) rather than the entire organism, which may in any case be a **mosaic**. (3) Nucleic acid sequences are said to be cloned when they are inserted into **vectors**

(e.g. **plasmids**, **yeast artificial chromosome**) and then copied along with them within host cells.

Club moss See **Lycophyta**.

Cnidaria Subphylum of the **coelenterata** containing hydroids, jellyfish, sea anemones and corals. Gut incomplete (one opening); ectoderm containing **cnidoblasts**. Two structural forms: (i) attached, sessile *polyp*, (ii) free-swimming *medusa*. Former is a cylindrical sac with mouth and tentacles at opposite end to the attachment; latter is umbrella-shaped, with flattened enteron, and mouth in middle of concave under-surface. Sometimes the phases alternate in a single life-cycle; sometimes only one phase occurs. Compare **Ctenophora**.

Cnidoblast (thread cell) Specialized stinging cell found only in **Cnidaria** and a few of their predators which incorporate them. Several kinds exist for, e.g., adhesion, penetration, injection. This cell produces an inert **organelle**, a *cnida* (nematocyst), commonly regarded as an *independent effector* (see **effector**). Compare **lasso cell**.

CNS See **central nervous system**.

CoA See **coenzyme A**.

Coacervate Inorganic colloidal particle (e.g. clay) on to which have been adsorbed organic molecules; maybe acted as an important concentrating mechanism in prebiotic evolution. See **origin of life** for discussion of illites and kaolinites.

Coated pit Specialized regions of most eukaryotic **cell membranes**, appearing as depressions in electron micrographs before pinching off as a **coated vesicle** to initiate an endocytic cycle. Clathrin-coated pits are the vehicles for **receptor**-mediated endocytosis.

Coated vesicles Membranous vesicles (*c.* 50 nm diameter) budded off endocytically from the plasma membrane and some other cell membranes. Either **clathrin**-coated or non-clathrin-coated; the former arise from coated pits during receptor-mediated **endocytosis** (e.g. in uptake of **cholesterol** as **LDL**), each vesicle containing a portion of the extracellular fluid and contributing to the bulk-phase aspect of endocytosis (*pinocytosis*); the latter (coat proteins include *β*-COP, recruited from a protein pool in the cytosol) carry cargo from the endoplasmic reticulum to and through the **Golgi apparatus**. Clathrin- and non-clathrin-coated vesicles are budded off from the *trans* Golgi network, the former carrying proteins to endocytic organelles and secretory granules. Coat proteins are normally jettisoned soon after vesicle formation and prior to fusion with the target membrane, when the cargo (glycoprotein, neurotransmitter, etc.) is released (see Fig. 26). See **cell membranes**, **cell locomotion**, **phagocytosis**.

Coccidia Sporozoan protozoa parasitic in guts of vertebrates and invertebrates. Probably ancestral to haemosporidians (e.g. malarial parasites). Give rise to diseases termed *coccidioses*.

Coccolith A calcified organic **scale** occurring in some members of the **Prymnesiophyta**.

Coccyx Fused tail vertebrae. In man comprises two to three bones.

Cochlea Diverticulum of the sacculus of

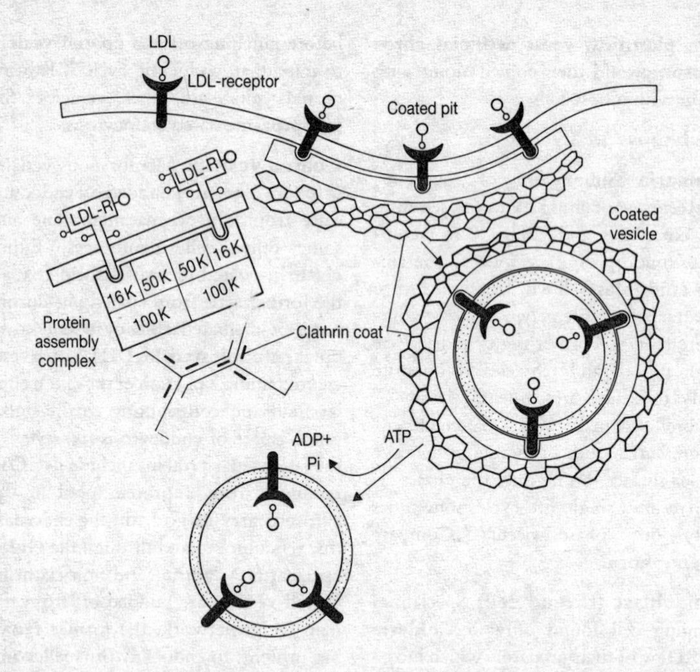

Fig. 26 *Involvement of a coated pit and coated vesicle in endocytosis of LDL via a cell-surface LDL-receptor. Details of the protein components of the assembly complex are given (K = kilodalton).*

inner ears of crocodilians, birds and mammals, usually forming a coiled spiral. Contains the organ of Corti, involved in sound detection and pitch analysis, a longitudinal mound of **hair cells** running the length of the cochlea supported on the *basilar membrane* with a membranous flap (*tectorial membrane*) overlaying the hair cells. Vibrations in the round window caused by vibrations in the **ear ossicles** are transmitted to the perilymph on one side of the basilar membrane, vibrating it and stimulating hair cells of the organ of Corti. The apex of the cochlea (spiral top) is most sensitive to lower frequency vibrations, the base to higher frequencies. See **vestibular apparatus**.

Cocoon Protective covering of eggs, larvae, etc. Eggs of some annelids are fertilized and develop in a cocoon. Larvae of many endopterygotan insects spin cocoons in which pupae develop (cocoon of silkworm moth is source of silk). Spiders may also spin cocoons for their eggs.

Codominant See **dominance**.

Codon Coding unit of **messenger RNA**, comprising a triplet of nucleotides which base-pairs with a corresponding triplet (*anticodon*) of an appropriate

transfer RNA molecule. Some codons signify termination of the amino acid chain (the *nonsense codons* UAG, UAA and UGA, respectively *amber, ochre* and *opal*). For role of AUG codon, see **protein synthesis**. See **genetic code**.

Coefficient of selection, s Proportionate reduction in contribution to the gene pool (at a specified time *t*) made by gametes of a particular genotype, compared with the contribution made by the standard genotype, which is usually taken to be the most favoured. If s = 0.1, then for every 100 zygotes produced by the favoured genotype only 90 are produced by the genotype selected against, and the genetic contribution of the unfavoured genotype is 1 − s. See **fitness**.

Coelacanthini Large suborder of **Crossopterygii**, mostly fossil (Devonian onwards). Freshwater, but with living marine representatives (*Latimeria*) in the Indian Ocean. Thought to have been extinct since the Cretaceous; but since 1938 several have been found. **Choanae** absent. Living coelacanth posterior dorsal and anal fins (both medial) resemble their (paired) pelvic fins in musculature, innervation and skeletal details. In this they differ from other sarcopterygians and other fishes, and the reasons are as yet unclear. Considerable debate surrounds whether coelacanth α- and β-haemoglobins and **mtDNA** resemble more closely those of other teleosts or those of larval amphibians.

Coelenterata Phylum of diploblastic and radially symmetrical aquatic animals, comprising the subphyla **Cnidaria** and **Ctenophora**. Ectoderm and endoderm separated by *mesogloea* (jelly-like matrix of variable thickness) and enclosing the gut cavity (*coelenteron*), with a single opening to the exterior. Peculiar cell types include **musculo-epithelial cells**, and either **cnidoblasts** (cnidarians) or **lasso cells** (ctenophores).

Coelom Main (secondary) body cavity of many triploblastic animals, in which the gut is suspended. Lined entirely by mesoderm. Principal modes of origin are either by separation of mesoderm from endoderm as a series of pouches which round off enclosing part of the archenteron (**enterocoely**), or *de novo* by cavitation of the embryonic mesoderm (**schizocoely**). Contains fluid (coelomic fluid), often receiving excretory wastes and/or gametes, which reach the exterior via ciliated funnels and ducts (*coelomoducts*). May be subdivided by septa into *pericardial*, *pleural* and *peritoneal* coeloms enclosing respectively the heart, lungs and gut. Reduced in arthropods, restricted to cavities of gonads and excretory organs, the main body cavity being the **haemocoele**, as in Mollusca.

Coelomate An animal with a **coelom**.

Coelomoduct Mesodermal ciliated duct (its lumen never intracellular) originating in the **coelom** and growing outwards from the gonad or wall of the coelomic cavity to fuse with the body wall. Sometimes conveys gametes to exterior (see **Müllerian duct**, **Wolffian duct**); sometimes excretory, e.g. kidneys of molluscs. Compare **nephridium**. See **kidney**.

Coenobium Type of algal colony where the number of cells is determined at its formation. Individual cells are incapable of cell division, are arranged in a

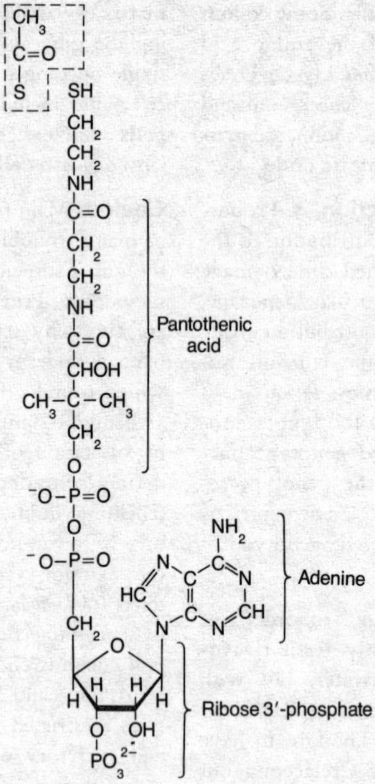

Fig. 27a *Structure of the* coenzyme A *molecule. The modification to form acetyl-coenzyme A is indicated at top left.*

specific manner, are coordinated and behave as a unit; e.g. *Volvox*, *Pandorina*, *Pediastrum*, *Hydrodictyon*.

Coenocyte (adj. coenocytic) Multinucleate mass of protoplasm formed by division of nucleus, but not cytoplasm, of an original cell with single nucleus; e.g. many fungi and some green algae. Compare **syncytium**. See **acellular**.

Coenospecies Group of related species with the potential, directly or indirectly, of forming fertile hybrids with one another.

Coenzyme Organic molecule (often a derivative of a mononucleotide or dinucleotide) serving as **cofactor** in an enzyme reaction, but, unlike a **prosthetic group**, binding only temporarily to the enzyme molecule. Often a recycled vehicle for a chemical group needed in or produced by the enzymic process, reverting to its original form when the group is removed – often by another

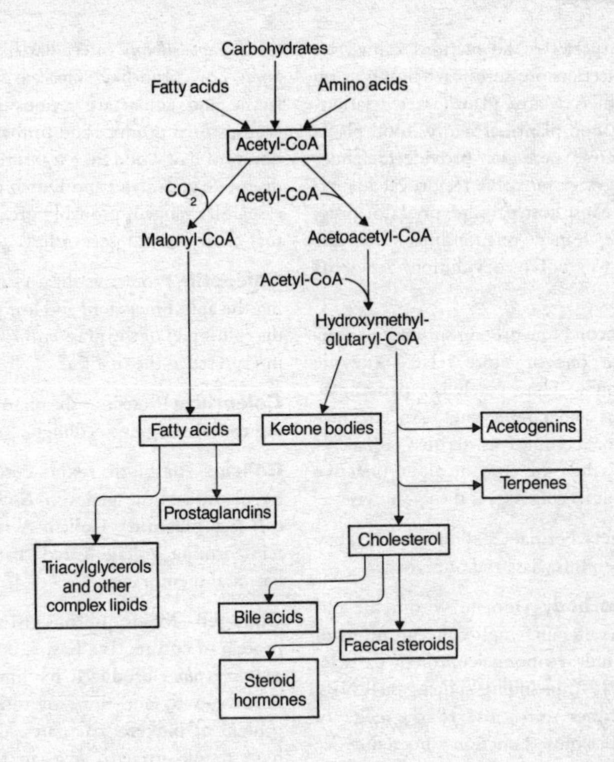

Fig. 27b *Some metabolic pathways in which acetyl-coenzyme A is involved.*

enzyme in a pathway. Removal by coenzymes of reaction products from the enzyme environment may be essential to prevent end-product inhibition of the enzyme. In heterotrophs they are frequently derivatives of water-soluble **vitamins**. See **coenzyme A**, **coenzyme Q**, **FAD**, **FMN**, **NAD**, **NADP**.

Coenzyme A (CoA, CoA-SH) Mononucleotide phosphate ester of pantothenic acid (a vitamin for vertebrates). Carrier of acyl groups in fatty acid oxidation and synthesis, pyruvate oxidation (see **Krebs cycle**) and various acetylations. When carrying an acyl group, referred to as acetyl coenzyme A (acetyl-CoA). See Figs. 27a and 27b.

Coenzyme Q (CoQ, ubiquinone) Lipid-soluble quinone coenzyme transporting electrons from organic substrates to oxygen in mitochondrial respiratory chains. Several forms; but all function by reversible quinone/quinol redox reactions. In plants, the related *plastoquinones* perform similar roles in photosynthetic electron transport. See **electron transport system**.

Coevolution Evolution in two or

more species of adaptations caused by the selection pressures each imposes on the other. Many plant/insect relationships (food plant/herbivore, food plant/pollinator, nest site provider/defender from grazers) involve reciprocal adaptations. Most host/parasite, predator/prey, cleaner/cleaned relationships, etc., are likely to involve coevolution. See **arms race**.

Cofactor Non-protein substance essential for one or more related enzyme reactions. They include **prosthetic groups** and **coenzymes**. An enzyme-cofactor complex is termed a *holoenzyme*, while the enzyme alone (inactive without its cofactor) is the *apoenzyme*.

Cohort Formal taxonomic category between Infraclass and Superorder.

Coiled body Domain within the nucleus containing high concentrations of small nulear ribonucleoprotein particles (snRNPs), including splicing snRNPs. Sometimes seen close to the edge of the nucleolus. Function uncertain. See **RNPs**.

Colchicine Antimitotic drug, binding to one tubulin molecule and preventing its polymerization. Depolymerization of tubulin then results, disappearance of the mitotic spindle blocking the cell mitosis. The drug taxol stabilizes microtubules, causing free tubulin to polymerize and for this different reason also arrests the cell in mitosis.

Coleoptera Beetles. Huge order of endopterygote insects. Fore-wings (elytra) horny, covering membranous and delicate hind-wings (which may be small or absent) and trunk segments. Biting mouthparts. Larvae may be active predators (*campodeiform*), caterpillar-like (*eruciform*), or grub-like (*apodous*). Many larvae and adults are serious pests of crops, stored produce and timber. Some borers of live wood may transmit fungal disease (e.g. *Scolytus* and Dutch elm disease). Size range is probably greater than that of any other insect order.

Coleoptile Protective sheath surrounding the apical meristem and leaf primordia (plumule) of the grass embryo; often interpreted as the first leaf.

Coleorhiza Protective sheath surrounding **radicle** in grass seedlings.

Colicins Antibiotic toxins determined by plasmid-borne genes in ***Escherichia coli*** (see **plasmid**). Colicin A is a protein forming voltage-gated channels in bacterial membranes.

Collagen Major fibrous (structural) protein of **connective tissue**, occurring as *white fibres* produced by fibroblasts. Forms up to one third of total body protein of higher vertebrates. Provides high tensile strength (e.g. in tendon), without much elasticity (unlike **elastin**). Collagen fibres are composed of masses of *tropocollagen* molecules, each a triple helix of collagen monomers. Yields *gelatin* on boiling.

Collar cell See **choanocyte**.

Collateral bundle See **vascular bundle**.

Collembola Springtails. Order of small primitively wingless insects (see **Apterygota**). Two caudal furcula fold under the abdomen and engage the hamula (another pair of abdominal appendages) prior to explosive release resulting in the spring. First abdominal segment carries ventral tube for adhesion. Compound eyes absent. No Malpighian tubules, and usually no tracheal system. Immensely abundant in soil, under bark, on pond surfaces, etc., forming vital link in detritus food chain.

Collenchyma Tissue composed of collenchyma cells, which provide mechanical support in many young growing plant structures (stems, petioles, leaves), but uncommon in roots. Collenchyma tissue occurs in discrete strands or as continuous cylinders beneath the epidermis, as well as bordering veins in dicotyledon leaves. Cells are typically elongated and contain unevenly thickened, non-lignified primary walls, and are living at maturity. Being primary walls are readily stretched, offering little resistance to elongation. Compare **sclerenchyma**.

Colloid A substance having particles of about 100–10,000 nm diameter, which remain dispersed in solution. Such *colloidal solutions* are intermediate in many of their properties between true solutions and suspensions. Brownian motion prevents colloid particles from sedimenting under gravity, and in lyophilic colloids (solvent-loving) such as aqueous protein solutions, each particle attracts around it a 'shell' of solvent forming a hydration layer, preventing them from flocculating. Competition for this solvent by addition of strong salt solution will precipitate the colloid.

Colon Large intestine of vertebrates, excluding narrower terminal rectum. In amniotes and some amphibians, but not fish, is clearly marked off from small intestine by a valve. In mammals at least, has essential water-absorbing role in preparation of faeces. Bacteria housed within it produce vitamins (esp. **vitamin K**).

Colony (1) Several plant and animal organizations where various more or less (and often completely) distinct individuals live together and interact in mutually advantageous ways. Sometimes, as in the alga *Volvox* and certain ciliates, the organization approaches multicellularity and may even have been a transitional stage in its attainment; but there is usually insufficient communication or division of labour between cells for full multicellular status (see **coenobium**). In colonial **Cnidaria** and **Ectoprocta** there may be considerable **polymorphism** between the individuals (termed **zooids**) with some associated division of labour. In all these there are good asexual budding abilities, and it is likely that the colonial habit originated by failure of buds to separate. Not so in colonial insects (e.g. **Hymenoptera**, **Isoptera**), although here too division of labour is the rule (see **caste**). Among vertebrates, several bird and mammal species live and/or breed in colonies, and many behavioural adaptations reflect this. (2) A group of microorganisms (bacteria, yeasts, etc.) arising from a single cell and lying on surface of food source – as in culture on agar.

Colony stimulating factors (CSFs) Glycoproteins stimulating mitosis in (i.e. they are mitogens of) bone marrow granulocyte and macrophage progenitor cells. They also maintain survival and promote differentiation in these cells and enhance phagocytosis by mature cells. Lung tissue is a major source. Nomenclature of CSFs is not standardized; e.g., multi-CSF is also called **inter-**

leukin 3 and stimulates multipotential stem cells to produce all the major blood cell types, while the CSF haemopoietin (see **haemopoiesis**) is also regarded as a **growth factor**. CSF-1 binds receptor tyrosine kinase of mononuclear phagocytes, initiating signal transduction with pleiotropic results due to expression of several different genes (e.g. some cyclin genes). Some can force leukaemic cells to differentiate and so stop dividing. Some CSFs form part of the **innate immune response**.

Colostrum Cloudy fluid secretion of mammary glands during the first few days after birth of young and before full-scale milk production. Important source of antibodies (passive immunity) too large to cross the placenta. Rich in proteins, but low in fat and sugar.

Columella (Bot.) (1) Dome-shaped structure present in sporangia of many Zygomycotina (fungi) of the order Mucorales (pin moulds); produced by formation of convex septum cutting off sporangium from hypha bearing it. (2) Sterile central tissue of moss capsule. (Zool.) The *columella auris*, or stapes. The **ear ossicle** of land vertebrates (often complex in reptiles and birds) homologous with the **hyomandibular** bone of crossopterygians, and transmitting air vibrations from the ear drum (to which it is primitively attached) directly to the oval window of the inner ear (amphibia, primitive reptiles). In mammals it no longer attaches to the ear drum, articulating with the incus at its outer edge.

Commensalism See **symbiosis**.

Commissure Nerve tissue tract joining two bilaterally symmetrical parts of the central nervous system. In arthropods and annelids, they connect ganglia of the paired ventral nerve cords, and the supraoesophageal ganglia (brain) with the suboesophageal. Commissures unite the vertebrate cerebral hemispheres (see **corpus callosum**).

Common (Of vascular bundles) passing through stem and leaf. Compare **cauline**.

Community Term describing an assemblage of populations living in a prescribed area or physical habitat, inhabiting some common environment. An organized unit in possessing characteristics additional to its individual and population components, functioning as a unit in terms of flow of energy and matter. The biotic community is the living part of the ecosystem. It remains a broad term, describing natural assemblages of variable size, from those living upon submerged lake sediments to those of a vast rain forest. See **association**, **consociation**, **society**.

Compaction See **cleavage**.

Companion cells Small cells characterized by dense cytoplasm and prominent nuclei, lying side-by-side with sieve tube cells in the phloem of flowering plants and arising with them by unequal longitudinal division of a common parent cell. One of their functions is to transport soluble food molecules into and out of sieve tube elements.

Comparative methods Species comparisons are the most commonly used technique for estimating whether and how organisms are adapted to their environments, providing **Darwin** with evidence for many of the arguments developed in his works *The Origin of Species* and *The Descent of Man*. Their popularity stems from our inability to do the experiments ideally suited to test evolutionary explanations, although proponents would argue that evolution provides its own natural experiments. Comparative methods essentially aim to determine whether or not different species living in the same environment predictably evolve the same characteristics, and nowadays greater weight is put on the number of times the same characteristic has evolved rather than simply on the number of species having it. In the last decade or so, statistical (and computerized) techniques have helped solve two problems inherent in comparative studies: evolutionary history and alternative explanations. Thus, monogamy and polygyny have evolved several times in birds and mammals; but monogamy (and egg-laying and feathers) predominates in birds and polygyny (and fur and viviparity) in mammals. It is unlikely that monogamy is causally associated with feathers or polygyny with fur, yet the correlations between these characters is highly significant. In order to explain adaptations we need to sift out the correlated but causally unrelated pairs of characters from the causally correlated, and in the case above we can do this only when we use information from polygynous birds and monogamous mammals. This way the merits of alternative phylogenies and adaptive explanations can be compared.

Compartments Anatomical regions in some (maybe many) animals, their boundaries well defined in development by cell lineage and expression of regulatory genes. In *Drosophila*, a compartment comprises cells forming more than one **clone** (or **polyclone**), whose growth respects a compartment boundary even when one of the component clones is a rapid-growing mutant. In vertebrates, the search is on for equivalents of such **homeotic** genes as *engrailed* and *wingless* in insects (see **segment polarity genes**), which delimit **parasegments**, and the ***Hox* gene** *Hox-2.9* seems to be the vertebrate homologue of *engrailed* and is involved as a regulatory gene in compartmentation of the hindbrain (see **rhombomere**). Fate of compartment polyclones is associated with the pattern of expression of homeotic genes.

Compensation point Light intensity at which rate of respiration by a photosynthetic cell or organ equals its rate of photosynthesis. At this intensity there is no net gain or loss of oxygen or carbon dioxide from the structure. Compensation points for most plants occur around dawn and dusk, but vary with the species.

Competence A cell able to take up DNA and be transformed is described as 'competent'.

Competent Describing embryonic cells while still able to differentiate into wide variety of cell types. Pluripotent, and undetermined. See **presumptive**, **determined**.

Competition The effect (result) of a common demand by two or more organisms upon a limited supply of resource, e.g. food, water, minerals, light, mates, nesting sites, etc. When *intraspecific*, it is a major factor in limiting population size (or density); when *interspecific*, it may result in local extinction of one or more competing species. An integral factor in Darwinian theory. See **competitive exclusion principle**, **density-dependence**, **natural selection**.

Competitive exclusion principle The empirical generalization, often regarded as axiomatic, probably first enunciated by J. Grinnell (1913) but generally attributed to G. F. Gause (1934), that as a result of interspecific competition, no two ecological niches can be precisely the same (see **niche**); i.e. niches are mutually exclusive, and mutual coexistence of two species will require that their niches be sufficiently different. See **character displacement**.

Complement Nine interacting serum proteins (beta globulins, C1–C9), mostly enzymes, activated in a coordinated way and participating immunologically in bacterial lysis and macrophage chemotaxis. Genetic loci responsible map in the S region of the **H-2 complex** in mice, and the HLA-B region of the MHC region in man. The principal event in the system is cleavage of the plasma protein C3, with subsequent attachment of the larger cleavage product (C3b) to receptors present on neutrophils, eosinophils, monocytes, macrophages and **B cells**. C3b combines with immune complexes and causes them to adhere to these white cells, promoting ingestion. C3 cleavage (*fixation*) terminates a **cascade**, itself initiated by immune complexes.

Complementary DNA See **cDNA**.

Complementary males Males, often degenerate and reduced, living attached to females; e.g. in barnacles (Cirripedia), ceratioid angler fish.

Complementation Production of a phenotype resembling wild-type (normal) when two mutations are brought together in the same cell in the *trans* configuration (on separate chromosomes); it occurs infrequently if the same functional unit (**cistron** or **gene**) is defective on both chromosomes. The ***cis-trans* test** is the rigorous method for defining the limits of cistrons, but often only the *trans* complementation test is made, the *cis* combination of individually recessive mutations almost always giving a wild phenotype.

Compost and composting See **decomposer**.

Compound leaf Leaf whose blade is divided into several distinct leaflets.

Concentric bundle See **vascular bundle**.

Conceptacle Cavity containing sex organs, occurring in groups on terminal parts of branches of thallus in some brown algae, e.g. bladderwrack, *Fucus vesiculosus*.

Condensation reaction Reaction in which two molecules are joined together (often by a covalent bond) with elimination of elements of water (i.e. one H_2O molecule). Occurs in all cells, during polymerization and many other processes. Contrast **hydrolysis**.

Conditioned reflex (conditional reflex) A **reflex** whose original unconditioned stimulus (US) is replaced by a novel conditioned stimulus (CS), the response remaining unaltered. Classically, the CS is presented to an animal either just prior to or in conjuction with the US; they are *associated* (presumably by formation of new neural pathways), and will each elicit the response when presented separately. Although Ivan Pavlov (1849–1936) is best remembered for conditioning dogs to salivate in response to the sound of a bell (CS) instead of the sight or smell of food (US), his dogs also became excited at the CS and moved to where it or the eventual food was delivered. A simple form of **learning**. See **conditioning**.

Conditioning Associative learning. There are two broad categories. (1) *Classical conditioning*, in which an animal detects correlations between external events, one of which is either *reinforcing* or *aversive*, modifying its behaviour in such a way that appropriate *consummatory responses* (usually) are elicited. (2) *Instrumental* or *operant* conditioning, in which aspects of consummatory responses are modified as an animal correlates and learns how variations in those responses affect its success in attaining a reinforcing, or avoiding an aversive, stimulus. Both are adaptive, since opportunities for successful instrumental conditioning may depend upon appropriate prior classical conditioning. See **consummatory act**, **learning**, **reinforcement**.

Condylarthra Extinct order of Paleocene and Eocene ungulate (hooved) mammals, possibly ancestral to most other ungulates and even to carnivores.

Condyle Ellipsoid knob of bone, fitting into corresponding socket of another bone. Condyle and socket form a joint, allowing movement in one or two planes, but no rotation; e.g. condyle at each side of lower jaw, articulating with skull; *occipital condyles* on tetrapod skull, fitting into atlas vertebra.

Cone (Bot.) *Strobilus*. Reproductive structure comprising a number of **sporophylls** more or less compactly grouped on a central axis; e.g. cone of pine tree. (Zool.) (1) Light-sensitive **receptor** of most vertebrate retinas, though not usually of animals living in dim light. The cone-shaped outer segment appears to be a modified **cilium**, with the 9 + 2 microtubule arrangement where the outer segment joins the rest of the cell. Pigment in the cone (*retinene*) requires bright light before it bleaches; three different classes of cone each contain a combination of retinene with a different genetically determined protein (opsin) bleaching at different wavelengths (red, green, blue). Cones are thus responsible for colour vision. There are about 6 million cones per human **retina**, mostly concentrated in the **fovea**. (2) Any of the cusps of mammalian molar teeth.

Conformation The three-dimensional shape (configuration) of a molecule. The final conformation adopted, preparatory to function, is the molecule's *native conformation*.

Congenital Of a property present at birth. In one sense a synonym of **heritable** (e.g. disorders arising from mutation such as **Down's syndrome**, **phenylketonuria**). Other infectious diseases (e.g. German measles) may be transmitted across the placenta from mother to foetus, as can some venereal diseases (e.g. congenital syphilis) which are otherwise transmitted by sexual intercourse. In some cases, however, the newborn may acquire a venereal disease by infection from the mother at birth; this would be congenital infection.

Conidiophore See **conidium**.

Conidium Asexual spore of certain fungi, cut off externally at apex of specialized hypha (conidiophore).

Conifer Informal term for **cone**-bearing tree. See **Coniferophyta**.

Coniferophyta The most widespread and numerous of the gymnosperms, having active cambial growth and simple leaves. Ovules and seeds are exposed; sperm non-flagellated. Comprise some 50 genera and 550 species, ranging from the giant redwood (*Sequoia sempivirens*), to the pines (*Pinus*), firs (*Abies*), spruces (*Picea*), etc. Conifer history extends back at least to the late **Carboniferous** period; the cordaites of that period were primitive conifers. Leaves of modern conifers possess drought-resistant features which may be related to the diversification of the division during the relatively dry, cold **Permian** period when increasing global aridity would have favoured structural adaptations like those of the conifer leaves. One genus, *Metasequoia*, was abundant during the **Tertiary** in Eurasia and North America; it was first described from fossil material in 1941, then in 1944 a live specimen was found in the Sichuan Province in south-western China. Subsequently, thousands more were found growing in China.

Conjugated protein Protein to which a non-protein portion (**prosthetic group**) is attached.

Conjugation (1) Union in which two individuals or filaments fuse together to exchange or donate genetic material. The process can involve a *conjugation tube* as in members of the Zygnematales (e.g. *Spirogyra*), or a *copulation tube* as seen in some **Bacillariophyta**, or no tube at all as in the Desmidiales. Process also occurs in fungi. In bacterial conjugation only a portion of the genetic material is transferred from the donor cell. Sex pili form (see **pilus**), followed by transmission of one or more plasmids from one cell to the other; less commonly the chromosome itself is transferred (see **F factor** for *Hfr* strain), when conjugation may be employed in **chromosome mapping**. Conjugation occurs in some animals; in ciliates (e.g. *Paramecium*) interesting varieties of the process occur. In the simplest case, after partial cell fusion, macronuclei disintegrate and each of the two micronuclei undergoes meiosis, after which three of the four nuclei from each cell abort. The remaining nucleus in each cell divides mitotically, one of the two nuclei from each cell passing into the other in reciprocal fertilization. The cells separate, their nuclei each divide mitotically twice, two nuclei reforming the macronucleus while the other two regenerate the micronucleus. The significance of this is obscure, but cultures of just one mating type seem to 'age' and die out sooner than those that can conjugate; **cytoplasmic inheritance** is probably involved. Compare **fertilization**.

Conjunctiva Layer of transparent epidermis and underlying connective tissue covering the anterior surface of the vertebrate eyeball to the periphery of the cornea, and lining the inner aspect of the eyelids.

Connective tissue A variety of vertebrate tissues derived from **mesen-**

chyme and the ground substance these cells secrete. A characteristic cell type is the **fibroblast**, producing fibres of the proteins *collagen* and *elastin*, providing tensile strength and elasticity respectively. Another protein, *reticulin*, is associated with polysaccharides in the **basement membranes** underlying epithelia and surrounding fat cells in **adipose tissue**. Loose connective tissue (*areolar tissue*) binds many other tissues together (e.g. in capsules of glands, meninges of the CNS, bone periosteum, muscle perimysium and nerve perineurium). Collagen fibres align along the direction of tension, as in **tendons** and **ligaments**. The viscosity of many connective tissues is due to the 'space-filling' *hyaluronic acid*. The **peritoneum**, **pleura** and **pericardium** (serous membranes) are modified connective tissues, as are the skeletal tissues **bone** and **cartilage**. Besides supportive roles, connective tissue is defensive, due largely to presence of **histiocytes** (macrophages), which may be as numerous as fibroblasts. These and **mast cells** constitute part of the **reticulo-endothelial system**. Connective tissues are frequently well vascularized, and permeated by tissue fluid.

Consensus sequences Conserved nucleotide sequences found in all examples of a regulatory region in DNA, such as **promoter** regions.

Consociation (Of plants) **Climax community** of natural vegetation, with an **association** dominated by *one* particular species; e.g. beech wood, dominated by the common beech tree.

Conspecific Of individuals that are members of the same species.

Constitutive enzyme An enzyme synthesized continuously, regardless of substrate availability. Compare **inducible enzyme**.

Constitutive mutant, constitutive phenotype Of mutants, initially studied in the context of the **lac operon** in *E. coli*, whose continuous and abnormal production of a cell product is caused by a mutation in the cistron encoding the **repressor** molecule of the appropriate **promoter** region, or from a mutation in the promoter region itself. For gain-of-function mutations, see **mutation**.

Consummatory act An act, often stereotyped, terminating a behavioural sequence and leading to a period of quiescence. Compare **appetitive behaviour**. Factors, sometimes very specific, which lead to termination of the sequence are sometimes referred to as *consummatory stimuli*.

Contact inhibition Phenomenon in which cells (e.g. fibroblasts) grown in culture in a monolayer normally cease movement at point of contact with another cell (see **adhesion**). When all available space is filled by cells, they also cease dividing – a phenomenon known as *contact inhibition of cell division*. Probably a normal self-regulatory device in tissue and organ size. It tends to be lost in transformed **cancer cells**.

Contact insecticide Insecticide whose mode of entry to the body is via the cuticle rather than the gut. DDT and dieldrin are notorious examples, their fat-solubility (often needed to penetrate waxy epicuticle) resulting in accumulation in fat reserves of animals in higher **trophic levels**.

Continental drift Theory, widely accepted since about 1953 but better described nowadays as *plate tectonics*, that crustal plates bounded by zones of tectonic activity, with the continents upon them, move slowly but cumulatively relative to one another. Helps explain distribution of many fossil and present-day forms formerly interpreted by invoking supposed land-bridges, and elevations and depressions of the sea bed. See **Gondwanaland**, **Laurasia**, **Pangaea**.

Continuous culture Procedure employed in **bioreactors** in which both the nutrients/substrates are added and the cells are harvested at a steady rate.

Contour feather See **feather**.

Contraception Deliberate prevention of fertilization and/or pregnancy, usually without hindering otherwise normal sexual activity. Includes: preventing sperm entry by use of a *condom*, a protective sheath over the penis; preventing access of sperm to the cervix by means of a *diaphragm* placed over it manually before copulation (later removed); preventing **implantation** by means of an *intra-uterine device* (*IUD*), a plastic or copper coil inserted under medical supervision into the uterus; and the **contraceptive pill**. Withdrawal of the penis prior to ejaculation (*coitus interruptus*) is *not* an effective method. *Abstention* during the phase in the **menstrual cycle** when fertilization is likely is a further method. Use of condoms is fairly effective, with the added advantage of reducing the risk of infection by microorganisms during intercourse. IUDs can cause extra bleeding during menstruation and may not be tolerated by some women. Male fertility control currently involves weekly injection of testosterone ethanoate, suppressing pituitary LH and FSH; sperm disappear from the ejaculate after 120 days. Recovery of fertility occurs about 100 days after injections cease. *Sterilization* must be regarded as irreversible. In women, this involves tubal ligation (preferably by endoscopy, e.g. laparoscopy), in which each oviduct is tied or blocked by diathermy (heating), clips or bands. In men, this involves vasectomy (diathermy of the vasa deferentia).

Contraceptive pill An oral pill for women, usually containing a combination of **oestrogen** and **progesterone**. Taken each day for the first 21 days after completion of menstruation, but not for the next seven. Resultant fall in blood oestrogen and progesterone levels allows menstruation to occur. Oral contraceptive pills may have unpleasant side effects (mainly due to oestrogens), and in some women contribute to thrombosis. Ovulation is inhibited through preventing normal gonadotropic effects of the **hypothalamus**. Alternative methods of contraception are often advisable. *Norplants* are tube-like pellets surgically implanted under the skin which release female contraceptive hormones for five years. They are cheaper than a supply of the standard oral pill, are removable, and have proved successful in parts of Scandinavia and may soon be employed in parts of the USA. *Depo-Provera* is a long-acting progestin (contains no oestrogen) injected into muscle tissue every six months. It maintains normal FSH activity and follicle development but prevents ovulation and fertilization by blocking LH release. It can cause bleeding, weight gain, depression and headaches. So-called *morning after pills* taken for a few days by women after intercourse (i.e. postcoitally) generally contain high oestrogen levels and speed up passage of the ovum down the oviduct, impairing its implantation and subsequent survival. Their routine use is ill-advised, but extenuating circumstances (e.g. in cases of rape) may override.

Contractile ring Bundle of filaments, largely **actin**, which assembles just beneath the plasma membrane at anaphase during animal cell divisions and in association with **myosin** generates the force pulling opposed membrane surfaces together prior to 'pinching-off' and completion of cytokinesis.

Contractile root Root undergoing contraction at some stage, causing a change in position of the shoot relative to the ground. Some corm-bearing plants produce large, thick, fleshy roots possessing few root hairs. These roots store large amounts of carbohydrate in the cortex, which may be rapidly absorbed by the plant. The cortex then collapses and the root contracts downwards, pulling the corm deeper into the soil.

Contractile vacuole Membrane-bound organelle of many protozoans (especially ciliates, sponge cells, and flagellated algal cells – vegetative cells, zoospores and gametes). In flagellated algae there are usually two anterior contractile vacuoles; however, they can be located posteriorly (e.g. in some **Chrysophyta**). Contractile vacuoles occur more frequently in freshwater than marine algae, suggesting that they maintain a water balance in cells, since cells in freshwater possess a higher concentration of dissolved substances in their protoplasm than in the surrounding medium, so that there is a net increase of water into the cells. Contractile vacuoles act to expel this excess water. A vacuole will fill with an aqueous solution (*diastole*), and then expel the solution outside of the cell (*systole*); this is repeated rhythmically, and if two contractile vacuoles are present, they usually fill and empty alternately. The process of filling is ATP-dependent, and therefore active. An alternative theory is that the vacuoles remove wastes from the cell. Dinoflagellates (**Dinophyta**) have a similar, but more complex structure called a **pusule**, which functions similarly to a contractile vacuole.

Control, controlled experiment See **experiment**.

Convergence (convergent evolution) The increasing resemblance over time of distinct evolutionary lineages, in one or perhaps several phenotypic respects, increasing their *phenetic* similarity but generally without associated genetic convergence. Usually interpreted as indicating similar selection pressures in operation. Structures coming to resemble one another this way are **analogous**. Convergence poses problems for any purely phenetic **classification**. Compare **parallel evolution**. See **cladistics**.

Copepoda Large subclass of **Crustacea**. No compound eyes or carapace, and most only a few mm long. Usually six pairs of swimming legs on thorax; abdominal appendages absent. Filter-feeders, using appendages on head. Some marine forms, e.g. *Calanus*, occur in immense numbers in plankton and are vital in grazing food-webs. About 4500 species; some (e.g. fish louse) parasitic.

Copia Transposable elements in *Drosophila* which resemble integrated retroviral proviruses. Extractable from eggs and cultured cells as double-stranded extrachromosomal circular DNA.

Copy number (1) The number of copies of a gene in a genome. (2) The number of copies of **plasmid** per host chromosome, typically in a yeast or bacterial cell.

Coracoid A cartilage bone of vertebrate shoulder girdle. Meets scapula at glenoid cavity, but reduced to a small process in non-monotreme mammals. See **pectoral girdle**.

Coral See **Actinozoa**.

Coralline A term referring to some members of the red algae (**Rhodophyta**) which become encrusted with lime (e.g. Corallinaceae).

Cordaitales Order of extinct Palaeozoic gymnophytes that flourished particularly during the Carboniferous. Tall, slender trees, with dense crown of branches bearing many large, simple, elongated leaves. Sporophylls distinct from vegetative leaves, much reduced in size and arranged compactly in small, distinct, male and female cones. Microsporophylls stamen-like, interspersed among sterile scales; megasporophylls similarly borne, each consisting of a stalk bearing a terminal ovule.

Corium See **dermis**.

Cork (phellum) Protective tissue of dead, impermeable cells, formed by the **cork cambium** which, as the diameter of young stems and roots increases, replaces the epidermis. During differentiation of the cork cells, their inner walls are lined with a relatively thick layer of a fatty substance **suberin**, which makes them impermeable to water and gases. The walls of cork cells may also become lignified. See **lignin**.

Cork cambium (Bot.) One of two lateral meristems in vascular plants which produces secondary tissues, which comprise the secondary plant body. Cork cambium is responsible for the formation of cork, which usually follows formation of both secondary xylem and secondary phloem. Repeated divisions of the cork cambium give rise to radial rows of compactly arranged cells, most of which are cork cells. These cells are formed toward the outer surface of the cork cambium, and the phellodem is formed toward the inner surface. Togerther, cork, cork cambium, and phelloderm comprise the periderm. See **cork**, **vascular cambium**.

Corm Organ of vegetative reproduction; swollen stem-base containing food material and bearing buds in the axils of scale-like remains of leaves of previous season's growth; food reserves not stored in leaves (compare **bulb**). Examples include crocus and gladiolus.

Cormophytes Refers to plants that possess a stem, leaf and roots (e.g. ferns, seed plants). Contrast **Bryophyta**.

Cornea Transparent exposed part of the sclerotic layer of vertebrate and cephalopod eyes. Flanked in former by conjunctiva and responsible for most refraction of incident light (a 'coarse focus'), the lens producing the final image on the retina. Composed of orderly layers of **collagen**, lacking a blood supply, its nutrients derived via aqueous humour from **ciliary body**.

Cornification (keratinization) Process whereby cells accumulate the fibrous protein *keratin* which eventually fills the cell, killing it. Occurs in the vertebrate epidermis, in nails, feathers and hair. See **cytoskeleton**.

Corolla Usually conspicuous, often coloured, part of a flower within calyx, consisting of a group of petals. See **flower**.

Corolla tube Tube-like structure resulting from fusion of petals along their edges.

Coronary vessels Arteries and veins of vertebrates carrying blood to and from the heart.

Corpora allata Small ectodermal endocrine glands in the insect head, connected by nerves to the **corpora cardiaca** (to which they may fuse) and producing juvenile hormone (neotenin) which is responsible for maintenance of the larval condition during moulting. Decreasing concentration of their product is associated with progressive sequence of larval stages. Their relative inactivity in final larval stage of **Endopterygota** brings about pupation, and their complete inactivity in the pupa is responsible for differentiation into the final adult stage. Removal is termed *allatectomy*.

Corpora cardiaca (oesophageal ganglia) Transformed nerve ganglia derived from the insect foregut, usually closely associated with the heart. Connected by nerves to, or fusing with, the **corpora allata** and producing their own hormones; but mainly storing and releasing brain neurosecretory hormones, particularly *thoracotrophic hormone* which stimulates thoracic (prothoracic) glands to secrete **ecdysone**, initiating moulting.

Corpus callosum Broad tract of nerve fibres (commissures) connecting the two **cerebral hemispheres** in mammals in the neopallial region.

Corpus luteum Temporary endocrine gland of ovaries of elasmobranchs, birds and mammals. In mammals develops from a ruptured **Graafian follicle**, and produces **progesterone** (as in elasmobranchs). Responsible in mammals for maintenance of uterine endometrium until menstruation, also during pregnancy. Its normal life (fourteen days in humans) is prolonged in pregnancy by *chorionic gonadotrophin* (hCG in humans). Its initial growth is due to **luteinizing hormone** from the anterior pituitary. See **oestrous cycle**.

Cortex (Bot.) In some brown and red algae, tissue internal to epidermis but not central in position; in lichens, compact surface layer(s) of the thallus; in vascular plants, parenchymatous tissue located between vascular tissue and epidermis. (Zool.) (1) Outer layers of some animal organs, notably vertebrate **adrenals**, **kidneys** and **cerebral hemispheres**. (2) Outer cytoplasm of cells (*ectoplasm*) where this is semi-solid (see **cytoskeleton**).

Corticosteroids (corticoids) Steroids synthesized in the **adrenal** cortex from **cholesterol**. Some are potent hormones. Divisible into *glucocorticoids* (e.g. **cortisol**, cortisone, corticosterone), and *mineralocorticoids* (e.g. **aldosterone**). Some synthetic drugs related to cortisone (e.g. prednisone) reduce inflammation (e.g. in chronic bronchitis, relieving airway obstruction).

Corticotropin See **ACTH**.

Cortisol (hydrocortisone) Principal glucocorticoid hormone of many mammals, humans included. (Corticosterone is more abundant in some small mammals.) Promotes **gluconeogenesis** and raises blood pressure. Low plasma cortisol level promotes release of *corticotropin releasing factor* (*CRF*) from the **hypothalamus**, causing release in turn of ACTH from the anterior **pituitary**. Prevents excessive immune response.

Corymb Inflorescence, more or less flat-topped and indeterminate.

Cosmid (cosmid vector) Hybrid **vectors** bearing the complementary single-stranded *cos* (cohesive) sites ('sticky ends') by which **lambda phage** circularizes, plus a standard plasmid replication origin and a gene for drug-resistance (e.g. tetracycline-resistance). Can carry up to 45 kb of DNA to be cloned – about three times that of phage vectors, but less than **yeast artificial chromosomes**. The sequences separately cloned are usually fragments which have some overlap with at least one other fragment in the set, enabling **chromosome walking**.

Cosmoid scale Non-placoid scale, having a thinner and harder outer layer composed of enamel-like material (*ganoine*) with hard non-cellular cosmine layer beneath. Growth is from the edge on the underside, since no living cells cover the surface. Characteristic of living and extinct lobefin fish (**Crossopterygii**), such as *Latimeria*. See **ganoid scale, placoid scale**.

Costa (1) In some members of the **Bacillariophyta** (diatoms), a ridge in the silica cell wall formed by well-defined siliceous ribs. (2) The midrib, or multilayered area, of a bryophyte leaf.

Costal Relating to ribs.

Cost of meiosis The disadvantage which most (amphimictic) sexual individuals seem to incur in contributing copies of only half their genomes to any of their offspring (through meiosis) whereas greater genetic fitness would seem to come from producing parthenogenetic offspring. See **sex**.

Cotyledon (seed leaf) Leaf, forming part of seed **embryo**; attached to embryo axis by hypocotyl. Structurally simpler than later formed leaves and usually lacking chlorophyll. Monocotyledons have one, dicotyledons two, per seed; the number varies in gymnophytes. Play important role in early stages of seedling development. In non-endospermic seeds, e.g. peas, beans, they are storage organs from which the seedling draws nutrients; in other seeds, e.g. grasses, compounds stored in another part of the seed, the **endosperm**, are absorbed by transfer cells on the outer epidermis of the cotyledons and passed to the embryo. Cotyledons of many plants (epigeal) appear above the soil, develop chlorophyll, and photosynthesize. See **scutellum**.

Cotylosauria (Mesosauria) The 'stem reptiles' of the late Palaeozoic and Triassic. Limbs splayed sideways from the body; superficially rather amphibian. Probably a heterogeneous (polyphyletic) order.

Countercurrent system System where two fluids flow in opposite directions, one or both along vessels so apposed to one another that exchange of contents, heat, etc., occurs resulting in the level dropping progressively in one fluid while it rises progressively in the other. It may involve active secretion, as in the *countercurrent multiplier* system in the loop of Henle in the vertebrate **kidney**, or be passive, as in the *countercurrent exchange* of respiratory gases in the teleost gill, the mammalian **placenta** and the **vasa recta**. See **rete mirabilis** and Fig. 45.)

Counterstaining See **staining**.

COV See **cross-over value**.

Coxa Basal segment of insect leg, linking trochanter and thorax.

Coxal bones See **pelvic girdle**.

Coxal glands Paired arthropod **coelomoducts**. In Arachnida, opening on one or two pairs of legs; in Crustacea, a pair of coelomoducts (antennal glands) on the third (antennal) somite, or on the somite of the maxillae; sometimes both. In Onychophora, a pair in most segments. Excretory.

cpDNA Chloroplast DNA. Larger than its mitochondrial counterpart mtDNA, but like it circular. Encodes the chloroplast's ribosomal RNAs and transfer RNAs, and part of ribulose bisphosphate (RuBP) carboxylase. Nuclear DNA encodes much of the rest of chloroplast structure. Many cpDNA mutations affecting the chloroplast are transmitted maternally, e.g. some forms of **variegation**. See **maternal inheritance**.

CpG island See **DNA methylation.**

C_3, C_4 plants See **photosynthesis**.

Cranial nerves Peripheral nerves emerging from brains of vertebrates (i.e. within the skull); distinct from **spinal nerves**, which emerge from the spinal cord. Dorsal and ventral roots of nerves from several segments are involved, but (unlike those of spinal nerves) these remain separate. Each root is numbered and named as a separate nerve; numbering bears little relation to segmentation, but does to the relative posteriority of emergence. There are 10 pairs of cranial nerves in anamniotes; 11 or 12 pairs in amniotes. Nerves I and II (*olfactory* and *optic* nerves) are largely sensory; III (*oculomotor*) innervates four of the six eye muscles; IV (*trochlear*) innervates the superior oblique eye muscle; V (*trigeminal*) is sensory from the head, but motor to the jaw muscles; VI (*abducens*) innervates the posterior rectus eye muscle; VII (*facial*) is partly sensory, but mainly motor to facial muscles in mammals; VIII (*vestibulocochlear*) is sensory from the inner ear; IX (*glossopharyngeal*) is mainly sensory from tongue and pharynx; X (*vagus*) is large, including sensory and motor fibres to and from viscera; XI (*accessory*) is a motor nerve accessory to the vagus; XII (*hypoglossal*) is motor, serving the tongue.

Craniata See **Vertebrata**.

Cranium The vertebrate skull. See **neurocranium**.

Crassulacean acid metabolism (CAM) A variant of the C_4 pathway of **photosynthesis** that evolved independently in many succulent plants (e.g. Cactaceae, Crassulaceae). Photosynthetic cells can fix carbon dioxide in the dark via phosphoenolpyruvate carboxylase forming malic acid which is stored in the vacuole. During the next light period, the malic acid is decarboxylated, and, with CO_2, is transferred to ribulose 1,5-bisphosphate (RuBP) of the **Calvin cycle** within the same cell. CAM plants, both C_4 and C_3 pathways (like C_4 plants); differ from C_4 plants in having a temporal separation of the two pathways in the CAM plants, rather than a spatial one as in C_4 plants.

CAM plants, then, are largely dependent upon nighttime accumulation of carbon dioxide for photosynthesis because stomata are closed during the day, which retards water loss. This is an adaptation to conditions of high light intensity and water stress; efficiency of water use can be many more times greater than C_3 or C_4 plants. CAM is widespread among vascular plants, more so than C_4 plants, being reported in at least 23 families of flowering plants.

Creatine Nitrogenous compound ($NH_2.C[NH].N[CH_3].CH_2.COOH$), derivative of arginine, glycine and methionine; reversibly phosphorylated to *phosphocreatine* (see Fig. 57), which transfers its phosphate to ADP via the enzyme *creatine kinase*. Found in muscle. Its anhydride breakdown product, *creatinine*, is excreted in mammalian urine. See **phosphagen, muscle contraction**.

Creatinine See **creatine**.

Creodonts Order of extinct mammals. Very varied, lasting into the Miocene, and ancestral to Fissipedia (dogs, cats, etc.). See **Carnivora**.

Cretaceous Geological period lasting from about 135–70 Myr BP. Much chalk deposited; anthophytes (flowering plants) and many groups of insects appeared, and became dominant. Large dinosaurs radiated, but along with ammonites and aquatic reptiles became extinct by the close. Climate tropical to subtropical throughout; Africa and South America separate. See **extinction**.

Crinoidea Feather stars; sea lilies. Primitive class of **Echinodermata**. Have long, branched, feathery arms; well-developed skeleton; tube feet without suckers; usually sedentary and stalked, with mouth upwards; microphagous; most modern forms free as adults. Long and important fossil history from Ordovician onwards (providing crinoid marble).

Crista (pl. cristae) See (1) **vestibular apparatus**, (2) **mitochondrion**.

Critical group Group of evidently closely related organisms, not easily categorized taxonomically. Used in the context of those apomicts which also reproduce by normal amphimictic means.

Crocodilia Sole surviving order of **archosaurs**. Alligators and crocodiles, appearing in the Triassic. Ancestors probably bipedal. Internal nares (**choanae**) open far back in the mouth owing to presence of long bony **false palate**. Some exhibit considerable parental care. Close affinities with birds.

Cro-Magnon man Earliest 'anatomically modern' humans. See ***Homo***.

Crop In vertebrates, distensible expanded part of oesophagus in which food is stored (esp. birds). In invertebrates, an expanded part of the gut near the head, in which food may be stored or digested.

Crop milk Secretion comprising sloughed crop epithelium of both sexes in pigeons, on which nestlings are fed.

Production influenced by **prolactin**, like mammalian milk.

Crop rotation The practice of growing different crops in regular succession to assist control of insect pests and diseases, increase soil fertility (especially when one season's growth includes nitrogen-fixing leguminous plants), and decrease erosion.

Cross The process or product of cross-fertilization. Contrast **selfing**.

Crossing-over Mutual exchange of sections of homologous chromatids in first meiotic prophase (see **homologous recombination** for diagram). Under the influence of enzymes (see **gene conversion**, **splicing**). Responsible for *chiasmata* visible in meiotic chromosomes, and for non-random **recombination**, its frequency and distribution are under genetic control (see **suppressor mutation**). Responsible for the **chiasmata** observed in first meiotic division. Non-homologous cross-overs, and cross-overs between sister chromatids in **mitosis** occur (see **parasexuality**, **twin spots**). See **cross-over value, DNA repair mechanisms, synaptonemal complex, transposable elements**.

Crossopterygii Order (sometimes superorder, or subclass) of **Osteichthyes**, in the heterogeneous subclass **Choanichthyes**. One known living form (*Latimeria*), the **coelacanth**; fossil forms included ancestors of land vertebrates. Bony skeleton; paired fins, with central skeletal axes; **cosmoid scales**, or derivatives. First appeared in Devonian around 400 Myr BP, the fossil record disappearing around 70 Myr BP. Differ from **Dipnoi** in having normal conical teeth. See **Rhipidistia**.

Cross-over value (COV, recombination frequency) The percentage of meiotic products that are recombinant in an organism heterozygous at each of two linked loci. In diploid organisms, most easily measured by crossing the double heterozygote to the double recessive. This value gives the **map distance** between the two loci, used in **chromosome mapping**. The percentage can never exceed 50%, which value would indicate absence of **linkage** between the loci. The **chi-squared test** can give the probability that this is so.

Cross-pollination Transfer of pollen from stamens to stigma of a flower of a different plant of the same species. Compare **self-pollination**.

Crown gall See ***Agrobacterium***.

Crozier formation (hook formation) (1) Similar to formation of **clamp connection** in Basidiomycotina, but occurring in dikaryotic cells of certain Ascomycotina; a hook develops in the ascogenous hypha where conjugate nuclear division takes place and is followed by cytokinesis; crozier formation may or may not immediately precede formation of an ascus. (2) In ferns and allies, the coiled juvenile leaf or stem.

Crustacea Class of **Arthropoda**, including shrimps, crabs, water fleas, etc. Mostly aquatic, with gills for gaseous exchange (and often nitrogenous excretion). Many segments with characteristic **biramous appendages**. Head bears single pairs of **antennules**, **antennae** (both primarily sensory), mandibles and maxillae (both for feeding). A pair of compound eyes common. Trunk composed of thorax and abdomen, often poorly distinct. Chitinous cuticle (**copepods** produce several million tons of **chitin** per year) often impregnated with calcium carbonate and excretory wastes (calcium reabsorbed prior to moulting). Small coelom, partly represented by 'kidneys' (antennal glands, maxillary glands) located in the head. Sexes usually separate; development usually via a **nauplius** larva. Includes hugely important microphagous filter-feeders in freshwater and marine food webs. Major subclasses are **Branchiopoda, Ostracoda, Copepoda, Cirripedia, Malacostraca**.

Cryobiology Study of effects of very low temperatures on living systems. Some organisms or their parts (e.g. corneas, sperm) can be preserved under these conditions.

Cryophytes Organisms growing upon ice and snow; mostly algae, but including some mosses, fungi and bacteria. Algae may be so abundant as to colour the substratum, as in 'red snow', due to the green alga *Chlamydomonas nivalis*. Algae can also be frozen into marine ice packs, possibly adapted to brine cell environments, as well as being found attached to the under surface of ice packs, where, for example, in the arctic they can form layers 2–3 cm thick, with diatoms being common along with small species of dinoflagellates and cryptophytes.

Crypsis A relational term, indicating that an individual organism in a particular environment setting tends to be overlooked by one or more potential predators, through its having some combination of size, shape, colour, pattern and behaviour. Such characters may be polymorphic, as in the banding and shell colour polymorphism of the snail genus *Cepaea*. To the extent that such a combination can be shown to improve the bearer's fitness by reducing attention from predators, it is regarded as cryptic. There are similarities between crypsis and **mimicry**. See **industrial melanism**, **search image**.

Cryptic species See **species**.

Cryptophyta Cryptophytes (cryptomonads). A small group of algal protists, comprising primarily flagellates that occur in both fresh and marine waters, where they appear to be light-sensitive, often forming their deepest living populations in clear **oligotrophic** lakes. They are found in the plankton of montane and north-temperate lakes throughout the winter. Motile cells possess two apically or laterally attached flagella (slightly unequal in length) at the base of a depression. Depending upon the species, there are two rows of microtubular hairs attached to the flagella. Small, 150 nm diameter organic **scales** are common on flagellar surfaces, and sometimes the cell body. Cells are dorsiventrally compressed in one plane, naked and surrounded by a periplast composed of the plasmalemma and a plate or series of plates directly under the plasmalemma. Chloroplast (with its own two envelope membranes) is surrounded by two membranes of **chloroplast Endoplasmic Reticulum**; and a **nucleomorph** is present. Starch grains – similar to those of potato, green algae and dinoflagellates (α-1,4 glucan; 30% amylose and amylopectin) – occur between the outer and inner membranes of chloroplast endoplasmic reticulum. A few, lacking chloroplasts, are heterotrophic. Major carotene and xanthophyll are β-

carotene and diatoxanthin respectively. Alloxanthin, unique to the cryptophytes, can be used in palaeolimnology to detect past fluctuations in their populations. **Eyespots** sometimes present. Cryptophytes possess projectiles called *ejectosomes* whose discharge results in movement of a cell in the opposite direction; function possibly as an escape mechanism. Some species form symbiotic relationships. For evolutionary origin, see **algae**.

Cryptorchid With testes not descended from abdominal cavity into scrotum.

CSF Acronym for any of **cerebrospinal fluid, colony stimulating factor** or **cytostatic factor**.

Ctenidia Gills of **Mollusca**, situated in the mantle cavity. Involved in gaseous exchange, excretion, and/or filter-feeding.

Ctenophora Comb-jellies, sea-gooseberries. Subphylum of **Coelenterata**; with **lasso cells**, but no cnidoblasts. Body neither polyp nor medusa; movement by cilia fused in rows (*combs* or *ctenes*); no asexual or sedentary phase. See **Cnidaria**.

Cultivar Variety of plant found only under cultivation.

Cusp Pointed projection on biting surface of mammalian molar tooth. Each cusp is termed a *cone*. Cusping pattern may have taxonomic value.

Cuticle Superficial non-cellular layer, covering and secreted by the epidermis of many plants and invertebrates (esp. terrestrial species). In plants, an external waxy layer covering outer walls of epidermal cells; in bryophytes and vascular plants comprises waxy compound, cutin, almost impermeable to water. In algae may contain other compounds. In higher plants cuticle is only interrupted by stomata and lenticels. Its function is to protect against excessive water loss as well as protecting against mechanical injury.

The arthropod cuticle contains alpha-**chitin**, proteins, lipids, and polyphenol oxidases involved in its tanning (see **sclerotization**). Normally subdivisible into an outer non-chitinous *epicuticle* and an inner chitinous *procuticle* (itself comprising an outer *exocuticle* and an inner *endocuticle*). The epicuticle (1 μm thick in insects) is composed of cemented and polymerized lipoproteins, and affords good waterproofing and resistance to desiccation (less so in Onychophora), having been of utmost importance in terrestrialization by insects. The endocuticles of decapod crustaceans are highly calcified, and often thick. In many arthropod larvae and some adults the cuticle remains soft and flexible (due largely to the protein *arthropodin*) but at some hinges (e.g. insect wing bases) another protein, *resilin*, enables greater flexibility still. Tanning hardens much of the arthropod cuticle, producing **sclerites**. Hardening is also brought about by water loss as the water-soluble arthropodin is converted to the insoluble protein sclerotin.

Cuticularization Process of **cuticle** formation.

Cutins Lipids found in the plant **cuticle**. The enzyme cutinase is secreted by pollen tubes during their initial growth on the stigma.

Cutinization Impregnation of plant cell wall with **cutin**.

Cutting Artificially detached plant part used in vegetative propagation.

Cuvier, Georges (1769–1832) Professor of vertebrate zoology at the *Musée d'Histoire Naturelle* in Paris (see **Lamarck**). Polymath, specializing in geology and palaeontology. Formed the premise that any animal is so adapted to its environment (or conditions or existence) that it can *function* successfully in that environment. All parts of the animal therefore had to interrelate to form a viable whole; but certain parts are relatively invariant between organisms, and may therefore have value in **classification**. His was not, however, an evolutionary system of classification, unlike Lamarck's; it is fairly clear that the two did not enjoy a cordial relationship. Cuvier had a genius for 'reconstructing' a whole vertebrate skeleton from a single bone.

Cuvierian duct Paired major (common cardinal) vein of fish and tetrapod embryos returning blood to heart from **cardinal veins** (under gut) in a fold of coelomic lining forming posterior wall of pericardial cavity. Becomes part of the superior vena cava of adult tetrapods.

C-value (constant-value) The amount of DNA per haploid **genome**. Among eukaryotes, varies from 0.009 pg in the yeast *Saccharomyces cerevisiae* to 700 pg in *Amoeba dubia*, being increased by transposable elements and non-coding DNA (1 pg = 10^{-12} g). The amount appears to determine nuclear volume.

Several unicellular protists contain much more DNA than mammals, which gives rise to the C-value paradox, compounded by the fact that many sibling species have very different C-values – with no comparable morphological differences. It is non-genic DNA which accounts for this, varying in eukaryotes from perhaps 3×10^6 base pairs to 1×10^{11} base pairs (a 100,000-fold range). See **repetitive DNA**, **selfish DNA**.

Cyanelle Endosymbiotic blue-green alga, usually in association with protozoa. See **Glaucophyta**.

Cyanobacteria (Cyanophyta, Myxophyta) Blue-green algae. Monerans, sharing general prokaryotic characteristics (related to **bacteria**), and including unicellular, colonial and filamentous forms. Viewed by phycologists and bacteriologists as algae and bacteria respectively. In the marine environment, blue--green algae trap and bind sediment as well as precipitating carbonate producing **stromatolites**. These are known from the **Precambrian**, when they freely populated large areas. Prior to 2000 Myr BP, no grazing or boring organisms had evolved; when they did, stromatolites declined abruptly. Blue-green algae were probably the first oxygen-producing organisms to evolve, responsible for early accumulations of atmospheric oxygen. Many extant species photosynthesize under aerobic and anaerobic conditions (oxygenic and anoxygenic photosynthesis). Under aerobic conditions, electrons for photosystem I are derived from photosystem II, but under anaerobic conditions, in the presence of sulphur, electrons are derived from reduction of sulphur (facultative phototrophic anaerobes). Blue-green algal photosynthetic apparatus includes chlorophyll *a*, carotenoids (including some specific to blue-green algae: myxoxanthophyll, oscillaxanthin), and phycobiliprotein pigments (water-soluble chromoproteins assembled in macromolecular aggregates, phycobilisomes), attached to the outer surface of thylakoid membranes scattered throughout the cell. Phycobiliproteins have two different roles: they provide a reserve of cellular nitrogen and collect light in photosynthesis. They also change in concentration in response to light quality and growth conditions, and play a role in chromatic adaptation. Blue-green algal cells contain myxophycean starch (glycogen granules), stored carbohydrate (similar to amylopectin), and cyanophycin granules which store protein in the form of polypeptides as well as polyphosphate bodies which contain phosphate.

A gelatinous sheath usually envelops the cells, and can be distinctly red or blue depending upon whether the cells are growing in highly acidic or alkaline environments respectively. Fimbriae or **pili** protrude from the walls of some blue-green algae and are probably involved in prokaryotic–eukaryotic cell interactions, e.g. symbiosis. Cell wall construction and chemistry resembles that of Gram-negative bacteria, but more complex; major wall constituent is peptidoglycan.

Species reproduce through cell division; some filamentous species form specialized vegetative fragments called hormogonia (singium) which serve in vegetative reproduction; they possess many gas vacuoles which aid buoyancy and dispersal. Many benthic (particularly epipelic) species are capable of moving (gliding) when in contact with a substratum without evident organs of locomotion. Gliding takes place within a mucilage sheath, with the sheath sticking to the substratum and being left behind by the advancing filament. Regularly arranged fibrillar extensions of the protoplasm moving in waves against the sheath may be responsible for the propulsive force in gliding. In some species, gliding is accompanied by rotation of the filament.

Endospores and exospores are formed by some genera by internal division of the protoplast; an irregular mass of spores (e.g. *Dermocarpa*), or a row of spores borne externally (e.g. *Chaemisiphon*). Resting spores (perennating structures) called akinetes are also produced by some species in response to stress. Other filamentous blue-green algae form heterocysts, which are larger than vegetative cells and are the production sites of the N-fixing enzyme nitrogenase (see **nitrogen fixation**). Evidence exists for bacterium-like sexual reproduction (conjugation) and genetic recombination.

Cyanobacteria are widely distributed, occurring in waters of varying salinity, nutrient status and temperature, with species found in both the benthos and plankton, as well as aerial habitats, including damp soil; they play an important role as primary colonizers in the establishment of a soil flora and in the accumulation of humus. Generally more abundant in neutral or slightly alkaline water but also in alkaline hot springs and in the marine environment, particularly the plankton where they comprise most of the picophytoplankton (tiny coccoid cells of *Synechococcus* and *Synechocystis*). In many nutrient-rich (**eutrophic**) lakes several species often form very large summer/early autumn populations which may reach 'bloom' proportions (e.g. *Anabaena* spp., *Microcystis aeruginosa, Aphanizomenon flos-aquae*). Blue-greens may also be obligate phototrophs, facultative chemoheterotrophs, or photo-heterotrophs.

Some strains of *Anabaena flos-aquae* and *Microcystis aeruginosa* produce toxins which can be poisonous to animals that drink the water. The ingested algae die in the digestive tract releasing toxins that include anatoxin A, B, C, and D (*Anabaena flos-aquae*) and saxitoxin (*Microcystis aeruginosa, Aphanizomenon flos-aquae*). The latter is the same **alkaloid** responsible for paralytic shell-fish poisoning.

Many species of blue-greens form mutualistic relationships with a diverse array of organisms, e.g. in forming **lichens** with fungi, with bryophytes, ferns, seed plants and, among animals, with sloths.

Cyanocobalamin (vitamin B_{12}) Cobalt- and nucleotide-containing vitamin, synthesized only by some micro-organisms. In vertebrates, carried across the gut wall by a glycoprotein (*intrinsic factor*) of gastric juice. Essential for erythrocyte maturation and nucleotide synthesis (through derivative, *coenzyme B_{12}*). Absence in diet or lack of intrinsic factor cause pernicious anaemia. See **vitamin B complex**.

Cyanome See **Glaucophyta**, **Rhodophyta**.

Cyanophychin granule Large bodies in the cells of blue-green algae (**Cyanobacteria**) composed of protein in the form of polypeptides, usually containing aspartic acid and arginine in the form of L-arginyl-poly (L-aspartic acid). Vary in appearance, but normally full of convoluted membranes. Amount of cyanophycin varies with the growth cycle being low in exponentially growing cells, but high in those in stationary growth phase. Polypeptide is produced by a non-ribosomal enzymatic process.

Cycadofilicales (Pteridospermae) Order of extinct, Palaeozoic gymnosperms that flourished mainly during the Carboniferous. Of great phylogenetic interest. Reproduced by seeds, but with fern-like leaves; internal anatomy combined fern-like vascular system with development of secondary wood. Micro- and mega-sporophylls little different from ordinary vegetative fronds; not arranged in cones. Compare **Cycadophyta**.

Cycadophyta Cycads. Extant and extinct seed plants; palm-like, found mainly in the tropics and subtropics; mostly fairly large plants, some with a distinct trunk densely covered with leaf bases of shed leaves. Exhibit secondary growth from a vascular cambium; central portion of the trunk comprises a large pith. Reproductive units are more or less reduced leaves with attached sporangia that are loosely or tightly clustered into a cone-like structure near the apex of the plant; pollen and seed are produced on different plants. Sperm is flagellated and motile, but is carried in a pollen tube to the vicinity of the ovule. Comprises two classes: the Cycadopsida, which includes the extant species that first appeared during the late **Carboniferous**; and the Cycadeoidopsida, which includes the fossil genera. These formed one of the dominant elements of mid-**Mesozoic** floras. Together with the cycads, they were so abundant that the Mesozoic era has been called the 'Age of Cycads'. Evolved in late **Carboniferous** or **Permian**, reaching a zenith of development in the **Jurassic**, declining dramatically to extinction in late **Creta-**

ceous. They appear to parallel, in a general manner, the rise and decline of the dinosaurs, which has led to the suggestion that cycadeoids and dinosaurs were to some degree interdependent; but no direct evidence exists. They resembled cycads with entire or pinnate leaves. Stems possessed a large pith and cortex, and a narrow cylinder of xylem, resembling that of extant Cycadophyta. Primary xylem was **endarch**; secondary xylem rays arranged in radial rows. Tracheids possessed bordered pits of a circular or scalariform design. Reproductive cones **bisporangiate**, the closed nature of which suggests a high degree of self-pollination, although insects and other animals may have chewed the cones, allowing parts to fall to the ground, where the pollen became windblown, hence disseminated.

Cyclic AMP (cAMP, cyclic adenosine monophosphate) Cyclic nucleotide **second messenger** produced from ATP by the soluble (or as in some receptor cells, membrane-bound) enzyme adenylate cyclase. cAMP-gated Ca^{2+} channels occur in vertebrate olfactory receptor ciliary membranes (see **cyclic GMP**). Can bind to regulatory subunits of inactive enzymes (e.g. protein kinase A) to liberate catalytic subunits. See **cascade**.

Cyclic GMP (cGMP, cyclic guanosine monophosphate) Cyclic nucleotide **second messenger** produced from GTP by the soluble (or in some receptor cells, membrane-bound) enzyme guanylate cyclase. Usually present in animal cells at about a tenth of cAMP concentration; but level can be raised by **inositol 1,4,5 triphosphate** pathway. Activates G-kinase, which in turn phosphorylates target proteins. Some membrane ion channels, e.g. in vertebrate photoreceptors, are specifically gated by cGMP (see **rhodopsin**); others (e.g. in vertebrate olfactory receptor ciliary membranes) are gated equally by cAMP and cGMP. See **G-protein**, **cyclic AMP**, **nitric oxide**.

Cyclic nucleotides See **cyclic AMP cyclic GMP**.

Cyclins Proteins of dividing eukaryotic cells, most of them accumulating and disappearing in waves during the **cell cycle** (see Fig. 15). Cyclin D (= Cln3) is an exception in not oscillating and may be an 'initiator cyclin' helping coordinate entry into the cell cycle with cell growth. Cyclins are sometimes grouped into 'mitotic cyclins' (A and B) and 'G1 cyclins' (C, D, E and F). Oscillations are engineered in part by transcriptional control and in part by their degradation by specific **ubiquitin** ligases, enzymes which selectively bind them to ubiquitin at precise points in the cell cycle. Extracellular influences such as reduced nutrient supply and removal of **growth factors** can prevent G1 cyclin breakdown and halt the cycle at G1 (see **CKI**s). Without degradation of mitotic cyclins, a cell is held in metaphase, breakdown returning the cell to interphase. The speed of cleavage cell cycles in frogs is affected by how fast cyclin accumulates; but this role cannot yet be universalized. Encoded by a diverse family of related genes, cyclins are often stored by way of maternal mRNA in the egg cytoplasm. They often exert their effects by binding to cyclin-dependent kinases (CDKs). See **maturation promoting factor**.

Cycloheximide Antibiotic, inhibiting **translation** of nuclear mRNA on cytoplasmic ribosomes. Prokaryotic translation unaffected, so used to distinguish mitochondrial/chloroplast from nuclear encoding of cell protein.

Cyclomorphosis A term that describes the seasonal **polymorphism** frequently observed in plankton organisms (e.g. dinoflagellates such as *Ceratium hirundinella*; rotifers; *Daphnia*). In its extreme form it appears to be exhibited primarily by organisms that reproduce throughout most of the year by asexual or parthenogenetic means, giving rise to genetically homogeneous clones.

Cyclosis Circulation of protoplasm (cytoplasmic streaming) in many eukaryotic cells, especially large plant cells. Sometimes restricted, as in small plant cells and animal cells, to jerky movement of organelles and granules in the cytoplasm. Some eukaryotic cells (e.g. squid giant axons) use both **actin**-based (**myosin**-dependent) and **tubulin**-based (**dynein**- or **kinesin**-dependent) cytoskeletal systems for membrane-bound organelle transport. One suggestion is that microtubules deliver organelles rapidly anywhere from the **centrosome** to the cell periphery, after which actin microfilaments route them to their specific terminals, such as the plasma membrane. Both mechanisms are ATP-dependent.

Cyclostomes (1) **Agnatha** whose sole living representatives are lampreys and hagfishes. Eel-like but jawless, with a sucking mouth and one nostril (lampreys), or two (hagfishes); without bone, scales or paired fins. Sometimes placed in a single order (**Cyclostomata**), but more usually regarded as two long-separated orders (*Petromyzontiformes*, lampreys, *Myxiniformes*, hagfishes) in subclasses Monorhina and Diplorhina respectively. Lampreys are ectoparasitic on vertebrates and **anadromous**; hagfishes are colonial burrowers, feeding largely on polychaete worms or corpses. Lamprey larva is the **ammocoete**. (2) A suborder of **Ectoprocta**.

Cyme Branched, flat-topped or convex **inflorescence** where the terminal flower on each axis blooms first.

Cypsela Characteristic fruit of Compositae (sunflower, daisy, etc.). Like an achene (and usually so described), but formed from an inferior ovary and thus sheathed with other floral tissues outside ovary wall. Strictly, a pseudo-nut, being formed from two carpels.

Cysticercus Bladderworm larva of some tapeworms. Some (e.g. *Echinococcus*) are asexual in an encapsulated cysticercus known as a *hydatid cyst*. Here, brood capsules form on the inner wall, each budding off scolices producing new hydatid cysts. Human liver may become infected with these cysts. See **polyembryony**.

Cystocarp Structure developed after fertilization in red algae; consisting of filaments bearing terminal **carpospores** produced from the fertilized carpogonium, the whole enveloped in some genera by filament arising from neighbouring cells.

Cystolith Stalked body, consisting of ingrowth of cell wall, bearing deposit of calcium carbonate; found in epidermal cells of certain plants, e.g. stinging-nettle. See **statolith**.

Cytochalasins Anti-cytoskeletal drugs binding reversibly to **actin** monomers.

Cytochromes System of electron-transferring proteins, often regarded as enzymes, with iron-porphyrin or (in *cyt c*) copper-porphyrin as prosthetic groups; unlike in haemoglobin, the metal atom in the porphyrin ring must change its valency for the molecule to function. Located in inner mitochondrial membranes, thylakoids of chloroplasts and endoplasmic reticulum. See **chemiosmotic theory**, **electron transport system**.

Cytogenetics Study of linking cell structure, particularly number, structure and behaviour of chromosomes (e.g. their rate of replication) to data from breeding work. Often provides evidence for phylogeny. See **taxonomy**.

Cytokines Soluble substances, such as **interleukins**, which mediate interactions between cells, often promoting cell growth and division. Typically proteins, acting in a **paracrine** fashion, and often pleiotropically. See **growth factors**, **interferons**, **tumour necrosis factors**.

Cytokinesis Division of a cell's cytoplasm, as opposed to its nucleus. Distinguished, therefore, from **mitosis** and **meiosis**. In eukaryotic cells, onset is usually in late anaphase, when the plasma membrane in middle of cell is drawn in to form a *cleavage furrow*, formed by a contractile ring of **actin** microfilaments; this enlarges and finally breaks through the remains of the spindle fibres to leave two complete cells.

Cytokinins (phytokinins) Group of plant **growth substances** recognized and named because of their stimulatory effect (requiring **auxin**) on plant cell division; but of diverse origin. Chemically identified as purines; first discovered (kinetin) was isolated from yeast, animal tissues and sweet corn kernels. Substances with similar physiological action occur in fruitlets, coconut milk and other liquid endosperms; also in microorganisms causing plant tumours, witches' brooms and infected tissues. Other growth-promoting influences include cell enlargement, seed germination, stimulation of bud formation, delay of senescence, and overcoming apical dominance. Effects are thought to involve increased nucleic acid metabolism and protein synthesis. Their mode of transport in plants is unresolved. See **gibberellins**.

Cytology Study of cells, particularly through microscopy.

Cytolysins Soluble products of macrophages and cytotoxic **T cells**, which kill **antigen-presenting cells**, sometimes (as with perforins) by perforating the cell's plasma membrane, sometimes as **tumour necrosis factors**.

Cytolysis Dissolution of cells, particularly by destruction of plasma membranes.

Cytoplasm All cell contents, including the plasma membrane, but excluding any nuclei. Comprises cytoplasmic matrix, or **cytosol**, in which **organelles** are suspended, some membrane-bound and some not, plus crystalline or otherwise insoluble granules of various kinds. In amoeboid cells, there is a distinction between a semi-solid outer *plasmagel* (ectoplasm) and a less viscous inner *plasmasol*. A highly organized aqueous fluid, where enzyme localization is of paramount importance. See **cell membranes**, **cytoskeleton**.

Cytoplasmic ground substance Equivalent to the **cytosol**.

Cytoplasmic inheritance (1) Eukaryotic genetics involving DNA lying outside the nucleus, often in organelles (see cp**DNA**, mt**DNA**, **plasmid**) or endosymbionts (see **kappa particles**). Patterns of inheritance from such a source characteristically fail to observe Mendelian ratios. (2) Inheritance of a cytoplasmic pattern, apparently independently of both nuclear and organelle DNA. Compare **maternal inheritance**; see **aberrant chromosome behaviour** (3).

Cytoplasmic male sterility (CMS) Trait of higher plants (e.g. maize) determined by either a mitochondrial **plasmid** gene or a mtDNA gene, and an example of **maternal inheritance**. Pollen production aborts in development and plants are therefore self-sterile. Important agronomically, since hybrid plant lines can be produced combining desirable characters from different inbred parents.

Cytoplasmic streaming See **cyclosis, microtubule**.

Cytosine A **pyrimidine** base found in the nucleic acids DNA and RNA, as well as in appropriate nucleotides and their derivatives.

Cytoskeleton Network of **actin** microfilaments, tubulin **microtubules** and **intermediate filaments**, much of it just beneath the plasma membrane, conferring upon a eukaryotic cell (especially an animal cell) its shape and generating the spatial organization within it, providing its attachment capabilities and enabling it to move materials both within it and out from it. It provides several structures involved in locomotion (e.g. in muscle contraction, amoeboid and ciliary locomotion). Concentration changes in such diffusible substrates as cyclic nucleotides and divalent ions (generated by interactions between a cell and its **extracellular matrix**) can alter polymerization of cytoskeletal proteins. In the myxomycotan *Dictyostelium*, phosphorylation of actin monomers by tyrosine kinase, causing depolymerization of subcortical cytoskeletal actin, occurs when starved cells round up and lose adhesion with the substratum on return to rich growth medium; dephosphorylation of actin, linked to assembly of filaments and formation of spikes, occurs when cells spread and extend pseudopodia when starved. In mammalian fibroblasts, assembly of focal adhesions and stress fibres is governed by the GTP-ase Rho (see **adhesion**, Fig. 1 **G-protein**). Actin is the major constituent in cell surface **microvilli**, **microspikes** and stereocilia, while tubulin is the main constituent in cilia and flagella. Beneath the cell surface, belt **desmosomes**, **stress fibres** and **contractile rings** all involve actin, sometimes in association with myosin or other protein filaments. Animal epithelial cells often have keratin-attachment sites (spot desmosomes) helping to link cells together. When such cells die, cross-linked keratinized cytoskeletons may form a protective surface layer, as in **hair** and **nails**. See **cell locomotion**, **cyclosis**, **intercellular junction**.

Cytosol The fluid and semi-fluid matrix of the cytoplasm, including the **cytoskeleton**, in which are suspended the organelles.

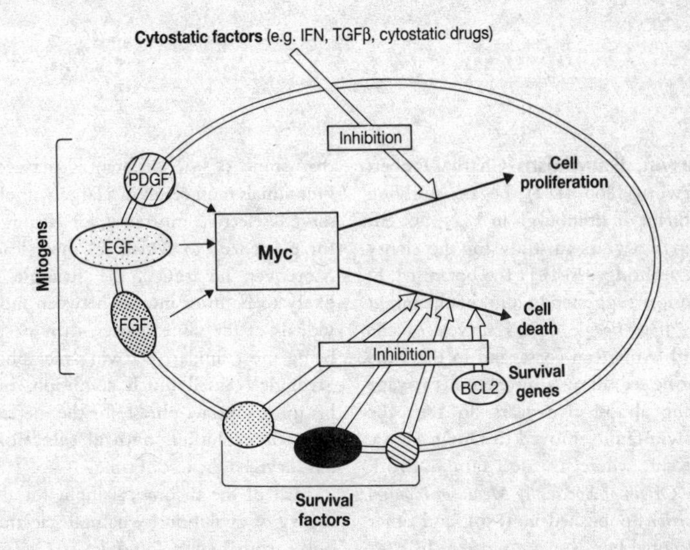

Fig. 28 *Diagram indicating how c-Myc protein can induce both cell proliferation and programmed cell death given appropriate conditions. See* **cytostatic factors** *and* **programmed cell death** *entries.*

IFN = interferon; TGFβ = transforming growth factor β; PGDF = platelet-derived growth factor; EGF = epidermal growth factor; FGF = fibroblast derived growth factor.

Cytostatic factors (CSFs) Any substance with the ability to stop cell growth, including interferons (IFNs) and transforming growth factor *β* (TGF*β*). However, the term 'cytostatic factor' is often reserved for the complex protein (which contains as a subunit the protein product of the vertebrate c-*mos* gene, c-Mos) which holds oocyte development at metaphase of the second meiotic division, possibly by preventing **cyclin** degradation and maintaining activity of **maturation promoting factor**. See Fig. 28 and the **cell cycle** diagrams.

Cytotoxic Poisonous, or lethal, to cells. See **T cell**.

Darwin, Darwinism Charles Robert Darwin (1809–1882) began studying medicine at Edinburgh in 1825, but left after two years to study for the clergy at Cambridge. In 1831 the botanist J. S. Henslow suggested to him that he might join HMS *Beagle* on its survey of the South American coast, and in the same year he set sail on a momentous voyage lasting almost five years. In 1842 the Darwin family moved to Down House in Kent, where his most famous work *The Origin of Species by Means of Natural Selection* (published in 1859), and other influential books, were written. In 1857 Darwin received a letter from Alfred Russel **Wallace** indicating that the two men were thinking along similar lines as to the mechanism of evolution – natural selection. After many years of ill health, Darwin died at Down House on 19 April 1882.

In essence, *Darwinism* is the thesis that species are not fixed, either in form or number; that new species continue to arise while others become extinct; that observed harmonies between an organism's structure and way of life are neither coincidental nor necessarily proof of the existence of a benevolent deity, but that any apparent design is inevitable given: (a) the tendency of all organisms to over-produce, despite the limited nature of resources (e.g. food) available to them; (b) that few individuals of a species are precisely alike in any measurable variable; (c) that some at least of this variance is heritable; and (d) that some of the differences between individuals must result in a largely unobserved selective mortality, or 'struggle for existence', to which (a) must lead. Moreover, he argued, the 'struggle' is likely to be most intense between individuals of the same species, their needs being most similar. Darwin's metaphor 'struggle' caused much confusion, but his more abstract phrase for the mechanism of evolution, **natural selection**, has survived the test of time.

Much of the theoretical input for the theory of evolution by natural selection came from Lyell's *Principles of Geology* (1830), which Darwin took on HMS *Beagle*, and Malthus's *Essay on the Principle of Population* (1798), which he read 'for amusement' in 1838. The chief empirical influences included observations on the fauna and flora of the Galapagos archipelago, so similar to – yet distinct from – those on the mainland, the fossil armadillos and ground sloths of Patagonia, and the effects of **artificial selection** on domesticated plants and animals. Darwinism argues for common genealogical links between living and fossil organisms (see also **comparative methods**). Darwin gave an unconvincing account of the origin and nature of phenotypic variation, so essential to his theory, and had rather little to say on the strict title of his major work; but Darwinism remains the single most powerful and unifying theory in biology. See **biogeography**, **evolution**, **neo-Darwinism**.

Daughter cells and nuclei Cells and nuclei resulting from the division of a single cell.

Day-neutral plants Plants that flower without regard to day-length. See **photoperiodism**.

Deamination Removal of an amino (-NH_2) group, frequently from an **amino acid**, by *transaminase* **enzymes**. In mammals, occurs chiefly in the liver, where the amino group is used in production of **urea**. The process is important in **gluconeogenesis**, where resulting carbon skeleton yields free glucose.

Decapoda (1) Order of **Malacostraca**, including prawns and lobsters (long abdomens), and crabs (reduced abdomens). Three anterior pairs of thoracic appendages are used for feeding; remaining five are for walking (hence name) or swimming, although first and second of these may bear pincers (chelae). Fused **cephalothorax** covered by a **carapace**, which may be heavily calcified. (2) Suborder of **Mollusca**, of the class Cephalopoda, having ten arms; squids, cuttlefish. Compare **Octopoda**.

Decerebrate rigidity See **reticular formation**.

Decidua Thickened and highly vascularized mucus membrane (endometrium) lining the uterus in many mammals (not in *ungulates*) during pregnancy. Some or all of the decidua comes away with the **placenta** at birth.

Deciduous (Of plants) shedding leaves at a certain season, e.g. autumn. Compare **evergreen**.

Deciduous teeth (milk teeth, primary teeth) First of the two sets of teeth which most mammals have; similar to second (permanent) set which replaces it, but having grinding teeth corresponding only to the **premolars**, not to the **molars**, of the permanent set.

Decomposer Any **heterotroph** breaking down dead organic matter to simpler organic or inorganic material. In some **ecosystems** the *decomposer food chain* is energetically more important than the *grazing food chain*. All release a proportion of their organic carbon intake as CO_2 and the heat release, as evidenced by compost heaps, can be considerable; a useful aspect is that most pathogenic bacteria and cysts, eggs, and immature forms of plant and animal parasites that may have been present, as in sewage, are killed.

An important factor in composting is the carbon to nitrogen ratio. A ratio of about 30 : 1 (by weight) is optimal: any higher and microbial growth slows. Because composting greatly reduces the bulk of plant wastes, it can be very useful in waste disposal.

Decussation Crossing of nerve tracts (e.g. some **commissures**) from one side of the brain to the other. Some fibres of the optic nerves and corticospinal tracts of vertebrates cross over in this way, so that parts of the visual field from each retina are transmitted to contralateral optic tecta; each side of the body is served by its contralateral cerebral hemisphere. See **optic chiasma**.

Deficiency disease Disease due to lack of some essential nutrient, in particular a **vitamin**, trace element or essential **amino acid**.

Degeneration Reduction or loss of whole or part of an organ during the course of evolution, with the result that it becomes **vestigial**. In the context of cells (e.g. nerve fibres) it usually implies their disorganization and death. Compare **atrophy**.

Dehiscent (Of fruits) opening to liberate the seeds; e.g. pea, violet, poppy.

Dehydration Elimination of water. (a) Prior to **staining**, usually achieved by soaking for up to 12 hours in successively stronger ethanol (ethyl alcohol), with at least two changes of 100%, in the preparation of tissues for microscopical examination. Failure to dehydrate properly leads to shrinkage and brittleness on embedding in paraffin, and deterioration of histological structure. See **clearing**. (b) Dry mass of soil and biological material is found after dehydration by heating to constant weight in an oven at 90–95°C.

Dehydrogenase See **oxidoreductase, enzyme**.

Deletion Type of chromosomal **mutation** in which a section of chromosome is lost, usually during **mitosis** or **meiosis**. Unlike some mutations, deletions are not usually reversible or correctable by a **suppressor mutation**. See **chromosome mapping**.

Deme, -deme Denotes a group of individuals of a specified taxon – the specificity given by the prefix used. The term deme on its own is not generally advocated, but where found usually refers to a group of individuals below, at or about the **species** level. Thus, groups of individuals of a specified taxon in a particular area (*topodeme*); in a particular habitat (*ecodeme*); with a particular chromosome condition (*cytodeme*); within which free gene exchange in a local area is possible (*gamodeme*), and of those in a gamodeme which are believed to interbreed more or less freely under specified conditions (*hologamodeme*). In zoology, the term deme without prefix tends to be used in the sense of gamodeme. See **infraspecific variation**.

Demography Numerical and mathematical analysis of populations and their distributions.

Denaturation Changes occurring to molecules of globular proteins and nucleic acid in solution in response to extremes of pH or temperature, or to urea, alcohols or detergents. Most visible effect with globular proteins is decrease in solubility (precipitation from solution), non-covalent bonds giving the molecule its physiological secondary and tertiary structures being broken but the covalent bonds providing primary structure remaining intact. Solutions of double-stranded DNA become less viscous as denaturation by any of the above factors results in strand separation by rupture of hydrogen bonds. This *melting* of DNA occurs with just very small increases in temperature, and may be reversed by the same temperature drop, the *re-annealing* being used in **DNA hybridization** techniques for assessing the degree of genetic similarity between individuals from different taxa.

Dendrite One of many cytoplasmic processes branching from the **cell body** of a nerve cell and synapsing with other neurones. Several hundred **boutons** may form synaptic connections with a single cell body and its dendrites.

Dendrochronology Use of isotopes

and annual rings of trees to assess age of tree; in fossils, used to date the stratum and/or make inferences concerning palaeoclimate. See **growth ring**.

Dendrogram Branching tree-like diagram indicating degrees of phenetic resemblance between organisms (a *phenogram*), or their phylogenetic relationships. In the latter, the vertical axis represents time, or relative level of advancement. A **cladogram** is a dendrogram representing phylogenetic relationships as interpreted by **cladistics**. See **taxonomy**.

Denitrification Process carried out by various facultative and anaerobic soil bacteria, in which nitrate ions act as alternative electron acceptors to oxygen during respiration, resulting in release of gaseous nitrogen. This nitrogen loss accounts in part for the lack of fertility of constantly wet soils that support nitrate-reducing anaerobes and for lowered soil fertility generally, products of denitrification not being assimilable by higher plants or most microorganisms. Bacteria such as *Pseudomonas*, *Achromobacter* and *Bacillus* are particularly important. See **nitrogen cycle**.

Density-dependence Widely observed and important way in which populations of cells or organisms are naturally regulated. One or more factors act as (a) increasing brakes on population increase with increased population density, and/or (b) decreasing brakes on population increase with decreased population density. There must be a *proportional* increase or decrease in the effect of the factor on population density as density rises or falls respectively. For example, the proportion of caterpillars parasitized by a fly must increase with increase in caterpillar density if the fly is to act as a density-dependent control factor. Since the caterpillar and fly may be regarded as a kind of **negative feedback** system, some have suggested that **ecosystems** might self-regulate by this sort of process (see **balance of nature**, **chaos**). **Contact inhibition** by cells is a form of density dependent inhibition of tissue growth.

Density gradient centrifugation Procedure whereby cell components (nuclei, organelles) and macromolecules can be separated by ultracentrifugation in caesium chloride or sucrose solutions whose densities increase progressively along (down) the centrifuge tube. Can be used to separate and hence distinguish different DNA molecules on the basis of whether or not they have incorporated heavy nitrogen (^{15}N) atoms. Cell components or macromolecules cease sedimenting when they reach solution densities that match their own buoyant densities.

Dental formula Formula indicating for a mammal species the number of each kind of tooth it has. The number in upper jaw of one side only is written above that in lower jaw of the same side. Categories of teeth are given in the order: incisors, canines, premolars, molars. The formulae for the following mammals are:

	i c pm m
hedgehog	$\frac{3 \cdot 1 \cdot 3 \cdot 3}{2 \cdot 1 \cdot 2 \cdot 3}$
grey squirrel	$\frac{1 \cdot 0 \cdot 2 \cdot 3}{1 \cdot 0 \cdot 1 \cdot 3}$

ruminants $\dfrac{0 . 0\text{–}1 . 3 . 3}{3 . 1 . 3 . 3}$

cats $\dfrac{3 . 1 . 2 . 0}{3 . 1 . 2 . 1}$

man $\dfrac{2 . 1 . 2 . 3}{2 . 1 . 2 . 3}$

Dentary One of the tooth-bearing **membrane bones** of the vertebrate lower jaw, and the only such bone in lower jaws of mammals, one on each side.

Denticle See **placoid scale**.

Dentine Main constituent of teeth, lying between enamel and pulp cavity. Secreted by **odontoblasts** (hence mesodermal in origin), and similar in composition to **bone**, but containing up to 70% inorganic material. Ivory is dentine. See **tooth**.

Dentition Number, type and arrangement of an animal's **teeth**. Where teeth are all very similar in structure and size (most non mammalian and primitive insectivore mammals) the arrangement is termed *homodont*, where there is variety of type and size of teeth, the arrangement is termed *heterodont*. Teeth which are replaced once only, as in mammals, are termed diphyodont; those replaced continuously, as in reptiles, are termed polyphyodont. See **deciduous teeth**, **dental formula**, **permanent teeth**.

Deoxyribonuclease See **DNase**.

Deoxyribonucleic acid See **DNA**.

Dephosphorylation See **phosphorylation.**

Dermal bone (membrane bone) Vertebrate bone developing directly from mesenchyme rather than from pre-existing cartilage (cartilage bone). Largely restricted in tetrapods to bones of **cranium**, **jaws** and **pectoral girdle**. See **ossification**.

Dermaptera Small order of orthopterous, exopterygote insects, including earwigs. Fan-like hind wings folding under short, stiff, forewings (resembling elytra); but wingless forms common; biting mouthparts; forceps-like cerci at end of abdomen.

Dermatogen See **apical meristem**.

Dermatophyte Fungus causing disease of the skin or hair of humans and other animals. Two of the most common diseases are athlete's foot and ringworm.

Dermis (corium) Innermost of the two layers of vertebrate skin, much thicker than the **epidermis**, and comprising **connective tissue** with abundant collagen fibres (mainly parallel to the surface); scattered cells including **chromatophores**; blood and lymph vessels and sensory nerves. Sweat glands and hair follicles project down from the epidermis into the dermis, but are not strictly of dermal origin. Responsible for tensile strength of skin. May contain **scales** or **bone**. See **dermal bone**.

Dermoptera Small order of placental mammals (one genus, *Cynocephalus*); so-called flying lemurs, although the Tertiary paromomyids provide the only fossils currently considered to lie within the order; a Chiroptera–Dermoptera clade is accepted by many. See **Archonta**. Also called colugos. The lower incisors are each divided and comb-like. They glide by means of the *patagium*, a hairy membranous skin fold stretching from neck to webbed finger tips and thence to webbed toes and tip of longish tail.

Desert A major **biome**, characterized by little rainfall and consequently little or no plant cover. Included are the *cold deserts* of polar regions, such as tundra and areas covered by permanent snow and ice. *Hot deserts* have very high temperatures, often exceeding 36°C in the summer months. Rainfall may be less than 100 mm per year. Deserts can be extensive: the African Sahara is the world's largest; Australian deserts cover some 44% of the continent. Annual plants (desert ephemerals) are most important in these conditions, both numerically and in kind. They have a rapid growth cycle which can be completed quickly when water is available, seeds surviving in desert soils during periods of drought. Perennials that do occur are mostly bulbous and dormant for most of the time. Taller perennials are either succulents (e.g. cacti) or possess tiny leaves that are leathery or shed during periods of drought. Many succulents exhibit **crassulacean acid metabolism**, absorbing carbon dioxide at night. See **osmoregulation**.

Desmid Informal taxon referring to two families of freshwater green algae (**Chlorophyta**) of the Class **Charophyceae**, Order Zygnematales. The *saccoderm desmids* (Mesotaeniaceae) are basically non-filamentous algae possessing non-porate walls, and do not form a new semicell during cell division. The nucleus is central in the cell and three chloroplast types can be found (axial and plate-like, stellate or spiral); one to several pyrenoids may be found. Sexual reproduction occurs via a conjugation tube. The *placoderm desmids* (Desmidiaceae) or 'true desmids' are distinguished from the saccoderm desmids in that each cell comprises two parts (semicells), with a much more complex, two-layered wall, perforated by a system of pores. The two semicells are usually separated by a median constriction (sinus), joined by a connecting zone (isthmus). Cells may be solitary, joined end-to-end in filamentous colonies, or united in amorphous colonies. Their taxonomy is complicated by polymorphism. Cells possess a single nucleus and two chloroplasts, one in each semicell, with one to several pyrenoids. Mucilage is secreted through pores in the cell wall and is responsible for movement and, when copiously secreted, for adherence to a substratum (commonly as epiphytes or epiliths). Barium sulphate crystals occur within a vacuole in each semicell. Sexual reproduction occurs via conjugation, but no tube is formed and the gametes are amoeboid. Ornamentation develops on the zygote walls, and meiosis precedes germination. Cell division is unique since it results in each semicell of the parent cell forming a new semicell, and becoming a new individual. As a group, desmids are **indicators** of relatively clean, unpolluted water, with low calcium and magnesium concentrations, and an acidic pH.

Desmin An **intermediate filament** protein characteristic of smooth, striated and cardiac muscle cells and fibres. In sarcomeric muscle it may help to link the Z-discs together; in smooth muscle it probably serves to anchor cells together. Also a prominent component of fibres on the cytoplasmic side of **desmosomes**.

Desmosomes One kind of **intercellular junction** (see Fig. 65) found typically where animal cells need firm attachment to one another against severe stress which would tear or shear tissues. A *belt desmosome* (zonula adhaerens) comprises a band of contractile **actin** filaments near the apical end of each epithelial cell, just under the cell membrane. A sheet of cells so united may roll up to form a tube by contraction of these filaments – as in neural tube formation. *Spot desmosomes* are sites of keratin filament attachment on the inner cell membrane surface, the filaments forming a network within the cell and connected to those in other cells, spot desmosomes being paired in adjacent cells. Hemidesmosomes are plasma membrane anchorage sites for **intermediate filaments**, establishing and maintaining strong adhesion between epithelial cells and the underlying basement membrane and connective tissue proteins. They contain one type of **integrin** heterodimer, and like spot desmosomes are rivet-like, probably transferring stress from the epithelium to underlying connective tissue via the **basal lamina**. See **cytoskeleton**

Desmotubule The tubule that traverses a plasmodesmatal canal, uniting the endoplasmic reticulum of two adjacent plant cells.

Determinate growth (Bot.) Growth of limited duration; characteristically seen in floral meristems and leaves of plants.

Determined Term applied to an embryonic cell after its fate has been irreversibly fixed. See **competence**, **organizer**, **positional information**, **presumptive**.

Detritus Organic debris from decomposing organisms and their products. The source of nutrient and energy input for the *detritus food chain*. See **decomposer**.

Deuteromycota (fungi imperfecti) Formerly Deuteromycotina (Deuteromycetes). Form-Division of Kingdom **Fungi**, comprising about 25,000 species, all lacking a sexual stage and thereby lying outside the remainder of fungal classification, heavily based as it is upon sexual reproduction. Some species may be secondarily sexual, others possibly never possessed a sexual stage or it became lost in the course of evolution. Generally, believed to be the non-sexual stages of fungi belonging to either the **Ascomycota** or **Basidiomycota**, with the largest number in the former division. Frequently, a sexual stage is discovered later, and both stages (anamorph and telomorph) are then transferred to the group to which the telomorph belongs. In general, classification within the Deuteromycota is based upon (a) the manner in which spores (conidia) are formed, and (b) conidial characteristics, producing an artificial or form group maintained primarily for convenience, comprising a diverse range of fungi. Deuteromycota are found in virtually every conceivable habitat and on every type of substratum. Many are saprotrophs in soil; others are plant and animal parasites (e.g. causing such diseases as ringworm and athlete's foot); some occur in flowing water. Many moulds are included, some commercially important (e.g. *Penicillium roque-*

fortii and *P. camembertii* in cheese production; *Aspergillus oryzae* in production of soy paste and in the brewing of sake; *A. oryzae* and *A. soyae* used together to produce soy sauce ('*shoyu*'). Strains of the genus *Penicillium* produce **antibiotics**, the first discovered by Sir Alexander Fleming, who noted in 1928 that a strain of *Penicillium* that had contaminated a culture of the bacterium *Staphylococcus* growing upon nutrient agar had completely halted the growth of the bacterium. Ten years later, Howard Florey and his colleagues of the University of Oxford purified penicillin. Citric acid is produced commercially in large amounts from colonies of *Aspergillus* grown under very acid conditions. Deuteromycotan moulds grow prolifically on artificial media, and are widely used in genetic, biochemical and nutritional research.

Deuterostomia That assemblage of coelomate animals (some call it an infragrade) in which the embryonic **blastopore** becomes the anus of the adult, a separate opening emerging for the mouth (see **stomodaeum**). It thus includes the **Pogonophora**, **Echinodermata**, **Hemichordata**, **Urochordata** and **Chordata**. Compare **Protostomia**.

Development Complex processes and events whereby a multicellular organism reaches its full size and form. Involves both genetic and environmental influences; but the distinction is somewhat arbitrary due to **epistasis**. Development in which **cleavage** of the egg produces blastomeres lacking the capacity to develop into entire embryos when isolated, even under favourable conditions, specific regions of the embryos being absent. Opposite of **regulative development.** Tunicates and echinoderms are among those animals with clear mosaic development. Those cleavage divisions which set up the mosaic condition are termed determinate cleavage divisions. May sometimes result from non-random distributions of cytoplasm in the egg. See **epigenesis**, **homeotic**, **long-germ**, **maternal effect**, **mosaic development**.

Devonian **Geological period** lasting from about 400–350 Myr BP. Noted for Old Red Sandstone deposits, for the variety of fossil fish, including **Actinopterygii, Choanichthyes** and primitive Amphibia (i.e. ichthyostegids). Primitive land plants (**Rhyniophyta, Zosterophyta, Trimerophyta**) became extinct. Lycopods (**Lycophyta**) first appeared; however, their diversity at this time suggests they evolved earlier, perhaps in the **Silurian** period. Progymnosperms first appeared in the Devonian, while the earliest seed to date is from the late Devonian, some 35 Myr after the first vascular plants appeared. Ecologically, sea covered most of the land, with mountains locally.

Dextrans Storage polysaccharides of yeasts and bacteria in which D-glucose monomers are linked by a variety of bond types, producing branched molecules.

Dextrin Polysaccharide formed as intermediate product in the hydrolysis of **starch** (e.g. to maltose) by **amylases**.

Dextrose Alternative name for **glucose**.

Diabetes (1) *Diabetes insipidus*. An uncommon disorder in which a copious urine arises usually owing to a person's inability to secrete **antidiuretic**

hormone. (2) *Diabetes mellitus*. The insulin-dependent form (IDDM, responding to insulin treatment) generally presents early in life and results from autoimmune destruction of insulin-producing β-cells in the pancreatic islets of Langerhans; a classic symptom is presence of glucose in the urine. Twin studies indicate multifactorial inheritance. Loci on human chromosome 6, linked to **MHC** loci, are implicated; but mouse studies suggest other loci as well are involved in that species. It is not inherited in a simple Mendelian way in either species. The non-insulin dependent form of the disorder, which presents late in life, appears from human twin studies to have high heritability.

Diacylglycerol A **second messenger**; cleavage product of phospholipase C activity upon phosphatidyl inositol bisphosphate (PIP2) of many mammalian (and maybe other) plasma membranes. Activates protein kinase C (C-kinase) when bound to Ca^{2+} and phosphatidylserine. Rapidly phosphorylated or cleaved to arachidonic acid. See **inositol 1,4,5 triphosphate**.

Diadelphous (Of stamens) united by their filaments to form two groups, or having one solitary and the others united; e.g. pea. Compare **monadelphous, polyadelphous**.

Diad-symmetry/asymmetry See **dyad-symmetry/asymmetry**.

Diageotropism Orientation of plant part by growth curvature in response to stimulus of gravity, so that its axis is at right angles to direction of gravitational force, i.e. horizontal; exhibited by rhizomes of many plants. See **geotropism, plagiotropism**.

Diaheliotropism See **phototropism**.

Diakinesis Final stage in the first prophase of **meiosis**.

Dialysis Method of separating small molecules (e.g. salts, urea) from large (e.g. proteins, polysaccharides) when in mixed solution, by placing the mixture in or repeatedly passing it through a semi-permeable bag or *dialysis tube*, e.g. made of cellophane, surrounded by distilled water (which itself may be removed and replaced). Small molecules will diffuse out of the mixture into the surrounding water, whereas large molecules are prevented by size from doing so. The principle underlies the design of artificial kidneys, which work by *renal dialysis*.

Diapause Term used to indicate period of suspended development in insects (and occasionally in other invertebrates). In insects, usually a true **dormancy**, implying a *condition* rather than a *stage* in morphogenesis. Insect diapause can occur at any stage in development, perhaps most commonly in eggs or pupae, but usually only once in any life cycle.

Diaphragm Sheet of tissue, part muscle and part tendon, covered by a serous membrane and separating thoracic and abdominal cavities in mammals only. It is arched up at rest, its flattening during inspiration reducing the pressure within the thorax, helping to draw air into the lungs. See **ventilation**.

Diaphysis Shaft of a long limb bone, or central portion of a vertebra, in mammals. Contains an extended **ossification** centre. See **epiphysis**.

Diapsid Vertebrate skull type in which two openings, one in the roof and one in the cheek, appear on each side. The feature is found in many reptiles (e.g. **archosaurs**) and all birds.

Diastase See **amylases**.

Diastema Gap in the tooth row of some (typically herbivorous) mammals separating cropping teeth (incisors/canines) from cheek teeth. Apes typically have diastemata ('monkey-gaps') in their upper jaws into which the lower canines fit. Almost half the specimens of the early hominid *Australopithecus afarensis* have small diastemata in the upper jaw, an increasingly uncommon trait in later hominids.

Diastole Brief period in the vertebrate **heart cycle** when both atria and ventricles are relaxed, and the heart refills with blood from the veins. The term may also be used of relaxation of atria and ventricles separately; in which case the terms *atrial diastole* and *ventricular diastole* are used, and are not to be confused with true diastole. Compare **systole**.

Diatom See **Bacillariophyta**.

Dichasium A **cyme** possessing two axes running in opposite directions; the type of cyme produced in those plants having opposite branching in the inflorescence.

Dichlamydeous (diplochlamydeous) (Of flowers) having perianth segments in two whorls.

Dichogamy Condition in which male and female parts of a flower mature at different times, ensuring that self-pollination does not occur. See **outbreeding**, **protandry**, **protogyny**. Compare **homogamy**.

Dichotomous venation Branching of leaf veins into two more or less equal parts, without any fusion after they have branched.

Dichotomy Branching, or bifurcation, into two equal portions. *Dichotomous keys* are employed in those identification manuals where one passes along a path dictated by consecutive decisions, each choice being binary (there being just two alternatives), one route involving the strict negation of the other. See **identification keys**.

Dicotyledonae Larger of the two classes of **Anthophyta** (flowering plants); distinguished from Monocotyledonae by presence of two leaves (**cotyledons**) in the embryo, by usually net-like leaf venation, by stem vascular tissue in the form of a ring of open bundles, and by flower parts in multiples of four or five. Pollen is usually tricolpate (having three furrows or pores), and commonly there is true secondary growth with a vascular cambium present. There are about 170,000 species, including many types of forest tree, potatoes, beans, cabbages, and such ornamentals as roses, clematis and snapdragon.

Dictyoptera Insect order containing the cockroaches and mantids, a group sometimes included in the **Orthoptera**. Largely terrestrial; wings often reduced or absent, and in general poor fliers; fore wings modified to form rather thick leathery tegmina (similar to elytra). Specialized stridulatory and auditory apparatus absent.

Dictyostele An amphiphloic **siphonostele**, comprising independent vascular bundles occurring as one or more rings. Present in stems of certain ferns.

Individual bundles here are termed *meristeles*. See **stele**, **protostele**.

Dictyostelium A genus of **myxomycotan** protists.

Diencephalon Posterior part of vertebrate forebrain. See **brain**.

Differentiation The process whereby cells or cell clones assume specialized functional biochemistries and morphologies previously absent. Such *determined* cells usually lose the ability to divide. Usually associated with the selective expression of parts of the genome previously unexpressed, brought on e.g. by cell contact, cell density, the extracellular matrix and molecules diffusing in it (e.g. see **growth factors**). Division of labour thus achieved is one evolutionary by-product of **multicellularity**.

Diffuse porous wood Secondary **xylem** in which the vessels (pores), when viewed in cross-section, are distributed fairly uniformly throughout a growth layer; or any change in size is gradual from early to late wood. Compare **ring porous wood**.

Diffusion Tendency for particles (esp. atoms, molecules) of gases, liquids and solutes to disperse randomly and occupy available space. Process is accelerated by rise of temperature, the source of movement being thermal agitation. Cells and organisms are dependent on the process at many of their surfaces and interfaces; on its own it is often inadequate for their needs. See **active transport**, **facilitated diffusion**, **osmosis**, **water potential**.

Digenea Order of the **Trematoda**. Includes those flukes which are usually vertebrate endoparasites as adults and mollusc endoparasites as sporocysts and rediae. Suckers simple. E.g. *Fasciola*, **Schistosoma**.

Digestion Breakdown by organisms, ultimately to small organic compounds, of complex nutrients that are either acted upon outside of the organism (e.g. by saprotrophs), or have entered some organelle (e.g. food vacuole), or organ (enteron) or gut specialized for the purpose. Often includes the physical events of chewing and emulsification besides chemical breakage of covalent bonds by mineral acids and enzymes. Food molecules are often too large to simply diffuse across **cell membranes**, and their digestion is first required. A gut forms a tube to confine ingested material, while extracellular enzymes hydrolyse it, given appropriate conditions (e.g. pH, temperature). Later stages of digestion may occur through enzymes located in the brush borders of intestinal epithelia (as with nucleotidases and disaccharides (see **Paneth cells**, **microvillus**). After digestion, molecules are incorporated (assimilated) into cells of the body.

Although plants lack guts, their cells can digest contained material (e.g., polysaccharides, lipids) (see **lysosome**). Carnivorous plants (pitcher plant, sundew, Venus fly trap) capture and digest insects with enzymes secreted by the plant, which absorbs the available nutrients (e.g. nitrogenous compounds and other organic compounds as well as minerals). Digestive enzymes are synthesized in glands. Until stimulation, these enzymes are stored in enlarged vacuoles and in the wall. Although both digestion and **respiration** are catabolic, and digestion like some respiration is anaerobic, digestion does not release significant amounts of energy.

Digit Finger or toe of vertebrate **pentadactyl limb**. Contains phalanges. May bear nails, claws or hooves.

Digitigrade Walking on toes, rather than on whole foot (**plantigrade**). Only the ventral surfaces of digits used. E.g. cat, dog.

Dihybrid cross A cross between two organisms or stocks heterozygous for the same alleles at the same two loci under study. A classic example was **Mendel's** crossing of F1 pea plants obtained as progeny from plants *homozygous* for different alleles at two such loci. This gave his famous 9 : 3 : 3 : 1 ratio of phenotypes (often called the *dihybrid* ratio); but by no means all dihybrid crosses give this ratio. The phrase *dihybrid segregation* is often used to describe the production of these ratios. *Dihybrid selfing* may be possible in some hermaphrodites (e.g. peas) and monoecious plants. Contrast **monohybrid cross**. See **linkage**.

Dikaryon (dicaryon) Fungal hypha or mycelium in which cells contain two haploid nuclei which undergo simultaneous division during formation of a new cell. This forms a third, dikaryotic, phase (*dikaryophase*) interposed between the haploid and diploid phases in a life cycle. Occurs in the **Ascomycota**, where it is usually brief, and in the **Basidiomycota**, where it is of relatively long duration. The paired nuclei may or may not be genetically similar. See **heterokaryosis, monokaryon**.

Dimorphism Of members of a species, structures, etc., existing in two clearly separable forms. E.g. *sexual dimorphism*, often very pronounced, between the two sexes; *heterophylly* in some plants (e.g. water crowfoot), in which leaves in two different environments have different morphologies. See **dosage compensation, polymorphism**.

Dinoflagellate See **Dinophyta**.

Dinophyta (Pyrrophyta) Dinoflagellates. Important freshwater and marine planktonic algae (Kingdom **Protista**). The oldest fossils date from the **Silurian**; they were probably the dominant primary producers in the **Palaeozoic** seas. A diverse group of unicells; some biflagellated (dinoflagellates proper), some non-motile (coccoid, filamentous, palmelloid, amoeboid), they possess chlorophylls *a* and c_1, β-carotene, and unique xanthophylls (peridinin). Most chlorophyll *a* and peridinin occur together in a water-soluble protein complex, the peridinin-chlorophyll *a*-protein. Chloroplasts are surrounded by the chloroplast envelope and typically one member of **chloroplast endoplasmic reticulum** (not continuous with the nuclear envelope); the few lacking this probably evolved from dinoflagellates with one. A typical motile dinoflagellate cell comprises two parts: the upper *epicone* and lower *hypocone*, separated by the transverse girdle or *cingulum*. Both epicones and hypocones are normally divided into several thecal plates. A longitudinal sulcus runs perpendicular to the girdle down one side of the hypocone. Both the transverse and longitudinal flagella emerge through the thecal plates in the area where the girdle and sulcus meet. Longitudinal flagellum projects out behind the cell; transverse flagellum is wavelike and closely appressed to the girdle. Fibrillar hairs may cover the longitudinal flagellum. The transverse flagellum is about 2–3 times as long as the longitudinal flagellum, but has a helical shape and comprises: (a) an axoneme, whose form approximates a helix; (b) a striated strand running parallel to the longitudinal axis of the axoneme but outside the loops of the coil; and (c) a flagellar sheath enclosing both axoneme and striated strands. Flagellum (left-hand screw) causes forward motion, while simultaneously rotating the cell; longitudinal flagellum acts as a rudder.

Pigmented and colourless forms exist. Some are heterotrophic, others parasitic, saprophytic, symbiotic or holozoic. Storage product is starch (similar to that of higher plants) and occurs in the cytoplasm. The nucleus is mesokayotic (see **mesokaryote**). Dinophyceae possess several different projectile organelles (e.g. trichocysts, cnidocysts, nematocyst-tacaiocysts, muciferous bodies; see **nematocyst**). **Eyespots**, some complex and lens-like, may or may not be present. Some species exhibit **bioluminescence**.

All produce resting cysts – smooth or spiny, with highly resistant cell walls (containing a chemical similar to sporopollenin); fossilized examples (cysts) were discovered independently by palaeontologists and classified under a separate taxonomic scheme and called *hystrichosphaerids* (hystrichospores), identical to many extant resting spores.

Some species, particularly those lodged in shellfish, produce dangerous toxins causing death of fish and shellfish during 'red tides', when blooms colour the water red. When humans eat contaminated shellfish, poisoning and even death result. A lethal product, the toxin's principal action is on the central and peripheral nervous systems.

Dinosaur See **Ornithiscia**, **Saurischia**.

Dioecious Unisexual, male and female reproductive organs being borne on different individuals. Compare **monoecious**, **hermaphrodite**. See **outbreeding**.

Diphyodont See **dentition**.

Diplanetism In some **Protista** (e.g. some **Oomycota**), the occurrence of a succession of two morphologically different zoospore stages separated by a resting stage. Contrast **monoplanetism**.

Diploblastic Level of animal organization in which the body is composed of two cell layers (germ layers), the outer **ectoderm** and inner **endoderm**. Found only in the **Coelenterata**, in which a jelly-like mesogloea separates the layers.

Diploid Nuclei (and their cells) in which the chromosomes occur as homologous pairs (though rarely *paired up*), so that twice the **haploid** number is present. Also applicable to appropriate tissues, organs, organisms and phases in a life cycle (see **sporophyte**). Most **somatic cells** of animals are diploid (see **male haploidy**), but some of the cells of the **germinal epithelium** engage in **meiosis**, giving haploid products. See **alternation of generations**, **dikaryon**.

Diploid apogamy See **apomixis**, **parthenogenesis**.

Diplont, Diplophase The **diploid** stage of a **life cycle**. See **alternation of generations**, **haplont**.

Diplopoda Class (or subclass) of **Arthropoda**, containing millipedes. Abdominal trunk segments fused in pairs to form *diplosegments*, each with two pairs of legs; exoskeleton calcareous; ocelli and one pair of club-like antennae present; young usually hatch with three pairs of legs, suggesting possible relations with the Insecta (see **neoteny**). Development gradual. See **Chilopoda**, **Myriapoda**.

Diplospory (Bot.) Form of *apomixis* in which a (diploid) megaspore mother cell gives rise directly to the embryo. See **parthenogenesis**.

Diplotene Stage in first prophase of **meiosis**.

Dipnoi (Sarcopterygii) The order of **Choanichthyes** including lungfishes.

Diptera Two-winged (or true) flies. Large order of the **Insecta**, with enormous specialization and diversity among its members. Endopterygote. The hind pair of wings is reduced to form balancing **halteres**. Head very mobile; compound eyes and ocelli present; mouthparts suctorial, usually forming a *proboscis* and sometimes adapted for piercing; larvae legless.

Disaccharide A carbohydrate comprisin two monosaccharide groups joined covalently by a *glycosidic bond*. The group includes **lactose**, **maltose** and **sucrose**.

Disc flower Actinomorphic tubular flowers (florets) composing the central portion of the flower head (capitulum) of most Asteraceae (Compositae); contrasted with the flattened, zygomorphic ray-shaped florets on the margins of the head.

Disinfectant Substance used particularly on inanimate surfaces to kill microorganisms, thus sterilizing them. Hypochlorites, phenolics, iodophores (complexes of iodine less staining, toxic and irritant than iodine solutions) and detergents all have disinfectant ability. The phenolic *hexachlorophene* is widely used in the food industry and hospital wards to reduce pathogenic staphylococci. See **antiseptic**.

Displacement activity Act expressive of internal *ambivalence* and seemingly irrelevant or inappropriate to the context in which it occurs. Tends to occur when an animal is subject to opposing motivations or when some activity is thwarted.

Disruptive colouration Colouration in animals tending to break up their outlines, thus avoiding visual predation.

Disruptive selection See **natural selection, polymorphism**.

Distal Situated away from; e.g. from place of attachment; from the head, along an antero-posterior axis; from the source of a gradient. Thus, the insect abdomen is *distal* to the head. Contrast **proximal**.

Diuresis Increased output of urine by kidney, as occurs after drinking much water or taking *diuretic* drugs. See **antidiuretic hormone**.

Diurnal rhythm See **circadian rhythm**.

Diversity (species diversity) See **biodiversity**.

Diverticulum Blind-ending tubular or sac-like outpushing from a cavity, often from the gut.

Division Major group in the Linnean hierarchy used in classifying plants. Includes closely related classes and is the taxonomic category between kingdom and class; equivalent to *phylum* in animal classification.

Dizygotic twins See **fraternal twins.**

DNA (deoxyribonucleic acid) The nucleic acid forming the genetic material of all cells, some organelles, and many viruses; a major component of **chromosomes** and the sole component of **plasmids**. A polymer (polynucleotide), which is formed in cells by enzymatic dephosphorylation and **condensation** of many (deoxyribo) **nucleoside triphosphates** (esp. dATP, dGTP, dCTP, dTTP). The product is a long chain of (deoxyribo)-**nucleotides**, bonded covalently by *phosphodiester bonds*. *Duplex DNA* comprises two such antiparallel strands (running in opposite directions; see Fig. 29) but complementary in base composition, and held together in a double helix by hydrogen bonds between the complementary bases, by electronic interactions between bases, and by hydrophobic interactions. Each strand comprises a *sugar-phosphate backbone* from which the bases project inwardly. The stacking of the complementary bases in the double helix produces two grooves in the sugar-phosphate backbone termed the major and minor grooves (see Fig. 29a, **DNA-binding proteins, B-form helix**). Each end of the double helix has one 5′-ending strand paired to a 3′-ending strand, where the 5′ and 3′ indicate which carbon atoms of the two terminal deoxyriboses are bonded to their terminal phosphate group. It is as duplex DNA that nuclear DNA is normally found, and this form is the ultimate store of *molecular information* for all cells (but single-stranded DNA **bacteriophages** occur). Unlike RNA, DNA is not hydrolysed by dilute alkali. Two of the bases abundant in the nucleotides of DNA are purines (adenine and guanine), that form hydrogen bonds with two common pyrimidines (thymine and cytosine respectively, see **base pairing**). This ability of one strand of duplex DNA to act as template for the other enables DNA to be replicated by DNA **polymerase** in a *semiconservative* way (see **DNA replication**). A combination of X-ray crystallographic and chemical data (see **base ratio**) led J. Watson and F. Crick in 1953 to propose the three-dimensional model of duplex DNA held today. See **denaturation**, **gene**, **gene manipulation**, **genetic code**, **protein synthesis**, **reverse transcriptase**, **RNA**.

DNA-binding proteins Proteins (prokaryotic and eukaryotic) with DNA-binding ability. The majority function as **transcription factors**, including steroid-receptor proteins, **TATA factor**, heat-shock proteins (see **chaperones**) and the **tumor necrosis factor**, p53. DNA-binding **domains** can fold correctly without the rest of the polypeptide chain, whereas particular *motifs* are conserved substructures that cannot fold independently. These proteins have various ways of inserting an α-helix into the major groove of DNA, and interactions with the sugar-phosphate backbone are critical for positioning so that protruding structures align correctly with specific base-pairs. A *zinc finger* motif occurs as a protrusion in the DNA-binding domain of some proteins, esp. among those binding steroids (i.e. steroid **nuclear receptors**) and those acting as transcription factors for

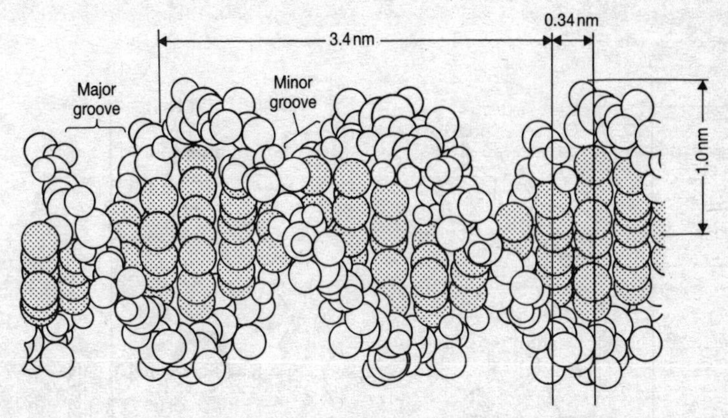

Fig. 29a *Diagrammatic section of duplex DNA, showing the minor and major grooves and approximate dimensions. The two outer helical sugar-phosphate backbones are lightly shaded and the base-pairs are darkly shaded.* (From *Introduction to Genetic Analysis* 5/e by Griffiths, Miller, Suzuki, Lewontin and Gelbert. Copyright © W. H. Freeman and Company, 1993. Reprinted with permission.)

mRNA promoters. It is a cysteine- and histidine-rich region, complexing a zinc atom. The DNA-binding *helix-turn-helix* motif is a feature of some bacterial regulatory proteins and the products of **homeogenes**. Replication protein A is a multi-subunit complex which binds single-stranded DNA and may initiate eukaryotic DNA replication (and p53 seems to inhibit this ability). See **recA**, **RNPs**, **homologous recombination**.

DNA cloning See **polymerase chain reaction**, **vector**.

DNA fingerprinting Technique, developed in 1984, which makes use of the hypervariable **repetitive DNA** sequences between (e.g. human) genes, there being strong similarities in these between relatives compared with a random outgroup. Using very small samples of DNA, these are first digested by a **restriction enzyme** then subjected to gel **electrophoresis**. After removal of the fragments to a nylon membrane a small radioactive double-stranded **DNA probe** (with specific hypervariable sequences it recognizes) is washed over the membrane (see **southern blot technique**). The bonded fragments show up in the X-ray film, looking rather like a bar code, and each individual has a unique pattern. The rigour of the technique has been disputed in law courts, not least in forensic contexts; however, fingerprinting provides invaluable information on mating systems in wild populations, sometimes making redundant such terms as 'monogamy', 'polygyny', 'promiscuity', etc. at the sexual level (however useful at the social level). See **RFLP**, **satellite DNA**.

DNA gyrase Enzyme capable of adding supercoils to circular **DNA** (topoisomerases remove the supercoils). See **DNA replication**.

DNA hybridization A technique often

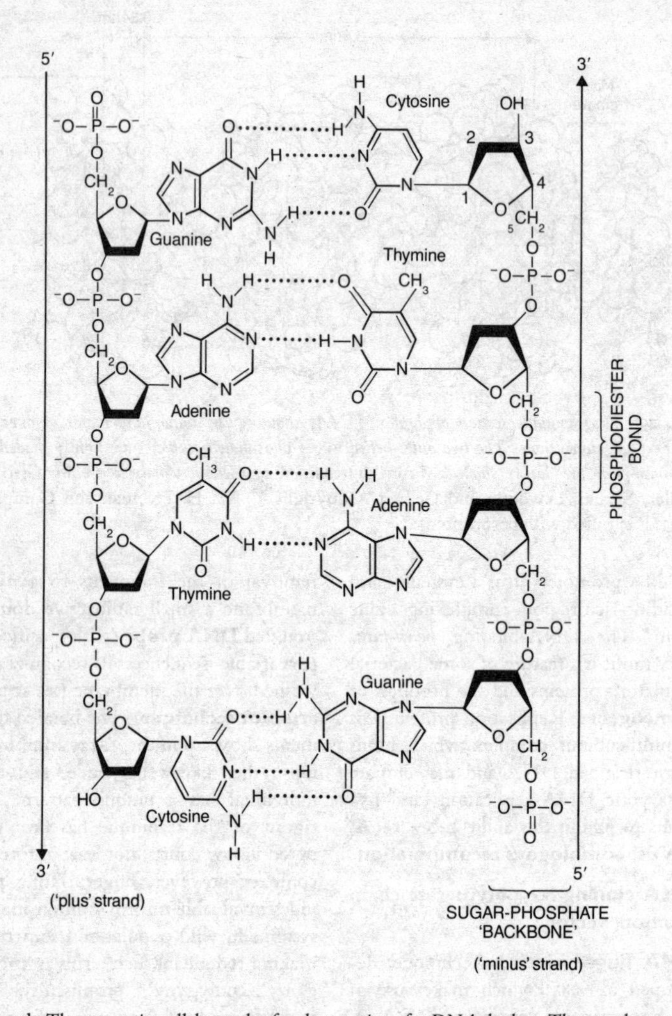

Fig. 29b *The two antiparallel strands of a short section of a DNA duplex. The strands actually twist round each other in a double helix. The dots linking the central base-pairs represent hydrogen bonds. Primed numbers (5′, 3′) indicate the carbon atom numbers on the deoxyribose moiety involved in the phosphodiester bonding of each chain. These carbon atoms are numbered in the deoxyribose moiety attached to the top cytosine. The left-hand chain runs from 5′ to 3′ top to bottom, the right-hand chain runs from 5′ to 3′ bottom to top.*

used in experimental taxonomy, in which a source of duplex DNA is 'melted' (see **denaturation**) by slight temperature rise in solution, and allowed to re-anneal with either a similarly treated sample of DNA from a different source, or else commonly a single-stranded RNA sample (e.g. messenger RNA). The time taken to re-anneal, or the thermal stability of the hybrid duplex, indicates the degree of complementarity of the original strands. This can be used to indicate whether a piece of duplex DNA codes for the polypeptide translated from the messenger RNA, or perhaps to estimate the degree of relatedness of the organisms providing the original duplex DNA' sources. See also **DNA ligase**, **gene manipulation**.

DNA ligase Enzyme which repairs 'nicks' in the DNA *backbone*; i.e. where the *phosphodiester bond* linking adjacent nucleotides has yielded 3′-hydroxyl and 5′-phosphate groups. Its role therefore overlaps that of some **DNA polymerases**. Valuable for hybridization (insertion) of DNA fragments with appropriate overlapping or 'sticky' ends. See **gene manipulation.**

DNA loop See **enhancer**.

DNA methylation Vertebrate DNA contains a covalently modified variety of cytosine,5-methylcytosine. Restriction enzyme Hpa II cleaves only sequence CCGG when the Cs are unmethylated indicating that, in 98% of the mammalian genome, the dinucleotide CpG occurs at about 25% of its expected frequency – and then in this methylated form. In the remaining 2% of the genome, CpG occurs at its expected frequency but is unmethylated and stable (except in the inactivated X chromosome in female placentals). These so-called 'CpG islands' are about 1 kb long and commonly occur at the *5′-ends of genes and serve to indicate* the proximity of a gene. Methylation can interfere with transcription, e.g. by inhibiting binding of ***trans*-acting control elements** to DNA and interfering with **nucleosome** formation. It is possible that methylation binds a gene into a non-transcribable state once trans-acting factors have initiated that state. See **DNA repair mechanisms**, **genomic imprinting**, **heterochromatin**.

DNA polymerase **Multienzyme complex** which incorporates appropriate nucleoside triphosphates into a DNA chain. Bacterial polymerases (DNA pols) and eukaryotic polymerases (DNA polymerase α–δ) are of several forms and require short complementary oligonucleotide RNA sequences, and several associated proteins, for *initiation*, after which they carry out chain *elongation*. See **DNA repair mechanisms**, **polymerase chain reaction**.

DNA probe A defined and fairly short DNA sequence, isotopically or otherwise labelled, which can be propagated by **gene manipulation** and introduced to DNA from the same or another individual (often from a different taxon) in order to detect complementary DNA sequences through **DNA hybridization**. DNA probes can be used to identify embryos carrying genes for human genetic disorders such as **sickle-cell anaemia** and **a-thalassaemia** using chorionic villus sampling (see **chorion**). See also **southern blot technique**.

DNA repair mechanisms General cellular DNA repair mechanisms involve enzymes in the removal of loose ends, filling-up of single-stranded regions and ligation of single-strand nicks. Some bacterial and eukaryotic DNA **polymerases** can replace a nucleotide they insert incorrectly. **DNA ligase** oithen seals the phosphodiester bond. To avoid removing the nucleotide from the wrong (i.e. error-free) strand, cells methylate DNA which has been formed some while; repair enzymes thus distinguish old from new DNA, and repair only the new strand error. Mutants lacking such repair mechanisms are likely to be more susceptible to irradiating sources and to express (somatic) mutations so induced. In *photoreactivation*, cells of many organisms (but apparently not of placental mammals) repair radiation-damaged DNA (as from UV light) using an enzyme that functions when exposed to strong visible light. The main damage products are *pyrimidine dimers* formed by linking adjacent pyrimidines in the same DNA strand. The photoreactivating enzyme monomerizes these dimers again. See **crossing-over**, **gene conversion**, **homologous recombination**.

DNA replication Almost universal biological processes, in which DNA duplexes are catalytically and *semiconservatively* replicated by a **DNA polymerase** (see **multienzyme complex**) at rates of between 50–500 nucleotides per second (but the polymerase requires a short complementary RNA primer). The duplex is first 'unzipped' by helicases (see Fig. 30) by breaking the hydrogen bonds holding base-pairs together. The resulting Y-shaped molecule is a *replication fork* and occurs first at *replication origins* (of which there may be a hundred or more) on the chromosome. Special **DNA-binding proteins** stabilize the two strands to prevent re-annealing. Completion of DNA replication is a key event in the control of the **cell cycle.** Appropriate DNA polymerases then move down the two single-stranded arms in a 5′-to-3′ direction (see **DNA**), incorporating nucleotides in accordance with **base pairing** rules. Energy is supplied by hydrolysis of substrate nucleoside triphosphates, also catalysed by the polymerase. There are usually several simultaneous replication forks on one replicating chromosome, and newly-synthesized sections are joined up by the DNA *ligase* component of the polymerase (see **DNA repair mechanisms**). Failure to replicate exactly results in a **mutation**.

DNase (DNAase, deoxyribonuclease) An enzyme (of which there are many forms) breaking down DNA by hydrolysis of the phosphodiester bonds of its sugar-phosphate backbone. Depending on the enzyme, it does this at either the 3′- or the 5′- end of the bond. As with peptidases, there are *endonucleases* and *exonucleases*, cleaving respectively terminal and non-terminal nucleotides from either a single strand or from both strands of the duplex, depending on the type of DNase. Pancreatic juice contains DNases. Valuable in **gene manipulation**. See **restriction endonuclease**.

DNA sequencing Determination of the sequence of nucleotides making up a length of DNA. **Restriction endonucleases** digest the strand; the fragments are isolated by gel **electrophoresis**, and

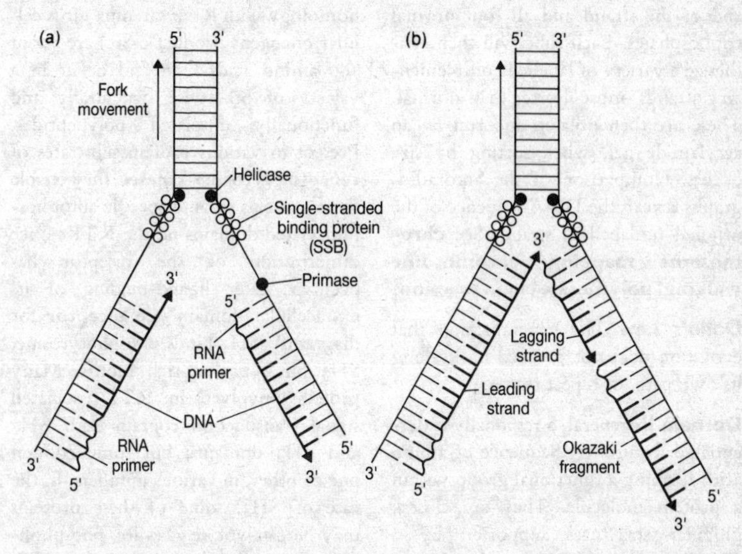

Fig. 30 (a) *Initiation of DNA synthesis by an RNA primer.* (b) *DNA synthesis proceeds by continuous synthesis on the leading strand and discontinuous synthesis on the lagging strand.* (From *Introduction to Genetic Analysis 5/e* by Griffiths, Miller, Suzuki, Lewontin and Gelbert. Copyright © W. H. Freeman and Company, 1993. Reprinted with permission.)

then the sequence can be determined by rendering the DNA single-stranded and using it as a template for **DNA polymerase** to resynthesize the complementary strand with labelled nucleoside triphosphates, or by chemical analysis of the fragments. See Fig. 31.

Two methods have dominated the field since 1975, both using single-stranded DNA. In the Maxam and Gilbert (base-destruction) method, duplex 3′ ends are labelled with ^{32}P, digesting the molecule. Strands of one length are then isolated, the two strands separated, from which one is isolated to form a homogeneous population. This is split into four, each part being subjected to selective destruction of a different one or two of the four bases present (G, A & G, T & C or C) in such a way that sequences of different lengths result. After separation on a gel, bands of labelled fragments are revealed by autoradiography, different for each of the four treatments. Comparison of the four bands displays successively shorter bands from top to bottom of the gel, and the presence or copresence of bands indicates which base was present at that place in the strand.

In the Sanger (dideoxy) method, four tubes of single-stranded DNA are given all necessary DNA-synthesizing requirements, but a small proportion of a different dideoxynucleoside triphosphate (which stops chain elongation) is given to each tube – in addition to a short labelled primer complementary to the end of the strand and all four normal triphosphates. Each tube will then synthesize a variety of labelled complementary strands, most shorter than normal. These are then isolated and run on an acrylamide gel, when sorting by size occurs. Comparison of the autoradiographs reveals the DNA sequence of the original (unlabelled) strand. See **chromosome mapping**, **chromosome walking**, **polymerase chain reaction**.

Dollo's Law The generalization that evolution does not proceed back along its own path, or repeat routes.

Domain In general, a regionally differentiated feature. (1) Sequence of amino acids forming a functional group within a protein molecule. They are closed, globular structures supported by a closed system of hydrogen bonds and built around a well-packed interior of hydrophobic groups. Domains include regulatory, catalytic and DNA-binding types and are the 'modules' whose exact linear assembly determines the precise characteristics of the protein. There may be several structural ways of achieving a functional domain; thus **DNA-binding proteins** have at least one DNA-binding domain, but this may take the form of any of several structural types, termed **motifs**. Crystallography shows that the domain organization of actin, heat-shock (chaperone) protein Hsc70 and hexokinase B are remarkably similar, yet they share little sequence similarity. SH2 and SH3 domains (Src homology 2 and 3 domains, named from their homology with Rous sarcoma virus cellular oncogene product c-Src) are about 100 amino acids long and occur in a variety of otherwise structurally and functionally unrelated polypeptides. Present in various protein substrates of **receptor tyrosine kinases**, they enable these proteins to bind specific autophosphorylated domains on the RTKs once dimerization of the receptor has occurred after ligand-binding of its extracellular domain (see **receptor** for diagrams). SH2 binds phosphotyrosine; SH3 binds proline-rich motifs. Many proteins involved in RTK-associated signal transduction contain both SH2 and SH3 domains; but some contain one or other, in various numbers. In the case of SH3, some of these proteins may be catalytic (e.g. c-Src, phospholipase Cγl and Ras-GAP), others not. Some cytoskeletal proteins (e.g. myosin-lB) also contain SH3. (2) For packaging domains in chromatin, see **position effect** variegation. (3) In syncytia, specialized domains may occur in which some nuclei may become transcriptionally distinct from the remainder, as with myofibre nuclei in the region of a neuronal synapse. (4) Plasma membrane domains refer to regions with different structure and function from the general cell surface. They may result from phospho- or glycolipid separation from other lipids; or from groupings of intrinsic and extrinsic membrane proteins and often involve the **cytoskeleton**. Nuclear domains are also being recognized. (5) An example of a gene

Fig. 31 *The Sanger ('dideoxy') method of* **DNA sequencing.** *dNTP stands for any of four deoxynucleoside triphosphates. ddATP stands for dideoxyATP.* (a) *The molecular process;* (b) *the resulting gel appearance.* (From *Recombinant DNA 2/E* by Watson, Gilman, Witkowski, and Zoller, 1992. Reprinted with permission of W. H. Freeman and Company.)

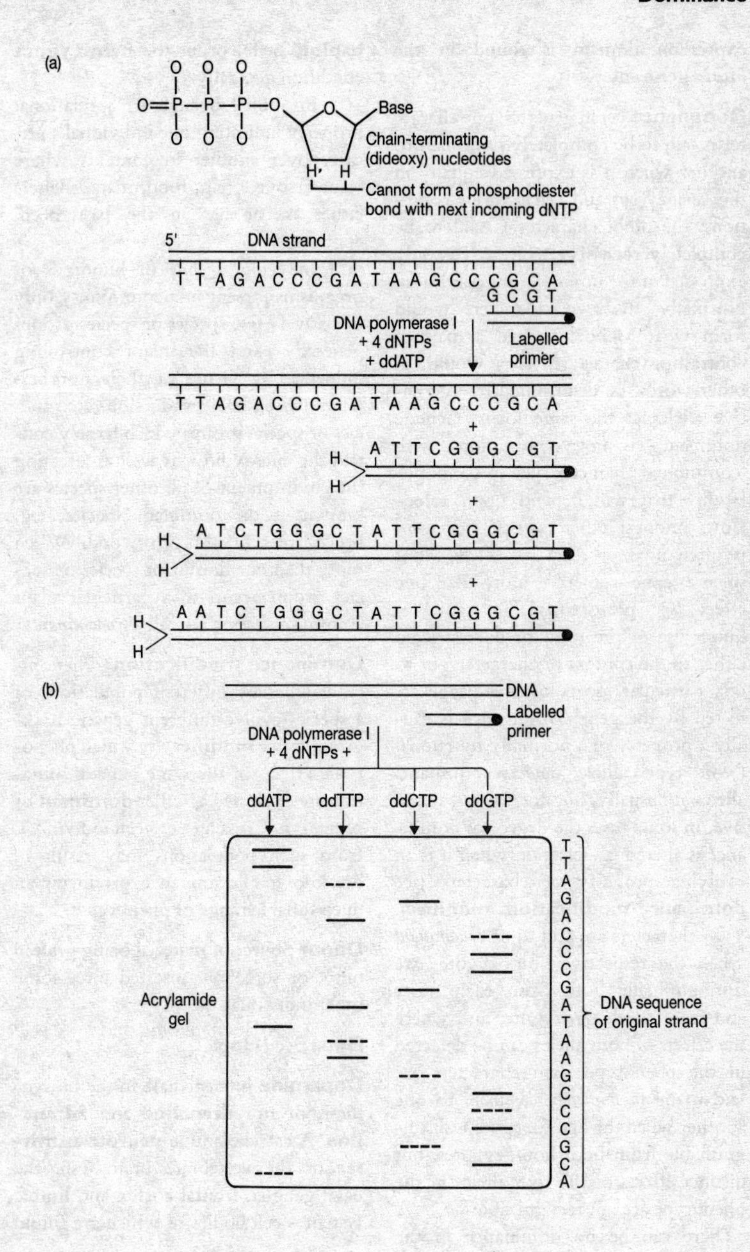
(a)
Base
Chain-terminating (dideoxy) nucleotides
Cannot form a phosphodiester bond with next incoming dNTP
5'
DNA strand
T T A G A C C C G A T A A G C C C G C A
G C G T
DNA polymerase I + 4 dNTPs + ddATP
Labelled primer
T T A G A C C C G A T A A G C C C G C A
A T T C G G G C G T
A T C T G G G C T A T T C G G G C G T
A A T C T G G G C T A T T C G G G C G T
(b)
DNA
Labelled primer
DNA polymerase I + 4 dNTPs +
ddATP
ddTTP
ddCTP
ddGTP
Acrylamide gel
T T A G A C C C G A T A A G C C C G C A
DNA sequence of original strand

expression domain is found in the ***bicoid* gene** entry.

Dominance (1) In genetics, one character is said to be completely dominant to another when it is expressed equally in the *homozygous* and *heterozygous* conditions; the other character is said to be completely **recessive** to it, and is only expressed in the homozygous condition. Normally, the two characters would form what Mendel termed a 'pair of contrasting characters'; they would, in other words, be determined by alternative alleles at the same locus. Genetic dominance is not synonymous with 'commonest character type in the population': that will depend upon **selection**, amongst other factors. The term is often used of *genes* (or alleles); but since a 'gene' can have more than one effect (see **pleiotropy**) accuracy requires use of the term to be restricted either to the context of characters, or to one particular aspect of phenotype affected by the gene. Dominance is usually a property of a normally functional ('wild type') allele; defective (mutant) alleles are usually, but not always, recessive. In some cases the degree of dominance is altered by selection when it is an evolving property of characters (see **dominance modification**, **modifier**). Two characters are said to be *codominant* when the respective homozygotes are distinguishable both from each other and from the heterozygote, and where the effects of both alleles can be detected in the phenotype; two characters are said to be *incompletely dominant* to one another when the heterozygote is distinguishable from both homozygotes, but distinct effects of the two alleles in the phenotype are not recognizable.

There can be no dominance in the **haploid** state, or in the **hemizygous** condition generally.

(2) In animal behaviour, a relational property indicating one individual's priority over another in contexts where some resource (e.g. food, mate, shelter) either is, or has in the past been, contested.

(3) In ecology, out of hundreds of organisms present in a community only a relatively few species or species groups generally exert the major controlling influence by virtue of their numbers (abundance), size, production, etc.; species or species groups which largely control the energy flow as well as affecting the environment of all other species are known as the dominant species, e.g. beech trees in a beech wood. When more than one dominant species or species group occurs in a particular plant community, they are called *codominants*.

Dominance modification Phenomenon whereby different populations of a species evolve different genetic backgrounds (see **modifier**) by which phenotypic effects of the same genetic mutation are expressed as either **dominant** or **recessive**. Crossing between individuals from such populations may result in breakdown of dominance, producing an unclassifiable range of phenotypes.

Donor Source of material being grafted onto, or somehow inserted into, some other individual.

Dopa See ***l*-dopa**.

Dopamine Intermediate in the biosynthesis of **noradrenaline** and **adrenaline**. A catecholamine **neurotransmitter** of the vertebrate brain, esp. the basal ganglia, frontal cortex and limbic system – cell bodies of which are found

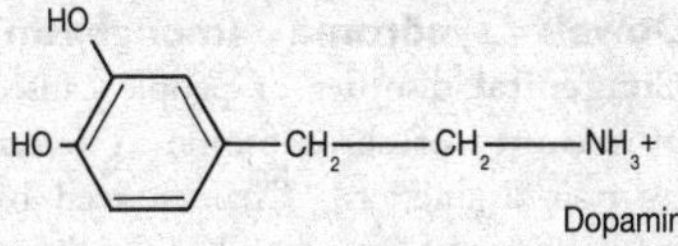

HO
HO–⟨benzene ring⟩–CH_2—CH_2—NH_3^+
Dopamine

largely in the substantia nigra; low levels in the human *caudate nucleus* produce symptoms of Parkinson's disease, treatable by administration of the precursor ***l*-dopa**. In schizophrenics there is increased sensitivity of dopamine receptors in the frontal cortex.

Dormancy (Bot.) See **dormant**. (Zool.) Term sometimes used of insect and other animal **diapause**.

Dormant In a resting condition. Alive, but with relatively inactive metabolism and cessation of growth. Dormancy may involve the whole organism (higher plants and animals) or be confined to reproductive bodies (e.g. resting spores such as statoblasts, fungal sclerotia, bacterial spores). May be due to unfavourable conditions, and end as these ameliorate. Many seeds (e.g. pea, wheat), though capable of germinating after harvesting, do not do so unless kept moist. On the other hand, a dormant period is part of an annual rhythm for most plants. Often has survival value (e.g. winter dormancy in deciduous trees). After vegetative growth and flowering in spring, many bulbs (e.g. snowdrop, daffodil) have a dormant period coinciding with conditions favourable to growth of other plants. This is common in plants of moist, tropical climates. Dormancy of seeds in conditions otherwise favourable to germination is common (e.g. hawthorn, the weed wild oats) and is associated with incomplete development of the embryo, impermeable seed coats, limiting entry of water and/or oxygen, inhibitors and absence of growth stimulators. Dormancy in some seeds and deciduous trees is regulated by photoperiod (see **phytochrome**). See **aestivation**, **diapause**, **hibernation**.

Dorsal (Zool.) Designating the surface of an animal normally directed away from the substrate; in chordates, the surface (posterior) in which the **neural tube** forms, lying closest to the eventual nerve cord. In flatfish, the apparent adult dorsal surface is in fact lateral. (Bot.) Also used of leaves; synonymous with **abaxial**.

Dorsal aorta See **aorta**.

Dorsal lip See **blastopore**, **organizer**.

Dorsal placentation Attachment of ovules to midrib of carpels in apocarpous gynoecia.

Dorsiventral (dorsoventral) Term generally used to indicate some gradient or morphological feature associated with the **axis** linking the upper and lower parts of an organism or its parts. As with leaves, it often indicates some difference in structure along the axis. Compare **isobilateral**.

Dosage compensation Some animals whose **sex determination** mechanism employs differences in the sex chromosome ratio compensate for the resulting dosage imbalances of some X-linked genes (and their products) in males and females. **Modifier** loci on the X chromosome (*dosage compensators*) act either to enhance gene expression in the heterogametic sex or repress such expression in the homogametic sex. Uniquely in mammals, dosage compensation is achieved by inactivation of one of the two X chromosomes of females in every cell early in development. This involves an 'X-inactivation centre' on the X chromosome from which an unknown signal spreads to achieve inactivation. Very early in eutherian development it is the paternally derived X chromosome which is inactivated while X-inactivation is random later. In marsupials, it is always the paternally derived X chromosome which is inactivated (see **Barr body**). **DNA methylation** does not appear to be the primary signal for inactivation.

Double circulation See **heart**.

Double fertilization The unique and probably universal condition in flowering plants (**Anthophyta**) whereby, from a single pollen grain, the two generative nuclei within the pollen tube fuse with different nuclei within the **embryo sac** of the ovule, one with the egg cell nucleus (the product of **automixis**) to form the zygote, the other with the diploid secondary endosperm nucleus to form the triploid primary endosperm nucleus. This appears to ensure that no nourishment (as endosperm) is laid down in the prospective seed until a zygote has been formed to take advantage of it.

Double helix See **DNA**.

Double recessive Individual or stock in which each of two loci involved in breeding work is homozygous for alleles bringing about expression of **recessive** characters. See **backcross**.

Doubling rate Time required for a population of a given size to double in number.

Down feather See **feather**.

Down's syndrome (mongolism) **Congenital** disorder of people caused by **trisomy** of chromosome 21 (often by non-disjunction). Characterized by mental retardation, *mongoloid* facial features, simian palm and reduced life expectancy. Has a frequency of about one per 700 live births. See **amnion**.

Drive Specific causal explanations are now sought for most animal activities, so *general drive theories* of motivation have been surpassed by investigation of the control of behaviour rather than its powering. Those specific causal influences promoting an action may be regarded as a part of that activity's *specific drive mechanism*.

Driving genes See **aberrant chromosome behaviour** (4), **selfish DNA/genes**.

Drosophila Genus of fruit flies (Diptera). Probably the best described animal genetically, and of enormous significance to studies of **linkage**, **cytogenetics**, **speciation** and, most recently, developmental biology (e.g. see **compartment, homoeotic gene**).

Drug Any substance, e.g. neurotransmitters and hormones, capable on administration of altering the behaviour and/or physiology of an organism. See **alkaloid**, **psychoactive drugs**, and (for detoxification of lipid-soluble drugs) **endoplasmic reticulum**.

Drupe Succulent **fruit** in which the wall (pericarp) comprises an outer skin (epicarp), a thick fleshy mesocarp, and a hard stony endocarp enclosing a single seed. Commonly called a stone-fruit; e.g. plum, cherry. Compare **berry**. In some drupes the mesocarp is fibrous;

e.g. in the coconut the pericarp has tough, leathery epicarp, thick fibrous mesocarp and hard endocarp enclosing the seed and forming with it the nut we buy. Compare **nut**.

Dryopithecine Term given to a heterogeneous group of hominoids of uncertain relationship to one another and to other forms, apparently representing a hominoid radiation occurring between 17 and 12 Myr BP. Three tribes have been proposed: Afropithecini (17–15 Myr BP; thickened molar enamel, enlargement of premolars and Proconsul-like postcrania), Kenyapithecini (15–14 Myr BP; even thicker molar enamel) and Dryopithecini (12–8 Myr BP; primitively thin molar enamel, limb bones with rounded shafts, distal humerus with rounded capitulum and deep olecranon fossa – adaptations for brachiation, greater brow ridge development). *Dryopithecus* itself remains the only European Miocene ape and may represent a sister group to the African apes/hominid clade. See ***Proconsul***, **Ponginae**.

Ductless gland See **endocrine gland**.

Ductus arteriosus (duct of Botallo) Vascular connection between pulmonary trunk (**aortic arch VI**) and **aorta** (aortic arch IV) in amniote embryos, serving as a bypass for most blood from the right ventricle past the lungs while they are deflated and functionless. When the pulmonary circuit opens at birth the ductus closes and atrophies.

Ductus Cuvieri See **Cuvierian duct**.

Duodenum Most anterior region of small intestine of mammals, its origin guarded by the pyloric sphincter. Receives the bile duct and pancreatic duct. Characterized by alkaline-mucus-secreting *Brunner's glands* in the submucosa. So-called because it is about 12 fingerbreadths, about 25 cm, long in man; site of active digestion and absorption; like the rest of the small intestine, its luminal surface has numerous villi.

Duplex Of a molecule composed of two chains or strands, usually held together by hydrogen bonds; e.g. double-stranded (duplex) DNA.

Duplication Chromosomal **mutation** in which a piece of chromosome is copied next to an identical section, increasing chromosome length. Can result from non-homologous **crossing over** in which two homologous chromosomes pair up imprecisely and a crossover transfers an abnormally large piece of one chromosome to its homologue, resulting in a **deletion** on one chromosome and a duplication on the other. See **gene duplication**.

Dura mater See **meninges**.

Dyad-symmetry/asymmetry (Of a DNA region) which has symmetry/asymmetry on either side about a central pair of base-pairs. Dyad-symmetric sequences tend to attract **DNA-binding proteins** with a leucine zipper motif composed of homodimeric (identical) subunits whereas dyad-asymmetric sequences tend to attract those with heterodimeric (non-identical) subunits.

Dye Alternative term for stain. See **staining**.

Dynamin A putative eukaryotic microtubule motor protein, found esp. in vertebrate brain tissue (up to 1.5% total extractable protein). Probably GTP-dependent in action; may function in membrane traffic.

Dynein A major eukaryotic motor protein. Its microtubule-activated ATPase is responsible for axoneme motility in **cilium** and **flagellum**, while cytoplasmic dynein moves membrane-bound organelles along microtubules in the opposite direction from those attached by **kinesin**, towards the minus (proximal) end. May also be involved in early stages of chromosome movements of nuclear division (see **kinetochore**). Addition of ATP or GTP prevents binding of cytoplasmic dynein to polymerizing microtubules.

Dysgenesis See **hybrid dysgenesis**.

Dystrophic Term applied to lakes and rivers having heavily stained brown water, through receipt of large amounts of exogenous organic matter with a high humic organic content. The colour originates from bog soils or from peat at the lake margin or in the catchment through which the streams or rivers flow. See **eutrophic**, **mesotrophic**, **oligotrophic**.

Ear, inner Membranous labyrinth. Vertebrate organ which detects position with respect to gravity, acceleration, and sound. Lies in skull wall (auditory capsule); impulses transmitted to brain via **auditory nerve**. Comprises the **vestibular apparatus** and the **cochlea**. See **lateral line system**.

Ear, middle Tympanic cavity. Cavity between eardrum and auditory capsule of tetrapod vertebrates (but not urodeles, anurans or snakes). Derived from a gill pouch (spiracle). Communicates with pharynx via eustachian tube, and is filled with air, ensuring atmospheric pressure is maintained on both sides of the eardrum. In it lie the **ear ossicles**.

Ear ossicles Bones in middle ear connecting eardrum to **inner ear** in tetrapod vertebrates. Instead of just the single auditory bone (stapes, see **columella**) of amphibia and primitive reptiles, mammals have in addition the incus and malleus. The first retains its original attachment to the oval window of the inner ear, but here articulates via the incus with the malleus which attaches to the eardrum (tympanum). These last two bones have evolved respectively from the quadrate and articular bones of mammal-like reptiles, in which they were involved in jaw suspension. By this articulation the pressure of the stapes on the oval window is amplified 22 times compared with that of the pressure waves on the tympanum: vibrations are damped, but produce larger forces.

Ear, outer (or external) That part of the tetrapod ear, absent from amphibians and some reptiles, external to the eardrum. Comprises a bony tube (*external auditory meatus*). In addition in mammals there is a flap of skin and cartilage (the pinna) at the outer opening which amplifies and focuses pressure waves upon the eardrum. Well developed in nocturnal mammals (e.g. bats).

Eardrum (tympanum, tympanic membrane) Thin membrane stretching across the aperture between skull bones at the surface of the head (most anurans and turtles) or within an external meatus (most reptiles, birds and mammals). Vibrates, often aperiodically, transmitting external air pressure changes to **ear ossicles** of middle ear cavity.

Ecad (Bot.) A habitat form, showing characteristics imposed by the habitat conditions and non-genetic; also called ecophenes or phenoecotypes. Compare **ecotype**.

Ecdysis Moulting in arthropods. Periodic shedding of the **cuticle** in the course of growth. In insects this includes much of the lining of the tracheal system. The number of larval moults varies (up to fourteen in **Apterygota**); in endopterygotes there is one pupal moult (producing adult), but among insects only apterygotans moult as adults. In most crustaceans it proceeds throughout adult life. Insect ecdysis is under the control of **ecdysone**.

Ecdysone (moulting hormone, growth-and-differentiation hormone) Hormone produced by insect *thoracic* (*prothoracic*) *glands*, and possibly also by the crustacean *Y-organ*. In insects its release is under the control of *thoracotropic hormone* produced by neurosecretory cells in the brain and released from the **corpora cardiaca**. In crustaceans the brain neurosecretion is produced in the X-organs and transported to the *sinus glands* of the eye-stalk. Its release inhibits release of moulting hormone by the Y-organs. In insects at least ecdysone induces 'puffing' of selected chromosome regions, the sequence being tissue-specific. This is associated with selective gene transcription, notably by the epidermis; but one of its major effects is to make appropriate cells sensitive to *juvenile hormone* from the **corpora allata**, with which ecdysone works to bring about moulting to the appropriate developmental stage. See **diapause**.

Ecesis Germination and successful establishment of colonizing plants; the first stage in succession.

Echidna Spiny anteater. See **Monotremata**.

Echinodermata Phylum of marine and invertebrate deuterostomes; typically with pentaradiate symmetry as adults; an internal skeleton of calcareous plates in the dermis; **tube feet**; nervous system typically one circular and five longitudinal nerve cords, lacking brain and ganglia; surface epithelium often ciliated, and sensory; coelom well developed, including peculiar **water vascular system**; no excretory organs; larvae typically pelagic, roughly bilaterally symmetrical, with tripartite coelom (*oligomerous*) and an often dramatic metamorphosis. Affiliations with **Hemichordata**. Includes classes Stelleroidea (including subclasses Asteroidea, the starfish, and Ophiuroidea, the brittlestars); Echinoidea (sea urchins, etc.); Holothuroidea (sea cucumbers); Crinoidea (crinoids); and the new class Concentricycloidea (sea daisies).

Echinoidea Sea urchins, heart urchins, etc. Class of **Echinodermata**; lacking separate arms; more or less globular in shape; mouth downwards; with rigid calcareous *test* of plates in dermis bearing spines and defensive **pedicellariae**; browsers and scavengers, often in enormous numbers, on sea bed.

Echolocation Method used by several nocturnal, cave-dwelling or aquatic animals for determining positions of objects by reflection of high-pitched sounds. Many bats and dolphins use it, as do oil birds and the platypus.

Ecodeme See **deme**.

Ecological memory The capacity of past states or 'experiences' to influence present or future responses of a community. Existence of memory in communities has always been implicit in ecology; it is an historical feature, and comprises all potential recruit-species that are not completely excluded because of spatial and temporal heterogeneity.

Ecological niche See **niche**.

Ecology Term deriving from the Greek οἶκος (house, or place to live); the study of relationships of organisms or groups of organisms to their environments, both animate and inanimate. Increasingly quantitative, employing modelling and computer simulations. See **ecosystem**, **trophic level**, **competition**, and cross-references included there.

Ecospecies Group within a species comprising one or more **ecotypes**, whose members can reproduce amongst themselves without loss of fertility among offspring. Approximates to a *hologamodeme* (see **-deme**), or to an ideal 'biological' **species**, and as a term is used more in botanical than in zoological contexts. See **infraspecific variation**.

Ecosystem Community of organisms, interacting with one another, plus the environment in which they live and with which they also interact; e.g. a lake, a forest, a grassland, tundra. Such a system includes all abiotic components such as mineral ions, organic compounds, and the climatic regime (temperature, rainfall and other physical factors). The biotic components generally include representatives from several **trophic levels**; primary producers (autotrophs, mainly green plants), macroconsumers (heterotrophs, mainly animals) which ingest other organisms or particulate organic matter, microconsumers (saprotrophs, again heterotrophic, mainly bacteria and fungi) which break down complex organic compounds upon death of the above organisms, releasing nutrients to the environment for use again by the primary producers. See **balance of nature**, **food chain**, **pyramid of biomass**.

Ecotone The transition between two or more diverse communities, as between forest and grassland. Zone which may have considerable length, yet be far narrower than adjoining communities.

Ecotoxicology The study of the fate and adverse effects of chemicals on ecosystems.

Ecotype Term, generally applied in botanical contexts, referring to a species population exhibiting genetic adaptation to the local environment, whose phenotypic expression withstands transplantation of the plant, or its offspring, to a new environment. Ecotypes can be distinguished by morphology, physiology and phenology, and are potentially infertile with other ecotypes of the same species. Most species comprise an assemblage of ecotypes, each ranging in size from a single population to a regional group of many populations; the wider the species' range, the more ecotypes it has. The terms race, genecotype and ecological race are sometimes used synonymously for ecotype. Random genetic variants (individuals, or groups of individuals) within ecotypes are called biotypes. See **infraspecific variation**, **ecospecies**. Compare **ecad**.

Ectexine See **exine**.

Ectoderm Outermost **germ layer** of metazoan embryos, developing mainly into epidermal and nervous tissue and, when present, **nephridia**.

Ectophloic siphonostele A siphonostele with phloem external to the xylem.

Ectoplasm (ectoplast) See cell **cortex**.

Ectoprocta (Polyzoa, formerly Bryozoa) Phylum of colonial and often polymorphic coelomates, retaining continuity by coelomic tubes (cyclostomes) or merely by a tissue strand (ctenostomes, cheilostomes). Feeding (polyp) individuals do so by microphagy using a **lophophore** of tentacles, and secreting a calcareous zooecium (together termed a **zooid**). See **statoblast**.

Ectothermy See **poikilothermy**.

Ectotrophic (Of mycorrhizas) with the mycelium of the fungus forming an external covering to the root and penetrating only between the outer cortical cells; e.g. in pine trees. See **mycorrhiza**; compare **endotrophic**.

Edaphic factors Environmental conditions determined by physical, chemical and biological characteristics of the soil.

Edentata (Xenarthra) Aberrant order of eutherian Mammalia, mainly of South American history and distribution. Includes tree sloths, anteaters, armadillos and extinct glyptodonts. Only anteaters are truly toothless (hence ordinal name), the others having molars at least.

Effector Cell or organ by which an animal responds to internal or external stimuli, often via the nervous system. Include muscles, glands, chromatophores, cilia. Cnidoblasts are often regarded as *independent effectors* in that they do not seem to require stimulation from other cells (e.g. of the nervous system) for their activity.

Efferent Leading away from; e.g. from the central nervous system (motor nerves), from the gills (blood vessels) or from a glomerulus (arteriole).

Egestion Removal of undigested material and associated microorganisms of the gut flora (up to 50% dry weight in man) from the anus. This material has never been inside body cells. A quite different process from **excretion**, with which it may be confused. The voided material is termed *egesta*.

Egg cell See **ovum**.

Egg membranes Few animal eggs, if any, have just a plasma membrane separating the cytoplasm from the external environment. Additional membranes are: (1) *Primary membranes*: the vitelline membrane, or thicker chorion. (2) *Secondary membranes*: consisting of or formed by the follicle cells around the egg. (3) *Tertiary membranes*: secreted by accessory glands, oviduct, etc., including albumen, shell membranes, egg 'jelly', etc. Protective against mechanical damage, desiccation.

EGTA A calcium-specific chelating agent.

Eicosanoids Collective term for derivatives of 20-carbon precursor fatty acids (eicosanoic acids, e.g. arachidonic acid), including **prostaglandins**, prostacyclins, thromboxanes and leukotrienes. Prostacyclins and thromboxanes tend to have opposite physiological effects. The former (formed in arterial walls) are powerful inhibitors of **platelet** aggregation, relaxing arterial walls and inducing drop in blood pressure; the latter are platelet-derived and induce platelet aggregation (see **blood clotting**), contraction of artery walls and raising of blood pressure. Leukotrienes promote **inflammation**, the strength depending on the fatty acid origin.

Elaioplast Colourless plastid (leucoplast) in which oil is stored; common in liverworts and monocotyledons.

Elasmobranchii Subclass of **Chondrichthyes**, appearing in the middle Devonian. Includes sharks (**Selachii**), skates and rays (Rajiformes) and angel sharks (Squatiniformes). Cartilaginous skeleton; dermal denticles probably the remnants of ancestral bony placoderm armour; upper jaws independent of braincase (hyostylic jaw suspension) or in some sharks with anterior attachment to braincase (amphistylic jaw suspension). Gills border gillslits (usually five); spiracle present. Internal fertilization, male having *claspers*, modified pelvic fins, acting as intromittant organs. Tail heterocercal; teeth in rows, replacing in turn those lost. The **Holocephali** (Chimaeras) form a second chondrichthyan subclass.

Elastin Principal fibrous protein of the yellow fibres of animal **connective tissue**. Numerous in lungs, walls of large arteries and in ligaments. Highly extensible and elastic. Compare **collagen**.

Elater (1) Elongated cell with wall reinforced internally by one or more spiral bands of thickening, occurring in numbers among spores in capsules of liverworts. Assist in discharge of spores by movements in response to humidity changes. (2) Appendage of spores of horsetails; formed from outermost wall layer, coiling and uncoiling as the air is dry or moist; possibly assisting in spore dispersal.

Electric organs Organs of certain fishes which produce electric currents by means of modified muscle cells (*electrocytes*) which no longer contract but generate ion current flow on nervous stimulation. Two basic kinds: those producing strong stunning current (e.g. electric eel, electric ray, electric catfish), and those producing currents of low voltage (e.g. in Mormyridae, Gymnotidae excepting electric eel) as a continuous series of pulses for locating prey and obstacles in muddy water, and for mate location.

Electrocardiogram (ECG) Record of electrical changes associated with the **heart cycle**, usually by means of electrodes placed on the patient's skin. Can also monitor foetal heart in the uterus.

Electroencephalogram (EEG) Record of changes in electrical potential ('brain waves') produced by the cerebral cortex; detected through the skull and picked up by electrodes placed on the scalp. The waves are then amplified. Four main types: *alpha* (produced when awake, but disappearing when asleep); *beta* (appear when nervous system is active – as in mental activity); *theta* (produced in children, and in adults in emotional stress situations); *delta* (occur in sleeping adults; in awake adults they indicate brain damage).

Electron microscope See **microscope**.

Electron transport system (ETS, electron transport chain) Functional chain of independent and mobile enzymes (mostly conjugated proteins, e.g. **cytochromes**) and associated **coenzymes** (e.g. **quinones**) in an ion-impermeable membrane. Essential to the oxidoreduction chemistry of aerobic respiration and photosynthesis (see Fig. 32). It was once thought that these components were structurally ordered; but it is now thought they achieve electron transfer from one member of the chain to another, down a **redox** gradient, through random collisions. In respiration, the cytochrome and quinol oxidases achieve a sustained flow of electrons (from the initial respiratory substrates) by catalysing efficient O_2-reduction. Through the exergonic electron transfer involved, energy release pumps protons across the membrane, providing a store of potential energy in the form of a proton gradient. As the protons return down their electrochemical gradient through specific channels in the membrane, they provide the *protonmotive force* needed for ATP synthesis by ATPase, itself associated with the proton channels. This process is common to the inner membranes of **mitochondria** and the thylakoid system of **chloroplasts**. In chloroplasts the electrons are first boosted to a high energy level by photons; in mitochondria they are derived from hydrogen atoms (also the proton source) covalently bonded in electron-rich respiratory substrates. Associated coenzymes, not all intrinsic to the membrane, may include NAD, NADP, FAD, flavoproteins, plastoquinone and ubiquinone. Similar ETSs occur in bacterial membranes (e.g. see **bacteriorhodopsin**). See **superoxides**.

Electron spin resonance See **radiometric dating**.

Electrophoresis Technique for separating charged molecules in buffer solution, particularly proteins, nucleic acids and their degradation products, based on different mobilities (caused by their different net charges at a given pH) in an electric field generated by direct current through the buffer. Substances for separation are usually allowed to move through a porous medium such as a gel (e.g. starch, agar, polyacrylamide) or paper (e.g. filter, cellulose acetate). Separated substances occur in bands on the medium and may be stained or identified by some labelling device, by fluorescence, by comparisons with knowns, or by removal and subsequent analysis. In *immunoelectrophoresis* antigens are placed in wells cut in agar gel. After separation of antigens by electrophoresis a trough is cut between the wells, filled with antibody, and diffusion allowed to take place. Where antigen meets appropriate antibody, arcs of precipitin form, allowing complex antigen mixtures to be compared.

Electroporation (electroporosity) Process by which an electrical potential applied across (typically) plant cells in culture disrupts their membranes sufficiently to create tiny pores through which DNA in the medium can be taken up – a small fraction of which may get incorporated into the cell's genome.

Elongation factors (EFs) Proteins which, once bound to GTP, bind the amino acid component of an aminoacyl-tRNA molecule enabling it to bind an empty **A-site** on a ribosome. Other EFs are involved in dissociation of GDP from, and binding of GTP to, the EFs described above. These latter dissociate from the tRNA once codon–anticodon pairing has occurred, enabling the amino acid to join the growing polypeptide chain. Compare **initiation factors**. See **ribosome**.

Emasculation Removal of stamens from hermaphrodite flowers before they

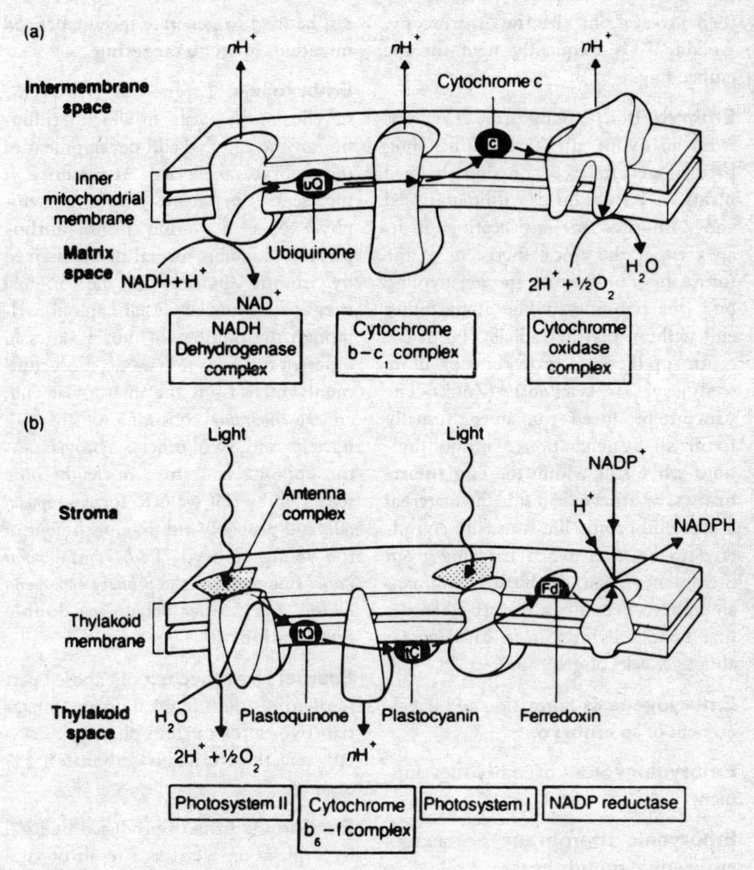

Fig. 32 *Hypothesized arrangements of electron-transporting molecules within (a) mitochondrial and (b) chloroplast membranes. Protons are extruded from mitochondria during activity but taken into thylakoids during the light reactions of photosynthesis.*

have liberated their pollen, usually as a preliminary to artificial hybridization.

Embedding Method employed in the preparation of permanent microscope slides of thin tissue sections. After **dehydration** and **clearing**, the material is put into molten paraffin wax (usually for 1–3 hours, with one or two changes of wax) which impregnates the tissue. After setting, the wax block is sectioned using a **microtome**. The wax is removed by xylene, itself removed by absolute alcohol, and gradual rehydration of the section is achieved by passing for a few minutes through progressively more dilute alcohols. Staining can then proceed. In electron microscopy, Araldite ® is frequently used for the embedding.

Embryo (Bot.) Young plant developed from an ovum after sexual (including parthenogenetic) reproduction. In seed plants, it is contained within the seed and comprises an axis bearing at its apex either the apical meristem of the future shoot or, in some species, a young bud (the plumule), while at the other end is the root (the radicle). From the centre of the axis grow one or more seed leaves (cotyledons). (Zool.) The structure produced from an egg (usually fertilized), by generations of mitotic divisions while still within the **egg membranes**, or otherwise inside the maternal body. Embryonic life is usually considered to be over when hatching from membranes occurs (or birth); in humans an embryo becomes a **foetus** when the first bone cells appear in cartilage (at about 7 weeks of gestation).

Embryogenesis Formation and development of an **embryo**.

Embryology Study of embryo development.

Embryonic membrane See **extra-embryonic membranes**.

Embryonic stem cells (ESCs) Pluripotent stem cells isolated directly from the **inner cell mass** of the mammalian blastocyst. They can be maintained *in vitro* as stem cell lines or made to differentiate in a variety of ways. When microinjected into a host blastocyst, and even earlier stages (e.g. the 8-cell embryo), they contribute to all the tissues of the resulting **chimaera**, including the **germ line**. ESCs are useful in studies of developmental regulation and can be used to generate specific defined mutations by **gene targeting**.

Embryo sac Large oval cell in the nucellus of the ovule, in which fertilization of the egg cell and development of the embryo take place. At maturity, it represents the entire female gametophyte of a flowering plant (**anthophyta**). Contains several nuclei derived by mitotic division of the original **megaspore** nucleus (itself haploid). Although the number of nuclei varies in different types of embryo sac, most commonly there is, at the micropylar end, an *egg-apparatus* consisting of the egg nucleus and two others, *synergids*. At the opposite end three nuclei become separated by cell walls to form *antipodal cells* and probably aid in nourishment of the young embryo. Two central *polar nuclei* fuse to form the *primary endosperm nucleus*. For further details, see **double fertilization**.

Enamel Hard covering of exposed part (crown) of tooth; 97% inorganic material (two thirds calcium phosphate crystals, one third calcium carbonate), 3% organic.

Enation Outgrowth produced by local hyperplasia on a leaf as a result of viral infection.

Endarch Type of primary xylem maturation, characteristic of most stems, where the oldest xylem elements (protoxylem) are closer to centre of axis than those formed later. Compare **exarch**.

Endemism Occurrence of organisms or taxa (termed *endemic*) whose distributions are restricted to a geographical region or locality, such as an island or continent. (2) Continual occurrence in a region of a particular (endemic) disease, as opposed to sporadic outbreaks of it (epidemics).

Endergonic (Of a chemical reaction) requiring energy; as in synthesis by green plants of organic compounds from water and carbon dioxide by means of solar energy. Compare **exergonic**. See **thermodynamics** for more detail.

Endexine Inner layer of **exine** of bryophyte spores and vascular plant pollen grains.

Endocarp Innermost layer of the carpel wall, or pericarp of fruit, in flowering plants. Frequently used to denote the 'stone' of drupes.

Endocrine gland (ductless gland) Gland whose product, one or more **hormones**, is secreted directly into the blood and not via ducts (compare **exocrine gland**). The gland may be a discrete organ, or comprise more scattered and diffuse tissue. Examples include: **adrenal**, **ovary**, **pancreas**, **pituitary**, **placenta**, **testis**, **thyroid**. See **endocrine system**.

Endocrine system Physiologically interconnected system of **endocrine glands** occurring within an animal body. Compared to neurotransmitters, the more diffuse hormonal outputs can take more time to reach effective concentrations, and therefore require a longer physiological half life (i.e. persistence in the body). **Hormones** generally exert effects over longer timescales, appropriate in growth, timing of breeding and control of blood and tissue fluid composition. Effects of peptide hormones depend as much on the distribution of membrane **receptors** on target cells as on the molecules secreted (see also **nuclear receptors**). See **nervous system** for a further comparison of roles, and **neuroendocrine coordination**, **neurohaemal organ**, **neurosecretion**.

Endocrinology Study of the structure and function of the **endocrine system.**

Endocytosis Collective term for **phagocytosis** and **pinocytosis**. An essential process in much eukaryotic **cell locomotion**. For recptor-mediated endocytosis, see **receptor** (2) and refs. there. See also **exocytosis**, **cell membranes**, **phagocyte**, **coated vesicle**, and Figs. 26, 33 and 75.

Endoderm (entoderm) Innermost **germ layer** of an animal embryo. Composed like mesoderm (when present) of cells which have moved from the embryo surface to its interior during **gastrulation**. Develops into greater part of gut lining and associated glands, e.g. where applicable, liver and pancreas, thyroid, thymus and much of the branchial system. Not to be confused with **endodermis**.

Endodermis Single layer of cells forming sheath around the vascular region (stele), most clearly seen in roots; in some stems identifiable by its content of starch grains (the *starch sheath*). Usually regarded as innermost layer of cortex. In roots, most characteristic feature of very young endodermis is band of impervious wall material, the Casparian strip, in radial and transverse walls of cells. With age, especially in monocotyledons, endodermis cells (except **passage cells**) may become further modified by deposition of layers of suberin over entire wall surface followed, particularly on the inner tangential wall, by a layer of cellulose, sometimes lignified. Endodermis is important physiologically in control of transfer of water and solutes between cortex and vascular cylinder, since these must pass through protoplasts of endodermis cells.

Endolymph Viscous fluid occurring within the vertebrate **cochlea** and **vestibular apparatus**. These are separated from the skull wall by **perilymph**.

Endometrium Glandular **mucous membrane** lining the uterus of mammals. Undergoes cyclical growth and regression or destruction during the period of sexual maturity. Receives embryo at **implantation**. See **oestrous cycle**, **menstrual cycle**, **placenta**.

Endomitosis (endoreduplication) Process whereby all the chromosomes of an **interphase** nucleus replicate and separate within an intact nuclear membrane (which does not divide). No spindle or other mitotic apparatus found. Resulting nuclei are **endopolyploid**, the degree of ploidy sometimes exceeding 2000. Compare **polyteny**, in which chromosomes do not separate after duplication.

Endonuclease See **DNAse.**

Endopelon Community of algae living and moving within muddy sediments. See **benthos**.

Endopeptidase Proteolytic enzyme hydrolysing certain peptide bonds in protein molecule, e.g. pepsin. Compare **exopeptidase**.

Endophyton Community of algae growing between cells of other plants, or in cavities within plants. Well known associations occur in some liverworts. See **benthos**.

Endoplasm That part of a cell's cytoplasm distinguished from the **ectoplasm** (if any) by greater fluidity; may be termed *plasmasol*.

Endoplasmic reticulum (ER) Eukaryotic cytoplasmic organelle comprising a complex system of membranous stacks (cisternae) and not unlike chloroplast thylakoids in appearance, but often being continuous with the outer of the two nuclear membranes and, like this membrane, bearing attached ribosomes on the cytosol side (when termed *rough ER*). A ribosome-free system of tubules (*smooth ER*), continuous with the cisternae, projects into the cytosol and pinches off transport vesicles. ER is not physically continuous with the **Golgi apparatus**, but is functionally integrated with it. A large rough ER is indicative of a metabolically active (e.g. secretory) cell.

ER seems to be the sole site of membrane production in a eukaryotic cell, membrane proteins and phospholipids being incorporated from precursors in the cytosol. Enzymes in the lipid bilayer pick up fatty acids, glycerol phosphate and choline and create **lecithin**, while protein components are fed into the ER lumen as they are produced at ribosomes bound to attachment sites on the cisternae. **Glycosylation** of newly synthesized proteins occurs within the cisternae through activity of *glycosyl transferase* located in the ER membrane. Only rough ER is involved in **protein synthesis**. Smooth (*transcisternal*) ER is generally a small component but from it **transport vesicles** (some of them **coated vesicles**) are budded off to carry protein and lipid to other parts of the cell. Some proteins are processed in the Golgi apparatus after the vesicles have fused there. Smooth ER contains enzymes involved in lipid metabolism and so is abundant in cells synthesizing cholesterol and those converting it to steroid hormones. Liver cells contain large amounts of smooth ER, in which enzymes of the cytochrome P450 family detoxify lipid soluble drugs and their metabolites (which otherwise remain in membranes) by direct reduction of carbonyl groups (>C=O) to hydroxyl groups (>HC–OH) and by conjugation of these with sulphate or glucuronic acid, rendering them water-soluble and excretable. Extra smooth ER made during a time of such drug administration seems to be removed afterwards by autophagosomes. See Figs. 14 and 51.

Endopodite See **biramous appendage**.

Endopolyploidy The result of **endomitosis**.

Endopterygota Insects with complete metamorphosis (pupal stage in life cycle) and with wings developing within the larva (see **imaginal disc**), although first visible externally in the pupa. Sometimes regarded as a subclass. Includes orders Neuroptera (lacewings); Coleoptera (beetles); Strepsiptera (stylopids); Mecoptera (scorpion flies); Siphonaptera (fleas); Diptera (true flies); Lepidoptera (butterflies, moths); Trichoptera (caddis flies); Hymenoptera (bees, ants, wasps). Often used synonymously with Holometabola, but see **Thysanoptera**. Compare **Exopterygota**.

End organ Structure at peripheral end of a nerve fibre; usually either a **receptor** or a motor end-plate (see **neuromuscular junction**).

Endorphins Peptide **neurotransmitters**, isolated from the **pituitary gland**, having morphine-like pain-suppressing effects. Also implicated in memory, learning, sexual activity, depression and schizophrenia. See **enkephalins**.

Endoskeleton Skeleton lying within the body. Vertebrate cartilage and bone provide support, protection and a system of levers enabling manipulation of the external environment; arthropods have internal projections of their cuticle (*apodemes*) for muscle attachment; echinoderms and annelids, among other invertebrates, use a *hydrostatic skeleton* to greater or lesser extent, and these too are endoskeletons. In vertebrates comprises both *axial skeleton* (bones/cartilages around body axis, esp. vertebral column, cranium, ribs, sternum, hyoid) and *appendicular skeleton* comprising **pectoral girdle** and **pelvic girdle** and skeletons of associated limbs/fins.

Endosome Membranous organelle (see Fig. 33) to which molecules taken up by **endocytosis** are transferred prior to the endocytic vesicle's becoming a **lysosome** by fusion of transport vesicles from the **Golgi apparatus**. Vesicles bud from it to return **receptor** molecules to the cell surface. There is evidence for at least two kinds of endosomal compartment: a 'housekeeping endosome' common to most if not all eukaryotic cells (e.g. fairly close to the basolateral membranes of epithelial cells and within the cell bodies and dendrites of neurons) which recycles, among others, low- density lipoprotein receptors (see **LDL**) and **transferrin** receptors, and a specialized

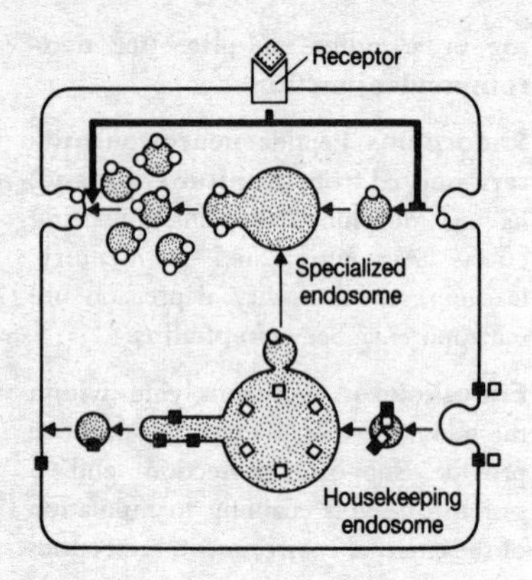

Fig. 33 *The cell-surface modification pathway and the involvement of distinct, but connected,* **endosome** *compartments.*

Dark squares = LDL-receptors, transferrin receptors, etc.; light squares = LDL, transferrin, etc.; circles = other (specialized) proteins, as in epithelial apical and synaptic vesicles; diamond = hormone or transmitter (e.g.).

endosome (near the apical membrane of epithelial cells and within axons and nerve termini of neurons) from which vesicles leave and target apical membrane and synaptic vesicle proteins (see **protein targeting**).

Endosperm Nutritive tissue surrounding and nourishing the embryo in seed plants. (1) In flowering plants (**Anthophyta**), formed in embryo sac by division of usually triploid endosperm nucleus after fertilization. In some seed plants (non-endospermic, exalbuminous), it is entirely absorbed by the embryo by the time seed is fully developed (e.g. pea, bean seeds); in other seeds (endospermic, albuminous), part of the endosperm remains and is not absorbed until seed germinates (e.g. wheat, castor oil). (2) Also applied to tissue of female gametophyte in conifers and related plants which is formed by cell division within the embryo sac before fertilization, outer layers persisting in the seed. Compare **perisperm**.

Endospore (1) Spore formed within a parent cell; in bacteria (mostly Gram-positive species, e.g. the fermenting *Clostridium* and aerobic *Bacillus*, but also a few Gram-negative sulphate-reducers, e.g. *Thiobacillus*); a thick-walled spore – usually one per cell – resistant to heat and harsh chemicals; in Cyanobacteria, a thin-walled spore. Term also used for inner layer of spore wall. See **Gram's stain**. (2) Structure within the elaters of spores of the sphenophyte *Equisetum*.

Endostyle Ciliated and mucus-secreting groove or pocket in ventral wall of pharynxes of urochordates, hemichordates, cephalochordates and ammocoete larvae of lampreys. The vertebrate thyroid is probably homologous with it.

Endosymbiosis Symbiotic association between cells of two or more different species, one inhabiting the other, the larger being host for the smaller. In *serial endosymbiosis*, one after another such symbiotic associations may occur telescoped within the largest cell. It is believed to account for the occurrence of eukaryotic chloroplasts (ancestor a cyanobacterium?), mitochondria (ancestor a purple photosynthetic bacterium?) and, some believe, cilia. See **Algae** (especially); also **Cryptophyta**, **Glaucophyta**, **Kappa particles**.

Endothelin A 21-amino acid peptide made by endothelial cells and ten times more potent in **vasoconstriction** than even angiotensin II.

Endothelium Single layer of flattened, polygonal cells lining vertebrate heart, blood and lymph vessels. Mesodermal in origin. See **plasminogen activators**.

Endothermic See **homeothermic**.

Endotoxin Glycolipids attached to cell walls of certain Gram-negative bacteria, giving them pathogenicity (e.g. *Salmonella typhi*, causing typhoid fever). Often complexed with protein. Released during autolysis. Compare **exotoxin**.

Endotrophic (Of mycorrhizas) with mycelium of the fungus within cells of root cortex; e.g. orchids (where it may be the sole means of nutrient support, host cells digesting the hyphae). See **mycorrhiza**; compare **ectotrophic**.

Endozoon Community of algae living within animals.

End-product inhibition (retro-inhibition, feedback inhibition) The inhibition of an **enzyme**, often the first in a metabolic pathway, by the product of the last enzyme in the pathway. Ensures against overproduction of the final product. See **allosteric**, **regulatory enzyme**. Compare **catabolite repression**.

Energy flow The passage of energy through an **ecosystem** from source (generally the sun), through the various **trophic levels** (within organic compounds), and ultimately out to the atmosphere as respiratory heat loss. There is about 90% loss of energy between one trophic level and the next in the grazing food chain. See **pyramid of biomass**.

Enhancer Site on eukaryotic DNA at which a protein may bind and turn on transcription of a particular gene (cistron), which may be either close-by or relatively distant (e.g. some tens of kilobases away) on the same chromosome. This may involve DNA looping brought about by regulatory DNA-binding proteins so that sites normally separated on a chromosome are brought into close contact. Some enhancers (e.g. glucocorticoid enhancer) bind steroid-nuclear protein complexes, enabling binding in turn by RNA polymerase II and the initiation of transcription. Enhancers exhibit tissue or species specificity; e.g. some viral enhancers only function in the species in whose cells the virus grows best. See Fig. 41.

Enkephalins Peptide **neurotransmitters** isolated from the thalamus and parts of the spinal cord and concerned with pain-related pathways. Morphine-like pain reducers. See **endorphins**.

Enrichment culture Microbiological technique allowing selection and isolation from a natural, mixed population of microorganisms of those having growth characteristics desired by the investigator. Involves culturing on a medium whose composition is adjusted for selective growth of desired organisms, by altering nutrients, pH, temperature, aeration, light intensity, etc. Employed in bacteriology, mycology and phycology.

Enterocoely Method of **coelom** formation within pouches of mesoderm budded off from embryonic gut wall. Develops this way in echinoderms, hemichordates, brachiopods and some other animals.

Enterokinase (enteropeptidase) Enzyme (peptidase) secreted by vertebrate small intestine, converting inactive trypsinogen to active trypsin. Removes a small peptide group. Component of succus entericus. See **kinase**.

Enteron (coelenteron) The gut (gastrovascular) cavity within the body wall of coelenterates, having a single opening serving as both mouth and anus. May be subdivided by mesenteries (as in sea anemones); sometimes receives the gametes (as in jellyfish). May serve as hydrostatic skeleton. See **archenteron**.

Entomogenous (Of fungi) parasitic of insects.

Entomology Study of insects.

Entomophagous Insect-eating.

Entomophily Pollination by insects.

Entoprocta Phylum of pseudocoelomate and mostly marine invertebrates, of uncertain relationships. Trochophore larva attaches by its oral surface; stolon grows out from the new aboral surface and produces a colony of adult individuals. These feed by ciliated tentacles which are simply folded away inside their protective cover, not withdrawn into a body cavity as in **Ectoprocta**. Excretion by protonephridia. Anus opens within tentacular ring.

Entrainment Synchronization of an endogenous rhythm with an external cycle such as that of light and dark. See **circadian rhythm**.

Envelope Term applied to the two membranes surrounding the nucleus, the membranous covering of those viruses which have them (i.e. enveloped viruses), and the complex of one or more membranes and peptidoglycan forming the bacterial cell surface (see **Gram's stain**).

Environment Collective term for the conditions in which an organism lives, both biotic and abiotic. Compare **internal environment**.

Enzyme A protein catalyst produced by a cell and responsible for the high rate and specificity of one or more intracellular or extracellular biochemical reactions. Enzyme reactions are always reversible. Almost all enzymes are globular proteins consisting either of a single polypeptide or of two or more polypeptides held together (in quaternary structure) by non-covalent bonds. By virtue of their three-dimensional configurations in solution, enzymes act upon other molecules (substrates), and thus catalyse one type of (but not necessarily just one) chemical reaction.

Their shapes provide them with one or more *active sites* (domains) which bind temporarily and usually non-covalently with compatible substrate molecules to form one or more *enzyme-substrate* (*ES*) *complexes*, catalysis occurring only during the brief existence of the complex. One or more *products* are then released as the active site is freed again to bind fresh substrate. Active sites have conformations and charge distributions which are substrate-specific and their component amino acids commonly alter their relative three-dimensional positions (termed an *induced fit*) as the substrate binds, enabling several sub-reactions involved in catalysis to proceed.

Enzymes do nothing but speed up the rates at which the *equilibrium positions* of reversible reactions are attained. In some poorly understood way, ultimately explicable in terms of **thermodynamics**, enzymes reduce the *activation energies* of reactions, enabling them to occur much more readily at low temperatures – essential for biological systems. When placed in low water concentrations, many enzymes catalyse the reverse of the reaction normally promoted in the biological system.

It is now known that RNA molecules can act as catalysts of reactions, sometimes involving themselves as substrates (see **splicing**). When they involve non-self RNA molecules as substrates, as some do, they can be regarded as enzymes in the full sense (see **ribozymes, telomere**).

In general, cells can do only what their enzymes enable them to do. During both evolution and multicellular development, cells come to look and function differently from each other because they come to have different biochemical capabilities. An enzyme's presence in a cell is dictated by the expression of one or more cistrons encoding it; thus **differentiation** is understood through molecular biology (see **gene expression**).

Because enzyme molecules are generally globular proteins, their shapes and functions may be affected by pH changes in their aqueous environments (see **denaturation**). Denaturation by extremes of pH is usually reversible; not so denaturation by heat. Temperature increase will raise the rate of collision of enzyme and substrate molecules, thus increasing the rate of ES complex formation and raising the reaction rate. This is opposed by increased enzyme denaturation as the *optimum temperature* for the reaction is exceeded. Eventually the reaction ceases, sometimes only at temperatures well in excess of 100°C (see **Archaebacteria**).

At any one instant, the proportion of enzyme molecules bound to substrate will depend upon the substrate concentration. As this is increased, the initial velocity of the reaction (V_o) on addition of enzyme increases up to a maximum value, V_{max} (see Fig. 34), at which substrate level the enzyme is said to be *saturated* (all active sites maximally occupied), and no further addition of substrate will increase V_o. The value of substrate concentration at which $V_o = \frac{1}{2}V_{max}$ is known as the **Michaelis constant** (K_m) for the enzyme–substrate reaction. Low K_m indicates high affinity of the enzyme for the substrate.

Some enzymes (e.g. aspartase) bind just one very specific substrate molecule; others bind a variety of the same kind (e.g. all terminal peptide bonds in the case of exopeptidases). The difference arises from the degree of *stereospecificity* of the enzyme. Many need an attached **prosthetic group** or a diffusible **coenzyme** for activity. In such enzymes the protein component is termed the *apoenzyme* and the whole functional enzyme-cofactor complex is termed the *holoenzyme*. Enzymes requiring metal ions are sometimes termed *metalloenzymes*, the commonest ions involved being Zn^{2+}, Mg^{2+}, Mn^{2+}, Fe^{2+} or Fe^{3+}, Cu^{2+}, K^+ and Na^+. These ions commonly provide a needed charge within an active site.

Some enzymes occur as part of a **multienzyme complex**. In nearly all cases, the shape of the enzyme alters as the ES complex forms, and this brings appropriate groups into such proximity that they are obliged to react. In so doing their electrostatic and hydrophobic bondings to the enzyme break, they fall away, and the enzyme returns to its original shape again. This *induced fit theory* is supported by X-ray crystallographic evidence. The suffix *-ase* often replaces the last few letters of a substrate's name to give the common name of the enzyme using it as substrate: thus sucrase digests sucrose. But an international code for enzymes recognizes six major categories of enzyme function, numbered as follows: 1. *oxido-reductases* (e.g. dehydrogenases), catalysing **redox reactions**; 2. *transferases*, transferring a group of atoms from one substrate to another; 3. *hydrolases*, catalysing hydrolysis reactions; 4. *lyases*, catalysing additions to double bonds (saturating them); 5. *isomerases*, performing isomerizations; 6. *ligases*, performing condensation reactions involving ATP cleavage.

Allosteric enzymes have, in addition to an active site, another stereo-specific site to which an *effector*, or *modulator* molecule can bind. When it does, the shape of the active site is altered so that it can or cannot bind substrate (allosteric stimulation or inhibition respectively). In this way the enzyme can be part of a fine control circuit, requiring the presence or absence of a substance – in addition to substrate presence – before enzyme activity proceeds. Some allosteric enzymes respond to two or more such modulators, permitting still finer control over timing of enzyme activity (see **regulatory enzyme**).

Feedback (or *retro-*) *inhibition* of a biochemical pathway is often achieved by *allosteric inhibition* of the first enzyme in the sequence by the final product. The product binds non-covalently to the modulator site on the enzyme, closing the active site allosterically.

Enzyme inhibition of a simpler kind is achieved in competitive inhibition, where an inhibitor substance competes with the substrate for the enzyme's active site. The binding is reversible so that the percentage inhibition for fixed inhibitor level decreases on addition of substrate. An extremely important example of this involves probably the most abundant enzyme, *ribulose bisphosphate carboxylase*, the CO_2-fixing enzyme in C_3 **photosynthesis**, in which O_2 molecules compete with CO_2 molecules for the active site (see **photorespiration**). In *uncompetitive inhibition* the inhibitor combines with the ES complex (one piece of evidence for the latter's existence), which cannot therefore yield normal product. In *non-competitive inhibition* (a form of allosteric inhibition) the inhibitor binds at a non-active site on the enzyme and ES complex so as to deform the active site and prevent ES breakdown, a process unaffected by increasing substrate concentration, being either reversible or irreversible.

Some enzymes are *constitutive*, being synthesized independently of substrate availability; others are *inducible* (e.g. many liver enzymes), being synthesized only when substrate becomes available. The molecular biology of this is to some extent explained in **gene expression**.

Some enzymes are located randomly in the cytosols of cells; others have very restricted distributions and may be attached to particular membranes or within the matrices of particular organelles. One effect of the latter restriction is that initial velocities of reactions (V_0) can be quite high for a substrate level that would be too low if the molecules were randomized over the whole cell. Another advantage is that incompatible reactions can be kept physically separated.

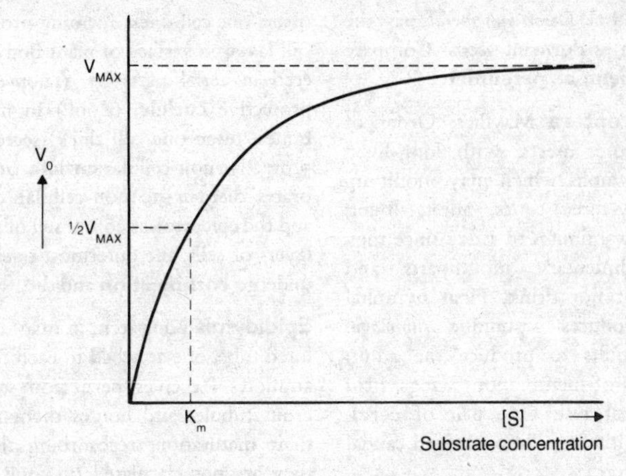

Fig. 34 *Effect of increasing substrate concentration on velocity of enzyme–substrate reaction.*

Methods are now available for attaching some enzymes, and even cells containing them, to insoluble support materials which *immobilize* them, holding them in place (e.g. for an industrially important reaction). The immobilizing medium may be a silica gel lattice, a collagen matrix or cellulose fibres; or enzymes can be encapsulated in beads of alginate or polymer microspheres. Advantages include recovery of the enzyme, lack of contamination of product by enzyme and sometimes greater enzyme stability at extremes of pH and temperature. Immobilization is of great service in continuous fermentations (see **bioreactor**).

Enzyme inhibition See **enzyme**.

Enzyme kinetics Study of the effects of substrate, inhibitor and modulator concentrations on the rate of an enzyme reaction, particularly on initial velocities (V_0). The interrelationships are normally expressed graphically, giving enzyme–substrate–inhibitor curve characteristics, one example being included in the entry for **enzyme**, Fig. 34.

Eocene Geological epoch of the **Tertiary** period lasting from about 54–38 Myr BP. Mammals and birds radiated extensively; initial formation of grass lands occurred. Australia separated from Antarctica, and India collided with Asia. In general, climate was mild to tropical.

Eosinophil One type of white blood cell. **Myeloid** cells with, in man, a bilobed nucleus, and cytoplasmic vesicles (granules) capable of fusing with the plasma membrane on appropriate stimulation and releasing a toxic protein against large targets, especially parasitic worms. Also release antihistamine, damping inflammatory responses. Migrate towards regions containing **T cell** products. Capable of limited phagocytosis.

Ephemeral Plant with a short life cycle (seed germination to seed production), having several generations in one year;

e.g. groundsel. *Desert ephemerals* pass the dry season as dormant seeds. Compare **annual**, **biennial**, **perennial**.

Ephemeroptera Mayflies. Order of exopterygote insects; with long-lived aquatic nymphs which may moult up to twenty-three times, adults living from a few minutes to a day since they have rudimentary mouthparts and neither eat nor drink. Final nymphal moult produces a unique *subimago*, which moults to produce the adult. Two pairs of membranous wings, held vertically at rest. One pair of **cerci**, with or without additional third caudal prolongation.

Ephyra Pelagic larval stage in life cycle of Scyphozoa (jellyfish); develops into adult medusa. Budded asexually from sessile scyphistoma.

Epiblast Cell layer of avian blastodisc and mammalian blastocyst; presumptive ectoderm. Compare **hypoblast**.

Epiboly Process, observed in amphibian and other vertebrate embryos, during which the region occupied by cells of the animal half of the blastula expands over the vegetal half. In amphibians the cells migrate and roll under through the **blastopore**, the vegetal cells remaining as just a plug filling the blastopore.

Epicotyl Upper portion of the axis of an embryo or seedling, above the cotyledons and below the next leaf or leaves. Compare **hypocotyl**.

Epidemic Large-scale temporary increase in prevalence of a disease due to a parasite or some health-related event. Compare **endemic**.

Epidermis Outermost layer of cells of a multicellular organism. (Bot.) Primary tissue, one cell thick, forming protective cell layer on surface of plant body, covered in aerial parts by a non-cellular protective **cuticle**. (Zool.) In invertebrates, often one cell thick, secreting a protective non-cellular **cuticle**. In vertebrates there is no non-cellular cuticle, and the epidermis is composed of several layers of cells; the outermost ones often undergo **cornification** and die.

Epididymis Long (6 m in man) convoluted tube, one attached to each testis in amniotes. Receives sperm from seminiferous tubules and houses them during their maturation, reabsorbing them if they are not ejaculated (in four weeks in man). Peristaltic contractions of the epididymides propel sperm into the sperm duct during ejaculation. Derived embryologically from the mesonephric (Wolffian) duct.

Epigamic (Of animal characters) attractive to the opposite sex and therefore subject to **sexual selection**. Often concerned with courtship and mating.

Epigeal (1) Seed germination in which the seed leaves (cotyledons) appear above the ground; e.g. lettuce, tomato. Compare **hypogeal**. (2) Of animals, inhabiting exposed surface of land, as distinct from underground.

Epigenesis Theory of reproduction and development deriving from Aristotle and espoused by William Harvey (1651) that the parts of an embryo are not all present and *preformed* at the start of development but arise anew one after the other during it. See **epistasis**, **preformation**.

Epigenetics Term introduced in 1947 by the geneticist C.H. Waddington for the branch of biology which studies those causal interactions between genes and their products which bring the phenotype into being. It has two main aspects: (i) changes in cellular composition (cell differentiation, or histiogenesis) and (ii) changes in geometrical form (morphogenesis).

Epiglottis Cartilaginous flap on ventral wall of mammalian pharynx. The glottis pushes against it during swallowing, preventing food, etc., from entering the trachea.

Epigynous See **receptacle**.

Epinasty (Bot.) More rapid growth of upper side of an organ. In a leaf, would result in a downward curling leaf blade. The growth substance **ethene** has been implicated. Compare **hyponasty**.

Epinephrine American term for **adrenaline**.

Epipelon Extremely widespread community of algae occurring in all waters where sediments accumulate onto which light penetrates. The species are almost all microscopic, living on and in the surface millimetres of the sediment, being unable to withstand long periods of darkness and anaerobic conditions. Motile species exhibit endogenous vertical migration rhythm. An important algal community, particularly in shallow ponds and lakes, as well as in highly transparent oligotrophic and montane lakes. See **benthos**.

Epipetalous (Of stamens) borne on the petals, with stalks (filaments) more or less fused with the petals and appearing to originate from them.

Epiphysis (1) Separately ossified end of growing bone, forming part of joint; peculiar to mammalian limb bones and vertebrae. Separated from rest of bone (**diaphysis**) by cartilaginous plate (epiphysial cartilage). Epiphysis and diaphysis fuse when growth is complete. See **bone**, **growth hormone**. (2) Synonym for **pineal gland**.

Epiphyte Plant attached to another plant, not growing parasitically upon it but merely using it for support; e.g. various lichens, mosses, algae, ivy, and orchids, all commonly epiphytes of trees.

Epiphyton Community of organisms living attached to other plants, sometimes in very large populations; well developed in aquatic habitats where algae attach to other plants.

Epipsammon Community of algae found living attached to sand grains in both freshwater and marine environments. It includes very small species of diatoms and blue-green algae, and includes both motile and non-motile species. Motile species, like those of the **epipelon**, exhibit an endogenous vertical migration rhythm; however the speed of movement is slower with epipsammic species.

Episome A genetic element (DNA) that may become established in a cell either autonomously of the host genome, replicating and being transferred independently, or else as an integrated part of the host genome, participating with it in recombination and being transferred with it. Term first applied to temperate **bacteriophage**, but includes **plasmids**. See **F factor**, **transposon**.

Epistasis Interaction between non-allelic genetic elements or their products, sometimes restricted to cases in which one element suppresses expression of another (*epistatic dominance*). Analogous to genetic **dominance**. Segregation of epistatic genes in a cross can modify expected phenotypic ratios among offspring for characters they affect. See **hypostasis, modifier, suppressor, mutation, polygenes, genetic variation**.

Epithelium (Zool.) Sheet or tube of firmly coherent cells (see **desmosome**) with minimal material between them, of ectodermal or endodermal origin, lining cavities and tubes and covering exposed surfaces of body; one surface of epithelium is therefore free, the other usually resting on a **basement membrane** over connective tissue. Its cells are frequently secretory, secretory parts of most glands being epithelial. Classified according to: height relative to breadth (e.g. *columnar*, *cuboidal*, or *squamous*, in order of diminishing relative height); whether the sheet is one cell thick (*simple*) or many (*stratified* or *pseudostratified*); and presence of cilia (*ciliated*). When morphologically identical tissue is derived from mesoderm, it is either **endothelium** or **mesothelium**. See **intercellular junction** (Fig. 65) for polarity in epithelial cells. (Bot.) Layer of cells lining schizogenously formed secretory canals and cavities, e.g. in resin canals of pine.

Epitope Antigenic determinant. See **antigen**.

Epitreptic behaviour Behaviour by one individual tending to cause the approach of a member of the same species (a conspecific).

Epizoite Non-parasitic sedentary animal living attached to another animal. Compare **epiphyte**.

Epizoon Community of algae living attached to the outer surfaces of animals, which may range from tiny aquatic invertebrates to fishes and whales.

Equatorial plate Plane in which the chromosomes of a cell lie during metaphase of mitosis and meiosis; the equator of the spindle.

Equilibrium potential Potential (voltage gradient) at which a particular ion type passes equally easily in either direction across a cell membrane. Different ions have different equilibrium potentials. See **membrane potential**.

Equisetales Horsetails. See **Sphenophyta**.

Ergot Several ascomycotan fungi are parasitic on higher plants. The disease of plants called ergot is caused by *Claviceps purpurea*, a parasite of rye (*Secale cereale*) and other grasses. Dark spur-shaped **sclerotia** develop in place of healthy grain in a diseased inflorescence. After the grain shatters, they overwinter in the soil. Activated by frost, they form several multicellular spore-bearing structures which contain abundant perithecia. Ascospores are shed when rye and other grasses are flowering and germinate among the flowers. The resulting mycelium gives rise to abundant conidia, which are embedded in a sticky liquid, and are spread further by insects. Fungus converts individual immature fruits into sclerotia. Ergot is a serious disease of rye because when eaten it can cause severe illness among domestic animals and humans. Ergotism, the toxic condition caused by eating grain infected with ergot, is often accompanied by gangrene, nervous spasms, psychotic delusions and convulsions. Occurred frequently during the Middle Ages, when it was known as St Anthony's Fire. Ergot contains the **alkaloid** lysergic acid amide (LSA), a precursor of lysergic acid diethylamide (LSD). Some compounds in ergot (e.g. ergotamine, ergonovine) are used medicinally.

Erythroblast Nucleated bone marrow cell which undergoes successive mitoses, develops increasing amounts of haemoglobin, and gives rise to a *reticulocyte*, and finally the fully differentiated **red blood cell**.

Erythrocyte See **red blood cell**.

Erythropoiesis Red blood cell formation. See **haemopoiesis**.

Escape Cultivated plant found growing as though wild, some detrimentally to the natural ecosystems (e.g. purple loosestrife, *Lythrum salicaria*), which in certain areas of the United States and Canada is rapidly taking over wetlands, causing them to become dry.

Escherichia coli (*E. coli*) Motile, Gram-negative, rod-shaped bacterium (Enterobacteriaceae) used most extensively in bacterial genetics and molecular biology. Normal inhabitant of the human colon; is usually harmless although some strains can cause disease. See **bacteria**, **chromosome**, **Jacob-Monod theory**, **Gram's stain**.

Essential amino acid See **amino acid**.

Essential fatty acid **Fatty acids** required in the diet for normal growth. In mammals, include linoleic and gamma-linolenic acids, obtained from plant sources, without which poor growth, scaly skin, hair loss and eventually death occur. Precursors of arachidonic acid and **prostaglandins**.

Essentialism The view, associated in particular with Aristotle, that for any individual there is a definitive set of properties, individually necessary and collectively sufficient, rendering it the kind of individual that it is. This approach has at times been adopted in the context of the taxa used in classification, sometimes rhetorically and in opposition to evolutionary theories. See **natural kind**, **nominalism**.

Etaerio (Of fruits) an aggregation; e.g. of achenes, in buttercup; of drupes, in blackberry.

Ethene See **Ethylene**.

Ethiopian Designating a zoogeographical region comprising Africa south of the Sahara. Sometimes Madagascar is treated as a separate region (the Malagasy Region).

Ethology Study of animal behaviour in which the overriding aim is to interpret behavioural acts and their causes in terms of evolutionary theory. The animal's responses are interpreted within the context of its actual environmental situation.

Ethylene (ethene) Simple gaseous hydrocarbon (C_2H_4) produced in small amounts by many plants (found in flowers, leaves, leafy stems, roots, and in some species of fungi), and acts as a plant hormone, or **growth substance**. The biosynthesis of ethylene begins with the amino acid methionine, which reacts with ATP to form a compound known as S-adenosylmethionine (SAM). SAM is then split into two different molecules, one of which contains a ring consisting of three carbon atoms (1-

aminocyclopropane-1-carboxylic acid, ACC). This compound is then converted into ethylene, carbon dioxide and ammonia by enzymes on the **tonoplast**. The reaction forming ACC is the step of the pathway that is affected by several treatments (e.g. high **auxin** concentration, air pollution damage, wounding). These stimulate ethylene production by plant tissues. Release of ethylene commonly inhibits auxin synthesis (negative feedback) and transport. It normally inhibits longitudinal growth, but promotes radial enlargement of tissues. The final shape and size of cells, as influenced by ethylene, are the result of its interaction not only with auxin, but also with gibberellic acid and cytokinin. Its effect upon fruit ripening has agricultural importance, e.g. to promote ripening of tomatoes picked green and stored in ethylene until marketed. Also used to ripen grapes. Ethylene promotes abscission of leaves, flowers and fruits in a variety of plant species, and is commonly used commercially to promote fruit loosening in cherries, grapes and blueberries. It is also used as a thinning agent in commercial prune and peach orchards. In **monoecious** flowers ethylene appears to play a major role in determining the sex of the flowers (e.g. in Cucurbitacaeae, ethylene is important in sex expression and is associated with the promotion of femaleness). See **climacteric**.

Etiolation Phenomenon exhibited by green plants when grown in darkness. Such plants are pale yellow because of absence of chlorophyll, their stems are exceptionally long owing to abnormal lengthening of internodes, and their leaves are reduced in size.

Eubacteriales Eubacteria; a large and diverse order of **bacteria**, lacking photosynthetic pigments. Simple, undifferentiated cells with rigid cell walls, either spherical or straight rods. If motile, move by peritrichous flagella. Thirteen recognized families. Includes the important genera *Azotobacter* and *Rhizobium* (both nitrogen-fixers), *Escherichia*, etc.

Eucarpic (Of fungi) with a mature thallus differentiated into distinct vegetative and reproductive portions. Compare **holocarpic**.

Eucaryote See **eukaryote**.

Euchromatin Eukaryotic chromosomal material (chromatin) staining maximally during metaphase and less so in the interphase nucleus, when it is less condensed. See **chromosome**, **heterochromatin**.

Eugenics Study of the possibility of improving the human **gene pool**. Historically associated with some extreme political tendencies and with encouragement of breeding by those presumed to have favourable genes and discouragement of breeding by those presumed to have unfavourable genes; nowadays the more humanitarian **genetic counselling** has largely replaced talk of eugenics.

Euglenophyta Euglenoids. Algal protists characterized by possession of chlorophylls a and b; one membrane of **chloroplast endoplasmic reticulum**; a **mesokaryotic** nucleus; flagella possessing fibrillar hairs in one row; no sexual reproduction and paramylon formed as the storage product in the cytoplasm. Mostly unicellular algae surrounded by a flexible or rigid pellicle found inside the plasmalemma. The pellicle is mostly proteinaceous, comprising a number of helically wound strips. When flexible, euglenoid cells can undergo a flowing (*euglenoid*) movement. Some are surrounded by rigid lorica (e.g. *Trachelomonas*). Muciferous bodies are also present in the cells.

Euglenoid flagellates are found in most freshwater habitats, particularly in water polluted by organic waste or decaying organic matter; also in marine or brackish water, the open sea, in tidal areas among seaweeds, and as sand inhabitants on beaches. Brackish epipelic species can colour estuarine sediment bright green at low tide.

Euglenoids probably arose from the ingestion of a green algal chloroplast by a protozoan in the Kinetoplastida (bodonids, trypanosomatids, and the closely related protozoan *Isonema*). Such an endosymbiotic event would have occurred when the food vacuole membrane of the protozoan became the single membrane of the chloroplast endoplasmic reticulum surrounding the two membranes of the chloroplast envelope. See **nucleolus**.

Eukaryote (eucaryote) Organism in whose cell or cells chromosomal genetic material is (or was) contained within one or more nuclei and so separated from the cytoplasm by two nuclear membranes. Some eukaryotic cells (e.g. mammalian erythrocytes, phloem sieve tubes) lose their nuclei during development; but all are distinguished from prokaryotic cells by having somewhat denser (80S) **ribosomes**, a greater variety of membrane-bound organelles, their generally much larger size and the presence of the proteins **actin, myosin, tubulin** and **histones**; though an actin-like protein has been found in *Escherichia coli*. Indeed, evidence that several protein domains have been conserved since before the bifurcation of prokaryotes and eukaryotes requires us to reconsider the distinction between these grades of organization at the molecular level. But no prokaryote engages in mitosis or meiosis, and **cell locomotion** in eukaryotes involves different **motor proteins**. Compare **prokaryote**. See **cell, nucleus, mesokaryote, urkaryote**.

Euphotic zone (photic zone) Uppermost zone of lakes, seas and rivers, with sufficient light for active photosynthesis. In clear water, may extend to 120 metres.

Euploid Term describing cells whose nuclei have an exact multiple of the **haploid** set of chromosomes, there being no extra or fewer than that multiple. Thus, **diploid**, **triploid**, **tetraploid**, etc., cells are all euploid. Compare **aneuploid**.

Euryhaline Able to tolerate a wide variation of osmotic pressure of environment. Compare **stenohaline, osmoregulation**.

Eurypterida Fossil subclass of the **Merostomata**, appearing in the Ordovician. Free-swimming, marine, brackish and freshwater forms; prosoma with six pairs of ventral appendages, the first being **chelicerae**, the others modified for grasping, walking and swimming. Larva resembled trilobite larva of king crab. Active predators, about two metres in length. See **Arachnida**.

Eurythermous Able to tolerate wide variations of environmental temperature. Compare **stenothermous**.

Eusocial Term applied generally to

certain colonial insects which exhibit cooperative brood care, overlap between generations, and reproductive **castes**. Includes termites, bees, ants and wasps.

Eusporangiate (Of sporangia in vascular plants) arising from a group of parent cells and possessing a wall of two or more layers of cells. Spore production greater than in the **leptosporangiate** type.

Eustachian tube Tube connecting middle ear to pharynx in tetrapod vertebrates. Allows equalization of air pressure on either side of eardrum. See **ear, middle**; **spiracle**.

Eustele Stele in which primary vascular tissues are arranged in discrete strands around a pit.

Eustigmatophyta Protistan algae; basically unicellular, living in freshwater or on damp soil, producing a small number of zoospores, most with a single emergent flagellum; but a second basal body is present, indicating a biflagellate ancestry. The emergent flagellum has microtubular hairs and is inserted subapically. Named after the large, orange-red **eyespot** at the anterior end of the zoospore, independent of the chloroplast (main difference from the **Xanthophyta**). Chlorophyll *a* and *β*-carotene are present, with two major xanthophylls (violaxanthin and vaucheriaxanthin). Violaxanthin is the major light-harvesting pigment in the Eustigmatophyta. Chloroplasts have three thylakoids per band with no girdle band beneath the chloroplast envelope. Two membranes of **chloroplast endoplasmic reticulum** surround the chloroplast; but there is no connection between this and the outer nuclear membrane. Vegetative cells normally possess a characteristic polygonal pyrenoid, although this is absent in the zoospores.

Eutheria (Placentalia) Placental mammals. Infraclass of the **Mammalia**, and the dominant mammals today. Most of the 3800 species occur within about six orders: Insectivora (e.g. shrews, hedgehogs), Chiroptera (bats), Rodentia (e.g. mice, rats), Artiodactyla (e.g. deer, pigs), Carnivora (e.g. cats, dogs, weasels) and Primates (e.g. lemurs, monkeys, apes, humans). Appear in Upper Cretaceous, at time of dinosaur extinction. Connection between embryo and uterus intimate and complex; amnion and chorion present; umbilicus links embryo to chorio-allantoic **placents**; scrotum posterior to penis. Gestation period of varying length; newborn young more advanced developmentally than in other mamals. Great **adaptive radiation** in early Cenozoic. **See Prototheria**, **Metatheria**.

Eutrophic (Of lakes and rivers) originally introduced to describe the phytoplankton assemblages characteristic of 'lowland lakes', which received a rich source of nutrients (e.g. phosphorus, nitrogen). Thus, the original use of the term implied variation in nutrient content. Nowadays, eutrophic is used in a more general sense to describe a lake which has a high concentration of nutrients; highly productive in terms of the organic matter produced. Compare **dystrophic**, **mesotrophic**, **oligotrophic**.

Eutrophication Usually rapid increase in the nutrient status of a body of water, both natural and occurring as a by-product of human activity. May be caused by run-off of artificial fertilizers from agricultural land, or by input of sewage or animal waste. May occur when large flocks of migrating birds collect around watering holes. Leads to reduction in species diversity as well as change in species composition, often accompanied by massive growth of dominant species. Excessive production stimulates respiration, increasing dissolved oxygen demand and leading to anaerobic conditions, commonly with accumulation of obnoxious decay and animal death. Artificial eutrophication can be slowed or even reversed by removal of nutrients at source, but may require costly sewage treatment plants. See **Cyanobacteria**.

Evergreen (Of plants) bearing leaves all year round (e.g. pine, spruce). Contrasted with **deciduous**.

Evocation Ability of an inducer to bring forth a particular mode of differentiation in a tissue which is *competent*. It has been suggested that the inducer brings about release of a substance (the *evocator*) which initiates the differentiation. See **induction**, **organizer**.

Evolution (1) *Microevolution*: changes in appearance of populations and species over generations. (2) *Macroevolution* or *phyletic evolution*: origins and **extinctions** of species and grades (see **speciation**).

Microevolution includes changes in mean and modal phenotype, morph ratios, etc. such as occur within populations from one generation to the next. When statistically significant changes in such variables (or the genes responsible for them) occur with time, a population may be said to evolve. Macroevolution includes large-scale phyletic change over geological time (e.g. successive origins of crossopterygian fish, amphibians, reptiles, birds and mammals), as well as extinctions of taxa within such groups. It is usually accepted that causes of evolutionary change include **natural selection** and **genetic drift**, and that macroevolutionary change can be explained by the same factors that bring about microevolution.

Debate has recently centred upon the rate of evolutionary change. Some biologists accept that evolution largely occurs by gradual **anagenesis**; others stress the role of **cladogenesis** and take the view that species persist unchanged for considerable periods of time, and that relatively rapid speciation events punctuate the fossil record (*punctuated equilibrium*). Darwin considered both to be possibilities. At the molecular level, controversy centres on the respective influences in evolution of random alterations in genetic material (the *neutralist* view) and of selective changes (the *selectionist* view). See **molecular clock**. Opposed to evolutionary explanations of the composition of the Earth's fauna and flora is the group of views termed '**special creationism**', which holds that there are no bonds of genetic relationship between species, past or present. See **origin of life**.

Although Anaximander (6th. cen. BC), Empedocles (5th. cen. BC) and Aristotle (4th. cen. BC) all held evolutionary views of some kind, they depended more on *a priorism* than on observation and testable theory. **Lamarck** is often considered the most influential evolutionary thinker prior to Charles **Darwin** and Alfred **Wallace** but his theory was very different from theirs. They themselves drew apart on the question of human origins and the role of sexual selection.

Evidence for common descent and the fact of evolution comes principally from molecular biology (see **DNA hybridization**, **electrophoresis**, **genetic code**, **molecular clock**), comparative biochemistry, comparative morphology (e.g. anatomy and embryology), geographical distributions of organisms and **fossil** records. The modern theory of evolution (**neo-Darwinism**) derives largely from the kind of genetical knowledge which Darwin lacked, principally the occurrence of Mendelian segregation, which helps explain how variations can be maintained in populations. Evidence for microevolution and Darwinian natural selection (amounting to his 'special theory of evolution') stems largely from population genetics (e.g. see **industrial melanism**), although Darwin himself drew heavily on the analogy of **artificial selection**. See **natural selection**.

Evolutionarily stable strategy (ESS) A heritable strategy (commonly but by no means always behavioural) which, if adopted by (expressed in) most members of a population, cannot be supplanted in evolution by an alternative (mutant) strategy. The strategy may be complex and involve a variety of different sub-responses in accordance with environmental changes, not least other organisms' behaviours. See **game theory**.

Evolutionary taxonomy A school of biological **classification** which makes use of both phenetic and phylogenetic data in classifying organisms. Because there is no theoretical guide as to when one approach should be used and when the other, this very influential school has been criticized by adherents of **cladistics**. See **parallel evolution**.

Evolutionary transformation series A pair of **homologous** characters, one derived directly from the other. See **plesiomorphous**, **apomorphous**, **cladistics**.

Evolution of sex See **sex** (2).

Exarch Type of maturation of primary xylem in roots, in which the oldest xylem elements are located closest to the outside of the axis. Compare **endarch**.

Excretion (1) Any process by which an organism gets rid of waste metabolic products. Differs from **egestion** in that wastes removed are products of the organism's cells rather than simply undigested wastes; and from **secretion** since substances produced would generally be harmful if allowed to accumulate, and as a rule have no intrinsic value to the organism. The simplest excretory method is passive diffusion, either through the normal body surface or across organs with enlarged surface areas (gills, lungs). These may be supplemented or replaced by internal excretory organs, particularly where the body surface cannot be used. Excretory organs typically remove metabolic products from interstitial fluids (e.g. lymph, blood plasma). The gut occasionally serves as a route for excretory products, but is not an excretory organ. Nitrogenous excretion is usually in the form of ammonia (aquatic environments), urea (terrestrial environments) or uric acid (environments where water is at a premium). Common invertebrate excretory organs include **flame cells**, **nephridia**, and **Malpighian tubules**, but in some cases (e.g. large crustaceans) excretion may be deposited in the exoskeleton, commonly to be lost during moulting. Vertebrate **kidneys** work by filtration and selective reabsorption, and, like some invertebrate excretory organs, also have roles in **osmoregulation**.

Excretion in plants includes **guttation** and removal by diffusion of excess oxygen produced by photosynthesis, since oxygen may inhibit that process. Leaf fall also removes a number of metabolic wastes. (2) A substance, or mixture of substances, excreted: *excreta*.

Exercise For effects of training, see **lipoprotein**, **muscle contraction**.

Exergonic (Of a chemical reaction) yielding energy. See **thermodynamics**.

Exine Outer layer of spores and pollen grains; usually divided into two main layers: an outer ectexine and an inner endexine. Often composed of **sporopollenin**.

Exocrine gland Any animal gland of epithelial origin which secretes, either directly or most commonly via a duct, onto an epithelial surface. See **gland**, **endocrine gland**.

Exocytosis Process whereby a vesicle (e.g. secretory vesicle), often budded from the **endoplasmic reticulum** or **Golgi apparatus**, fuses with the plasma membrane of the cell, with release of vesicle contents to exterior. Common process in **secretion**. When restricted to anterior region of cell it is an important stage in much eukaryotic **cell locomotion**. Compare **endocytosis**. See **synaptic vesicles**.

Exodermis Layer of closely fitting cortical cells with suberized walls, replacing the withered piliferous layer in older parts of roots.

Exogamy See **outbreeding**.

Exons Used in two senses: (i) any sequence of DNA represented by its RNA equivalent in mRNA (i.e. after **RNA processing**); (ii) any DNA sequence encoding and giving rise to a translated polypeptide sequence (often a protein **domain**). Exons alternate with **introns** in most eukaryotic, and some prokaryotic, genes. In *exon shuffling*, recombination occurs between different intron sequences at the DNA level (as opposed to the splicing which occurs during **RNA processing**) to generate novel exon sequences.

Exonuclease Enzyme which removes nucleotides one by one from the end of a polynucleotide chain. See **DNase**.

Exopeptidase Proteolytic enzyme which removes amino acids one by one from the end of a protein molecule. Compare **endopeptidase**.

Exopodite See **biramous appendage**.

Exopterygota (Heterometabola) Winged insects with incomplete metamorphosis; sometimes regarded as a subclass of the **Insecta**. No pupal stage. Wings develop outside the body; successive larvae (nymphs) become progressively adult-like. Includes palaeopteran orders Ephemeroptera and Odonata; orthopteroid orders Plecoptera, Grylloblattoidea, Orthoptera, Phasmida, Dermaptera, Embioptera, Dictyoptera, Isoptera and Zoraptera; and hemipteroid orders Psocoptera, Mallophaga, Siphunculata, Hemiptera and Thysanoptera. See **Endopterygota**.

Exoskeleton Skeleton covering the outside of the body, or located in the skin. In arthropods (see **cuticle**), secreted by the epidermis; in many vertebrates, e.g. tortoises, armadillos, the exoskeleton consists of bony plates beneath the epidermis. Many primitive jawless vertebrates (ostracoderms) and primitive jawed vertebrates (placoderms) had body armour comprising bony skin plates and scales. The scales and denticles of modern fish are remnants of this.

Exotoxin Toxin released by a microorganism into surrounding growth medium or tissue during *growth phase* of infection. Generally inactivated by heat and easily neutralized by specific antibody. Produced mainly by Gram-positive bacteria, such as the agents of botulism, diphtheria, *Shigella* dysentery and tetanus. The alga *Prymnesium parvum* forms a potent exotoxin that causes extensive fish mortalities in brackish water conditions in many countries in Europe and in Israel. Compare **endotoxin**.

Experiment The intentional manipulation of material conditions so as to elicit an answer to a question, often posed in the form: what is the effect of x on y? The aim of the experimenter is to isolate x as the only free variable, keeping constant all other variables which might affect the value of y. Values of x can then be paired off with values of y, when changes in x are said to be the cause of any changes in y. A similar approach compares the results of two experimental situations differing in just one initial condition, which often has zero value in one of the experimental situations (called the *control*) but is allowed free range over its values in the other experimental situation (called the *experiment*). The effects of this free- ranging variable are then compared with the effect of its absence (zero value), and since it is the only independent variable, any differences in effect can be said to have been caused by changes in its value. Controlled experiments must have this comparative element. The rationale is to eliminate all possible alternative causes of effects save the one under investigation. Without such controlled experiments, the material causes of phenomena could never be ascertained. It is often assumed, not always with justification, that methods used to study biological material do not themselves affect the properties being studied.

Explanation A phenomenon may be said to have been fully explained when all its component parts can be formally deduced as consequences of sets of actual initial conditions (the minor premises) satisfying the terms of whichever general law (the major premise) represents our most inclusive summary of the relevant experimental data to date. Attempts to explain biological phenomena solely in terms of the language employed in physics and chemistry exemplify what is termed *reductionism*. Most people believe this can only be achieved if terms peculiar to biology can be 'paired off' by identity or equivalence relations to terms in the physical sciences. It is highly contentious whether this can be achieved, even in principle.

Explantation See **tissue culture**.

Exponential growth Growth of cells, populations, etc., in which rate of increase is dependent only upon the number of individuals and their potential net reproductive rate. In other words, no competition occurs between individuals for resources nor is there any other detrimental effect of individuals upon one other. Such a situation is characteristic of the initial growth phase of microorganisms in cultures, or of organisms introduced into regions where food is not limiting and where natural controls (e.g. predators, parasites) are absent. The exponential growth curve can be defined by the equation:

$$N_t = N_o e^{(b-d)t}$$

where t is a very short time interval
N_t is the number of individuals after time t
N_o is the number of individuals at the beginning of the time interval
b is the 'birth' rate during time t
d is the 'death' rate during time t
e is a constant, taken for convenience to be the base of Napierian logarithms, 2.718 (the exponential constant).

Expression signals The molecular signals which may need to be spliced upstream of a gene (or its cDNA equivalent) for it to be properly transcribed (initiated and terminated) and translated (i.e. expressed) in a different organism. These may take the form of an 'expression casette' in a plasmid **vector** containing all appropriate signals, such as a strong **promoter** (often patented) and sequences to achieve high copy number, proper ribosome binding and correct targeting of product (see **protein targeting**). Such signal sequences are needed when, e.g., a product protein has to be secreted from the host cell so that it can be easily extracted from the culture (e.g. yeast) in cell-free solution; but these signals may need subsequent removal (if not done by the host), e.g. to avoid immunogenicity. Eukaryotic genes for expression in prokaryotic hosts need a prokaryotic promoter and must be in the form of cDNA to avoid problems of **RNA processing**. See **yeast artificial chromosome**.

Expression system The system used for the expression of a piece of recombinant DNA and from which the desired product is released or retrieved. Common expression systems are *Escherichia coli*, *Bacillus subtilis*, various yeasts, Chinese hamster ovary (CHO) and mouse myeloma cells. Transgenic sheep and cattle are increasingly used in specialized protein production. Baculovirus vector expression within insects is used in **vaccine** production and where complex glycosylation and other post-translational modification of proteins are required. Retroviral vectors are commonly used in mammalian cells where expression of **oncogenes** and anti-oncogenes (retarding cell growth) are being studied. See **vector**.

Expression vector A vector in which a piece of foreign DNA is not only cloned, but expressed as well. See **expression signals**, **vector**.

Expressivity The level to which a given genotype is expressed in an individual's phenotype. Compare **penetrance**. See **genomic imprinting**, **modifier**.

Extensor Muscle or tendon straightening a joint, antagonistic to **flexor**.

Exteroceptor A **receptor** detecting stimuli emanating from outside an animal. Compare **interoceptor**.

Extinction Termination of a genealogical lineage. Used most frequently in the context of species, but applicable also to populations and to taxa higher than species. Agents of 'background rate' extinction include competition, predation (e.g. see **species flock**) and disease, alteration of habitat and random fluctuations in population size.

There have been four periods of so-called *mass extinction*, during which the Earth's fauna has suffered extinction rates far higher than the normal background rate. These occurred in the Ordovician, the late Devonian, the late Permian (225 Myr BP), the late Triassic (190 Myr BP) and the late Cretaceous/Tertiary (K/T, 57 Myr BP). Possible causes of greater than normal extinction rates include evolutionary competition, geological (e.g. volcanic, deep-sea warming) and climatic change and cometary or other impact. Victims of the K/T extinction included the dinosaurs and 60–75% of all marine species, and evidence (high iridium levels and soot in clays at the K/T boundary) suggests that cometary impact could have resulted in large-scale fire. This could have released huge volumes of oxides of nitrogen into the atmosphere, causing severe acid rain and reducing surface temperatures. One likely genetic factor in extinction as population size decreases is *inbreeding depression*. See **punctuated equilibrium**.

Extracellular In general, occurring outside the plasma membrane; but where a **cell wall** is present, often refers to the region surrounding this. See **glycocalyx**.

Extracellular matrix (ECM) Complex network of macromolecules lying between cells where these form tissues and colonies, and in cell walls and cuticles. Comprises mainly locally secreted proteins including, in animals, structural collagens and elastin; adhesive **fibronectin** and **laminins** (see **integrins**) and polysaccharides (e.g. **proteoglycans**, **glycosaminoglycans**). Forms **basal lamina** in animals between epithelium and underlying **connective tissue**. Houses many local signalling molecules profoundly influencing cell division, differentiation, growth (see **growthfactors**) and **programmed cell death**.

Extrachromosomal inheritance Inheritance of genetic factors not forming part of a chromosome. Examples include **plasmid**, mitochondrial and chloroplast inheritance. Inheritance of a variety of intracellular symbionts may also be regarded as extrachromosomal. See **cytoplasmic inheritance**, **episome**.

Extraembryonic coelom In amniote development, the space lying between the mesoderm layers lining inner surface of the chorion and outer surface of the amnion.

Extraembryonic membranes The **yolk sac**, **chorion**, **amnion** and **allantois** of amniote vertebrates; membranes derived from the zygote but lying outside the epidermis of the embryo proper. Have played a major part in evolution of vertebrate terrestrialization. See Fig. 35.

Eye Sense organ responding to light. In invertebrates, either a simple scattering of light-sensitive pigment spots in the general epithelium but more often comprising an optic cup of receptor cells with screening pigment cells (each functional unit an *ocellus*), lacking a refractive surface so that no image can be formed. Although a lens may be present, most ocelli can only differentiate between light and dark. Nonetheless, this enables orientation with respect to light direction and intensity.

The basic unit of the arthropod compound eye is the *ommatidium*, comprising a cornea lens, crystalline cone, a group of usually 7–8 sense (retinula) cells radially arranged around a central rhabdome formed from their innermost fibrillar surfaces (rhabdomeres composed of microvilli), in which the light-sensitive pigment is located, each rhabdome extending into a nerve fibre distally.

Higher molluscan (i.e. cephalopod) eyes (e.g. of *Octopus*) resemble those of vertebrates in complexity (see **convergence**); however, there is an ommatidium-like organization in the retina. For details of the vertebrate eye, see diagram and entries for structures labelled. See also **tapetum**.

Eye muscles (a) *Extrinsic* (outside eyeball). In vertebrates six such muscles rotate the eyeball: a pair of anterior oblique and four, more posterior, rectus muscles. Supplied by cranial nerves III, IV and VI. (b) *Intrinsic* (inside eyeball); see **iris, ciliary body**.

Eyespot (stigma) (1) Light-sensitive pigment spots of some invertebrates. See **eye**. (2) Rather a misnomer for orange to red-coloured lipid droplets or globules close to or within the chloroplasts of some eukaryotic flagellates. The colour is imparted by carotenoid pigments. In the green algae (**Chlorophyta**), for example, the eyespot is always within the chloroplast situated anteriorly near the flagellar bases, and the lipid droplets (one to several layers) are contained in the stroma between the chloroplast envelope and the outermost thylakoids. In other algae (e.g. **dinoflagellates**) the lipid droplets may be in the cytoplasm, and not surrounded by a membrane or in a plastid-like structure, while the most complex eyespot comprises a lens mounted in front of a pigment cup (e.g. the dinoflagellate *Nematodinium armatum*). In flagellates, it is highly probable that the 'eyespot' casts a shadow on the presumably light-sensitive swelling at the flagellar bases, the flicker frequency indicating the angle of the cell rotation with respect to the light source. See **signal transduction**.

Extraembryonic membranes

(a)

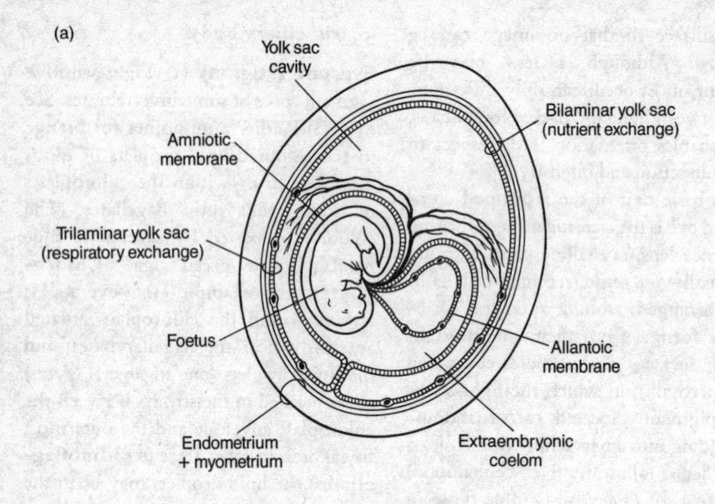

(b)

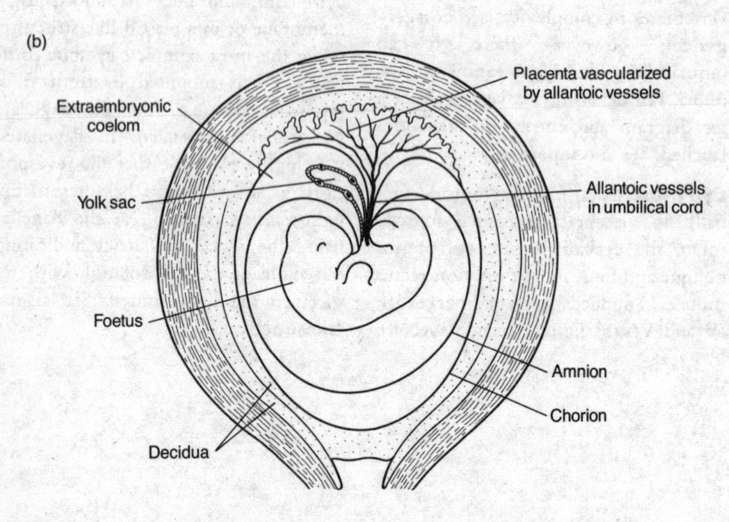

Fig. 35 *Extraembryonic membranes during development of (a) wallaby (marsupial) and (b) human. The uterus wall is outermost in both.*

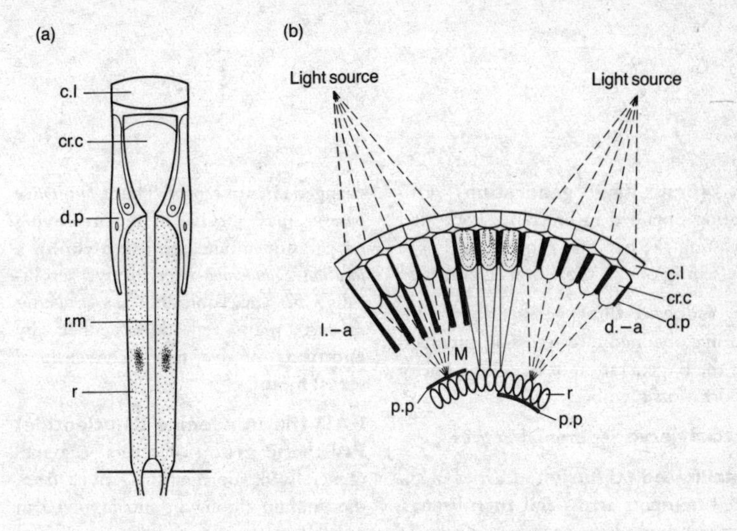

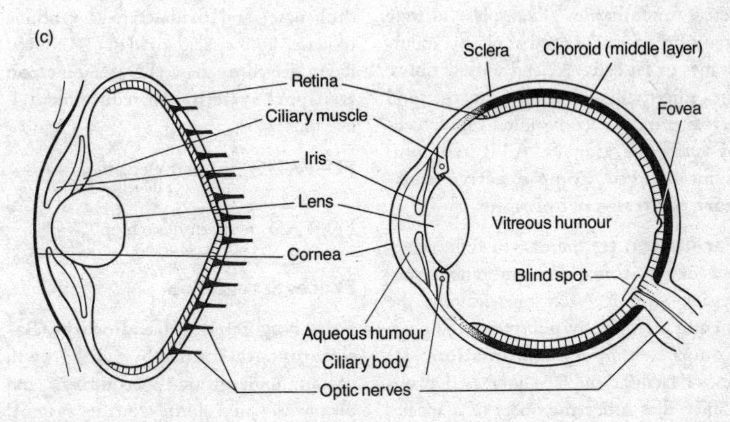

Fig. 36 *An ommatidium from an insect compound apposition eye (b) A superposition compound eye, light-adapted (l.-a) and dark-adapted (d.-a). The central three ommatidia show the proposed laminated structure of the cones; c.l = corneal lens, cr.c = crystalline cone, d.p = distal pigment cell, p.p = proximal pigment cell, r = retinula cell (photoreceptor), r.m = rhabdome. (c) Comparison of cephalopod eye (left) with human eye (right).*

F_1 (first filial generation) Offspring obtained in breeding work after crossing the parental generation (P_1) or by selfing one or more of its members.

F_2 (second filial generation) Offspring obtained after crossing members of the F_1 generation or by selfing one or more of its members.

Facial nerve See **cranial nerves**.

Facilitated diffusion Carrier-mediated transport across **cell membranes**, the transported molecule never moving against a concentration gradient. Only speeds up rate of equilibrium attainment across membrane. Examples include transport of glucose across plasma membranes of fat cells, skeletal muscle fibres, the microvilli of ileum mucosa and across proximal convoluted tubule cells of vertebrate kidneys. ATP hydrolysis is not involved. Compare **active transport**. See **transport proteins**.

Facilitation (1) Increase in responsiveness of a postsynaptic membrane to successive stimuli, each one leaving the membrane more responsive to the next. Compare temporal **summation**. (2) Social facilitation. The increased probability that other members of a species will behave similarly once one member has acted in a certain way. See **nervous integration**.

Facultative Indicating the ability to live under altered conditions, or to behave adaptively under markedly changed circumstances. Thus a *facultative parasite* may survive in the free-living or parasitic mode (see **mixotroph**); a *facultative anaerobe* may survive aerobically or anaerobically; a *facultative apomict* may reproduce either by **apomixis** or by more conventional sexual means.

FAD (flavin adenine dinucleotide) **Prosthetic group** of several enzymes (generally flavoproteins). Derived from the vitamin riboflavin and involved in several **redox reactions**, e.g. as catalysed by various dehydrogenases (e.g. NADH dehydrogenase, succinate dehydrogenase) and oxidases (e.g. xanthine oxidase, amino acid oxidase). Reduced flavin dehydrogenase (FD, see **electron transport system**) can reduce methylene blue:

$$\text{FD–FADH}_2 + \underset{\text{(blue)}}{\text{methylene blue}} =$$

$$\text{FD–FAD} + \underset{\text{(colourless)}}{\text{methylene blue}_{\text{red}}}$$

Faeces See **egestion**.

Fairy ring Circle of **basidiocarps** (**Basidiomycota**) formed by radial growth of an underground secondary (and dikaryotic) mycelium from its original starting point marking the form and extent of the underground mycelium. Some are estimated to be up to 500 years old. The grass immediately inside the ring is often stunted and a lighter green than outside it.

Fallopian tube In female mammals, the bilaterally paired tube with funnel-shaped opening just behind ovary, leading from perivisceral cavity (coelom) to uterus. By muscular and ciliary action it conducts eggs from ovary to uterus. Is frequently the site of fertilization. Represents part of **Müllerian duct** of other vertebrates.

False annulus Discrete grouping of thick-walled cells on the jacket of some fern sporangia, not directly influencing dehiscence.

Family Category employed in biological **classification**, below order and above genus. Typically comprises more than one genus. Familial suffixes normally end-aceae in botany and-idae in zoology. Term also employed in the wider taxonomy of proteins, e.g. the protein family encoded by ***cdc* genes**, and the families of **protein kinases**.

Fascia Sheet of connective tissue, as enclosing muscles.

Fasciation Coalescing of stems, branches, etc., to form abnormally thick growths.

Fascicle (1) Bundle of pine leaves or other needle-like leaves of gymnophytes. (2) Now obsolete term, formerly applied to vascular bundle.

Fascicular cambium Cambium that develops within a vascular bundle.

Fat (neutral fat) Major form of **lipid** store in higher animals and some plants. Commonly used synonymously with **triglyceride**, which not only stores more energy per gram than any other cell constituent ($2\frac{1}{2}$ times the ATP yield of glycogen) but, being hydrophobic, requires less water of hydration than polysaccharide and is therefore far less bulky per gram to store. **Adipose tissue** is composed of cells with little besides fat in them. Hydrolysed by lipases to yield fatty acids and glycerol. See **chylomicron**.

Fat body (1) Organ in abdomen of many amphibia and lizards containing **adipose tissue**, used during hibernation. (2) In insects, diffuse tissue between organs, storing fat, protein, occasionally glycogen and uric acid.

Fate map Diagram showing future development of each region of the egg or embryo. A series of such maps indicates the trajectories of each part from egg to adult. Construction may involve vital staining, cytological and genetic markers. Most easily constructed in cases of highly **mosaic development**.

Fatty acid Organic aliphatic and usually unbranched carboxylic acid, often of considerable length. Condensation with glycerol results in ester formation to form mono–, di–, and triglycerides (fat). Commonly a component of other **lipids**. Free fatty acids are transported in blood plasma largely by albumin (but see **lipoproteins**). Saturated fatty acids have no double bonds; monounsaturated have one and polyunsaturated more than one. Classification of polyunsaturated fatty acids into families is based on how many carbon atoms the last double bond is from the methyl ($-CH_3$) end of the molecule: thus *n*-3, *n*-6 and *n*-9 describe families where the last double bond is 3, 6 and 9 carbon atoms from the methyl end respectively. **Fatty acid oxidation** makes an important contribution to a cell's energy release.

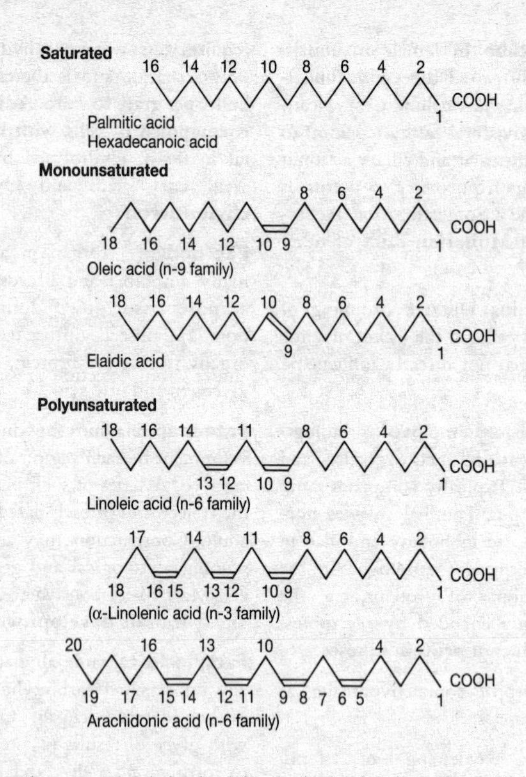

Fig. 37 *Fatty acid structures (methyl end to left). Double bonds indicated by parallel pairs of lines; carbon atoms numbered.* (From *Human Nutrition and Dietetics* (9th edn) by J. S. Garrow and W. P. T. James. Copyright © Churchill Livingstone, 1993.)

Some types of unsaturated fatty acids stored in a cell's membranes can be released and transformed into local cell-signalling **eicosanoids**. A diet lacking in **essential fatty acids** results in changes in the fatty acid compositions of cell membranes leading to their malfunction. Pregnant women should ensure their intake of linoleic and α-linolenic acids is sufficient to allow proper foetal brain cell production (50% of brain mass is due to lipids), and subsequently for proper milk production. These two fatty acids are required for production of the small amounts of the long-chain polyunsaturated fatty acids arachidonic and docosahexanoic acids present in human milk and required for active neonatal brain development. Saturated fatty acids include *palmitic acid*, $CH_3(CH_2)_{14}COOH$ and *stearic acid*, $CH_3(CH_2)_{16}COOH$; unsaturated fatty acids include *oleic acid*, $CH_3(CH_2)_7CH:CH(CH_2)_7COOH$. See **fatty acid oxidation, lipase**.

Fatty acid oxidation (beta-oxidation) Prior to oxidation, fatty acids undergo a complex activation in the cytosol followed by transport across the mitochondrial membranes, whereupon the acyl group binds to **coenzyme A** to form a fatty acyl-CoA thioester. The terminal two carbon atoms are removed enzymatically (forming acetyl CoA, for entry into the **Krebs cycle**) while another CoA molecule is bound to the remaining fatty acid chain. Each sequential 2-carbon removal is accompanied by dehydrogenation and production of reduced NAD for entry into the **electron transport system** of mitochondria and ATP production. See **vitamin E**.

Feather Elaborate and specialized epidermal production characteristic of birds as a class, providing thermal insulation, colouration and, generally, lift and thrust of wings during flight. Develops from *feather germ*, a minute projection from the skin, within which longitudinal ridges of epidermal cells (*barb ridges*) form early on. On each ridge further cells of appropriate shape and position form the *barbules* after keratinization, some cell processes becoming *barbicels*, or hooks. A deep pit in the epidermis, the *feather follicle*, surrounds the bases of feathers. The first feathers are *down feathers* in which the quill is very short, a ring of barbs with minute and non-interlocking barbules sticking up from its top edge giving a soft and fluffy texture. Some follicles produce down feathers throughout life, but most are pushed out by new feathers during moulting. The adult (*contour*) feathers grow in definite tracts on the skin, with (except in penguins and **ratites**) bare patches between.

Other feather types include: *intermediate feathers*, showing a combination of features of contour and down feathers; *filoplumes* (*plumulae*), which are hair-like, usually lacking vanes; *vibrissae*, stiff and bristle-like, often around the nares; and *powder down*, which is soft downy material giving off dusty particles used in feather cleaning.

The amount of keratin required to make a new set of contour feathers may cause the timing of moulting to be under strong selection pressure. Feathers may be pigmented or have a barbule arrangement which produces interference colours. See Fig. 38.

Fecundity Reproductive output, usually of an individual. Number of offspring produced. See **fitness**.

Feedback inhibition See **end-product inhibition**.

Femur (pl. femora) (1) Thigh-bone of tetrapod vertebrates. (2) The third segment from the base of an insect leg.

Feral Of domesticated animals, living in a wild state. See **escape**.

Fermentation Enzymatic and anaerobic breakdown of organic substances (typically sugars, fats) by microorganisms to yield simpler organic products. Pasteur showed in about 1860 that microorganisms were responsible (contrary to view of Liebig). Kuhne called the 'active principle' an **enzyme** in 1878, and Buchner first isolated a fermentative cell-free yeast extract in 1897.

The term is often used synonymously with anaerobic **respiration**. Classic examples include alcohol production by yeasts, and the conversion of alcohol to vinegar (acetic acid) by the bacterium

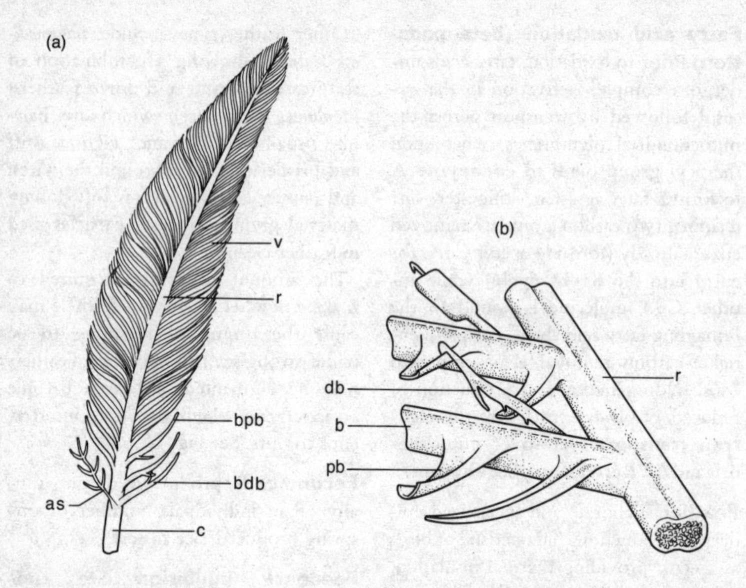

Fig. 38 (*a*) *Generalized contour feather; as = aftershaft, bdb = barb with distal barbule, bpb = barb with proximal barbule, c = calamous or quill, r = rachis, v = one side of vane.* (*b*) *Enlarged view of part of rachis of contour feather seen from dorsal side. Three proximal barbules (pb) have been cut short. These and the distal barbules (db) are less widely spaced than shown here. b = barb.*

Acetobacter aceti, a process commonly called *acetification*. Lactic acid production by animal cells is another example. In all cases the final hydrogen acceptor in the pathway is an organic compound. See **biotechnology**.

Fermenter See **bioreactor**.

Ferns See **Pterophyta**.

Ferredoxins (iron-sulphur proteins) Proteins containing iron and acid-labile sulphur in roughly equal amounts; extractable from wide range of organisms, where they are components of the **electron transport systems** of mitochondria (involved in aerobic respiration) and of chloroplasts, where they undergo reversible Fe^{2+}/Fe^{3+} transitions. Ferrodoxin has several roles in the chloroplast, including donation of electrons to enzymes involved in amino and fatty acid biosynthesis.

Ferritin Iron-storing protein (esp. in spleen, liver and bone marrow). The iron (Fe^{3+}) is made available when required for haemoglobin synthesis, being transferred by **transferrin**.

Fertilization (syngamy) Fusion of two **gametes** (which may be nucleated cells or simply nuclei) to form a single cell (*zygote*) or fusion nucleus. Commonly involves cytoplasmic coalescence (*plasmogamy*) and pooling of nuclear material (*karyogamy*). With **meiosis** it forms a fundamental feature of most eukaryotic sexual cycles, and in general the gametes that fuse are **haploid**. When both are motile, as primitively in plants, fertilization is *isogamous*; when they differ in size but are similar in form it is *anisogamous*; when one is non-motile (and usually larger) it is termed *oogamous*. This is the typical mode in most plants, animals and many fungi. In many gymnophytes and all anthophytes neither gamete is flagellated, and a **pollen tube** is involved in the fertilization process. In animals, *external fertilization* occurs (typically in aquatic forms) where gametes are shed outside the body prior to fertilization; *internal fertilization* occurs (typically as an adaptation to terrestrial life) where sperm are introduced into the female's reproductive tract, where fertilization then occurs. After fertilization the egg forms a *fertilization membrane* to preclude further sperm entry. Sometimes the sperm is required merely to activate the egg (see **activation**, **parthenogenesis**, **pseudogamy**). See also **acrosome**, **double fertilization**.

Feulgen method Staining method applied to histological sections, giving purple colour where DNA occurs.

F factor (F plasmid, F particle, F element, sex element, sex factor) One kind of **plasmid** found in cells of the bacterium *E. coli*, and playing a key role in its sexuality (i.e. inter-cell gene transfer). It encodes an efficient mechanism for getting itself transferred from cell to cell, like many drug-resistance plasmids. Rarely, the F plasmid integrates into the host chromosome (forming an *Hfr* cell), when the same transmission mechanism results in a segment of chromosome adjacent to the integrated F plasmid being transferred from donor cell to recipient. In this condition it behaves very like a λ **prophage**, replicating only when the host chromosome does. When this cell conjugates with a cell lacking an F particle, a copy of the *Hfr* chromosome passes along the conjugation canal, the F particle entering last (see **chromosome mapping**, **transduction**), if at all. The resulting diploid or partial diploid cells do not remain so for long since recombination (hence *high frequency recombinant*, *Hfr*, strain) between the DNA duplexes occurs and the emerging clones are haploid.

Fibre (Bot.) See **sclerenchyma**. (Zool.) Term applied to thin, elongated cell (e.g. nerve fibre, muscle fibre), or the characteristic structure adopted by molecules of collagen, elastin and reticulin. See also **filament**.

Fibril (1) Submicroscopic thread comprising cellulose molecules, in which form cellulose occurs in the plant cell wall; (2) thread-like thickening on the inner faces of large hyaline cells in the leaf or stem cortex of the moss *Sphagnum*.

Fibrin Insoluble protein meshwork formed on conversion of fibrinogen by thrombin. See **blood clotting**.

Fibrinogen Plasma protein produced by vertebrate liver. See **blood clotting**.

Fibrinolysis One of the homeostatic processes involved in **haemostasis**. As in **blood clotting** the major inactive participant is a plasma protein, here *plasminogen*, which is converted to the **serine protease** enzyme *plasmin* by a variety of activators. Plasmin dissolves blood clots and removes fibrin which may otherwise build up on endothelial walls.

Fibroblast Characteristic cell type of vertebrate connective tissue, responsible for synthesis and secretion of extracellular matrix materials such as tropocollagen, which polymerizes externally to form **collagen**. Migrate during development to give rise to mesenchymal derivatives. See **filopodium**.

Fibroin A major protein component of silk; rich in β-pleated sheets.

Fibronectin An important cell **adhesion** molecule which also seems to guide cell migration in embryos. Cultured human T cells require adhesion to fibronectin in order to divide.

Fibrous root Root system comprising a tuft of adventitious roots of more or less equal diameter arising from the stem base or hypocotyl and bearing small lateral roots; e.g. wheat, strawberry. Compare **tap root**.

Fibula The posterior of the two bones (other is **tibia**) in lower part of hindlimb of tetrapods. Lateral bone in lower leg of human.

Filament (1) Stalk of the **stamen**, supporting the anther in flowering plants. (2) Term used to describe thread-like thalli of certain algae and fungi. (3) Term used of many long thread-like structures or molecules. Viruses whose coat proteins produce a rod-shaped virion (e.g. **TMV**) are described as *filamentous*. See **actin**, **gill**, **myosin**.

Filarial worms Small parasitic nematode worms of humans and their domestic animals, typically in tropical and semitropical regions. *Filariasis* is caused by blockage of lymph channels by *Wuchereria bancrofti* (up to 10 cm long), the young (microfilariae, 200 μm long) accumulating in blood vessels near the skin. Transmitted by various mosquitoes. Cause gross swellings of legs: *elephantiasis*. Another filarian, *Onchocerus volvulus*, transmitted by blackflies (*Simulium* spp.), causes *onchoceriasis* (river blindness). The flies need water to breed, and inject the worms when they bite humans. These cause fibrous nodules under the skin and inflammation of the eye, leading to blindness. See **superspecies**.

Filicales Order of **Pterophyta** (ferns), including the great majority of existing ferns and a few extinct forms. Perennial plants with a creeping or erect rhizome, or with an erect aerial stem several metres in height (e.g. tropical tree ferns). Leaves are characteristically large and conspicuous. Sporophylls either resemble ordinary vegetative leaves, bearing sporangia on the under surface, often in groups (sori), or else are much modified and superficially unlike leaves (e.g. royal fern). Generally homosporous, prothalli bearing both antheridia and archegonia; but includes a small group of aquatic heterosporous ferns. See **life cycle**.

Filoplume (plumule) See **feather**.

Filopodium Dynamic extension of the cell membrane up to 50 μm long and about 0.1 μm wide, protruding from the surfaces of migrating cells, e.g. **fibroblasts**, or growing nerve axons. Grow and retract rapidly, probably as a result of rapid polymerization and depolymerization of internal actin filaments, which have a paracrystalline arrangement. Possibly sensory, testing adhesiveness of surrounding cells. Smaller filopodia, up to 10 μm long, are termed *microspikes*. See **cell locomotion**.

Filter feeding Feeding on minute particles suspended in water (**microphagy**) which are often strained through mucus or a meshwork of plates or lamellae. Water may be drawn towards the animal by cilia, or enter as a result of the animal's locomotion. Very common among invertebrates, and found among the largest fish (basking and whale sharks) and mammals (baleen whales).

Finger domain An amino acid sequence within a protein which binds a metal atom, producing a characteristic 'finger-like' conformation within the protein. Such domains tend to bind nucleic acid and may be found repeated tandemly as in *multifinger loops* (see **transcription factors**). Compare **homoeobox**. See **nuclear receptors**, **ubiquitin**.

Fingerprinting See **DNA fingerprinting**.

Fin ray See **fins**.

Fins (1) Locomotory and stabilizing projections from the body surface of fish and their allies. Include unpaired medial **agnathan** *fin-folds* with little or no skeletal support; but the term generally refers to the medial and paired ray fins of the **Chondrichthyes** and **Osteichthyes**, in which increasingly extensive skeletal elements (*fin rays*) articulate with the vertebrae, and **pectoral** and **pelvic girdles**. The pectoral and pelvic fins are paired and are used for steering and braking. The dorsal, anal and caudal fins are unpaired and medial, opposing yaw and roll. The caudal fin (tail) is generally also propulsive (see **heterocercal**, **homocercal**). Fins are segmented structures, seen clearly in the muscle attachments of the ray fins of **Actinopterygii**. (2) Paired membranous and non-muscular stabilizers along the sides of arrow worms (Chaetognatha). (3) Horizontal and muscular fringe around the mantle of cephalopods such as the cuttlefish (*Loligo*) by means of which its gentler swimming is achieved.

Fish General term, covering **Agnatha** (jawless fish), **Chondrichthyes** (cartilaginous fish) and **Osteichthyes** (bony fish).

Fission Form of vegetative reproduction, involving the splitting of a cell into two (*binary fission*) or more than two (*multiple fission*) separate daughter cells. See **cell division**.

Fissipedia Suborder of **Carnivora**, including all land carnivorous mammals. Canines large and pointed; jaw joint a transverse hinge (preventing grinding); carnassial teeth often present. Includes cats (Felidae), foxes, wolves and dogs (Canidae), weasels, badgers, otters (Mustelidae), civets, genets and mongooses (Viverridae), hyaenas (Hyaenidae), racoons and pandas (Procyonidae), and bears (Ursidae).

Fitness (selective value) Factor describing the difference in reproductive success of an individual or genotype relative to another. Usually symbolized by w. Often regarded as compound of survival (longevity) and annual fecundity.

(1) Of individuals. Lifetime reproductive success; either 'lifetime reproductive output' (the lifetime fecundity), or the number of offspring reaching reproductive age. Both omit information on the reproductive output of these offspring, and hence on the number of grandchildren reaching reproductive age, or the number of great grandchildren doing so, and so on. Fecundity alone is therefore only one component of fitness: an individual may leave more descendants in the long term by producing fewer total offspring but by ensuring a greater probability of their survival to reproductive age (e.g. by provisioning fewer seeds with more food reserves). Likewise, natural selection will favour any heritable factor that improves the chances of a gene's representation in subsequent generations. Thus an individual may promote future representation of its own genes, even if it leaves no offspring itself, by contributing to the fitness of close relatives. Any actions which do so contribute improve an agent's *inclusive fitness*, calculated from that individual's reproductive success plus its effects upon the reproductive success of its relatives, each effect weighted by the relative's coefficient of **relatedness** to the agent. Likewise, an individual's fitness may be improved by the effects of its relatives. See **helper**, **unit of selection**.

(2) Of genotypes. Usually applied to a single locus, where the fitness value, w, of a genotype such as Aa is defined as $1 - s$, where s is the *selection coefficient* against the genotype. The existence of fitness differences between genotypes creates selection for the evolution of the genetic system itself. See **coefficient of selection**.

Fixation (1) In microscopy, the first step in making permanent preparations of organisms, tissues, etc., for study. Aims at killing the material with the least distortion. Solutions of formaldehyde and osmium tetroxide often used. Some artifacts of structure usually produced.

(2) Of genes. The spread of an allele of a gene through a population until it comes to occupy 100% of available sites (i.e. until it is the only allele found at that locus). It is then *fixed* in the population.

(3) Of elements, e.g. carbon, nitrogen. Conversion of an inorganic source of the element to an organic source. C-fixation occurs in photosynthesis; N-fixation occurs in soils, ponds, etc., through the action of prokaryotes (e.g. bacteria, blue-green algae).

Flagellata See **Mastigophora**.

Flagellum (1) Extension of the cell membrane of certain eukaryotic cells, comprising an internal axoneme of nine doublet microtubules surrounding two central doublets. Upon entering the cell body, the two central microtubules end at a double plate, whereas the nine peripheral doublets continue into the cell, usually picking up extra structures transforming them into triplets. Between the microtubules in the flagellum's basal region is the basal body, attached to which are the fibrillar roots. Thus, the structure is identical to that of a **cilium** (see Fig. 23), but more variable in length and generally longer. The flagellar membrane may be smooth, possessing no hairs on its surface (smooth, whiplash or acronematic flagellum); or it may have hairs on its surface (hairy tinsel or pantonematic flagellum). Two types of

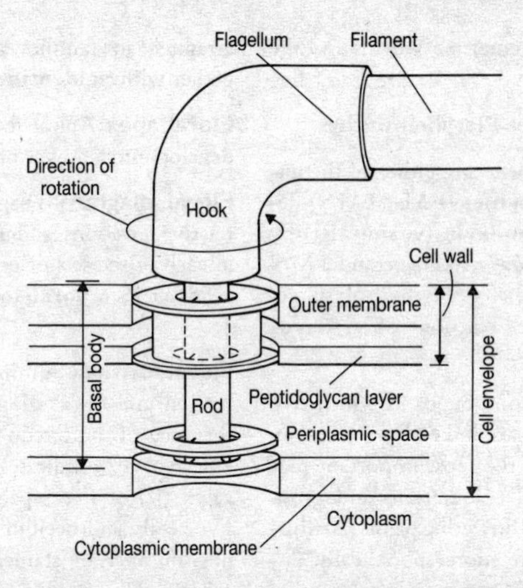

Fig. 39 *The wheel-like rotating basal body complex of the flagellum of* E. coli, *encoded by about thirty-five genes.* (Adapted from *The Sensing of Chemicals by Bacteria*, by Julius Adler. Copyright © Scientific American, Inc., 1976. All rights reserved.)

hairs occur: (a) *fibrous hairs* or *flimmer filaments* (5 nm thick), composed of glycoprotein, which are solid and wrap around the flagellum, increasing its surface area and efficiency of propulsion; (b) *tubular hairs* or *mastigonemes* (20 nm thick, 1 μm long), composed of proteins and glycoprotein, which are tripartite having a tapering base, a microtubular shaft and terminal filaments. Able to reverse the thrust of the flagellum. In addition to hairs, several types of scales, both inorganic and organic, can be found on the surfaces of flagella.

Flagella beat in wave-like undulations, unlike cilia, whose down-beat power-stroke is followed by an up-stroke offering less resistance. In some algae and fungi, flagella have a locomotory role, propelling organisms through the water, allowing them to undergo diurnal vertical phototactic rhythms. In plants, such as mosses, liverworts, ferns, cycads and ginkgophytes, and some algae, (e.g. *Chara* and *Fucus*), flagella are only found on gametes.

(2) In some prokaryotes, a hollow membrane-less filament, 3–12 mm long and 10–20 mm in diameter, composed of helically arranged subunits of the protein *flagellin*. The attachment of the flagellum is by *hook*, *bearing* and *rotor*. The flagellum is thus in the form of a fixed helix, several often rotating in unison, powered by *proton motive force* (see **bacteriorhodopsin**). Involved in chemotactic responses by the cell. See **locomotion**, **pili**.

Flame cell (solenocyte) Cell bearing a bunch of flickering flagella (hence name) and interdigitating with a *tubule cell* (which forms a hollow tube by wrapping itself around the extracellular space). Combined, they form the excretory units (*protonephridia*) of the **Platy-**

helminthes, nemertine worms and the **Entoprocta**.

Flatworms See **Platyhelminthes**.

Flavin Term denoting either of the nucleotide **coenzymes** (FAD, FMN) derived from **riboflavin** (vitamin B_2) by the enzymes *riboflavin kinase* and FMN *adenylyltransferase*. ATP hydrolysis accompanies the reactions. See **flavoproteins**.

Flavonoids Compounds in which two 6-carbon rings are linked by a 3-carbon unit. They are the most important pigments in floral colouration, and probably occur in all flowering plants (**Anthophyta**), but are more sporadically distributed among the members of other groups of vascular plants; rare in algae and animals. They function in blocking far-ultraviolet radiation, which is highly destructive to nucleic acids and proteins, usually selectively admitting blue-green and red wavelengths, important in photosynthesis. One major class of flavonoids is the **anthocyanins**. Another is the flavonols which are commonly found in leaves and flowers, and contribute to the ivory or white colours of certain flowers.

Flavoproteins A group of conjugated proteins in which one of the *flavins* FAD or FMN is bound as prosthetic group. Occur as dehydrogenases in **electron transport systems**.

Flexor Muscle or tendon involved in bending a joint; antagonizes extensors.

Flimmer See **flagellum** (1).

Flora (1) Plant population of a particular area or epoch. (2) List of plant species (with descriptions) of a particular area, arranged in families and genera, together with an **identification key**.

Floral apex Apical meristem that will develop into a flower or inflorescence.

Floral diagram Diagram illustrating relative positions and number of parts in each of the sets of organs comprising a flower. See **floral formula** and Fig. 40.

Floral formula Summary of the information in a **floral diagram**. The floral formula of buttercup (*Ranunculaceae*), $K_5C_5A\infty G\underline{\infty}$, indicates a flower with a calyx (K) of five sepals, corolla (C) of five petals, androecium (A) of an indefinite number of stamens and a gynoecium (G) of an indefinite number of free carpels. The line below the number of carpels indicates that the gynoecium is superior. The floral formula of the campanula (Campanulaceae), $K_5C_{(5)}A_5G_{\overline{(5)}}$, shows that the flower has five free sepals, five petals united () to form a gamopetalous corolla, five stamens, and five carpels united () to form a syncarpous gynoecium. The line above the carpel number indicates that the gynoecium is inferior (see **receptacle**).

Floral tube Cup or tube formed by fusion of basal parts of sepals, petals and stamens, often in flowers possessing inferior ovaries. (See **receptacle**.)

Floret One of the small flowers making up the composite inflorescence (Compositae), or the spike of grasses. In the former, *ray florets* are often female while *disc* florets are often hermaphrodite. See **gynomonoecious**.

Floridean starch Polysaccharide storage product occurring in red algae

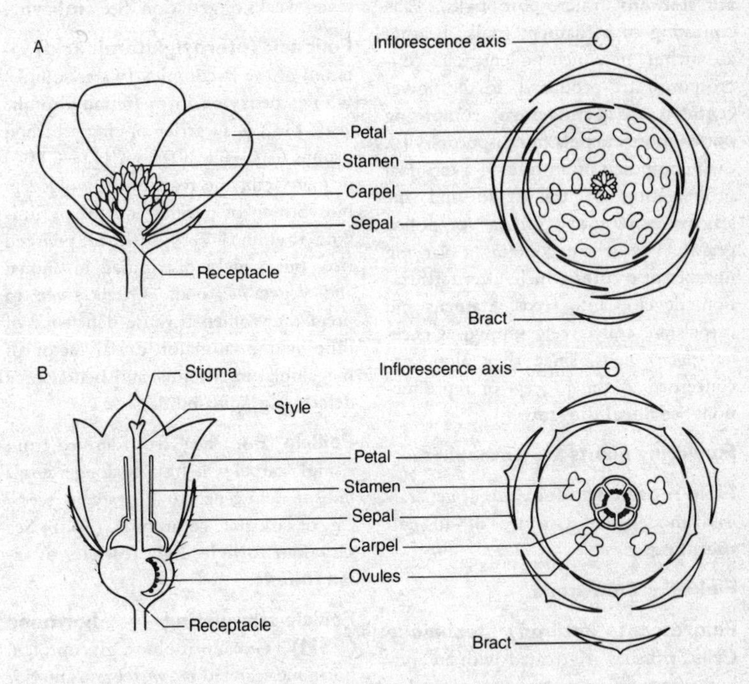

Fig. 40 *Diagrams illustrating flower structure; half-flower (median vertical section) and floral diagram (right).* ***A*** *Buttercup,* ***B*** *Campanula.*

rhodophyta); somewhat similar to amylopectin.

Florigen Hypothetical plant 'hormone' (see **growth substance**), invoked to explain transmission of flowering stimulus from leaf, where it is perceived, to growing point.

Floristics Study of composition of vegetation in terms of species (**flora**) present in a particular region. Floristics aims to account for all plants of the region, with keys, descriptions, ranges, habitats and phenology, and to offer analytical explanations of the flora's origin and geohistorical development.

Flower Specialized, determinate, reproductive shoot of flowering plants (**Anthophyta**), consisting of an axis (**receptacle**) on which are inserted four different sorts of organs, all evolutionarily modified leaves. Outermost are **sepals** (the *calyx*, collectively), usually green, leaf-like, and enclosing and protecting the other flower parts while in the bud stage. Within the sepals are petals (the *corolla*, collectively), usually conspicuous and brightly coloured. Calyx and corolla together constitute the *perianth*. They are not directly concerned in reproduction and are often referred to as *accessory flower parts*. Within the petals are **stamens** (microsporophylls), each consisting of a filament (stalk) bearing an **anther**, in which pollen grains (microspores) are produced. In the flower centre is the **gynoecium**, comprising one or more **carpels** (megasporophylls), each composed of an **ovary**, a terminal prolongation of the **style** and the **stigma**, a receptive surface for pollen grains. The ovary contains a varying number of **ovules** which, after fertilization, develop into seeds. Stamens and carpels are collectively known as *essential flower parts*, since they alone are concerned in the process of reproduction. See **floral diagram**.

Flowering plants See **Anthophyta**.

Fluid mosaic model Current generalized model for structure of all **cell membranes**.

Fluke See **Trematoda**.

Fluorescent antibody technique Cells or tissues are treated with an antibody (specific to an antigen) which has been labelled by combining it with a substance that fluoresces in UV light and can thereby indicate the presence and location of the antigen with which it combines.

FMN (flavin mononucleotide) A **flavin**; derivative of riboflavin. Prosthetic group of some **flavoproteins**.

Focal contacts See **adhesion**.

Foetal membranes See **extra-embryonic membranes**.

Foetus In mammals, the stage in intra-uterine development subsequent to the appearance of bone cells (osteoblasts) in the cartilage, indicating the onset of **ossification**. In humans, this occurs after seven weeks of gestation. See **embryo**.

Folic acid (pteroylglutamic acid) Vitamin of the B-complex (water-soluble) whose coenzyme form (tetrahydrofolic acid, FH_4) is a carrier of single carbon groups (e.g. $-CH_2OH$, $-CH_3$, $-CHO$) in many enzyme reactions. Involved in biosynthesis of purines and the pyrimidine thymine. Very little in polished rice, but widely distributed in animal and vegetable foods. Often given to pregnant women because deficiency of folic acid is a major causal factor of megaloblastic anaemia and neural tube defects (e.g. spina bifida).

Follicle (Bot.) Dry fruit derived from a single carpel which splits along a single line of dehiscence to liberate its seeds; e.g. of larkspur, columbine. (Zool.) See **Graafian follicle**, **hair follicle**, **ovarian follicle**.

Follicle-stimulating hormone (FSH) Gonadotrophic glycoprotein hormone secreted by vertebrate anterior **pituitary** gland. Stimulates growth of follicular cells of **Graafian follicles** in the ovary and formation of spermatozoa in testis. See **maturation of germ cells**, **menstrual cycle**.

Follicular phase Phase in mammalian **oestrous** and **menstrual cycles**, in which Graafian follicles grow and the uterine lining proliferates due to increasing oestrogen secretion.

Fontanelle Gap in the skeletal covering of the brain, either in the chondrocranium or between the dermal bones, covered only by skin and fascia. Present in new-born babies between frontal and parietal bones of the skull; closes at about eighteen months.

Food chain A metaphorical chain of organisms, existing in any natural community, through which energy and matter are transferred. Each link in the chain feeds on, and hence obtains energy from, the one preceding it and is in turn eaten by and provides energy for the one succeeding it. Number of links in the chain is commonly three or four, and seldom exceeds six. At the beginning of the chain are green plants (autotrophs). Those organisms whose food is obtained from green plants through the same number of links are described as belonging to the same **trophic level**. Thus green plants occupy one level (T^1), the **producer** level. All other levels are **consumer** levels: T^2 (herbivores, or primary consumers); T^3 and T^4 (secondary consumers; the smaller and larger carnivores respectively). At each trophic level, much of the energy (and carbon atoms) is lost by respiration and so less biomass can be supported at the next level. Bacteria, fungi and some protozoa are consumers that function in decomposition of all levels (see **decomposer**). All the food chains in a community of organisms make up the **food web**. See **pyramid of biomass**.

Food vacuole Endocytic vesicle produced during **phagocytosis**. See **endocytosis**.

Food web The totality of interacting **food chains** within a community of organisms. See **ecosystem**.

Foramen Natural opening (e.g. **foramen magnum** of skull, **foramen ovale** of foetal heart). Foramina in bones permit nerves and blood vessels to enter and leave.

Foramen magnum Opening at back of vertebrate skull, at articulation with vertebral column, through which spinal cord passes.

Foramen ovale Opening between left and right atria of hearts of foetal mammals, normally closing at birth (failure to do so resulting in 'hole' in the heart). While open, it permits much of the oxygenated blood returning to the foetal heart from the placenta to pass across to the left atrium, thus bypassing the pulmonary circuit (which in the absence of functional lungs is largely occluded). From there blood passes via the left ventricle to the foetal body.

Foraminifera Order of mainly marine protozoans whose shells form an important component of chalk and of many deep sea oozes (e.g. *Globigerina ooze*). Shells may be calcareous, siliceous, or composed of foreign particles. Thread-like pseudopodia protrude through pores in the shell and may or may not exhibit cytoplasmic streaming. See **Heliozoa, Radiolaria**.

Forebrain (prosencephalon) Most anterior of the three expansions of the embryonic vertebrate brain. Gives rise to *diencephalon* (thalamus and hypothalamus) and *telencephalon* (cerebral hemispheres). Also the origin of the eyestalks. Associated originally with olfaction.

Form (Bot.) Smallest of the groups used in classifying plants. Category within species, generally applied to members showing trivial variations from type, e.g. in colour of the corolla. See **infraspecific variation**. (Zool.) Used more or less synonymously with *morph*, to indicate one of the forms within a dimorphic or polymorphic species population.

Formation See **biome**.

Fossil Remains of an organism, or direct evidence of its presence, preserved in rock, ice, amber, tar, peat or volcanic ash. Animal hard parts (hard skeletons) commonly undergo *mineralization*, a process which also turns sediment into hard rock (both regarded as diagenesis). The aragonite (a form of $CaCO_3$) of molluscs and gastropods may recrystallize as the common alternative form, calcite; or it may dissolve to leave a void. This mould may then be filled later by *replacement*, involving precipitation of another mineral (possibly calcite or silica). Partial replacement and impregnation of the original hard parts in both plants and animals by mineral salts (*permineralization*) may occur, especially if the material is porous – as are wood and bone. Fossils may occur *in situ*, or else (derived fossils) be released by erosion of the rock and subsequent reposition in new sediments. Sometimes, the fossil imprints of different locomotory styles (gaits) of the same individual animal are given different taxonomic names. Fossils may be dated by various methods, and provide direct evidence for **evolution**, as well as telling us about past conditions on Earth. See **geological periods**. For fossils of mankind's ancestors, see **hominid**.

Fossorial Of animals adapted to digging, burrowing.

Founder effect Effects on a population's subsequent evolution attributable to the fact that founder individuals of the colonizing population have only a small and probably non-representative sample of the parent population's **gene pool**. Subsequent evolution may take a different course from that in the parent population as a result of this limited genetic variation. Likely to occur where colonization is a rare event, as on oceanic islands, and where the colonizer is not noted for mobility. May be combined with effects of **genetic drift**.

Fovea (Bot.) Pit in the wall of palynomorphs, such as spores, pollen, or dinoflagellate cysts. (Zool.) Depression in retina of some vertebrates, containing no rod cells but very numerous cone cells. May lie in a circular region termed the **macula**. Blood vessels absent, and no thick layer of nerve fibres between cones and incoming light as in rest of inverted retina. It is a region specialized for acute diurnal vision. Found in diurnal birds, lizards and primates, including man.

Fractionation See **cell fractionation**.

Fraternal twins Dizygotic twins who develop as a result of simultaneous fertilization of two separate ova. Such twins are no more alike genetically than other siblings. See **monozygotic twins**.

Freemartin Female member of unlike-sexed twins in cattle and occasionally other ungulates. Sterile, and partially converted towards hermaphrodite condition by hormonal (or possibly H–Y **antigen**) influence of its twin brother reaching it through anastomosis of their placentae.

Free nuclear division Stage in development in which unwalled nuclei result from repeated division of primary nucleus.

Free nuclear endosperm Endosperm in which there are many nuclear divisions without cell division (cytokinesis) before cell walls start to form.

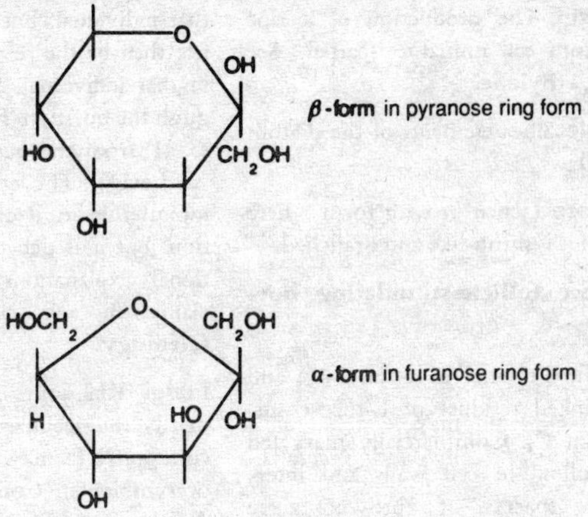

β-form in pyranose ring form

α-form in furanose ring form

Two isomers of D-fructose

Freeze drying Method of preserving unstable substances by drying when deeply frozen.

Freeze-etching Technique used in electron **microscopy** for examining the outer surfaces of membranes.

Freeze-fracture Technique used in electron **microscopy** for examining the inner surfaces of membranes.

Frequency-dependent selection Form of **selection** occurring when the advantage accruing to a character trait in a species population is inversely proportional to the trait's frequency in the population. When rare, it will be favoured by selection; when common, it will be at a disadvantage compared with alternative traits. Two or more traits determined by the same genetic mechanism (locus, or loci) may thus coexist in the population in a condition of **polymorphism**.

Frond Term applied to leaf of a fern as well as divided leaves of other plants (e.g. palm).

Frontal bone A **membrane bone**, a pair of which covers the front part of the vertebrate brain (forehead region in man). Air spaces (*frontal sinuses*) extend from nasal cavity into frontal bones of mammals.

Frontal lobe Major part of the **cerebral cortex** of the primate brain, including human's. Behind frontal bone. Has numerous connections with many parts of the brain.

Fructification Reproductive organ or fruiting structure, often used in the context of fungi, myxomycetes and bacteria.

Fructose A ketohexose reducing sugar, $C_6H_{12}O_6$. In combination with glucose, forms sucrose (non-reducing). The sweetest of sugars.

Fruit Ripened ovary of the flower, enclosing seeds.

Fruit set The production of a ripe fruit from an immature carpel. See **auxins**, **ethylene**.

Frustule Silica elements of the diatom cell wall.

Fruticose Lichen growth form where the thallus is shrub-like and branched.

FSH See **follicle-stimulating hormone**.

Fucoidin Polymer of α-1,2, α-1,3 and α-1,4 linked residues of L-fucose sulphated at C_4. Commercially marketed phycocolloid in cell walls and intercellular spaces of brown algae (**Phaeophyta**).

Fucosan vesicles Refractive vesicles, usually around the nucleus, containing a tannin-like compound in the brown algae (**Phaeophyta**). Also called *physodes*.

Fucoserraten A sexual attractant (gamone) produced by macrogamete (egg cell) in the brown algal genus *Fucus*.

Fucoxanthin Xanthophyll pigment present with chlorophylls (and other pigments) in various algal groups; e.g. **Phaeophyta**, **Chrysophyta**, **Prymnesiophyta**, **Bacillariophyta**.

Function In one sense, the function of a component in an organism is the contribution it makes to that organism's **fitness**. Therefore it may also be the ultimate reason for that component's existence in the organism, having been selected for in previous generations. This does not exclude the possibility that a component may arise by mutation in an individual and have immediate selective value (and hence function) in that individual; but its function would not then be the reason for its existence in that individual. We need to distinguish the question 'for what end does X exist?' from the question 'what caused X to exist?'. The latter requires an etiological (historical and causal) explanation; but it is debatable whether functional explanations justify etiological claims in any clear-cut way. See **teleology**.

Fungi Kingdom of eukaryotic, primarily multicellular organisms lacking chlorophyll (hence saprotrophs, parasites or symbionts). Comprise the divisions **Ascomycota**, **Basidiomycota**, **Zygomycota**, and the form-division **Deuteromycota**. Traditionally, fungi have also included heterotrophic protists (**Oomycota**, **Chytridiomycota**); however, little evidence exists for direct relationship. No protist group seems ancestral to fungi. The oldest fossils resembling fungi occur in strata about 900 Myr old (among the oldest eukaryotes); the oldest identified with certainty are from the **Ordovician** period (500–450 Myr BP). The first fungus was probably unicellular, from which coenocytic forms evolved. The organizational unit in fungi is the *hypha* (matted to form a *mycelium*); the most complex structures are reproductive bodies involved in spore production. With oomycotes and chytrids excluded from the kingdom, no motile cells (e.g. **zoospores**) are formed at any stage in the life cycle. Cell walls are composed primarily of **chitin**. Glycogen is the primary storage polysaccharide; lipids also serve an important storage function. Saprotrophic and parasitic fungi often form specialized hyphae such as *rhizoids*, or *haustoria*, both absorbing nutrients. Some fungi have complicated life cycles involving several spore types, and peculiar genetic mechanisms (e.g. **clamp connection**, **crozier formation**, **dikaryon**). See **myxomycota**, **yeasts**.

Fungicide A compound destructive to fungi.

Fungi imperfecti See **Deutero** mycota.

Funiculus (Bot.) Stalk attaching ovule to the placenta in an ovary.

Furcula See **wishbone**.

Fusiform initial (Bot.) Meristematic cells of the vascular cambium. Vertically elongated cells, much longer than their width, appearing flattened or brick-shaped in transverse section. Produce xylem and phloem cells having their long axes orientated vertically. Comprise the axial system of the secondary vascular tissues. See **ray initials**.

G_0, G_1, G_2 phases See **cell cycle**.

GABA (γ-aminobutyric acid) Inhibitory **neurotransmitter** in the (human) brain, potentiated by benzodiazepine tranquillizers. Derivative of **glutamic acid**. See **synapse**.

G-actin See **actin**.

Gain-of-function mutation See **mutation**.

Galactose An aldohexose sugar; constituent of **lactose**, and commonly of plant polysaccharides (many gums, mucilages and pectins) and animal **glycolipids** and **glycoproteins.**

Gall bladder Muscular bladder arising from **bile duct** in many vertebrates, storing bile between meals. Bile is expelled under influence of the intestinal hormone **cholecystokinin.**

Gametangial contact Form of **conjugation** in which, following growth and contact of the gametangia, nuclei are transferred from the antheridium through a fertilization or copulation tube; e.g. in Oomycete fungi.

Gametangium (Bot) Gamete-producing cell; most commonly in the contexts of algae and fungi. However, more complex antheridia, oogonia and archegonia are sometimes cited as examples too. Compare **sporangium**.

Gamete (germ cell) Haploid cell (sometimes nucleus) specialized for **fertilization**. Gametes which so fuse may be identical in form and size (*isogamous*) or may differ in one or both properties (*anisogamous*). The terms 'male' and 'female' are often applied to gametes, but serve only to indicate the sex of origin, for gametes do not have sexes. Where they differ in size it is customary to refer to the larger gamete as the *macrogamete*, and to the smaller as the *microgamete*. Sometimes plasmogamy is absent in fertilization, in which case the nuclei which fuse may be regarded as gametes. See **autogamy**, **automixis**, **maturation of germ cells**, **ovum**, **parthenogenesis**, **sperm**.

Game theory In biology, denotes all approaches to the study of 'decision making' by living systems (usually lacking conscious overtones) in which an organism's responses to its conditions are viewed as *strategies* whose (evolutionary) goal is maximization of the organism's **fitness**. Often convenient to regard each organism as having at any time a decision procedure for responding to future circumstances in such a way as to maximize any possible payoff to itself while minimizing the payoffs to others (but see **inclusive fitness**). Decision procedures which cannot be superseded by rival procedures will be **evolutionarily stable strategies.** See **arms race**, **optimization theory**.

Gametocyte Cell (e.g. oocyte, spermatocyte) undergoing meiosis in the production of gametes. Primary gametocytes undergo the first meiotic division; secondary gametocytes undergo the second meiotic division. See **maturation of germ cells**.

Gametogenesis Gamete production. Frequently, but by no means always, involves **meiosis**. In eukaryotes there are often haploid organisms and stages in the life cycle where gametes can only be produced by **mitosis**. See **spermatogenesis, oogenesis**.

Gametophore In bryophytes, a fertile stalk bearing gametangia.

Gametophyte In plants showing **alternation of generations**, the haploid (n) phase; during it, gametes are produced by mitosis. Arises from a haploid spore, produced by meiosis from a diploid **sporophyte.** See **life cycle.**

Gamma globulins (immune serum globulins) Class of globular serum proteins. Includes those with **antibody** activity, and some without.

Gamone Compound involved in bringing about fusion of gametes in some brown algae (Phaeophyta); e.g. ectocarpan (in *Ectocarpus*), multifidin and aucanten (in *Cutleria*), fucoserraten (in *Fucus*).

Gamopetalous (sympetalous) (Of a flower) with united petals; e.g. primrose. Compare **polypetalous**.

Gamosepalous (Of a flower) with united sepals; e.g. primrose. Compare **polysepalous**.

Ganglion Small mass of nervous tissue containing numerous **cell bodies** with synapses for integration. **Central nervous systems** of many invertebrates contain many such ganglia, connected by nerve cords. In vertebrates the CNS has a different overall structure, but ganglia occur in the peripheral and **autonomic nervous systems**, where they may be encapsulated in connective tissue. Some of the so-called nuclei of the vertebrate brain are ganglia.

Ganglioside Type of glycolipid common in nerve cell membranes.

Ganoid scale Scale characteristic of primitive **Actinopterygii**. Outer layer is hard inorganic enamel-like ganoine, thicker than in otherwise similar **cosmoid scale**. Grows in thickness by addition of material both above (ganoine) and below (laminated bone). Found today in e.g. *Polypterus*, *Lepisosteus* and sturgeons.

Gap genes A class of *Drosophila* segmentation genes (e.g. *Krüppel* (*Kr*), *hunchback* (*hb*) and *knirps* (*kni*)), mutants of which delete several adjacent segments and create gaps in the anterior-posterior pattern. As a result of signals laid down by maternal genes in oogenesis (see ***bicoid* gene**, ***oskar* gene**) they are the first zygotic genes to be expressed in *Drosophila* development. Most encode **transcription factors** whose transient gradients constitute **positional information** in the embryo and determine patterns of expression of **homeotic genes** and other patterning genes (e.g. **pair-rule genes** and **segment-polarity genes**), although mutual interactions complicate the picture. The *Drosophila* gap gene *knirps* encodes a hormone receptor-like protein of the steroid-thyroid superfamily essential for abdominal segmentation, similar to ligand-dependent DNA-binding proteins of vertebrates (see **receptor**

proteins). The *Krüppel* product (K_r) and *hb* product (Hb) bind to different DNA sequences upstream of the two *hb* promotors.

Gap junction See **intercellular junction**.

GAPs (GTPase-activating proteins) Proteins stimulating the GTP-hydrolysing activity of specific GTPases (e.g. **Ras protein**-related GTPases). The GTP p190 appears to be involved in coupling signalling pathways involving Ras and Rho. See **rho proteins, GEFS G-protein, receptor**.

GAP protein See **GAPs**.

Gas bladder (swim bladder, air bladder) Elongated sac growing dorsally from anterior part of gut in most of the **Actinopterygii**. In fullest development (in **Acanthopterygii**) acts as hydrostatic organ; but may also act as an accessory organ of gaseous exchange, as a sound producer, or as a resonator in sound reception. Opinions differ as to whether the gas bladder or the vertebrate lung is the ancestral structure; they are certainly homologous.

Gastric Of the stomach. *Gastric juice* is a product of vertebrate *gastric glands*, and contains hydrochloric acid, proteolytic enzymes and mucus.

Gastrin Hormone secreted by mammalian stomach and duodenal mucosae in response to proteins and alcohol. Stimulates gastric glands of stomach to secrete large amounts of gastric juice. Relaxes pyloric sphincter and closes cardiac sphincter. Oversecretion may result in gastric ulcers. See **secretin**, **cholecystokinin**.

Gastropoda Large class of the **Mollusca**. Marine, freshwater and terrestrial. Head distinct, with eyes and tentacles; well-developed, rasping tongue (radula). Foot large and muscular, used in locomotion. Visceral hump coiled, and rotated on the rest of body (torsion) so that the anus in the mantle cavity points forward; some forms undergo a secondary detorsion. Visceral hump commonly covered by a single (univalve) shell. Often a trochosphere larva. Includes subclasses Prosobranchia (e.g. limpets), Opisthobranchia (e.g. sea hares) and Pulmonata (e.g. snails, slugs).

Gastrotricha Class of the **Aschelminthes** (or a phylum in its own right), probably closely related to nematode worms. Composed of a small number of cells, these minute aquatic invertebrates have an elastic cuticle but unlike nematodes have a ciliated but acellular hypodermis. Hermaphrodite or parthenogenetic. No larval stage. See **Rotifera.**

Gastrula Stage of embryonic development in animals, succeeding **blastula,** when the primary **germ layers** are laid down as a result of the morphogenetic processes of **gastrulation.**

Gastrulation Phase of embryonic development in animals during which the primary **germ layers** are laid down; its onset is characterized by the morphogenetic movements of cells, typically through the **blastopore**, forming the **archenteron**. Movements may result in **epiboly**, but frequently also *emboly* in which cells invaginate, involute and ingress.

Gas vacuole Structure aiding dispersal/buoyancy comprising gas vesicles (hollow, cylindrical tubes with conical ends) found in the cytoplasm of all orders of **Cyanobacteria** (blue-green algae), except the Chamaesiphonales. Gas vacuoles do not possess a true protein-lipid membrane but rather protein-ribs or spirals arranged like the hoops of a barrel, a form of membrane which is quite rigid (gas inside is at 1 atmosphere pressure) and permeable to gases. Inner membrane surface is hydrophobic, preventing condensation on it of water droplets, restraining them by surface tension, water seeping through the pores. The outer surface is hydrophilic to maximize interfacial tension, which would otherwise collapse the gas vesicle. It functions in light shielding and/or buoyancy.

Gated channels **Transport proteins** of membranes, not constitutively (permanently) open to the passage of molecules, but capable of closure. *Ligand-gated channels*, such as those responding to **neurotransmitters**, open only in response to an extracellular ligand; *voltage-gated channels* (e.g. the **sodium pump** of nerve and muscle fibres) are dependent for opening and closure upon an appropriate membrane potential. Others may only open when concentrations of certain ions in the cell are appropriate. See **impulse**, **muscle contraction**.

GEFs (guanine-nucleotide-exchange factors) Proteins activating **Ras proteins** through exchange of bound GDP for free GTP.

Gel Mixture of compounds, some commonly polymeric, having a semisolid or solid constitution. See **chromatography**, **electrophoresis**.

Gel electrophoresis See **electrophoresis**.

Gel filtration See **chromatography**.

Gemma Organ of vegetative reproduction in mosses, liverworts and some fungi. Consists of a small group of cells of varying size and shape that becomes detached from the parent plant and develops into a new plant; often formed in groups, in receptacles known as gemmae-cups.

Gemmule (1) Of sponges, a bud formed internally as a group of cells, which may become free by decay of the parent and subsequently form a new individual. Freshwater sponges overwinter in this way. (2) See **pangenesis**.

Gene Usually regarded as the smallest physical unit of heredity encoding a molecular cell product; commonly considered also to be a **unit of selection**. The term gene (coined by W. Johannsen in 1909) may be used in more than one sense. These include: a) **allele**, b) **locus**, and c) **cistron**. What **Mendel** treated as algebraic units ('factors') or 'atoms of heredity' obeyed his laws of inheritance and were considered to be the physical determinants of discrete *phenotypic* characters. This may be called the classical gene concept (see **genetics**). In 1903, W. S. Sutton pointed out that the segregation and recombination of Mendelian factors studied in heredity found a parallel in the behaviour of chromosomes revealed by the microscope. Through the work of T. H. Morgan in the period 1910–20, chromosomes came to be regarded as groups of linked genes (or, more abstractly, of their loci), and the positions of loci and their representative alleles were first mapped on the chromosomes of *Drosophila* in this period. Morgan found that alternative genes (alleles) at a locus could mutate from one to another.

The importance of genes in enzyme production first emerged through work on the chemistry and inheritance of eye colour in *Drosophila*, and through work on **auxotrophic** mutants of the mould *Neurospora crassa* by G. W. Beadle and E. L. Tatum (1941). It heralded the modern phase of genetics and molecular biology.

Genes soon became accepted as the heritable determinants of enzymes (one gene: one enzyme). However, the correspondence between the nucleotide composition of a gene and the amino acid composition of its encoded product was first revealed in variants of haemoglobin (a non-enzymic protein) and its genes, and their precise sequential correspondence (*colinearity*) was first established in detailed studies of the bacterial enzyme *tryptophan synthetase* and its gene. Great precision was by now being achieved in fine genetic mapping of **bacteriophage** chromosomes using the ***cis*-trans test**, and genes were soon regarded as nucleic acid sequences, mappable geographically on a chromosome, each encoding a specific enzyme, or (as in the polypeptide subunits of haemoglobin) non-enzyme protein. Subsequent work on *tryptophan synthetase* of the bacterium *E. coli* showed that two genes were required to encode this enzyme, and that their different polypeptide products associated to give the *quaternary structure* of the functional enzyme (see **protein**). The *functional gene concept* thus denoted a nucleic acid sequence encoding a single polypeptide chain. Nowadays, the term 'gene' is used to indicate the length of nucleic acid encoding any molecular cell product, be it a polypeptide, transfer RNA or ribosomal RNA molecule, and can usually be equated with **cistron** (but see **allelic complementation**). Class I genes (in the nucleolar organizer) encode the large rRNAs and 5.8S RNA; Class II genes include those encoding polypeptides and some snRNAs (see **RNPs**); Class III genes encode tRNA, 5S RNA and some snRNAs. The initial RNA transcripts of all three classes usually undergo **RNA processing**, Class II transcripts in particular having their **exons** spliced out. The manner in which these get excised from pre-RNA often dictates the precise cell product encoded by the gene, particularly when an intron contains a stop codon. This need not cause serious problems for gene definition, since just one promoter is usually employed – even though the final gene products may differ. But occasionally, true gene overlap does occur (see **genetic code**) through a shift in reading frame. One stretch of DNA (or RNA) may then be part of two genes; but even this leaves the functional gene concept unscathed. When specified by name, genes are given italics (e.g. *cdc2*, *osk*, etc.); but their protein products may be given first-letter or full capitals and are unitallicized. See **chromosome**, **protein synthesis** and genetic references below.

Gene amplification Process in which a small region of the **genome** of a cell is selectively copied many times while the rest remains unreplicated. Occurs in some specialized cell lines where large quantities of a particular cell product are needed rapidly. In rRNA cistrons, up to 1000 extra nucleoli may arise in

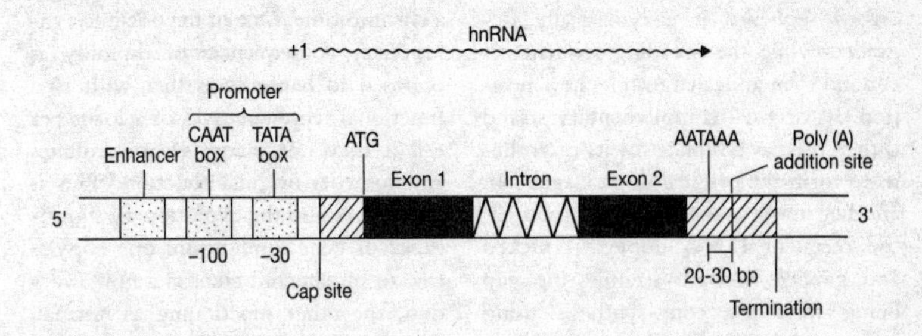

Fig. 41 *Generalized diagram of a eukaryotic Class II gene with a single intron. Enhancers may occur within or downstream of the gene. Transcription factor binding sites may also occur at the 5′-end, and occasionally the CAAT and TATA consensus sequences are absent. The region encoding hnRNA is indicated above. The cap site indicates where post-transcriptional modification occurs to the hnRNA by addition of a 'cap' (see* **RNA capping**).

amphibian oocytes in this way, with consequent large-scale ribosome production. Cistrons for rRNA are amplified in all cells with nucleoli. Gene amplification is associated with some kinds of dry resistance in cell cultures; amplification of cellular **oncogenes** is a fairly common feature of tumour cells. The whole phenomenon of selective DNA amplification is rich in theoretical interest. A powerful method of gene amplification *in vivo* (and of selective DNA in general) involves the *rolling circle*, whereby an extrachromosomal circular copy of the DNA sequence is produced which in turn produces many copies containing tandem repeats of the sequence. If these linearize and re-integrate into the chromosome, the enlarged genome will contain these identical repeat sequences, with consequent increase in copy number. Gene amplification is an excellent marker of genetic instability and commonly occurs in **neoplasms** (see **tumour necrosis factors**). See **gene duplication**, **nucleolus**, **polymerase chain reaction**, **polyteny**.

Gene bank Alternative term for **gene library**.

Genecology Study of population genetics with particular reference to ecologies of populations concerned.

Gene conversion Phenomenon, in eukaryotes occurring mainly at synapsis during meiosis, whereby a donor DNA sequence, a few hundred bases or perhaps a kilobase in length, is transferred from one gene to another having substantial sequence homology (usually between homologous loci, but sometimes between related sequences at non-homologous loci, notably those of dispersed **multigene families**). The donor sequence is repaired back to its original form. It may be responsible for much of the diversity in some mammalian immunoglobulin production (see **antibody diversity**). In one model this involves 'nicking' (cutting) of a single-strand invading DNA sequence and melting (unzipping) of the invaded duplex DNA so that *heteroduplex* base pairing between the two can occur. The ousted sequence is enzymatically degraded while the invading sequence is cut and then annealed into its new position. Its original complementary strand is then used as template for its resynthesis to form the original duplex again. In another model, increasingly favoured, the recipient DNA duplex is nicked and gapped in both strands, the gap being filled by copy-synthesis using both strands of the donor duplex as templates. This would generate heteroduplexes only in regions flanking the gap. In a heterozygote of the yeast *Saccharomyces*, where conversion occurs at a rate of several per cent per gene per meiosis, each allele can usually convert the other with about equal frequency, but examples of strongly biased conversion are known which could, in principle, lead to fixation of the favoured allele (see **mating type**).

Initiation of the cutting, and hence of the recombination, seems in one form of the process to occur within a gene promoter region. There is growing support for the view that the sites of heteroduplex formation (*Holliday junctions*) are responsible for much of eukaryotic crossing-over: there is about 30 – 50% association of gene conversion with crossing-over. Gene conversion in prokaryotes involves similar processes, although the initial alignment of homologous duplexes is less highly organized (see ***recA***, ***recB*** and ***recC***). See **recombination**.

Gene dosage Effective number of copies of a gene in a cell or organism. See **dosage compensation**, **gene amplification**.

Gene duplication Mechanisms resulting in tandem duplication of loci along a chromosome. One of the possible evolutionary consequences of diploidy as opposed to haploidy is that with two functional representatives of a locus per cell it may not matter if one mutates and loses its original function. This is very likely also the evolutionary significance of gene duplication: one copy is free to mutate and take on a new function, the other functioning as normal. The enzymes of the glycolytic pathway may have arisen this way from a common ancestral gene sequence, as most certainly do the various types of globin in haemoglobin. Controversy surrounds whether random fixation or positive selection of mutations occurs in the duplicated gene(s). At least one piece of evidence indicates selection operates. Non-homologous **crossing-over** is one mechanism for producing gene duplication. See **multigene families**, **gene amplification**, **duplication**.

Gene expression/gene regulation The effect of those mechanisms which dictate whether or not a particular genetic element is transcribed (acts as a template for mRNA synthesis) at any particular time. In prokaryotes it may best be explained by some variant of the **Jacob-Monod theory**; in eukaryotes too this theory may find application, although different **chromosome** structure here raises fresh problems. It is common in eukaryotes to recognize two classes of regulatory phenomena involving gene expression: short-term (reversible) regulation, and long-term (often irreversible) regulation. Short-term regulation often relates to a cell's production of inducible and repressible enzymes. Steroid hormones ('effectors', see **ecdysone**) frequently bind to receptor proteins (see **nuclear receptors**) in the cell prior to entry into the nucleus and activate transcription of selected genes (see **transcription factors**). Long-term eukaryote regulation includes those processes involved in: a) rendering a cell **determined**, prior to differentiation, b) **maternal effects**, c) the origins of facultative and constitutive **heterochromatin**. The precise roles of **nucleosomes** and of nucleosome-free regions of eukaryotic chromosomes in the control of gene expression have still to be clarified. See **chromosomal imprinting**, **DNA methylation**, **enhancers**.

Gene flow The spread of genes through populations as affected by movements of individuals and their propagules (e.g. spores, seeds, etc.), by **natural selection** and **genetic drift**. See **vagility**, **panmixis**.

Gene frequency Frequency of a gene in a population. Affected by **mutation**, **selection**, emigration, immigration and **genetic drift**. See **Hardy-Weinberg equilibrium**.

Gene library (gene bank) Term given to the collection of DNA fragments resulting from digestion of a genome by a **restriction endonuclease**. Each fragment is clonable by inserting it into an appropriate phage **vector** and introducing it into an appropriate host cell (e.g. *E. coli*) for copying. cDNA libraries are libraries of those genes being transcribed in a system. See **chromosome walking**.

Gene manipulation (gene modification) Set of procedures by which selected pieces (genes) of one genome (e.g. human) can be enzymatically cut out from it, spliced into a vector (e.g. a **plasmid**) and inserted into an appropriate host microorganism (e.g. *E. coli*, yeast, etc.) in which it is replicated and passed on to all daughter cells of the microorganism forming a clone. A cloned gene may then be modified by site-directed mutagenesis (see **mutagen**). Various enzymes used include **restriction endonucleases**, **DNA ligases**, etc. If the aim is mass production of the substance encoded by the transposed gene, then insertion is so arranged that transcription of the gene occurs within the clone of cells, producing large amounts of the desired substance (e.g. insulin). Clones can be screened for their activities and selected appropriately. The recombinant DNA thus artificially produced might be hazardous unless properly contained, and strict precautions are applied in such work. Thus only non-pathogenic strains are used as hosts, or else strains that can only grow in laboratory conditions. See **biotechnology**, **gene library**, **vector**.

Gene overlap See **genetic code**.

Gene pool Sum total of all genes in an interbreeding population (gamodeme) at a particular time. See **deme**.

Generative cell (Bot.) In the pollen grain, the cell of the male **gametophyte** which divides mitotically to produce two generative nuclei (gametes). See **double fertilization**.

Generative nucleus See **double fertilization, generative cell.**

Generator potential (**receptor potential**) Initial depolarization of the membrane of an excitable cell (receptor, nerve or muscle) by stimulus or transmitter, which triggers an **action potential**

when threshold depolarization is reached.

Generic (Adj.) Of **genus**.

Genet The genetic individual. Particularly employed in the context of vegetative (clonal) reproduction and growth. Compare **ramet**.

Gene targeting Techniques enabling us to create mutations in (i.e. place foreign DNA in) specific genes at will. Used to inactivate (create **null mutation**) in mammalian genes, often by **homologous recombination** using a targeting vector containing a disrupted gene construct; but far easier to achieve in yeast, with its smaller genome size and simpler recombination machinery. Valuable tool in study of gene function. Introduced into animal **embryonic stem cells**, they could be transmitted to the germ line and hence to the gene pool. In the future, *gene transplacement* might be possible in mammals, a gene being precisely replaced by an engineered alternative (currently possible with yeasts).

Gene therapy See **human gene therapy**.

Genetic Concerned with genes, or their effects. Compare **hereditary**.

Genetic assimilation Phenomenon involving conversion of an acquired character (resulting maybe from transference of individuals from one environment to another) into one with greater **heritability** than it had before, where the causal mechanism involved is selection acting on the genotypes of the transferred population. The significance of the process in evolutionary terms is debatable: there is no assurance that the initial acquired character will be adaptive in the conditions bringing it about. See **mutation**, **phenocopy**.

Genetic code Table of correspondence between (a) all possible triplet sequences (codons) of messenger RNA and (b) the amino acid which each triplet causes to be incorporated into protein during **protein synthesis**. In addition, certain triplets cause termination of polypeptide chain synthesis and occur regularly at the 3′-ends of polypeptide-encoding sequences (open reading frames). They may also arise as motivating mutations within an encoding sequence (see **codon**). DNA is sometimes spoken of as a 'code', but this is shorthand. Indeed, the genetic code itself is really a *cipher*, since the amino acids in a polypeptide correspond to the letters of an alphabet rather than to words. Since more than one triplet may encode some amino acids, the code is said to be degenerate. The translation of mRNA into protein is only possible because of the specificity of transfer RNA molecules for particular amino acids – itself the result of specificity of the enzymes which activate amino acids and bind them to appropriate tRNAs.

The code is remarkably uniform from prokaryotes to eukaryotes; but in mitochondria there are fewer codons and slightly different reading rules; in *Mycoplasma capricolum* the codon UGA is read as tryptophan rather than as a stop codon. Until the 1970s it was thought that the reading frame for translating mRNA into polypeptide never overlapped: that there were unequivocal initiation sites for any mRNA molecule. However, overlapping reading frames (overlapping genes) are widespread in viruses (e.g. φX174), or-

1st position (5′ end) ↓	2nd position U	C	A	G	3rd position (3′ end) ↓
U	Phe	Ser	Tyr	Cys	U
	Phe	Ser	Tyr	Cys	C
	Leu	Ser	STOP	STOP	A
	Leu	Ser	STOP	Trp	G
C	Leu	Pro	His	Arg	U
	Leu	Pro	His	Arg	C
	Leu	Pro	Gln	Arg	A
	Leu	Pro	Gln	Arg	G
A	Ile	Thr	Asn	Ser	U
	Ile	Thr	Asn	Ser	C
	Ile	Thr	Lys	Arg	A
	Met	Thr	Lys	Arg	G
G	Val	Ala	Asp	Gly	U
	Val	Ala	Asp	Gly	C
	Val	Ala	Glu	Gly	A
	Val	Ala	Glu	Gly	G

Fig. 42 *Diagram of the amino acids encoded by* **RNA** *triplets (codons). The triplets run from the 5′-end to the 3′-end of the* **RNA**. *The three nucleotides for any triplet are found by taking one from the left column, one from the horizontal row and one from the right column. The amino acid indicated at their intersection is that encoded by that triplet. Thus CCC encodes proline. Three stop codons are also indicated.*

ganelles and bacteria; both same-strand and complementary strand overlaps exist. See **wobble hypothesis**.

Genetic counselling Service, generally provided by specialists in human genetic disorders. Seeks to explain to parents with children already affected by genetic disorders the nature of those disorders, and the probability of their (and their children) having further affected offspring, and helps families to reach decisions and take appropriate action in the light of this information. Many carriers of genetic disorders (i.e. heterozygous for the condition, but not expressing it) can be diagnosed through screening procedures (see **RFLP**, **polymerase chain reaction**); e.g., several inherited mutations predisposing patients to cancer can now be detected early in life. The probability of someone being a carrier can often be ascertained from information about the occurrence of the disorder in close relatives and, to an increasing extent, by DNA analysis. See **amnion**, **chorion**, **human gene therapy**.

Genetic drift (Sewall Wright effect) Statistically significant change in population gene frequencies resulting not from selection, emigration or immigration, but from causes operating randomly with respect to the fitnesses of the alleles concerned. Such random sampling error might for example occur if, in a population of beetles, a disproportionately large number of those killed by a wandering elephant happened to be heterozygous for a recessive eye colour. The frequency of this, and maybe other, alleles could now alter significantly even though no selection had taken place. Genetic drift is expected to be of significance only in small populations, where alleles may easily go to extinction or fixation by chance alone. In large populations, effects of sampling error are usually considered to be small in comparison with those of **natural selection**. See **founder effect**.

Genetic engineering See **gene manipulation.**

Genetic fingerprinting See **DNA fingerprinting**.

Genetic immunization See **vaccines**.

Genetic load The proportionate decrease in average fitness in a population relative to that of the fittest or optimal genotype, as caused principally by mutation (mutational load) and segregation (segregational load). Mutational load is the degree to which mutation impairs the average fitness of a population. J. B. S. Haldane (1937) argued that when all loci are considered, this equals the gametic mutation rate multiplied by a factor of 1 or 2 (depending on dominance) and that since partial dominance is usual the mutation load is twice the total rate per gamete. So, for mutations with large heterozygous effect, if the total mutation rate per gamete for these loci is 0.01, population fitness would be lowered by 0.02. However, adjustments have to be made to allow for mutations of minor effect. Segregational load is the loss of fitness due to segregation in polymorphic populations, where genes are maintained by **heterozygous advantage** and, as in the case of **sickle-cell anaemia**, may be high. An alternative view is that many or most polymorphisms involve selectively neutral allelic differences (see **neutralism**).

Genetic mapping See **chromosome mapping**.

Genetic marker Genes, DNA sequences, heterochromatic regions or any other distinguishing feature of genotypes, chromosomes or karyotypes which may be used to keep track of specific chromosomes, cells or individuals. Almost by definition they must exist in at least two alternative and (preferably) readily identifiable forms; when the rarest has a frequency of 1–2%, the marker is described as polymorphic (see **polymorphism**). Genetic markers enabled **Mendel** to score discrete phenotypes and develop his theories of segregation and random assortment and provided subsequent workers with the tool to develop the principles of **linkage** and the chromosomal theory of inheritance. Markers are now routinely used to identify transgenic cells or organisms carrying a particular **vector**. Expression of the marker gene, carried on the same vector as another introduced gene of unrelated but useful effect, indicates which cells or organisms also harbour this other gene. If the marker gene carries drug resistance, positive selection of the cells can be achieved by administration of the drug. Marker genes closely linked to as yet unisolated genes for human genetic disorders are of enormous help in identifying cells, embryos and individuals carrying these deleterious alleles. Once isolated, these genes will be locatable by **DNA probes**. Other markers include **RFLPs** and some **satellite DNA**s. In developmental biology, cell markers are frequently used. These are cells whose phenotype indicates the occurrence of some genetic event, e.g. mitotic recombination or onset of gene expression, as a guide to some important developmental feature (e.g. a compartment boundary). Good cell markers do not damage their cells and are autonomous: their expression should be restricted to the cells carrying them. They should not, e.g., cause other cells to express the marker phenotype simply through leakage of product.

Genetics Study of heredity and variation in biological systems. The origin of the modern, particulate, theory of inheritance is marked by the work of **G. Mendel**, with many other contributors to this classical phase of genetics in which phenotypic ratios in breeding tests were ultimately explained in terms of chromosome behaviour. Post-classical work, largely with microorganisms and phages, directed attention towards a biochemical understanding of genetics, isolation of DNA and determination of its structure, eventually enabling **gene manipulation**. Population (ecological) genetics attempts to quantify the roles of selection and genetic drift in shaping the **genetic variation** within populations. See **gene**.

Genetic variation Occurrence of genetic differences between individuals, most commonly studied in species populations. Upon such differences, when expressed, can natural selection act. **Mutation** is the ultimate source of genetic variation; but in most sexual populations **meiosis** then results in recombination both between and within parentally-derived chromosomes, generating enormous genetic diversity among gametes and, at least potentially, among offspring phenotypes. The **breeding system** of the population is an important consideration, affecting the relative levels of free and potential variability. *Free variability* is genetic variation which is expressed phenotypically and approximates to the proportion of homozygotes in the population; *potential variability* is genetic variation which does not express itself phenotypically and approximates to the proportion of heterozygotes in the population. See Fig. 43.

However, two or more loci often affect the same character (polygenic inheritance), and in the simplest case (where both loci are additive with respect to effect on phenotype) may generate the same phenotypes in the double heterozygote, GgHh, as in both the double homozygotes, GGhh and ggHH, thus 'protecting' the latter from selection, even though their *homozygous potential variation* contributes to the total

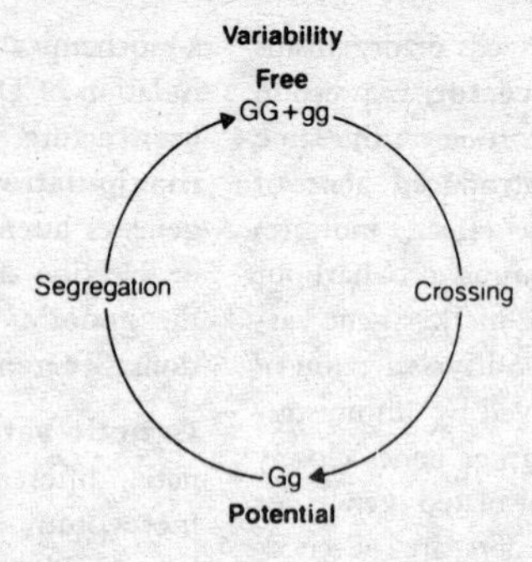

Fig. 43a *Free and potential variability. Free variability (open to direct selection) is represented by differences in phenotype between GG and gg genotypes and is converted to potential variability of Gg by crossing and is released again by segregation.*

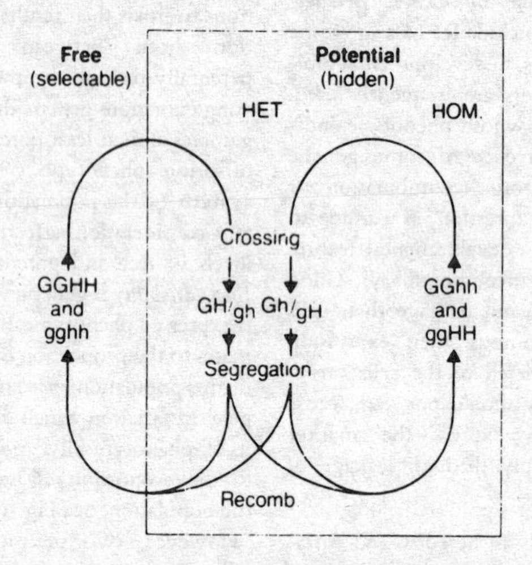

Fig. 43b *The states of variability in a system where the G and H loci are additive in effect on phenotype, so that GgHh and GGhh are phenotypically identical. The extremes of phenotype (free variability) are expressed by the double dominant and double recessive.*

potential variation in the population. Alleles always exert their effects against a 'genetic background' which can modify their expressions.

Only through crossing, with its subsequent segregation, can a major part of the potential variability in a population be freed and become available for selection. Inbred populations, comprising all possible types of homozygotes, will harbour almost as much potential variability as outbred ones with the same genes in the same frequencies; but this variability is never freed in inbred populations and they tend to be evolutionarily static. See **variation**, **polymorphism**, **heritability**.

Genitalia (Zool.) External reproductive organs. In many arthropods, especially insects, structures of male and female genitalia are often species-specific and serve as a prezygotic mechanism preventing hybridization.

Genome The total genetic material within a cell or individual, depending upon context. A bacterial genome comprises a circular chromosome containing an upper limit of about 0.01 pg (1 picogram = 10^{-12} g) of DNA; a haploid mammalian genome contains from about 3 to 6 pg of DNA, while some amphibia and psilopsid plants may contain well over 100 pg of DNA per haploid genome. The DNA content may or may not correlate with chromosome number, the fit being generally better in plants than in animals. Much eukaryote DNA is not part of structural genes coding for detectable cell products. The evolution of genome size is much debated. See **C - value.**

Genome mapping See **human genome project** for some of the methods being employed there. It should be remembered that two sorts of map are available: genetic maps based largely on linkage data (see **chromosome mapping**) and physical maps based largely on cloned DNA segments (see **vector**, **yeast artificial chromosomes**).

Genomic imprinting The heritable modification of expressivity and/or **penetrance** of genes in offspring determined by sexual provenance – viz., which parental sex they are inherited from. Maternally derived genes experience oogenesis whereas paternally derived genes experience spermatogenesis; consequently, maternal and paternal diploid embryos are not equivalent (diploid embryos with two egg or two sperm genomes generally fail to develop). Can be both developmentally stage-specific and tissue-specific. Implicated in an increasing number of human genetic disorders, but the molecular mechanisms are only just becoming clear. Sometimes (at least) involves **DNA methylation** (more methylation, less expression). See **chromosomal imprinting**.

Genotype Genetic constitution of a cell or individual, as distinct from its **phenotype**.

Genus Taxonomic category, between **family** and **species**, which may include one or more examples of species. See **binomial nomenclature**, **classification**.

Geological periods, epochs See Table 5. The fossil record commences in pre-Cambrian times (see **Archaean, Proterozoic)** with organisms resembling bacteria and blue-green algae (see **stromatolites**) in deposits 3 billion years old. A few green algae have been identified from an upper Precambrian formation about 1 billion years old. In Cambrian rocks, in addition to algae and bacteria, a diverse range of aquatic invertebrate animals is found (e.g. brachiopods, trilobites and the onychophoran *Aysheaia*), including many which apparently left no descendants. Fungi and spores with wall markings like those of various land plants were also present. In the Ordovician period there was

Geological periods, epochs

Era	Period	Epoch	*Millions of years (Myr) since start of period*
Cenozoic	Quaternary	Holocene	(11,000yr)
		Pleistocene	2–3
	Tertiary	Pliocene	12
		Miocene	25
		Oligocene	40
		Eocene	57
		Palaeocene	70
Mesozoic	Cretaceous		135
	Jurassic		195
	Triassic		225
Palaeozoic	Permian		270
	Carboniferous		350
	Devonian		400
	Silurian		440
	Ordovician		500
	Cambrian		600

Table 5 *Main fossil-bearing geological periods and approximate time-scales since the beginnings of the periods.*

sufficient land vegetation to support such animals as burrowing worms, while the first fish-like vertebrates (ostracoderms) appeared. In this and the Silurian period, land vegetation seems to have included plants with bryophyte-like attributes, while primitive land vascular plants appeared and expanded in the late Silurian and early Devonian (420–390 Myr BP). Terrestrial arthropods have been found in early Silurian (Pennsylvania) and Upper Silurian of England (411 Myr BP): the arthropod invasion of land occurred prior to the Devonian radiation of vascular plants. But by the late Devonian both insects and vertebrates (amphibians) had appeared on land. Club mosses, horsetails and ferns (many of them tree-like) were now abundant there, providing an enormously rich flora in the Carboniferous period. Tremendous accumulations of remains of Carboniferous plants, partially decayed and subjected to intense pressure, formed the coal seams of today. The Mesozoic era was the age of the great dinosaurian reptiles, while gymnophytes were the dominant land plants. Flowering plants appeared in the Jurassic period and radiated during the late Cretaceous.

However, despite an increase in species numbers, it seems that at this time they played a subordinate role in vegetation that was largely fern-dominated. The ascendancy of mammals also began with the end of the Mesozoic era. See **continental drift**, and individual periods, epochs and groups of organisms.

Geophytes Class of **Raunkiaer's life forms**; plants possessing perennating buds below the soil surface on a corm, bulb, tuber or rhizome.

Geotaxis **Taxis** in which the stimulus is gravitational force.

Geotropism (gravitropism) (Bot.) Orientation of plant parts under stimulus of gravity. Main stems, *negatively geotropic*, grow vertically upwards and, when laid horizontally, exhibit increased elongation of cells in growth region on lower side at tip of stem, which turns upwards and resumes its vertical position. Main roots, *positively geotropic*, grow vertically downwards and if laid horizontally exhibit increased elongation of cells on upper side of growth region, the root turning down again as a result. In shoots placed in a horizontal position, differences in **gibberellin** and **auxin** concentrations develop between upper and lower sides. Together, these cause lower side of the shoot to elongate more than upper side, giving the observed upward growth. When it eventually resumes vertical growth, lateral asymmetry in growth substance concentrations disappears, and growth continues vertically. In roots, growth substance asymmetries are less well understood. There is evidence that gradients of these substances within the root cap are brought about by relatively tiny movements of starch-containing plastids (amyloplasts), for when a plant is placed in a horizontal plane, these plastids move from the transverse walls of vertically growing roots and come to rest near what were previously vertically orientated walls. Then, after several hours, the root curves downwards and these plastids return to their original positions. It is not yet clear how these plastid movements translate into growth substance gradients. There is little evidence for the role of auxin (IAA) in roots; some have been unable to find it in the root cap at all. Instead, **abscisic acid** is found there, and has been shown to be redistributed from the cap to the root itself, and to act as an inhibitor on cells in the region of elongation on lower side of a horizontally positioned root. See **diageotropism**, **plagiotropism**.

Germ cell Any of the cells forming a germinal epithelium, plus its cell products, the **gametes**. In vertebrates, *primordial germ cells* migrate from the early gut or yolk sac to sites of the eventual genital ridges in ovaries or testes where, after rounds of mitosis, some of their daughter cells will undergo meiosis and differentiate into gametes. See **germ plasm**, **ovary**, **testis**.

Germinal epithelium See **ovary**, **testis**.

Germinal vesicle Enlarged oocyte nucleus formed during diplotene of first meiotic prophase. See **meiotic arrest**.

Germ layer One of the main layers or groups of cells distinguishable in an animal embryo during and immediately after gastrulation. Diploblastic animals have two such layers: ectoderm (outermost), and endoderm (innermost). Triploblastic animals have a third layer, mesoderm, situated between these two. Roughly speaking, ectoderm gives rise to epidermis and nerve tissue; endoderm gives gut and associated glands; mesoderm gives blood cells, connective tissue, kidney and muscle. Cartilage derives from more than one lineage. The derivations of various organs and tissues in vertebrates are indicated in Fig. 44.

Germ line That cell line which, early in development of many animals, becomes differentiated (determined) from the remaining *somatic cell line*, and alone has the potential to undergo **meiosis** and form gametes. See **germ plasm**, **polar plasm**.

Germ plasm, continuity of The germ plasm theory (**Weismann**, 1892) held that the nuclei of an individual's germ cells, unlike those of its body cells, are qualitatively identical to the nucleus of the zygote from which the individual developed. Weismann held that cellular differentiation was preceded by some loss of material from the nucleus of a cell, but that heredity from one generation to another was brought about by transference of a complex nuclear substance (germ plasm), itself part of the germ plasm of the original zygote which was not used up in the construction of the animal's body, but was reserved unchanged for the formation of its germ cells. See **aberrant chromosome behaviour** (3).

Gestation period Period between conception and birth in viviparous animals. Where there is no conception (e.g. in some cases of **parthenogenesis**), it would be the interval between egg maturation and birth.

Giant chromosome See **polyteny**.

Giant fibres (giant axons) Nerve axons of very large diameter (e.g. 1 mm in squids). Occur in many invertebrates (e.g. annelids, crustacea, and nudibranch and cephalopod molluscs) and some vertebrates where rapid conduction is required (e.g. for escape) and achieved through reduced electrical resistance of the axoplasm with larger axon diameter. May be either a single enormous cell or the result of fusion of many cells. There are fewer synaptic barriers in these nerves than is usual, increasing speed of conduction.

Gibberellins Class of plant **growth substances**, originally isolated from the fungus *Gibberella fujkoroi* when it caused abnormal elongation of infected rice plants in the 1930s. The best-studied is *gibberellic acid*, GA_3, which has the following structure:

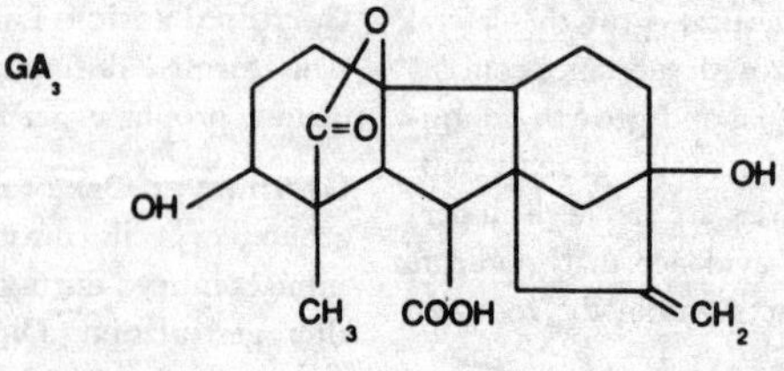

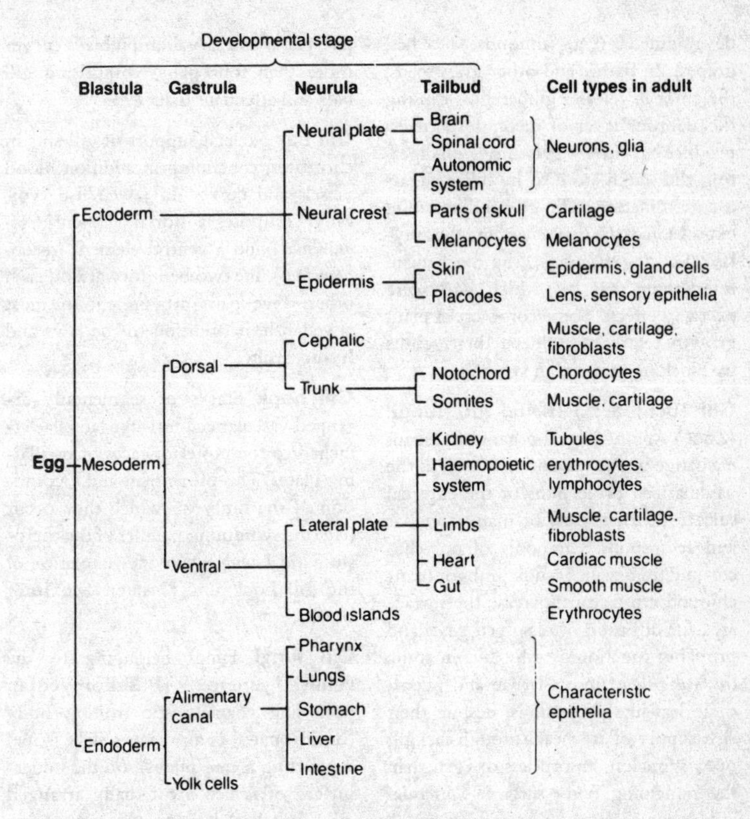

Fig. 44 *Times of production and* **germ layer** *origins of the basic tissues and organs of a vertebrate. The diagram represents a decision-making plan for differentiation although many more decision points are actually involved. Some cell types, e.g. cartilage, have more than one derivation.*

Gibberellins have spectacular effects upon stem elongation in certain plants; they cause dwarf beans to grow to the same height as tall varieties, are involved in seed germination, and will substitute in many species (e.g. lettuce, tobacco, wild oats) for the dormancy-breaking cold or light requirement, promoting early growth of the embryo. Specifically, they enhance cell elongation, making it possible for the root to penetrate growth-restricting seed coat or fruit wall, which has practical application in ensuring uniformity of germinating barley in production of barley malt used in the brewing industry. Can be used to promote early seed production of biennial plants; stimulate pollen germination and growth of pollen tubes in some species; and can promote fruit development (e.g. almonds, peaches, grapes). In barley and other grass seeds, the embryo releases gibberellins causing the aleurone layer of the endosperm to produce enzymes (e.g. α-amylase), digesting the starch store to mobilize sugars for germination. The gibberellin causes expression of the gene encoding the amylase (i.e. de-repression), the mechanism resembling that by which **ecdysone** exerts its effects. Some commercial plant growth retardants achieve their results by blocking gibberellin synthesis.

Gill (Bot.) See **lamella**, **gill fungi**. (Zool.) Any of several organs of gaseous exchange in aquatic animals, such as the vascularized projections of the external surface (*external gills*) of many annelids and arthropods. Parapodia of polychaetes and thin-walled trunk limbs of branchiopod crustaceans increase the surface area for diffusion of dissolved gases and probably function as gills; but in some *tracheal gills* of insect larvae and pupae, these leaf-like projections, despite their rich supply of tracheae (though lacking open spiracles), absorb less oxygen than the remaining body surface. *Spiracular gills* occur in some aquatic insect pupae where one or more pairs of spiracles is drawn out into long processes and generally supplied with a **plastron**. Vertebrate gills are either *external* (ectodermal) or *internal* (endodermal). The former occur in larvae of a few bony fish (*Polypterus* and some lungfish) and of amphibians, and are almost always soon lost or replaced by internal gills. Fish gills may serve as organs of **osmoregulation**. See **ventilation** and Fig. 45.

Gill arches Visceral arches between successive gill slits in jawed fish (typically five pairs) and larval amphibians (never more than four pairs) comprising gill bars and attendant tissues.

Gill bar Skeletal support of gill slits in chordates, containing in addition blood vessels and nerves. In jawed fish typically comprises a dorsal element (*epibranchial*) and a ventral element (*ceratobranchial*), the two bent forward on each other. Five pairs are present in most jawed fish, in addition to the jaws and **hyoid arch**.

Gill book Stacks of segmentally arranged vascularized leaf-like lamellae attached to the posterior faces of oscillating plates. This movement and locomotion of the limbs on which they occur (e.g. the swimming paddles of the merostomatid *Limulus*) provide ventilation of the gill book under water. See **lung book**.

Gill fungi Fungi belonging to the family Agaricaceae (**Basidiomycota**); possessing characteristic fruiting body (**basidiocarp**) comprising a stalk (stipe) supporting a cap (pileus), on the undersurface of which are radially arranged gills (lamellae) bearing the hymenium; e.g. mushrooms, toadstools.

Gill pouch Outpushing of side-wall of pharynx towards epidermis in all chordate embryos. Precursor of gill slit in fish and some amphibia; in terrestrial vertebrates breaks through to exterior only temporarily or (as in humans) not at all. See **spiracle**.

Gill rakers Skeletal projections of inner margins of gill bars of fish, particularly elongated in those which strain incurrent water for food particles.

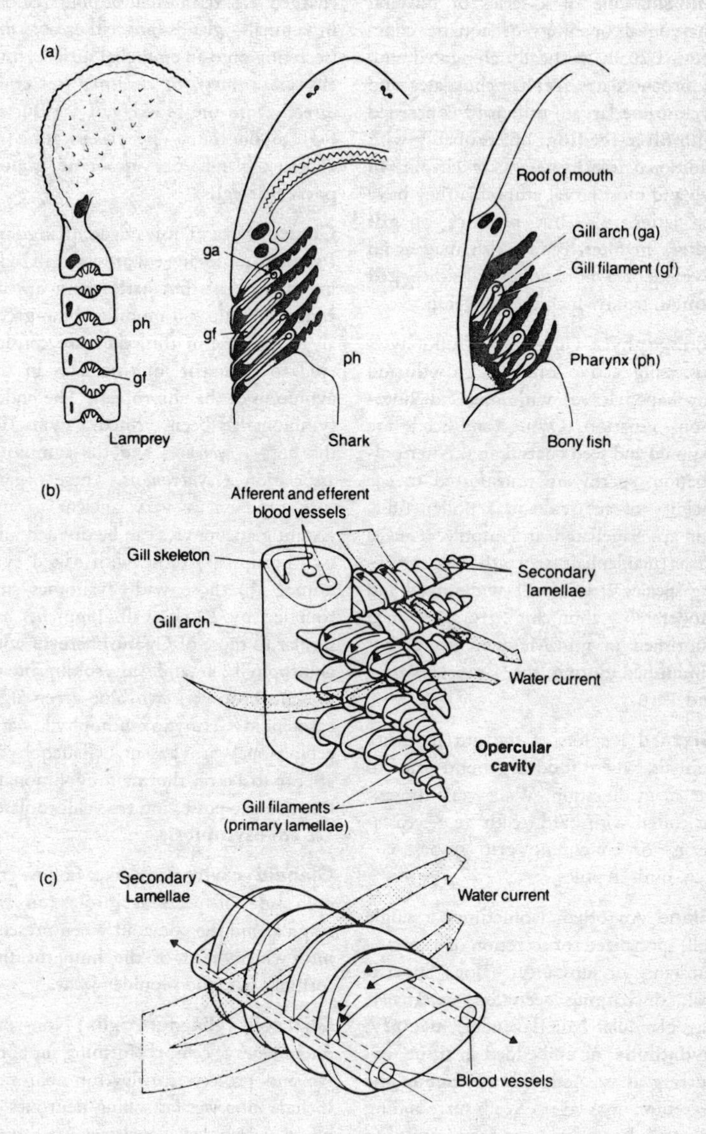

Fig. 45 *(a) The arrangement of gills in three types of fish. (b) The arrangement of secondary gill lamellae in a bony fish. (c) The countercurrent flow of blood through the lamellae.*

Gill slit One of a series of bilateral pharyngeal openings of aquatic chordates. Usually vertically elongated and, in urochordates, cephalochordates and cyclostome larvae, primarily concerned with **filter-feeding**, but probably with additional role in gaseous exchange. In fish and most larval amphibia they have the latter role; but presence of **gill rakers** in filter-feeding fish may again give them a nutritional role. See **gill pouch**, from which they develop.

Ginkgophyta Ginkgos. Gymnophytes possessing active cambial growth and fan-shaped leaves with open, dichotomous venation. Ovules and seeds are exposed and seed coats fleshy. After pollination, sperm are transported to the vicinity of an ovule in a pollen tube, but are flagellated and motile. *Ginkgo bilboa* (maidenhair tree) is the sole surviving species of this once widespread and moderately abundant group, which flourished in mid-Mesozoic times but diminished during the later Mesozoic and Tertiary.

Gizzard Region of the gut in many animals, where food is ground prior to the main digestion. Walls very muscular, often with hard 'teeth' (e.g. crustaceans) or containing grit, stones, etc. (e.g. birds, reptiles).

Gland An organ (sometimes a single cell) specialized for secretion of a specific substance or substances. (Bot.) Superficial, discharging secretion externally, e.g. glandular hair (lavender), **nectary**, **hydathode**, or embedded in tissue, occurring as isolated cells containing the secretion, or as layer of cells surrounding intercellular space (secretory cavity or canal) into which secretion is discharged; e.g. resin canal of pine. (Zool.) In animals, glands are either *exocrine* (secreting onto an epithelial surface, usually via a duct), or *endocrine* (secreting directly into the blood, not via ducts). Fig. 46 illustrates one classification of exocrine glands. See **apocrine gland**, **paracrine cells**.

Glaucophyta Those algae (Kingdom **Protista**) possessing eukaryotic cells lacking chloroplasts but harbouring instead endosymbiotic and modified blue-green algae. Because of this, they are considered to represent intermediates in the evolution of the chloroplast. The endosymbiont has been termed a **cyanelle**, the host a *cyanome*, and the symbiotic association a *syncyanosis*. These organisms represent a very ancient group. Extant glaucophytes can be divided into two groups: (a) those with naked cyanomes; (b) those with cyanomes surrounded by a cell wall. Pigments are similar to those of **Cyanobacteria** with chlorophyll *a* and phycobiliproteins present; however, two blue-green algal carotenoids (myxoxanthophyll and echinenone) are absent. Glaucophytes appear to be on the main evolutionary line to algae possessing true chloroplasts. See **endosymbiosis**.

Glenoid cavity Cup-like hollow on each side of pectoral girdle (on the scapula, and the coracoid when present) into which head of the humerus fits, forming tetrapod shoulder-joint.

Glial cells (glia, neuroglia) Non-conducting nerve cells, performing supportive and protective roles for neurones. Include *astrocytes* (attaching neurones to blood vessels), *oligodendrocytes* (forming myelin sheaths of axons of central nerv-

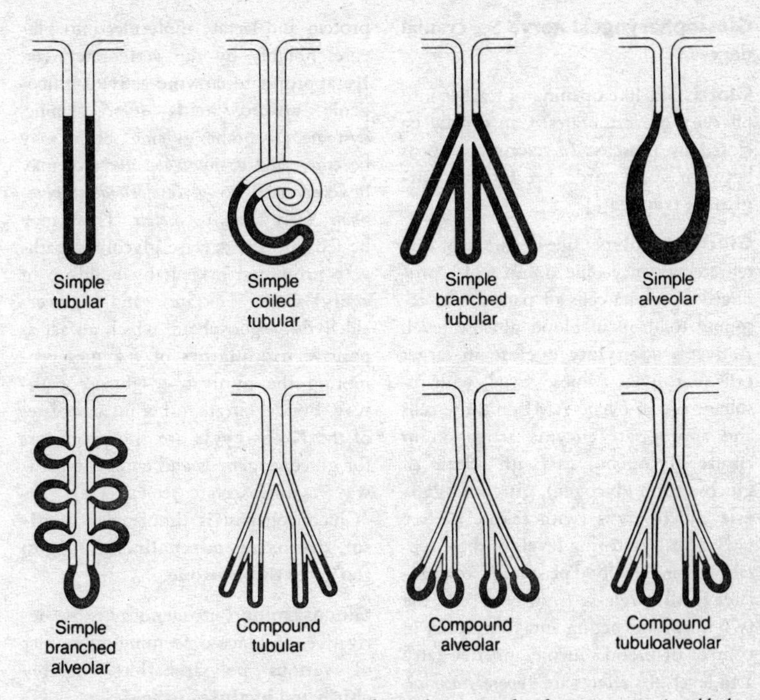

Fig. 46 *Diagram illustrating the major types of animal exocrine* **gland**; *secretory portions black.*

ous system), **Schwann cells**, *microglia* (phagocytic) and *ependyma* cells (lining ventricles of the brain and cerebrospinal canal).

Globigerina ooze Calcareous mud, covering huge areas (about one third) of ocean floor. Formed mainly from shells of **foraminifera**, *Globigerina* being an important genus.

Globulins Group of globular proteins, soluble in aqueous salt solutions and of wide occurrence in plants (e.g. in seeds) and animals (e.g. in vertebrate blood plasma). In humans, **electrophoresis** can separate alpha$_1$-, alpha$_2$-, **beta-** and **gamma-globulins**. Normal individuals belong to one of three genetic types, separable by electrophoretograms of their plasma proteins. Some globulins are involved in lipid and iron transport. See **transferrin**.

Glomerulus (1) Small knot of capillaries covered by basement membrane and surrounded by Bowman's capsule, forming part of a **nephron** of vertebrate **kidney**. Through it small solute molecules (i.e. not cells or plasma proteins) are filtered under pressure from the blood to form the *glomerular filtrate*. (2) *Caudal glomeruli*; small tissue masses containing **retia mirabilia** in mammalian tails, some involved in heat conservation.

Glossopharyngeal nerve See **cranial nerves**.

Glottis Slit-like opening of trachea into pharynx of vertebrates. Can usually be closed by muscles. In mammals, opens between vocal cords. See **larynx**, **epiglottis**, **trachea**.

Glucagon Polypeptide hormone of vertebrates (twenty-nine amino acids), produced by alpha-cells of pancreas in response to drop in blood glucose level. Activates **adenylate cyclase** in target cells (e.g. liver, adipose tissue) with resultant rise in cyclic AMP in those cells and appropriate enzyme activation to ensure glycogenolysis (with release of glucose from glycogen), **gluconeogenesis** and lipolysis (with release of free fatty acids), restoring levels of these metabolites in the blood plasma. Also stimulates **insulin** release from beta-cells, the two hormones acting antagonistically in control of blood glucose and free fatty acid levels. Its effects are *hyperglycaemic*.

Glucocorticoids Steroid adrenal cortex hormones (principally **cortisol** and cortisone) concerned with normal metabolism and resistance to stress conditions (e.g. long-term cold, starvation) by promoting deposition of glycogen in liver, **gluconeogenesis** and release of fatty acids from fat reserves. They render blood vessels more sensitive to vasoconstrictors, thereby raising blood pressure. Stabilize lysosomal membranes, thus inhibiting release of inflammatory substances (and are hence anti-inflammatory). Undersecretion results in *Addison's disease*, oversecretion in *Cushing's syndrome*. See **ACTH, corticosteroids**.

Gluconeogenesis Conversion of fat, protein and lactate molecules into glucose, notably by the vertebrate liver. By appropriate enzyme activity glucogenic **amino acids** (e.g. alanine, cysteine, threonine, glycine, serine) may be converted to *pyruvate*; glycerol may be converted to *glyceraldehyde–3–phosphate*; fatty acids to *acetate*. These may be fed into the reverse glycolytic pathway promoted in cells by build-up of **acetyl–CoA**, citrate and glyceraldehyde–3–phosphate, which all act as positive **modulators** of enzymes promoting the pyruvate-to-glucose pathway. Besides citrate, other intermediates of the **Krebs cycle** are also precursors for gluconeogenesis and enter the pathway via oxaloacetate (see Fig. 47a).

Gluconeogenesis is stimulated by **cortisol**, **thyroxine**, **adrenaline**, **glucagon** and **growth hormone**.

Glucosamine Nitrogenous hexose derivative of glucose forming monomer of various polysaccharides, notably **chitin** and **hyaluronic acid.**

Glucose (dextrose) The most widely distributed hexose sugar (*dextrose* in its dextrorotatory form). Component of many disaccharides (e.g. sucrose) and polysaccharides (e.g. starch, cellulose, glycogen). An *aldohexose* reducing sugar, and (as glucose-1-phosphate) the initial substrate of **glycolysis** for most, if not all, cells. When completely oxidized (in combined glycolysis and aerobic respiration) sufficient energy is released per glucose molecule under intracellular conditions for the generation of thirty-six molecules of ATP from ADP and P_i, in the overall equation:

$$C_6H_{12}O_6 + 6O_2 = 6CO_2 + 6H_2O$$

Glume (serile glume) Chaffy bract,

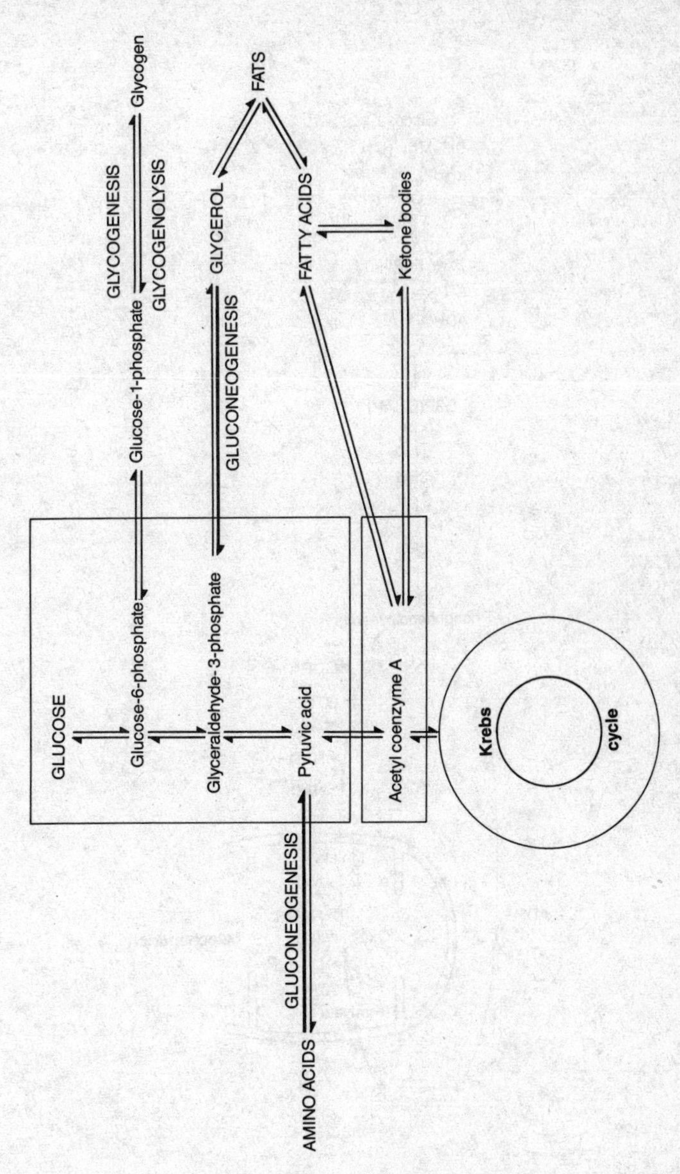

Fig. 47a *Diagram to illustrate where inputs to the glycolytic pathway occur for resynthesis of glucose.*

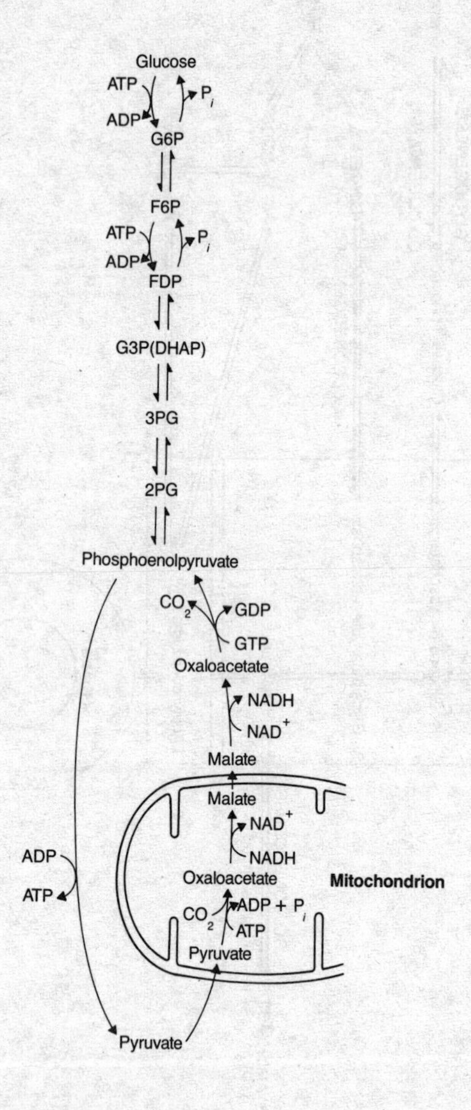

Fig. 47b *Diagram indicating role of mitochondria in reversal of glycolytic pathway during gluconeogenesis.*

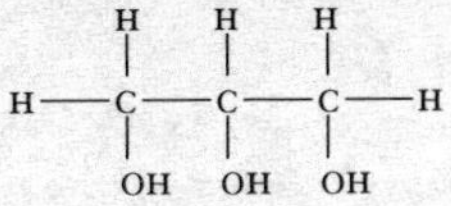

(a) **glycerol**

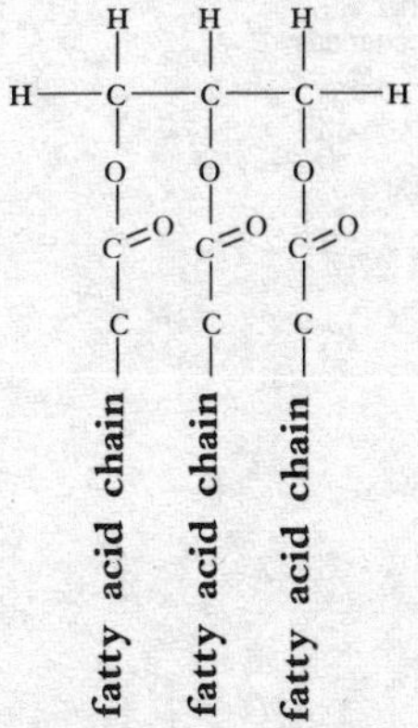

(b) **triglyceride** (neutral fat)

a pair of which occurs at the base of grass spikelet, enclosing it.

Glutamic acid (Glu) Acidic amino acid (typical pK value = 4.4), with side chain nearly always negatively charged at physiological pH (hence commonly called glutamate). Glutamine is an uncharged derivative, with a terminal amine replacing carboxylate; proline and arginine are also derivatives. An important **neurotransmitter**, not least of the mammalian brain.

Gluten Protein occurring in wheat giving firmness to risen dough in bread-making. Those allergic to gluten suffer from *coeliac disease*, resulting in poor absorption of dietary components and in the consequent *malabsorption syndrome*. Gluten-free products are available for such people.

Glycan Synonym of **polysaccharide**.

Glyceride Fatty acid ester of glycerol. When all three –OH groups of glycerol are so esterified the result is a *triglyceride*. See **fat**.

Glycerol A trihydric alcohol and component of many lipids, notably of glycerides. See **diacylglycerol**.

Glycocalyx (cell coat) Carbohydrate-rich region at surfaces of most eukaryotic cells, deriving principally from oligosaccharide components of membrane-bound **glycoproteins** and **glycolipids**, although it may also contain these substances secreted by the cell. Role of the cell coat is not properly understood yet. See **cell membranes**, **phospholipids**, **extracellular matrix**.

Glycogen The chief polysaccharide store of animal cells and of many fungi; often called 'animal starch'. Also found in many bacteria. It resembles **amylopectin** structurally in being an α-[1,4]-linked homopolymer of glucose units, although it is more highly branched. It can be isolated from tissues by digesting them with hot KOH solutions. As with amylopectin, it gives a red-violet colour with iodine/KI solutions. Its hydrolysis is termed *glycogenolysis*. Like starch it is osmotically inactive and therefore a suitable energy storage compound. See **gluconeogenesis**, **glycolysis**, **glucagon**, **insulin**.

Glycolipid Lipid with covalently attached mono- or oligosaccharides; found particularly in the outer half of phospholid bilayers of plasma membranes. Considerable variation in composition both between species and between tissues. All have a carbohydrate polar head end. Range in complexity from

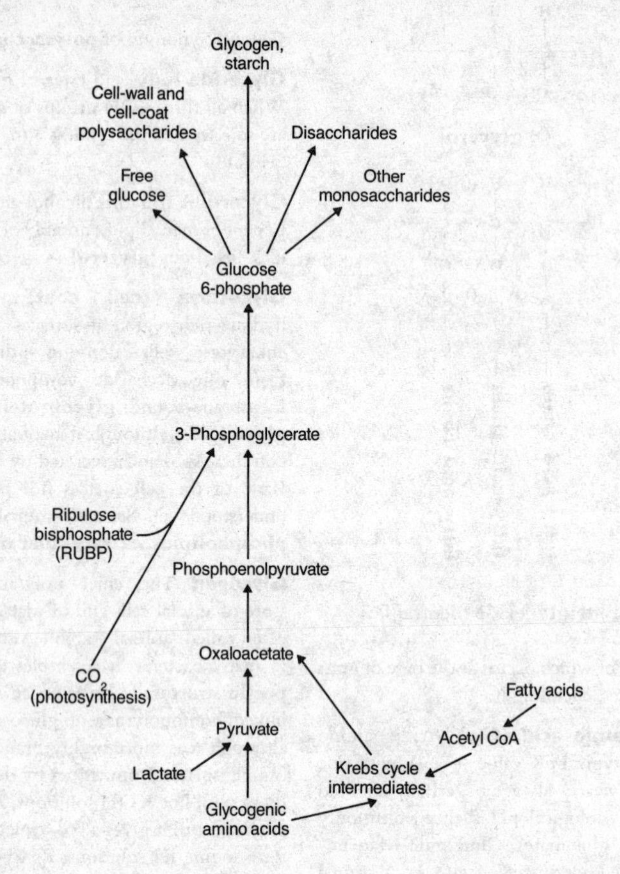

Fig. 48 *Diagram of biochemical pathways linking some non-carbohydrates to carbohydrates. Glucose-6-phosphate acts as a key branch-point.*

relatively simple galacto-cerebrosides of the Schwann cell **myelin sheath** to complex *gangliosides*. May be involved in cell-surface recognition. See **glycosylation**, **glycoprotein**, **cell membranes**.

Glycolysis Anaerobic degradation of glucose (usually in the form of glucose-phosphate) in the cytosol to yield pyruvate, forming initial process by which glucose is fed into aerobic phase of **respiration**, which usually occurs in **mitochondria**. Cells without mitochondria (and those prokaryotes without a **mesosome**) rely on glycolysis for most of their ATP synthesis, as do facultatively anaerobic cells (e.g. striated muscle fibres) when there is a shortage of oxygen. The pathway is illustrated in Fig. 49. It generates a net gain of two molecules of ATP per molecule of glucose used, plus reducing power in the form of two $NADH_2$ molecules. The $NADH_2$ is available for reduction of pyruvate to lactate, or of acetaldehyde to alcohol, or for fatty acid and steroid synthesis from acetyl coenzyme A, as occurs in liver cells.

Most significant in glycolysis is the hydrolysis of each fructose 1,6–bisphosphate molecule into two triose phosphate molecules, the remaining steps in the pathway thereby effectively occurring twice for every initial glucose-phosphate molecule used. Conversion of fructose–6–phosphate to fructose 1,6–bisphosphate is the main rate-limiting step in glycolysis, and phosphofructokinase, the enzyme involved, is a **regulatory enzyme**, modulated by the ratio in the levels of (AMP + ADP):ATP in the cytosol so that high ATP levels inhibit glycolysis. Enzymes involved in the glycolytic pathway appear to have evolved from one ancestral enzyme by a process involving **gene amplification**. See **Pasteur effect**, **pentose phosphate pathway**. For *aerobic glycolysis*, see **cancer cell**.

Glycoprotein Protein associated covalently at its n-terminal end with a simple or complex sugar residue. In **proteoglycans** the carbohydrate forms the bulk of the molecule, with numerous long and usually unbranched **glycosaminoglycans** bound to a single core protein. These important extracellular components contrast with cell surface glycoproteins, which generally comprise short but often complex non-repeating oligosaccharide sequences bound to an integral membrane protein. Proteins become glycosylated (i.e. have their sugar residues added) in the **endoplasmic reticulum** and **Golgi apparatus**. See **cell membranes**, **glycosylation**, **glycolipids**.

Glycosaminoglycans (GAGs) Long, unbranched polysaccharides (formerly called mucopolysaccharides) of repeated disaccharide units, one member always an amino sugar (e.g. N-acetylglucosamine, N-acetylgalactosamine). They comprise varying proportions of the extracellular matrices of tissues, where they are often numerously bound to a core protein to become *proteoglycans*, e.g. **hyaluronic acid**, **chondroitin**, **heparin**. See **glycocalyx**, **cell membranes**.

Glycoside Substance formed by reaction of an *aldopyranose* sugar, such as glucose, with another substance such that the aldehyde moiety in the sugar is replaced by another group. Glycosidic bonds form the links between monosaccharide units in the formation of polysaccharides. Some plant glycosides, termed *cardiac glycosides*, alter the excitability of heart muscle and may be defensive; examples include *ouabain* and *digitalin*. See **anthocyanins**, **tannins**, **alkaloids**.

Glycosylation Bonding of sugar residue to another organic compound. **Glycoproteins** are formed in the lumen of rough endoplasmic reticulum, but may be subsequently modified in the lumen of the **Golgi apparatus**, where other amino acids of the protein may become glycosylated. Nucleotides may be glycosylated, UDP-glucose being an important coenzyme in transport of glucose, most probably in cell

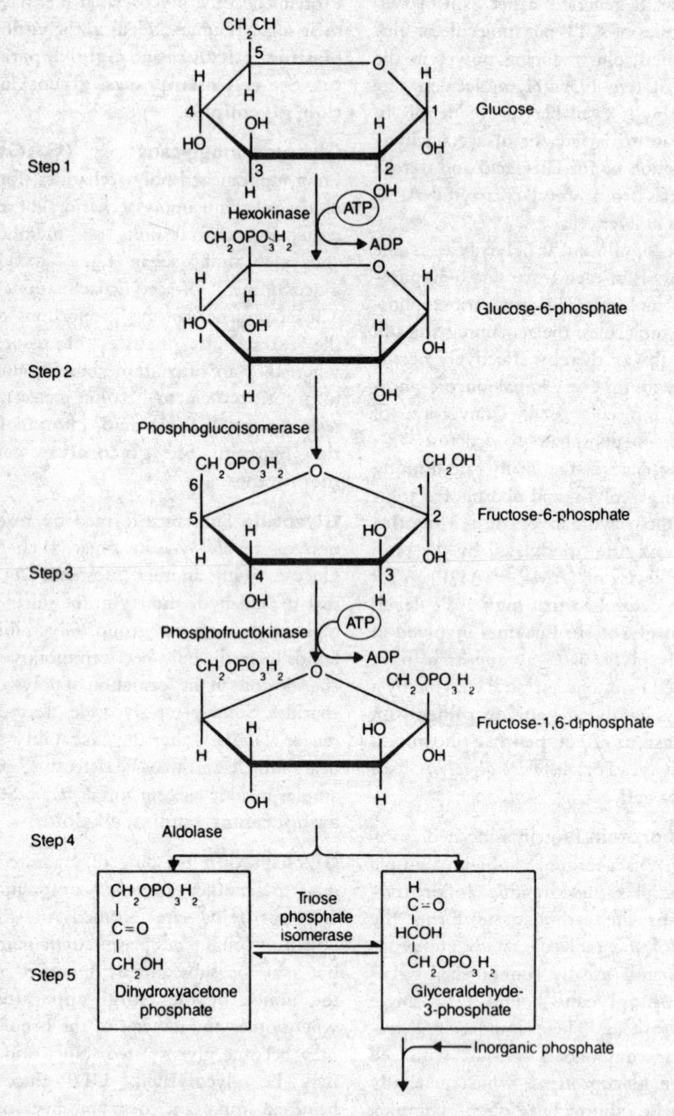

Fig. 49 *The stages of vertebrate* **glycolysis** *and the enzymes involved. Because of the hydrolysis at Step 5 all later stages are represented by two molecules for every original glucose molecule.*

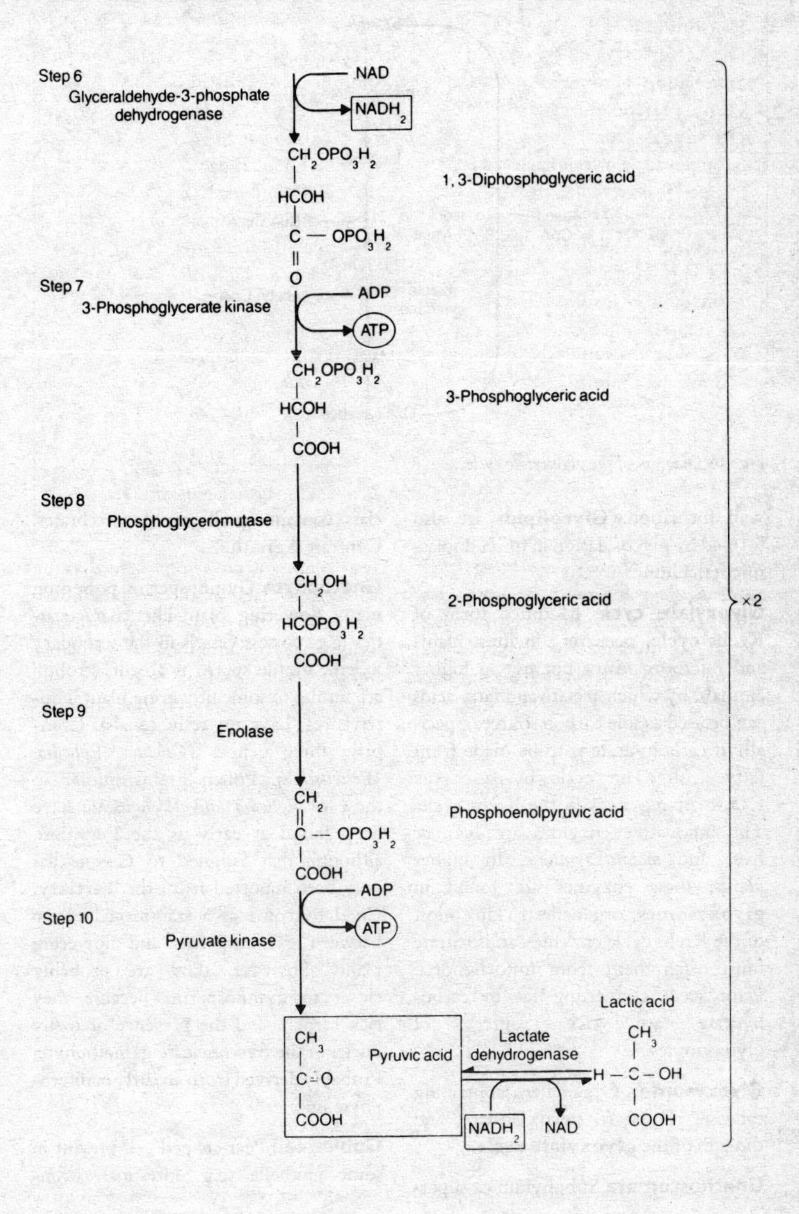
Step 6
Glyceraldehyde-3-phosphate dehydrogenase
NAD
NADH2
CH2OPO3H2
HCOH
C — OPO3H2
O
1, 3-Diphosphoglyceric acid
Step 7
3-Phosphoglycerate kinase
ADP
ATP
CH2OPO3H2
HCOH
COOH
3-Phosphoglyceric acid
Step 8
Phosphoglyceromutase
CH2OH
HCOPO3H2
COOH
2-Phosphoglyceric acid
×2
Step 9
Enolase
CH2
C — OPO3H2
COOH
Phosphoenolpyruvic acid
ADP
ATP
Step 10
Pyruvate kinase
CH3
C=O
COOH
Pyruvic acid
Lactate dehydrogenase
NADH2
NAD
Lactic acid
CH3
H – C – OH
COOH

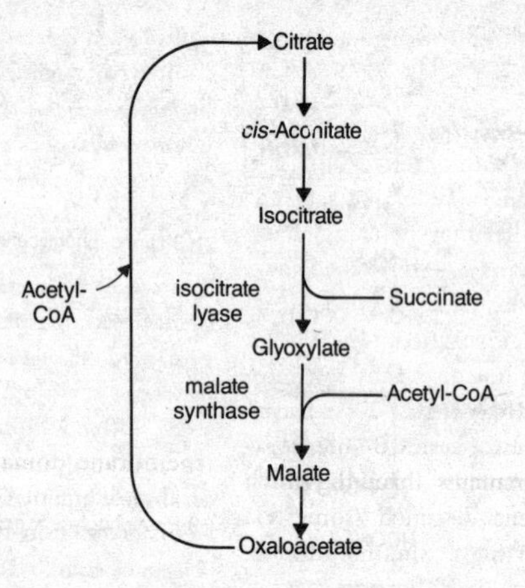

Fig. 50 *Diagram of the glyoxylate cycle.*

wall formation. **Glycolipids** are also formed by glycosylation in the endoplasmic reticulum.

Glyoxylate cycle Modified form of **Krebs cycle**, occurring in most plants and microorganisms but not in higher animals, by which acetate and fatty acids can be used as sole carbon source, especially if carbohydrate is to be made from fatty acids. The cycle by-passes the CO_2-evolving steps in the Krebs cycle. The innovative enzymes are *isocitrate lyase* and *malate synthase*. In higher plants, these enzymes are found in **glyoxysomes**, organelles lacking most of the Krebs cycle enzymes; so isocitrate must reach them from mitochondria. Plant seeds converting fat to carbohydrate are rich sources of glyoxysomes.

Glyoxysome Organelles containing catalase, related to **peroxisomes**, and the sites of the **glyoxylate cycle**.

Gnathostomata Subphylum or superclass containing all jawed vertebrates. Contrast **Agnatha**.

Gnetophyta Gymnosperms possessing many flowering plant-like characteristics (e.g. possess vessels in the secondary xylem, motile sperm is absent, stroboli are similar to some flowering plant inflorescences, have no resin canals). Comprise three genera (*Gnetum*, *Ephedra*, *Welwitschia*). Pollen grains similar to those of *Ephedra* and *Welwitschia* have been found as early as the **Permian**, although that assigned to *Gnetum* has only been reported from the **Tertiary**. Cited by some as a transitional group between gymnosperms and flowering plants; however, they are probably closer to gymnosperms because they lack carpels, and the presence of many nuclei in the free-nucleate gametophyte. Probably derived from an early coniferophyte line.

Goblet cell Pear-shaped cell present in some epithelia (e.g. intestinal, bronchial), and specialized for production of **mucus**.

Golgi apparatus (Golgi body, Golgi complex) A dynamic eukaryotic organelle, comprising a system of stacked and roughly parallel interconnecting flattened sacs (cisternae) sandwiched between two complex networks of tubules (*cis* and *trans* Golgi networks), the whole situated close to, but physically separate from, the **endoplasmic reticulum** (ER). It provides a series of membranous subcompartments through which move components destined (some via **endosomes**) for the plasma membrane, secretory vesicles and lysosomes (this requires **protein targeting**). Stacks of cisternae can move towards the 'minus' (organizing) ends of **microtubules**, maintaining the organelle's central position in the cell needed for its role in membrane traffic. Numbers vary from one to hundreds per cell; tend to be interconnected in animal but not plant cells, with up to 30 cisternae (normally about 6) per Golgi body. Each cisterna has a *cis* surface (towards the nucleus) and a *trans* surface (away from the nucleus). Transport vesicles (including non-clathrin-coated) from the ER arrive and fuse with the *cis* Golgi network, adding their membrane material to the cisternae and depositing **glycoproteins** for processing within the cisternal lumina. Some of the oligosaccharide of the glycoprotein may be removed, while other sugar units are added to yield mature glycoproteins of different kinds – possibly vesicle-specific (see **major histocompatibility complex**). Many of these will be retained within the membrane during modification. Vesicular carriers engage in membrane traffic between components of the Golgi complex, apparently budding from one compartment and fusing with the next, *en route* to the *trans* Golgi network, where two types of vesicles are budded off: **coated vesicles** (about 50 nm in diameter), and larger secretory vesicles (about 1000 nm in diameter). Much remains to be learned about these movements, which seem to involve **phospholipid transfer proteins**. Possibly, different 'membrane scaffolds' are associated with budding and non-budding membrane domains, restricting the lateral movement of 'resident' Golgi proteins from non-budding regions. Non-clathrin coated vesicles are characterized by several types of COP (coatomer protein), uncoating of which is a prerequisite for fusion of vesicle with an appropriate acceptor membrane – in turn governed by soluble 'fusion proteins' which link the two membranes and small Ras-like GTPases. The fungal product brefeldin A causes uncoating of COPs and leads to disintegration of the whole organelle.

At the onset of mitosis, exocytic and endocytic membrane traffic ceases and the Golgi body fragments into tubular clusters and small vesicles (compare **nucleus**). See **lysosome**, **cell membranes** and Fig. 51.

In plant cells, the Golgi apparatus is involved in secretion; e.g. it synthesizes cell wall polysaccharides, which collect in vesicles pinched off from the cisternae. These secretory vesicles migrate and fuse with the plasmalemma; vesicles discharge their polysaccharide contents to the exterior, where they become part of the cell wall. In some algae, scales (both organic and inorganic) are formed in vesicles cut off from the Golgi apparatus before being transported to the cell periphery. In diatoms (**Bacillariophyta**), the Golgi apparatus again gives rise to translucent vesicles which collect beneath the plasmalemma, where they fuse to form the **silica deposition vesicle** within which the siliceous cell wall is synthesized.

Gonad Animal organ producing either sperm (**testis**) or ova (**ovary**). See **ovotestis**.

Gonadotrophins (gonadotropins, gonadotropic hormones) Group of vertebrate glycoprotein hormones, controlling production of specific hormones by gonadal endocrine tissues. Anterior pituitaries of both sexes produce **follicle-stimulating hormone** (FSH) and **luteinizing hormone** (LH, or interstitial cell stimulating hormone (ICSH) in males); but their effects in the two sexes are different. **Human chorionic gonadotrophin** (HCG) is an embryonic product whose presence in maternal urine is usually diagnostic of pregnancy. Release of FSH and LH is controlled by hypothalamic *gonadotrophic-releasing factors* (GnRFs). **Prolactin** is also gonadotrophic. See **menstrual cycle**.

Gondwanaland Southernmost of the two Mesozoic supercontinents (the other being **Laurasia**) named after a characteristic geological formation, the *Gondwana*. Comprised future South America, Africa, India, Australia, Antarctica and New Zealand (the last breaking earliest from the supercontinent, with present-day examples of a relict Gondwana-like flora and fauna). Rifts between Gondwanaland and Laurasia were not effective barriers to movements of land animals until well into the Cretaceous. By the dawn of mammalian radiation Gondwanaland had largely split into its five major continental regions, each being the nucleus of radiation for its inhabitant fauna and flora. Flora was characterized by *Glossopteris*; podocarps and tree ferns still persist in New Zealand, as do the reptile *Sphenodon* (see **Rhynchocephalia**), giant crickets and flightless birds (e.g. kiwi and, up to 5000 years ago, moas). See **continental drift**, **zoogeography**.

Gonochorism Condition of having sexes separate; individuals having functional gonads of only one type. Bisexual.

G_1, G_2 phase Phases of the **cell cycle**.

G-protein Three-subunit eukaryotic protein, coupling light or hormonal activation of membrane receptor to activation of a target protein (e.g. **adenylate cyclase**), thereby amplifying the signal, or membrane ion channel, mediated by dissociation of a G-protein subunit bound to GTP. Different G-proteins enable sorting out of signals from membrane receptors to effector molecules within the cell. Although they are present in plant cells, it is not clear what role they play in signalling. See **cytoskeleton**, **GTP**, **Ras protein**, **receptor**, **second messenger**.

Graafian follicle Fluid-filled spherical vesicle in mammalian ovary containing **oocyte** attached to its wall. Growth is under the control of **follicle-stimulating hormone** of anterior pituitary, its rupture (*ovulation*) also being a gonadotrophic effect (see **luteinizing hormone**). After ovulation the follicle collapses, but theca and granulosa cells grow and proliferate forming the **corpus luteum**. Androgen precursors

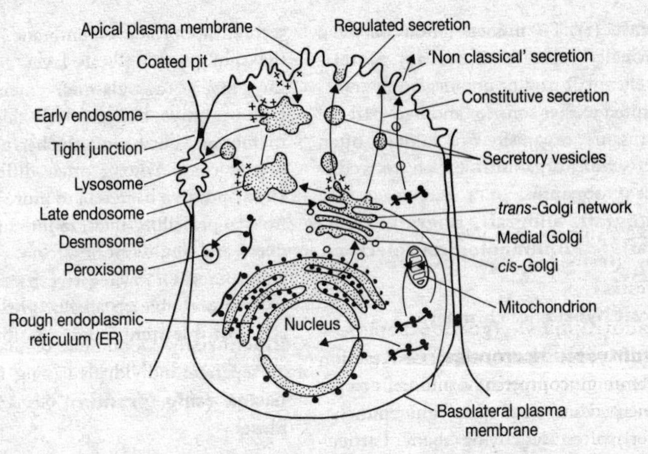

Fig. 51a *Involvement of the Golgi apparatus in some of the protein targeting pathways in a mammalian epithelial cell. The stippled compartments are topologically equivalent to one another and to the outside of the cell.*

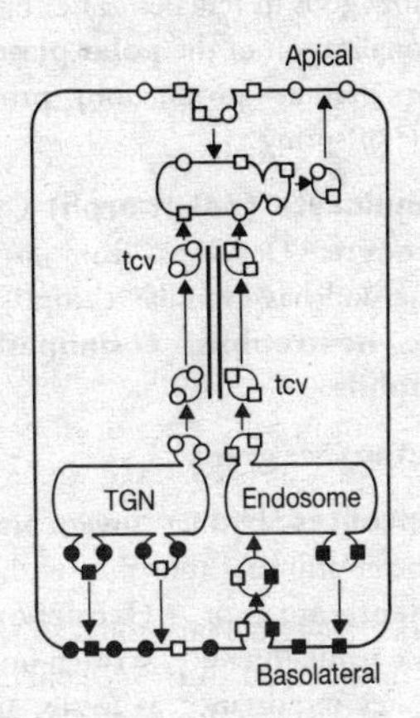

Fig. 51b *Protein targeting examples in an epithelial cell. Newly synthesized proteins of the basolateral membrane (dark squares and circles) go straight there from the* trans *Golgi network (TGN); proteins (dark squares) that also have an endosomal targeting sequence go to basolateral endosomes. Newly synthesized proteins destined for the apical membrane either go directly there (open circles) or go first to the basolateral membrane (open squares) prior to entry into the basolateral endosome, whence it is sorted into a transcytotic vesicle (tcv) and transported along microtubules (parallel lines) to the apical membrane. See* **endosome**.

are made by the *theca cells* of the Graafian follicle, and aromatized to oestrogens by the *granulosa cells*; in primates *theca lutein cells* of the corpus luteum make oestrogen precursors. See **menstrual cycle, ovary**.

Grade A given level of morphological organization sometimes achieved independently by different evolutionary lineages, e.g. the mammalian grade. See **clade**.

Graft (1) To induce union between normally separate tissues. (2) A relatively small part of one organism transplanted either on to another part of the same organism (in animals often the whole organism) or on to a different organism, or a part of it. See **autograft**, **allograft**, **isograft**, **xenograft**. See **immunological tolerance**, **MHC**.

Graft hybrid See **chimaera**.

Graft-versus-host response Reaction of immunocompetent donor cells to recipient tissues (e.g. skin, gut epithelia, liver), often destroying them. Particularly problematic in bone transplants.

Gramicidin See **ionophore**.

Gram-negative bacteria See **Gram's stain**.

Gram's stain Stain devised by C. Gram in 1884 which differentiates between bacteria which may be otherwise similar morphologically. To a heat-fixed smear containing the bacteria is added crystal violet solution for 30 s which is then rinsed off with Gram's iodine solution; 95% ethanol is applied and renewed until most of the dye has been removed (20 s – 1 min). Those bacteria with the stain retained are *Gram-positive* (e.g. *Bacillus subtilis*, *Staphylococcus aureus*); those without are *Gram-negative* (e.g. *Escherichia coli*, *Agrobacterium tumefaciens*). A counterstain (e.g. eosin red, saffranin, brilliant green) is then applied, colouring the Gram-negative bacteria but not the Gram-positive ones. Differentiation reflects differences in amount and ease of access of **peptidoglycan** in the bacterial envelope. Gram-negative bacteria have a second lipoprotein membrane outside the thin peptidoglycan layer covering the inner (cytoplasmic) membrane. Gram-positive bacteria lack this outer membrane, but have a thicker peptidoglycan coat. Among other differences, Gram-positive bacteria are more susceptible to penicillin, acids, iodine and basic dyes, and some species form resistant endospores; Gram-negative bacteria are more susceptible to alkalis, **antibodies** and **complement**, and do not form endospores. See Fig. 52.

Grana (sing. granum) See **chloroplasts**.

Grandchildless (gs) Adjective describing a group of at least eight **maternal effect** genes of *Drosophila* which when homozygous in females causes failure of regionalization of the **polar plasm** (and polar granule production) producing sterile offspring.

Granulocyte (polymorph) Granular **leucocyte**. Develops from **myeloid tissue** and has granular cytoplasm. Include **neutrophils**, **eosinophils** and **basophils**.

Granum See **grana**.

Graptolites Extinct invertebrates of doubtful affinities (possibly with either **Coelenterata** or **Hemichordata**), whose name (literally, written on stone) indicates importance as fossils, notably of shales. Upper Cambrian–Lower Carboniferous; used to subdivide the Ordovician and Silurian.

Grassland (prairie, steppe) Includes a wide variety of plant communities. Some intergrade with savannas, others with deciduous woodland or desert. Unlike savannas, grasslands lack trees –

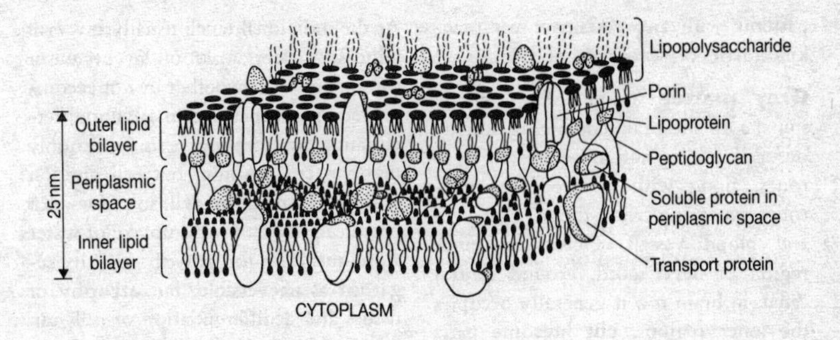

Fig. 52 *The arrangement of the double membrane envelope of Gram-negative bacteria such as* E. coli.

except along rivers and streams. They are generally characterized by cold winters and occur over large areas in the interior portions of continents. Annual precipitation determines distribution of the grass species; e.g. typical short-grass species occur where precipitation is low, while taller species dominate in moister areas. Grassland regions have been used extensively for agriculture, particularly cultivation of cereal crops and for pasture. They were once inhabited by herds of grazing mammals associated with large predators. Such herds were widespread during Pleistocene glaciations.

Gravitropism See **geotropism**.

Great chain of being (scala natura) View proposed by Aristotle and incorporated by Leibniz in his metaphysics and by **Buffon** (for whom it was a scale of degradation, from man at the top): that there is a linear and hierarchical progression of forms of existence, from simplest to most complex, lacking both gaps and marked transitions.

Greenhouse effect Effect in which short wavelength solar radiation entering the Earth's atmosphere is re-radiated from the Earth's surface in longer infrared wavelengths, and then reabsorbed by components of the atmosphere to become an important factor in heating the total atmosphere. Effect resembles heat reflection by greenhouse glass. Oxygen, **ozone**, carbon dioxide, methane and water vapour all absorb in the infrared wavelengths, and increasing amounts of carbon dioxide from the combustion of fossil fuels, the destruction of tropical forest cover and the ploughing of soil are growing factors in raising the average atmospheric temperature. Methane is a much more effective greenhouse gas than is carbon dioxide and its increasing level is causing concern. Predictions include that the increase in temperature will cause an increase in the great deserts of the world, and that increased photosynthetic activity will result from increased carbon dioxide concentrations in the atmosphere. The phenomenon is alarming because the consequences are not known, existing global climate models being too unsophisticated to predict future climate patterns with any certainty; yet mankind carries on seemingly obliviously.

Grey matter Tissue of vertebrate spinal cord and brain containing numerous cell bodies and dendrites of neurones, along with unmyelinated neurones synapsing with them, glial cells and blood vessels. Occurs as inner region of nerve cord, around central canal; in brain too it generally occupies the inner regions, but in some parts (e.g. **cerebral cortex** of higher primates), some cell bodies of grey matter have migrated outwards to form a third layer on top of the white, axon zone.

Ground meristem Primary **meristem** in which procambium is embedded and which is surrounded by protoderm; matures to form the **ground tissues**.

Ground tissue (Bot.) All tissues except the epidermis (or periderm) and the vascular tissues; e.g. those of the cortex and pith.

Group selection Postulated evolutionary mechanism whereby characters disadvantageous to individuals bearing them, but beneficial to the group of organisms they belong to, can spread through the population countering the effects of selection at the individual level. Invoked to explain apparent reproductive restraint by individuals when environmental resources are scarce, and other cases of apparent **altruism** which may have a simpler explanation in terms of **natural selection**. Some hold that group selection may have played a part in the evolution of **sex**. See **unit of selection**.

Growth Term with a variety of senses. At the individual level, usually involves increase in dry mass of an organism (or part of one), whether or not accompanied by size increase, involving differentiation and morphogenesis. Commonly involves cell division, but cell division without increase in cell size does not produce growth. Nor is uptake of water alone sufficient for growth. Usually regarded as irreversible, but **atrophy** of tissues and dedifferentiation of cells can occur. Algal and plant growth forms may be *diffuse*, where most cells are capable of division; *apical*, where a single terminal cell gives rise to cells beneath; *trichothallic* where a cell divides forming a hair above the thallus below; *promeristematic* where a non-dividing apical cell controls a large number of smaller dividing cells beneath it; *intercalary* where a zone of meristematic cells occurs forming tissue above and below the meristem; *meristodermic*, when a layer of usually peripheral cells divides parallel to the thallus surface to form tissue below the meristoderm (usually cortex). In most higher plants, growth is restricted to **meristems**. In animals, growth is more diffuse. The terms 'lytic growth' and 'lysogenic growth' are often used in the context of **bacteriophage** 'life cycles'. See **growth curve**, **exponential growth**, **allometry**, **Ras proteins**, **auxin**, **growth hormones**.

Growth cone The expanded tip of a nerve axon, whose dynamic structure resembles that of a motile cell, which leads the extending axon along its appropriate path during growth.

Growth curve Many general features of the growth of a population may be indicated by the growth curve of unicellular organisms under optimal condi-

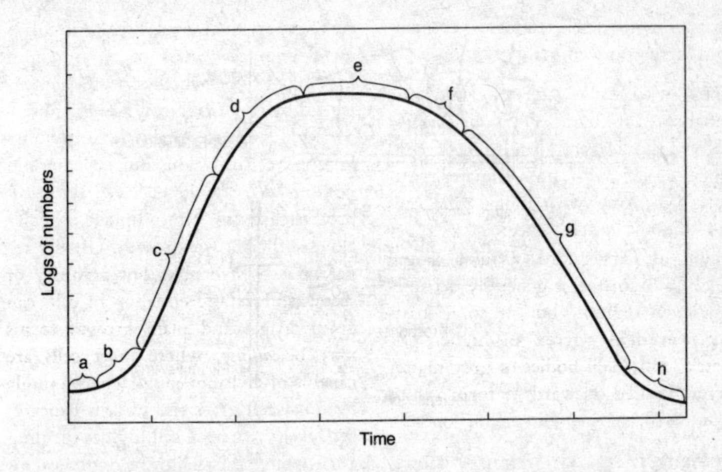

Fig. 53 *The* **growth curve** *of unicellular organisms under optimal growth conditions. Phases a to h are explained in the entry.*

tions for growth. (See Fig. 53.) Stages (a) to (e) represent a logistic growth curve (see **exponential growth**).

(a) *Lag Phase*: latent phase, in which cells recover from new conditions, imbibe water, produce ribosomal RNA and subsequent proteins. Cells grow in size, but not in number. (b) *Phase of Accelerated Growth*: fission initially slow, cell size large. During this phase, rate of division increases and cell size diminishes. (c) *Exponential or Logarithmic Phase*: cells reach maximum rate of division. Characteristic of this phase that numbers of organisms, when plotted on a logarithmic scale, generate a straight-line slope. (d) *Phase of Negative Growth Acceleration*: food (e.g.) begins to run out, waste poisons accumulate, pH changes and cells generally interfere with one another. Increase in number of live cells slows as rate of fission declines. (e) *Maximum Stationary Phase*: number of cells dying balances rate of increase, resulting in a constant total viable population. (f) *Accelerated Death Phase*: cells reproduce more slowly and death rate increases. (g) *Logarithmic Death Phase*: numbers decrease at unchanging rate. (h) *Phase of Readjustment and Final Dormant Phase*: death rate and rate of increase balance each other, and finally there is complete sterility of the culture. See **J-shape growth form**.

Growth factors (GFs) Diffusible molecules variously signalling the division, locomotion, survival and/or differentiation of cells; e.g. **colony stimulating factors** and some **interleukins**. All bind **receptors** of the cell surface or **extracellular matrix (ECM)**, which are often linked to **signal transduction** pathways via such intracellular enzymes as **protein kinases**. *T cell growth factor* (TCGF, interleukin 2) is required for proliferation of T cells, TCGF receptors appearing on

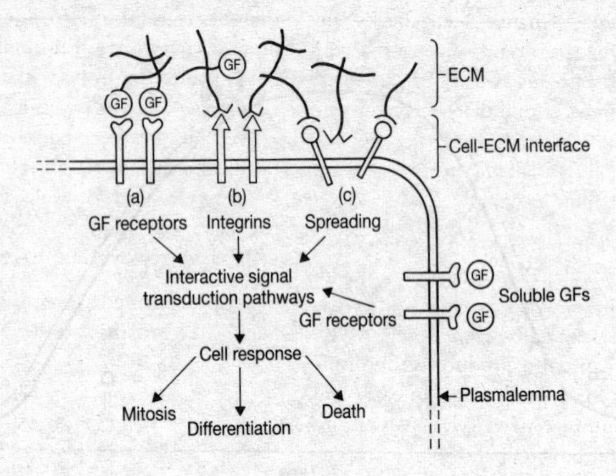

Fig. 54 *Possible interactions between* **growth factors**, *the extracellular matrix (ECM) and receptors (shown variously as structures crossing the plasmalemma).*

the T cell surface in response to mitogen (antigen); *platelet-derived growth factor* (PDGF) is a chemoattractant and stimulates connective tissue and neuroglial cells (see **platelets**); *epidermal growth factor* (EGF) stimulates many cell types to divide; *insulin-like growth factors* (ILGF-I and ILGF-II) collaborate with PDGF and EGF to stimulate fat and other connective tissue cells, causing permanent exit from the **cell cycle**; *transforming growth factor β* (TGF-*β*, or activin) prepares many cells to respond to other GFs, stimulates the synthesis of ECM components and reversibly decreases the rate of transit through the cell cycle – a feature shared with *nerve growth factor* (NGF), which promotes axon growth and longevity of sympathetic and some sensory and CNS neurons. See **tumour necrosis factors**, Fig. 54, **programmed cell death**.

Growth hormone (GH, somatotrophin) Polypeptide (and most abundant) hormone produced by somatotroph cells of anterior **pituitary**. These cells proliferate and secrete on binding growth hormone-releasing factor (GRF) at their **G-protein**-linked receptors, a signal which is transduced via cyclic AMP. Regulates deposition of collagen and chondroitin sulphate in bone and cartilage; promotes mitosis in osteoblasts and increase in girth and length of bone prior to closure of **epiphyses**. Transported in plasma by a **globulin** protein. Its release is prevented by hypothalamic **somatostatin**. Induces release from liver of the hormone **somatomedin**, which mediates its effects at the cell level. Low levels in humans result in pituitary dwarfism (epiphyseal growth plates close before normal height attained); high levels prior to closure of the growth plates results in gigantism (increase in long bone length) and after closure of growth plates to acromegaly (thick bones of hands, feet, cheeks and jaws). Dwarfism and gigantism syndromes in humans have been explained by mutations in one or other genes encoding the variety of molecules regulating the hypothalamic-pituitary-target tissue signalling system.

Growth ring (Bot.) Growth layer in secondary xylem or secondary phloem, as seen in transverse section. The periodic and seasonally related activity of vascular cambium produces growth increments, or growth rings. In early and middle summer new xylem vessels are large and produce a pale wood, contrasting with the narrower and denser vessels of late summer and autumn, which produce a dark wood. Width of these rings varies from year to year depending upon environmental conditions, such as availabilities of light, water and nutrients and temperature. See **dendrochronology**.

Growth substance (Bot.) Term used in preference to plant hormone, to include all natural (endogenous) and artificial substances with powerful and diverse physiological and/or morphogenetic effects in plants. Sites of natural production are often diffuse and rarely comprise specialized glandular organs – often just a patch of cells, commonly without physical contact. The minute quantities produced, and the notorious synergisms in their modes of action, pose profound research problems. See **auxins**, **abscisic acid**, **cytokinins**, **ethene**, **gibberellins**.

gs See **grandchildless**.

GTP (guanosine 5′-triphosphate) Purine nucleoside triphosphate. Required for coupling some activated membrane receptors to **adenylate cyclase** activity. A GTP-binding protein (**G-protein**) hydrolyses the GTP and keeps the enzyme's activity brief. Involved in tissue responses to various hormones (e.g. **adrenaline**), a precursor for **nucleic acid** synthesis and essential for chain elongation during **protein synthesis**. See **cyclic GMP**.

Guanine Purine base of **nucleic acids**. See **GTP**, **inosinic acid**.

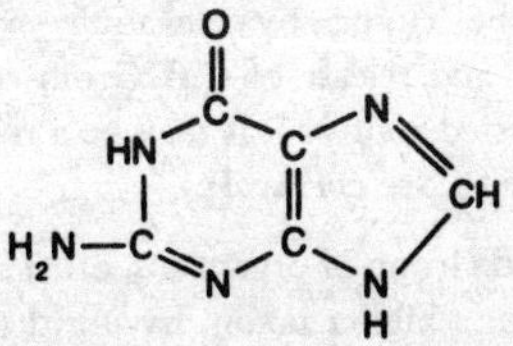

Guanosine. Purine nucleoside. Comprises D-ribose linked to guanine by a beta-glycosidic bond. See **GTP**.

Guard cells (Bot.) Specialized, crescent-shaped, unevenly thickened epidermal cells in pairs surrounding a stomatal pore (see **stoma**). Changes in shapes of guard cells, due to changes in their turgidities, control opening and closing of the stomata, and hence affect rate of loss of water vapour in **transpiration** and amount of gaseous exchange. It is now generally accepted that turgor changes in the guard cells are caused principally by salt uptake and extrusion, although the presence of starch is held to play a role, too. Guard-cell cell membranes can extrude protons using an ATP-driven proton pump, potassium ions (K^+) replacing them down an electrical gradient, thereby reducing water potential and generating turgor, while chloride ions (Cl^-) enter to maintain pH. Protons are produced within the cell by conversion of starch to malate. Guard cell chloroplasts appear not to fix much CO_2, and contain starch at night rather than during daytime. The growth substance **abscisic acid** is believed to act on the guard cell membrane and bring about opening of calcium (Ca^{2+}) channels there so that Ca^{2+} enters the cell. Once inside, Ca^{2+} binds a **calmodulin** and through its action as a second messenger, changes the cell's turgidity in some as yet unspecified way. Stomata can be closed by elevating external Ca^{2+} and release of Ca^{2+} from cells of a stressed root or leaf may be a relevant message. See **cell wall**.

Guild Group of organisms, often species within a higher taxon, having the same broad feeding habits; e.g. phytophages (plant- feeders, themselves divisible into chewers and suckers), parasitoids, scavengers, etc.

Gustation The sense of taste. See **receptor**, **signal transduction**.

Gut See **alimentary canal**.

Guttation Excretion of water drops by plants through **hydathodes**, especially in high humidity, due to pressure built up within the xylem by osmotic absorption of water by roots. See **root pressure**.

Gymnosperm A term no longer used in the formal schemes of classification, but still used informally. The term literally means 'naked-seeded plants', ovules and seeds lying exposed on their sporophyll (or analagous) surfaces, this being one of the principal characteristics of non-anthophyte seed plants. The four divisions with living species today are **Coniferophyta**, **Cycadophyta**, **Ginkgophyta** and **Gnetophyta**. See also **Progymnospermophyta**.

Gynandromorph Animal, usually an insect, which is a genetic **mosaic** in that some of its cells are genetically female while others are genetically male. Loss of an X-chromosome by a stem cell of an insect which developed from an XX zygote thus produces a clone of 'male' tissue. Sometimes expressed bilaterally, one half of the animal being phenotypically male, the other female. Also occurs in birds and mammals. See **intersex**.

Gynandrous (Bot.) (Of flowers) having stamens inserted on the gynoecium.

Gynobasic (Bot.) (Of a style) arising from base of ovary (due to infolding of ovary wall in development); e.g. white dead nettle.

Gynodioecious Having female and hermaphroditic flowers on separate plants; e.g. thymes (*Thymus*). Compare **androdioecious**, **gynomonoecious**.

Gynoecium (pistil) Collective term for the carpels of a flower; i.e. the female components of the flower. Compare **androecium**.

Gynogenesis Condition whereby a female animal must mate before she can produce parthenogenetic eggs. In some triploid thelytokous animals (the salamanders *Ambystoma*, the fish *Poeciliopsis*) the sperm penetrates the egg to initiate cleavage but contributes nothing genetically.

Gynomonoecious (Bot.) With female and hermaphroditic flowers on the same plant; e.g. many Compositae. Compare **andromonoecious**, **gynodioecious**.

Gyrogonite Lime encrusted, fossilized oogonia and encircling sheath cells (nucule) of the **Charophyceae**. Earliest fossils occur in the Upper **Silurian**. Gyrogonites of extant species date from the Upper **Carboniferous**.

Gyrus A fold (convolution) on the surface grey matter of the **cerebral cortex**, between which occur deep grooves, or sulci. Both gyri and sulci are functionally differentiated and identified by names.

Habitat Place or environment in which specified organisms live; e.g. sea shore. Compare **niche**.

Habituation **Learning**, in which an animal's response to a stimulus declines with repetition of the stimulus at the same intensity. Needs to be distinguished experimentally from sensory **adaptation** and muscular fatigue.

Haem (heme) Iron-containing **porphyrin**, acting as **prosthetic group** of several pigments, including **haemoglobin**, **myoglobin** and several **cytochromes**.

Haemagglutination Clumping of red blood cells due to cross-linking of cells by antibody to surface antigens. Not to be confused with **blood clotting**.

Haemagglutinin Glycoprotein product of influenza virus, associated, like neuraminidase, with the encapsulating host cell membrane and involved in attachment to host cells. Antigenic variation (shift) in haemagglutinin is responsible for new epidemics of the virus, against which earlier antibodies are ineffective.

Haemerythrin Reddish-violet iron-containing respiratory pigment of sipunculids, one polychaete, priapulids and the brachiopod *Lingula*. Prosthetic group is not a porphyrin, the iron attaching directly to the protein. Always intracellular.

Haemocoel Body cavity of arthropods and molluscs, containing blood. Continuous developmentally with the **blastocoele**. Unlike the **coelom**, it never communicates with the exterior or contains gametes. Often functions as a hydrostatic skeleton. Contains haemolymph.

Haemocyanin Copper-containing protein (non-porphyrin) respiratory pigment occurring in solution in haemolymph of malacostracan and chelicerate arthropods and in many molluscs. Blue when oxygenated, colourless when deoxygenated.

Haemoglobin Protein respiratory pigment with iron(Fe^{2+})-containing porphyrin as prosthetic group. Tetrameric molecule, comprising two pairs of non-identical polypeptides associated in a quaternary structure, binding oxygen reversibly, forming *oxyhaemoglobin*. Occurs intracellularly in vertebrate **erythrocytes**, but when found in invertebrates (e.g. earthworms) is usually in simple solution in the blood. Also found in root nodules of leguminous plants (as *leghaemoglobin*), but only if *Rhizobium* is present. Scarlet when oxygenated, bluish-red when deoxygenated. The ability of the haemoglobin molecule to pick up and unload oxygen depends on its shape in solution, which varies allosterically with local pH (see **buffer**). This in turn is a function of the partial pressure of CO_2 (see **Bohr effect**). Haemoglobins are adapted for maximal loading and unloading of oxygen within the

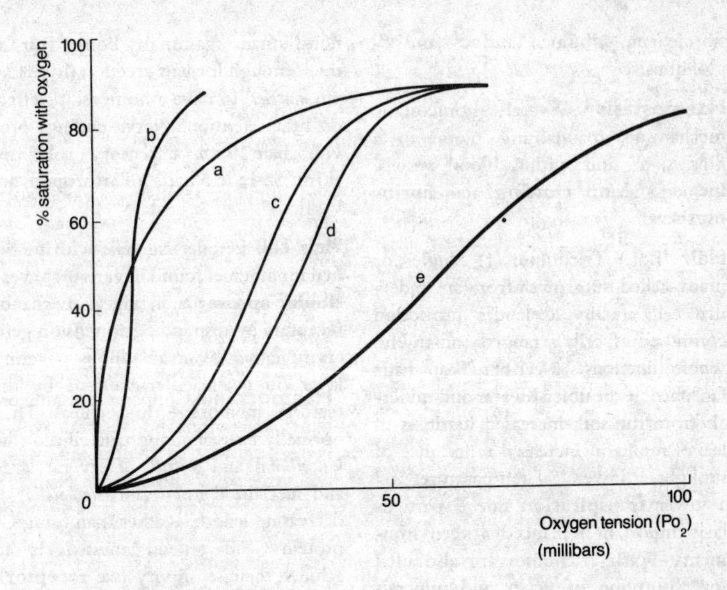

Fig. 55 *Oxygen equilibration curves for: (a) myoglobin, (b) Arenicola Hb, (c) human foetal Hb, (d) adult human Hb, (e) pigeon Hb.*

oxygen tension ranges occurring in their respective organisms. See Fig. 55.

Normal adult human haemoglobin (HbA) contains two α- and two β-globin chains; foetal haemoglobin (HbF) contains two α- and two γ-globin chains. The transition from foetal to adult haemoglobin production begins late in foetal life and is completed in early infancy. Some carbon dioxide is carried by haemoglobin as *carbamino compounds*, and one consequence of inhaling cigarette smoke is that carbon monoxide produced binds irreversibly to haemoglobin, forming *carboxyhaemoglobin*. See **myoglobin**, **sickle-cell anaemia**, **thalassaemias**.

Haemolysis Rupture of red blood cells (e.g. through osmotic shock) with release of haemoglobin.

Haemophilia Hereditary disease, sex-linked and recessive in humans, in which blood fails to clot owing to absence of Factor VIII (see **blood clotting**).

Haemopoiesis Blood formation; in vertebrates includes both plasma and cells. In anoxia, the kidney produces a **colony stimulating factor**, *erythropoietin*, stimulating red cell production in red bone marrow (see **myeloid tissue**). In vertebrate embryos, erythropoiesis occurs commonly in yolk sac, liver, spleen, lymph nodes and bone marrow; in adults is restricted to red bone marrow (and yellow bone marrow in long bones under oxygen stress). Leucocytes also originate in myeloid tissue but from different stem cells from erythrocytes. Much of the plasma protein is formed in the liver, notably fibrineogen, albumen and α- and β-globulins.

Haemostasis Several homeostatic mechanisms maintaining blood in a fluid state, and within blood vessels. Includes **blood clotting** and **fibrinolysis**.

Hair (Bot.) Trichome. (1) Single- or many-celled outgrowth from an epidermal cell; usually a slender projection composed of cells arranged end-to-end whose functions are various. Root hairs facilitate absorption of water and minerals from the soil; increased hairiness of leaves results in increased reflectance of sunlight, a lower leaf temperature, and a lower **transpiration** rate – particularly important in plants of arid environments. Foliar trichomes are also used for absorption of water and minerals (e.g. bromeliads); those of the saltbush (*Atriplex*) secrete salt from leaf tissue, preventing toxic accumulation in the plant; others may act as a defence against insects (hooked hairs of some species, e.g. *Phaseolus vulgaris*, may impale insects and larvae; glandular hairs may provide a chemical defence). (2) A simple filament of the **Cyanobacteria**, devoid of a sheath. (3) A flagellar appendage, either a fibrous solid hair or a tubular hair (see **flagellum**). (Zool.) Epidermal thread protruding from mammalian skin surface, composed of numerous cornified cells. Each develops from the base of a **hair follicle** (its 'root') in the undersurface of which (the hair bulb), cells are produced mitotically. Hair colour depends upon the amount of **melanin** present, and with age an increasing presence of air bubbles results in total reflection of light, making hair appear white. In most non-human mammals body hair is thick enough for hair erection (by *erector pili muscles*) to have a homeostatic effect on heat retention. Nerve endings provide hair with a sensory role (see **skin**). So-called 'hairs' of arthropods are bristles.

Hair cell Ectodermal cells with modified membranes found in vertebrate **vestibular apparatus**, acting as mechanoreceptors by responding to tension generated either by a gelatinous covering layer (in maculae, cristae), or by the tectorial membrane (in cochlea). They normally bear one long true cilium (the *kinocilium*) and a tuft of several large and specialized microvilli (*stereocilia*) of decreasing length. Rather than using G-proteins and second messengers to achieve their sensitivity (see **receptor**), an ion-channel gating 'lid' is located near the tip of each pivoting stereocilium so that when the channel opens to calcium (distortion of 0.1 nm is sufficient) there is a marked dip in the stiffness-displacement curve for these organelles and a receptor potential is generated.

Hair follicle Epidermal sheath enclosing the length of a hair in the skin. Surrounded by connective tissue serving for attachment of erector pili muscles. May house a **sebaceous gland**. See **skin**.

Hairpin (foldback) loops Palindromic duplex DNA sequences (**inverted repeat sequences**) which can fold back on themselves and form hairpin double-stranded structures on renaturing after being denatured. Single-stranded RNAs may also form such loops. Sometimes the degree of base-pairing required for hairpins to form is not great.

Haldane's rule J. B. S. Haldane (1922) proposed the rule that 'when in the F_1 offspring of two different animal races one sex is absent, rare or sterile, that sex is the heterogametic sex'.

Halophyte Plant growing in and tolerating very salty soil typical of shores of tidal river estuaries, saltmarshes, or alkali desert flats.

Hallux 'Big toe'; innermost digit of tetrapod hind foot. Often shorter than other digits. Turned to the rear in most birds, for perching. Compare **pollex**.

Haltere Modified hind wing of **Diptera** (two-winged flies) concerned with maintenance of stability in flight. Comprises basal lobe closest to thorax, a stalk and an end knob. Halteres are like gyroscopes; their combined nervous input to the thoracic ganglion enables adjustment of wings to destabilizing forces.

Hamilton's rule Prediction that genetically determined behaviour which benefits another organism, but at some cost to the agent with the allele(s) responsible, will spread by **selection** when the relation $(rb - c) > 0$ is satisfied; where r = degree of **relatedness** between agent and recipient, b = improvement of individual **fitness** of recipient caused by the behaviour and c = cost to agent's individual fitness as a result of the behaviour. See **altruism, kin selection.**

Haplochlamydeous See **monochlamydeous**.

Haplodiplontic (Of a **life cycle**) in which both haploid and diploid mitoses occur.

Haploid (Of a nucleus, cell, etc.) in which chromosomes are represented singly and unpaired. The haploid chromosome number, *n*, is thus half the **diploid** chromosome number, 2*n*. The amount of DNA per haploid genome is its **C-value**. Haploid cells are commonly the direct product of **meiosis**, but haploid mitosis is relatively common too. No haploid cell can undergo meiosis. Diploid organisms generally produce haploid gametes. In humans, $n = 23$. See **alternation of generations**, **male haploidy**, **polyploidy**.

Haploid parthenogenesis (Bot.) Development of an embryo into a haploid sporophyte from an unfertilized egg on a haploid gametophyte. See **parthenogenesis**.

Haplo-insufficiency See **mutation**.

Haplont Organism representing the **haploid** stage of a **life cycle**. Compare **diplont**.

Haplontic (Of a **life cycle**) in which there is no diploid mitosis, but in which haploid mitosis does occur.

Haplostele Solid cylindrical **stele** in which a central strand of primary xylem is sheathed by a cylinder of phloem.

Haplotype A haploid genotype. The gametes produced by a normal outbred diploid individual will be of a variety of haplotypes.

Hapten Molecule binding specifically to an antigen-binding site (epitope) on an antibody molecule or lymphocyte receptor, but without inducing any immune response.

Hapteron (holdfast) Bottom part of some algae, attaching the plant to the substratum; may be discoid or root-like in structure.

Haptonema Filamentous appendage arising near eukaryotic **flagellum**, but thinner and with different properties and structure. Occurs in many of the **Prymnesiophyta** (haptophytes). Function not fully understood; but can serve as temporary attachment to a surface and has been implicated in acquisition of food. Structurally variable; e.g. may be reduced, short and bulbous, or a long whiplash. In transverse section it comprises three concentric membranes surrounding a core containing seven microtubules; there is no contact between the core microtubules and the outer portion of the haptonema. Space between innermost and middle membranes is a vesicle, continuous over the tip of the core. Commonly covered with small scales; most are capable of coiling.

Haptotropism (thigmotropism) (Bot.) A **tropism** in which the stimulus is a localized contact, e.g. tendril in contact on one side with solid object such as a twig; response is curvature in that direction producing coiling around the object.

Hardy-Weinberg theorem (H-W law, principle or equilibrium) Theorem predicting for a normal amphimictic population the ratios of the three genotypes (e.g. AA:Aa:aa) at a locus with two segregating alleles, A and a, given the frequencies of these genotypes in the parent population. The theorem assumes: random (i.e. non-assortative) mating, no **natural selection**, **genetic drift** or **mutation**, or immigration or emigration. Its utility is that once the parental population's genotypic ratios have been determined, their predicted ratio in the next generation can be checked against the observed values, and any departures from expectation tested for significance (e.g. by **chi-squared test**). If significant, and if all assumptions other than selection can be discounted, then this is *prima facie* evidence that selection caused the departures from expected values. If in the parental generation the frequencies of alleles A and a are p and q respectively, then the theorem states that the genotypic ratios in the next and all succeeding generations will be:

AA : Aa : aa
$p^2 : 2pq : q^2$

Hatch-Slack pathway See **photosynthesis**.

Haustorium (pl. **haustoria**) Specialized penetrative food-absorbing structure. Occurs (1) in certain fungal parasites of plants, at the end of a hyphal branch within a living host cell; (2) in **lichens**, commonly penetrating the algal cells; (3) in some parasitic plants (e.g. dodder), withdrawing material from the host tissues.

Haversian system (osteon) Anatomical unit of compact **bone**. Comprises a central *Haversian canal*, which branches and anastomoses with those of other Haversian systems and contains blood vessels and nerves, surrounded by layers of bone deposited concentrically by osteocytes and forming cylinders. Blood is carried from vessels at the bone surface to the Haversian system by *Volkmann's canals*.

H-2 complex Mouse major histocompatibility complex. See **MHC**.

HDL High-density **lipoprotein**.

Heart Muscular, rhythmically contracting pump forming part of the cardiovascular system and responsible for blood circulation. All hearts have valves to prevent back-flow of blood during contraction. Initiation of heart beat may be by extrinsic nerves (*neurogenic*), as in many adult arthropods, or by an internal pacemaker (*myogenic*), as in vertebrates and some embryonic arthropods.

Vertebrate heart muscle (**cardiac muscle**) does not fatigue, and is under the regulation of nerves (see **cardio-acceleratory/inhibitory centres**) and hormones (e.g. **adrenaline**). The basic S-shaped heart of most fish comprises four chambers pumping blood unidirectionally forward to the gills. This *single circulation* has the route:

body → heart → gills → body.

Replacement of gills by lungs in tetrapods was associated with the need for a heart providing a *double circulation* in order to keep oxygenated blood (returning to the heart from the lungs) separate from deoxygenated blood (returning to the heart from the body). The blood route becomes:

body → heart → lungs → heart → body.

In amphibians, two atria return blood from these two sources to a single ventricle, and separation is limited; reptiles have a very complex ventricle with a septum assisting separation; birds and mammals have two atria and two ventricles, one side of the heart dealing with oxygenated and the other with deoxygenated blood (see **heart cycle**). In annelids, the whole dorsal aorta may be contractile with, in addition, several vertical contractile vessels, or 'hearts'. The insect heart is a long peristaltic tube lying in the roof of the abdomen and perforated by paired segmental holes (*ostia*) through which blood enters from the haemocoele. There may be accessory hearts in the thorax. Blood is driven forwards into the aorta, which opens into the haemocoele. A similar arrangement occurs in other arthropods. The basic molluscan heart comprises a median ventricle and two atria. See **pericardium**.

Heart cycle (cardiac cycle) One complete sequence of contraction and relaxation of heart chambers, and opening and closing of valves, during which time the same volume of blood enters and leaves the heart. Chamber contraction (*systole*) is followed by its relaxation (*diastole*) when it fills again with blood. In mammals and birds ventricular diastole draws in most of the blood from the atria; atrial systole adds only 30% to ventricular blood volume. The **pacemaker** and its associated fibres ensure that the two atria contract simultaneously just prior to the two ventricles. Atrioventricular valves open when atrial pressure exceeds ventricular and close when ventricular pressure exceeds atrial. See **cardio-acceleratory centre**.

Heartwood Central mass of xylem tissue in tree trunks; contains no living cells and no longer functions in water conduction but serves only for mechanical support; its elements frequently blocked by **tyloses**, and frequently dark-coloured (e.g. ebony), impregnated with various substances (tannins, resins, etc.) that render it more resistant to decay than surrounding sapwood.

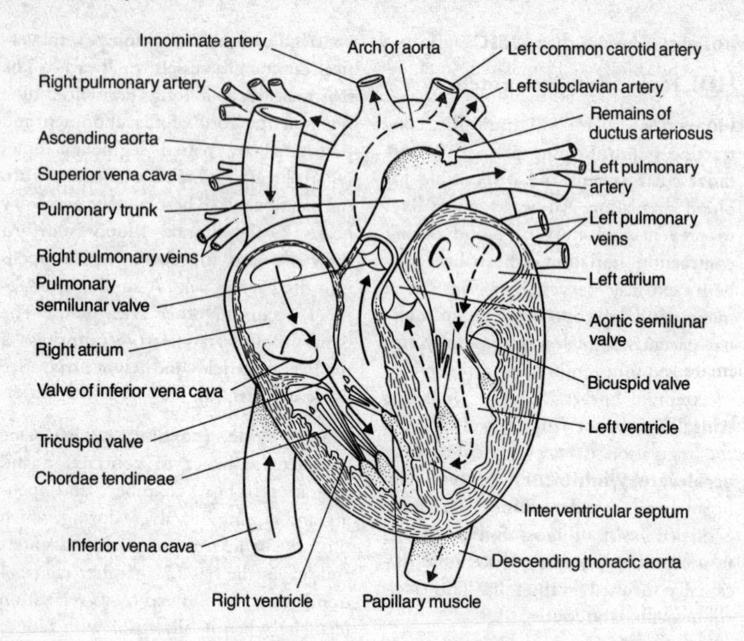

Fig. 56 *Mammalian* **heart** *showing direction of oxygenated (--->) and deoxygenated (→) blood.*

Heat-shock proteins See **chaperones**.

HeLa cell Cell from human cell line widely used in study of cancer. Original source was Helen Lane, a carcinoma patient, in 1952.

Helicase An ATP-dependent enzyme which, at a replication fork, breaks the hydrogen bonds holding the two strands of duplex DNA together. Special proteins then bind each of the DNA strands and prevent re-annealing.

Heliophytes Class of **Raunkiaer's life forms.**

Heliotropism The diurnal movement of the leaves and flowers of many plants which are orientated either perpendicular (*diaheliotropism*) or parallel (*paraheliotropism*) to the sun's direct rays (also known as solar tracking). Unlike stem **phototropism**, leaf movement of heliotropic plants does not result from asymmetric growth; in the majority of cases, movement involves **pulvini** at the bases of the leaves; sometimes the whole petiole possesses pulvinal-like properties.

Heliozoa Sarcodina of generally freshwater environments, without shell or capsule, but sometimes with siliceous skeleton, and usually very vacuolated outer protoplasm. Locomotion by 'rolling', successive pseudopodia pulling the animal over in turn. Food is caught by cytoplasm flowing over axial supports of pseudopodia. Flagellated stage common. Some are autogamous; binary fission usual.

Helix-turn-helix genes Genes encoding those **DNA-binding proteins** whose structural motif is the helix-turn-helix. They include **homeobox** genes as a subset.

Helminth Term usually applied to parasitic flatworms, but occasionally to nematodes also.

Helper An animal which helps rear the young of a conspecific to which it is not paired or mated. Commonly there are genetic bonds between the helper and its beneficiary 'family'. Many of the studies are on communal nesters in birds. Of considerable theoretical interest. See **Hamilton's rule**, **inclusive fitness**, **altruism**.

Helper cell See **T cell** and Fig. 41.

Hemicelluloses Heterogeneous group of long-chain polysaccharides, (mostly $\beta 1 \rightarrow 4$-linked) and composed entirely of a single monomer, be it arabinose, xylose, mannose or galactose. Hydrogen-bonded to cellulose in plant **cell walls**, especially in lignified tissue. More soluble and less ordered than cellulose; may function as food reserve in seeds (e.g. in endosperm of date seeds).

Hemichordata Group of marine organisms of disputed phylogenetic relationships. Either a distinct phylum, or a subphylum of the **Chordata**. Includes the pterobranchs, and the acorn worms (enteropneusts) such as *Balanoglossus*. Lack both **endostyle** and any homologue of **notochord**. Possess *proboscis pore* (detectable also in cephalochordates and craniates) and gill slits. Nerve cord(s) usually solid. Development indirect; the enteropneust larva is the *tornaria*.

Hemicryptophytes Class of **Raunkiaer's life forms**.

Hemidesmosome See **desmosome**.

Hemimetabola See **Exopterygota**.

Hemiptera (Rhyncota) Large order of exopterygotan insects. Includes aphids, cicadas, bed bugs, leaf hoppers, scale insects. Of enormous economic importance. Usually two pairs of wings, the anterior pair either uniformly harder (*Homoptera*) or with tips more membranous than the rest of the wing (*Hemiptera*). Mouthparts for piercing and sucking. Many are vectors of pathogens.

Hemizygous Term applied to cell or individual where at least one chromosomal locus is represented singly (i.e. its homologue is absent), in which case the locus is hemizygous. Sometimes a chromosome pair bears a non-homologous region (as in the **heterogametic sex**), or all chromosomes are present singly (as in **haploidy**).

Hensen's node See **primitive streak**, **retinoic acid**.

Heparin **Glycosaminoglycan** product of **mast cells**; an anticoagulant, blocking conversion of prothrombin to thrombin. Reduces **eosinophil** degranulation. Stored with **histamine** in mast cell granules, and hence found in most connective tissues.

Hepatic (Adj.) Relating to the **liver**.

Hepaticae See **Hepaticopsida**.

Hepaticopsida (Hepaticae) Liverworts. Class of **Bryophyta**, whose sporophytes develop capsule maturation and undergo meiosis before the seta elongates. Consist of a thin prostrate, or creeping to erect body (thallus), a central stem with three rows of leaves, attached to the substratum by rhizoids. Sex organs antheridia and archegonia, variously grouped; microgametes flagellated and motile. Fertilization is followed by development of a capsule containing spores, which germinate on being shed to form most usually a short thalloid **protonema** from which new liverwort plants arise. Includes leafy and thallose species. Generally occur in moist soils, on rock, or epiphytically. Rarely aquatic. See **life cycle**.

Hepatic portal system System of veins and capillaries conveying most products of digestion (not **chylomicrons**) in cephalochordates and vertebrates from the gut to the liver. Being a portal system, it begins and ends in capillaries.

Herb Plant with no persistent parts above ground, as distinct from shrubs and trees.

Herbaceous Having the characters of a herb.

Herbarium Collection of preserved and diverse plant specimens, usually arranged according to a classificatory scheme. Used as a reference collection for checking identities of newly collected specimens, as an aid to teaching, as a historical collection, and as data for research.

Herbivore Animal feeding largely or entirely on plant products. See **food web**, **carnivore**, **omnivore**.

Hereditary (Adj.) Of materials and/or information passed from individuals of one generation to those of a future generation, commonly their direct genetic descendants. Hereditary and genetic material are not identical: an egg cell, e.g., contains a great deal of cytoplasm that is non-genetic; material passed from mother to embryo across a placenta might also be termed hereditary but not genetic.

Heritability Roughly speaking, the degree to which a character is inherited rather than attributable to non-heritable factors; or, that component of the variance (in the value) of a character in a population which is attributable to genetic differences between individuals. Estimation of heritability is complex. May be regarded as the ratio of additive genetic variance to total phenotypic variance for the character in the population, where *additive genetic variance* is the variance of breeding values of individuals for that character, and where *breeding value* (which is measurable) is twice the mean deviation from the population mean, with respect to the character, of the progeny of an individual when that individual is mated to a number of individuals chosen randomly from the population.

Hermaphrodite (bisexual) (Bot.) (Of a flowering plant or flower) having both stamens and carpels in the same flower. Compare **unisexual**, **monoecious**. (Zool.) (Of an individual animal) producing both sperm and ova, either simultaneously or sequentially. Does not imply self-fertilizing ability, but if self-compatible such individuals would probably avoid the **cost of meiosis**. Commonly, but not exclusively, found in animals where habit makes contact with other individuals unlikely (e.g. many parasites) or hazardous. Rare in vertebrates. See **ovotestis**, **parthenogenesis**.

Heterocercal Denoting type of fish tail (caudal fin) characteristic of **Chondrichthyes**, in which vertebral column extends into dorsal lobe of fin, which is larger than the ventral lobe. Compare **homocercal**.

Heterochlamydeous (Of flowers) having two kinds of perianth segments (sepals and petals) in distinct whorls. Compare **homochlamydeous**.

Heterochromatin Parts of, or entire, chromosomes which stain strongly basophilic in interphase. Such regions are transcriptionally inactive and highly condensed. *Facultative heterochromatin* (as in inactivated X-chromosomes of female mammals) occurs in only some somatic cells of an organism and appears not to comprise repeat DNA sequences. The resulting animal may thus be a **mosaic** of cloned groups of cells, each with different heterochromatic chromosome regions. *Constitutive heterochromatin* (e.g. **telomeres**, around human **centromeres**, and **chromocentres** of e.g. *Drosophila*) comprises condensed chromosome regions that occur in all somatic cells of an organism and often consist of DNA repeat sequences. Can cause **position effect** variegation.

Heterochrony Changes during **ontogeny** in the relative times of appearance and rates of development of characters which were already present in ancestors. Sometimes regarded as inclusive of two distinct processes: *progenesis*, in which development is cut short by precocious sexual maturity; and *neoteny*, in which somatic development is retarded for selected organs and parts. See **allometry**.

Heterocyst Specialized cell of some filamentous genera of **Cyanobacteria**. Larger than vegetative cells and dependent on them nutritionally (via microplasmodesmata), with a higher respiration rate but only half the phycobilisomes; they lack photosystem II of **photosynthesis**, have no short wavelength form of chlorophyll a (670 nm) but a higher concentration of P_{700} in photosystem I. Unlike vegetative cells, they appear empty and not granular under light microscopy, but develop from them by dissolution of thylakoids and storage products and production of new internal membranes and a **multilayered structure** outside the cell wall. Heterocysts are involved in **nitrogen fixation**, producing the enzyme nitrogenase. They divide only exceptionally, have a limited physiological life (vacuolizing when senescent) and usually break off from the filament, causing a fragmentational form of vegetative reproduction.

Heterodimer A protein (e.g. an **integrin**) composed of two different polypeptide chains (e.g. α and β) held together in quaternary structure. In homodimers, the two polypeptides are identical. See Fig. 110.

Heterodont See **dentition**.

Heteroduplex The double helix (duplex) formed by annealing of two single-stranded DNA molecules from different original duplexes so that mispaired bases occur within it. Heteroduplex regions are likely to occur as a result of most kinds of **recombination** involving breakage and annealing of DNA, and may be short-lived, for when heteroduplex DNA is replicated any mispaired bases should base-pair normally in the newly synthesized strands. Furthermore, most organisms have **DNA repair mechanisms** of greater or lesser efficiency for correcting base-pair mismatches by excision/replacement. See **DNA hybridization**, ***recA***.

Heteroecious (Of **rust fungi**) having certain spore forms of the life cycle on one host species, and other forms on an unrelated host species; e.g. *Puccinia graminis* (wheat rust). Compare **autoecious**.

Heterogametic sex The sex producing gametes of two distinct classes (in approx. 1 : 1 ratio) as a result of its having **sex chromosomes** that are either partially **hemizygous** (as in XY individuals) or fully hemizygous (as in XO individuals). This sex is usually male, but is female in birds, reptiles, some amphibia and fish, Lepidoptera, and a few plants. Sometimes the XY notation is restricted to organisms having male heterogamety, female heterogamety being symbolized by ZW (males here being ZZ). See **homogametic sex**, **sex determination**, **sex linkage**.

Heterograft See **xenograft**.

Heterokaryosis Simultaneous existence within a cell (or hypha or mycelium of **fungi**) of two or more nuclei of at least two different genotypes to produce a '*heterokaryon*'. These nuclei are usually from different sources, their association being the result of plasmogamy between different strains. They retain their separate identities prior to karyogamy. Heterokaryosis is extremely common in coenocytic filamentous fungi. **Ascomycota** and **Basidiomycota** have a dikaryotic phase in their life cycles (see **dikaryon**). See **parasexuality**.

Heterokont Any organism producing, at some stage in its life-cycle, cells with two anteriorly attached undulipodia (see **flagellum**) of different lengths and/or structure. The three algal taxa with such cells (**Xanthophyta**, **Chrysophyta**, **Phaeophyta**) are sometimes grouped into a single phylum, Heterokonta.

Heteromerous (Of lichens) where the thallus has algal cells restricted to a specific layer, creating a stratified appearance.

Heterometabola See **Exopterygota**.

Heteromorphism Occurrence of two or more distinct (heteromorphic) morphological types within a population, due to environmental and/or genetic causes. Examples include genetic **polymorphism**, **heterophylly**, and the phenotypically distinct phases of those life cycles, particularly in the algae (e.g. the green alga *Derbesia*), with **alternation of generations**.

Heterophylly Production of morphologically dissimilar leaves on the same plant. Many aquatic plants produce submerged leaves that are much dissected, while floating leaves of the same plant are simple and entire. Juvenile stages of some plants have leaves that are morphologically different from 'adult' forms. See **heteromorphism**, **phenotypic plasticity**.

Heteroptera Suborder of **Hemiptera.**

Heteropycnosis The occurrence of **heterochromatin**. Heterochromatic regions were formerly called heteropycnotic.

Heterosis See **hybrid vigour**.

Heterosporous (Bot.) Of individuals or species producing two kinds of spore, microspores and megaspores, that give rise respectively to distinct male and female gametophyte generations. Examples are found in some club mosses and ferns and all seed plants. See **alternation of generations, life cycle**.

Heterostyly Condition in which the length of style differs in flowers of different plants of the same species, e.g. pin-eyed (long style) and thrum-eyed (short style) primroses (*Primula*). Anthers in one kind of flower are at same level as stigmas of the other kind. A device for ensuring cross-pollination by visiting insects. Compare **homostyly**. See **polymorphism, supergene**.

Heterothallism (Of algae, fungi) the condition whereby sexual reproduction occurs only through participation of thalli of two different **mating types**, each self-sterile. In fungi, includes *morphological heterothallism*, where mating types are separable by appearance, and *physiological heterothallism*, where interacting thalli (often termed plus and minus strains) offer no easily recognizable differences by which to distinguish them. Compare **homothallism**.

Heterotrichous (Of algae) having a type of thallus comprising a prostrate creeping system from which project erect branched filaments. Common in filamentous forms.

Heterotrophic (organotrophic) Designating those organisms dependent upon some external source of organic compounds as a means of obtaining energy and/or materials. All animals, fungi, and a few flowering plants are *chemoheterotrophic* (chemotrophic), depending upon an organic carbon source for energy. Some autotrophic bacteria (purple non-sulphur bacteria) are *photoheterotrophic*, using solar energy as their energy source but relying on certain organic compounds as nutrient materials. Both these groups contrast with those organisms (*photoautotrophs*) able to manufacture all their organic requirements from inorganic sources, and upon which all heterotrophs ultimately depend. Thus all herbivores, carnivores, omnivores, saprotrophs and parasites are heterotrophs. See **decomposer**.

Heterozygous Designating a locus, or organisms, at which the two representatives (alleles) in any diploid cell are different. Organisms are sometimes described as *heterozygous for* a character determined by those alleles, or *heterozygous at* the locus concerned. Thus, where two alleles *A* and *a* occupy the *A-locus*, of the three genotypes possible (*AA*, *Aa*, *aa*), *Aa* is heterozygous while the other two are **homozygous**. See **dominance**.

Heterozygous advantage Selective advantage accruing to heterozygotes in populations and which may be responsible for some balanced **polymorphisms**. Mutations in diploids arise in the heterozygous condition and must therefore confer an advantage in that state if they are to spread. Theory has it that if a mutation is advantageous, selection will make its advantageous heterozygous effects dominant and its deleterious effects recessive, thus giving heterozygotes a higher **fitness** than homozygotes. There are remarkably few well-documented examples in which the evidence for heterozygous advantage is conclusive. See **sickle-cell anaemia**. Compare **balanced lethal system**.

Hexokinase Enzyme phosphorylating free glucose within cell, producing glucose-phosphate, which cannot pass out across the plasma membrane. See **insulin**.

Hexose Carbohydrate sugar (monosaccharide) with six carbon atoms in its molecules. Includes glucose, fructose and galactose. Combinations of hexoses make up most of the biologically important disaccharides and polysaccharides.

Hibernation Far-reaching physiological adaptation of some homeothermic animals to prolonged cold. Marked drops in body temperature (to perhaps 1°C above ambient temperature) and basal metabolic rate (down to 1% of normal) are associated with the ability to elevate these again if conditions become dangerously cold (contrast **diapause** in insects). Common in mammalian orders Insectivora, Chiroptera and especially Rodentia. See **adipose tissue**, **aestivation**.

High-energy phosphates Group of phosphorylated compounds transferring chemical energy required for cell work. Depends upon their tendency to donate their phosphate group to water (to be hydrolysed) as is indicated by their **standard free energies** of hydrolysis (the more negative it is, the greater the tendency to be hydrolysed). *Phosphate-bond energy* (not *bond energy*, the energy required to *break* a bond) indicates the difference between free energies of reactants and products respectively before and after hydrolysis of a phosphorylated compound.

Fig. 57 indicates tendencies for phosphate groups (shown by arrows) to be transferred between commonly occurring phosphorylated compounds of cells. High-energy phosphate bonds arise because of *resonance hybridity* between single and double bonds of the phosphorus/oxygen atoms, which renders them more stable (i.e. they have less free energy) than expected from structure alone. All common phosphorylated compounds of cells, including ADP and AMP, are resonance hybrids at the phosphate bond; but the terminal phosphate of AMP has a standard free energy less than half as negative as those of ADP and ATP.

High-mobility group (HMG) proteins Acid-soluble, non-histone nuclear **DNA-binding proteins** with an HMG domain capable of bending or looping DNA in such a way as to form DNA supercoils. Some may have a role in immunoglobulin gene maturation.

Hill reaction Light-induced transport of electrons from water to acceptors such as potassium ferricyanide (*Hill reagents*) which do not occur naturally, accompanied by the release of oxygen. During it, electron acceptors become reduced. Named after R. Hill, who studied the process in 1937 in isolated chloroplasts. See **photosynthesis**.

Hilum Scar on seed coat, marking the point of former attachment of seed to funicle.

Hindbrain Hindmost of the three ex-

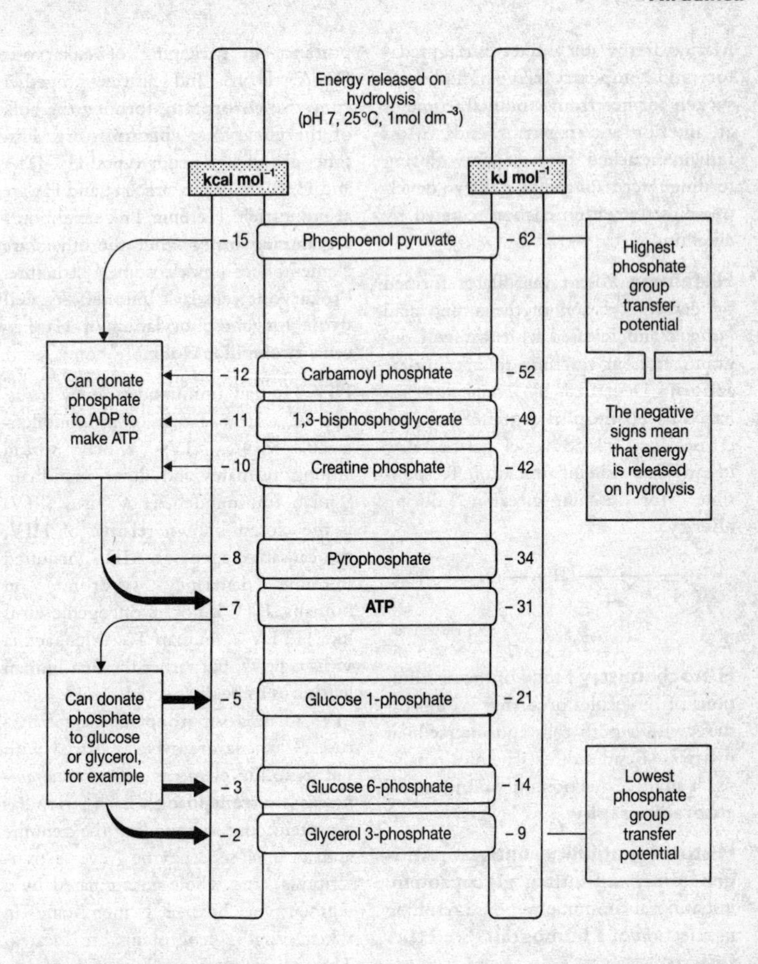

Fig. 57 *The flow of phosphate groups from substances with high transfer potential to those with low-transfer potential, assuming molar concentrations of reactants and products.* See **high energy phosphates**.

panded regions of the vertebrate brain as marked out during early embryogenesis, developing into the cerebellum and medulla oblongata. See **brain**.

Hip girdle See **pelvic girdle**.

Hippocampus Part of the cerebral cortex, with a special role in learning (in which long-term **potentiation** is involved).

Hirudinea Leeches. Class of **Annelida**.

Marine, freshwater and terrestrial predators and temporary ectoparasites with suckers formed from modified segments at anterior and posterior ends. Most remain attached to host only during feeding. Hermaphrodite; embryo develops directly within cocoon secreted by clitellum.

Histamine Potent vasodilator formed by decarboxylation of the amino acid *histidine* and released by **mast cell** degranulation in response to appropriate antigen. Degraded by *histaminase* released by **eosinophil** degranulation. Increases local blood vessel permeability in early and mild inflammation. Responsible for itching/sneezing during **allergy**.

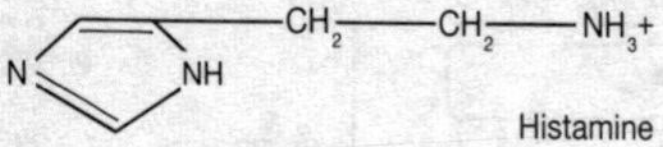

$-CH_2-CH_2-NH_3^+$
N NH

Histamine

Histochemistry Study of the distributions of molecules occurring within tissues, within both cells and intercellular matrices. Besides direct chemical analysis, it involves sectioning, **staining** and **autoradiography**.

Histocompatibility antigen Antigen (often cell-surface **glycoprotein**) initiating an immune response resulting in rejection of a **homograft**. See **HLA system**.

Histogenesis Interactive processes whereby undifferentiated cells from major **germ layers** differentiate into tissues.

Histology Study of tissue structure, largely by various methods of **staining** and **microscopy**.

Histones Basic proteins of major importance in packaging of eukaryotic DNA. DNA and histones together comprise **chromatin**, forming the bulk of the eukaryotic **chromosome**. Histones are of five major types: H1, H2A and H2B are lysine-rich; H3 and H4 are arginine-rich. H1 units link neighbouring **nucleosomes** while the others are elements of nucleosome structure. Prokaryotic cells lack histones. See **cell cycle** for phosphorylation of H1 histone; **molecular clock**.

HIV Human immunodeficiency virus; one of a large group of immunodeficiency viruses (IVs) widely spread among primates and other mammals. Simian immunodeficiency virus (SIV) is the closest known relative of HIV, the causative agent of AIDS (acquired immune deficiency syndrome) in humans. HIV is not an oncogenic virus like HTLV-1 (human T cell leukaemia virus type 1), but rather the first human lentivirus to be discovered.

IVs form a subgroup of the retroviruses – whose life cycle is shared with the genomic elements called *retrotransposons* (see **transposon**). Each virion has a protein core surrounding the genome and an enclosed enzyme (reverse transcriptase), the whole encapsulated by a segment of host cell membrane in which viral glycoproteins are located. This glycoprotein (gp120) recognizes and binds to the **accessory molecule** CD4, so HIV infects any $CD4^+$ cell, including T-helper (T_H) cells (see **T cell**). Once bound, the membranes of HIV and the host cell fuse, releasing the infective virus core within the host cell.

HIV's genome comprises an RNA molecule (two copies per virion), 5–10 kb in length, housing at least three genes (*gag*, encoding core proteins; *pol* encoding viral enzymes; *env*, encoding envelope glycoproteins). On entry into the host cell, viral reverse transcriptase creates a cDNA copy of the RNA genome which is then converted into double-stranded DNA capable of inserting into the human genome and existing as a provirus for long periods; however, there is no evidence that HIVs integrate into the germ line. Eventually, productive virus synthesis occurs and new HIV particles leave the host cell encapsulated in its modified membrane. T cells producing HIV no longer divide, and eventually die. Also, because of the gp120-CD4 binding, HIV-infected cells bind to uninfected $CD4^+$ cells to produce syncytia, at which point the uninfected T cells lose their immune capacity and die. Immunodeficiency results.

The virus is transmitted in sexual fluids and blood plasma (hence on contaminated hypodermic needles) as well as across the placenta and probably in breast milk. Perinatal infection (at the time of birth) through haemorrhage is also possible. Those receiving blood transfusions, haemophiliacs receiving blood products, and those using non-sterile needles for intravenous drug injection may all at times risk contamination, although blood products are increasingly screened/treated for HIV. Travellers to the 'Third World' should preempt such risks by obtaining medical insurance enabling them to fly home in emergency.

There are numerous strains of HIV that cross-react minimally or not at all with neutralizing antibodies targeting other strains and it is not known how many strains would be needed in a **vaccine** providing broad anti-HIV protection. Testing a live-attenuated HIV vaccine on humans would be risky, but with a projected 30–40 million people infected with HIV by year 2000 (90% in developing countries), all avenues deserve consideration and some products have already entered safety trials with uninfected volunteers. Clinical trials using the drug GEM-91 (see **oligonuceotide**) are taking place.

HLA system (human leucocyte antigen system) The most important human **major histocompatibility complex**, located as a gene cluster on chromosome 6 and probably involving several hundred gene loci.

hnRNA (pre-mRNA) Heterogeneous nuclear RNA. See **RNA processing**.

hnRNP See **RNA processing**.

Holarctic Zoogeographical **region** almalgamating the Palaearctic and Nearctic regions.

Holliday structure (Holliday junction) Alternative term for the cross-strand exchange structure involved in some forms of **homologous recombination** events.

Holoblastic Form of **cleavage**.

Holocarpic (Of fungi) having the mature thallus converted in its entirety to a reproductive structure. Compare **eucarpic**.

Holocene (recent) The present, post-Pleistocene, epoch (system) of the **Quaternary** period.

Holocentric Of chromosomes with diffuse **centromere** activity, or a large numberof centromeres. Common in some insect orders (Heteroptera,

Lepidoptera) and a few plants (*Spirogyra, Luzula*).

Holocephali Subclass of the **Chondrichthyes**, including the ratfish *Chimaera*. First found in Jurassic deposits. Palatoquadrate fused to cranium (*autostylic* jaw suspension). Grouped with elasmobranchs because of their common loss of bone.

Holocrine gland Gland in which entire cells are destroyed with discharge of contents (e.g. sebaceous gland). Compare **apocrine gland**, **merocrine gland**.

Holoenzyme Enzyme/cofactor complex. See **enzyme**, **apoenzyme**.

Hologamodeme See **deme**.

Holometabola Those insects with a pupal stage in their life history. See **Endopterygota**. Compare **Thysanoptera**.

Holophytic Having plant-like nutrition; i.e. synthesizing organic compounds from inorganic precursors, using solar energy trapped by means of chlorophyll. Effectively a synonym of photoautotrophic. Compare **holozoic**; see **autotrophic**.

Holostei Grade of **Actinopterygii** which succeeded the chondrosteans as dominant Mesozoic fishes. Oceanic forms became extinct in the Cretaceous, but living freshwater forms include the gar pikes, *Lepisosteus*, and bowfins, *Amia*. Superseded in late Triassic and Jurassic by **Teleostei**. Tendency to lose **ganoine** covering to scales.

Holothuroidea Sea cucumbers. Class of **Echinodermata**. Body cylindrical, with mouth at one end and anus at the other; soft, muscular body wall with skeleton of scattered, minute plates; no spines or pedicillariae; suckered **tube feet**; bottom-dwellers, often burrowing; tentacles (modified tube feet) around mouth for feeding. Lie on their sides.

Holotype (type specimen) Individual organism upon which naming and description of new species depends. See **neotype**, **binomial nomenclature**.

Holozoic Feeding in an animal-like manner. Generally involves ingestion of solid organic matter, its subsequent digestion within and assimilation from a food vacuole or gut, and egestion of undigested material via an anus or other pore. Compare **holophytic**.

Homeobox Conserved DNA sequence motif of *c.* 180 base pairs, encoding a conserved domain (*homeodomain*) of 60 amino acids occurring in certain genes from probably all multicellular eukaryotes which, in animals at least, form a set of homologous gene families. First identified in 1984 within several **homeotic genes** of *Drosophila*, the homeobox product confers the helix-turn-helix motif upon a protein, giving it its DNA-binding properties (see **DNA-binding proteins**, **regulatory gene**). In eukaryotes these proteins therefore tend to be localized within nuclei, acting as **transcription factors**. See **homeogenes**, **compartment**.

Homeodomain See **homeobox**.

Homeogenes **Homeobox**-containing genes; of wide (possibly universal) occurrence in animals, from cnidarians to vertebrates, but also present in other eukaryotes (plants, fungi and slime moulds, at least). One subset, the *Hox* (Antennapedia-like) homeogene family (see **homeotic genes**), is involved in encoding the relative positions of structures along the antero-posterior body axis of (possibly all) animals and has recently been suggested as the defining character (synapomorphy) of the kingdom (see **zootype**). The same colinear order of expression of *Hox* gene clusters and their homologues is found in embryos as distant phylogenetically as those of fruit flies and mice: gene expression at the 3' end of each cluster on the chromosome starts in anterior embryonic regions, each next gene in the cluster in the 5' direction along the chromosome being expressed in progressively more posterior embryonic parts. Genes in a *Hox* cluster are numbered according to their positions within it. The **antennapedia complex** and **bithorax complex** of *Drosophila* form a single (split) *Hox* gene cluster. See **chromosomal imprinting**, **GAP genes**.

Homeostasis Term given to those processes, commonly involving *negative feedback*, by which both positive and negative control are exerted over the values of a variable or set of variables, and without which control the system would fail to function.

(1) *Physiological*. Various processes which help regulate and maintain constancy of the internal environment of a cell or organism at appropriate levels. Each process generally involves: (a) one or more sensory devices (misalignment detectors) monitoring the value of the variable whose constancy is required; (b) an input from this detector to some effector when the value changes, which (c) restores the value of the variable to normality, consequently shutting-off the original input (negative feedback) to the misalignment detector. In unicellular organisms homeostatic processes include osmoregulation by contractile vacuoles and movement away from unfavourable conditions of pH; in mammals (homeostatically sophisticated) the controls of blood glucose (see **insulin**, **glucagon**), CO_2 and pH levels, and its overall concentration and volume (see **osmoregulation**), of ventilation, heart rate and body temperature provide a few examples. (2) *Developmental*. Mechanisms which prevent the **fitness** of an organism from being reduced by disturbances in developmental conditions. The phrase *developmental canalization* has been used in this context. (3) *Genetic*. Tendency of populations of outbreeding species to resist the effects of artificial **selection**, attributable to the lower ability of homozygotes than heterozygotes in achieving *developmental homeostasis*. (4) *Ecological*. Several ecological factors serving to regulate population density, species diversity, relative biomasses of trophic levels, etc., may be thought of as homeostatic. See **arms race**, **balance of nature**, **chaos**, **density dependence**.

Homeothermy Maintenance of a constant body temperature higher than that of the environment. Involves physiological **homeostasis**. Characteristic of mammals and birds. Some fish are able to keep some muscles considerably warmer than the surrounding water, and there is considerable evidence that many large extinct reptiles were homoiothermic. See **poikilothermy**.

Homeotic genes Term describing a *control gene* which, by either being tran-

Hominid classifications

Old scheme	*New scheme*
Superfamily Hominoidea	Suborder Hominoidea
Family Hylobatidae	Family Proconsulidae
Family Pongidae	Genus *Proconsul*
Subfamily Dryopithecinae	Family Hylobatidae
Genus *Proconsul*	Family Hominidae
Genus *Dryopithecus*	Subfamily Dryopithecinae
Subfamily Ponginae	Tribe Sivapithecini
Genus *Pongo*	Genus *Sivapithecus*
Subfamily Gorillinae	Tribe Pongini
Genus *Gorilla*	Genus *Pongo*
Genus *Pan*	Subfamily Homininae
Family Hominidae	Tribe Gorillini
Genus *Australopithecus*	Genus *Gorilla*
Genus *Homo*	Genus *Pan*
	Tribe Hominini
	Genus *Australopithecus*
	Genus *Homo*

scribed or remaining silent during development (according to decisions between alternative pathways of **development)**, can profoundly affect the developmental fate of a region of a plant or animal's body. As yet found only in insects, one nematode and a few plants. A hierarchical sequence of binary decisions could provide clones of cells with 'genetic addresses' for differentiation. In the insect wing **imaginal disc**, such a decision sequence appears to be: anterior/posterior, dorsal/ventral, proximal/distal. A *homeotic mutation* is a DNA sequence change comprehensively transforming its own and its descendant cells' morphologies into those of a different organ, in insects, normally one produced by a different imaginal disc. In *Drosophila*, examples include *engrailed, antennapedia* and *bithorax*. Implications of homeosis for **homology** are controversial. *Antennapedia* mutants have all or parts of their antennae converted into leg structures. Since antennae are regarded as the paired appendages of the second embryonic somite, their homology with and evolutionary development from paired ambulatory appendages receives some support from this source. The same applies to the segmentally-arranged mouthparts.

(2) Of organs whose positions are altered as a result of homeotic mutation. See **compartment**, **homeobox**, **imaginal disc**.

Home range That part of a habitat that an animal habitually patrols, commonly learning about it in detail; occasionally, identical to the animal's total range. Differs from a **territory** in that it is not defended, but may be geographically identical to it if it becomes a territory at some part of the year.

Hominid Of Family Hominidae (Order Primates, Suborder Hominoi-

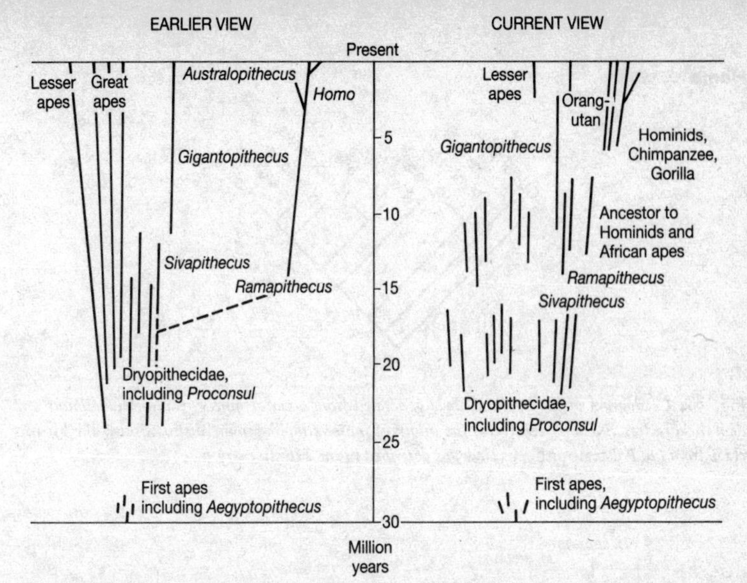

Fig. 58 *Changing views on hominoid relationships. Unbroken lines indicate fossil finds; broken lines indicate possible ancestry for which no fossils have provided evidence.*

dea). For long only **australopithecines** and ***Homo*** were included within the family, the great apes lying in a distinct family (Pongidae). Current taxonomy includes the **dryopithecines**, **Ponginae** and **Homininae** within the Hominidae, molecular evidence indicating that the old Pongidae/Hominidae distinction is unwarranted as African apes and humans are more closely related to one another than any of them is to the orang-utan. See schemes opposite.

Homininae Hominid subfamily (members are hominines). Includes the African ape and human clades. See **hominid**, **Hominini**.

Hominini Tribe within the hominid subfamily Homininae (members are hominins). Most distinctive attribute is bipedalism – upright walking. Short face (lacking specialized brain-cooling muzzles of baboons), small incisors and canines and tendency towards parabolic shape of dentary also characteristic, as is the elaboration of material culture. Probably originating *c.* 6–5 Myr BP, earliest fossils date from *c.* 3.5 Myr BP. Includes ***Homo***, and the **australopithecines** – whose detailed relationships with *Homo* require further fossil material. Premolecular data were once thought to suggest *Ramapithecus* (see **Ponginae**) as earliest fossil ape to date with human affinities; postmolecular data suggest a much more recent separation of the human lineage from apes, at perhaps as little as 5 Myr BP. Some even place African anthropoid apes (*Pan*, *Gorilla*) and hominins in the same clade, with sivapithecines in the sister clade.

Hominoid Member of the primate superfamily Hominoidea (see Fig. 58). Includes gibbons (Hylobatidae), the great apes (Pongidae) and **hominids**.

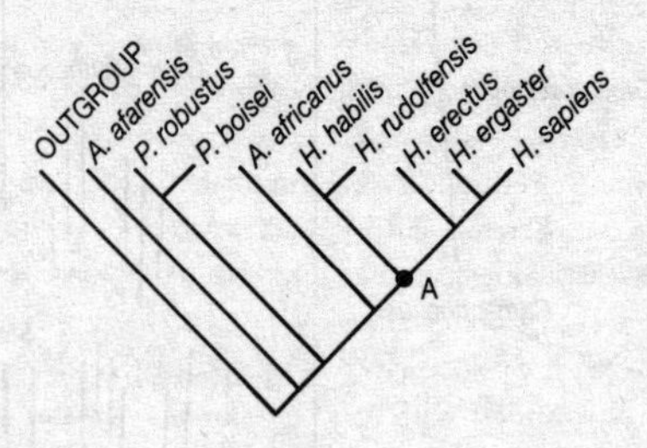

Fig. 59a *Cladogram of the hominid clade generated from a set of ninety cranial, mandibular and dental characters. Node A represents the origin of those autapomorphies distinguishing the* Homo *clade from the* Australopithecus *clade, as described in the* **Homo** *entry.*

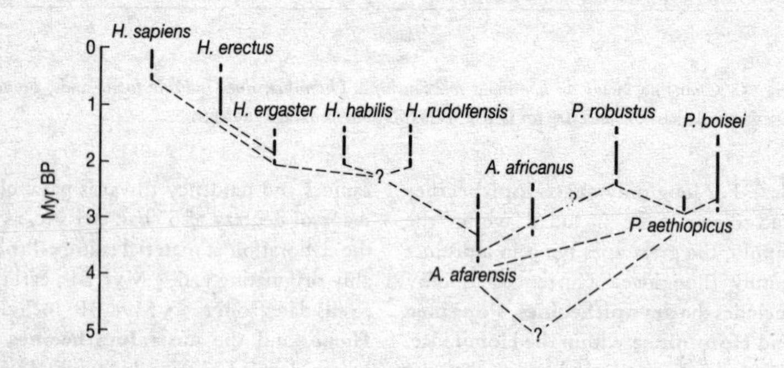

Fig. 59b *Phylogenetic scheme for later hominid evolution. The horizontal axis corresponds roughly to relative and absolute postcanine tooth size. The three taxa furthest to the right form a putative* Paranthropus *clade, and, as indicated, there are alternative views on its relationship with* Australopithecus, *in which it is often subsumed.*

Distinguished from other **catarrhine** primates by widening of trunk relative to body length, elongated clavicles, broad iliac blades and broad, flat back. Normally no free tail after birth, when the spinal column undergoes curvature as an adaptation to partial or complete bipedalism. See **Anthropoidea**.

Homo Primate genus within the subfamily Homininae (see **hominid**), including mankind and its immediate relatives. The three species regularly (but not universally) recognized are *H. habilis, H. erectus* and *H. sapiens*, only the last of which, modern man, exists today.

The genus is defined cladistically by the autapomorphies distinguishing it from *Australopithecus*, including increased cranial vault thickness and height, reduced postorbital constriction, more anteriorly situated foramen magnum, reduced lower facial prognathism, narrower tooth crowns (esp. mandibular premolars) and reduction in length of the molar tooth row. Additional to these autapomorphies, cranial capacity (brain volume) is greater than 700 cm^3, rising to in excess of 1600 cm^3; dental arcade is evenly rounded with no diastema in most individuals; pelvic girdle and limb skeletons are fully adapted for bipedalism; and hands are capable of precision grasp. The stock immediately ancestral to *Homo* is generally thought to have been **australopithecine**, possibly a descendant of *A. afarensis*.

In 1964 an early (1.9–1.6 Myr BP) fossil hominin (Tribe **Hominini**) was reported from Olduvai, Tanzania, with brain smaller than *H. erectus* but larger than *Australopithecus*. Attributed to a new species, *H. habilis*, it had advanced jaw and dental features but was australopithecine postcranially. Subsequent more varied fossils from Lake Turkana (1.9–1.8 Myr BP) suggest there may have been more than one early hominin present. The controversial taxon *H. rudolfensis* has been proposed for some Lake Turkana and Lake Malawi fossils, combining a relatively large brain and *Homo*-like postcranial features with a face and dentition resembling the more robust australopithecines ('*Paranthropus*'). The present date for *H. rudolfensis* is 2.4–1.8 Myr BP, the taxon possibly arising during the climatic cooling event in E. Africa at 2.5 Myr BP. Culturally, all these fossils belong in the traditional Lower **Palaeolithic**, with characteristic Oldowan tool culture comprising stone choppers and crude scrapers; some tools may also have been made from bone, teeth, etc., and possibly from wood.

Various fossils are thought to form intermediates between *H. habilis* and *H. erectus*. The earliest *H. erectus* fossils come from Lake Turkana, date from at least 1.7 Myr BP and indicate close temporal proximity to *H. habilis*. *H. erectus* is distinguished by larger cranial capacity (750–1250 cm^3), flattened skull vault, large supraorbital torus and marked postorbital constriction, and early African fossils (1.7 Myr BP, sometimes given specific status, *H. ergaster*) indicate close temporal proximity of *H. erectus* and *H. habilis*. Evidence of Eurasian *H. erectus* now extends back to 1.8 Myr BP (Dmanisi, Georgia). The earliest evidence of the Acheulian culture associated with *H. erectus* is African at *c.* 1.4 Myr BP, and it appears that the species moved with its culture into Eurasia prior to the extensions of the polar ice sheets (0.9–0.7 Myr BP).

Debate surrounds the autapomorphies of *H. sapiens*, continuity occurring between fossils of *H. sapiens* and *H. erectus* at sites with large samples. *H. sapiens* is normally distinguished by increased cranial capacity (early forms average 1166 cm^3), reduction in size of posterior and increase in size of anterior teeth, and reduced muscular robusticity, esp. of lower limbs; but overlap in all these occurs. Studies using electron spin resonance (ESR, see **radiometric dating**) on the Chinese Jinniushan skull (*H. sapiens*) at 200 Kyr BP raise questions about the possible coexistence of *H. sapiens* and *H. erectus*.

The taxonomic status of Neanderthal man (*c.* 230–36 Kyr BP) is problematic. It is generally held that they derive from

Middle Pleistocene populations which were either late *H. erectus* or a descendant of that species (*H. heidelbergensis* or possibly 'archaic' *H. sapiens*), possibly before 400 Kyr BP. Fossils come principally from Europe and the Middle East, mass spectrometry ($^{230}Th/^{234}U$) and ESR studies on Israeli specimens indicating Neanderthals (at Tabun) and modern *H. sapiens* (at Skhul & Qafzeh) were approximately coeval at 100 Kyr BP (± 5 Kyr). Some French fossils suggest Neanderthals and modern humans coexisted until *c.* 36 Kyr BP, although it is not clear whether they were reproductively isolated (see **mitochondrial Eve**). European Neanderthals varied anatomically, but compared with modern humans had the following: cranial capacity 1300–1640 cm^3; enlarged occiput (esp. during Wurm glaciation); cranium with lower, flatter, crown; chin receding or absent; cheek bones large; large supraorbital torus; large nasal cavity and nasal prognathism; larger (esp. broader) incisors and canines. Postcranially, Neanderthals were short, powerfully built people; but the popular preconceptions of stupidity combined with brute animal strength require overhauling. These people had fire, but it seems that until about 60 Kyr BP did not build elaborate hearths, the most complex to date being scoops in a cave floor. After that time, well-built hearths in open sites do appear, indicating use of temporary or semi-permanent huts and cabins. The discovery of quite complete Neanderthal bodies in burial-like position might indicate only the absence of large carnivores from the relevant caves.

The precise origins of anatomically modern humans prior to 40 Kyr BP await clarification, but an African origin at *c.* 100 Kyr BP has received recent fossil support. Controversy surrounds the origins of such symbolic behaviour as art and ornament and whether their origins were gradual or sudden. It is customary to regard the Upper Palaeolithic Aurignacian, at > 43 Kyr BP in Eastern Europe, as the first such 'true' industry, coinciding as it did with the spread of anatomically modern humans into Europe. In France, however, flake-based Mousterian culture and Neanderthals themselves persisted after this date. From this time on, open-air (as opposed to cave) burials and impressive camp-sites become common. Anatomically modern humans probably originated in Africa at perhaps 130 Kyr BP, arrived in the Middle East by *c.* 100 Kyr BP (Qafzeh & Skhul) and reached Australia by *c.* 55 Kyr BP. The antecedents of Aurignacian culture may have been in Southeast Asia. This model, lacking continuity between *H. erectus* and *H. sapiens* anywhere outside Africa, stands in contrast to the largely abandoned view that different modern populations arose at different times from different *H. erectus* stocks (the 'multiregional continuity' model). In this context, the Jinniushan skull's date might be controversial.

Homocercal Designating those outwardly symmetrical fish tails (caudal fins) in which the upper lobe is approximately of the same length as the lower. Typical of modern **Actinopterygii**; evolved from the **heterocercal** fin.

Homochlamydeous (Of flowers) having perianth segments of one kind (sepals) in two whorls. Compare **heterochlamydeous**.

Homodont See **dentition**.

Homoeobox See **homeobox**

Homogametic sex The sex producing gametes which are uniform with respect to their **sex-chromosome** complement. In mammals, this is the female sex (XX); in birds, reptiles and lepidopterans it is the male sex (ZZ). See **heterogametic sex**, **sex determination**.

Homogamy Condition in which male and female parts of a flower mature simultaneously. Compare **dichogamy**.

Homograft See **allograft**.

Homokaryon (homocaryon) Cell, fungal hypha or mycelium with more than one genetically identical nucleus in its cytoplasm; in fungi such nuclei commonly are haploid. See **coenocytic**.

Homologous (1) For homologous characters, see **homology**. (2) Homologous chromosomes: those capable, at least potentially, of pairing up to form **bivalents** during first prophase of **meiosis**, having approximately or exactly the same order of loci (but not normally of alleles). See **aneuploid**, **polyploidy**, **hemizygous**.

Homologous recombination Process whereby, after a nuclease has cut two strands with the same polarity, these cut DNA segments search for and recognize homologous sequences (through mediation of proteins like RecA and single-stranded **DNA-binding proteins**) and form cross-strand exchange structures (**Holliday structures**), after which both crossing strands are cut and finally ligated. Easily employed in yeast **gene targeting**. See Fig. 60.

Homology A controversial term. In evolutionary biology denotes common descent. Two or more structures, developmental processes, DNA sequences, behaviours, etc., usually occurring in different taxa, are said to be *homologous* if there is good evidence that they are derivations from (or identical to) some common ancestral structure, developmental process, etc. Very often a structure, etc., serving one function in one taxon has come, often with modification, to serve a different function in another. Two of the **ear ossicles** of mammals are homologous with the articular and quadrate bones of ancestral reptiles; the vertebrate **thyroid** gland is almost certainly homologous with the **endostyle** of urochordates, etc. The term may be applicable even when, as with the vertebrate **pentadactyl limb**, comparable structures do not always arise from the same embryonic segment. Where structures are repeated along the organism with little or no modification, as occurs in **metamerism**, they are termed *serially homologous* structures. Much evidence for homology is likely to come from work on **homeotic** mutants. See **analogous**.

In **phylogenetics**, or **cladistics**, two characters, etc., are homologous if one (the *apomorphic character*) is derived directly from the other (the *plesiomorphic character*). The relationship is often termed special homology. Some workers in cladistics equate homology with **synapomorphy**. In molecular biology, the term often indicates a significant degree of sequence similarity between DNA or protein sequences.

Homonomy Characters of two or more taxa which have the same development, are found on different parts of the organism, and whose developmental pathways have a common evolutionary origin. E.g. each mammalian hair is *homonomous* with all other mammalian hair.

Homoplasy In **cladistics**, the term used to denote parallel or convergent evolution of characters.

Homoptera See **Hemiptera**.

Homosporous Having one kind of spore giving rise to gametophytes bearing both male and female reproductive organs, e.g. many ferns. Compare **heterosporous**. See **life cycle**.

Homostyly Usual condition in which flowers of a species have styles of one length, as opposed to **heterostyly**.

Homothallism (Of algae, fungi) the condition whereby thalli are morphologically and physiologically identical, so that fusion can occur between gametes produced on the same thallus. Compare **heterothallism**.

Homozygous Any **locus** in a diploid cell, organism, etc., is said to be homozygous when the two **alleles** at that locus are identical. Organisms are said to be homozygous for a character when the locus determining that character is homozygous. Homozygous mutations are normally expressed phenotypically, unless the genetic background dictates otherwise (see **penetrance**). Characters which are **recessive** are only expressed in the homozygous (or **hemizygous**) condition. See **heterozygous**.

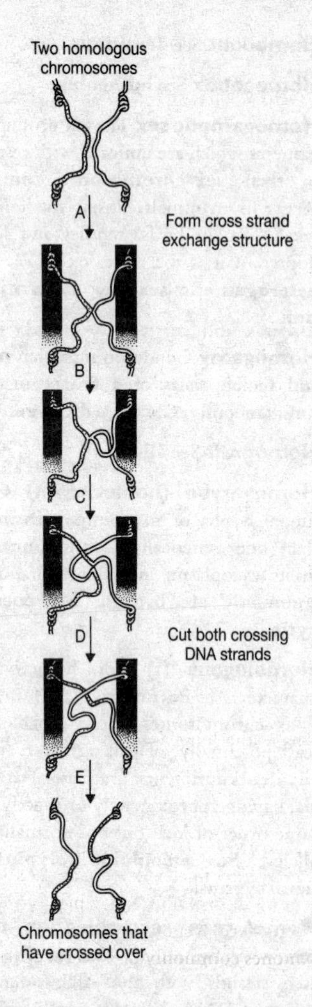

Fig. 60 *The general enzymatic recombination event between two homologous chromosomes leading to crossing-over.*

Hormogonium Short piece of filament, characteristic of some filamentous blue-green algae, that becomes detached from the parent filament and moves away by gliding, eventually developing into a separate filament. Several may develop from one filament.

Hormone Term once applied in both botanical and zoological contexts, but now restricted to the latter (see **growth substance**). Denotes any molecule, usually of small molecular mass, secreted directly into the blood by ductless glands and carried to specific target cells/organs by whose response they bring about a specific and adaptive physiological response. The term *chemical messenger* is still sometimes used in this context, **second messengers** being molecular signals produced within, but not exported by, a cell. Neurotransmitters and neurosecretions customarily fall outside the compass of the term hormone, a distinction blurred by **neurohaemal organs**. Hormones tend to be either water-soluble peptides and proteins, or lipid-soluble steroids, retinoids, thyroid hormones and vitamin D_3. The latter have the longer physiological half-lives and are hydrophobic, being rendered soluble by binding to specific transport proteins (see **nuclear receptors, transcription factors**). In this form they may enter nuclei to bring about selective **gene expression** and typically mediate long-term responses. Water-soluble hormones commonly bind to **receptor** sites on cell membranes (see **adenylate cyclase**) and tend to mediate short-term responses. Examples of hormones include **adrenaline**, **ecdysone**, **gastrin**, **thyroxine**, **insulin**, **testosterone** and **oestrogen**. See **endocrine system**, **pituitary gland**, **prostaglandins**, **cascade**.

Horn Matted hair or otherwise keratinized epidermis of mammal, surrounding a knob-like core arising from a dermal bone of the skull. Neither the core nor the keratinized sheath is ever shed, nor do they ever branch, unlike **antlers**.

Horsetails See **Sphenophyta**.

Host (1) Organism supporting a **parasite** in or on its body and to its own detriment. A *primary* (*definitive*) *host* is that in which an animal parasite reproduces sexually or becomes sexually mature; a *secondary* (*intermediate*) *host* is that in which an animal parasite neither reproduces sexually nor attains sexual maturity, but which generally houses one or more larval stages of the parasite. (2) Organism supporting (e.g. housing) a non-parasitic organism such as a commensal. See **symbiosis**.

Host race See **infraspecific variation**, **speciation**.

Host restriction See **phage restriction**.

***Hox* genes** A subset of **homeogenes**, encoding **positional information**.

HTLV (human T-leukaemia virus) See **HIV**.

Human Term indicating any **hominid**.

Human chorionic gonadotrophin (hcg) Peptide hormone produced by developing human blastocyst and **placenta**, prolonging until four months of pregnancy the period of active secretion of oestrogens and progesterone by the **corpus luteum**, which otherwise atrophies, inducing menstruation. Its presence in urine is usually diagnostic of early pregnancy (see **immunoassays**). See **menstrual cycle**.

Human gene therapy Treatment of human disease by gene transfer. The first such trial, begun in 1990, involved transfer of the adenosine deaminase gene into lymphocytes of a patient whose defect in this enzyme (leading to immune deficiency) would otherwise have proved fatal. Most approved gene therapy trials involve use of retroviral vectors for gene transfer into cultured human cells that would be administered to patients; but trials employing liposomes in the direct transfer of plasmid DNA into patients have occurred in some cases of cancer and cystic fibrosis (CF). Retroviral vectors are retroviruses lacking all functional viral genes, so no viral protein is produced in infected cells. Viral replication is achieved using 'packaging cells' that produce all the viral proteins but no infective virus, so that when the DNA form of the retrovirus is introduced into these cells virions carrying vector RNA are produced that can infect target cells in the patient but cannot spread thereafter, so rather than 'infection', the term 'transduction' is used to describe this process. This results in efficient and stable targeting of cells, the main drawback being the inability of the vectors to infect non-dividing cells. There is also the risk of insertional mutagenesis (e.g. activation of **oncogenes**). Adenoviruses are also being employed as vectors, notably for transferring the normal gene (*CTFR*) in CF patients, with the lung as initial target. Proposals for human gene therapy have to pass several levels of review to ensure safety, not least ensuring there is no transmission of transgenes to the **germ line**. See **genetic counselling**.

Human genome project By 1987, a genetic map with an average resolution of about 10 cM (see **centimorgan**) between markers had been achieved. The human genome is estimated to contain about 3×10^9 base pairs. The human genome project is an international research programme, launched in 1988 and expected to be completed by 2005 (at an estimated cost of $3 billion), whose objective is to map all human genes precisely to their respective positions on chromosomes and to identify their DNA sequences. Techniques involved include the use of **genetic markers** (e.g. **RFLPs** and **satellite DNA**), **chromosome jumping, chromosome walking**, **DNA sequencing**, and production of **gene libraries**. It is anticipated that by the mid-1990s marker spacing will be close to 1–2 cM. Such high resolution maps will make easier the study of diseases involving **multifactorial inheritance**.

Human placental lactogen (hPL, hCS) Hormone produced by human placenta after about five weeks of pregnancy. Major effect is to switch maternal metabolism from carbohydrate to fat utilization. **Insulin** antagonist.

Humerus Bone of tetrapod fore-limb adjoining **pectoral girdle** proximally, and both radius and ulna distally. See **pentadactyl limb**.

Humoral Transported in soluble form, particularly in blood, tissue fluid, lymph, etc. Often refers to hormones, antibodies, etc. See **humoral immunity**.

Humoral immunity Immunity due to soluble factors (in plasma, lymph or tissue fluid). Production of **antibody** constitutes *humoral response* to an antigen. Contrasted with *cellular response* (see **immunity**).

Humus Complex organic matter resulting from decomposition of dead organisms (plants, animals, decomposers) in the soil giving characteristic dark colour to its surface layer. Colloidal (negatively charged), improving cation absorption and exchange and preventing leaching of important ions, thus acting as a reservoir of minerals for plant uptake; water-retention of soil also improved. See **soil profile**.

Hyaluronic acid Non-sulphated **glycosaminoglycan** of D-glucuronic acid and N-acetylglucosamine; found in extracellular matrices of various connective tissues.

Hyaluronidase Enzyme hydrolysing hyaluronic acid, decreasing its viscosity. Of clinical importance in hastening absorption and diffusion of injected drugs through tissues. Some bacteria and leucocytes produce it. Reptile venoms and many sperm **acrosomes** contain it.

H-Y antigen (H-W antigen) Minor **histocompatibility antigen** encoded by locus on Y sex chromosome of most vertebrates (W sex chromosome in birds), and responsible for rejection of tissue grafted on to animal of opposite sex. Not now thought to be a product of the gene for testis-determining factor (TDF). See **sex determination**, **sex reversal gene**.

Hybrid In its widest sense, describes progeny resulting from a cross between two genetically non-identical individuals. Commonly used where the parents are from different taxa; but the term also has wider applicability as with *inversion hybrids* (inversion heterozygotes) where offspring are simply heterozygous for a chromosome **inversion**. Where parents of a hybrid have little chromosome homology, particularly where they have different chromosome numbers, hybrid offspring will be sterile (e.g. *mules*, resulting from horse × donkey crosses) through failure of chromosomes to pair during **meiosis**, although one offspring sex may be partially or completely fertile. Hybrid sterility is one factor maintaining species boundaries, and selection against hybrids is a major factor in theories of **speciation**. Mitosis in hybrid zygotes is unlikely to be affected by lack of parental chromosome homology but development may be thwarted by imbalance of gene products. Sometimes hybridization (especially between inbred lines of a species) may produce **hybrid vigour**. See **allopolyploidy**.

Hybrid dysgenesis (HD), or dysgenesis Infertility and other defects arising from crossing of certain genetic strains, notably in *Drosophila* where high sterility and increased chromosome mutability occur among offspring from crosses between laboratory female and wild male stocks. The non-reciprocal nature of these results and the discovery of chromosomal mutational 'hot-spots' housing extra DNA implicated chromosomal insertions, and transposable elements are now known to be responsible. These HD insertions are integrated into wild stock genomes, encoding repressor molecules which inhibit transposition and hence mutagenesis. Repressor concentration would thus be high in wild eggs, preventing transposition of elements donated by wild males. Absence of HD insertions from laboratory strains means that incoming wild HD elements find a repressor-free environment in the egg, transpose readily and cause chromosome damage and consequent sterility in offspring. See **P element**.

Hybridization (1) Production of one or more **hybrid** individuals. (2) Molecular hybridization (see **DNA hybridization**). (3) See **cell fusion**.

Hybridoma Clone resulting from division of hybrid cell resulting from artificial fusion of a normal antibody-producing **B cell** with a B cell tumour cell. Technique involved in production of *monoclonal antibodies*. See **antibody**.

Hybrid swarm Continuum of forms resulting from hybridization of two species followed by crossing and backcrossing of subsequent generations. May occur when habitat is disturbed or newly colonized, as with the oaks *Quercus ruber* and *Q. petraea* in Britain. See **introgression**.

Hybrid vigour (heterosis) Increased size, growth rate, productivity, etc., of the offspring resulting usually from a cross involving parents from different inbred lines of a species, or occasionally from two different (usually congeneric) species. Possibly results from **heterozygous advantage** or, probably more generally, from fixation of different deleterious recessives in the inbreds.

Hybrid zone Area (zone) between two populations normally recognized as belonging to different species or subspecies and occupied by both parental populations and their phenotypically recognizable hybrids. Existence of a narrow hybrid zone may indicate that the parent populations are distinct evolutionary species; a wide zone may indicate that they are geographical variants of the same evolutionary species. Not to be confused with **introgression**. See **semispecies**.

Hydathode Water-excreting gland occurring on the edges or tips of leaves of many plants. See **guttation**.

Hydatid cyst Asexual multiplicative phase of some tapeworms (e.g. *Echinococcus*) within the secondary host (e.g. man, sheep, pig) in which a fluid-filled sac produces thousands of secondary cysts (brood capsules), each of which buds off a dozen or so retracted scolices. In humans the cysts may become malignant and send *metastases* around the body with sometimes fatal results. An example of **polyembryony**.

Hydranth See **polyp**.

Hydrocortisone See **cortisol**.

Hydrogen bond Electrostatic attraction forming relatively weak non-covalent bond between an electronegative atom (e.g. O, N, F) and a hydrogen atom attached to some other electronegative atom. Responsible in large measure for secondary, tertiary and quaternary **protein** structures, for **base pairing** between complementary strands of nucleic acid, for the cohesiveness and high boiling point of water.

Hydroid Member of the **Hydrozoa**, in its polyp form.

Hydrolase Enzyme catalysing addition or removal of a water molecule. See **hydrolysis**.

Hydrolysis Reaction in which a molecule is cleaved with addition of a water molecule. Some of the most characteristic biochemical processes (e.g. digestion, ATP breakdown and other dephosphorylations such as those in respiratory pathways) involve hydrolysis reactions. Chemically it is the opposite of a **condensation reaction**.

Hydrophily Pollination by means of water.

Hydrophyte (1) Plant whose habitat is water, or very wet places; characteristically possessing **aerenchyma**. Compare **mesophyte**, **xerophyte**. (2) Class of **Raunkiaer's life forms**.

Hydroponics System of large-scale plant cultivation developed from water-culture methods of growing plants in the laboratory. Roots are allowed to dip into a solution of nutrient salts, or else plants are allowed to root in some relatively inert material (e.g. quartz sand, vermiculite) irrigated with nutrient solution. The external environment is commonly kept artificially constant.

Hydrosere **Sere** commencing in water or otherwise moist sites.

Hydrotropism **Tropism** in which the stimulus is water.

Hydroxyapatite Crystalline calcium phosphate. Mineral component of **bone**. Used in column chromatography for eluting proteins with phosphate buffers.

5-Hydroxytryptamine See **serotonin**.

Hydrozoa Class of **Cnidaria** containing hydroids, corals, siphonophores, etc. Usually there is an **alternation of generations** in the life cycle between a sessile polyp (hydranth) phase and a pelagic medusoid phase; but one or other may be suppressed. Most polyp forms (but not *Hydra*) are **colonial**, showing division of labour between feeding and reproductive individuals. Gonads ectodermal, unlike those of **Scyphozoa**.

Hymenium Layer of regularly arranged spore-producing structures in the fruit bodies (**ascocarps** and **basidiocarps**) of many fungi (e.g., **Ascomycota**, **Basidiomycota**). May or may not be exposed.

Hymenoptera Large and diverse order of endopterygote insects, including sawflies (Symphyta), bees, ants, wasps and ichneumon flies (Apocrita). Fore wings coupled with hind wings by hooks. Mouthparts typically for biting, but sometimes for lapping or sucking (as in bumble bees). Ovipositor used, besides egg-laying, for sawing (sawflies), piercing or stinging. Abdomen often constricted to form a thin waist, its first segment fused with metathorax. Larvae generally legless. Bees, ants and wasps often **eusocial**.

Hyoid arch Vertebrate **visceral arch** next behind jaws. Dorsal part forms **hyomandibular**; ventral part in adults forms *hyoid bone*, usually supporting tongue. Contains facial nerve; gives rise to many face muscles.

Hyomandibular Dorsal element (bone or cartilage) of hyoid arch, taking part in jaw attachment in most fish (see **hyostylic**). Becomes columella auris (stapes) in tetrapods (see **ear ossicles**).

Hyostylic Method of jaw suspension of most modern fishes. Upper jaw has no direct connection with the braincase and the jaw is supported entirely by the hyomandibular. This widens the gape. See **autostylic, amphistylic**.

Hypermutation The introduction of point mutations at high rate into variable (*V*) region genes of **antibody**-producing B cells in germinal centres of lymph nodes, leading to **affinity maturation** of antibodies during an immune response.

Hyperparasite Organism living parasitically upon another parasite. Provoked Jonathan Swift's doggerel expressing supposed infinite regress.

Hyperplasia Increase in amount of tissue resulting from cell division (each remaining the same size) as opposed to increase in cell volumes *per se* (hypertrophy).

Hypertonic Relational term expressing the greater relative solute concentration of one solution compared with another. The latter is *hypotonic* to the former. A hypertonic solution has a lower **water potential** than one hypotonic to it, and has a correspondingly greater **osmotic pressure.** See **isotonic, euryhaline**.

Hypertrophy Often interchangeable with **hyperplasia**. Sometimes used of enlargement of individual components of tissues, organs, etc., without increase in cell division. See **regeneration**.

Hypervariable regions See **antibody diversity, DNA fingerprinting**.

Hypha (Of fungi) a tubular filament or thread of a thallus, often vacuolated. Increases in length by growth at its tip; but proteins are synthesized throughout the **mycelium** and transported to hyphal tips by cytoplasmic streaming. May be septate (with cross-walls) or non-septate. New hyphae arise by lateral branching. See **coenocyte**.

Hypoblast Cell layer of avian blastodisc and mammalian blastocyst; presumptive endoderm. Compare **epiblast**.

Hypocotyl Part of a seedling stem, below the cotyledon(s).

Hypodermis Layer of cells immediately below the epidermis of leaves of certain plants, often mechanically strengthened (e.g. in pine), forming an extra protective layer, or forming water-storage tissue.

Hypogeal (Of cotyledons) remaining underground when the seed germinates; e.g. broad bean, pea. Contrast **epigeal**.

Hypoglossal nerve See **cranial nerves**.

Hypogynous See **receptacle**.

Hyponasty (Bot.) More rapid growth of lower side of an organ than upper side; e.g. in a leaf, resulting in upward curling of leaf-blade. Compare **epinasty**.

Hypophysis See **pituitary gland**.

Hypostasis Suppression of expression of a (*hypostatic*) gene by another non-allelic gene. Compare **recessive**. See **suppressor mutation**, **epistasis**.

Hypothalamus Thickened floor and sides of the third ventricle of the vertebrate forebrain (diencephalon). Its nuclei control many activities, largely homeostatic. It integrates the **autonomic nervous system**, with centres for sympathetic and parasympathetic control; receives impulses from the viscera. Ideally situated to act as an integration centre for the endocrine and nervous systems, secreting various **releasing factors** into the **pituitary** portal system and neurosecretions into the posterior pituitary. Releases substances inhibiting release of releasing factors (e.g. see **somatostatin**). Contains control centres for feeding and satiety – the latter inhibiting the former after feeding. In higher vertebrates, is a centre for aggressive emotions and feelings and for psychosomatic effects. Contains a thirst centre responding to extracellular fluid volume; helps regulate sleeping and waking patterns; monitors blood pH and concentration and, in homeotherms, body temperature. See **neuroendocrine coordination** and Fig. 13.

Hypothallus Thin, shiny, membranous adherent film at base of a slime mould (**Myxomycota**) fruitification.

Hypothesis A temporary working explanation or conjecture, commonly based upon accumulated data, which suggests some general principle or relationship of cause and effect; a postulated solution to a problem that may then be tested experimentally. See **null hypothesis**.

Hypotonic Of a solution with a lower relative solute concentration (higher **water potential**) than another. See **hypertonic**, **euryhaline**.

Hypoxanthine Base present in small amounts in certain transfer RNA molecules, at the 5′-end of the anticodon and capable of base-pairing with U, C and A in the corresponding mRNA codon. Hydrolytic product of adenine. See **wobble hypothesis**.

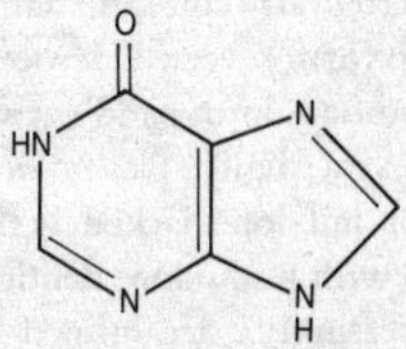

Hypoxanthine

IAA Indole-3-acetic acid. Most common **growth substance** in plants, produced in apical meristems of shoots and tips of coleoptiles. One of several **auxins**.

IAN Indole-3-acetonitrile. Natural plant **auxin**.

I-band See **striated muscle**.

Ichthyosauria Extinct reptilian order (subclass Euryapsida), as fossils from Triassic to Cretaceous. They were not dinosaurs (see **Archosaurs**), but were contemporaries. Vertebral column curved downwards to form reverse heterocercal tail; legs modified into paddles, with addition of extra digits. Fleshy dorsal fin and upper tail lobe lacked skeletal support. Jaws with homodont dentition. Became increasingly streamlined for aquatic locomotion; convergence with porpoises, etc.

ICSH Interstitial-cell stimulating hormone. See **luteinizing hormone**.

Identical twins See **monozygotic twins**.

Identification keys Keys used in discovering the name of a specimen are commonly constructed so as to lead the investigator through a sequence of choices between mutually exclusive character descriptions, so chosen as to eliminate all but the specimen under observation. The format is commonly **dichotomous**. A disadvantage arises when not all characters of the dichotomous key are observable, as through damage or incompleteness. A *polyclave* overcomes this, placing reliance only upon characters observable in the specimen to hand. It commonly comprises a set of punch-cards, each representing a different form or state that a character can take. Each species within the set dealt with by the cards is located on a master sheet and is given a unique number representing its set of character-states. When sufficient cards with character descriptions appropriate to the specimen are held up together only one punch-hole remains open, and the number corresponding to that hole is the number in the master sheet which identifies the species (or other taxonomic unit being identified).

Idioblast (Bot.) Cell clearly different in form, structure or contents from others in the same tissue; e.g. cystolith-containing parenchyma cells.

Idiotype Antigenic constitution of the variable (V) region of an immunoglobulin molecule. See **antibody**, **antibody diversity**.

IgA Monomeric or polymeric immunoglobulin, often dimeric (composed of two polypeptides). Most abundant in seromucous secretions such as saliva, milk, and such as occur in urogenital regions. See **antibody**.

IgD Low titre immunoglobulin found attached to **B cell** membranes. Of uncertain function.

IgE Low titre immunoglobulin located on basophil and mast cell surfaces. Possibly involved in immunity to helminths; also involved in asthma and hay fever hypersensitivity.

IGF Insulin-like growth factor. See growth hormone.

IgG Class of monomeric immunoglobulin proteins with four subclasses (IgGl–4), accounting for at least 70% of human immunoglobulin titre. Each molecule contains two heavy and two light chains. The sole antitoxin class, and major antibodies of secondary immune responses (see **B cell**). The only antibody class to cross the mammalian **placenta**. See **antibody**.

IgM Class of large pentameric immunoglobulin molecules (five linked subunits), largely confined to plasma. Produced early in response to infecting organisms, whose surface antigens are often complex.

Ileum Region of mammalian small intestine closest to colon and developing from region occupied by embryonic yolk sac; not anatomically distinct from **jejunum**. Its numerous villi, intestinal glands (crypts of Lieberkühn) with their characteristic **Paneth cells** and **goblet cells**, enable it to serve as the major digesting and absorbing region of the mammalian alimentary canal.

Ilium Paired bone forming dorsal part of tetrapod **pelvic girdle** (present in rudimentary form even in fishes) and articulating with one or more sacral vertebrae.

Imaginal disc Organ-specific **primordia** of holometabolous insects, derived from blastoderm and distributed mainly in larval thorax. Composed of imaginal cells. There are nineteen such discs in *Drosophila* larvae, which evaginate at metamorphosis and differentiate largely into adult epidermal structures: eyes and antennae from one pair; front legs from another pair; wings and halteres from two more pairs; genitals from a midline unpaired disc, and so on. Most discs are **determined** in the larval insect and remain so through artificial subculture; but *transdetermination* of cells may occur. They tend to differentiate according to their original position in embryo regardless of transplantation to new locations. Much remains to be learnt of the evolution of this mode of development. Cues for differentiation of discs into adult tissues are apparently hormonal; but the origin of the determined state appears to depend upon **positional information**, expressed as genetic addresses. See **chromosomal imprinting**, **compartment**, **homeotic gene**.

Imago Sexually mature adult insect.

Imbricate (Of leaves, petals, etc.) closely overlapping.

Immobilization See **enzyme**.

Immune tolerance (immunological tolerance) Acquired inability to react to particular self- or non-self-antigens. Both **B cells** and **T cells** display tolerance, generally to their specific antigen classes. The concentration of antigen required to induce tolerance in neonatal B cells is 100-fold less than for adult B cells. Responsible for suppression of transplant rejection. First noticed in non-identical twin cattle which shared foetal circulations (i.e. were synchorial). See **thymus**.

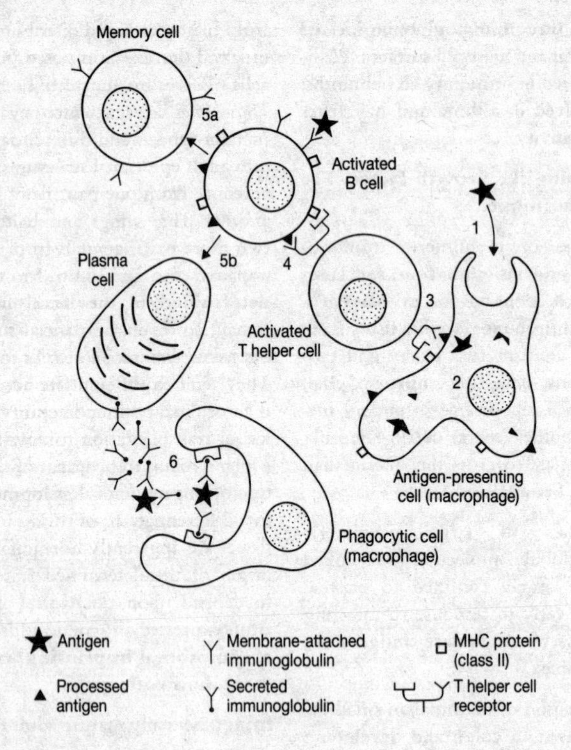

Fig. 61 *The humoral (B-cell-mediated) immune response to an infection. (1) The antigen (the infecting agent) is taken up by a cell such as a macrophage. (2) This cell processes the antigen, breaking it down into components that are then displayed on the surface of the cell. (3) The antigen is recognized by a T helper cell, which in turn (4) activates B cells that are also carrying pieces of the antigen. The activated B cells either (5a) differentiate into memory cells, which respond in future infections caused by the same agent, or (5b) become plasma cells that secrete antibody. (6) The antibodies bind to the antigen, thereby forming a complex that is destroyed by macrophages.* (Adapted from *The Molecules of the Immune System*, by S. Tonegawa. Copyright © Scientific American, Inc., 1985. All rights reserved).

Immunity Ability of animal or plant to resist infection by parasites and effects of other harmful agents. Essential requirement for survival, since most of these organisms are perpetually menaced by viruses, bacteria and fungi, or parasitic animals.

In animals there are two functional divisions of the immune system: *innate* (*non-specific*) *immunity* and *adaptive* (*specifically acquired*) *immunity*. The former includes several barriers to pathogen entry (e.g. lysozyme, mucus, intact skin/cuticle, sebum, stomach acid, ciliary respiratory lining and commensal gut competitors) as well as non-specific cellular responses, e.g. production of interferons and colony-stimulating factors and the activities of non-lymphocytic **leucocytes** of the reticulo-endothelial system and of neutrophils). Adaptive immune responses, unlike innate immune responses, differ in quality and/or quantity of response on repeated exposure to antigen: the primary response to antigen takes longer to achieve significant antibody titre than does the secondary response. They include *active natural immunity*, in which the animal's **memory cells** respond to a secondary natural contact with antigen by multiplication and specific antibody release; and *active induced immunity*, in which a **vaccine** (see also **inoculation**) initially sensitizes memory cells. *Passive immunity* may be either natural, as by acquisition of antibodies via the placenta or colostrum in mammals, or induced, usually via specific antibodies injected intravenously. There is sometimes a distinction made between *cell-mediated* (lymphocytic and phagocytic) responses and *humoral* (antibody) responses in immunity, but it is never clear-cut: cells are involved in initiation of antibody responses and cell-mediated responses are unlikely in the complete absence of antibody. Cell-mediated immunity tends nowadays to refer to any immune response in which antibodies play a relatively minor role. For distinctions between *primary* and *secondary immune responses*, see **B cell**. See **antigen-presenting cell**, **T cell**, **inflammation**, and Fig. 61.

Immunity in plants is due to structural features, such as a waxy surface preventing wetting and consequent development of pathogens, thick cuticles preventing entry of germ-tubes of fungal spores; or immunity may be protoplasmic, the protoplast being an unfavourable environment for further development of the pathogen (see **phytoalexins**); or it might be acquired immunity (in context of viral diseases) when the plant recovers from an acute disease, or when resistance to virulent strains is conferred by presence of avirulent ones. In the latter (non-sterile) cases, active virus persists in the recovered or protected plant. Freedom from a second attack of an acute disease, or protection from the effects of virulent strains, persists only as long as the plants are infected. Plants are not known to produce antibodies. See **resistance genes**.

Immunization Process rendering an animal less susceptible to infection by pathogens, to toxins, etc. May confer active **immunity** through use of **vaccine**; or passive immunity through injection of appropriate antibodies (in antiserum). See **inoculation**.

Immunoassays Methods for quantitative analysis of the amount of protein or other antigen present. *Solid-phase immunoassay* involves fixing antibody to the antigen on a polyvinylchloride sheet, putting on a drop of serum (or urine) and washing off after the antigen–antibody complex has had time to form and then adding a second labelled or fluorescent antibody, this time specific to a different epitope of the antigen. The amount of the second antibody which binds is proportional to the amount of antigen present. *Enzyme-linked immunosorbant assay* (ELISA) is similar, but an enzyme (e.g. alkaline phosphatase) instead of the label is attached to the second antibody. This can convert a colourless substance to a coloured product, or non-fluorescent to fluorescent, when added. Sensitivities of both methods can be extremely high. ELISA is used to assay human urine for **human chorionic gonadotrophin** in a common pregnancy test.

Immunoelectrophoresis See **electrophoresis**.

Immunofluorescence Use of antibodies, with a fluorescent marker dye attached, in order to detect whereabouts of specific antigens (e.g. enzymes, glycoproteins) by formation of antibody-antigen complexes which show up on appropriate illumination.

Immunogen Any substance that can induce an immune response. Compare **antigen**.

Immunoglobulin (Ig) Those members of the **immunoglobulin superfamily** with antibody activity. See **IgA**, **IgD**, **IgE**, **IgG**, **IgM**.

Immunoglobulin (Ig) superfamily A large glycoprotein superfamily including **antibodies** and their membrane-bound isotypes; **T cell receptors**, lymphocyte Fc receptors, the **CD2, CD4** and **CD8 accessory molecules** and **MHC** molecules. All contain one or more homologous immunoglobulin-like **domains** about 100 amino acids long, containing two antiparallel β-pleated sheets.

Immunological distance A unit indicating the closeness of relationship between two organisms, based on such evidence as **immunoassays** and immunoprecipitation. The smaller the value, the greater the cross-reaction and the closer the relationship.

Immunological memory See **B cell**, **memory cells**.

Immunological tolerance See **immune tolerance**.

Immunosuppression Clinical procedures suppressing a patient's rejection of an organ transplant, including total body irradiation, selective drugs (e.g. the fungal derivative cyclosporin A, which suppresses T cells causing rejection, but not B cells, allowing the body to mount an immune response), or such anti-lymphocyte monoclonal antibodies as anti-lymphocyte globulin or anti-CD4. It is likely that certain TH cell-derived cytokines (e.g. interferon IFN_{ψ}) suppress the production of IgE.

Implantation (nidation) Attachment of mammalian **blastocyst** to wall of uterus (endometrium) prior to further development, placenta formation, etc. In humans the blastocyst is small and penetrates the endometrium, passing into the subepithelial connective tissue (*interstitial implantation*). This involves breaking of **junctional complexes** between endometrial cells, and proteolytic enzymes may be secreted by the **trophoblast** to achieve this.

Imprinting (1) Form of **learning**, often restricted to a specific *sensitive period* of an animal's development, when a complex stimulus may appear to elicit no marked response at the time of reception but nonetheless comes to form a model whose later presentation (or something appropriately similar) elicits a highly significant response. Particularly prevalent in birds. *Filial imprinting* involves narrowing of preferences in social companion (e.g. to mother, or to artificial object, in ducklings); *sexual im-*

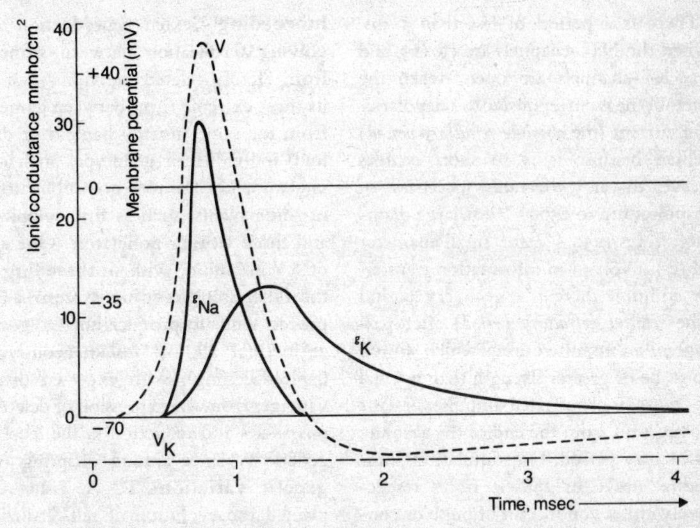

Fig. 62 *The action potential curve (V) resulting from changes in sodium and potassium conductances (gNa, gK) across an axon membrane at a point on its surface during propagation of an* **impulse**.

printing involves the preferential directing of sexual behaviour towards individuals similar to those encountered early in life. (2) See **chromosomal imprinting, genomic imprinting**.

Impulse An **all-or-none response** comprising an **action potential** propagated along the plasmalemma of an excitable tissue cell, such as a nerve axon (between **nodes of Ranvier** in myelinated axons) or muscle fibre. Impulses are initiated at synapses by depolarizations of the postsynaptic membrane's resting potential (usually internally negative by some 70 mv), generally brought about by release of an excitatory neurotransmitter molecule (for general details see **acetylcholine**). This opens up Na^+- and K^+-*ligand-gated channels* in the postsynaptic membrane and allows influx of Na^+ and efflux of K^+ along their electrochemical gradients; Cl^- channels remain closed if the transmitter is excitatory. Enzymic degradation of the transmitter restores these channels to their closed state, but current resulting from ion flow opens *voltage-gated* Na^+- and K^+-channels in the adjacent membrane, and the flow of ions which results causes depolarization, further current flow along the membrane and further depolarization.

Voltage-gated channels close again when depolarization in their region reaches its peak. These combined causes and effects result in propagation of an action potential away from the site of original depolarization. Because the K^+-channels open later than the Na^+-channels and stay open longer, the action potential has the characteristics shown in Fig. 62.

There is a period of less than 1 ms, when the Na^+-channels are closing and the K^+-channels are open, when the membrane is unresponsive to a depolarizing current (the *absolute refractory period*) which because it is so short enables nerves to carry the rapid succession of impulses (up to 2500 s^{-1} for large diameter fibres, 250 s^{-1} for small diameter fibres) involved in information transfer. In addition there is a recovery period (the *relative refractory period*) after passage of an impulse during which stimuli must be of greater strength than normal to cause a propagated impulse. It lasts about 2 ms from the end of the absolute refractory period. On stimulation, individual nerve or muscle fibres respectively either conduct an impulse or contract, or they do not. There are no partial conductions or contractions because of the statistical way in which their membrane ion channels open: only when sufficient ligand-gated channels are open (the threshold level) does sufficient depolarization occur to open adjacent voltage-gated channels (the *all-or-none rule*). Their depolarizing effect in turn causes adjacent voltage-gated channels to open in a reiterated fashion (*accelerating positive feedback*). Unmyelinated nerves conduct at from 0.5–100 m. s^{-1}, increasing with diameter; myelinated axons conduct at around 120 m. s^{-1}. Rise in temperature up to about 40°C increases conduction rate.

Depolarization of muscle *sarcolemma* occurs through release of acetylcholine at **neuromuscular junctions**. Impulses are then propagated along the sarcolemma in just the same way as along nerves, but are carried inwardly to myofibrils by transverse tubules. See **resting potential**, **muscle contraction**, **gated channels**.

Inbreeding Sexual reproduction involving fertilization between gametes from closely related individuals, or in its most extreme form between gametes from the same (usually haploid or diploid) individual or genotype. Such *selfing* is not uncommon, even obligatory, in some plants, such as first colonizers and those lacking pollinators. One end of a continuum, with **outbreeding** at the other (see **breeding system**). The process tends to produce homozygosity at loci (at all loci instantaneously in haploid selfing), with expected disadvantages from the expression of deleterious alleles and reduction in the level of genetic variance among offspring (see **genetic variation**). R. A. Fisher explained the evolution of self-fertilizing plant polulations from cross-fertilizing ones on the basis that in a stable population they will contribute three gametes rather than two to the succeeding generation: one megagamete, one self-fertilizing microgamete and one outcrossing microgamete. If the reduction in fitness due to segregation of recessive lethals is more than offset by the 50% excess in gene transmission, the 'gene for self-fertilization' will tend to increase. He reasoned that most plant species are either mainly selfing or mainly outcrossing, with few examples having a mixture of both types, and this appears to be the case. Self-fertilizers do evolve despite their being less fit than outcrossers; but they lack **heterozygous advantage** and are likely to become extinct in competition with them. However, many plant populations which do outbreed and inbreed (e.g. *Viola*, violets, and gynodioecious species) may, by regular exposure to selection of rare alleles with recessive deleterious effects, be 'purged' of two such alleles for each death resulting from their expression. See **assortative mating**, **genetic load**, **inbreeding depression**.

Inbreeding depression Increase in proportion of debilitated or inviable offspring consequent upon **inbreeding**.

Incisor Chisel-edged tooth of most mammals, occurring at the front of the dentary. Primitively three on each side, of both upper and lower jaws. Gnawing teeth of rodents (which grow continuously) and tusks of elephants are modified examples. Used for nipping, gnawing, cutting and pulling. See **dental formula**.

Inclusion granule Microscopically visible bodies produced in the cytoplasms of many plant and animal cells, sometimes in the nucleus, as a result of viral infection. Often consist largely of virions, which may form crystals.

Inclusive fitness See **fitness**.

Incompatibility (Bot.) (1) In flowering plants, the failure to set seed (i.e. failure of fertilization and subsequent embryo development) after either self- or cross-pollination has occurred. It is due to the inability of the pollen tubes to grow down the style. In tobacco, *Nicotiana alata*, this involves degradation of RNA in incompatible pollen; in the cabbage genus *Brassica*, a membrane-spanning **protein kinase** may represent the initial transduction step in signalling between pollen and stigma which destroys incompatible pollen. Incompatibility (*S* locus) alleles probably encode such kinases whose extracellular domains are glycoproteins. (2) In physiologically heterothallic organisms, it is the failure to reproduce sexually in single or mixed cultures of the same mating type. Genetically determined in both cases, it prevents the fusion of nuclei alike with respect to alleles at one or more loci, thus preventing inbreeding. It is analogous to negative **assortative mating**. (3) In horticulture, inability of the scion to make a successful union with the stock. (Zool.) The cause of rejection of a graft by the host organism through an immune response. See **immunity**.

Incomplete dominance See **dominance**.

Incomplete flower Flower which lacks one or more of the kinds of floral parts, i.e. lacking sepals, petals, stamens or carpels.

Incus One of the mammalian **ear ossicles**, homologous with the quadrate bone of other vertebrates.

Indehiscent (Of fruits) not opening spontaneously to liberate their seeds; e.g. hazel nuts.

Independent assortment (1) See **Mendel's Laws**. (2) Events occurring in normal diploid **meiosis** which cause one representative from each non-homologous chromosome pair to pass together into any gamete randomly, irrespective of the eventual genetic composition of the gamete. Results in random **recombination** and is an important source of **genetic variation** in eukaryotic populations. See **aberrant chromosome behaviour** (*meiotic drive*).

Indeterminate growth Unrestricted or unlimited growth; continues indefinitely.

Indeterminate head Flat-topped

inflorescence possessing sterile flowers with the youngest flowers in the centre.

Indicator, indicator species Species whose ecological requirements are well understood and which, when encountered in an area, can provide valuable information about it. In palaeolimnology, for example, certain diatoms, chrysophytes and ostracods are invaluable indicator species enabling inferences to be made about past lake environments. Absence of an indicator species (e.g. a lichen) from an area where it might be expected to occur could be symptomatic of pollution or some other environmental impoverishment.

Indigenous Indicating an organism native to a particular locality or habitat.

Indoleacetic acid See **IAA**.

Inducible enzyme Enzyme synthesized only when its substrate is present. See **enzyme**, **gene regulation**, **Jacob-Monod theory**.

Induction In embryology, the process resulting from combined effects of **evocation** and competence (see **competent**); results in production by one tissue (the *inducing tissue*) of a new cellular property in a dependently differentiating second tissue where the inducing tissue neither exhibits the resulting property nor alters its developmental properties as a result of the interaction. *Primary induction* events take place early in development; *secondary inductions* take place later in development.

Indusium Membranous outgrowth from undersurface of leaves of some ferns, covering and protecting a group of developing sporangia (a sorus).

Industrial melanism Occurrence, common in insects, of high frequencies of dark (*melanic*) forms of species in regions with high industrial pollution, where surfaces on which to rest are darkened by soot and where atmospheric SO_2 levels are high enough to prevent crustose lichen growth. A mutation darkening an individual will tend to be selected for (and hence come to predominate) in polluted regions since it will decrease the bearer's risk of falling prey to a visual predator; but in non-polluted parts of the species range the non-melanic form will be advantageous and occur with higher frequency. In the peppered moth, *Biston betularia*, heterozygous mutants collected in the mid-nineteenth century were paler than they are today, providing evidence in support of the theory that **dominance** is an evolving property of characters in populations of species. Industrial melanism provides one of the best examples of evolution within species and of selection resulting in **polymorphism**; but not all melanism is necessarily adaptive against visual predation. *Thermal melanism* has been suggested in one ladybird (*Adalia bipunctata*), in which dark forms absorb more energy in regions where atmospheric soot lowers levels of incident solar radiation. They warm up earlier in the season, and gain a reproductive benefit by being mobile sooner than non-melanics. The precise roles of migration and predation on gene frequencies in industrial melanism have yet to be elucidated.

Infarction See under **sclerosis**.

Inflammation Local response to injury in vertebrates; also involved in **allergy**. Involves vasodilation and increased permeability of capillaries in damaged area due largely to release of **histamine** and serotonin from **mast cells**. White blood cells, nutrients and fibrinogen enter and neutrophils are followed into the area by monocytes which become transformed into wandering **macrophages** for engulfing dead tissue, dead neutrophils, bacteria, etc. Fibrin forms from fibrinogen leaked into the tissues from blood, creating an insoluble network localizing and trapping invading pathogens, forming a fibrin clot preventing haemorrhage while isolating any infected region. Most inflammatory responses are usually down-regulated by circulating steroids, including **cortisol**, but are activated by interleukin-1 (see **interleukins**) and are dependent upon the sequential migration of the above-mentioned effector cells from the peripheral blood to the site of injury in response to the release of various **chemokines**. *Pus* usually results following inflammation and comprises dead and living white blood cells and cell remains from damaged tissues. See **eicosanoids**.

Inflorescence Collective term for specific arrangement of flowers on an axis, grouped according to the method of branching, into: a) indefinite, or racemose; b) definite, or cymose. In a), branching is monopodial, inflorescences consisting of a main axis which increases in length by growth at its tip, giving rise to lateral flower-bearing branches. These open in succession from below upwards or, if the inflorescence axis is short and flattened, from the outside inwards. The following are recognized (see Fig. 63): **raceme**, whose main axis bears stalked flowers; **panicle**, compound raceme, such as oat; **corymb**, raceme with flowers borne at the same level due to elongation of the stalk (pedicel) of lower flowers, e.g. candytuft; **spike**, raceme with sessile flowers, e.g. plantain; **spadix**, spike with fleshy axis, e.g. cuckoo pint; **catkin**, spike of unisexual, reduced and often pendulous flowers, e.g. hazel, birch; **umbel**, raceme in which the axis has not lengthened, the flowers arising at the same point to form a head with the oldest outside and youngest at the centre, e.g. carrot, cow parsley; **capitulum**, where the axis of the inflorescence is flattened and laterally expanded, with growing point in centre, and bearing closely crowded sessile flowers (florets), the oldest at the margin and youngest at the centre, e.g. dandelion.

In b), branching is sympodial and the main axis ends in a flower, further development taking place by growth of lateral branches, each behaving in the same way. The **cyme** is described as a **monochasium** when each branch of the inflorescence bears one other branch (e.g. iris), and as a **dichasium** when each branch produces two other branches (e.g. stitchwort). Inflorescences are often *mixed* (part indefinite, part definite): a raceme of cymes.

Infraspecific variation Variation within **species**. It takes several different forms. Clarification depends on population structure, **breeding system** and effectiveness of gene flow of the particular case. *Clines* refer to variable phenotypic characters whose distributions display gradients mappable geographically on to gradients in environmental conditions. Morph ratio clines occur when the ratios of different morphs change in a similarly graded way. Phenotypic plasticity may be the cause but where genetic fixation is involved the variation may be closer to ecotypic. **Ecotypes**

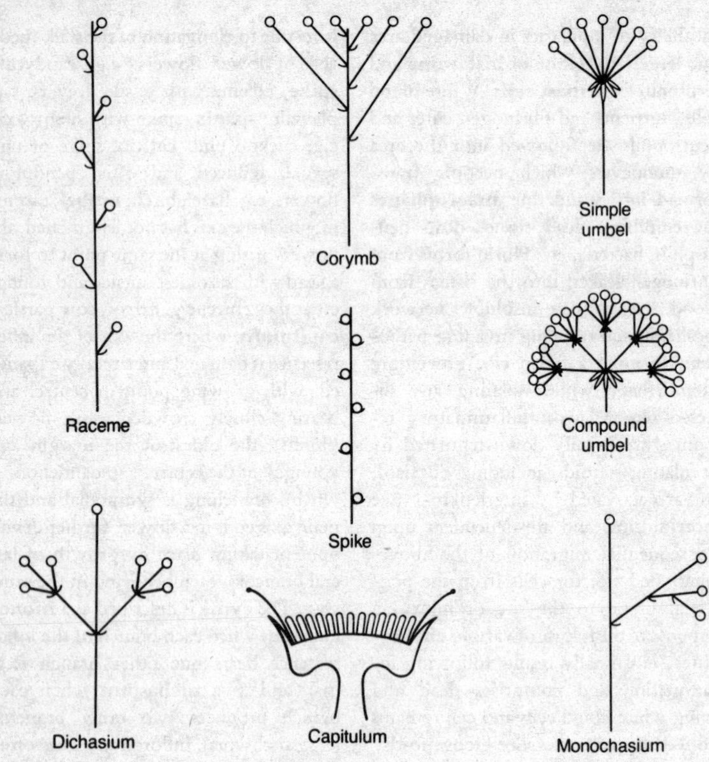

Fig. 63 *Diagrams of different types of* **inflorescence**.

(botanical equivalent of races in **zoology**; see later) involve adaptation of populations to local edaphic, climatic or biotic influences. A *form* in botany is the category within the species generally applied to members showing trivial variations from normal (e.g. in petal colour). In zoology the term is often synonymous with morph (see below), or else a seasonal variant, or used as a neutral term when it is unclear whether a species, subspecies or lesser category is appropriate. A *morph* is one form of a polymorphic species population (see **polymorphism**).

A *race* is a non-formal category used chiefly in zoological contexts. *Geographical races* approximate to subspecies (see later). Host races are those species populations with the same favoured hosts (if parasites) or food plants (when egg-laying, feeding, etc.); such preferences may involve various genetic and non-genetic influences. A *subspecies* is a formal taxonomic category used to denote various forms (types), commonly geographically restricted, of a polytypic species. It should ideally be used of evolutionary lineages rather than mere phenetic subdivisions of a species. Most easily applied when a population is geographically isolated from other populations of the species (e.g. on an island, mountain top). Subspecific status is often conferred on populations which are really part of a clinal series for the characters used but where intermediate populations have not been studied. In some groups (e.g. Diptera) taxonomists have dispensed with the category. A *variety* is a formal category in botany below the level of subspecies and is used of groups which differ, for various reasons, from other varieties within the same subspecies.

Infundibulum (1) Outpushing, or stalk, from floor of vertebrate forebrain attaching the **pituitary** to the hypothalamus. Its terminal swelling produces the posterior pituitary (neurohypophysis); its tissues combine with those growing up from the embryonic mouth to form the combined pituitary organ. (2) Anterior end of ciliated funnel of vertebrate oviduct.

Infusoria Term formerly applied to rotifers, protozoa, bacteria, etc., found in cooled suspensions of boiled hay, etc.

Inguinal Relating to the groin.

Inhibition (Nervous). Prevention of activation of an effector through action of nerve impulses. Some inhibitory **neurotransmitter** molecules hyperpolarize rather than depolarize postsynaptic membranes at synapses thus reducing the probability of a propagated action potential at a synapse (see **inhibitory postsynaptic potential**). Alternatively, an inhibitory neurone may, by its activity, reduce the amount of excitatory neurotransmitter released by another neurone stimulating an effector. See **summation**.

Inhibitory postsynaptic potential (IPSP) Hyperpolarization of a postsynaptic membrane at a synapse; brought about usually by release of a **neurotransmitter** from the presynaptic membrane which fails to open ligand-gated Na^+- or K^+-channels, but instead opens Cl^--channels, making the inside of the cell more negative (polarized) than it was during the resting potential. Tends to inhibit formation of an **action potential** at a synapse. See **summation**.

Initial(s) (Bot.) Cell, or cells, from which tissues develop by division of differentiation, as in **apical meristems**; or a cell from which an antheridium develops in bryophytes.

Initiation factors (IFs) (1) Soluble proteins (often quaternary) initiating transcription or translation of RNA during **protein synthesis**. In transcription, IFs enable an RNA polymerase to locate its consensus **promoter** sequence. In initiating translation, they bind to the small ribosomal subunit, or to the initiator tRNA, and enable mRNA and the initiator tRNA to join the *initiation complex* prior to arrival of the large ribosomal subunit. In some cells the rate of translation is controlled by IFs, which are generally removed prior to binding of the large ribosomal subunit. GTP is required at the same stage as the initiation factors. Compare **elongation factors**. (2) Some **transcription factors** are also described as initiation factors.

Innate immune response See **immunity**.

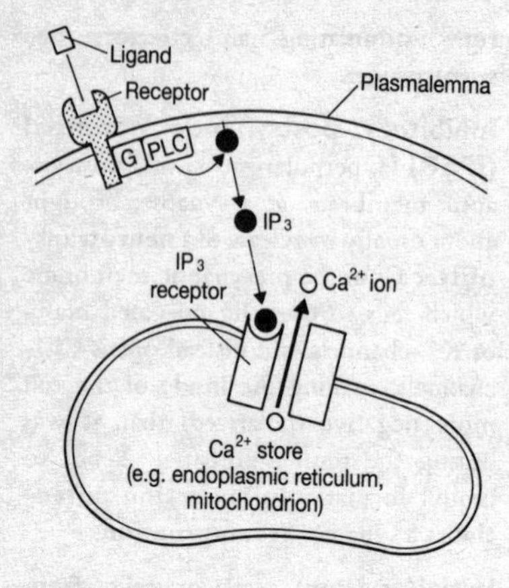

Fig. 64 *The signal pathway thought to cause release of Ca^{2+} ions into the cytosol from intracellular stores, involving* **phospholipase C** *(PLC). See* **inositol 1,4,5-triphosphate**.

Inner cell mass Group of cells formed after sinking inwardly from outer layer of the mammalian morula (blastocyst); determined by the 64-cell stage to become the future embryo rather than trophoblast.

Inner ear See **ear, inner**.

Innervation Nerve supply to an organ.

Innominate artery Short artery arising from aorta of many birds and mammals and giving rise to right subclavian artery (to fore-limb) and right carotid artery (to head).

Innominate bone Each lateral half of the **pelvic girdle** when pubis, ilium and ischium are fused into a single bone as in adult reptiles, birds and mammals.

Inoculation Injection of living or otherwise mildly infective pathogen into a person or domestic animal followed usually by a mild but non-fatal infection which results in the patient's immunity to the virulent pathogen. Nowadays rarely used, immunization by non-infective agents being preferred. See **immunity**.

Inoperculate Opening of a **sporangium** or **ascus** by an irregular tear or plug to liberate spores.

Inosinic acid (IMP) Purine nucleotide precursor of AMP and GMP. Also a rare monomer in nucleic acids where, being similar to guanine, it normally pairs with cytosine. Where it occurs at the 5′-end of an anticodon it may pair with adenine, uracil or cytosine in the 3′-end of the codon. See **wobble hypothesis**.

Inositol Water-soluble carbohydrate (a sugar alcohol) required in larger amounts than vitamins for growth by some organisms.

Inositol 1,4,5-triphosphate (IP3, InsP3) A **second messenger** produced by phospholipase C activity as a breakdown product of the minor cell membrane phospholipid phosphatidylinositol. Hydrophilic, IP3 diffuses into the cytosol and initiates calcium ion (Ca^{2+}) release from the endoplasmic reticulum and from intact plant cell vacuoles and initiates stomatal closure when released from its bound form within guard cells (see **calmodulin**). Another second messenger, **diacylglycerol**, is a product of the same phospholipase C activity. IP3 is the common product of antennal pheromone stimulation in various insect species, and IP3-gated ion channels have been located in olfactory receptors of vertebrates and arthropods. See **cyclic GMP**, **phospholipid transfer proteins**. See Fig. 64.

Inquilinism See **symbiosis**.

Insecta (Hexapoda) Class of **Arthropoda** whose members have a body with distinct head, thorax and abdomen (see **tagmosis**). Head bears one pair of antennae and paired mouthparts (mandibles, maxillae and a single fused labium); thorax bears three pairs of legs and frequently either one or two pairs of wings on second and/or third segments; abdomen bears no legs but other appendages may be present (e.g. see **cerci**). Found as fossils from Devonian onwards. Most have a tracheal system with spiracles for gaseous exchange, and excretion by means of **Malpighian tubules**. **Metamorphosis** either effectively lacking (**Apterygota**), partial (**Exopterygota**) or complete (**Endopterygota**). More numerous in terms of species and individuals than any other metazoan class.

Insectivora Order of placental mammals (e.g. moles, shrews, hedgehogs); a primitive insect-eating or omnivorous group resembling and probably phylogenetically close to Cretaceous ancestors of all placentals. Have small, relatively unspecialized teeth (but incisors tweezer-like). Tree shrews and elephant shrews tend nowadays to be placed in separate orders, Macroscelidia and Scandentia respectively.

Insertion sequence (IS, i. element) One sort of **transposable element** capable of inserting into bacterial chromosomes using enzymes they encode. Their ends form **inverted repeat sequences**. IS's may also occur in **plasmids** (e.g. **F factor**). Can mediate integration of plasmids into main bacterial chromosome by recombination, but cannot self-replicate; are therefore only inherited when integrated into other genomes which do have DNA replication origins. Can mediate a variety of **deletions**, **inversion** and self-excision. See **auxotroph**.

Instar Stage between two ecdyses in insect development or the final adult stage.

Instinct Behaviour which comprises a stereotyped pattern or sequence of patterns; typically remains unaltered by experience, appears in response to a restricted range of stimuli and without prior opportunity for practice. Distinction between this and learnt behaviour has been blurred by research in the last two decades: attention has focused upon developmental pathways of different behavioural responses. Even learnt behaviour presumably has some heritable component; the **heritability** of some behavioural patterns is high, and such behaviour tends still to be termed instinctive.

Insulin Protein hormone comprising 51 amino acids in two chains held together by disulphide bridges. Secreted by β-cells of vertebrate pancreas in response to high blood glucose levels, e.g. after a meal, as monitored by the β-cells themselves, and active only after removal of two amino acid sequences (pre-proinsulin gives proinsulin which gives insulin). Promotes uptake by body cells (esp. muscle, liver, adipose cells) of free glucose and of amino acids by muscle; essentially anabolic in its action. Is thus a *hypoglycaemic* hormone – the only one in most vertebrates – reducing blood glucose. Increases rate of fusion of glucose transporter-bearing vesicles with plasma membrane of sensitive cells; also activates their transporter-mediated glucose uptake and causes synthesis of hexokinase, which phosphorylates glucose on entry to cell, preventing its diffusion out. Low insulin levels cause the glucose transporters to accumulate once again in intracellular vesicles. The cell-surface receptor for insulin is a **tyrosine receptor kinase**. Opposed in its action by **glucagon**, the two hormones together regulating and maintaining blood glucose at appropriate levels (about 100 mg glucose/100 cm^3 blood in humans) through negative feedback via the pancreas. See **diabetes**.

Insulin-like growth factor (IGF, ILGF) See **growth factors**.

Integrins A versatile family of cell-surface transmembrane receptor proteins, binding ligands of the extracellular matrix (ECM) and proteins on other cells' surfaces, thereby performing **adhesion** functions. Each is non-covalently heterodimeric ($\alpha\beta$), the specific ECM ligand bound depending upon the particular combination of the 15 α- and 8 β-subunits from which an integrin can be assembled. There are at least four β-subfamilies, the $\beta1$ subfamily alone including receptors for such ECM components as fibronectin, laminin and collagen. Once ligand-bound, integrins are thought to transduce information about the extracellular environment to the **cytoskeleton** (and possibly to the nucleus). But signals from the cytosol also affect avidity of cell-surface adhesion molecules; so integrins may be involved in information flow both ways across the plasmalemma. The **growth factor** TGF-β up-regulates all integrins.

Integument (Bot.) (Of seed plants) outer cell layer or layers of ovule covering nucellus (megasporangium) and ultimately forming the seed coat. Most flowering plants have two integuments, an inner and an outer. (Zool.) Outer protective covering of an animal, such as skin, cuticle.

Intercalary (Of a meristem) situated between regions of permanent tissue, such as at bases of nodes and leaves in many monocotyledons, or at junctures of stipes and blade in some brown algae.

Intercalated disc See **cardiac muscle**.

Intercellular Occurring between cells. Often applied to the matrix or ground substance secreted by cells of a tissue, as in **connective tissues**. For *intercellular fluid*, see **tissue fluid**. See **extracellular**, **interstitial**, **intracellular**.

Intercellular junction Any of a variety of cell–cell **adhesion** mechanisms, particularly abundant between animal epithelial cells, the three commonest of which are (1) **desmosomes**, which are principally adhesive, (2) *gap junctions*, involved in intercellular communication, and (3) *tight junctions*, occluding the intercellular space, thereby restricting movements of solutes. Gap junctions consist of cylindrical channel proteins with a channel diameter of 1.5 nm, coupling cells electrically (as in electrical synapses and cardiac muscle) and in all probability metabolically (see **plasmodesmata**). Tight junctions perform a selective barrier function in cell sheets, preventing diffusion of ions, etc., from one side of the sheet to the other through intercellular spaces. This is essential to proper functioning of epithelia such as intestinal mucosae and proximal convoluted tubules of vertebrate kid-

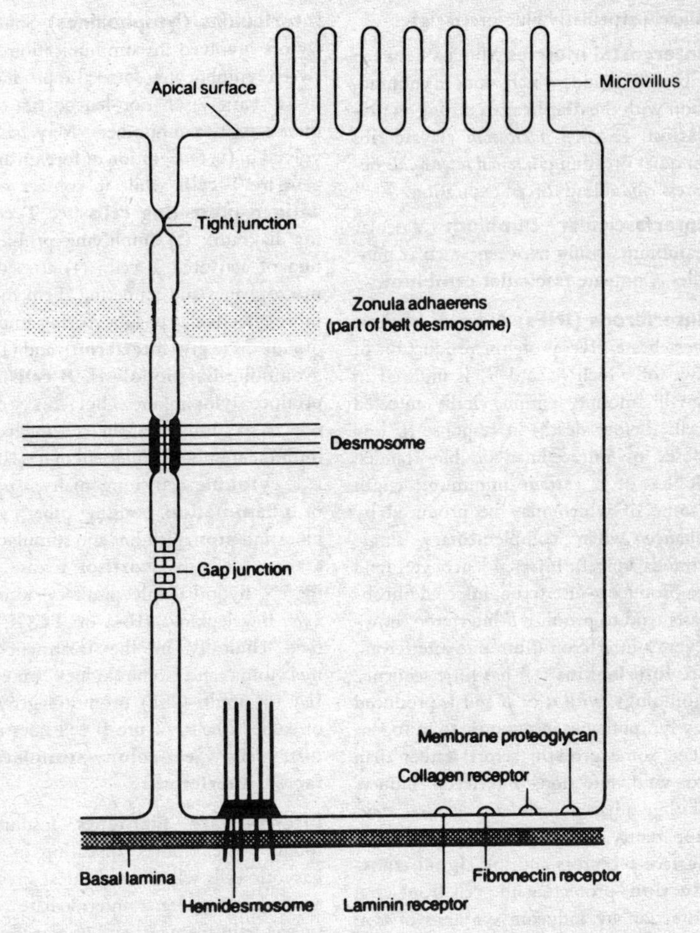

Fig. 65 *Some of the cell–cell adhesion mechanisms of epithelia. See* ***intercellular junction***.

neys. Tight junctions also limit diffusion of membrane proteins and lipids in the outer (not the inner) lipid layers of adjacent membranes. Proteins destined for the apical and basolateral portions of polarized epithelial cells linked by tight junctions are sorted differently, the *trans* Golgi network sometimes, perhaps often, playing a part in this. See Fig. 65.

Intercellular space (Bot.) In plants, air-filled cavities between walls of neighbouring cells (e.g. in cortex and pith) forming internal aerating system. Spaces may be large, making tissue light and spongy as in **aerenchyma**, occasionally harbouring algae, particularly blue-green algae.

Intercostal muscles Muscles between ribs of tetrapods which work in conjunction with the **diaphragm** during **ventilation**. *External intercostals* elevate ribs in quiet breathing; *internal intercostals* depress ribs aiding forced expiration.

Interfascicular cambium Vascular cambium arising between vascular bundles. Compare **fascicular cambium**.

Interferons (INFs) A group of three vertebrate glycoproteins, production of two of which (α and β) is induced in small amounts within virally infected cells (before death) in response to low doses of intracellular double-stranded RNA of a certain minimum length (some of which may be produced by chance when complementary single strands anneal). Infected leucocytes tend to produce α-interferon; infected fibroblasts tend to produce β-interferon; however, γ-interferon ('immune interferon', see **interleukins** (3)) has little sequence homology with α or β and is produced by lymphocytes in response to mitogens (e.g. some growth factors) rather than to viral infection. Interferon induces within adjacent cells an *antiviral state*, apparently binding to specific cell-surface receptors and, by **signal transduction**, protecting the cell from viral infection by inducing synthesis of two enzymes which, activated by double-stranded RNA, inhibit the viral production cycle. One of these digests any single-stranded RNA in the cell, thereby killing the cell. Some viruses (e.g. adenoviruses), by producing high levels of short double-stranded RNA, circumvent the antiviral state.

Intergradation zone See **hybrid zone**.

Interleukins (lymphokines) Soluble factors involved in communication between lymphocytes; some also produced by a variety of non-leucocytic cells (hence term a misnomer). May be involved in (1) recognition of foreign antigens by **T cells** while in contact with **antigen-presenting cells** (see **T cells** for diagram); (2) amplifying proliferation of activated T cells; (3) attracting macrophages and rendering them more more effective at phagocytozing micro-organisms (e.g. γ-**interferon**); and (4) in promoting **haemopoiesis. B cells** can produce lymphokines, but this seems not to be important in cell-mediated immune response. Interleukin-1 (IL-1) is a **cytokine** activating many aspects of **inflammation**, boosting other cytokines and prostaglandins and stimulating a more defensive **cortisol** release via the hypothalamic–pituitary–adrenal axis. Interleukin-2 (IL-2 or TCGF) is used clinically in the treatment of melanoma and some kidney cancers. IL3 (or multi-CSF) promotes growth of some T cells, of pre-B cell lines and mast cells. See **colony-stimulating factor**, **interferons**.

Intermediate filaments Insoluble, tough protein fibres appearing in eukaryotic cells where mechanical stress is applied. Diameters intermediate between actin filaments and microtubules. Help keep Z-discs of adjacent muscle sarcomeres in line. See **cytoskeleton**.

Interneurone (internuncial neurone, relay neurone) Neurone synapsing between sensory and motor neurones in a typical spinal **reflex** arc. Vertebrate interneurones are confined to **grey matter** of central nervous system. Afford cross-connections with other neural pathways, enabling **integration** of reflexes, and learning.

Internode Part of plant stem between two successive **nodes**.

Interoceptor Receptor detecting stimuli within the body, in contrast with exteroceptor. Includes **baroreceptors**, **proprioceptors**, pH-receptors and receptors sensitive to concentrations of dissolved O_2 or CO_2 (see **carotid body**, **carotid sinus**).

Interphase Interval between successive nuclear divisions, usually mitotic in proliferative cells; also preceding, and occasionally following, meiosis. Somewhat misleading term suggesting a quiescent or resting interval in the **cell cycle** (indeed the nucleus is often termed a resting nucleus during it). On the contrary, it is the period during which most components of the cell are continuously made. Cell mass generally doubles between successive mitoses. See Fig. 15b (**cell cycle**) for some of the molecular details.

Intersex Individuals, often sterile and usually intermediate between males and females in appearance; sometimes hermaphrodite; resulting from failure of the mechanism of **sex determination**, often through chromosomal imbalance. A **freemartin** is an example where hormonal causes are involved. The discovery of intersexes in *Drosophila* led to an understanding of its balance mode of sex determination. See **gynandromorph**, **testicular feminization**.

Interspecific Between species; as in *interspecific* **competition**.

Interstitial Lying in the spaces (interstices) between other structures, *interstitial cells* of vertebrate gonads lie either between the ovarian follicles or between the seminiferous tubules of the testis; the latter cells secrete the hormone **testosterone.** See **luteinizing hormone**.

Interstitial cell stimulating hormone (ICSH) See **luteinizing hormone**.

Interstitial fluid See **tissue fluid**.

Intestine That part of the **alimentary canal** between stomach and anus or cloaca. Responsible for most of the digestion and absorption of food and (usually) formation of dry faeces. In vertebrates, former role is often performed by the anterior *small intestine* (see **duodenum**, **jejunum**, **ileum**), which commonly has a huge surface area brought about by a combination of (i) folds of its inner wall, (ii) **villi**, (iii) **brush borders** to the epithelial mucosal cells, (iv) **spiral valves**, if present, (e.g. elasmobranchs) and (v), in herbivores especially, its considerable length. The more posterior *large intestine*, or *colon*, is usually shorter and produces dry faeces by water reabsorption. The junction between the two intestines is marked in amniotes by a valve, and often a **caecum**. Products of digestion are absorbed either into capillaries of the submucosa or, in **chylomicrons**, into lacteals of the lymphatic system. Such digestion as occurs in the caecum and large intestine is largely bacterial.

Intracellular Occurring within a cell, which generally means within and including the volume limited by the plasma membrane. Contents of food vacuoles and endocytotic vesicles, although geographically within the cell, are not strictly intracellular until they have passed through the vacuole or vesicle membrane and into the cytosol. See **endosymbiosis**, **glycocalyx**.

Intraspecific Within a species. See **deme, infraspecific variation**.

Introgression (introgressive hybridization) Infiltration (or diffusion) of genes of one species population into the gene pool of another; may occur when such populations come into contact and hybridize under conditions favouring one or the other, the hybrids and their offspring backcrossing with the favoured species population. See **hybrid, hybrid swarm**.

Intromittant organ Organ used to transfer semen and sperm into a female's reproductive tract. Include **claspers, penis**.

Intron DNA sequence (or its RNA transcript) lying within a coding sequence, but usually not itself encoding cell product, and resulting in so-called 'split-genes'; almost universal in eukaryotic genes (but not **interferon** genes), where on average there are ten per gene; but few examples of prokaryotic introns. Usually spliced out from RNA (pre-mRNA) during **RNA processing** to avoid translation into missense protein (see **spliceosome**). Many of the vertebrate genes cloned to date contain several or many introns. DNA from T4 phage and from fungal and plant organelles contains *self-splicing* introns, but eukaryote introns are not self-splicing. At least one intron has been shown to cause mitochondrial recombination (see **mitochondrion**). Until recently, it was widely held (the 'introns early' view) that the progenote ancestor of all living groups of organisms had introns in its DNA, for they had been found in **Archaebacteria** but not in eubacteria (which were thought to have lost them). Now, however, they have been found in the Cyanobacteria (putative ancestors of chloroplasts) and those purple bacteria believed to have given rise to mitochondria (see **endosymbiosis**). This favours the alternative 'introns late' view that introns are essentially selfish elements (see **selfish DNA/genes**) which have invaded eukaryotic genes subsequent to the evolution of the nucleus. It now looks as though eukaryotic nuclear (spliceosomal) introns evolved from group II self-splicing introns (absent from the nucleus) which moved by retroposition from the mitochondrion, whence they had come by way of purple bacteria (see **spliceosome**). Subsequent fragmentation of one of these could then have produced the genes for small nuclear (sn) RNAs (see **RNPs**) able to splice all group II introns present in the nuclear genome (in *trans*), making intron autocatalysis redundant.

Introrse (Of anther dehiscence) towards the centre of a flower, promoting self-pollination. Compare **extrorse**.

Intussusception (Bot.) Insertion of new cellulose fibres and other material into an existing and expanding **cell wall**, increasing its surface area. Cellulose microfibrils are interwoven among those already existing, as opposed to being deposited on top of them. Compare **apposition**.

Inulin Soluble polysaccharide, composed of polymerized fructose molecules, occurring as stored food material in many plants, such as members of the Compositae and in dahlia tubers. Absent from animals.

Invagination Intucking of a layer of cells to form a pocket opening on to the original surface. Common in animal development, as during **gastrulation**.

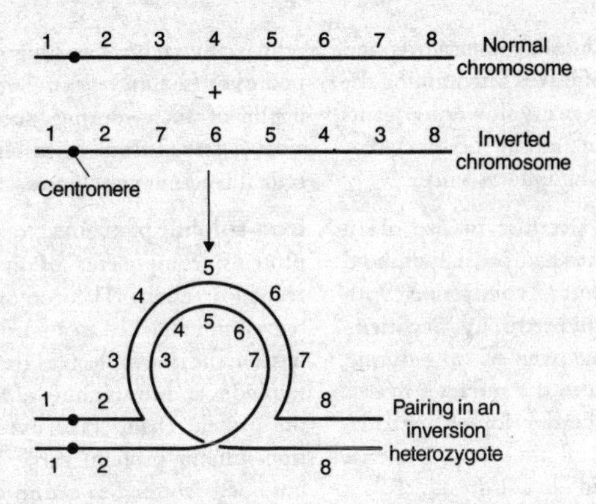

Fig. 66 *Diagram showing the effect of a large* **inversion** *upon pairing of a bivalent during meiotic prophase. Cross-overs within such an inversion hybrid (heterozygous for the inversion) result in acentric fragments and reduced fertility. Numbers indicate positions of loci.*

Inversion (1) A type of chromosome **mutation** in which a section of chromosome is cut out, turned through 180° and rejoined, or spliced back, to the chromosome upside down. If long enough, it results in an *inversion loop* in **polytene** or meiotic cells heterozygous for the inversion. The inverted region is stretched in order to pair up homologously with its partner, as shown in Fig. 66.

If **crossing-over** occurs within the inversion loop then the chromosomes which result are nearly always abnormal, having either deletions, duplications, too many centromeres or none at all (acentric). This normally results in reduced fertility. See **supergene**. (2) Hydrolysis of sucrose by **invertase** to equimolar concentrations of glucose and fructose.

Invert sugar Equimolar mixture of glucose and fructose, usually resulting from digestion of sucrose by **invertase**.

Invertase (sucrase) See **sucrose**.

Invertebrate Term designating any organism that is not a member of the **Vertebrata**. There are many invertebrate chordates.

Inverted repeat sequence DNA sequences, often lying on either side of **transposable elements** and, when single-stranded, run in opposite directions along the chromosome (i.e. are palindromic) and may form a **hairpin loop** by folding back and base-pairing. Double-stranded inverted repeats are also found.

In vitro 'In glass' (Latin). Biological process occurring, usually under experimental conditions, outside the cell or organism; e.g. in a test-tube.

In vivo Biological process occurring within a living situation, e.g. in a cell or organism.

Involucre A protective investment. (1) In thalloid liverworts, scale-like upgrowth of the thallus overarching the archegonia; (2) in leafy liverworts and mosses, groups of leaves surrounding the sex organs; (3) in many flowering plants (e.g. Compositae), group of bracts enveloping the young inflorescence.

Involution (1) Decrease in size of an organ, e.g. thymus and other lymphoid tissue after puberty, contrasting with hyperplasia and hypertrophy. See **atrophy**. (2) Rolling over of cells during **gastrulation**, from the surface towards the interior of the developing gastrula. (3) Production of abnormal bacteria, yeasts, etc. (e.g. in old cultures).

Ion channels Membrane-spanning water-filled pores composed of protein, allowing ions across (highly selectively in eukaryotic cells, less so in prokaryotic). Not continuously open, but have 'gates' which open briefly on appropriate stimulation. Voltage-gated, ligand-gated, mechanically gated and **G-protein**-gated forms exist. Sodium and potassium channels (see **impulse, resting potential**) are examples. See **receptor** (2), and references there.

Ionophore One of a range of small organic molecules facilitating ion movement across a cell membrane (usually the plasma membrane). They either enclose the ion and diffuse through the membrane (e.g. valinomycin-K^+) or form pore-channels in the lipid bilayer (e.g. gramicidin), in which case water molecules are allowed through too. Some are products of microorganisms and may have adverse effects upon cells of competing species.

IP3 See **inositol 1,4,5-triphosphate**.

Iris Pigmented, muscular diaphragm whose reflex opening and closing causes varied amounts of light to fall upon the retinas of vertebrate and higher cephalopod **eyes** (the *iris reflex*). Contributes to depth of focus during **accommodation**. Derived from fused **choroid** and retinal layers in vertebrates.

Iron-sulphur proteins Non-haem iron proteins; components of the **electron transport chain**. The iron of these proteins is not attached to a porphyrin ring; instead, the iron molecules are attached to sulphides and to sulphurs of cysteines of the protein chain. Like **cytochromes**, iron-sulphur proteins carry an electron but not a proton. See **quinone**.

Ischaemia (adj. ischaemic) Reduced blood supply. See **nicotine**.

Ischium Ventral, back-projecting, paired bones of vertebrate **pelvic girdle**. They bear the weight of a sitting primate.

Isidium Rigid protuberance of upper part of a lichen thallus which may break off and serve for vegetative reproduction.

Islets of Langerhans Groups of endocrine cells scattered throughout the vertebrate **pancreas**; some of these cells secrete **insulin**, some **glucagon**.

Isoantigen See **alloantigen**.

Isobilateral (Of leaves) having the same structure on both sides. Characteristic of leaves of monocotyledons (e.g. irises), where leaf-blade is more or less vertical and the two sides are equally exposed. Compare **dorsiventral**.

Isodiametric Having equal diameters; used to describe cell shape when length and width are essentially equal.

Isoelectric point The pH of solution at which a given protein is least soluble and therefore tends to precipitate most readily. At this pH the net charge on each of the protein molecules has the highest probability of being zero, and as a result they repel each other least in solution. They also tend not to move in an electric field, e.g. during **electrophoresis**.

Isoenzymes (isozymes) Variants of a given enzyme within an organism, each with the same substrate specificity but often different substrate affinities (see **Michaelis constant**); separable by methods such as **electrophoresis**. E.g. lactate dehydrogenase occurs in five different forms in vertebrate tissues, the relative amounts varying from tissue to tissue.

Isogamy Fusion of gametes which do not differ morphologically, i.e. are not differentiated into macro- and microgametes. Compare **anisogamy**; see **sex**.

Isogeneic (syngeneic) Having the same genotype.

Isograft (syngraft) Graft between isogeneic individuals, such as identical twins, or mice of the same pure inbred line. Unlikely to be rejected. See **graft**.

Isokont (Bot.) Motile cell or spore possessing two flagella of equal length. Compare **heterokont**.

Isolating mechanisms Mechanisms restricting gene flow between species populations, sometimes of the same but usually of different species. Sometimes classified into *prezygotic mechanisms*, including any process (including behaviour) tending to prevent fertilization between gametes from members of the two populations, and *postzygotic mechanisms*, which prevent development of the zygote to maturity or render it partially or completely sterile. These mechanisms are likely to arise during geographical isolation (a non-biological isolating mechanism) of populations of the same species but they may be reinforced by selection during subsequent sympatry. Their role in sympatric speciation is under investigation. See **speciation**.

Isomerase Any enzyme converting a molecule to one of its isomers, commonly a structural isomer.

Isomorphic Used of **alternation of generations**, particularly in algae, where the generations are vegetatively identical. Compare **heteromorphic**.

Isopoda Order of the crustacean subclass **Malacostraca**, containing such forms as aquatic waterlice (e.g. *Asellus*) and terrestrial woodlice (e.g. *Oniscus*). No carapace; body usually dorso-ventrally flattened. Females have a brood pouch in which young develop directly. Little division of labour between appendages. Many parasitic forms. About 4000 species.

Isoptera Termites (white ants). Order of **eusocial** exopterygote insects, with an elaborate system of **castes**, each colony founded by a winged male and female; wings very similar, elongated, membranous and capable of being shed by basal fractures. Numerous apterous forms.

Isotonic Of solutions having equal solute concentration (indicated by their *osmotic pressure*). See **hypertonic**, **hypotonic**, **water potential**.

Isotype (1) See **antibody diversity**. (2) Duplicate of type specimen, or **holotype**.

Isozyme See **isoenzyme**.

Iteroparous Breeding more than once in its life. In botany, polycarpic. Compare **semelparous**.

Jacob-Monod theory An influential theory of prokaryotic **gene expression**, of some value in the explanation of eukaryotic gene expression. Its basic concept is that of the *operon*, a unit of **translation**, comprising a group of adjacent structural genes (**cistrons**) on the chromosome, headed by a non-coding DNA sequence (the *operator*) whose conformation binds a **DNA-binding protein** (the *repressor*, or *regulator*) encoded by a regulator gene elsewhere on the chromosome (see Fig. 67). In one model, the repressor binds the operator preventing the enzyme RNA polymerase from gaining access to an adjacent DNA sequence (the *promoter*), which it must do if any of the structural genes of the operon are to be transcribed. The repressor–operator complex is stable and only broken if another molecule, the *inducer*, binds to it, in which case the inducer–repressor complex loses its affinity for the operator and transcription of the operon's cistrons results. This serves to explain prokaryotic enzyme induction, where the presence of substrate, acting directly or indirectly as inducer, is required for an enzyme to be produced by a cell. In one model of enzyme repression, the repressor does not bind to the operator until it has itself bound to some other molecule, the *corepressor* (e.g. enzyme product, or other gene product). Only then will transcription of the operon be inhibited. Mutants in the repressor gene, affecting repressor shape so that it cannot bind the operator or the corepressor, will result in **constitutive** production of the operon's cistron products. One observation which the theory helped explain was that in the bacterium *Escherichia coli* the enzymes encoded by what is now referred to as the *lac* operon were either all produced together or not produced at all. Other **regulatory genes**, especially in eukaryotes, encode **transcription factors**, which bind upstream of genes and assist binding of RNA polymerase to the promotor. The operon concept seems inapplicable in eukaryotes, where polycistronic mRNA is rare or non-existent. See Fig. 113.

Jaws Paired (upper and lower) skeletal structures of **Gnathostomata** almost certainly deriving from the third pair of **gill arches** of an ancestral jawless (agnathan) vertebrate. Upper jaw (see **maxilla**) varies in its articulation with the braincase (see **autostylic**, **amphistylic** and **hyostylic jaw suspension**). Progressive reduction in number of skeletal elements in lower jaw (see **mandible**) during vertebrate evolution, only the dentaries remaining in mammals. Tooth-bearing. See **dentition**.

Jejunum Part of the mammalian small intestine succeeding the duodenum and preceding the ileum. Has larger diameter and longer villi than the rest of the small intestine, from which it is not anatomically distinct.

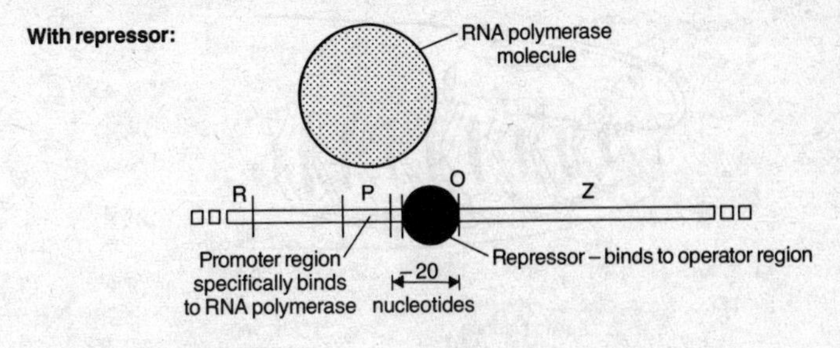

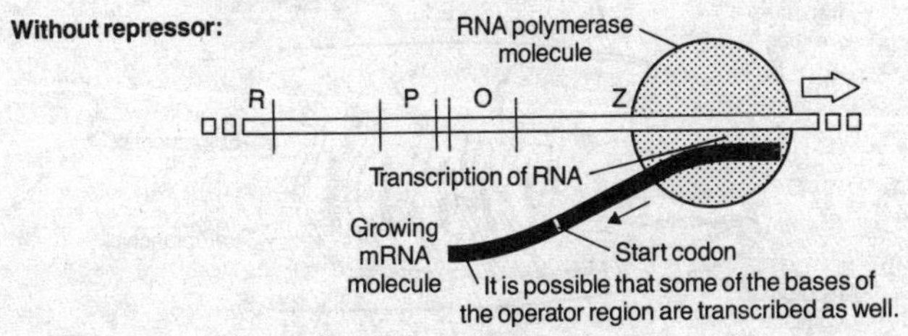

Fig. 67. *Simplified diagram indicating the effect of a repressor molecule (above) in inhibiting transcription of gene Z by RNA polymerase compared with loss of this inhibition (below) when it is removed. Gene R encodes the repressor.* © J. D. Watson: *Molecular Biology of the Gene* 3rd ed. (1976), Fig. 14-12(a). Pub. Benjamin/Cummings.

Jellyfish See **Scyphozoa**.

Joints Articulations of animal endoskeletons or exoskeletons, the former definitive of arthropods, the latter characteristic of vertebrates where they may be either immovable (e.g. between skull bones), partly movable (e.g. between vertebrae) or freely movable (e.g. hinge joints, ball-and-socket joints). Commonly a feature of lever systems employing antagonistic muscles.

J-shape growth form Type of population growth form, in which density increases rapidly in an exponential manner and then stops abruptly as environmental resistance or other limit takes effect more or less suddenly. See **growth curves**, **exponential growth**.

Jugulars Major veins in mammals and related vertebrates returning blood from the head (particularly the brain) to the superior vena(e) cava(e). Usually in the form of paired interior and exterior jugulars fusing on each side to form common jugulars before joining the subclavian veins and ultimately draining into the venae cavae.

Junctional complex See **intercellular junction**.

Jurassic **Geological period** of the **Mesozoic** era, extending from about 195–135 Myr BP. A major part of the age of reptiles, during which the **Archosaur** radiation, begun in the **Triassic**, continued. Plesiosaurs and ichthyosaurs also flourished. Mammal-like reptiles

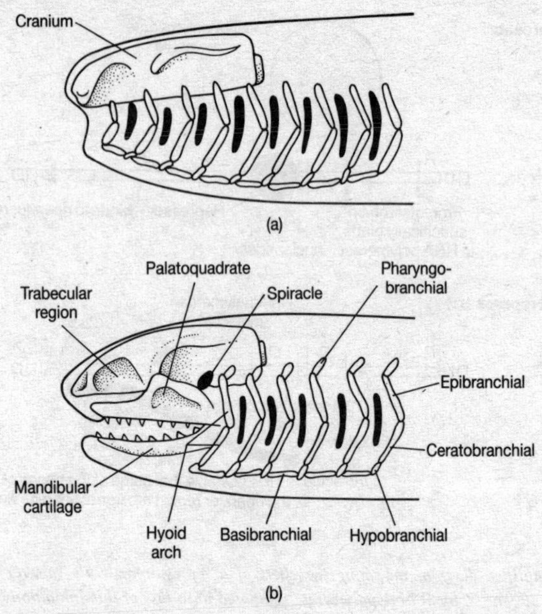

Fig. 68 *Proposed evolution of vertebrate jaw from anterior gill arches of agnathan ancestor. (a) agnathan, (b) gnathostome. The first gill arch may have had no descendant structure; the second probably gave rise to the trabecular cartilage, while a more posterior arch gave rise to the pair of jaws, for which the hyoid arch acts as a support in (b). The gill slit between jaws and hyoid arch was reduced to the spiracle seen in modern sharks.*

(e.g. therapsids) dwindled; true mammals were scarce. Earliest fossil birds discovered (e.g. ***Archaeopteryx***) were deposited in the Upper Jurassic. Climate was generally mild, continents low with large areas covered by seas. Gymno sperms, especially cycads, prominent.

Juvenile hormone (neotenin) Hormone produced by insect **corpora allata**. See also **ecdysone**.

K See **Kd**.

Kappa particles Gram-negative bacterial species present endosymbiotically (as commensals) within cytoplasm of ciliate *Paramecium aurelia*; their maintenance requires activity of some nuclear genes of host cell. May be transferred from one host cell to another during **conjugation**, and may produce toxins (e.g. *paramecin*) which kill sensitive *Paramecium* strains but not the producer cells (termed *killer cells*). Inheritance of the killer trait is an instance of extrachromosomal or **cytoplasmic inheritance**.

Karyogamy Fusion of nuclei or their components. Feature of eukaryotic sexual reproduction. See **fertilization**, **heterokaryosis**. Compare **plasmogamy**. For possible evolutionary origin, see **sex**.

Karyotype Characteristics of the set of chromosomes of a cell or organism (their number, sizes and shapes). A photograph or diagram of chromosomes, generally arranged in pairs and in order of size, is termed a *karyogram*.

kb Kilobase; 10^3 bases.

Kd (K) Kilodalton. Having a relative molecular mass of 1000.

Keel (carina) (Zool.) Thin medial plate-like projection from sternum (breast-bone) of modern flying birds (*carinates*) and bats providing attachment for wing muscles. Absent from **ratites** and many flightless birds. (Bot.) (1) Ridge alongside a fold applied to coalescent lower petals of a papilionaceous corolla of the pea family. (2) In some pennate diatoms, summit of a ridge bearing the raphe, where the valve is sharply angled at the raphe.

Kelp Common name given to brown algae (**Phaeophyta**) of Order Laminariales, the main group of brown algae having important economic uses (e.g. alginic acid, green manure, fucoidin, food).

Keratin Tough fibrous sulphur-rich protein of vertebrate epidermis forming resistant outermost layer of skin. See **cornification**, **cytoskeleton**.

Ketone bodies Substances such as acetoacetate and hydroxybutyrate produced mainly in the liver from acetyl coenzyme A, itself an oxidation product of fatty acids, released for use by peripheral tissues as fuel. The metabolic pathway is termed *ketogenesis*.

Kidney Major organ of nitrogenous **excretion** and **osmoregulation** in many animal groups of little or no homology. Its elements usually open directly to the exterior in invertebrates, but usually via a common excretory duct in vertebrates. Functional units in vertebrates, kidney tubules or *nephrons*, were probably originally paired in each trunk segment and drained through a pair of ducts, one on each side of the body (see **Wolffian duct**). In higher vertebrates, anterior tubules (forming what remains of the *pronephros*) are embryonic and transitory, kidney function being normally dominated by the opisthonephros (*mesonephros* and *metanephros*), whose segmental organization is all but lost in the adult, a new excretory duct (the *ureter*) draining from the mass of nephrons into the bladder. The mesonephros is the functional kidney in adult fish and amphibians. In embryonic amniotes the two kidneys are initially mesonephric, lying in the trunk, their nephrons having glomeruli and coiled tubules; but the whole structure loses its urinary role during development and in males becomes invaded by the vasa efferentia. Each mesonephric duct gives off *ureteric buds* which grow into the intermediate mesoderm and develop into the collecting ducts, pelvis and ureter – in due course draining the *metanephros*. Each bud gives rise to a cluster of capillaries, a *glomerulus*, and a long tubule differentiating into *Bowman's capsule*, *proximal convoluted tubule*, *loop of Henle* and *distal convoluted tubule*, joining the collecting duct. Units developing from the cap tissue are termed *nephrons*, most of whose components lie in the kidney *cortex*, only the loops of Henle lying in the *medulla*, where they join the collecting ducts. Over a million nephrons may occur in each mammalian metanephros.

Hydrostatic pressure in the blood forces water and low molecular mass solutes (not proteins) out of the glomeruli into the Bowman's capsules. In mammals 80% of this glomerular filtrate is then reabsorbed across the cells of the proximal convoluted tubule by **facilitated diffusion** and **active transport** into capillaries of the **vasa recta** draining the kidney. The descending and ascending limbs of the loop of Henle form a **countercurrent system** whose active secretion and selective permeabilities result in a high salt concentration in the interstitial fluid of the medulla, enabling water to be drawn back into the medulla osmotically from the collecting ducts if these are rendered permeable by **antidiuretic hormone**. Urea is also reabsorbed, but never against a concentration gradient, half being excreted on each journey through the kidneys. Kidneys play a major role in osmoregulation and help regulate blood pH by controlling loss of HCO_3^- and H^+. The remnants of glomerular filtrate from all collecting ducts comprise the **urine**. See **renin**.

Kinase Enzyme transferring a phosphate group from a **high-energy phosphate** compound to a recipient molecule, often an enzyme, which is thereby activated and able to perform some function. Opposed by phosphatase activity, which removes the transferred phosphate group. See **phosphorylase kinase**, **tyrosine kynase**, **enterokinase**, **cascade**.

Kinesin One of any eukaryotic cell's three major motor proteins (**myosin** and **dynein** being the other two). Able to bind separately to microtubules and organelles and, through its ATPase activity, generate the force required to move the latter along the former (towards the *plus*, or distal, end). In contrast to myosin and dynein – also transducing proteins – binding of kinesin to its partner protein (e.g. tubulin of a microtubule) is promoted by ATP's presence, ATP being hydrolysed only while kinesin is bound. Once released, kinesin can take another step in the walk along the microtubule. Kinesin-dependent organelle movement, unlike dynein-dependent, is usually centrifugal. Plays a major role in transport within axons of neurons. See **cyclosis**.

Kinesis Movement (as opposed to growth) of an organism or cell in response to a stimulus such that rate of locomotion or turning depends upon intensity but not direction of the stimulus. Compare **taxis**, **tropism**.

Kinetin A purine, probably not occurring naturally, but acting as a **cytokinin** in plants.

Kinetochore Structure developing on **centromere** of chromosome, usually during late mitotic and meiotic prophases. **Microtubules** appear to embed on it (seemingly involving **dynein**) and possibly grow out from it.

Kinetoplast Organelle present in some flagellate protozoans (the Kinetoplastida, e.g. *Trypanosoma, Leishmania*) and containing sufficient DNA for this to be visible under light microscopy when suitably stained. Apparently a modified mitochondrion; commonly situated near the origin of a flagellum.

Kinetosome See **centriole**.

Kingdom Taxonomic category with the greatest generality commonly employed, inclusive of divisions (phyla). Controversy has occurred over the number of kingdoms to employ. Today, most recognize five kingdoms: **Monera** (prokaryotic organisms), **Protista**, **Fungi**, **Animal** and **Plantae**.

Kinin See **cytokinin**, **bradykinin**.

Kinorhyncha Class of minute marine **Aschelminthes** (or a phylum in its own right) with superficial metameric appearance but without true segmentation. Share a syncytial hypodermal structure with **Gastrotricha** but unlike them lack external cilia. Cuticle covered with spines. Muscular pharynx similar to that of nematode worms. Nervous system a ring around the pharynx with four longitudinal nerve cords. Usually dioecious.

Kin selection Selection favouring genetic components of any behaviour (in its broadest sense) by one organism having beneficial consequences for another, and whose strength is proportional to the **relatedness** of the two organisms. Contrast **group selection**. See **Hamilton's rule**.

Klinefelter's syndrome Syndrome occurring in men with an extra X-chromosome giving genetic constitution XXY. Usually results from **non-disjunction** and expresses itself by small penis, sparse pubic hair, absence of body hair, some breast development (gynaecomastia), small testes lacking spermatogenesis. Long bones often longer than normal. See **Turner's syndrome**, **testicular feminization**.

K_M-value See **Michaelis constant**.

Kranz anatomy (K. morphology) Wreath-like arrangement (*Kranz* being German for wreath) of palisade mesophyll cells around a layer of bundle-sheath cells, forming two concentric **chloroplast**-containing layers around the vascular bundles; typically found in leaves of C_4 plants, such as maize and other important cereals. See **photosynthesis**.

Krebs cycle

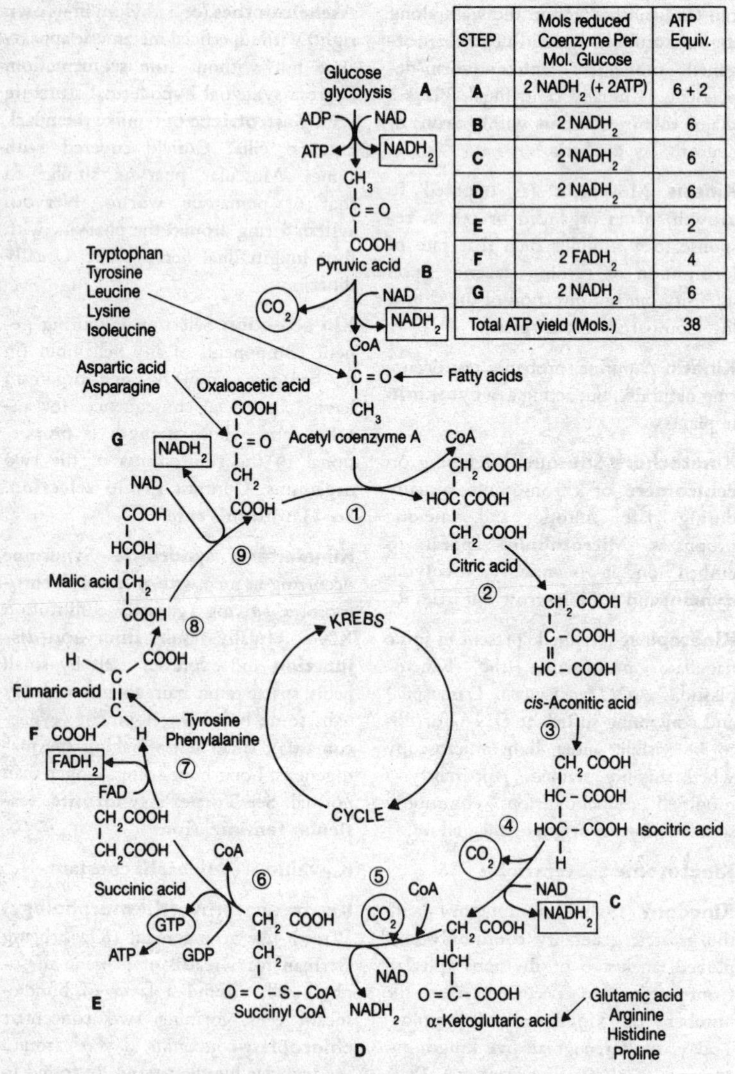

STEP	Mols reduced Coenzyme Per Mol. Glucose	ATP Equiv.
A	2 $NADH_2$ (+ 2ATP)	6 + 2
B	2 $NADH_2$	6
C	2 $NADH_2$	6
D	2 $NADH_2$	6
E		2
F	2 $FADH_2$	4
G	2 $NADH_2$	6
Total ATP yield (Mols.)		38

Fig. 69 *Diagram showing the* **Krebs cycle** *and its relationship to glycolysis and to amino acid and fatty acid input. The table indicates how one glucose molecule can provide the energy to release thirty-eight ATP molecules once the reduced coenzymes have passed their hydrogen atoms on to the respiratory chain and electron transport system.*

Krebs cycle (citric acid cycle, tricarboxylic acid cycle, TCA cycle) Cyclical biochemical pathway (see Fig. 69) of central importance in all aerobic organisms, prokaryotic and eukaryotic. The pathway is cyclical since the 'end-product' (oxaloacetic acid) is also an initial substrate for the next round; and being cyclical, the pathway does not suffer from **end-product inhibition**. The reactions themselves contribute little or no energy to the cell, but dehydrogenations involved are a source of electrons for **electron transport systems** via which ATP is produced from ADP and inorganic phosphate. In addition, some GTP is produced directly during the cycle. Cells with mitochondria perform the cycle by means of enzymes within the matrix bounded by the inner mitochondrial membrane; in prokaryotes these enzymes are free in the cytoplasm. Bulk of substrate for the cycle is acetate bound to coenzyme A as acetyl CoA (see **pantothenic acid**), but intermediates of the cycle can act as substrates. Acetate is usually derived from pyruvate produced by **glycolysis**, amino acid oxidation or fatty acid oxidation, a decarboxylation and oxidation within the mitochondrion then generating carbon dioxide, acetate and reduced NAD. See **ATP**.

Krummholz Region between the alpine and tree lines, where trees are dwarfed and deformed owing to severe environmental conditions, particularly wind.

K-selection Selection for those characteristics which enable an organism to maximize its **fitness** by contributing significant numbers of offspring to a population which remains close to its **carrying capacity**, *K*. Such populations are characterized by intensely competitive and density-dependent interactions among adults, with few opportunities for recruitment by young. Adults invest heavily in growth and maintenance, have a small reproductive commitment and generally have delayed maturity. See **density-dependence**, ***r*-selection**.

Kupffer cell See **reticulo-endothelial system**.

Labelling Variety of indispensable techniques for detecting presence and/or movement of certain isotopic atoms both *in vitro* and *in vivo*. Isotopes used are either radioactive (**radioisotopes**) or differ in their atomic masses without being radioactive. Commonly used radioactive isotopes include ^{14}C, ^{35}S, ^{32}P and ^{3}H; non-radioactive 'heavy' isotopes include ^{15}N and ^{18}O. Material containing the unusual isotope is often administered briefly (pulse-labelling), and 'chased' by unlabelled material. The time taken for label to pass through the system and the route it takes contribute to our understanding of the dynamics of biological systems and processes (e.g. cell membranes, photosynthesis, aerobic respiration, DNA and protein synthesis) as well as of molecular structure. See **autoradiography, southern blot technique, DNA hybridization**.

Labial Referring to **labium**, or to lip-like structure.

Labiate process Tube or opening through the valve wall of a diatom (**Bacillariophyta**) with an internal flattened tube or longitudinal slit often surrounded by two lips.

Labium (Bot.) (1) Lower lip of flowers of the family Labiatae. (2) The lip subtending the ligule in the lycopod *Isoetes*. (Zool.) (1) An insect mouthpart forming the lower lip and comprising a single structure formed from a pair of fused second maxillae and bearing a pair of palps distally. Compare crustacean **maxilliped**. (2) One member of two pairs of skin folds (*labia*) in female mammals protecting the exterior opening of the vagina.

Labrum Plate of exoskeleton hinged to front of the head in some arthropod groups and serving to enclose a space (cibarium) in front of the actual mouth. Found in insects and crustaceans. Trilobites had a similar structure, but placed ventrally on the head.

Labyrinthodontia Subclass of extinct amphibians found as fossils from Upper Devonian to Upper Triassic (see **geological periods**). Generally believed to have had rhipidistian ancestors.

Labyrinthulales Net slime moulds. Order of colonial organisms; cells naked, spindle-shaped, secreting an extracellular membrane-limited matrix in the form of a network of tubes (slimeways) through which they move. Aquatic; mostly marine. Affinities uncertain.

lac operon DNA sequence in the genome of the bacterium *Escherichia coli* comprising an *operator*, *promoter*, and three structural genes, transcribed into a single mRNA, encoding enzymes involved in lactate uptake and metabolism. See **Jacob-Monod theory**.

Lactation Process of milk production by mammary glands. Involves hormone activity, notably of **prolactin**, itself released by hypothalamic *prolactin releasing factor* reflexly secreted during sucking at nipples. Oestrogens and progesterone promote prolactin secretion but inhibit milk secretion; however after delivery the female sex hormone levels in the blood drop abruptly and this inhibition is removed. Sucking at the nipples reflexly releases **oxytocin** from the posterior pituitary causing muscle contraction in the breast alveoli and milk letdown. See **human placental lactogen**.

Lacteal Lymph vessel draining a **villus** of vertebrate small intestine. After digestion, reconstituted fats are released into the lacteal as **chylomicrons**.

Lactic acid A carboxylic acid, $CH_3CHOHCOOH$, produced by reduction of pyruvate during anaerobic respiration (see **glycolysis**). Many vertebrate cells can produce lactate, notably muscle and red blood cells. After transport in the blood it may be converted back by the liver to glucose during **gluconeogenesis**. Probably responsible for sensation of muscle fatigue. See **oxygen debt**. Also formed in metabolism of many bacteria (e.g. from lactose in souring of milk), as well as in fungi (e.g. Deuteromycotina), commonly in association with alcohol or acetaldehyde.

Lactose Disaccharide found in mammalian milk. Comprises galactose linked to glucose by a β.1–4 glycosidic bond. A reducing sugar.

Lacuna (Zool.) Any small cavity, such as those containing bone or cartilage cells.

Lagomorpha Order of placental mammals including pikas, rabbits and hares. Gnawing herbivores differing from rodents chiefly in having two pairs of incisors as opposed to one pair in the upper jaw, the second pair being small and functionless.

Lamarck, Lamarckism Jean Baptiste de Lamarck (1744–1829) was a French natural philosopher who united a wide range of scientific interests under general principles. Coined the term *biology* in 1802, and worked in the newly-created National Museum of Natural History as professor of the zoology of the lower animals, although his previous work had been largely botanical. First to classify animals into invertebrates and vertebrates and to use such taxa as Crustacea and Arachnida. Lamarck's evolutionary views owed something to **Buffon**'s interest in modification of organisms by changes in their environments – changes which for Lamarck altered an organism's needs and habits. He held (as did Geoffroy St Hilaire) that continuous spontaneous generation was required to restock the lowest life forms which had evolved into more complex ones.

For Lamarck the mechanism of evolutionary change was environmental: organs which assist an organism in its altered conditions are strengthened (e.g. the giraffe has acquired its high shoulders and long neck by straining to reach higher and higher into trees for leaves); others progressively atrophy through disuse. Such acquired characters, he thought, were then inherited. But there were and are no clear examples of such inheritance; nor does this theory account satisfactorily for evolutionary stability (stasis). It would require a theory of inheritance completely at variance with that receiving experimental support today. See **Cuvier**, **Darwin**, **Weismann**, **neo-Lamarckism**, **mutation**.

Lambda (λ) phage and repressor A temperate phage (see **bacteriophage**) which can undergo either lytic growth or enter lysogenic growth. The decision to move to lysogeny after infection is enhanced by λ repressor concentration, while transfer from lysogenic growth to the lytic cycle depends on repressor destruction. The λ repressor is a dimeric protein encoded by λ phage which prevents entry of further λ and whose transcription is activated by phage cII protein. It prevents binding of RNA polymerase and shuts off early phage promoters, repressing phage genes involved in the lytic phase, switching development into the lysogenic mode. The λ repressor promoter region is itself blocked by Cro protein, when the phage lytic cycle commences. See **vector**.

Lamella Any thin layer or plate-like structure. (Bot.) (1) One of the spore-bearing gills in the fruiting body of a mushroom or related fungus, attached to the underside of the cap (pileus) and radiating from centre to margin. (2) One of a series of double membranes (thylakoids) within a chloroplast which bear photosynthetic pigments. (3) In bryophytes, a thin sheet of flap-like plates of tissue on dorsal surface of the thallus or leaves. (Zool.) One of the concentric layers of hard calcified material of compact bone forming part of a **Haversian system**.

Lamellibranchia See **Bivalvia**.

Lamellipodium Sheet-like extension, or flowing pseudopodium, of the leading edges of many vertebrate cells during locomotion, some forming attachments with the substratum, others carried back in waves (ruffling, producing a *ruffled border*). Many give rise to microspikes. See **cell locomotion**, **lectin**.

Lamina (1) Sheet or plate; flat expanded portion of a leaf or petal. (2) In brown algae (**Phaeophyta**), expanded leaf-like portion of the thallus.

Lamina propria Loose connective tissue of a **mucous membrane** (e.g. of gut mucosa), binding the epithelium to underlying structures and holding blood vessels serving the epithelium.

Laminarin Storage polysaccharide product of the brown algae (**Phaeophyta**), occurring as oil-like liquid contained in a membrane-bound sac surrounding the pyrenoid but outside the **chloroplast**. Comprises β-1,3 linked glucans containing sixteen to thirty-one residues.

Laminar placentation Attachment of ovules over surface of carpel.

Laminins Proteins of the eukaryotic **extracellular matrix** containing many epidermal growth factor (EGF) repeats and implicated, like other such proteins, in developmental processes and stimulation of cell division.

Lamins See **nucleus**.

Lampbrush chromosome **Bivalents**

during diplotene in some vertebrate (notably amphibian) oocytes in which long chromatin loops, which are transcriptionally very active, form at right angles to the chromosome axis and become covered with newly transcribed RNA. Not certain that much of this RNA acts as mRNA in protein synthesis; but if it is functional, the high transcriptional activity is presumably an adaptation to serving a relatively large cell from a single nucleus. See **chromosome**, **gene amplification**.

Lampreys See **Cyclostomes**.

Lamp shells See **Brachiopoda**.

Land use ecology The study of the role and impact of ecological (or environmental) factors on land use patterns, productivity and management, in the past as well as the present.

Large intestine See **colon**.

Larva Pre-adult form in which many animals hatch from the egg and spend some time during development; capable of independent existence but normally sexually immature (see **paedogenesis**, **progenesis**). Often markedly different in form from adult, into which it may develop gradually or by a more or less rapid **metamorphosis**. Often dispersive, especially in aquatic forms. In insects especially, the phase of greatest growth in the life cycle. Examples include **ammocoete**, **caterpillar**, **leptocephalus**, **nauplius**, **tadpole**, **trochophore**, **veliger**.

Larynx Dilated region at upper end of tetrapod trachea at its junction with the pharynx. 'Adam's Apple' of humans. Plates of cartilage in its walls are moved by muscles and open and close glottis. In some tetrapods, and most mammals, a dorso-ventral and membranous fold (vocal cord) within the pharynx projects from each side wall, vibrations of these producing sounds. Movement of the cartilage plates alters the stretch of the cords and alters pitch of sound. See **syrinx**.

Lasso cell Cell type characteristic of **Ctenophora**, whose tentacles are armoured with these sticky thread-cells for capturing prey. Do not penetrate prey, unlike nematocysts of cnidarians.

Latent period (reaction time) Time between application of a stimulus and first detectable response in an irritable tissue.

Lateral line system (acoustico-lateralis system) Sensory system of fish and aquatic and larval amphibians whose receptors are clusters of sensory cells (*neuromast organs*) derived from ectoderm, found locally in the skin or within a series of canals, or grooves on head and body. Neuromasts resemble cristae of the **vestibular apparatus** of higher vertebrates, having a gelatinous cupula but lacking otoliths; they are probably homologous structures. Pressure waves in the surrounding water appear to distort the neuromasts, sending impulses via the vagus nerves on either side, where they associate with the *lateral lines* themselves – an especially pronounced pair of these sensory canals, one running the length of each flank. The head canals are served by the facial nerves.

Lateral meristems Meristems giving rise to secondary tissue; the vascular cambium and cork cambium.

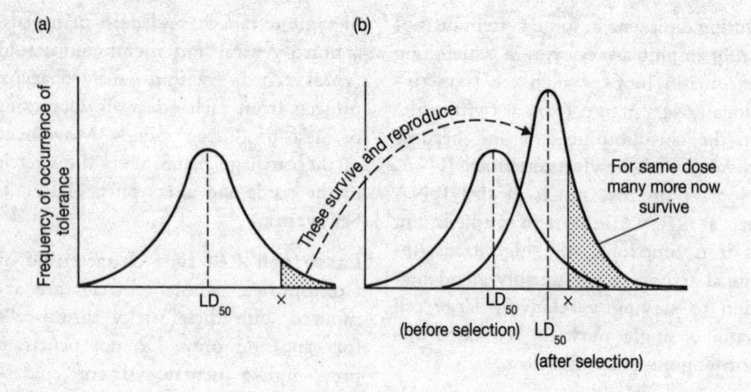

Fig. 70 *Artificial selection and the development of insecticide-resistance.* (a) *Tolerance distribution before selection;* (b) *tolerance distribution after selection (unwittingly) by agricultural practices, showing differences between parental and progeny populations.*

Lateral plate See **mesoderm**.

Latex Fluid product of several flowering plants, characteristically exuding from cut surfaces as a milky juice (e.g. in dandelions, lettuce). Contains several substances, including sugars, proteins, mineral salts, alkaloids, oils, caoutchouc, etc.; rapidly coagulates on exposure to air. Function not clearly understood, but may be concerned in nutrition and protection, as well as in healing wounds. Latex of several species is collected and used in manufacture of several commercial products, the most important being rubber.

Latimeria See **Coelacanthini**.

Laurasia One of two great Upper Carboniferous land masses, the other being **Gondwanaland**, formed by the breakup of Pangaea. Comprised what are now North America, Greenland, Europe and Asia. Originally straddling the Equator, it gradually moved northwards by continental drift. For much of the Jurassic and Cretaceous much of Europe was covered by the Tethys and Turgai Seas, the latter only drying up to link Europe and Asia about 45 Myr BP. Separation of Gondwanaland and Laurasia was completed by early Cretaceous (130 Myr BP) with the result that much of the later radiation of dinosaurs took place in Laurasian continents but not in Gondwanaland.

Laver General name given to the edible dried preparation made from the red alga *Porphyra* (Rhodophyta).

LD_{50} (Of pesticide, insecticide, etc.) the dose required to kill 50% of the pest population. It is generally found that after a pesticide's application, owing to selection for greater pesticide resistance, the population's LD_{50} increases and the dose required to achieve the same effect as on initial application has to be raised. (See Fig. 70.)

LDL (low-density lipoprotein) Spherical complexes (density 1.019–

1.063 g/cm^3) of phospholipid and specialized protein (apolipoproteins, or apoproteins) found in blood plasma. Because the apoproteins are amphipathic (one surface contains hydrophilic amino acid residues and the other hydrophobic ones) they can bind phospholipid by hydrophobic interactions. The core of the particle contains non-polar **cholesterol** esters and **triglyceride**. LDL binds cell surface LDL receptors, where they cluster in coated pits (see Fig. 26, **coated vesicles**). As a result, cholesterol is provided to cells throughout the body and excess is delivered to the liver for recycling or excretion as bile acids. LDL receptors are synthesized in response to a fall in the free cholesterol within a cell. Endothelial cells of the artery oxidise LDLs as they are transported into the artery wall and this oxLDL may transform monocytes into macrophages (see **sclerosis** (2)). See **lipoprotein**, **receptors**.

L-dopa (l-dihydroxyphenylalanine) Intermediate in the pathway from phenylalanine to noradrenaline and immediate precursor of the brain neurotransmitter **dopamine**. Dopamine is deficient from the caudate nuclei of the brain in patients with Parkinson's disease. While neither oral nor intravenous dopamine reaches the brain, *l*-dopa does so and is of widespread clinical use in treating parkinsonism, at least in the short term. Transplantation of embryonic adrenal medulla (production site for *l*-dopa) into the caudate nuclei of sufferers from parkinsonism is under clinical test but raises ethical issues.

Leaf Major photosynthesizing and transpiring organ of bryophytes and vascular plants; those of the former are simpler, non-vascular and not homologous with those of the latter, which consist usually of a leaf stalk (*petiole*), attached to stem by **leaf base**, the leaf blade typically lying flattened on either side of the main vascular strand, or *midrib*. The lamina is often lobed or toothed, possibly reducing the mean distance water travels from main veins to sites of evaporation, helping to cool the leaf. Hairy leaves probably reduce insect damage, as may latex channels, resin ducts, essential oils, tannins and calcium oxalate crystals. Some leaves produce hydrogen cyanide when damaged. During evolution, leaves have become modified to serve in reproduction, either sexually (as **sporophylls**), or asexually (as producers of propagating buds). Leaves of higher plants usually have a bud in their axils and are important producers of **growth substances**. See **cotyledon**, **cuticle**, **leaf blade**, **phyllotaxis**.

Leaf base (phyllopodium) Usually the expanded portion of the leaf, attached to the stem.

Leaf blade (lamina) Thin, flattened and flexible portion of a **leaf**; major site of **photosynthesis** and **transpiration**, for which it is admirably adapted. May be simple (comprising one piece) or compound (divided into separate parts, leaflets, each attached by a stalk to the petiole). A typical dicotyledon leaf presents a large surface area, photosynthesizing cells being arranged immediately below the upper epidermis allowing maximum solar energy absorption. The blade is provided with a system of supporting veins bringing water and mineral nutrients and removing photosynthetic products. Has a system of intercellular spaces opening to the atmosphere through **stomata**, permitting regulated gaseous exchange and loss of water vapour; its thinness reduces diffusion distances for gases, keeping their concentration gradients steep, so increasing diffusion rates. Its large surface area and transpiration rate help cooling. Evergreen leaves generally have blades twice as thick as deciduous ones, but neither can offer too great a wind resistance. See **Kranz anatomy**, **mesophyll**.

Leaf gap Localized region in vascular cylinder of the stem immediately above point of departure of **leaf trace** (leaf trace bundle) where the parenchyma rather than vascular tissue is differentiated. In some plants where there are several leaf traces (leaf trace bundles) to a leaf, these are associated with a single leaf gap. Leaf gaps are characteristic of ferns, gymnosperms and flowering plants (anthophytes).

Leaflet Leaf-like part of a compound leaf.

Leaf scar Scar marking where a leaf was formerly attached to the stem.

Leaf sheath Base of a modified leaf, forming a sheath around the stem (e.g. in grasses, sedges).

Leaf trace (1) Vascular bundle extending between vascular system of stem and leaf base; where more than one occurs, each constitutes a leaf trace. (2) Vascular supply extending between vascular system of stem and leaf base, consisting of one or more vascular bundles, each known as a *leaf trace bundle*.

Learning Acquisition by individual animal of behaviour patterns, not just as an expression of a maturation process but as a direct response to changes experienced in its environment. Various forms of learning include **conditioning**, **habituation** and **imprinting**. Insight learning, in which an animal uses a familiar object in a new way to solve a problem creatively, may be a form of instrumental conditioning: actions may come to be selected because their consequences form part of a route to obtaining a goal. Once clearly contrasted with **instinct**, but it is now realized that these are not mutually exclusive categories of process. Learning ability is a clear example of adaptability, and of the adaptiveness of behaviour. See **melanocyte-stimulating hormone**.

Lecithin See **phospholipid**.

Lectin Proteins and glycoproteins cross-linking cell-surface carbohydrates and other antigens, often causing cell clumping (see **agglutination**). May act as antigens themselves, as do the mitogenic lymphocyte-stimulators *phytohaemagglutinin* (*PHA*) and *concanavalin A* (*ConA*). In plants, may provide toxic properties of seeds. Are also involved in artificial **capping** of specific cell surface components.

Lectotype Specimen or other component of original material, selected to serve as a nomenclatural type when no **holotype** was designated at the time of publication, or as long as it is missing. See **isotype**, **neotype**, **syntype**.

Leghaemoglobin See **haemoglobin**.

Legume (1) A pod; fruit of members of the Family Leguminoseae (peas, beans, clovers, vetches, gorse, etc.). Dry fruit formed from single carpel that liberates its seeds by splitting open along sutures into two parts. (2) Used by agriculturists for a particular group of fodder plants (clovers, alfalfa, etc.) belonging to the Leguminoseae. Important in crop rotation, having symbiotic nitrogen-fixing root nodule bacteria.

Lemma Lower member of pair of **bracts** surrounding grass flower, enclosing not only the flower but also the other bract (**palea**).

Lemurs See **Dermoptera** (flying lemurs); and **Primates**.

Lens Transparent, usually crystalline, biconvex structure in many types of eye, serving to focus light on to light-sensitive cells. In vertebrate eyes, constructed of numerous layers of fibres of the protein *crystallin*, arranged like layers of an onion, normally enclosed in connective tissue capsule and held in position by suspensory ligaments, absorbing potentially damaging light of wavelength <400 nm. See **accommodation**, **eye**.

Lentic (Of freshwaters) where there is no continuous flow of water, as in ponds, lakes. Compare **lotic**.

Lenticel Small raised pore, usually elliptical, developing on woody stems in portions of the peridium in which **cork cambium** is more active than elsewhere. Lenticel tissue possesses numerous intercellular spaces which begin forming during the development of the first peridium, and in the stem, typically appear below stomata. Their function is to allow for gaseous exchange between the interior of the stem and the atmosphere. They are also formed upon the surfaces of some fruits (e.g. apples, pears) and roots.

Lentivirus Any of a group of 'slow' retroviruses inducing neurological impairment and chronic pneumonias. Include visna virus of sheep and **HIV**.

Lepidoptera Butterflies and moths. Endopterygote insect order. Two pairs of large membranous wings, with few cross-veins, and covered with scales; larva a **caterpillar**, usually herbivorous and sometimes a defoliator of economic importance. Adults feed on nectar using highly specialized and often coiled proboscis formed from grooved and interlocked maxillae. There is no simple way to distinguish all moths from all butterflies, but in Europe any lepidopteran with club-tipped antennae, flying in the day, and capable of folding its wings vertically over its back, is a butterfly.

Lepidosauria Dominant subclass of living reptiles; possess **diapsid** skulls and overlapping horny scales covering the body. Includes the orders Rhyncocephalia (*Sphenodon*, the tuatara) and Squamata (lizards, snakes, amphisbaenids).

Leptocephalus Oceanic larva of European eel. Migrates over 2000 miles across Atlantic from breeding site near West Indies (Sargasso Sea) to European fresh waters, where it becomes adult. Transparent.

Leptoma Thin area in wall of gymnophyte pollen, through which the pollen tube emerges.

Leptome (leptoid) Photosynthate-conducting cell, approaching phloem in structure and function, found in bryophytes (esp. Bryales).

Leptosporangiate (Of vascular plant sporangia) arising from a single parent cell and possessing a wall of one layer of

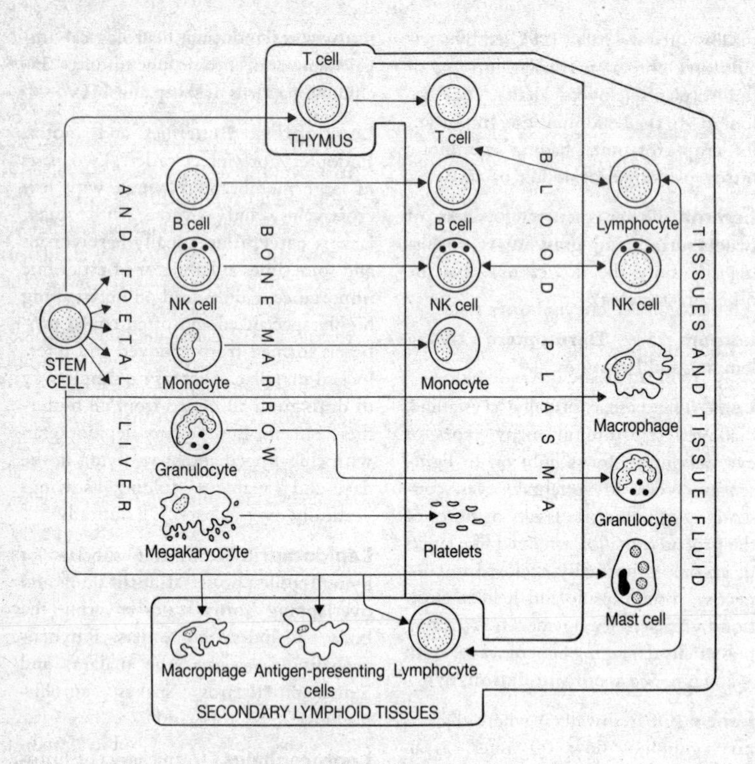

Fig. 71 *Origins of cells of the immune system, all derived from the haemopoietic stem cell. Granulocytes include neutrophils, basophils and esinophils. NK = natural killer cell. For details, see separate entries and* **lymphoid tissue, myeloid tissue.**

cells. Spore production is low in comparison with **eusporangiate** type.

Leptotene Stage in first prophase of **meiosis** during which chromosomes are thin and attached at both ends to the nuclear membrane. DNA has already replicated but each chromosome appears as one thread, sister chromatids being closely apposed.

Leucocyte (white blood cell) Nucleated blood corpuscle lacking haemoglobin. Includes *granulocytes* (neutrophils, eosinophils, basophils), with granules in their cytoplasm, developing typically from **myeloid tissue**, and *agranulocytes* (lymphocytes and monocytes), lacking cytoplasmic granules and developing typically from **lymphoid tissue**. Lymphocytes are of two kinds: T cells and B cells (see also **immunity**). The monocytes in tissue fluid are called *wandering macrophages*, and these and neutrophils are the major phagocytic leucocytes. Natural killer cells may also be of lymphoid origin. See specific cell types, and Fig. 71.

Leucoplast Colourless plastid found in cells of plant tissues not normally exposed to light. Includes **amyloplasts**, storing starch, **elaioplasts**, storing oil, and **aleuroplasts**, storing protein.

Leucosin See **chrysolaminarin**.

Leukaemia Malignant overproduction by **myeloid tissue** of white blood cells, crowding out normal red cell- and platelet-producing lines (leading to poor blood clotting) and resulting in a lack of mature and normal white cells. Death can result not so much directly from the cancer as from these indirect effects. Treatment by X-rays and chemotherapy may result in partial or complete remission; myeloid transplants sometimes required.

Leydig cells (interstitial cells) Groups of relatively scanty cells in the interstices between seminiferous tubules of the vertebrate testis responsible for steroid production (especially **testosterone**). Have unusually large smooth endoplasmic reticulum. Testosterone output is synergistically enhanced by **prolactin** and **luteinizing hormone**. See **maturation of germ cells**.

LH See **luteinizing hormone**.

Libriform fibre Xylem fibre having thick walls and greatly reduced pits.

Lice See **Siphunculata** (sucking lice), **mallophaga** (biting lice), and **Psocoptera** (book lice).

Lichen Within Kingdom **Fungi**. Symbiotic association between a fungus and an alga, developing into a unique morphological form quite distinct from either partner. The fungal partner (*mycobiont*) is usually a member of the **Ascomycota**, and sometimes the **Basidiomycota**; the algal partner (*phycobiont*) is most frequently either a green alga (**Chlorophyta**) or a blue-green alga (**Cyanobacteria**). About 90% of all lichens have as the phycobiont species of the green algal genera *Trebouxia*, *Pseudotrebouxia*, and *Trentepohlia*, and the blue-green alga *Nostoc*. Most lichens have only one species of alga, although there are a few with two. Lichen thalli may be (1) *foliose*, having several well-defined stratified layers including an upper cortex, an algal layer, a medulla and a lower cortex. They have two growth forms, either lobed and leaf-like and attached to the substratum by numerous rhizines or (less common) circular in outline and attached by single central cord; (2) *crustose*, having an upper cortex, an algal layer and a medulla in direct contact with the substratum via rhizoidal hyphae. They form a thin, flat crust on the substratum; (3) *squamulose*, having a structure intermediate between foliose and crustose thalli, where it comprises numerous small lobes or squamules. Internally there are no lower cortex or **rhizines**; (4) *fruticose*, having a radially symmetrical, cylindrical-to-flattened thallus in cross-section, hollow or solid, and either erect or pendulous. The outer cortex is quite thick, the algal layer thin and scattered, and neither rhizines nor a lower cortex are present (*Cladonia* has a squamulose primary thallus but a more conspicuous secondary fruticose thallus called a podetium); (5) *gelatinous*, having no stratified layers other than a thin upper and lower cortex and a medullary layer of loosely interwoven hyphae and scattered algae.

Non-sexual reproduction occurs through fragmentation of the thallus, as well as through production of *soredia*, which are microscopic structures comprising small clusters of algal cells together with some fungal hyphae and originating from the upper cortex (where they are numerous they form conspicuous powdery clumps), and *isidia*, which are minute outgrowths of the upper cortex, occurring over the entire surface of the thallus and consisting of protuberances of the upper cortex, algal and medullary tissues. The fungus may also reproduce non-sexually by conidia which may form in a **pycnidium**. Reproduction of algal cells is by simple cell division (zoospores and aplanospores have been reported only very rarely); when isolated in culture they readily form such asexual spores.

Sexual reproduction is confined to the fungus; structures formed commonly being **ascocarps** (apothecia or perithecia). Ascospores are formed in asci and discharged. On germination new lichen individuals are formed if the algal partner happens to be present; but in its absence the fungus dies.

In the lichen relationship, the fungus obtains organic carbon from the alga. Those that have blue-green algae (e.g. *Nostoc*) capable of fixing atmospheric nitrogen, also obtain nitrogen products. Lichen fungi form a close network of hyphae around the algal cells, and commonly produce **haustoria** which penetrate the algal cells; other fungi produce specialized organs called *appresoria* that lie along the surface of algal cells penetrating them by means of specialized pegs. Growth of many lichens is very slow (0.1–10 mm yr^{1} increase in colony diameter). Some mature lichens may approach 4,500 years or more. Lichens dominate the flora in large areas of mountain and arctic regions, where few other plants can exist. They occur on bare soil, rocks, tree-trunks, and fence posts. One species, *Verruccaria serpuloides*, is a permanently submerged marine species. Lichens play an important role in the primary colonization of bare areas. In the Arctic, certain lichens are a valuable food source (e.g. Iceland moss, reindeer moss). Others produce dyes (e.g. *Rocella* provides litmus), and are sources of medicines, poisons, cosmetics and perfumes. Many are sensitive to polluted air (e.g. air polluted with sulphur dioxide), probably due to the sensitivity of the phycobiont, and can be used as **indicators** of pollution and have been used to monitor fallout from nuclear tests.

Life Complex physico-chemical systems whose two main peculiarities are (1) storage and replication of molecular information in the form of nucleic acid, and (2) the presence of (or in viruses perhaps merely the potential for) enzyme catalysis. Without enzyme catalysis a system is inert, not alive; however, such systems may still count as biological (e.g. all viruses away from their hosts). Other familiar properties of living systems such as nutrition, respiration, reproduction, excretion, irritability, locomotion, etc., are all dependent in some way upon their exhibiting the two above-mentioned properties.

Living systems also have an evolutionary history. Whatever the **origin of life** may have been, all *existing* life forms derive from living antecedents.

The earliest living system would have been very different from any modern life form, particularly so in their genetic systems (modes of storage and implementation of molecular information).

Life cycle Progressive series of changes undergone by an organism or a lineal succession of organisms from fertilization to death of the stage producing the gametes beginning an identical series of changes. As do phenotypes, they display adaptation to the external environment; and within a species there is often more than one way in which a generation may be completed, not least through variation in the modes of reproduction. In eukaryotes, this sometimes involves an **alternation of generations** (*metagenesis*) between sexual and asexual stages. The simplest form of life cycle (found in both prokaryotes and eukaryotes) is purely vegetative, involving repeated binary fission, budding, vegetative cell division or cell division coupled with fragmentation of the filament or colony. Next in complexity are **asexual** cycles (again found in both prokaryotes and eukaryotes), involving production of spores (e.g. aplanospores, conidia, zoospores) which do not engage in fertilization and were formed other than by meiosis.

With the evolution of eukaryotic sexual reproduction, the stage was set for the evolution of diploidy – probably when two haploid cells combined to form a diploid zygote. Presumably, this zygote divided by meiosis to restore the haploid condition; however, some zygotes divided mitotically instead, with meiosis occurring later. In animals, this delayed meiosis results in production of gametes. In eukaryotes sexual life cycles may be classified according to whether or not haploid mitosis and/or diploid mitosis occur (see Fig. 72). Such life cycles may either be (a) *haplontic*, in which only haploid cells divide mitotically, the diplophase being represented by a single nucleus (e.g. some green algae such as *Spirogyra*, *Chlamydomonas*; fungi such as *Mucor*); (b) *diplontic*, in which all haploid cells are the immediated products of meiosis (e.g. most animals; the brown alga *Fucus*), or (c) *haplodiplontic* (or *diplohaplontic*), in which both haploid and diploid mitoses occur. Haplophase and diplophase may be more or less equally prominent (e.g. the green alga *Ulva*), or one may be dominant (haplophase in mosses and liverworts; diplophase in ferns and seed plants). In most haplodiplontic cases (yeasts are somewhat variable) gametes are produced by mitosis, and it is a mitotic product of the zygote that undergoes meiosis.

It is usual with plants and algae to refer to the haplophase and diplophase as the *gametophyte* and *sporophyte* respectively. Multicellular haploid organisms that appear alternately with the diploid forms are found in plants, and among the **Protista** in some brown, red and green algae, in two closely related genera of the **Chytridiomycota** and in one or more other groups. Such organisms display alternation of generations. Even in mosses and ferns these phases are practically distinct individuals. The gametophyte has become increasingly restricted during evolution, represented in flowering plants (**Anthophyta**) by just the microspore, megaspore and their greatly reduced mitotic products. Although the development of gametophytic heterospory, begun in ferns, has been taken far further in flowering plants, it is the sporophyte that has become the dominant phase from ferns to seed plants, indicating a probably greater robustness and adaptability of the diploid plant in terrestrialization.

Most animal life cycles are diplontic; but in some forms (e.g. polychaetes) the products of meiosis may undergo mitosis to produce gametes. In most cases of **male haploidy**, sperm are formed mitotically, although in bees they are formed by unipolar meiosis. Life cycles of animal parasites often involve **polyembryony** as well as two or more hosts; similarly some fungal parasites may require more than one host and produce several spore types (e.g. rust fungi), while life cycles of periodical cicadas (Hemiptera) may take as long as seventeen years to complete, nymphs taking this long to reach maturity. See **parthenogenesis**.

Ligament Form of vertebrate **connective tissue** joining bone to bone. *Yellow elastic ligaments* consist primarily of elastic fibres and form relatively extensible ligaments joining vertebrae, and true vocal cords; *collagenous ligaments* by contrast consist largely of parallel bundles of collagen and resist extension.

Ligase **Enzyme** catalysing condensation of two molecules and involving hydrolysis of ATP or another such triphosphate. **DNA ligase** is much used in **gene manipulation**, as well as forming part of **DNA repair mechanisms**.

Lignin Complex polymeric molecule composed of phenylpropanoid units associated with **cellulose** (as *lignocellulose*) in cell walls of sclerenchyma, xylem vessels and tracheids, making them strong and rigid. After cellulose, lignin is the most abundant plant polymer, forming 20–30% of the wood of trees.

Lignocellulose Major chemical component of wood. Valuable resource, not least of energy as when converted to methane or alcohol in techniques of **biotechnology**. See **cellulose**, **lignin**.

Ligule (1) Membranous outgrowth arising (a) from junction of leaf blade with leaf sheath in many grasses, (b) from base of leaves of certain lycopods. (2) Flattened corolla of ray flower in Compositae.

Limiting factor Any independent variable, increase in whose value leads to increase in the value of a dependent variable. Ideally, values of other independent variables should be held constant while this relationship is examined. In plots of dependent against independent variables, the latter are *limiting* only while there is a linear or near-linear relationship to the plot. Thus in the plot of initial velocity of an enzyme reaction (V_o) against substrate concentration, S,

Fig. 72. *(i) Scheme indicating the various ways in which mitosis and meiosis can be included within a eukaryotic sexual life cycle. Both fertilization and meiosis may or may not be followed by mitosis. The situation in which no mitosis at all occurs (the central route) is rarely if ever found: cell number would be increased solely by meiosis. (ii) The typical haplontic life cycle (e.g.* Mucor, Spirogyra, Chlamydomonas*). (iii) The typical haplodiplontic life cycle (e.g. mosses, liverworts, ferns, conifers, flowering plants). Spores may be produced homosporously or heterosporously. (iv) The typical diplontic life cycle (e.g.* Fucus, *most animals).*

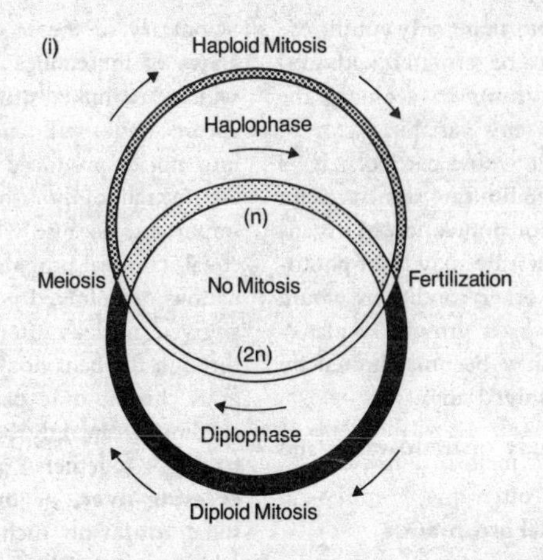
(i)
Haploid Mitosis
Haplophase
(n)
Meiosis
No Mitosis
Fertilization
(2n)
Diplophase
Diploid Mitosis

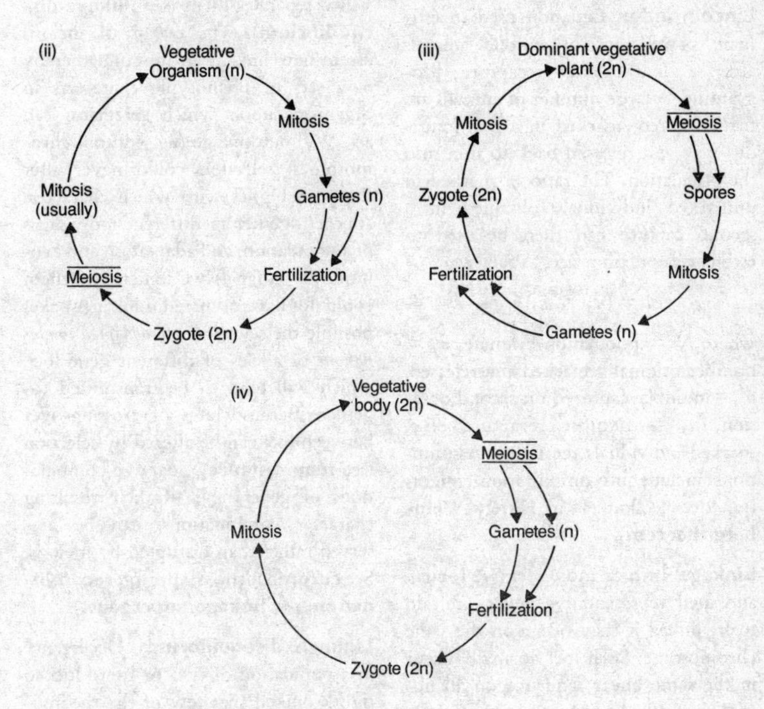
(ii)
Vegetative Organism (n)
Mitosis
Gametes (n)
Fertilization
Zygote (2n)
Meiosis
Mitosis (usually)
(iii)
Dominant vegetative plant (2n)
Meiosis
Spores
Mitosis
Gametes (n)
Fertilization
Zygote (2n)
Mitosis
(iv)
Vegetative body (2n)
Meiosis
Gametes (n)
Fertilization
Zygote (2n)
Mitosis

the latter is limiting only until the enzyme begins to be saturated with substrate (see **enzyme**). In ecology, the term applies to any variable factor of the environment whose particular level is at a given time limiting some activity of an organism or population of organisms; e.g. temperature may limit photosynthesis when other conditions would favour a higher rate; growth of planktonic diatoms may become limited by depletion of dissolved silica.

Limnology Study of fresh waters and their biota.

Limulus See **Merostomata**.

Lincoln index Equation used to estimate populations of mobile animals after a mark–release–recapture programme. A large number of animals are first captured, marked in a non-injurious way and released back to mix into the population. The ratio of marked to unmarked individuals obtained in a second capture can then be used to estimate population size, as follows:

$$N = (n_1 \times n_2)/n_3$$

where N = population estimate, n_1 = number originally marked and released, n_2 = number captured on second occasion, n_3 = number recaptured (i.e. marked individuals recaught). Assumptions include no immigration/emigration or selection (as in **Hardy-Weinberg theorem**).

Linkage Two or more loci (see **locus**), and their representative genes, are said to be linked if they occur on the same chromosome. Such loci normally occur in the same linear sequence on all homologous chromosomes, so that chromosomes form *linkage groups*.

Contrary to the second of **Mendel's Laws** of inheritance, eukaryotic genes which are linked (forming part of a chromosome) will tend to pass together into nuclei produced by **meiosis** and are not randomly assorted. This is very important, because whereas there is an equal statistical probability of all combinations of unlinked genes ending up in a given nucleus after meiosis (simply through the behaviour of non-homologous chromosome pairs), genes which are linked can only be prevented from passing together into nuclei by **crossing-over**, or by some chromosome **mutation** such as translocation, which separates them (see **linkage disequilibrium**). The events of meiosis create new linkage groups and thereby new sets of phenotypic characters in organisms upon which **selection** can act. Without linkage (i.e. without chromosomes) selection could never alter the probabilities with which characters appear together in different individuals of a population, and adaptation and evolution in so far as we understand them could not have occurred. Linkage makes possible the selection of *coadapted combinations* of alleles of different gene loci, which will tend to be transmitted together. Because rates of crossing-over between loci can be altered by selection (see **map distance**), adaptive combinations of genes, and of their resulting character combinations, can be preserved rather than disrupted by meiosis. See **chromosome mapping**, **recombination**, **sex linkage**, **supergene**.

Linkage disequilibrium Occurrence in a population of two or more loci so tightly linked that few of the theoretically possible gene combinations are found. E.g., if loci A and B are represented by just two alleles each in the population (*A* and *a*; *B* and *b*) yet only individuals with genotypes *AABB* and *aabb* are found in any frequency, then there is linkage disequilibrium between the two loci. Without crossing-over, and without selection for and against any of the genotypes, one expects linkage equilibrium ($AB + ab = Ab + aB$) to be achieved eventually. See **linkage**, **supergene**.

Linkage map See **chromosome map**.

Linkage mapping Genetic mapping. See **chromosome mapping**.

Linnaeus, Carolus (Adj. *Linnean*) Swedish naturalist (1707–88), physician and originator of modern system of **binomial nomenclature**. In his *Systema Naturae Fundamenta Botanica* he assigned to every known product of nature (including minerals) its place on one great system of classification; indeed, discovery of the natural method of such classification was for him the naturalist's task. He thought he had achieved this for species and genera but only partially for classes and orders. Nature was for him imbued with divine *economy*: nothing happened unnecessarily, nor did any created form ever become extinct, although fossil evidence clearly forced him to consider this possibility. He was a little perplexed by the discovery of hybridization between species, and in later works considered that new species might arise in this way; but fundamentally his thinking was by today's standards conservative and non-evolutionary. His chief critic was **Buffon**. See **great chain of being**.

Linnean hierarchy Arrangements of organisms into **taxa**; forming nested sets, with **kingdom** the highest taxonomic category and **species** the lowest.

Linoleic acid Essential fatty acid of mammals, making up to 20% of the total fatty acid content in their trigylcerides.

Lipase Enzyme hydrolysing fatty acid esters; it converts triglycerides (fats) into fatty acids and glycerol. Present in pancreatic juice and succus entericus of vertebrates. *Lipoprotein lipase* on surfaces of adipose tissue cells is largely responsible for hydrolysis of triglycerides within **chylomicrons** in blood plasma.

Lipids Wide and heterogeneous assemblage of organic compounds having in common their solubility in organic solvents such as alcohol, benzene, diethyl ether, etc. Include fats and waxes, steroids and sterols (e.g. cholesterol), glycolipids, phospholipids, terpenes, fat-soluble vitamins, prostaglandins, carotenes and chlorophylls. See **micelle**.

Lipogenesis Fatty acid synthesis.

Lipolysis Hydrolysis of lipids, particularly of triglycerides. See **adipose tissue, lipase**.

Lipoprotein Micellar complex of protein and lipids. *Apolipoproteins*, protein components of these complexes, transport otherwise insoluble lecithin, triglyceride and cholesterol in vertebrate blood plasma. The transport lipoprotein particle comprises an outer lipid bilayer with specific conjugated protein components, within which the transported molecules are either free or esterified to bilayer fatty acids. There is a complex turnover

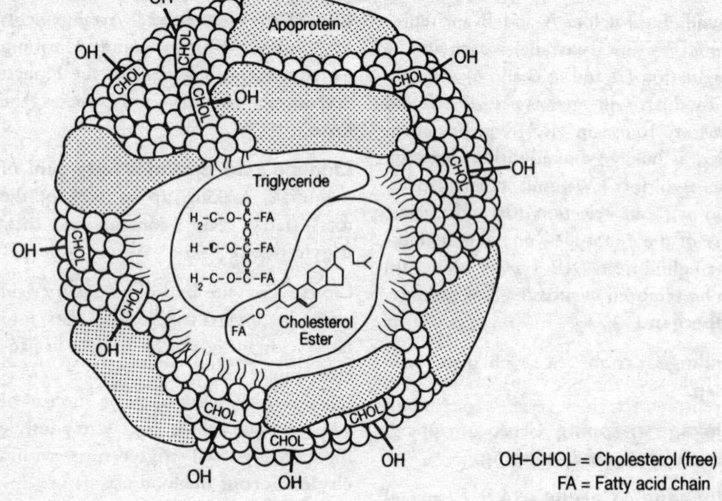

Fig. 73 *A human plasma* **lipoprotein**. *Diameter depends on the class of lipoprotein, but excepting larger chilomicrons would be ~ 40 nm.*

and interaction of these particles, **chylomicrons**, low density lipoprotein (**LDL**), high-density (**HDL**) and very high-density (**VHDL**) forms increasing relative to the amount of protein in the particle. Regular exercise (physical conditioning) raises blood HDL, which seems to counteract impact of **cholesterol** in heart disease, decrease triglyceride levels and improve lung function. See **lipase**, **adipose tissue**, **liver**, **coated pit**.

Liposome Artificially produced spherical lipid bilayers, 25 nm or more in diameter, which can be induced to segregate out of aqueous media. Their selective permeabilities to organic solutes and the subsequent reactions which can occur within them have led some to propose a similar structure as a cell prototype. See **coacervate**, **chylomicron**, **human gene therapy**.

List Cellulose extension of cell wall in some armoured dinoflagellates, usually extending out from the cingulum and/or sulcus.

Lithophyte Plant found growing on rocks.

Lithosere Sere originating on exposed rock surface.

Lithotroph Autotrophic bacterium (e.g. nitrifying or sulphur bacterium) obtaining its energy from oxidation of inorganic substances (e.g. sulphur, iron) by inorganic oxidants: the terminal hydrogen acceptor in respiration is always inorganic.

Littoral Inhabiting the bottom of sea or lake, near the shore. From shore to 200 m in the sea; to 6–10 m in lakes, depending upon the extent of rooted vegetation.

Liver Gland, usually endodermal in

origin and arising as a diverticulum of gut. Livers in different phyla are not homologous. In vertebrates its main glandular function is production of **bile**, which leaves via the hepatic duct for storage in the gall bladder. Has a wide capability enzymatically, much of which is inducible. Structural unit is the *lobule*, a roughly hexagonal block of cuboidal cells (*hepatocytes*) supplied at its corners with products of digestion via factors of the hepatic portal vein, and with oxygen by branches of the hepatic artery. Blood leaves the lobule at its centre through a vessel leading to the hepatic vein. Bile leaves through *bile canaliculi*, also at corners of lobules. As a homeostatic organ it is second to none, involved in production of **glycogen** from monosaccharides and fat and its subsequent storage; absorption and release of blood glucose, absorption of **chylomicrons**, deamination of amino acids, urea production, **gluconeogenesis**, raising of blood temperature, production of prothrombin, fibrinogen, albumin and other plasma proteins, storage of vitamins A, D, E and K, red blood cell breakdown, detoxification of some poisons and steroid hormone conversion to cholesterol.

Liverworts See **Hepaticopsida**.

Locule Compartment, cavity or chamber. In **Ascomycota**, chambers containing **asci**; in flowering plants (**Anthophyta**), cavity of the ovary where ovules occur; in diatoms (**Bacillariophyta**), a chamber within the frustule having a constricted opening on one side and a **velum** on the opposite side.

Loculicidal (Bot.) Describing dehiscence of multilocular capsule by longitudinal splitting along a dorsal suture (midrib) of each carpel; e.g. iris. Compare **septicidal**.

Locus (gene locus) Position on homologous chromosomes occupied, normally throughout a species population, by those genes which determine the state of a particular phenotypic character (see **gene**). E.g., a locus for eye colour in *Drosophila*. This position may be determined by **chromosome mapping**, various occupants of a locus in a given population being referred to as **alleles**. Relative positions of loci within an individual's cells may be altered by some chromosome **mutations**. See **linkage**.

Lodicules Reduced perianth of grass flowers; two small scale-like structures below ovary which at time of flowering swell up, forcing open enclosing bracts (pales), exposing stamens and pistil.

Log phase (logarithmic phase) See **growth curves**.

Lomasome Membranous evagination of the plasmalemma of a fungal cell or hypha, occurring singly or in groups and situated between the rest of the plasmalemma and the wall material. Consists of membranous tubules, vesicles or parallel sheets lying in a matrix. May play a role in normal development of either the plasmalemma or wall.

Lomentum Type of leguminous fruit, constricted between seeds and breaking into one-seeded portions when ripe; e.g. in bird's foot trefoil.

Long-day plants Plants that flower when exposed to dark periods less than a critical length and therefore flowering primarily during the summer. See **photoperiodism**, **short-day plants**, **day-neutral plants**, **phytochrome**.

Long-germ Mode of **development** in higher insects (e.g. Diptera) in which all body segment primordia are established at the blastoderm stage, segments becoming visible along the entire embryo soon after gastrulation. Compare **short-germ**.

Long shoot Main branch in some gymnophytes (gymnosperms) bearing short dwarf shoots; e.g. in *Pinus*, *Ginkgo*.

Looped domain See **chromosome**.

Loop of Henle See **kidney**.

Lophophore Hollow ring of ciliary feeding tentacles surrounding the mouth and (strictly) containing an extension of the coelom, as in Ectoprocta, Brachiopoda and Phoronidea. Applied loosely to any ring of oral tentacles, as in some polychaete worms and entoproctans.

Lorica Envelope surrounding some flagellated (e.g. *Trachelomonas*, *Dinobryon*) and ciliate (e.g. *Cothurnia*) protists, fitting loosely enough to allow the cell to move fairly freely – unlike shells and tests. Generally proteinaceous, of glycosaminoglycan (generally described as 'chitin', 'pseudochitin' or 'tectin') or cellulose.

Loss-of-function mutation See **mutation**.

Lotic Of freshwaters where there is continuous flow of water, e.g. streams, rivers. Compare **lentic**.

Luciferase, luciferin See **bioluminescence**.

Lumbar vertebrae Bones of the lower back region, lacking rib attachments and situated between thoracic and sacral vertebrae.

Lumen (1) Cavity within tube (e.g. within blood vessel, gut) or within sac. (2) Cavity within cell wall of plant cell, from which the protoplast has been lost.

Lung Sac-like organ of gaseous exchange, invariably with moist inner surface. (1) In vertebrates, they arise as a diverticulum of the pharynx (see **gas bladder**) and were present in fish prior to the rise of amphibians, serving probably as an adaptation to drought and/or poorly oxygenated water (see **Dipnoi**). Generally paired, and in most higher tetrapods internally subdivided into bronchi, bronchioles and alveoli, they lie surrounded by coelomic membranes that in mammals will later form fluid-lubricated pleural cavities separating lungs from thorax (see **mediastinum**). Lungs are relatively small in birds, where **air sacs** are the major respiratory surfaces. Lungs lack muscles of their own and are ventilated by rib muscles of the trunk and by the diaphragm (in mammals). Only in birds does air actually circulate through lungs; elsewhere air is tidal and terminates in richly vascularized alveoli, providing a huge surface area for gaseous exchange. The lung's (pulmonary) blood circulation has evolved in tetrapods towards the **double circulation** of birds and mammals. Stretch receptors in a lung's connective tissue walls feed back to **respiratory centres** controlling ventilation. (2) In molluscs, the lung is most advanced in pulmonate gastropods where it is a specialization of the mantle cavity and opens to the air via a valved pneumostome. Muscles ventilate the chamber.

Lung book Organ of gaseous exchange in some air-breathing arachnids (e.g. in scorpions and, in conjunction with tracheae, some spiders). Consists of leaf-like projections sunk into pits opening through a narrow pneumostome to the air. Four pairs occur in scorpions, two pairs or one pair in some spiders, none in others. Gaseous exchange is by diffusion.

Lungfish See **Dipnoi**.

Lutein A xanthophyll pigment.

Luteinizing hormone (LH, interstitial cell-stimulating hormone, ICSH) A glycoprotein **gonadotrophin**, produced by the anterior **pituitary** under influence from hypothalamic releasing factor (GnRF). Its main function in males is to stimulate interstitial cells of the testis to produce **testosterone**, which in turn shuts off GnRF release. In females, brings about ovulation (see **menstrual cycle**), and transforms the ruptured Graafian follicle into a corpus luteum with subsequent progesterone production.

Lycophyta Club mosses. There are five living genera: *Lycopodium*, *Selaginella*, *Isoetes*, *Phylloglossum* and *Stylites*, giving about 1000 living species, representatives of an evolutionary lineage extending back to the Devonian period. Lycopods split very early on into two major groups. The first remained herbaceous and is still represented in today's flora; the second, the lepidodendrids (or tree ferns), became woody and tree-like and were the dominant plants of coal-forming forests of Carboniferous period (e.g. *Lepidodendron*). Became extinct in the Permian period (about 280 Myr BP). Living lycopods are small, evergreen plants with upright or trailing stems that bear numerous small leaves. Sporangia are borne singly in axils of leaves, sporophylls occurring in groups at intervals along the stem or forming terminal cones. They are either homosporous (e.g. *Lycopodium*), with small mycotrophic prothalli, wholly or partly subterranean; or heterosporous (e.g. *Selaginella*), with reduced prothalli remaining largely enclosed by the spore wall. See **life cycle**.

Lymph Clear fluid within vessels of **lymphatic system** and derived from **tissue fluid** and resembling it in composition, except in those lymph vessels between the gut and venous system after a meal, when it is generally 'milky' due to presence of **chylomicrons**. It returns proteins from the tissue fluid to the blood. Lymphocytes enter as it passes through **lymph nodes**.

Lymphatic (lymphoid) system System of **lymph**-containing vessels and the organs producing and accumulating lymphocytes linked by these vessels in vertebrates (see Fig. 74). The 'second' circulatory system in these animals. Comprises a blindly-ending meshwork of highly permeable endothelial *lymph capillaries* (resembling blood capillaries, but having non-return valves) permeating most body tissues (not the nervous system) and joining to form larger vessels (usually not larger than 2–3 mm diameter), resembling veins, with non-return valves but thinner connective tissue walls. Finally joins the venous system, usually near the heart. **Tissue fluid** drains into lymph capillaries and is slightly concentrated by loss of some water and electrolytes as it passes along the system, under the same forces as achieve venous return. Most of the large *lymph trunks* enter the left *thoracic duct*

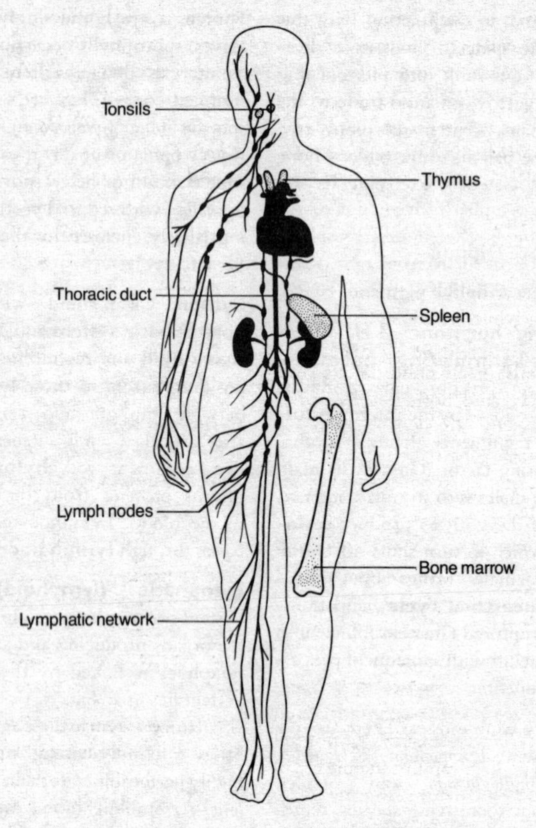

Fig. 74 *Diagram of the human lymphatic system and its associated lymphoid organs and tissues (labelled).*
Adapted from N. Staines, J. Brostoff and K. James, *Introducing Immunology*, Gower Medical Publishing, 1985, courtesy of the authors and publishers.

which opens into the venous system close to or at the left subclavian vein, but some also enter from a right thoracic duct near the right subclavian vein. See **lymph node, lymphoid tissue**.

Lymph heart Enlarged part of lymphatic vessel with muscular pulsating wall. Present in many vertebrates, but not in birds or mammals.

Lymph nodes Ovoid structures on lymph vessels of mammals and to a lesser extent of birds (but not other vertebrates); up to 25 mm long in humans, comprising connective tissue framework of capsule and inner extensions (trabeculae) supporting successive internal parenchyma tissues: (a) the cortex of **B cells**, (b) the paracortex of **T cells** and (c) the medulla of T and B cells. Afferent lymphatic vessels join the capsule, where valves ensure that lymph passes progressively towards the medulla, where an efferent lymph vessel rejoins the lymphatic circulation. As it passes through the node, lymph is processed by fixed macrophages of the **reticulo-endothelial system** lining the sinuses, while lymphocytes respond to antigens from adjacent **antigen-presenting cells**. Each node has its own blood supply, and both B cells and T cells respond by clonal expansion in a way dependent upon the type of antigenic stimulation. Compare **spleen**. See **lymphoma**.

Lymphocyte One of two kinds of vertebrate white blood cell (**leucocyte**), confined to the blood system. Most are *small lymphocytes* (diameter 6–8 μm), agranular with high nuclear:cytoplasmic ratio; but *large lymphocytes* are larger (diameter 8–10 μm) and granular, with lower nuclear:cytoplasmic ratio and with azurophilic granules (staining with azure dyes) in the cytoplasm; some may serve as **natural killer cells**. Small lymphocytes (**B cells** and **T cells**) are involved in both humoral and cell-mediated **immunity**. Lymphocytes can move by amoeboid locomotion but most are non-phagocytic. They develop in **lymphoid tissue**. See Fig. 71.

Lymphoid tissue Vertebrate tissue in which **lymphocytes** develop (e.g. Bursa of Fabricius in birds, thymus in mammals). Lymphocytes are produced in *primary lymphoid tissue* (thymus, embryonic liver, adult bone marrow) and migrate to *secondary lymphoid tissue* (spleen, lymph nodes, unencapsulated lymphoid regions of gut submucosa, respiratory and urogenital regions), where **antigen-presenting cells** and mature **T cells** and **B cells** occur. The kidney is a major secondary lymphoid organ in lower vertebrates. Tissues in animals without close chordate affinities may nonetheless form regional aggregates of phagocytes which are loosely termed lymphoid. See **myeloid tissue**.

Lymphokines Any soluble factors produced by lymphocytes; those produced by **T cells** often termed **interleukins**.

Lymphoma Any tumour composed of lymph tissue. *Hodgkin's disease* is a malignant lymphoma of reticulo-endothelial cells in lymph nodes and other lymphoid tissues.

Lysigenous (Of secretory cavities in plants) originating by dissolution of secreting cells; e.g. oil-containing cavities in orange peel. Compare **schizogenous**.

Lysis Destruction of cells through damage to or rupture of plasma membrane, e.g. by osmotic shock. In bacteria, may be brought about by infection by **bacteriophage**. See **hydrolysis**.

Lysogenic bacteria, lysogeny See **bacteriophage**.

Lysosomes Diverse membrane-bound vacuolar organelles, forming integral part of eukaryotic intracellular digestive system (see Fig. 75). Contain large variety of hydrolytic enzymes. Substances for digestion entering the cell's geography by endocytosis (heterophagy) become part of the lysosomal system when they fuse with *primary lysosomes* (0.5 μm diameter), which have not been

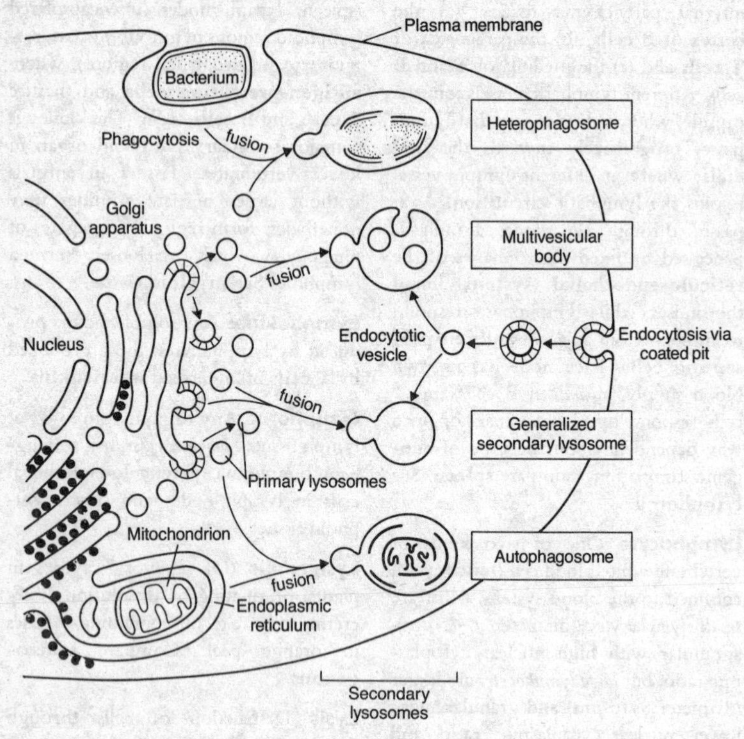

Fig. 75 *Diagram showing the source of primary lysosomes and their fusions with other organelles to form secondary lysosomes. Ladder-like rings indicate coated vesicles.*

involved in hydrolytic activity, budded from the **Golgi apparatus**. The vacuoles (*heterophagosomes*) formed by this fusion are sites of digestion analogous to the gut, and similar digestion products diffuse through specific channels or carriers into the cytosol through the lysosomal membrane. Undigested remains may persist in 'residual bodies' for some time. Autophagy often involves organelles being wrapped up in a membranous vacuole, probably originating from smooth endoplasmic reticulum, and its subsequent fusion with primary lysosomes to form *autophagosomes*. These are characteristic of cells involved in hormone-related developmental reorganization. Autophagosomes and heterophagosomes are types of *secondary lysosome*, and may be considerably larger than primary lysosomes. Some extracellular enzyme secretion results from fusion of primary lysosomes directly with the plasma membrane. See Fig. 51.

Lysozyme Class of **enzyme** catalysing hydrolysis of (glycosaminoglycan) walls of Gram-positive bacteria, leading to rupture and death of remaining protoplast. Secreted by skin and mucous membranes and found in tears, saliva and other body fluids of mammals; also in egg white. Provides one innate immune response to bacteria. Lysozyme was among the first proteins whose three-dimensional molecular structure was elucidated.

Macrocyst See **Myxomycota**.

Macrogamete (megagamete) See **gamete, anisogamy**.

Macromolecule Molecule of very high molecular weight, characteristic of biological systems, e.g. proteins, nucleic acids, polysaccharides, and complexes of these.

Macronucleus (meganucleus) See **nucleus**.

Macronutrients Substances required in large amounts for plant growth; e.g. nitrates, phosphates, sulphates, calcium and magnesium.

Macrophage Phagocytic cell of vertebrate connective tissue but not typically of the blood itself. Included here are *wandering macrophages* derived from monocytes (see **leucocyte**), and more static macrophages (*histiocytes*) dispersed throughout connective tissue but capable of migrating towards a site of infection. Phagocytes of the **reticulo-endothelial system** are rather more specialized macrophages, but all are important as **antigen-presenting cells**, capable of collecting immune intelligence about pathogens and activating a sensitized helper **T cell**.

Macrosporangium See **megasporangium**.

Macrospore See **megaspore**.

Macrosporophyll See **megasporophyll**.

Macula (1) See **fovea**. (2) Small receptor in walls of the utricle and saccule of tetrapod **vestibular apparatus** providing information about position of the head in relation to gravity when the animal is moving or at rest. Lying in planes perpendicular to one another, they comprise an epithelium containing support cells and **hair cells** whose stereocilia are embedded in a gelatinous layer on which lie *otoliths* of calcium carbonate which move the gelatinous layer, so pulling on the stereocilia. Impulses generated travel along **cranial nerve** VIII to the **medulla**. Otoliths have an inertia which makes them stay at relative rest or movement as the body respectively moves or comes to rest. See **statocyst**.

Major histocompatibility complex (MHC) Mammalian gene complex of several highly polymorphic linked loci encoding glycoproteins involved in many aspects of immunological recognition, both between lymphoid cells and between lymphocytes and antigen-presenting cells. Initially detected as the region encoding antigens involved in graft rejection, the most important region in humans (the HLA system) is located on the sixth chromosome pair; in mice the comparable region (the H-2 complex) is located on the seventeenth chromosome pair. *Class I* MHC molecules occur in low levels on most body cells, but abundantly on **T cells**; *Class II* MHC molecules are constitutive of lymphoid dendritic and **B cell** mem-

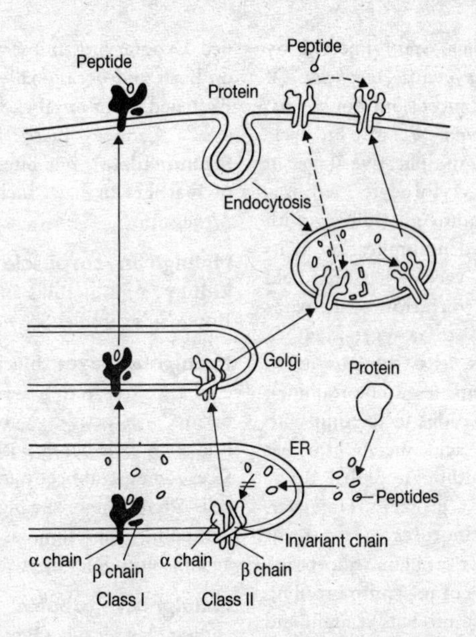

Fig. 76 *Two pathways for antigen presentation. The Class Iα and β_2m chains (shaded) are inserted into the endoplasmic reticulum (ER) where they associate non-covalently and bind peptide. Class I-peptide complexes are exported via the Golgi apparatus to the cell surface – most of the peptides being cleaved from proteins in the cytosol. Class IIα and β chains are inserted into the ER but associate with each other and the invariant chain which blocks peptide-binding until it is removed after processing in the Golgi apparatus and reach the cell surface, where they bind extracellular peptides for recognition by appropriate T cells.*

branes and inducible on macrophages. Both Class I and Class II molecules initially insert into the endoplasmic reticulum. There, the former bind intracellular antigenic peptide fragments prior to export to the cell surface via the Golgi apparatus; but the latter enter a type of Golgi vesicle destined to fuse with endocytic vesicles and bind antigenic peptides of extracellular origin prior to export to the cell surface for T cell surveillance. Several genes in the MHC appear to encode proteins involved in supply of peptide fragments to Class I and Class II molecules. See **accessory molecules**, **immunity**.

Malacostraca Largest subclass of **Crustacea**, including prawns and shrimps (Amphipoda), crabs and lobsters (Decapoda) and woodlice (Isopoda). Thorax with eight segments, covered by carapace (not in isopods or amphipods); abdomen usually with six. Abdominal appendages on all segments, usually functioning in swimming. Thoracic appendages serve for locomotion and sometimes for feeding, and often bear gills. Compound eyes stalked.

Malaria A widespread and debilitating human disease, caused by a protozoan parasite (*Plasmodium* spp.) injected by mosquitoes of the genus *Anopheles*. Of many forms, the most commonly fatal is due to *P. falciparum*. Parasite life cycle includes asexual multiplicative stages in human liver and erythrocytes, fertilization in the mosquito gut lumen, with subsequent asexual multiplication in its wall. Work on a vaccine has been held up by **antigenic variation** of the parasite. Mosquitoes are now resistant to many insecticides, as is the parasite to prophylactic drugs (e.g. chloroquine); but drug combinations interfering with plasmodial folic acid metabolism has been effective, although global drug-resistance is now a problem. However, use of **oligonucleotides** in treatment may side-step this problem. Short-lasting oil on surfaces of mosquito breeding pools, use of mosquito nets at night and mosquito-repellant creams are all of some help. Visitors to malarial areas need to seek advice and take appropriate precautions. See **DNA probe**, **sickle-cell anaemia**, **thalassaemia**.

Male haploidy Arrhenotoky. Condition in which males arise from unfertilized eggs, females from fertilized ones. Males have no father; females have two parents, but only three grandparents. Males are cytologically haploid, producing gametes by mitosis (see **life cycle**). Universal in hymenopteran insects; found also in some scale insects (Coccoidea) and mites, among other animal groups. Form of **parthenogenesis**. See **Hamilton's rule**, **thelytoky**.

Male sterility See **cytoplasmic male sterility**.

Malleus See **ear ossicles**.

Mallophaga Biting lice; bird lice, feather lice. Exopterygotan insects; ectoparasitic on birds and occasionally on mammals. Flattened, with tarsal claws and reduced eyes. Cannot pierce skin (unlike **Siphunculata**), but bite small particles of **feathers** or hair. Include hen louse, *Menopon*.

Malpighian corpuscle In vertebrate **kidney**, a glomerulus and its associated Bowman's capsule. See **nephron**.

Malpighian layer Innermost layer of epidermis of vertebrate skin, next to dermis. An active region of **mitosis**, non-stem cells being pushed to the surface, ageing and becoming cornified. Its cells often contain the pigment melanin, darkening the skin as a protection against ultraviolet light.

Malpighian tubules Long, blind-ending, slender tubes lying in the haemocoeles of most myriapods and terrestrial insects and arachnids (where independently evolved), but not Onychophora. Involved in excretion and osmoregulation. In insects, open into intestine near junction of hindgut and midgut. Water, nitrogenous waste and salts enter tubule lumen, probably by active secretion. The hindgut then selectively reabsorbs water and useful metabolites, leaving a precipitated concentrate rich in uric acid salts which then leaves with the faeces. See **coxal glands**.

Maltose Disaccharide sugar composed of two glucose molecules bonded in an α[1,4]glycosidic linkage. Formed by the enzymatic degradation of starch by **amylases**. Occurs e.g. in germinating seeds such as barley, and during digestion in animals.

Mammalia Vertebrate class originating

in Triassic; present forms distinguished by presence of body hair and **milk** secretion, generally via mammary glands. Homoiothermic; with a diaphragm used in lung ventilation. Lower **jaw** composed of a single pair of bones (dentaries), unlike therapsid reptilian ancestors, presumed also to have been homoiotherms. **Dentition** heterodont. Only the left systemic arch remains (see **aorta, dorsal**). Three bones form the **ear ossicles** in each middle ear. Mostly of small size until the Tertiary. Three extant subclasses: oviparous **Prototheria** (monotremes, e.g. spiny anteaters, duck-billed platypus); viviparous **Metatheria** (marsupials, e.g. opossums, wombats, koalas, kangaroos) and **Eutheria** (placentals, in sixteen extant orders). Several extinct orders. One posible phylogeny is illustrated in Fig. 77. See **mammal-like reptiles**.

Mammal-like reptiles (Therapsida) Reptilian subclass radiating from the mid-Permian and becoming extinct at the end of the Triassic. Sluggish herbivores and active carnivores, the latter with elbow and knee swung towards (tucked into) the body allowing rapid four-footed gait. The herbivores may have formed herds. Small Mesozoic therapsids gave rise to the earliest mammals, contemporaries of the ruling archosaurs for several million years. See **Reptilia**.

Mammary glands Milk-producing glands, peculiar to the ventral surfaces of female viviparous mammals (similar structures are present in monotremes). Rudimentary in males, unless abnormal hormonally. Develop from epidermis and resemble sweat glands, consisting of a branching series of ducts terminating in secretory alveoli during pregnancy. See **milk, prolactin**.

Mandible (1) Of vertebrates, the lower **jaw**. (2) Of insects, crustacea and myriapods, one of the first pair of **mouthparts** ('jaws'), usually involved in biting and crushing food and heavily sclerotized. See **maxilla**.

Mandibular arch See **visceral arches**.

Mannan Polysaccharide, composed of mannose units and occurring in cell walls of some algae and yeasts.

Mannitol A 6-carbon sugar alcohol universally present in the brown algae (**Phaeophyta**) in which D-mannitol is the accumulation product of photosynthesis (up to about 25% of the dry weight of some *Laminaria* species in the autumn).

Mannoglycerate Saccharide storage product in some red algae (**Rhodophyta**).

Mantle (Bot.) (1) A dense mass of fungal hyphae surrounding a root. (2) That part of the diatom valve that bends away at 90°. (Zool.) Surface layer of visceral hump of **Mollusca**, secreting shell. Flaps of the mantle enclose the mantle cavity. Similar structure is found in brachiopods.

Map distance A measure of the frequency of **crossing-over** between two linked chromosome marker loci, equal to the percentage recombination in meiotic products, provided there are no double or multiple cross-overs between the markers. The further apart geographically two loci are on a chromosome the more will their apparent distance (as derived from cross-over values)

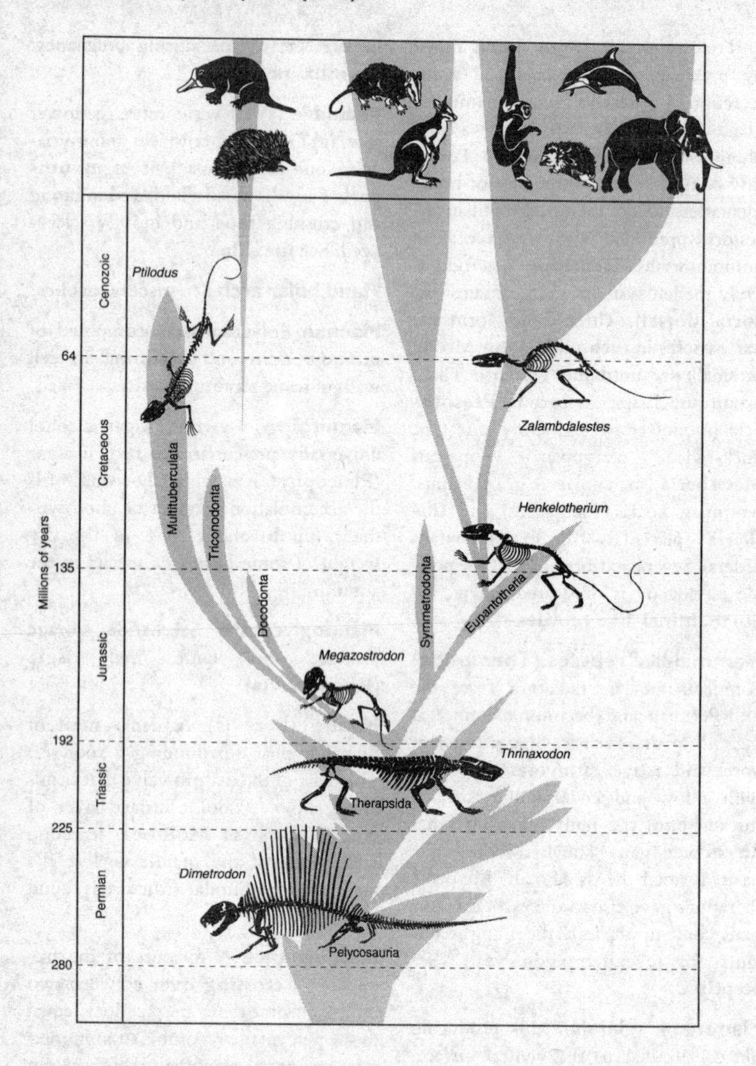

Fig. 77 *One possible phylogeny of the three extant mammalian subclasses (top): Prototheria (left), Metatheria (centre), Eutheria (right). Illustration by Lucrezia Beerli-Bieler, reproduced with the kind permission of Prof. R. D. Martin.*

be foreshortened by double and multiple cross-overs between them. Map distances between loci are reduced (underestimated) to the extent that such cross-overs occur between them, so that long map distances are best derived from summation of short distances, using intervening markers. Suppression of all crossing-over between two loci (see **suppressor mutation**) would effectively render them indistinguishable from a single complex locus. This might have selective advantages. See **supergene**.

Marginal meristem **Meristem** located along margin of leaf primordium and forming the blade.

Marginal placentation Attachment of ovules to carpel margin.

Marker See **genetic marker**.

Marsupial See **Metatheria**.

Marsupium Pouch of many marsupials and spiny anteater (*Echidna*, **Prototheria**). Fold of skin supported by epipubic bone of pelvic girdle, forming pouch containing mammary glands or similar structures, into which newborn (or eggs in echidnas) are placed. Young marsupials attach there to teats, or lick milk in the case of teatless *Echidna*.

Masseter muscle Muscle running from inside the zygomatic arch and inserting on the outside and rear of the mammalian lower jaw, pulling jaws forward and up (protraction) and assisting in side-to-side motions. Compare **temporalis muscle**.

Mass extinction See **extinction**.

Mass flow Hypothesis for explaining **translocation** of material in phloem of vascular plants.

Mast cell Granular **leucocyte** derived from **myeloid tissue**, often associated with mucosal epithelia. When not in connective tissue, dependent upon **T cells** for reproduction. Granules in cytoplasm contain **serotonin**, **heparin**, **histamine** and **tumour necrosis factor-**α; released when allergen cross-links the IgE molecules bound to plasma membrane. Involved in many allergies, but also in immunity to parasites. See **basophil**.

Mastigoneme See **flagellum**.

Mastigophora Often employed as a class of **Protozoa**, but including both holophytic, holozoic and facultative forms. Usually refers to flagellated protozoans, both *zoomastigophorans* such as trypanosomes and *phytomastigophorans* such as euglenoids. Close affinities of flagellated and pseudopodial protozoans have resulted in some authorities uniting them in the Class Sarcomastigophora. Likely that all these are grades rather than clades. See **flagellum**.

Mastoid process Part of mammalian auditory capsule (periotic bone) containing air spaces communicating with middle ear.

Maternal effect (m. influence) Instances in which some aspect of phenotype of normal sexually produced offspring is controlled more by the genotype of maternal parent than by its own. Distribution of cytoplasmic molecules in the egg, under the control of maternal genes, is one notable example in that it may have a marked effect upon **cleavage** and later development (see

cyclins, ***string*** gene, ***oskar*** gene). For instance, during oogenesis of the *Drosophila* egg, maternal mRNA products of the ***bicoid*** **gene** are retained at the anterior pole of the egg, and the eventual protein translation product disperses some way along the egg, forming the kind of gradient required of a positional signal in the **positional information** theory of development. See **maternal inheritance**, **germ plasm**, **grandchildless**.

Maternal gene Term sometimes used specifically for a gene responsible for a **maternal effect**.

Maternal inheritance Characters inherited through the female line only. Not all genes are transmitted via nuclear chromosomes (see cp**DNA**, mt**DNA**, **plasmids**). Evolution of anisogamy means that the female parent may contribute more genetically to its offspring than does the male parent. See **maternal effect**, **cytoplasmic inheritance**.

Mating type In, e.g. algae and fungi, used to designate a particular genotype with respect to compatibility in sexual reproduction. Gametes are identical in appearance and referred to as plus (+) and minus (−), rather than as male and female. In the budding yeast *Saccharomyces cerevisiae*, a single locus (*MAT*) determines whether a haploid cell is of mating type **a** or *α*. Each new generation switches its mating type, and it turns out that this involves a form of **gene conversion** in which unexpressed copies (so-called *cassettes*) of the two alternative mating type alleles (i.e. **a** and *α*) lie one on either side of the expressed form and that the expressed form is degraded each generation and replaced by the alternative allele, which persists as a flanker (see **conjugation** and **mitochondrion**). In the fission yeast *Schizosaccharomyces pombe*, cells of the two mating types (P and M) differ only in their mating potential and respond to pheromone of the opposite type by extending a conjugation tube towards the source. **Receptors** for the pheromones **a**-factor and *α*-factor in *S. cereviseae* and *P*-factor and *M*-factor in *S. pombe* all belong to the **rhodopsin** family of receptors. See **heterothallism**, **sex**.

Maturation of germ cells Processes normally taken to include the cell division and differentiation involved in the production of functional gametes from their germ mother cells.

During *oogenesis* in humans, oocyte development is arrested at first meiotic prophase in the newborn female (see **cytostatic factor**), most oocytes being surrounded by follicle cells, some already organized in developing follicles; but these degenerate before puberty. At puberty, **follicle stimulating hormone** restarts follicle development, the surrounding thecal cells secreting oestrogens which reduce FSH production but increase the number of FSH receptors on the thecal cells, so that as FSH level drops only that follicle with most receptors can bind the remaining FSH; the others die. Follicle cells contribute small molecules to the oocyte via gap junctions, enabling it to synthesize macromolecules; but **cyclic AMP** is also transferred and inhibits resumption of meiosis by phosphorylating key oocyte proteins (see **maturation promoting factor**). The mid-cycle surge

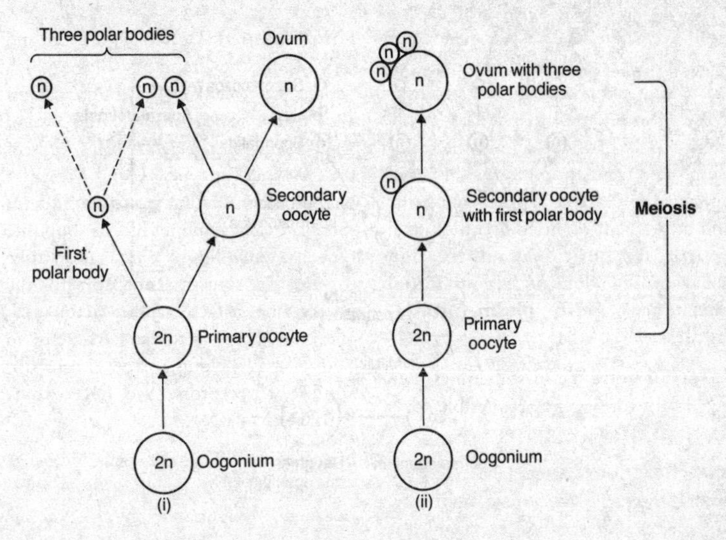

Fig. 78a *Diagram illustrating vertebrate oogenesis (i) with polar bodies shown distinct from oocyte; (ii) with polar bodies remaining attached to oocyte, as normally occurs.*

of pituitary **luteinizing hormone** disrupts the gap junctions causing a drop in cAMP and onset of the second meiotic division arrested at second meiotic metaphase, in which state the oocyte may be fertilized. The follicle enlarges and ruptures in response to LH and internal prostaglandin, releasing the secondary oocyte (see Fig. 78a).

Spermatogenesis in the **testis** (Fig. 78b) is also under pituitary control, LH (= ICSH) being the more important hormone, but acting synergistically with FSH. ICSH causes **Leydig cells** to release testosterone, some of which probably stimulates the germinal epithelium of the seminiferous tubules to divide. ICSH also affects the **Sertoli cells**, influencing the rate at which spermatozoa are released into their tubule lumen. Spermatogonia remain attached to each other after mitosis, as do their primary and secondary spermatocytes and spermatids (forming a *syncytium*). Meiosis takes about 24 days to complete in man and most sperm differentiation (*spermiogenesis*, or *spermateliosis*) occurs after meiosis is completed, spermatids losing excess cytoplasm and developing an **acrosome** and, from one of the two centrioles, a flagellum – all the while surrounded and protected from immune attack by their Sertoli cell (gametes are antigenically unlike body cells). Spermiogenesis takes about 9 weeks in man. Sperm normally cannot fertilize an ovum until they have been in the epididymis, nor until they have undergone **capacitation** in the female reproductive tract. See **menstrual cycle**.

Maturation promoting factor (MPF, M-phase promoting factor, growth-associated H1 kinase) A mitosis-inducing protein kinase govern-

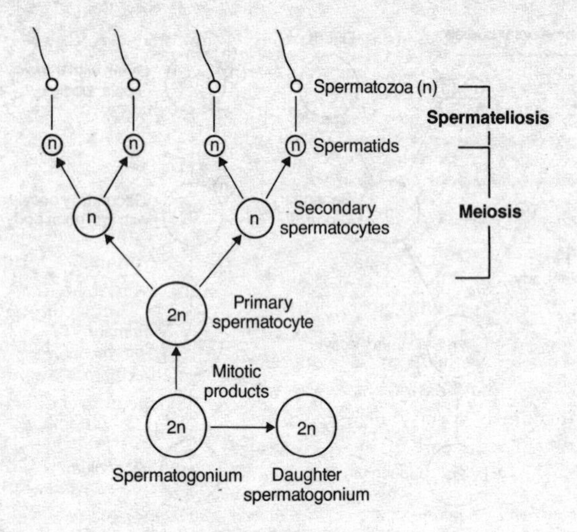

Fig. 78b *Diagram illustrating vertebrate spermatogenesis.*

ing the G2-M transition in the **cell cycle** and responsible for resumption of meiosis (i.e. ending meiotic arrest brought about by **cytostatic factor**) in ovulated amphibian and other vertebrate oocytes. Several of its major subunits are **cyclins**, but its catalytic subunit is the phosphoprotein pp34, identified as the product of the *cdc2* gene ($p34^{cdc2}$) in the fission yeast *Schizosaccharomyces pombe* and of the *CDC28* gene in the budding yeast *Saccharomyces cerevisiae* (see ***cdc* genes, yeasts**). Probably ubiquitous in eukaryotes, $p34^{cdc2}$ has an ATP-binding domain and is inhibited by phosphorylation of a tyrosine residue in that domain by *weel* kinase, among other tyrosine kinases. MPF is activated by the tyrosine phosphatase product of the (fission yeast) *cdc25* gene, when it triggers chromosome condensation, breakdown of the nuclear lamina and formation of the mitotic spindle (see Fig. 15 in the **cell cycle** entry). Some of these events may result directly from MPF-activation, others from subsequent regulatory **cascades**.

Maxilla (1) One of the **dermal bones** of the vertebrate upper jaw (and of the face in man) carrying all the upper teeth, except incisors. (2) Sometimes used for the whole vertebrate upper jaw. See **premaxilla**. (3) One of a pair of **mouthparts** in insects, crustaceans and myriapods, lying beneath the mandibles and articulating with the head capsule by a ball-and-socket joint. Sclerotized. In insects, bears a jointed sensory palp. In crustaceans, tends to be a phyllopodium, and may serve in feeding and gaseous exchange (see **biramous appendages**).

Maxillule Paired crustacean mouthpart, lying behind the mandibles and in front of the maxillae. Usually a phyllopodium (see **biramous appendages**), passing food to the mouth.

Meatus A passage; e.g. external auditory meatus (see **ear, external**).

Mechanoreceptor See **receptor**.

Meckel's cartilage Paired bar of cartilage, one forming each side of lower jaw of gnathostome embryo, and of adult elasmobranch. In most other vertebrates it becomes reduced, and partly ossified, as the *articular bone* (see **ear ossicles**), which in non-mammals forms hinge of lower jaw. Represents part of third **visceral arch**.

Meconium Contents of mammalian foetal intestine; derived from glands discharging into gut, and from swallowing of amniotic fluid during process of excretion.

Median Situated in or towards the plane dividing a bilaterally symmetrical organism or organ into right and left halves.

Mediastinum Space between the two pleural cavities containing the heart in its pericardium, aorta, trachea, oesophagus, thymus, etc.

Medulla Central part of an organ. (Bot.) Central core of usually parenchymatous tissue in stems where vascular tissue is in cylindrical form; functions in food storage. May also occur in some roots where the central tissue develops into parenchyma instead of xylem. Also refers to innermost region of the thallus in lichens and in some brown and red algae. (Zool.) (1) Central part of an animal organ, typically where outer part is termed the *cortex*. See **adrenal gland**, mammalian **kidney**, **lymph node**. (2) Abbreviation of **medulla oblongata**.

Medulla oblongata (medulla) Most posterior part of vertebrate brain. See **brainstem** and Fig. 13.

Medullary ray Thin vertical plate of parenchyma tissue, one to several cells wide, running radially through the **stele**. May be either *primary*, passing from pith (medulla) to cortex between primary vascular bundles, or *secondary*, formed from cambium during secondary thickening, ending blindly in secondary xylem and phloem. Since these latter have no connection with pith (medulla), they are sometimes termed *vascular* (*phloem* or *xylem*) rays. The function of medullary rays is in storage and radial translocation of synthesized (organic) materials.

Medullated nerve fibre A myelinated **nerve fibre**.

Medusa Free-swimming form of the **Cnidaria**. Shaped like bell or umbrella, swimming by rhythmic contractions of circular muscle in the rim and/or subumbrella producing a jet-propulsion effect through the water and involving concentrations of nerve cells (ganglia) and formation of nerve tracts or rings (two in jellyfish). Relatively thick mesogloea. Produced by budding, they themselves are sexual. Absent from life cycles of corals, sea anemones and some hydrozoans (e.g. *Hydra*).

Megagamete See **gamete**, **anisogamy**.

Megagametophyte In heterosporous plants, the female gametophyte; located within the ovule of seed plants.

Megakaryocyte Large, highly polyploid bone marrow cell giving rise to **platelets** by a kind of pinching-off, or cellular autotomy.

Meganucleus (macronucleus) See **nucleus**.

Megaphyll Type of leaf possessing a branched system of veins and with a **leaf trace** generally associated with one or more **leaf gaps** in **stele** of stem. Usually associated with siphonostelic vascular system in stem. Characteristic of ferns and seed plants. Compare **microphyll**.

Megasporangium (macrosporangium) **Sporangium** in which **megaspores** are formed; i.e. a meiosporangium of heterosporous plants, producing usually one to four megaspores. In seed plants, an **ovule**.

Megaspore (macrospore) Larger of two kinds of spore (meiospore) produced by heterosporous ferns; the first cell of the female gametophyte generation of these and of seed plants. It becomes the **embryo sac** in flowering plants (**Anthophyta**).

Megaspore mother cell Diploid cell (meiosporocyte) in which **meiosis** will occur, contained within the megasporangium (within nucellus of ovule); produces one or more megaspores. See **double fertilization**.

Megasporophyll (macrosporophyll) Leaf or modified leaf, bearing **megasporangia**. In flowering plants, the carpel. Compare **microsporophyll**.

Meiosis (reduction division) Process whereby a nucleus divides by two divisions into four nuclei, each containing half the original number of chromosomes, in most cases forming a genetically non-uniform haploid set (see Fig. 79). A necessary aspect of eukaryotic sexual reproduction, for without it fertilization would usually double the chromosome number every generation. Meiosis ensures that all gamete nuclei from diploid parents contain a haploid set of chromosomes. It also ensures, in sexually outbred populations at least, wide **genetic variation** between offspring. This results from genetic **recombination**, both random and non-random.

The first meiotic division is initiated by DNA replication (S phase), so that each chromosome comes to comprise two sister chromatids, as in mitosis; but at the start of prophase the sister chromatids of each chromosome remain tightly together, their early appearance (*leptotene*) giving the impression of unreplicated chromosomes. They remain attached to the nuclear membrane. Chromosomes then pair up homologously (*zygotene*), as a synaptonemal complex develops between them holding them together (*synapsis*). Each such pair is termed a *bivalent*; but where there is little homology (e.g. between sex chromosomes as in the heterogametic sex of most vertebrates) synapsis is only partial. Completion of synapsis is followed by shortening and thickening of the bivalents (*pachytene*), during which interval, often lasting for days, **crossing-over** occurs between non-sister chromatids, producing recombination between homologous chromosomes. The synaptonemal complex then dissolves (*diplotene*) and the two homologous chromosomes fall apart, except where crossing-over holds them together at one or more visible *chiasmata*. Chromosomes begin to unwind (decondense) and may commence RNA transcription again. At *diakinesis*, any RNA synthesis ceases (but see **lampbrush chromosomes**); bivalents shorten, thicken and detach from the nuclear membrane. For the first time they appear as four distinct chromatids, linked at their centromeres and chiasmata.

After first meiotic prophase, the nuclear membrane disintegrates and at *first metaphase* bivalents lie on the midline of the cell, between the two poles. To commence *first anaphase* the spindle fibres, attached to kinetochores of the centromeres, then pull the two members of each bivalent to opposite poles, the sister chromatids appearing somewhat more 'splayed out' than at mitotic anaphase. Chance governs which pole each chromosome of a bivalent moves to, ensuring random recombination between non-homologous chromosomes. Each of the two sets of chromosomes produced is **haploid**, although each dyad has chromatids of mixed parental origin. A short *first telophase* and *interphase* may follow; or the second meiotic division proceeds at once. Nuclear membranes normally reform, chromosomes decondensing, then recondensing, at a brief *second prophase*. The nuclear envelope breaks down, a spindle forming either parallel to (e.g. plant megaspores), or at right angles to, the first. *Second metaphase*, *anaphase* and *telophase* pass quickly, resembling mitotic phases except that non-identical sister *chromatids* separate, lying on the metaphase plate until the centromeres holding them together separate at anaphase. Again it is a matter of chance which of two poles a chromatid moves to. Nuclear membranes form around the four haploid chromosome sets to complete meiosis. See **life cycle**, **spermatogenesis**, **oogenesis** and **maturation of germ cells** for vertebrate gametogenesis. See **cost of meiosis**, **polyploid**.

Meiospore Spore produced by meiosis.

Meiotic arrest See **maturation promoting factor**.

Meiotic drive See **aberrant chromosome behaviour**.

Meissner corpuscles Moderately rapid-acting touch (velocity) receptors in dermis of vertebrate hairless skin.

Melanin Dark brown pigment of many animals and product of tyrosine metabolism, giving brown and yellow colouration to skin, hair, etc. Often located in melanophores. See **chromatophore**, **Malpighian layer**.

Melanism See **industrial melanism**.

Melanocyte-stimulating hormone Peptide hormone of the *pars intermedia* of the vertebrate anterior **pituitary gland** causing darkening of the skin in lower vertebrates by contraction of melanin in melanophores (see **chromatophore**). In humans it appears to raise the general excitability of neurones in the central nervous system, probably affecting learning. Release inhibited by hypothalamic factor.

Melanophore See **chromatophore.**

Melatonin See **pineal gland**.

Membrane See **cell membranes.**

Membrane bone See **dermal bone**.

Membrane potential Electrical

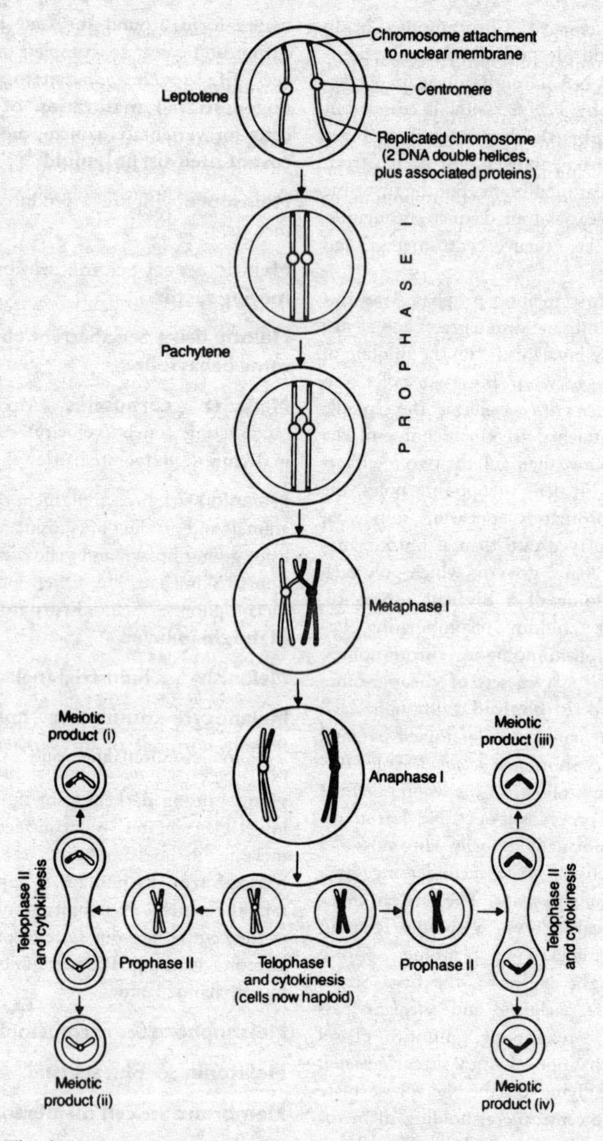

Fig. 79 *The diagram shows the behaviour of one pair of homologous chromosomes during* **meiosis**. *One cross-over results in a chiasma, visible at metaphase I. From this stage onwards the two chromosomes are shown black or white to indicate origins of segments.*

potential across a cell membrane. All plasma membranes have such voltage gradients, the inside negative with respect to the outside. Most pronounced in animal excitable tissues. See **resting potential**.

Membranous labyrinth Vertebrate inner ear, comprising the **vestibular apparatus** and, in higher vertebrates, the **cochlea**.

Memory cell A kind of mature **B cell** capable of clonal expansion when appropriately triggered (*sensitized*) by antigen, typically after initial antigen-presentation. See **antigen-presenting cell**.

Mendel, Johann (Gregor). Austrian experimental biologist (1822–84) of peasant stock. After study at Olomouc University, he entered Brno monastery in 1843 as it provided opportunities for continued academic work. His experimental work involved production of pea stocks (*Pisum* is self-fertile) pure-breeding for one or more characters; stocks which could be selfed and crossed in large numbers, reciprocally when necessary, whose seeds and offspring could be scored for several pairs of contrasting characters used. He carried out both monohybrid and dihybrid crosses. Since, as is now realized, the pairs of characters studied were determined by unlinked loci, he was able to obtain offspring in ratios enabling the subsequent formulation of laws of inheritance. His greatest conceptual innovation was to regard heritable factors determining characters as atomistic and material particles which neither fused nor blended with one another – a conclusion inescapable in the light of the experimental results he obtained. Mendel was elected Abbot of Brno monastery in 1868. His experimental results were confirmed in the first decade of the present century, after the 'rediscovery' of his laws.

Mendelian heredity (M. inheritance, Mendelism) The view, expressed here in modern terminology, that in eukaryotic genomes alleles segregate (separate into different nuclei) during **meiosis**, after which any member of a pair of alleles has equal probability of finding itself in a nucleus with either of the members of any other pair (if the loci are unlinked). As a result of chromosome behaviour during meiosis and fertilization, and of dominance and recessiveness among characters, ratios of characters among offspring phenotypes are predictable, given knowledge of the parental genotypes.

Mendel was unaware of the genetic role of chromosomes (see **gene**, **Weismann**), and studied inheritance of variation determined by unlinked allelic differences of major effect. He was also unaware of **polygenic inheritance** and **linkage**, both of which are liable to cause departures from Mendelian ratios in breeding work. Linkage provides a clear exception to the law of independent assortment (see **Mendel's Laws**). Mendelian ratios may also be distorted by **mutation**, **sex-linkage**, **meiotic drive**, **cytoplasmic inheritance**, **maternal effect**, **epistasis** and by **selection** among embryos or gamete types. **Male haploidy** will also result in distortions; but even here, as with sex-linkage, alleles behave in a basically Mendelian way. It is simply that the genetic system produces non-Mendelian ratios in breeding work.

Mendel's Laws This account of the

laws uses terminology which Mendel did not employ himself.

First Law, of Segregation: during meiosis, the two members of any pair of alleles possessed by an individual separate (segregate) into different gametes and subsequently into different offspring, neither having blended with nor altered the other in any way while together in the same cell (but see **gene conversion**). The law asserts that alleles retain their integrities (barring mutation) during replication from generation to generation. None of the cells normally produced by meiosis contains two alleles from any locus.

Second Law, of Independent Assortment: asserts that during meiosis all combinations of alleles are distributed to daughter nuclei with equal probability, distribution of members of one pair having no influence on the distribution of members of any other pair. It holds, in effect, that during meiosis random reassortment occurs between alleles at different loci. Thus if one locus is represented by the two alleles *A* and *a*, while another locus is represented by alleles B and b, then all four haploid nuclei *AB*, *aB*, *Ab* and *ab* will be formed in equal frequency by meiosis. Mendel's second law is refuted by **linkage.** Genes at linked loci tend to retain their linear sequences on chromosomes during meiosis and therefore tend to be inherited as blocks (but see **cross-over value**).

Mendel's first law is a consequence of the behaviour of all chromosomes during meiosis. His second law is a consequence of the independent behaviour of non-homologous chromosomes during meiosis. See **Mendelian heredity**.

Meninges Three membranous coverings of the vertebrate brain, spinal cord, and spinal nerves as far as their exits from between the vertebrae. Innermost is the vascular *pia mater*, separated from the *arachnoid* by the fluid-filled arachnoid spaces in which a fine web of fibres and villi reabsorb the **cerebrospinal fluid** back into the venous system. Outermost membrane, the *dura mater*, is a thick, fibrous membrane lining the skull and separated from the arachnoid below it by the dural sinus draining blood from the brain. The pia mater supplies capillaries to the ventricles of the brain.

Menstrual cycle See Fig. 80. Modified oestrous cycle of catarrhine primates; characterized by sudden breakdown of uterine endometrium, producing bleeding (*menstruation*), and by absence of a period of heat (oestrous) in which the animal is particularly sexually receptive. Controlled and coordinated nervously and hormonally, in particular by the hypothalamus and pituitary gland and the gonads. A mature human cycle has the following main sequence of events: a hypothalamic releasing factor (FSH-RF) causes release of **follicle-stimulating hormone** from the pituitary, initiating growth of one **Graafian follicle** in the ovaries and its consequent increased production of oestrogens, especially oestradiol. High oestrogen levels eventually inhibit release of both hypothalamic FSH-RF and LH-RF, decreasing FSH and LH output; but about two days prior to *ovulation* a marked rise in output of pituitary **luteinizing hormone** occurs. Through a positive feedback mechanism, rising levels of oestrogens induce LH release (via LHRF) from the pituitary, which has increased

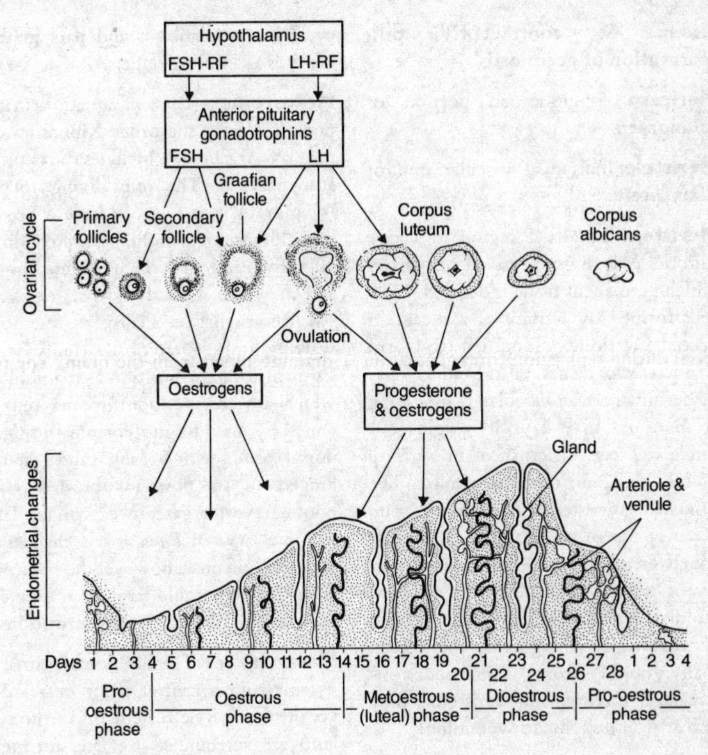

Fig. 80 *Diagram illustrating the hormonal and endometrial changes in the human menstrual cycle when no pregnancy occurs.*

sensitivity at this time. A mid-cycle surge of LH induces rupture of the Graafian follicle (ovulation) with release of the egg and development of the Graafian follicle into a **corpus luteum**, which secretes oestradiol and increasing amounts of progesterone, which maintains the vascularization of the uterine endometrium begun by oestrogens. Progesterone is secreted only so long as small amounts of pituitary LH maintain the corpus luteum, about 10–12 days in women unless pregnancy occurs. With atrophy of the corpus luteum, progesterone and oestradiol levels drop sharply, causing *menstruation* – the shedding of the uterine lining each month in the absence of pregnancy. If there is no pregnancy, the cycle repeats immediately as hypothalamic FSH-RF is released again, having been inhibited by the negative feedback effect of steroid sex hormones. If pregnancy occurs, maintenance of the corpus luteum is sustained for a short period by LH, then by hCG (**human chorionic gonadotrophin**) secreted by the implanted blastocyst and later by the placenta. See **contraceptive pill**, **maturation of germ cells**.

Mericarp Single-seeded portion of **schizocarp**.

Meristele Individual vascular unit of **dictyostele**.

Meristem Localized region of active mitotic cell division in plants, from which permanent tissue is derived. New cells formed by activity of a meristem become variously modified to form characteristic tissues of the adult (e.g. epidermis, cortex, vascular tissue, etc.). A meristem may have its origin in a single cell (e.g. in ferns), or in a group of cells (e.g. in flowering plants). The principal meristems in latter group occur at tips of stems and roots (**apical meristems**, or growing points), between xylem and phloem of vascular bundles (**cambium**) in cortex (**cork cambium**), in young leaves and (e.g. in many grasses) at bases of internodes (intercalary meristems). Meristems may also arise in response to wounding.

Meristematic activity State of active mitosis in a **meristem.**

Meristoderm Outer meristematic cell layer in certain brown algae.

Meroblastic See **cleavage**.

Merocrine gland Gland whose cells secrete their product while remaining intact: no portion is pinched off (see **apocrine gland**), nor do cells have to disintegrate in order to release their product (see **holocrine gland**). Examples include vertebrate salivary glands and exocrine cells of the pancreas.

Meroplanktonic Term describing organisms that spend part of their life cycle in the plankton and part in the benthos as a resting stage.

Merostomata Class of aquatic arthropods (formerly the order Xiphosura of the class Arachnida). Includes the extinct **Eurypterida**. The only living forms (king crabs, e.g. *Limulus*) have a broad cephalothorax (prosoma) covered dorsally by carapace which covers limbs and in which is located a pair of eyes. Chelicerae chelate. Pedipalps resemble walking limbs. Gnathobases (spiny basal segments of legs) function as mandibles, which are absent. Opisthosoma represented by fused tergites forming a single dorsal plate, with a long caudal spine hinged to its posterior border. Gill books serve for gaseous exchange. The trilobite larva of *Limulus* has chelicerae and lacks antennae; however the chelicerate arthropods could have been derived from a pro-trilobite. See **Arthropoda**.

Mesarch Type of maturation of primary xylem from a central point outwards; i.e. the oldest xylem elements (protoxylem) are surrounded by later-forming metaxylem.

Mesencephalon See **midbrain**.

Mesenchyme Embryonic mesoderm comprising widely scattered tissue giving rise to connective tissue, blood, cartilage, bone, etc.

Mesentery (1) Double-layered extension of the peritoneum attaching stomach and intestines to the dorsal body wall. Contains blood, lymph and nerve supply to these organs. (2) Vertical partitions of body wall of anthozoans (sea anemones) forming compartments within the enteron.

Mesocarp Middle layer of mature

ovary wall, or pericarp; between the exocarp and endocarp. See **fruit**.

Mesoderm Middle germ layer of triploblastic animals, coming to lie between ectoderm and endoderm from gastrulation onwards (see **germ layer** for mesodermal derivatives). In animals with a large coelom it is separable into an inner *splanchnic mesoderm* forming outer covering of digestive tract and its diverticula, and an outer *somatic mesoderm* from which, when present, develop the somatic skeleton and musculature, kidneys and gonads. Part of the mesoderm commonly becomes divided during development into a series of blocks, or *somites*, which form the basic developmental units of segmented coelomates. In vertebrates the somites, which are dorsal and on either side of the neural tube, each undergo differentiation into lateral *dermatome* and medial *myotome* with *sclerotome* around the neural tube and notochord. These will give rise respectively to dermis, striated muscle and vertebral column and adjacent rib components. Vertebrate mesoderm undivided into somites forms unsegmented *lateral plate mesoderm*, giving rise to splanchnic and somatic mesoderm. The *extraembryonic mesoderm* of amniotes is mesoderm that has spread out to cover the trophoblast and forms the **chorion**.

Mesogene development Development of stomata where **guard cells** and subsidiary cells share a common parental cell.

Mesogloea Layer of jelly-like material between ectoderm and endoderm of coelenterates. Merely a non-cellular collagenous membrane in hydrozoans such as *Hydra*, but much enlarged, thickened and fibrous in jellyfish (scyphozoans) and in these and some other groups comes to contain cells derived from the two tissue layers. It is not itself a germ layer.

Mesokaryote (dinokaryote) Term used to describe the nucleus of the **Dinophyta** and **Euglenophyta**, where the chromosomes remain condensed throughout the mitotic cycle. Nucleolus persistent, dividing by pinching in two; nuclei large, the nuclear membrane remaining intact during the entire mitotic cycle; chromosomes present attach to the nuclear membrane and not the spindle microtubules. In Dinophyta, few basic proteins (histones) are associated with the DNA. The term is used to denote the evolutionary position of dinoflagellates and euglenoids between prokaryotes and eukaryotes.

Mesonephros See **kidney**.

Mesophilic (Of microorganisms, e.g. *Escherichia coli*) with optimum growth temperature (measured by generations per hour) in the 30–40°C range. Compare **psychrophilic**, **thermophilic**.

Mesophyll Internal tissue of **leaf blade**; differentiated into upper *palisade* and lower *spongy* mesophyll, the latter cells with generally fewer chloroplasts and separated by large air spaces. See **leaf**.

Mesophyte Plant found growing under average conditions of water supply. Compare **hydrophyte**, **xerophyte**.

Mesosome Infolded region of plasma membrane of bacterial cell, containing electron transport system, attaching to the circular chromosome and involved in initiation (and termination?) of chromosomal replication. Generally located near newly-forming cell wall in binary fission. Very probably homologous with mitochondrial cristae. See **lomasome**.

Mesothelium Epithelium-like layer covering vertebrate **serous membranes**; flattened squamous cells derived from mesoderm, and therefore not strictly epithelial.

Mesotrophic (Of lakes and rivers) which are neither nutrient-rich nor nutrient-deficient (i.e. between **oligotrophic** and **eutrophic**). Category based upon so-called phytoplankton indices which involves comparing ratios of eutrophic to oligotrophic species. Compare **dystrophic**.

Mesozoic Geological era extending from 225–70 Myr BP; includes Triassic, Jurassic and Cretaceous periods. See **geological periods**.

Messenger RNA (mRNA) Single-stranded RNA molecule, translated on ribosomes into a polypeptide. Produced (transcribed) by RNA polymerase, only one of the two DNA strands being read by the polymerase and acting as template (see **protein synthesis**). The transcription product (hnRNA in eukaryotes) generally contains some base sequences not coding for any part of the eventual polypeptide, these being cut out enzymatically (see **RNA processing, intron**) before the processed mRNA is attached to a ribosome. The amino acid sequence of the polypeptide reflects the mRNA nucleotide sequence, as dictated by the **genetic code**. Most mRNA molecules are short-lived in the cell, but some eukaryotic cells lacking nuclei have relatively long-lived mRNA, as do some egg cells.

Metabasidium Cell in which meiosis occurs in members of the **Basidiomycota**. Compare **probasidium**.

Metabola See **Pterygota**.

Metabolic pathway Sequence of reactions, each catalysed by a different enzyme, leading to the formation of one or more functional products. Pathways may be linear (e.g. **glycolysis**) or cyclical (e.g. **Krebs cycle**), and a by-product of one may serve as a substrate in another. See **regulatory enzymes**.

Metabolism Sum of the physical and chemical processes occurring within a living organism; often intended to refer only to its enzymic reactions. May be regarded as comprising *anabolism* (build-up of molecules) and *catabolism* (breakdown of molecules). Frequently used in context of a particular class of compounds within an organism, e.g. fat metabolism. The term *metabolic rate* is often used rather loosely, more or less synonymously with respiratory rate, the level of oxygen consumption in aerobes being an indicator of the general metabolic activity of the organism. See **metabolite**.

Metabolite Substance participating in **metabolism**. Some are intermediary compounds of biochemical pathways; others are taken in from the environment. Primary metabolites are those required for or produced by metabolic reactions essential to the life of the organism; **secondary metabolites** are those required for, or produced by, reactions not essential to continuance of the organism's life.

Metacarpal bones Rod-like bones of fore-foot of tetrapods articulating with carpals (wrist-bones) proximally and finger bones (phalanges) distally. Usually one corresponding to each digit. Compare **metatarsal bones**. See **pentadactyl limb**.

Metacarpus Region of tetrapod fore-foot containing metacarpal bones. Palm region of man.

Metacentric See **centromere**.

Metachromatic granule (polyphosphate body, volutin granule) Spherical granule containing stored phosphate found in cells of blue-green algae (**Cyanobacteria**). Absent from young cells or cells grown in phosphate-deficient medium; prominent in older cells.

Metachromatic stain See **staining**.

Metachronal rhythm Pattern of beating adopted by groups of cilia, segmented parapodia of polychaetes and arthropod limbs, in which each unit (cilium, parapodium, limb) is at a slightly different stage in the beat cycle from those on either side of it. Result is a smooth progression of waves of beating along the units. Often occurs in forms of locomotion, microphagous feeding and exchange of water over a respiratory surface.

Metagenesis See **alternation of generations**.

Metameric segmentation (metamerism) See **segmentation**.

Metamorphosis Process during, and as a result of, which an animal undergoes a comparatively rapid change from larval to adult form. Under hormonal control, it is most notable in the life histories of many marine invertebrates, the majority of insects (especially **Endopterygota**) and of **Amphibia**. Often requires destruction of much larval tissue (see **lysosomes**) and changes in gene expression. Of enormous evolutionary and ecological significance. See **corpora allata**, **thyroxine**.

Metanephros See **kidney**.

Metaphase Stage in **mitosis** and **meiosis**.

Metaphloem Primary **phloem** formed after the protophloem.

Metaphyton Algae present between epiphytic algae, occurring as a loose collection of non-motile or weakly motile forms, lacking ready means of attachment to a substratum.

Metaplasia Transformation of one kind of adult cell into another not usually found in that part of the body. May occur during tumour formation.

Metarterioles Minute branches of arterioles (8–18 μm in diameter). Generally give rise to capillary beds which may be opened and closed by precapillary sphincters. See **capillary**.

Metastasis Movement of infectious microorganisms or malignant **cancer cells**, usually via blood stream or lymph, from one focus of growth to another part of the body where they set up further foci, in the case of cancer cells often by attachment to the endothelium of a capillary.

Metatarsal bones Rod-like bones of tetrapod hind-foot (forming metatarsus), usually one corresponding to each digit, articulating with ankle bone (tarsal) proximally and toe bones (phalanges) distally. See **pentadactyl limb**.

Metatheria Marsupials. Subclass or infraclass of **Mammalia** containing just one order (Marsupialia). Originated in early Cretaceous; short gestation period and very undeveloped offspring at birth; a protective pouch (marsupium) in most forms investing the mammary glands of the female and housing the young while they complete development. Restricted to Australian region (including New Guinea) and the New World; fossil marsupials appear to be absent from Europe and Mongolia. Decline in diversity during late Cretaceous. Many resemble placental mammals of various orders, exhibiting **convergence**. See **placenta**.

Metaxylem Primary xylem formed after the protoxylem. Cell elongation is generally complete, or nearly so, by the time of its production.

Metazoa Multicellular animals; a subkingdom of the **Animalia**. Characterized by differentiation of cells enabling division of labour, organization of at least tissue grade, development including recognizable larva and intercellular communication. Sponges (**Parazoa**) are sometimes regarded as non-metazoan since they lack a nervous system providing communication between cells; but this is achieved in other remarkable ways.

Methanogen An organism (e.g. see **Archaebacteria**) producing methane. Methanogenic bacteria are obligate anaerobes, most using CO_2 as their terminal electron acceptor in fermentation, although methanol, formate, methyl mercaptans, acetate and methylamines are also used. Major environments for methane production include marshes, swamps, rice paddies and the rumen.

MHC See **major histocompatibility complex**.

MHC restriction See **T cell**.

Micelle Particle of colloidal size, normally spherical, which in aqueous medium has a hydrophilic exterior and hydrophobic interior. Commonly consists of a phospholipid monolayer, sometimes with associated protein, within which non-polar materials may be housed and transported. Several storage polysaccharides (e.g. amylose, amylopectin) form hydrated micelles rather than true solutions within cells. See **bile**, **chylomicron**, **lipoprotein**, **liposome**.

Michaelis constant (Michaelis-Menten constant) For a given enzyme–substrate reaction, the value (K_M) of the substrate concentration at which the initial velocity of the reaction (V_O) is half maximal. A low value indicates high affinity of enzyme for substrate. Allosteric enzymes do not strictly have K_M values. See **enzyme**.

Microaerophilic (Of organisms, e.g. microorganism) requiring gaseous oxygen, but at a level lower than atmospheric (e.g. *Rhizobium*; see **nitrogen fixation**).

Microbe Microscopic organism of any type, whether prokaryotic or eukaryotic. Many are pathogenic.

Microbiology Branch of biology dealing with **microorganisms**.

Microbodies Term often used to indicate **peroxisomes** and glyoxysomes (see **glyoxylate cycle**).

Microdissection Technique used in operating upon small organisms whereby the specimen is dissected while viewed through a microscope. Instruments are normally manipulated by mechanical means. Micromanipulation is often used where living cells or their nuclei are removed or inserted.

Microfilament See **actin**, **cytoskeleton**.

Microfossil Microscopic **fossil**; includes spores, pollen grains, algae, tracheids, pieces of plant cuticle, animals, etc.

Microgamete See **gamete**, **anisogamy**.

Microgametophyte One of two types of gametophyte produced by **heterosporous** plants; develops from microspore. In ferns, comprises a *prothallus* (confined to the microspore in clubmosses like *Selaginella*); in seed plants, comprises microspore within wall of pollen grain, plus its mitotic products, including pollen tube and its nuclei.

Micrograph Photograph of an image obtained during either light or electron microscopy.

Micrometre (micron) Unit of length often used in microscopy. Symbol μm; 10^{-3} mm; 10^{-6} m. Many bacterial cells are approximately 1 μm in length. 1 μm = 1000 nm = 10 000 Å.

Micron See **micrometre**.

Micronucleus See **nucleus**.

Micronutrient Substance required by an organism from its environment for healthy growth, but only in minute amounts. Includes **trace elements** and **vitamins**.

Microorganism Microscopically small organism. Includes unicellular plants and animals, bacteria and many fungi. Viruses are commonly included as well.

Microphagy Methods employed by animals in feeding upon particles which are small in relation to their own size. Such feeding tends to occur continually, frequently by sieving of particles from water by such devices as baleen plates (baleen whales), by ciliary-mucus devices in pharyngeal gill slits (urochordates, cephalochordates) and other gills (bivalve molluscs); by lophophores; by trapping particles in bristle-fringed trunk limbs (many crustacea), or even cytoplasmic nets (heliozoans, foraminiferans).

Microphyll Type of leaf usually but not always small, possessing very simple vascular system comprising single branched vein. **Leaf trace** is not associated with a **leaf gap** in stele of stem. Associated with **protostelic** vascular system. Characteristic of club mosses (**Lycophyta**), horsetails (**Sphenophyta**) and related forms. Compare **megaphyll**.

Micropyle Canal formed by extension of integument(s) of ovule beyond apex of nucellus; recognizable in a mature seed as a minute pore in seed coat through which water enters at start of germination.

Microsatellite DNA See **satellite DNA**.

Microscope, microscopy Microscopes are instruments employing lenses to produce magnified images, and hence fine detail, of objects too small to observe clearly with the naked eye. Earliest *light microscopes* (visible light as the transmitted medium) appeared in about 1590 in Holland and had two glass lenses. In these *compound microscopes* one short focal length lens, the eye-piece lens, produces a magnified apparent image of a real image produced by a second short focal length lens, the objective lens, placed closer to the object. Total magnification is the product of the magnifications produced by these two lenses. Hooke (1665) first recorded cells in cork, and Leeuwenhoek first noted infusoria (1676) and bacteria (1683), which were first stained by von Gleichen (1778). In the 1870s E. Abbe and C. Zeiss produced the first oil-immersion objective lens enabling a good image of up to ×1500 magnification. Achromatic and aplanatic objective lenses (the finest available to this day) were developed by J. J. Lister, those of 1886 from Zeiss being of a very high quality.

The thinner the material being observed, the greater the clarity of image; hence the improvement of embedding, cutting and sectioning methods to match the evolution of the microscope. Light reflected from a point object cannot be recombined again to form another true point, but only a disc of light. When discs representing adjacent object points overlap detail is lost. The *resolving power* of a microscope, its ability to distinguish fine detail, is proportional to the wavelength of the transmitted medium. Visible light has a wavelength of about 0.5 μm and the best resolving power (even using visible light of the shortest wavelength) is about 0.45 μm. Objects closer together will not be resolved as more than one object. Where an object (e.g. a typical cell) is transparent, with features differing only in refractive index, light rays will emerge with different phase relations depending on the paths they have taken. The resulting image has uniform brightness, but the technique of *phase contrast microscopy* makes use of its phase differences so as to produce the image that would have been seen had these been amplitude differences.

Electron microscopes use wave properties of electrons fired through the object held in a vacuum. In transmission electron microscopes, electrons are accelerated through the microscope by a large voltage (up to 1000 kV), those passing through the object hitting a screen, which fluoresces giving an image. The electrons are focused by electron magnets. The wavelength of electrons is inversely related to the voltage used, but even in low voltage apparatus is about 0.005 nm – four orders of magnitude (10^4) less than that of a light wave. Resolving power of the order of 1 nm for biological material (less for crystals) is achievable. This is about 200 times better than in light microscopy. Materials for observation are commonly first fixed and then embedded in Araldite® prior to sectioning by an ultramicrotome, giving sections 20–100 nm thick. The dangers of so generating artefacts are well known and are avoided as far as possible. Biological material contains few heavy atoms and consequently its electron-scattering ability is poor, but can be improved by soaking the object in, or spraying it with, a salt of a heavy metal ('staining' it). In *scanning electron microscopy* the electron beam causes the object to emit its own electrons. These can be used to produce an image which is built up as the electron beam scans the specimen (rather in the way a television picture is produced). Scanning tunnelling electron microscopy (STM) enables visualization of atomic-scale images and even chemical modification at the nanometre level. Micrographs produced have a three-dimensional quality, but lower resolution than in transmission microscopy. One technique often employed for looking at hydrophobic interiors of membranes is *freeze-fracture electron microscopy*, in which cells are frozen in ice to the temperature of liquid nitrogen (−196°C) and then fractured, the plane of fracture tending to pass through the middles of lipid bilayers. Exposed fracture surfaces are then shadowed with platinum and carbon followed by digestion of the organic content (*freeze-etching*), leaving a platinum 'replica' which can be examined with the electron microscope. This is particularly useful for detecting where proteins are located in membranes. In electron microscopy, the specimen is inevitably in a vacuum and exposed to high temperature electron bombardment; hence it is impossible to view living material, specimens requiring cooling during viewing.

Microsomes Products of homogenization of endoplasmic reticulum. Rough microsomes are closed vesicles with ribosomes attached to their outer surfaces; smooth microsomes lack ribosomes and may derive from plasma membrane as well as from smooth endoplasmic reticulum.

Microspike Small **filopodium.**

Microsporangium Sporangium in which microspores are formed; borne on a microsporophyll. In seed plants, a *pollen sac*.

Microspore Smaller of two kinds of spore produced by heterosporous plants (e.g. ferns, seed plants); first cell of microgametophyte generation in these plants, developing after meiosis from a *microspore mother cell* (microsporocyte). In seed plants, becomes a pollen grain. Compare **megaspore**.

Microsporophyll Leaf, or modified leaf, that bears microsporangia. In flowering plants, a **stamen**. Compare **megasporophyll**.

Microtome Machine for cutting extremely thin sections of tissue, etc. For light microscopy the material is first fixed and either frozen or embedded in paraffin wax, sections cut with a steel knife usually being 3–20 μm thick; for electron microscopy, fixing is followed by embedding in a resin such as Araldite®, sections cut with the glass or diamond knife of an *ultramicrotome* being 20–100 nm thick.

Microtubular root Microtubules attaching to **basal body** of flagellum within a motile cell.

Microtubule One of the essential protein filaments of the **cytoskeletons** of probably all eukaryotic cells, and of their cilia, flagella, basal bodies, centrioles and mitotic and meiotic spindles. Microtubules originate at **MTOC**s. Each microtubule consists of a hollow cylinder, about 25 nm in diameter, made up of thirteen protofilaments of the protein **tubulin**. Each protofilament consists of globular tubulin molecules polymerized together. Like **actin** filaments, microtubules grow and depolymerize at different rates at their two ends. Microtubule roles include guiding organelle and chromosome movement

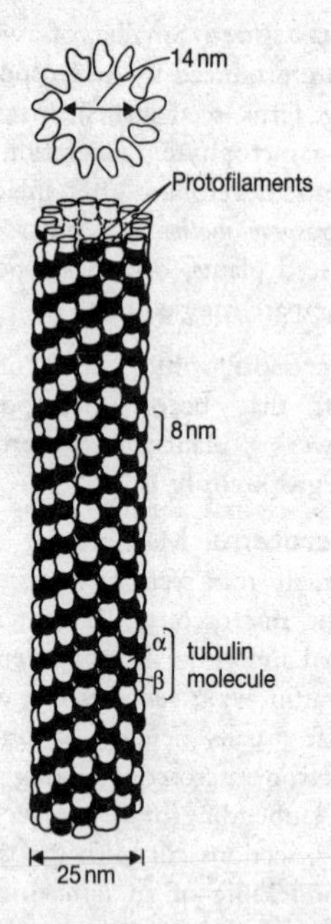

Fig. 81 *Arrangement of tubulin monomers (α and β globular protein molecules) to form a* **microtubule**, *cross section (*top*) and side view (*below*). Compare diameter with width of a bacterial envelope (see* **Gram's stain**).

in the cell, causing cell elongation by their own elongation and involvement in ciliary and flagellar beating. Various associated proteins play modifying roles in the behaviour of these microtubules. See Fig. 81, **spindle**.

Microvillus Minute finger-like projection from the surfaces of many eukaryotic cells, particularly animal epithelia involved in active uptake (e.g. small intestine, kidney proximal convoluted tubule) where, several thousand strong, they constitute the *brush borders* observed in electron micrographs. Each microvillus is about 1 μm long and 0.1 μm in diameter. About forty **actin** microfilaments run along its length, supported by accessory proteins (e.g. alpha-actinin, fimbrin). They may be capable of retraction and extension, possibly through sliding of actin over myosin (see **muscle contraction**). Brush borders may increase area of plasma membrane available for absorption 25-fold. *Stereocilia* are long, thick microvilli, about 4 μm in length (see **hair cell**), and the rhabdomeres of arthropod compound eyes are also composed of microvilli (see **eye**). See **intercellular junctions** for diagram.

Midbrain See **brain**.

Middle ear See **ear, middle**.

Middle lamella Layer of intercellular material cementing together the primary walls of adjacent cells. Contains calcium pectate. See **cell wall**.

Midgut (1) In vertebrate development, a somewhat arbitrary division of the endodermal layer in contact with remnants of the yolk sac and giving rise to part of duodenum, rest of small intestine and upper part of large intestine. (2) Cylinder of endodermal epithelial cells of the arthropod gut (mesenteron). Not lined by cuticle; secretes a *peritrophic membrane*, enclosing food and separating and protecting the epithelium from abrasion. Most digestion occurs here, with glycogen storage and secretion of urate (see **Malpighian tubules**). Commonly bears one or more pairs of diverticula (midgut or hepatic caeca), increasing its surface area.

Mildew (1) Plant disease caused by a fungus, producing superficial, powdery or downy growth on host surface. (2)

Fungus causing such a disease. (3) Often used synonymously with **mould**.

Milk (1) Complex aqueous secretion of **mammary glands**, with which mammals suckle their young. Composition varies, but usually rich in suspended fat, in protein (mostly *casein*) and sugar (mainly *lactose*). In addition, minerals, vitamins and antibodies (including antibacterial IgA in human milk, but not in cow's milk) and (again in humans but not in cows) the important iron-binding protein *lactoferrin* inhibiting bacterial growth in the baby's intestine. Milk production is under **prolactin** control, and its ejection involves **oxytocin** release during the *sucking reflex*. See **pasteurization**, **parturition**. (2) See **crop milk.**

Milk teeth See **deciduous teeth**.

Millipedes See **Diplopoda**.

Mimicry Relational term, indicating that the signal component of some mutually beneficial evolved signal–response pairing between two organisms (one signalling, the other responding) has been simulated or employed by a third party to its own advantage, often to the detriment of one or both of the original parties.

In *Müllerian mimicry* (after F. Müller) two or more organisms independently derive protection from predation, for example by tasting repellent; but they also benefit mutually through convergent evolution of similar warning (aposematic) colouration and pattern, predators learning by association to avoid both after tasting one. In *Batesian mimicry* (after H. W. Bates) an edible or otherwise relatively defenceless organism (the *mimic*) gains protection from predation by resemblance to a distasteful, poisonous or harmful organism, termed the *model*. In this case, the benefit gained will depend in part upon the ratio of availabilities of models and mimics, predators tending to require reinforcement of the learnt association of colour/pattern and disagreeableness. Some species (e.g. certain butterflies) may be Batesian mimics of more than one model (see **polymorphism**). In *aggressive mimicry*, the mimic feeds on one of the original parties in the original signal–response pairing. Some instances of **crypsis** (e.g. stick or leaf insects) may appear to overlap the criteria of mimicry given above; but it can be argued that crypsis does not involve manipulation of a true signal (as opposed to background noise) in the service of a new function. Many flowering plants, notably several orchids, achieve pollination through mimicry of signals (e.g. visual, olfactory) attracting specific insect pollinators. Mimicry has provided much support for Darwinian evolution. See **search image**.

Minimal medium Medium for growing microorganisms, containing ample quantities of all minimal nutritional requirements of the wild-type organism. See **prototroph**, **auxotroph**.

Minisatellite DNA See **satellite DNA**.

Miocene Geological epoch; subdivision of **Tertiary** period. Forests receded, grasslands spread along with radiation of grazing animals and apes. Climate was moderate, while extensive glaciation occurred in the Southern Hemisphere.

Miracidium Ciliated larval stage of endoparasitic trematode platyhelminths (flukes). Commonly bores into snail, where it develops into a sporocyst. See **trematoda**.

Mites See **Acari**.

Mitochondrial Eve Human mitochondrial DNA (see mtDNA) is mostly or entirely inherited via the female line (matrolinearly) and mutates at about ten times the rate of nuclear DNA. Attempts have been made to estimate the times of divergence of different human populations by assuming a direct proportionality between the sum of the differences in their mtDNA and the time since they diverged. But human mtDNA from a variety of geographical populations (so-called 'races') is remarkably uniform, although more diverse in African populations, indicating recency of common origin from an original ancestress, probably African, dubbed 'mitochondrial Eve'. Controversy surrounds the statistical methods employed to analyse mtDNA data, and mitochondrial Eve's date currently ranges from 400–60 Kyr BP. It is even disputed that she was of African origin. However, it is anticipated that greater knowledge of human mtDNA and more refined analytic techniques will lead to greater precision in these matters. There is no evidence from mtDNA for genetic input from Neanderthals to modern humans, but this does not preclude their hybridization (see ***Homo***).

Mitochondrion (chondriosome) Cytoplasmic organelle of all eukaryotic cells engaging in aerobic respiration, and the source of most ATP in those cells. Vary in number from just one to several thousand per cell. Most electron micrographs of mitochondria show them as either cylindrical (up to 10 μm long and 0.2. μm in diameter) or roughly spherical (0.5–5 μm in diameter). Have two membranes, the outer smooth and generally featureless, the inner invaginating to produce *cristae*, generally at right angles to long axis of the mitochondrion. Mitochondria from very active tissues have large numbers of tightly-packed cristae, those from relatively anoxic cells have few cristae. The matrix within the inner membrane is often granular and contains ribosomes (of prokaryotic type), several copies of a circular DNA molecule (see **mtDNA**) and other protein-synthesizing components, several enzymes (e.g. those for the **Krebs cycle**, part of the **urea** cycle and for **fatty acid oxidation**) and variable amounts of calcium and phosphate ions. Enzymes for haem synthesis (for cytochromes and haemoglobin) occur in the mitochondrial matrix.

Inner membrane is the site of the **electron transport system**, the ATP-synthesizing complex and enzymes involved in fatty acid synthesis. Enzymes in outer membrane include those involved in oxidation of adrenaline and serotonin, and others engaged in phospholipid metabolism. The two membranes differ in their permeabilities: outer freely permeable to small and medium-sized molecules and ions; inner permeable only to small uncharged molecules such as O_2 and undissociated H_2O and impermeable to glucose and $NADH_2$, but with various **transport proteins** permitting exchange of ATP and ADP, Krebs cycle intermediates and the accumulation of pyruvate, calcium and phosphate ions against concentration gradients, all coupled to electron transport (see **active trans-**

port). The ATP *synthetase* complex ($F_0F_1ATPase$) is also embedded in inner membrane, comprising at least nine different polypeptides in two separable units: five make a spherical head ($F_1ATPase$) which can catalyse ATP synthesis; remainder comprise a proton channel (F_0) through which protons re-enter the mitochondrion after ejection by the electron transport system (see **bacteriorhodopsin** and Fig. 32a).

New mitochondria arise by growth and division of existing mitochondria (resembling binary fission in bacteria). Many of their proteins are encoded by genes in the nuclear genome, in particular the enzymes of the Krebs cycle and fatty acid oxidation, much of the electron transport system, the outer membrane proteins and DNA and RNA polymerases. In the slime mould *Physarum*, mitochondrial fusion and genome recombination can occur between mif^+ (plasmid-containing) and mif^- strains. Most derived mitochondria contain the plasmid responsible. Yeast mitochondrial genomes may engage in recombination, intron $\omega+$ being commonly responsible, and nearly all mitochondria end up $\omega+$. Both the *mif* plasmid and $\omega+$ intron appear to be examples of **selfish DNA/genes**. Compare **F factor**. Mitochondrial **genetic code** differs from both 'nuclear' and bacterial codes in that the triplet UGA codes for tryptophan and is not a stop codon. Other codons may also have different meanings, even between mitochondria from different organisms.

Mitochondria of most C_3 plants of temperate latitudes do not engage in aerobic respiration during daylight, this being restricted to the hours of darkness. Instead these plants engage in the seemingly wasteful process of **photorespiration**, which is greatly reduced or absent in C_4 plants of tropical origin.

Mitochondria are generally regarded as having originated by **endosymbiosis** from a purple photosynthetic bacterium.

Mitosis (karyokinesis) Method of nuclear division (M-phase of the **cell cycle**) which produces two daughter nuclei, genetically identical to each other and to the original parent nucleus. Commonly accompanied by division of parental cytoplasm (cytokinesis) around the daughter nuclei to produce two daughter cells (cell division). It is the usual method by which nuclei are replicated during the growth, development and repair of multicellular organisms and coenocytic fungi. Prokaryotes lack nuclei and have no mitosis, replicating and segregating their DNA by a different process. Body cells (and nuclei) not undergoing mitosis are typically in **interphase.** The first stage of mitosis (*prophase*) commences with the first appearance of condensed chromosomes, each consisting of two identical sister chromatids held together by a **centromere**. In contrast to their behaviour in first meiotic prophase, homologous mitotic chromosomes do not normally pair up (see Fig. 82).

During *prophase* a spindle of **microtubules** assembles outside the nucleus. If centrioles are present, they have replicated in S-phase (see **cell cycle**) and form foci for origins of spindle fibres. The two **asters** get pushed apart by growth of microtubules. In *metaphase* the nuclear membrane disintegrates and microtubules growing from the **kinetochores** of the chromosome centromeres attach to those of the spindle, the resulting agitation making all chromosomes come to lie in one plane (the metaphase plate) half way between the two spindle poles. Each chromosome lies with its long axis at right angles to the axis of the spindle. Metaphase may be quite lengthy. *Anaphase* begins when kinetochores separate, the two sister chromatids of a chromosome being pulled towards the opposite spindle poles, which move further apart. Each chromatid is now a chromosome in its own right. A few minutes later, once chromosomes have reached the poles, a new nuclear membrane appears around each group of chromosomes, which decondense. Nucleoli reappear, constituting *telophase*. Mitosis is now complete and two daughter cells are normally produced by cytokinesis. Mitosis normally takes between a half and three hours. See **fission**, **cleavage**. Compare **meiosis**.

Mitospore Spore produced as a result of mitosis. Contrast **meiospore**.

Mitotic recombination See **recombination**, **twin-spots**.

Mitral valve Valve comprising two membranous flaps and their supporting tendons between the atrium and ventricle on left side of the hearts of birds and mammals. See **heart**.

Mixed tissue Tissue containing more than one cell type, all originating from same group of stem cells in development. **Xylem** and **phloem** are examples.

Mixotroph **Photoautotroph** capable of utilizing organic compounds in the environment; *facultative heterotroph* or *facultative parasite* may also refer to a mixotroph.

Mobile element See **transposable element**.

Modification Methylase activity preventing or reversing host restriction (see **phage restriction**). Specific bases of the phage DNA within target site of the restriction endonuclease are methylated. Non-heritable change.

Modifier Gene capable of altering expression of another gene at a different locus in the genome. Selection for such modifiers can establish a genetic background within which a new mutation at a locus will consequently be expressed to a greater or lesser extent. Modifiers are thus strong candidates for control of **dominance** and recessiveness of characters. See **dosage compensation**, **epistasis**, **genomic imprinting**, **penetrance**.

Modulator (Of enzyme). See allosteric **enzyme**, **ATP**.

Molar Crushing back tooth of mammal, without milk tooth predecessor (unlike premolar). Usually has several roots and complicated pattern of ridges and projections on grinding surface. See **dentition**.

Mole See **Insectivora**.

Molecular activity (turnover number) Number of substrate molecules catalysed per minute by a single enzyme molecule (or active site). Carbonic anhydrase has one of the highest molecular activities: 36×10^6 min^{-1} $molecule^{-1}$.

Molecular clock The more important it is for the amino acid sequence of a protein to be conserved, the more

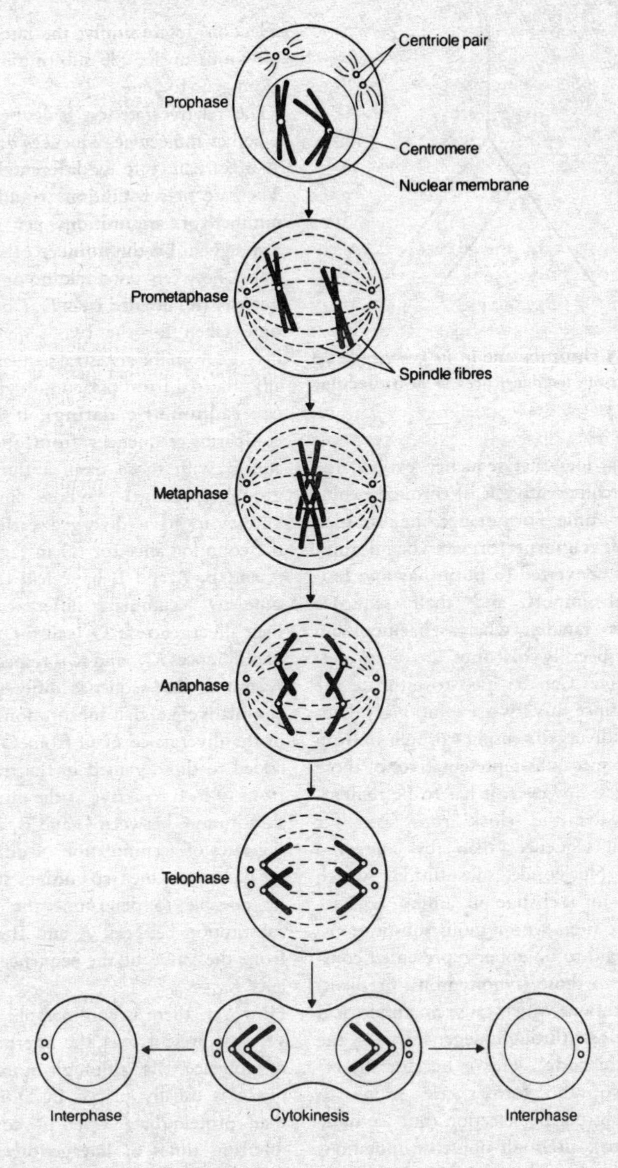

Fig. 82 *Diagrams showing the behaviour of one pair of homologous chromosomes in animal mitosis.*

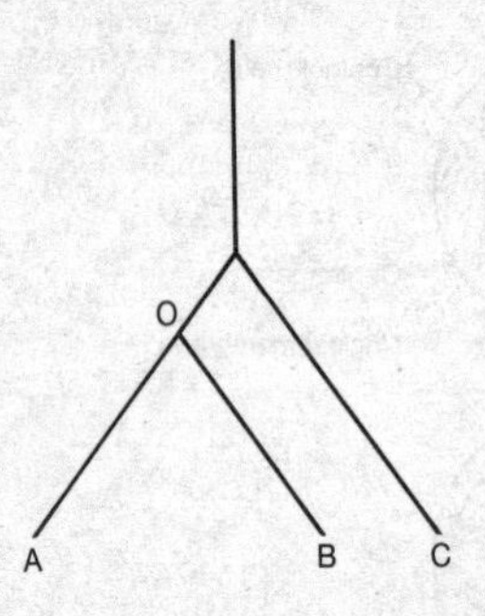

Fig. 83 *Figure illustrating the tree used in the relative-rate test as described in the* **molecular clock** *entry.*

slowly does that sequence evolve (the more conservative it is) through evolutionary time. For instance, the discarded peptide fragments formed when fibrinogen is converted to fibrin have no biological import and their sequence evolves rapidly, whereas haemoglobin and especially **histones** are very conservative. Can be used to estimate the time since any two evolutionary lineages diverged, using proteins from living species as representatives of those lineages; however, it has to be remembered that the 'clock' runs faster for neutral sequences than for conserved ones. Nucleotide substitutions which result in a change in amino acid sequence (non-synonymous substitutions) will tend to be under-represented compared to those (synonymous, or silent) substitutions which cause no amino acid alteration (through degeneracy of the **genetic code**). This is because the effects of the former are prone to weeding-out by selection (but see **neutralism**), although nonsense mutations (synonymous, but producing start or stop codons) would also be subject to selection. Interestingly, the rate of synonymous nucleotide substitution varies from gene to gene.

The 'relative-rate test' is devised to test whether molecular sequences do evolve at a constant rate in different lineages. The rate of substitution would be the number of substitutions per site per year, given by the number of substitutions between two homologous sequences (K) divided by $2T$, T being the time taken for the two sequences to diverge from the ancestral sequence, usually inferred from palaeontological data (see **radiometric dating**). It involves comparing sequences from species A and B with those from a third (outgroup) species, C, whose lineage is known to have diverged earlier than the common ancestor (O in Fig. 83) of A and B. A and B have had the same time to accumulate differences since their divergence at O (call the number of differences K_{OA} and K_{OB} respectively). Assuming that sequence differences are cumulative, i.e. that substitutions gained in the divergence of C from O can be added to those gained in the transition from O to B to arrive at the number of substitutions between C and B, and that the rates of accumulation of differences are the same in the two lineages, it should be possible to determine the rate of substitution between A and B directly from the value of the sequence difference K_{AC}–K_{BC}.
However, there is considerable controversy at present over the interpretation of sequence data, although more confidence is usually placed in DNA data than protein data when it comes to inferring times of lineage divergences since DNA sequences contain more information than protein sequences. This

is because replacement of an amino acid by another may require more than one non-synonymous substitution at the DNA level. See **mtDNA**, **mitochondrial Eve**, **mutation rate**.

Mollusca Phylum of bilaterally symmetrical, unsegmented invertebrates. Largely aquatic; coelom restricted to spaces around heart and within kidneys and gonads. Characteristically soft-bodied, with anterior head (rudimentary in Bivalvia); commonly with rasping tongue (radula), large muscular ventral foot (modified to 'arms' in Cephalopoda); with viscera usually in a hump dorsal to the foot. Outer layer of visceral hump called the *mantle*, usually covered by a shell. Mantle typically enclosing a *mantle cavity* in which usually lie the gills (ctenidia), anus and opening of kidney and reproductive ducts. Heart typically present; circulation often includes both open (haemocoele) and closed components. Nervous system of cords and ganglia (see **central nervous system**). Development mostly by spiral **cleavage**, often with trochophore larva. Includes Monoplacophora, Amphineura (chitons), Gastropoda (snails, slugs), Scaphopoda, Bivalvia (clams, oysters, etc.) and Cephalopoda (octopus, cuttlefish, squid). Enormous radiation among the phylum. About 80 000 species living; numerous fossils back to the Cambrian.

Monadelphous (Of stamens) united by their filaments to form tube surrounding the style; e.g. in lupin, hollyhock. Compare **diadelphous**, **polyadelphous**.

Monera The **kingdom** including all **prokaryotes**.

Mongolism See **Down's syndrome**.

Monocarpic (semelparous) (1) Having one carpel per flower, as in Leguminoseae. (2) (Of plants) reproducing only once and then dying. Contrast **polycarpic**.

Monochasium. See **inflorescence**.

Monochlamydeous (haplochlamydeous) (Of flowers) having only one whorl of perianth segments.

Monoclonal antibody (mAb) See **antibody**.

Monocolpate Referring to a pollen grain with one furrow or groove through which pollen tube will emerge.

Monocotyledonae Smaller of the two classes into which flowering plants (**Anthophyta**) are divided; distinguished from the **Dicotyledonae** by the presence of a single seed leaf (cotyledon) in the embryo, and by other structural features, such as parallel-veined leaves, stem vascular tissue in the form of scattered closed bundles, flower parts usually in threes or multiples of three. A few monocotyledonous plants are large (e.g. palm trees), but majority are small. Includes many important food plants; e.g. cereals, fodder grasses, bananas, palms, also ornamentals, e.g. orchids, lilies, tulips.

Monocyte Largest of the kinds of vertebrate **leucocyte**. Can differentiate into **macrophages** (see **LDL**) and are the source of numerous **growth factors**, notably PDGF and interleukin-1. See also **myeloid tissue** and Fig. 71.

Monoecious (1) (Of plants), having both male and female reproductive organs on the same individual; in flowering plants, having unisexual, male and female, flowers on the same plant, e.g.

hazel. See **dioecious**, **hermaphrodite**. (2) (Of animals) see **hermaphrodite**.

Monokaryon (monocaryon) A fungal cell, hypha or mycelium in which there are one or more genetically identical haploid nuclei. Characteristic of the early phase in life cycles of many of the **Basidiomycota**. Compare **dikaryon**, **heterokaryon**.

Monophyletic Of a taxon or taxa originating from and including a single stem species (either known or hypothesized) and either including the whole **clade** so derived (a holophyletic taxon), or else excluding one or more smaller clades nested within it (a paraphyletic taxon).

Monoplanetism In some **Protista** (e.g. some oomycota), where a zoospore of only one type is produced. Contrast **diplanetism**.

Monopodium An axis produced and increasing in length by apical growth; e.g. trunk of a pine tree. Compare **sympodium**.

Monosaccate Refers to a type of pollen grain with a single air bladder.

Monosaccharides Simple sugars, with molecules often containing either five carbon atoms (pentoses such as ribose, $C_5H_{10}O_5$) or six (hexoses such as glucose and fructose, $C_6H_{12}O_6$). Some, notably glucose and its amino derivatives, are monomers of biologically important **polysaccharides** and **glycosaminoglycans**. As carbohydrates their empirical formula approximates to $C_x(H_2O)_x$. Trioses (3-carbon sugars) such as phosphoglyceraldehyde and phosphoglyceric acid are important intermediates of many biochemical pathways (e.g. see **glycolysis**, **photosynthesis**).

Monosomy Abnormal chromosome complement, one chromosome pair in an otherwise diploid nucleus being represented by just a single chromosome. An instance of **aneuploidy**. May arise by **non-disjunction**. The norm in the heterogametic sex of those animals with the XO/XX mode of **sex determination**.

Monosulcate Refers to a pollen grain with one furrow or groove on distal surface.

Monotremata Monotremes. The only order of the mammalian subclass **Prototheria**.

Monozygotic twins Twins developing from two genetically identical cells, cleavage products of a single fertilized egg which become completely separate (failure to do so may give rise to Siamese twins). They will be of the same sex. The nine-banded armadillo regularly produces a genetically identical litter and this amounts to a form of asexual reproduction. See **fraternal twins**, **freemartin**, **polyembryony**.

Morph One form of **infraspecific variation**; one of the *forms* of a species population exhibiting **polymorphism**, generally of a markedly obvious kind.

Morphactins Group of synthetic compounds derived from fluorenecarboxylic acid, that reduce and modify plant growth. Internally, the orientation of the **spindle** axes of dividing cells is altered; most striking external effect is development of dwarf, bushy habit due to shortening of the internodes and loss of apical dominance (see **auxins**). Other effects include inhibition of phototropism and geotropism, of seed germination, and of lateral root development.

Morphogen Substance responsible for some aspect of **morphogenesis**. Controversy surrounding their putative existence seems to be resolving in their favour, largely through work on the development of the chick limb bud (see **retinoic acid**) and on developmental genetics in *Drosophila* and other arthropods (see ***bicoid*** **gene** and references cited there). It is now clear that some morphogens are **transcription factors** or transcription factor m**RNA** (see **maternal effect**).

Morphogenesis The generation of form and structure during development of an individual organism. *Morphogenetic movements* involve displacements and migrations of large numbers of cells during ontogeny, being particularly pronounced during **gastrulation**. One aspect of such movements, if they are to result in appropriate cell location, is **adhesion** between cells.

Considerable interest surrounds the theory that an animal's form develops through a hierarchical series of decisions determining the fates of cells and clones derived from them as cell number increases (see **phylotypic stage**, **zootype**). The cues enabling these decisions to be made are believed to result in **pattern formation**, and take the form of molecules (*morphogens*) activating or suppressing gene expression. See **compartment**, **positional information**.

Morphology (1) Study of the form, or appearance, of organisms (both internal and external). Anatomy is one aspect of morphology. *Comparative morphology* is important in evolutionary study. (2) The actual 'appearance' of an organism, including such diverse features as behaviour, enzyme constitution and chromosome structure. Any and all detectable features of an organism.

Morula Animal embryo during **cleavage**, at the stage when it is a solid mass of cells (blastomeres). Stage prior to the **blastula**. Human embryos implant at this stage.

Mosaic (1) Organism comprising clones of cells with different genotypes derived, however, from the same zygote (unlike **chimaeras**). See **heterochromatin**, **ovotestis**. (2) (Bot.) Symptom of many virus diseases of plants; patchy variation of normal green colour, e.g. tobacco mosaic.

Mosaic development See **development**.

Mosses See **Bryopsida**.

Motif See **DNA-binding protein**. Compare **domain**.

Motor end-plate See **neuromuscular junction**.

Motor neurone (motor nerve) Neurone carrying impulses away from central nervous system to an effector (usually a muscle or gland). Efferent neurone. In vertebrates, motor neurones leave spinal cord via the ventral horn of a spinal nerve.

Motor protein A protein which, through its ability to hydrolyse an appropriate energy-rich molecule (e.g. ATP, GTP), is able to generate the force required to move objects within cells, or the cells themselves – acting as a molecular transducer. Examples in

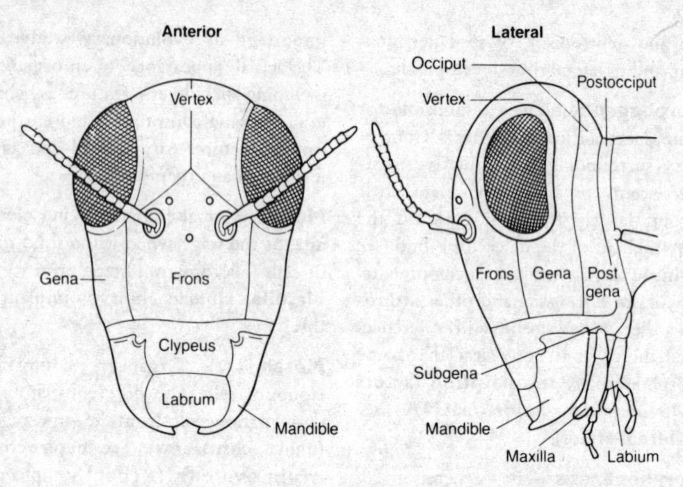

Fig. 84 *Anterior and lateral views of a locust head showing positions of mouthparts and head areas.*

eukaryotes are **myosin**, **dynein** and **kinesin**.

Motor unit One motor neurone, its nerve terminals, and the skeletal muscle fibres innervated by them. There may be 200 or more muscle fibres in each vertebrate motor unit in large muscles of the leg or trunk.

Mould (1) Any superficial growth of fungal mycelium. Frequently found on decaying fruit or bread. (2) Popular name for many fungi.

Moulting (1) For moulting in arthropods, see **ecdysis**. (2) The periodic shedding of **feathers** in birds, and of hair in mammals.

Mouthparts Paired appendages of arthropod head segments, surrounding the mouth and concerned with feeding and sensory information. Evolved from walking limbs. Similarities in mouthparts between different arthropod groups are often due to convergence rather than homology. Arachnids have a pair of **chelicerae** and **pedipalps** but no mandibles; crustaceans a pair each of **mandibles**, **maxillules**, **maxillae** and **maxillipeds**; insects a pair of mandibles and maxillae, a **labium** and a **labrum**; centipedes a pair of mandibles and two pairs of maxillae; millipedes a pair of mandibles and maxillae. Mouthparts show most adaptive radiation in insects, commonly being modified for piercing and/or sucking (see **proboscis**). See Fig. 84.

M-phase See **cell cycle**.

mtDNA Mitochondrial DNA. The mitochondrial genome almost always consists of a circular duplex, always less than 100 kilobases in length (16.5 kb in humans). In animals, generally contains less than twenty structural genes, including some for mitochondrial ribosomal RNAs, transfer RNAs and perhaps for some subunits of a few of the respiratory enzymes. Mitochondrial DNA evolves rapidly, with a mutation rate up to 10 times that of nuclear DNA. Once thought to exhibit strict **maternal inheritance**, but some paternal inheritance (esp. in mussels, *Mytilus*) has been shown. See **cytoplasmic male sterility, mitochondrial Eve, mitochondrion, plasmid, species flocks**.

MTOC (microtubule-organising centre) The cell's nucleation centre for **microtubules**, duplicated along with nuclear DNA replication during S-phase of the **cell cycle**. In most animals the **centrosome** is the MTOC; however, the microtubules are not directly nucleated by the centriole itself. It is not currently known how they do originate, although **cyclins** A and B are involved. Immunofluorescence work will be of value here. The fungal MTOC is the spindle pole body.

Muciferous body Organelle found in some algal cells, e.g. **Chrysophyta, Prymnesiophyta, Raphidophyta, Dinophtya**, containing granular material bound by a single membrane. On discharge, the contents of the vesicle often form a fibrous network outside the cell. In the **Euglenophyta** cells are permanently coated with a thin slime layer produced by the muciferous body, and those species with large muciferous bodies eject their contents upon irritation and produce a copious slime layer around the cell, which in *Euglena gracilis* comprises glycoprotein and polysaccharides.

Mucilage Slimy fluid containing complex carbohydrates; secreted by many plants and animals. Swells in water, and is often involved in water retention.

Mucilage canal Elongate cells in cortex of some brown algae (Order Laminariales) and cycads that may be involved in conduction of mucilage.

Mucilage hairs Specialized mucilage-producing hairs of bryophytes, occurring most commonly near leaf axils and growing points of gametophores.

Mucins (mucoproteins) Jelly-like, sticky or slippery **glycoproteins**. Formed by complexing a **glycosaminoglycan** to a protein, the former contributing most of the mass. Some provide an intercellular bonding material, others lubrication. See **peptidoglycan**.

Mucopeptides Major constituents of cell walls of blue-green algae (**Cyanobacteria**) and bacteria, constituting up to 50% of the dry weight. They are **glycosaminoglycans**, with chains of alternating N-acetylglucosamine and N-acetylmuramic acid residues linked by peptides sometimes including diaminopimelic acid; not found in eukaryotes.

Mucopolysaccharide See **glycosaminoglycan**.

Mucoproteins See **mucins**.

Mucosa See **mucous membrane**.

Mucous membrane General name for a moist epithelium and, in vertebrates, its immediately underlying connective tissue (*lamina propria*). Applied particularly in context of linings of gut and urogenital ducts. Epithelium is usually simple or stratified, often ciliated;

usually contains goblet cells secreting mucus.

Mucus Slimy secretion containing **mucins**, secreted by goblet cells of vertebrate mucous membranes. Some invertebrates produce similar sticky or slimy fluids

Müllerian duct Oviduct of jawed vertebrates (i.e. excluding Agnatha). Generally paired, except in birds. A mesodermal tube, arising embryologically from mesothelium of coelom in close association with **Wolffian duct**, and opening at one end by a ciliated funnel into coelom, joining the cloaca (or its remnant in placental mammals) at the other. Muscular and ciliary movements pass eggs down the tube; where fertilization is internal, sperms pass up it. In marsupials and placentals, gives rise to *Fallopian tube*, *uterus* and *vagina*. In all placentals, posterior parts of the pair of tubes fuse to produce a single, median, vagina and, as in humans, a single median uterus. The duct is just a remnant at most in males. See **Müllerian inhibiting substance**.

Müllerian inhibiting substance (Müllerian inhibitor) In male mammalian embryos, this is one of the earliest products of the foetal testis. It causes regression of the female **Müllerian duct** system (so no oviducts or uterus are formed) prior to production of androgens and development of the **Wolffian duct** system. See **testis**.

Müllerian mimicry See **mimicry**.

Müller's ratchet Process, first recognized by H. J. Müller, whereby in asexual populations no line can ever evolve to contain less mutational load than is present in the line currently with the least mutational load. The ratchet is supposed to move on one notch as each least mutationally loaded line in turn becomes extinct through genetic drift, leaving a more mutationally loaded line as the new least loaded one as a new mutant allele drifts to fixation. It has been suggested that one advantage of sex (recombination) may have been to slow the ratchet and provide time for back and compensatory mutations to appear.

Multiaxial Having an axis with several apical cells that give rise to nearly parallel filaments; found in many red algae (**Rhodophyta**).

Multicellularity Occurrence of organisms composed not merely of more than one cell, but of cells between which there is (a) considerable division of labour brought about by cell **differentiation**, and (b) a fairly high level of intercellular communication. Increase in cell number must be accompanied by intercellular **adhesion**, and by itself is of relatively little evolutionary significance. However it provides opportunities for division of labour and specialization between cells, which may enable further increase in size, providing scope for improved competitiveness in a variety of respects, and exploitation of fresh niches. Sponges (**Porifera**) display considerable communication between cells (see **amoebocytes**), although this does not take the form of a nervous system. Occasionally used rather loosely of groups of cells between which there is little differentiation and for which the terms colonial (e.g. *Volvox*) or filamentous (e.g. *Spirogyra*) are more appropriate. Social organization of prokaryotic

Myxobacteria and eukaryotic **Myxomycota** may well be termed multicellular. See **acellular**, **colony**, **coenobium**.

Multienzyme complex Complex and compound molecule usually containing several enzyme subunits performing different stages in a biochemical pathway along with intrinsic cofactors (prosthetic groups). The DNA replicating enzymes (DNA *polymerase complex*) of cells, including those coded by viral genomes, form such complexes, as do *pyruvate dehydrogenases* of mitochondria. Often of sufficient size to be visible in electron micrographs, e.g. **proteasomes** and **ribosomes**.

Multifactorial inheritance Any pattern of inheritance where variation in a particular aspect of phenotype is dependent upon more than one gene locus (in this sense synonymous with *polygenic*) but often also, especially in some human clinical disorders, upon particular environmental conditions which tip the balance in favour of expression: enough genes may predispose one to having the disease, but exposure to some environmental factor may be necessary before actual symptoms of the disease are presented. Allelic differences involved are usually of varying but small individual effect. Sex differences may also be involved. Examples include schizophrenia, diabetes and hypertension. See **human genome project**.

Multifinger loop See **transcription factors**.

Multigene families Genes with considerable base sequences in common and thought to have descended from a single ancestral gene through **gene duplication** and modification. The common base sequences are believed to be homologous and conserved. It is thought such multigene families are represented by genes encoding globins, immunoglobulins and **nuclear receptors**, by **homoeobox-** and **paired box**-containing genes and by genes whose products contain **multifinger loops**.

Multilayered structure (Of flagella) comprising a more or less rectangular body attached to the anterior end of the single broad band of microtubules in the **Charophyceae** and in the spermatozoids of lower land plants. It lies directly beneath the basal bodies of the flagella and comprises four layers, the layer closest to the plasmalemma containing microtubules of the root. Beneath are two electron-dense layers, the bottom-most layer comprising small microtubules. May be present in the Micromonadophyceae (**Chlorophyta**) but absent in the Chlorophyceae and Ulvophyceae.

Multinucleate Cell or hypha containing many nuclei. See **acellular**, **coenocyte**, **syncytium**.

Multiple allelism Simultaneous occurrence within a species population of more than two alleles (*multiple alleles*) at a given gene locus. A common phenomenon, contributing not only to continuous variation of a character in the population but sometimes also, as in inheritance of human **blood groups**, to discontinuous variation. Common at loci responsible for **incompatibility** in plants. See **supergene**.

Multipolar neurone See **neurone**.

Multituberculata Extinct mammalian order within the **Prototheria**. Mainly of late Jurassic–Cretaceous times, but by extending into the Cenozoic its span far exceeds that of any other mammalian order. Had multicusped teeth (as opposed to the three or fewer of therian mammals). Probably the first herbivorous mammals; somewhat rodent-like in form and dentition. Some were medium-sized. See **Pantotheria**.

Multivalents Associations of more than two chromosomes, joined by chiasmata, at meiosis in **polyploids** (seen most clearly at first metaphase). Where homology extends to more than just pairs of chromosomes (as it does particularly in **autopolyploid** cells) synapsis will probably result in crossing-over between any homologous regions present. When such multiply chiasmate chromosomes move apart at anaphase the result may be a chain or ring of chromosomes, and since multivalents do not always disjoin regularly, autopolyploids often have a proportion of meiotic products with unbalanced (aneuploid) chromosome numbers giving loss of fertility. See **allopolyploid**.

Mureins Group of **mucins** found in bacterial cell walls.

Muscarine An **alkaloid** which, like nicotine, mimics acetylcholine action at certain cholinergic junctions (*muscarinic junctions*). Affects target organs of parasympathetic nervous system (e.g. vagus nerve); its effects being blocked by atropine, those of nicotine by curare.

Musci See **Bryopsida**.

Muscle Any of a spectrum of animal tissues of mesenchymal origin, from fibroblast-like cells through to skeletal muscle fibres. Smooth and striated muscle are found in all animal phyla from the Coelenterata upwards. Contractile role of muscle is attributable to a proliferation of protein microfilaments from cell cortex to cell interior. See **striated muscle**, **smooth muscle**, **cardiac muscle**.

Muscle contraction The force-generating response of muscle to stimulation. May be *isotonic* (when muscle shortens during contraction) or *isometric* (when there is no change in muscle length during contraction). Molecular mechanism of contraction is probably similar in smooth, cardiac and striated muscle but is best understood in the last. Myofibrils in a resting striated muscle fibre consist of very precise arrangements of filaments of proteins **actin** and **myosin** (see **striated muscle**). In order for contraction to occur, several conditions must be met. Currently accepted account (*sliding filament hypothesis*) holds that force generation can only occur when actin and myosin filaments make contact and form the complex **actomyosin**. This they can only do in the presence of intracellular calcium ions (Ca^{2+}). But the tubular sarcoplasmic reticulum has **calcium pumps** which accumulate Ca^{2+}, keeping it scarce. These pumps are shut off when **action potentials** reach the terminal cisternae of the sarcoplasmic reticulum where they make contact with inwardly folded transverse tubules (T-tubules) of the fibre's outer membrane (see Fig. 107). Only then does Ca^{2+} flow from the cisternae, surrounding the myofilaments. These ions then bind to *troponin* molecules attached to actin filaments and in turn cause a shift in positions of *tropomyosin* molecules also attached to

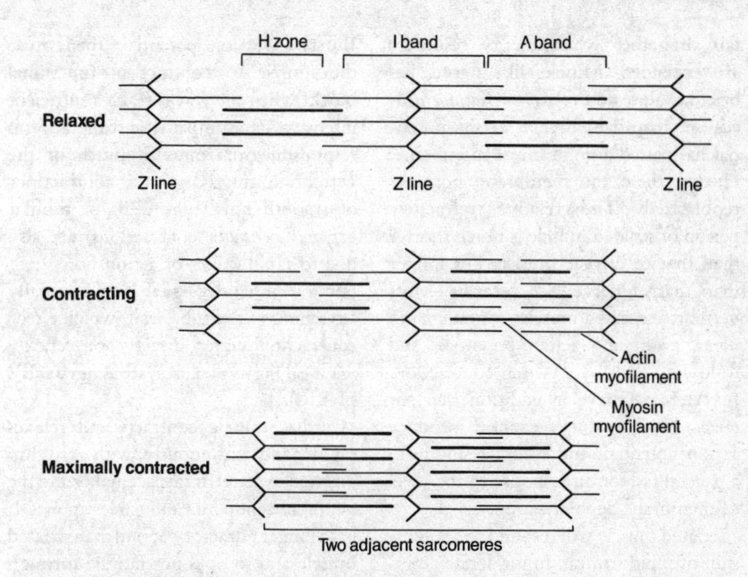

Fig. 85. *Diagram illustrating the sliding filament hypothesis of striated* **muscle contraction**.

actin. These unmask sites on actin to which myosin can bind, forming acto-myosin. But the 'heads' on myosin filaments will only bind to actin after they have been primed to do so by hydrolysis of ATP, which each head catalyses through its own ATPase activity. Only when ATP has been hydrolysed (i.e. before contraction occurs) is the myosin head ready to attach to an actin filament, changing its own shape (allosterically) so as to pull the actin filament past it. It can then release its bound ADP and inorganic phosphate, hydrolyse another ATP molecule (which requires magnesium ions, Mg^{2+}) and swing back ready for another power stroke. Repeated cycles of this sequence result in the actin filament sliding along relative to myosin; and since in muscle cells actin filaments are bound to Z-discs, the H-zone in the middle of the contracting sarcomere will narrow and darken and the I-bands disappear: the sarcomere has then fully contracted.

Striated muscle contracts with an **all-or-none response**. It relaxes only if ATP is available to free myosin from actin and if Ca^{2+} ions are actively removed from sarcoplasm by terminal cisternae, which they are in the absence of action potentials along the sarcolemma (and transverse tubules). In the absence of ATP the actomyosin complex persists and *rigor mortis* occurs

There is limited availability of ATP in sarcoplasm, and the immediate source of energy for its resynthesis is another **high-energy phosphate** compound, the **phosphagen** *creatine phosphate* (*phosphocreatine*). If muscle is forced to contract repeatedly the amplitude of contractions will decrease and ultimately fail altogether when energy resources are expended. Muscle, like nerve, has brief absolute and relative *refractory periods* (see **impulse**) after an action potential has passed a point on its membrane. During these the membrane becomes repolarized. The relative refractory period of striated muscle is much shorter than that of cardiac muscle, but longer than that of nerve. See **tetanus**. Most vertebrate striated muscles contain *twitch fibres*, producing action potentials and following each by a twitch contraction. Invertebrates have in addition *slow*, or *tonic*, *fibres*, which cannot produce action potentials and contract slowly in a graded fashion in response to the level of membrane depolarization.

Striated muscle will respond to a succession of rapid stimuli. In moderate activity oxygen supply to muscles is adequate to prevent build-up of lactic acid, but if work rate exceeds oxygen supply an **oxygen debt** arises. Prior to this, the oxygen stored in **myoglobin** will have been used up. *Muscle tone* (*tonus*) is the sustained low-level reflexly controlled background muscle contraction without which muscle flabbiness and poor responsiveness result. **Muscle spindles** maintain this tone, which is variable. The effect of *training* on muscles is mainly to increase capillary circulation and hence diffusion rates of materials to and from muscle fibres. Endurance training results in hypertrophy of muscle tissue (see **lipoprotein**) and mitochondrial number per fibre.

Smooth muscle contractions are either *tonic* or *rhythmic*. Some smooth muscle is self-exciting (see **myogenic**). In tonic contractions slow waves of electrical activity pass across adjacent cells, which are often fused to form a syncytium. Bursts of action potentials then cause the muscle to contract in functional blocks when the waves reach a sufficient intensity. Parasympathetic stimulation is responsible for tonus in much of the vertebrate gut. Rhythmic contractions of smooth muscle in walls of tubular organs (*peristaltic* contractions) are also due to rapid spikes of action potentials; but the initiating signal here is usually stretching of the tube wall, which elicits contraction either directly or reflexly via local nerve plexuses (see **Auerbach's plexus**).

Cardiac muscle contracts and relaxes rapidly and continuously, with a rhythm dictated by its intrinsic pacemaker and the neural and endocrine influences upon it. It has a longer refractory period than striated muscle and so does not fatigue through build up of lactic acid.

Muscle spindle Stretch receptor (proprioceptor) of vertebrate muscle. When muscle is stretched, sensory nerve endings in its spindles fire and send impulses via the spinal cord to the cerebellum and cerebral cortex. When muscle contracts these impulses are inhibited. Responsiveness of the spindle can be altered by central control from the reticular formation of the midbrain: each spindle can be made to contract even when its muscle is not contracting, thus enabling regulation of the rate of muscle contraction in response to different loads. Increased spindle firing due to muscle stretch will result in reflex muscle contraction and reduced spindle firing. The nerve circuitry is quite complex and involves inhibitory and excitatory loops.

Mushroom Common name for the edible fruiting bodies of fungi belonging to the Agaricaceae (**Basidiomycota**).

Mutagen Any influence capable of increasing **mutation** rate. Usual effect is chemical alteration, addition, substitution or dimerization of one or more bases or nucleotides of the genetic material (DNA or RNA). Mutagens include **alkylating agents**, acridine dyes, **ultraviolet irradiation** (e.g. producing thymidine dimers), **X-ray irradiation** and beta and gamma irradiation. *In vitro* mutagenesis is now routinely used to modify existing gene products. In site-directed mutagenesis, a synthetic oligonucleotide is used to achieve **gene conversion** in a plasmid prior to introduction into its *Escherichia coli* host. Alternatively, wild-type sequences can be removed from a plasmid and the desired mutant sequence (cassette) ligated in instead. See **Ames test**, **carcinogen**.

Mutant An individual, stock or population expressing a **mutation**.

Mutation Alteration in the arrangement, or amount, of genetic material of a cell or virus. They may be classified as either *point mutations*, involving minor changes in the genetic material (often single base-pair substitutions), or *macromutations* (e.g. deletions), involving larger sections of chromosome. Mutations often occur during DNA replication, may involve **transposable elements** and **transposons** and commonly involve enzyme activity. Precise definition of the term is difficult if one wishes to exclude such phenomena as **crossing-over** and other forms of recombination (see, e.g., **exons**, **antibody diversity**).

The effects of macromutations on chromosome structure are often visible during mitosis and meiosis and include **inversion**, **translocation**, **deletion**, **duplication** (see **transposable element**) and **polyploidy**. **Non-disjunction** produces chromosome imbalance (e.g. aneuploidy). Whereas point mutations commonly cause amino acid substitutions in the polypeptide encoded by the gene and often have minor effects on gene function, macromutations (especially if they involve deletions) commonly lead to syndromes of abnormalities which seriously reduce **fitness**, and are more often lethal. Many point mutations fail to result in amino acid substitutions because the **genetic code** is degenerate. Others, though altering amino acid sequence, either have little effect or only partially inactivate gene function (as in *leaky* auxotrophs of fungi and bacteria). But a single amino acid replacement in a critical position can abolish an enzyme's activity. One type of radical effect that can follow from a single base-pair substitution is the creation of **stop codon** within the open reading frame, usually rendering the translation product useless. Mutations creating stop codons (*nonsense* mutations) can often be phenotypically suppressed by compensating mutations in the anticodon sequence of tRNA molecules (*suppressor mutations*). *Frameshift mutations* are nucleotide additions and deletions not involving a multiple of three base-pairs, which move the 'reading frame' of tRNA to the left or right during the translation phase of **protein synthesis**. These usually have drastic effects on gene function. Most mutations lead to partial or complete loss of functional gene product and are recessive to wild-type alleles, normal gene product normally being produced in amounts sufficient for function even when the allele is present in single dose.

However, *loss-of-function mutations* are those causing the normal allele (now present in single dose) to produce inadequate product for normal phenotype (*haplo-insufficiency*, or *haplo-lethality*) and therefore tend to have a dominant effect on phenotype. *Gain-of-function mutations*, such as constitutively active alleles, tend either to cause overproduction of normal gene-product or production of novel or toxic gene products and tend also to have dominant effects on phenotype – esp. if key regulatory genes. See **dosage compensation**.

A distinction may be made in some multicellular organisms between mutations occurring in body cells (somatic mutations), and those occurring in **germ line** cells. Only the latter can normally be inherited and play a part in evolution (but see **polyploidy**).

By themselves, mutations do not normally direct the path of evolution: there is no 'mutation pressure'; rather, they provide the heritable variation upon which selection may act. It is usually accepted that mutations are random with respect to requirements of the cells or organisms in which they arise; but debate is developing as to whether or not some bacterial populations can respond to selection pressure by increasing the frequency of mutation rates for favourable genes. See **mutagen**, **gene conversion**, **genetic variation**, **molecular clock**, **replica plating**.

Mutation rate The frequency with which **mutations** arise in populations of organisms or in tissue culture. Experiments suggest that an average of about one base-pair changes 'spontaneously' per 10^9 base-pair replications. If proteins are, on average, encoded by about 10^3 base-pairs then it would take about 10^6 cell generations before the protein contained a mutation. See **molecular clock**, **mutagen**.

Mutualism See **symbiosis**.

Mycelium Collective term for mass of fungal hyphae constituting vegetative phase of the fungus; e.g. mushroom spawn. Some bacteria (e.g. Actinobacteria) have mycelial vegetative forms.

Mycobiont Fungal partner of a lichen; usually member of **Ascomycota**. Compare **phycobiont**.

Mycology Study of fungi.

Mycophage Fungal **phage**.

Mycoplasmas General name for group of prokaryotes distinguished by possession of the smallest known cells, as little as 0.1 μm in diameter. Cells lack cell walls, but possess ribosomes and all other protein-synthesizing machinery, encoded by a very small genome of fewer than 650 genes (about 20% of a bacterial genome). Some are saprotrophic; others are pathogens of plants and animals, including humans. Known by a variety of names, including *pleuropneumonia-like organisms* (PPLOs), after the disease caused in cattle by the first member to be described. Six genera to date, about 50 of the 60 species belonging to *Mycoplasma*. Their ribosomal RNA sequences suggest they form a natural group, probably of bacterial origin. *Mycoplasma-like organisms* (MLOs) have been isolated from over 200 plant species and have been implicated in more than 50 plant diseases, often with symptoms of yellowing or stunting; they appear to be confined to sieve-tubes and to be passed passively from one sieve-tube member to another through sieve-plate pores.

Mycorrhiza Symbiotic relationship between a fungus and the root of a higher plant. Mycorrhizae are of common occurrence, and can be found in most groups of vascular plants. Two main types of mycorrhizae occur: (a) *Endomycorrhizae*, which are most common, occurring in about 80% of all vascular plants; the fungal component is a species of the **Zygomycota**. These relationships are not highly species-specific. The fungal hyphae are found within the cortex of the plant root, where they form coils, swellings or minute branches in the cortical cells. The hyphae also extend into the surrounding soil. Two of the characteristic intracellular swellings are termed vesicles and arbuscles. Such mycorrhizae are particularly important in the tropics, where soils tend to retain phosphorus so tightly that it is available only in extremely small amounts. (b) *Ectomycorrhizae*, which are species-specific relationships characteristic of certain trees and shrubs, including beech, willow and pine families, as well as a few tropical trees. Trees growing at the timberline invariably have ectomycorrhizae, in which the fungus surrounds the root. The hyphae do not penetrate individual cells. The mycelium is extensive, spreads out into the surrounding soil and plays a highly important role in transferring organic carbon to the plant, replacing the function of root hairs, which are frequently absent. Species of the **Basidiomycota** most often form ectomycorrhizal relationships, but some involve species of the **Ascomycota**.

Mycosis Disease of animals caused by fungal infection; e.g. ringworm.

Mycotrophic (Of plants) having **mycorrhizas**.

Myelin sheath Many layers of membrane of **Schwann cell** (in peripheral nerves) or of oligodendrocyte (central nervous system) wrapped in a tight spiral round a nerve axon forming a sheath preventing leakage of current across the surrounded axon membrane except at *nodes of Ranvier*. Cell membranes comprising the myelin sheath contain large amounts of the glycolipid galactocerebroside. Such *myelinated neurones* conduct faster than equivalent non-myelinated ones as current 'jumps' from one node to the next. See **impulse**.

Myeloid tissue Major site of vertebrate **haemopoiesis**, restricted except in embryo to red bone marrow (e.g. in ribs, sternum, cranium, vertebrae, pelvic girdle). Mesodermal in origin. Myeloid stem cells give rise to cell lineages distinct from those of **lymphoid tissue**. Products include **monocytes** and descendant **macrophages**, **mast cells**, **basophils**, **neutrophils**, **eosinophils**, **megakaryocytes** and their **platelets**. In embryos, liver and spleen are principal myeloid sites.

Myeloma Malignant cancers of **myeloid tissue**; causing anaemia, especially in the middle-aged and elderly. See **cancer cell**.

Myoblast Precursor cells of vertebrate **skeletal muscle** fibres. Can divide but eventually fuse to form typical multinucleate syncytia of differentiated skeletal muscle tissue.

Myofibril Structural unit of **striated muscle** fibres, several to each fibre.

Myogenic (Of muscle tissue) capable of rhythmic contraction independently of external nervous stimulation as a result of presence of **pacemaker**. **Cardiac muscle** always has such a pacemaker and **smooth muscle** may have.

Myoglobin Conjugated protein of vertebrate muscle fibres. Single polypeptide, whose iron-haem prosthetic group binds molecular oxygen. Its oxygen equilibrium curve lies well to the left of **haemoglobin's**, enabling it to load and unload its oxygen at lower oxygen partial pressures. It gives up its oxygen only when the muscle is under anoxic conditions. See Fig. 55.

Myosin A protein found in the majority of eukaryotic cells. At least two classes of myosin exist, a single-headed tail-less variety (*myosin I*) involved in **cell locomotion**, and a two-headed, tailed variety (*myosin II*) involved in **muscle contraction**. Each filament of myosin II is thicker than those of **actin**, comprising a tail composed of two heavy α-helices of about 134 nm in length by which it interacts with other myosin filaments, and two pairs of light chains forming a pair of knob-like heads at each end containing actin-binding ATPase sites essential for **actomyosin** formation, as in muscle contraction. Its presence in **striated muscle** is responsible for A-bands. Myosin I molecules, on the other hand, are single-headed and instead of the tail have a non-α-helical domain housing a second actin-binding site which is ATP-independent (i.e. lacks ATPase activity). Molecules of cytoplasmic myosin (myosin I) can assemble into dynamic thick filaments, in contrast to the more static thick filaments formed on assembly of myosin II. Cytoplasmic myosins establish the cleavage furrow before onset of cytokinesis and generate the force required for constriction of the contractile ring – cells lacking thick cytoplasmic myosin filaments become large and multinucleate. Phosphorylation of cytoplasmic myosin by cdc2 kinase appears to inhibit its activity until the appropriate stage of the **cell cycle** (after nuclear division) and cdc2 kinase is itself inhibited. Cytoplasmic myosins can apparently interact with phospholipid bilayers of membranes and may mediate membrane-dependent cell movements which occur in the absence of myosin II. They have been located in microvilli and *Drosophila* eye rhabdomeres.

Myotome See **mesoderm**.

Myriapoda Class of **Arthropoda** including two important subclasses: carnivorous Chilopoda (centipedes), and herbivorous **Diplopoda** (millipedes). Have long bodies of many segments, and distinct heads bearing one pair of antennae, mandibles and at least one pair of maxillae (see **mouthparts**). Terrestrial, with tracheal system. Centipedes have one pair of legs per segment and flattened bodies; millipedes have two pairs of legs per segment (the result of segment fusion) and cylindrical bodies.

Myxamoeba Naked cell characteristic of vegetative phase of slime fungi (**Myxomycota**) and some simple fungi; capable of amoeboid locomotion.

Myxobacteria Small group of rod-shaped bacteria, distinguished by gliding movement in contact with a solid surface and by delicate, flexible, cell wall. In many, vegetative cells mass together (particularly when starved) to form minute fruiting bodies in which cells differentiate into durable spores. Compare eukaryotic **Myxomycota**.

Myxomycetes See **Myxomycota**.

Myxomycota Slime moulds (Kingdom **Protista**), forming a naked mass of protoplasm called the *plasmodium*, which is motile: As it moves bacteria, yeast cells, fungal spores and small particles of decaying plant and animal material are engulfed, and subsequently digested. Slime moulds appear to have no direct relationship to the cellular slime moulds (**Acrasiomycota**) or any other group. The plasmodium is multinucleate and aseptate and all nuclei divide by synchronous mitosis. The moving plasmodium is typically fan-shaped, with flowing protoplasmic tubules thicker at their base, and then spreading out, branching, and becoming thinner toward the periphery. Plasmodial growth continues until the food supply is exhausted. Then internal differentiation occurs and a fruiting body (sporangium) is produced. The fruiting structure may comprise a stalk and sporangium. The stalk and spore cells may become coated in a gelatinous cellulose wall, or the entire plasmodium may develop into a *plasmodiocarp* which may develop into an *aethalium* (when the plasmodium forms a large mound that is essentially a single, large sporangium). Spores germinate under favourable conditions to produce either unicellular myxamoebae or biflagellate cells which commonly behave as gametes. Amoeboid and flagellated cells are readily interconvertible. If food sources become limiting, an amoeba may secrete a thin wall and form a microcyst which can remain viable for a year or more. Meiosis occurs prior to spore formation, thus the plasmodium, which forms after plasmogamy and karyogamy, is diploid. Slime moulds are widely distributed in damp environments as long as there is an adequate food supply.

Myxophycean starch Storage polysaccharide of the **Cyanobacteria**, having a similar structure to glycogen; occurs as granules (α-granules) whose shape varies among species from rod-shaped granules, to 25-nm particles, to elongate 31–67 nm bodies.

Myxophyta See **Cyanobacteria**.

NAD (nicotinamide adenine dinucleotide, coenzyme I) See Fig. 86. Dinucleotide **coenzyme**, derivative of **nicotinic acid**, required in small amounts in many **redox reactions** where oxidoreductase enzymes transfer hydrogen (i.e. carry electrons), as in **Krebs cycle** and **electron transport system** (**ETS**) in respiration.

In the Krebs cycle the oxidized form of the coenzyme (NAD^+) receives one hydrogen atom from isocitrate (under the influence of isocitrate dehydrogenase) to become NADH, while another hydrogen atom from the isocitrate becomes a proton:

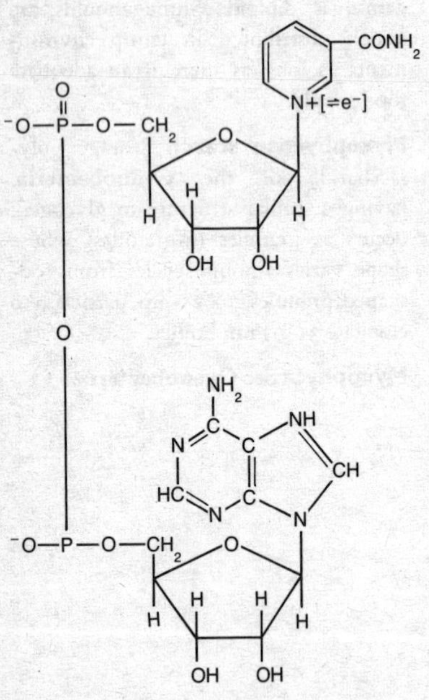

isocitrate + NAD^+ =
ketoglutarate + CO_2 + NADH + H^+

NAD dehydrogenase, which contains the flavoprotein FMN and is an important part of mitochondrial ETS, then transfers both the proton and the hydrogen from reduced NAD to its FMN component, releasing NAD^+ for reuse:

$$NADH + H^+ + FMN = NAD^+ + FMNH_2$$

By this means, electrons can be collected as NADH from a variety of sources and funnelled into the ETS. So NADH is an energy source for ATP synthesis, through its link to the ETS, and as the mitochondrion is impermeable to NADH, its electrons are first transferred to glycerol 3-phosphate, which crosses the outer mitochondrial membrane and is reoxidized to dihydroxyacetone phosphate on the inner membrane, electrons passing there to FAD (so entering the ETS) while dihydroxyacetone phosphate diffuses back to the cytosol.

Fig. 86 *Structure of* **NAD** *indicating where reversible reducton occurs, giving NADH.*

NADP (nicotinamide adenine dinucleotide phosphate, coenzyme II) A dinucleotide coenzyme; a phosphorylated NAD, and like it involved in several dehydrogenase-linked redox reactions, but usually acting as an electron donor. Not as abundant in animal cells as NAD. Involved more in synthetic (anabolic) reactions than in breakdown. Major role in electron transport during **photosynthesis**.

Nail Hard, keratinized epidermal cells of tetrapods covering upper surfaces of tips of digits, forming flattened structures, often for arboreal grasping locomotion. *Claws* are generally tougher still, narrowing and curving downwards at their tips. The mitotic epithelium of the *nailbed* is protected by a fold of skin. See **cytoskeleton**.

Nanometre Unit of length (nm). 10^{-9} metres; 10 Ångstroms; one thousandth of a **micrometre**. Formerly called a millimicron.

Nanoplankton See **plankton**, **phytoplankton**.

Nares (sing. naris) Nostrils of vertebrates. *External nares* open on to surface of head; *internal nares* are the **choanae**. Usually paired.

Nasal cavity Cavity in tetrapod head containing olfactory organs, communicating with mouth and head surface by internal and external **nares** respectively. Lined by mucous membrane. See **palate**.

Nastic movement (In plants) response to a stimulus that is independent of stimulus direction. May be a growth curvature (e.g. flower opening and closure in response to light intensity), or a sudden change in turgidity of particular cells causing a rapid change in position of an organ (e.g. leaf). These movements are classified according to nature of the stimulus; e.g. *photonasty* is a response to alteration in light intensity, *thermonasty* to change in heat intensity, *seismonasty*, to shock. Compare **tropism**.

Natural killer cell (NK cell) Type of **leucocyte** of higher vertebrates capable of recognizing alterations to surfaces of virally infected and cancerous cells, of binding to them and then killing them. Activated by **interferon**. Component of the **innate immune system**.

Natural kind A term filling a need experienced by some who seek to justify a classificatory system on objective, non-arbitrary grounds. Such people hold that objects would 'sort themselves' into categories, or kinds, if all facts about them were known. At such a time one could predict the kind to which an organism belonged given knowledge only of those facts about it that *materially caused* it to be a member of that kind. This predictability presupposes a degree of theoretical sophistication which might be unattainable given the epigenetic nature of biological systems. So, even if natural kinds do exist in biology, we might never be in a position to know which they are. See **classification**.

Natural selection (selection) Most widely accepted theory concerning the principal causal mechanism of evolutionary change ('descent with modification'); propounded by Charles **Darwin** and Alfred Russel **Wallace**. The theory asserts that, given diversity (both genetic and phenotypic) among individuals making up a species population, not all individuals in the population at time t_0 will contribute equally to the make-up of the population at a subsequent time t_1. To the extent that this is due to the effects of heritable differences upon individuals, natural selection has occurred.

Confusion arises over the use of Herbert Spencer's phrase 'the survival of the fittest'. Individual organisms do not survive through geological time (unlike some evolutionary lineages), but what they inherit and pass on does: that is, **genes** (see **unit of selection**). The theory of natural selection asserts that the genetic composition of an evolutionary lineage will change through time by non-random transmission of genes from one parental generation to the next, a non-randomness ('selection') due solely to the fact that not all gene combinations are equally suited to a given environment, and that consequently indivividuals differ in their biological (Darwinisn) **fitness**. Constraints upon phenotype from the environment, which produce this differential gene transmission, are termed '*selection pressure*'. It is commonly assumed that all regular components of a species' phenotype have been favoured by natural selection, but evolution may sometimes result from causes other than natural selection (see **genetic drift**).

When a character, especially a **Polygenic** one, is under *directional* selection (*orthoselection*) in a population, it undergoes an incremental or decremental shift in its mean value with time. When alternative phenotypes (e.g. red-eye and white-eye) at one character mode (here eye-colour) are each favoured in the same population, selection is *disruptive* and the population may become polymorphic for that character mode. When a particular (mean) character state is favoured, selection is *stabilizing*. See **sexual selection**, **arms race**.

Natural taxon Taxon comprising organisms more closely related to one another than to organisms in any other taxon at the same taxonomic level.

Nauplius Larval form of many crustaceans. Oval, unsegmented, bearing three pairs of appendages. It approximates to the 'head' of the eventual adult, successive segments being added in an anterior direction from the rear as it develops.

Nautiloid Designating those cephalopod molluscs resemblong and including the pearly nautilus, *Nautilus*. Numerous fossil forms, first appearing in the Cambrian. Have coiled or straight chambered shell.

Neanderthal man See ***Homo***.

Nearctic Zoogeographical region consisting of Greenland, and North America southwards to mid-Mexico.

Necridium (separation disc) Cell in a blue-green alga filament (**Cyanobacteria**), whose death results in formation of a **hormogonium**.

Necrosis Relatively uncontrolled process of cell/tissue death usually folowing gross perturbation to the cellular environment and loss of plasma membrane integrity causing water and ions to pass down their respective osmotic and chemical gradients with cell and organelle swelling followed by cell rupture. The causes are typically pathological. Contrast **apoptosis**.

Necrotrophic (Of organisms) feeding on dead cells and tissues.

Nectary Fluid-secreting gland of flowers. *Nectar* contains sugars, amino acids and other nutrients attractive to insects, especially in insect-pollinated flowers.

Negative feedback See **homeostasis**.

Negative staining Those methods employed in both light and electron **microscopy** by which only the background is stained, the unstained specimen showing up against it.

Nekton Swimming animals of pelagic zone of the sea or lake. Includes fishes and whales. See **benthos**, **plankton**, **pelagic**.

Nematocyst (Bot.) Structurally complex organelle found in members of the **Dinophyta**, fired out of the cell when irritated, resulting in a sudden movement of the cell in the opposite direction from the discharge. Include *trichocysts*, which discharge straight, tapering rods many times longer than the charged trichocyst (the actual benefit is obscure, but they could be a mechanism for quick escape or might spear a naked intruder); *cnidocysts*, which are explosive projectiles somewhat similar to nematocysts of coelenterates (anemones, *Hydra*). More complex is a *nematocyst–taeniocyst* complex occurring in the dinoflagellate *Polykrikos*; on discharge, this complex has a long thread attached to a conical structure at one end; probably used to capture prey which is then digested by the dinoflagellate. (Zool.) Inert stinging capsule produced by a cnidarian **cnidoblast**.

Nematoda Roundworms (eelworms, threadworms). Abundant and ubiquitous animal phylum (or class of the **Aschelminthes**). Unsegmented, triploblastic and pseudocoelomate. Circular in cross-section, with a characteristic undulating, or thrashing movement. Elastic **cuticle** of collagen acts as antagonist of the unique longitudinal muscles (no circular muscle) during swimming. Muscle cells have contractile bases adjacent to cuticle and non-contractile 'tails' lying in the vacuolated parenchymatous pseudocoelom. Gut with suctorial pharynx; anus terminal. Nervous system of simple nerve cords, ganglia and anterior nerve ring. Excretory system intracellular, consisting of two longitudinal canals. Cilia absent; sperm amoeboid. Sexes usually separate. Includes free-living and parasitic forms. **Filarial worms** of man and domestic animals, and root eelworms (e.g. of potato) are of great economic importance. Each species has its own number of cells, and there are usually four moults.

Nematomorpha Small phylum (or aschelminth class) of thin, elongated worms resembling nematodes but lacking excretory canals and with a brain linked to a single ventral nerve cord. Their larvae bore into insects. Like nematodes, have only longitudinal muscles in body wall. See **aschelminthes** for more general properties.

Nemertina (Nemertea, Rhynchocoela) Ribbon, or proboscis, worms. Small phylum of mostly marine worms with platyhelminth-like characteristics. Differ in having tube-like gut with mouth and anus (gut entire), peculiar proboscis, simpler reproductive system and a circulatory system.

Neo-Darwinism Brand of **Darwinism**, current since early decades of 20th century, which combines Darwin's theory of evolution by natural selection with **Mendelian heredity** and post-

Mendelian genetic theory. Accounts, more successfully than Darwin was able, for the origin and maintenance of variation within populations. The combination has to some extent resolved problems surrounding the nature and origins of species (see **speciation**).

Neogea See **Neotropical Region.**

Neogene Collective term for the Miocene, Pliocene and Pleistocene epochs. See **geological periods.**

Neognathae Largest superorder of **neornithes**, including all non-fossil and non-ratite species. Many orders. See **Palaeognathae**.

Neo-Lamarckism View, generally discredited, that acquired characters may be inherited. Notoriously espoused by the Stalinist biologist T. D. Lysenko (mainly for crops and domesticated animals). In another episode this century, Austrian biologist P. Kammerer tried to demonstrate the phenomenon in midwife toads and sea squirts. More recently still, genetic transfer of acquired immunity in rats has been alleged. Experimental support for the view has generally been inconclusive, and sometimes even fabricated. See **Lamarck**, **genetic assimilation**, **mutation**.

Neolithic (New Stone Age) Phase of human history, commencing approximately 10,000 years BP, and succeeding the **Palaeolithic**, during which domestication of animals and cultivation of plants first occurred. Production of sophisticated stone tools (sometimes by mass-production), arrowheads, fine bone ornaments, etc., took place. Farming began to supersede hunter-gathering ecologically. First detected in Mesopotamia, from which it spread.

Neopallium See **cerebral cortex.**

Neoplasm Tumour, or cancerous growth. Malignant if invasion or **metastasis** occurs, or is likely to occur; otherwise *benign*. See **cancer cell.**

Neornithes See **Aves**.

Neoteny Retardation of somatic development. Form of **heterochrony**, often confused with **progenesis**. Involves a slowing in rate of growth and development of specific parts of the body relative to (especially) the reproductive organs, although frequently accompanied by delayed onset of sexual maturity. Usually results in retention of juvenile features by otherwise adult animal. Formerly used in broad sense of **paedomorphosis**. Examples include the amphibian *Ambystoma* (axolotl); development of combat and display structures in some social mammals (e.g. mountain sheep, giraffe, African buffalo); increased brain size in slowly-growing mammals with small litter sizes and intense parental care, e.g. some cat species and, arguably, humans. A correlation exists between presence of neotenic attributes and conditions favouring **K-selection**.

Neotropical Region Zoogeographical region consisting of South and most of Central America. Compare **Nearctic.**

Neotype Specimen chosen as a replacement of the **holotype** when that is lost or destroyed.

Nephridium Tubular organ present in several invertebrate groups (platyhelminths, nemerteans, rotifers, annelids, some molluscan larvae and *Amphioxus*), developing independently of the **coelom** as an intucking of the ectoderm. Lumen often formed by intracellular hollowing-out of nephridial cells, closing internally (*protonephridium*) or acquiring an opening, the nephrostome, into the coelom (*metanephridium*). Protonephridia end internally either in **flame cells** or **solenocytes**. Adult annelids usually have metanephridia, often replacing larval protonephridia, and sometimes these have ciliated funnels and resemble **coelomoducts** (which they are not). Nephridia open either to outside of the body (via *nephridiopores*), or into the gut. Carry excretory products, wafted by cilia or flagella; may have an osmoregulatory role; occasionally carry gametes.

Nephron Functional unit of vertebrate **kidney**.

Neritic Waters over the continental shelf. Depth varies from just zero at low tide to as much as 200 m at the outer edge of the shelf. Much of the region is less than 80 m deep, and is well within the **photic zone** (i.e. euphotic). These waters are far more productive than the open ocean; it is over the continental shelves that the great fish and mammal migrations occur. Neritic waters differ from oceans in being less constant in chemical and physical features; the further inshore, the more variable these become.

Nerve (Bot.) Narrow thickened strip of tissue found running the length of the middle of a moss leaf (costa). (Zool.) Bundle of motor and/or sensory **neurones** and **glial cells**, with accompanying connective tissue, blood vessels, etc., in a common connective tissue sheath, or *perineurium*. Each neurone conducts independently of its neighbours. *Mixed nerves* contain both sensory and motor neurones. Nerves may be nearly as long as the whole animal and contain thousands of (usually myelinated) neurones. See **cranial nerve**, **spinal nerve**, **nervous system**.

Nerve cell See **neuron**.

Nerve cord Rod-like axis of nervous tissues forming, usually, a longitudinal through-conduction pathway and integration centre for both sensory and motor information and forming a major element of the **central nervous system**. Usually linked anteriorly to the **brain**. In segmented animals especially, a major route for **reflex arcs**. In invertebrates the nerve cord or cords are often *nerve chains*, composed of linked ganglia. In vertebrates it forms the **spinal cord**. Nerve cords in annelids and arthropods are usually ventral and paired; in chordates they are single, dorsal and usually hollow (see **neurulation**).

Nerve ending Structure forming either the sensory or motor end of a peripheral neurone. If the former, may comprise free nerve endings or a **receptor** end organ; if the latter, usually consists of a motor end-plate (see **neuromuscular junction**).

Nerve fibre Axon of a **neurone**, and its **myelin sheath** if present. Diameters vary from 1–20 μm in vertebrates up to 1 mm in **giant fibres** of some invertebrates. Nerve fibres commonly branch towards their termini into small-diameter 'twigs'.

Nerve impulse See **impulse**.

Nerve net Network of neurones, often diffusely distributed through tissues, making up all or most of the nervous system of coelenterates and echinoderms and a large component of peripheral nervous systems of hemichordates. Cells may fuse to form a syncytium or form specialized synapses capable of transmitting in both directions. In some coelenterates, especially motile ones, there is often division of labour between two nerve nets, one (*through-conduction net*) conducting faster and more unidirectionally than the other. Nerve nets characteristically conduct away slowly in all directions from a point of stimulation, temporal **summation** and **facilitation** involving an increasing area of excitation with increased stimulus strength. Found in gut walls of some arthropods, molluscs and vertebrates.

Nerve plexus Diffuse network of neurones and/or ganglia. In vertebrates, *brachial plexi* and *sacral plexi* are associated with limb movements and consist of anastomosing spinal nerves. *Solar plexus* is the collective term for a number of ganglia in the coeliac/anterior mesenteric region connected to the sympathetic nerve chain by the splanchnic nerves.

Nervous integration Process whereby sensory inputs, often from more than one source and modality, either give rise to unified motor responses or are stored under some principle of association. **Synapses** are the basic physical units of integration, providing for **summation** of excitatory and inhibitory potentials. Non-synaptic membranes of the postsynaptic cell may integrate through **adaptation**, **accommodation** or by their refractory periods (see **impulse**). Ganglia, nerve 'nuclei' and brains all depend on synaptic connexions for integration. Reflex arcs are major sites of nervous integration, *relay neurones* often communicating sensory information to the brain where even quite complex behaviour is often reflexly coordinated. Quite simple muscular activity often requires fairly complex nervous integration, in vertebrates employing feedback from **muscle spindles**, after which the **cerebellum** coordinates this information and initiates appropriate motor outputs. In segmented invertebrates this role is often performed by segmental ganglia, largely independently of the brain. All forms of **learning** involve integration of nervous information, as do long- and short-term memory. See **hypothalamus**, **neuroendocrine coordination**.

Nervous system Complement of nervous tissue (**neurones**, **nerves**, **receptors** and **glial cells**) serving to detect, relay and coordinate information about an animal's internal and external environments and to initiate and integrate its effector responses and activities. Present in all animals except sponges, developing from the ectodermal **germ layer**. Characteristic mode of information carriage lies in patterns of nerve **impulses** transmitted along neurones, **neurotransmitters** relaying the impulse pattern at **synapses**. Presence of synapses enables some nervous integration to occur; this is more marked the more advanced the nervous system. Nervous systems may be relatively simple, as in **nerve nets**, but even here there is a tendency towards *through-conduction pathways* enabling rapid transfer of impulses from one region to specific, and often distant, parts. Thus invertebrate and vertebrate **nerve cords** mediate local reflexes, via *segmental* and *spinal nerves* respectively. Further, ganglia often serve as integration centres. The nerve cord(s) and **brain** make up the **central nervous system** (CNS), the *peripheral nervous system* comprising most of the nerves conducting impulses towards and away from it (in vertebrates, these include the spinal and **cranial nerves**). The **autonomic nervous system** forms an additional visceral motor circuit in vertebrates, integrated anatomically and functionally with the CNS.

Compared with **endocrine systems**, nervous systems provide for *reception* of more specific environmental information and, through integration centres simultaneously receptive to inputs from a wide variety of sources (see **synapse**), can elicit responses (generally more complex, preprogrammed and integrated) far more quickly. There are no structures equivalent to sense organs or integration centres in endocrine systems, which rely upon unidirectional transport of solutes via the blood system (see **hormones**). It is the distribution patterns of receptor sites on the membranes of target cells which are instrumental in responses to hormonal messages, diffusely broadcast as they are when compared to the pin-point accuracy of nerve impulses. Nervous systems function via rapid, multi-directional and highly integrated through-conduction pathways. Different endocrine cells use different hormones to signal to target cells; but many nerve cells respond to the same transmitter yet still convey specific information. Hormones are much more diluted than transmitters, so must be effective at lower titres. Not surprisingly, rapid responses to environmental changes are integrated via the nervous system, whereas seasonal and circadian responses often involve the endocrine system. See **nervous integration**, **neuroendocrine coordination**.

Neural arch Arch of bone resting on centrum of each **vertebra**, forming tunnel (neural canal) through which spinal cord runs.

Neural crest Band of embryonic vertebrate ectoderm on both sides of the developing neural tube, giving rise to dorsal root ganglia, **chromaffin cells**, **Schwann cells**, and other cell types indicated in Fig. 44 for **germ layer**. Neural crest cells often attain their final positions after lengthy migrations.

Neural plate Flat expanse of chordate ectodermal tissue, the first-formed embryonic rudiment of the nervous system. Will sink and round up to form neural tube.

Neural spine Spine of bone which may be produced from top of **neural arch**, running up between dorsal muscles of each side and serving for their attachment. Successive spines are usually bound by ligaments.

Neural tube Hollow dorsal tube of chordate embryonic nerve tissue formed by rolling up of neural plate, *neural folds* so produced fusing in the mid-dorsal line, a process termed *neurulation* (see **desmosomes**). The epidermis then fuses above the neural tube. Failure to roll up and fuse gives rise to *spina bifida* and, in its most extreme form, *anencephaly*. The tube expands in front to form the brain and its ventricles, the narrower more posterior part forming the spinal cord; thereby produces the central nervous system and peripheral motor neurones.

Neuroblast Embryonic and presumptive nervous tissue cell.

Neurocranium The part of the skull surrounding brain and inner ear, as distinct from the part composing **jaws** and their attachments, the splanchnocranium.

Neuroendocrine cells See **neurosecretory cells**.

Neuroendocrine coordination Combined and integrated activities of the nervous and endocrine systems; involved in many physiological and behavioural responses to internal and external signals in multicellular animals.

Moulting (**ecdysis**) in some insects is initiated nervously but is mainly hormonally controlled; copulation and ovulation in many female mammals are integrated so that hormonal influences bring about **oestrous** behaviour (involving nervous control); the ensuing cervical stimulation initiates nervous release of gonadotrophic releasing factors (GnRFs), bringing about ovulation and increasing the probability of fertilization. The vertebrate hypothalamus and pituitary illustrate the close anatomical and physiological links between the nervous and endocrine systems. For instance, a sudden fall in air temperature around a mammal induces shivering. This involves sensory input from skin to the hypothalamus, whose output (along vagus nerve) causes adrenaline release from adrenals followed by rapid rhythmic contractions in appropriate body muscles and rise in body temperature. See **neurohaemal organ**, **neurosecretory cells**.

Neuroglia See **glial cell**.

Neurohaemal organ Organ lying outside the nervous system and storing secretion from numbers of **neurosecretory cells**, releasing it into the blood. Particularly widespread in arthropods. Insect **corpora cardiaca**, and *sinus glands* in the eyestalks of some crustaceans are examples.

Neurohumour See **neurotransmitter**.

Neurohypophysis See **pituitary gland**.

Neuromast organ See **lateral line system**.

Neuromuscular junction Area of membrane between a motor neurone and the muscle cell membrane forming a **synapse** between them. The area of muscle membrane under the nerve is the *motor endplate*. Each **skeletal muscle** fibre commonly receives just one terminal branch of a neurone. Each nerve impulse releases a 'jet' of acetylcholine, and resulting small depolarizations of the endplate summate in a graded way to a critical threshold level (about -50 mV internal negativity) at which an action potential is generated and travels along the muscle (see **impulse**, **muscle contraction**).

Neuron (neurone, nerve cell) Major cell type of nervous tissue, specialized for transmission of information in the form of patterns of **impulses**. Nucleus and surrounding cytoplasm comprise the *cell body* (*perikaryon*, or *soma*),

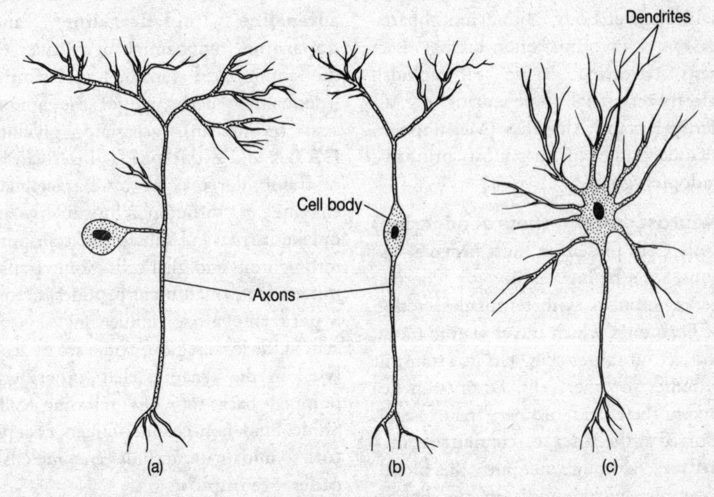

Fig. 87 *Schematic diagrams of three types of* **neuron**. (a) *Monopolar (many motor neurons and interneurons of higher vertebrates and many sensory neurons);* (b) *bipolar neuron (many sensory neurons);* (c) *multipolar neuron (most vertebrate interneurons and motor neurons). Impulses are normally considered as travelling from dendrites to axons, although this has not always been shown.*

which may be the site of multiple synaptic connexions with other nerve fibres. In non-receptor cells numerous projections from the cell body, *dendrites*, provide a large surface area for synaptic connexions with other neurones, while one or more other regions of the cell body (*axon hillocks*) extend into long thin *axons* carrying impulses away from the cell body to other neurones or to effectors, making contacts via **synapses** and usually through secretion of **neurotransmitters**. See Fig. 87.

The membrane of the axon hillock region commonly has the lowest threshold for production of **action potentials** and is often the site of their origin. Most neurones have one axon (*monopolar*); others have two (*bipolar*), or several (*multipolar*). Axons are also termed *nerve fibres*, and may or may not be covered in a **myelin sheath**. Myelinated fibres conduct faster than unmyelinated ones. Unmyelinated fibres, lacking the electrical insulation of myelinated fibres, are normally located in association centres of the nervous system (see **grey matter**).

Neurophysin See **neurosecretory cells**.

Neuropil(e) Mass of interwoven axons and dendrites within which neurone cell bodies of the central nervous system are embedded.

Neuroptera Order of **Endopterygote** insects including alderflies and lacewings. Two similar pairs of membranous wings which, when at rest, are held up over body. Biting mouthparts. Larvae carnivorous, often taking insect pests (lacewing larvae eat aphids). Mostly terrestrial; some aquatic (e.g. alderfly larvae). Alderflies (Megaloptera) include some of the most primitive endopterygotes.

Neurosecretory (neuroendocrine) cells Cells present in most nervous systems, combining ability to conduct nerve impulses with terminal secretion of hormones which travel via the blood and act on target cells. Do not transmit impulses to other cells. Distinction between these and ordinary neurones is blurred as the latter secrete **neurotransmitters**, although not into the blood. Examples include cells in the brains, **corpora cardiaca**, **corpora allata** and **thoracic glands** of insects, and cells in the vertebrate **hypothalamus**. Hormones produced are transported inside the axon, and then usually in the blood, by proteins termed *neurophysins*. See **neurohaemal organ**, **chromaffin cell**.

Neurotransmitters Low molecular mass substances released in minute amounts at interneural, neuromuscular and neuroglandular **synapses**. May be excitatory (depolarizing postsynaptic membrane) or inhibitory (hyperpolarizing postsynaptic membrane). Neuropeptides (e.g. **endorphins**, **enkephalins**, corticotropin releasing factor and **cholecystokinin**) are often derivatives of **polyproteins** and comprise the largest family. One widespread transmitter is **acetylcholine**, but L-glutamate (**glutamic acid**) serves at the majority of mammalian excitatory synapses and at insect and crustacean neuromuscular junctions. Other transmitters include **adrenaline**, **noradrenaline** and **dopamine** (monoamine derivatives of the amino acid tyrosine), **serotonin** (monoamine derivative of the amino acid tryptophan), aspartate, glycine, **GABA** and **nitric oxide**. Inactivation of transmitter may be by extracellular enzyme or through removal by an uptake carrier and subsequent transport into neurons and glial cells. Some transmitters (e.g. small brain peptides) act in a **paracrine** mode, influencing several cells in the local region. Some are hydrolysed in the synaptic cleft; others are pumped back into the releasing cell. Some bind non-channel-linked **receptors** to initiate intracellular enzyme **cascades**. See **impulse**.

Neurula Stage of vertebrate embryogenesis after most gastrulation movements have ceased, manifested externally by presence of **neural plate**. Stage ends when **neural tube** is complete. See **primitive streak**.

Neuter Organism lacking sex organs, but otherwise normal.

Neutralism The thesis that gene frequencies in populations owe more to chance than to **natural selection**. There are several molecular and mathematical studies which suggest that any selective forces bringing about changes in molecular structure (e.g. DNA and protein sequences) are at most exceedingly weak, suggesting that most evolution at this level takes place by random **genetic drift**. This contrasts with the view that selection is the major cause of allelic substitutions. See **molecular clock**.

Neutrophil Type of **leucocyte**. Phagocytic **polymorph**.

Newt See **Urodela**.

Niche (ecological niche) Originally (J. Grinnell, 1914) considered to be the spatial and dietary conditions, biotic and abiotic, within which a particular type of organism is found. It was appreciated these were complex and possibly different for different populations of the same species. A later approach (C. Elton, 1927) was to regard a niche as a role within a community enacted equivalently by different species in different communities and defined for animals largely by feeding habits and size. Both approaches view a niche as an immutable 'place' (in a broad sense) within a community, and neither identifies it with the organism occupying that place. A much more abstract approach (G. E. Hutchinson, 1944) took a niche to be the totality of environmental factors (the n-dimensional hyperspace) acting on a species (or species population). If temporal considerations are included (e.g. nocturnality/diurnality), then the **competitive exclusion principle** may be expressed thus: realized niches do not intersect. This approach defines a niche strictly with respect to (in terms of) its occupant, and with respect to a set of continuous (as opposed to discrete) axes, i.e. axes upon which all other niches are also defined. Modern niche theory is concerned particularly with resource competition between species. See **species**.

Nicking (Of DNA) cutting of a strand of DNA duplex by an endonuclease.

Nicotinamide adenine dinucleotide See **NAD**.

Nicotinamide adenine dinucleotide phosphate See **NADP**.

Nicotine Psychoactive **alkaloid** present in tobacco – cigarettes, snuff, chewing tobacco and some gums. Induces both physiological and psychological dependence (addiction) and, as with all stimulant drugs, depression can follow use. Appears to stimulate some of the receptors to **acetylcholine**, particularly those in ganglia of the **autonomic nervous system**, raising blood pressure, heart rate, release of **adrenaline** and tone and activity of the gastrointestinal tract. Reduces sensory output from muscle spindles, reducing muscle tone (partial cause of feeling relaxed). A general stimulant of the central nervous system, raising behavioural activity; may cause vomiting and nausea. In large doses, induces tremors and convulsions. Induces release of **antidiuretic hormone**, causing fluid retention. Reduces weight gain. With the carbon monoxide present in cigarette smoke, increases atherosclerosis (see **sclerosis**) and thrombosis (clotting) in coronary arteries; but if these are already atherosclerotic, causes cardiac *ischemia* (reduced blood supply) when oxygen supply fails to meet demand brought on by its stimulatory effect on heart muscle, inducing angina or myocardial infarction (heart attack). There is a 5 to 19-fold increased risk of death from coronary heart disease in smokers compared with non-smokers – worse still in diabetic smokers or those with already high blood pressure. Apart from the effects of nicotine in cigarette smoke, smoking is also the major cause of lung cancer in men and women and is a major cause of bladder cancer. Intake of alcohol while smoking exacerbates the risk of mouth cancers. At least 23 of the more than 2000 compounds in cigarette tar are carcinogenic. Nicotine crosses the placenta and the carbon monoxide a mother inhales in cigarette smoke reduces oxygen supply to the foetus. Smoking increases spontaneous abortion and stillbirth and probably results in reduced average body mass of neonates – with potentially irreversible effects on the intellectual and physical abilities of the child. See **caffeine**.

Nicotinic acid (**niacin**) **Vitamin** of the B-complex, lack of which is part of the cause of *pellagra* in man (symptoms being dermatitis and diarrhoea). Yeast, fortified white bread and liver are good sources, but it can also be synthesized in the body from the amino acid tryptophan, of which milk, cheese and eggs are good dietary sources. Contributes structural components to coenzymes **NAD** and **NADP**. Synthesized by many microorganisms.

Nictitating membrane Transparent membranous skinfold (third eyelid) of many amphibians, reptiles, birds and mammals, lying deeper than the other two; often very mobile. Moves rapidly over the cornea independently of the other two, if these move at all, cleaning it and keeping it moist. Often used under water.

Nidation See **implantation**.

Nidicolous Of those birds which hatch in a relatively undeveloped state (naked, blind) and stay in the nest, being tended, for some time after hatching. Compare **nidifugous**.

Nidifugous Of those birds which hatch in a relatively advanced and mobile state and are capable of leaving the nest immediately and of searching for food, often assisted by one (usually) or both parents. Compare **nidicolous**.

Nit Egg of human louse; cemented to hair.

Nitric oxide (NO) Neurotransmitter and muscle relaxant produced by the enzyme NO synthetase (NOS) on deamination of arginine. Binds avidly to haem groups, e.g. haemoglobins. Raises a cell's **cyclic GMP** level. When produced by damaged endothelial cells, promotes vasodilation and oxidizes low-density lipoprotein (see **LDL**, **sclerosis** (2)).

Nitrification Conversion of ammonium ions (NH_4^+) to nitrite and nitrate ions (NO_2^-, NO_3^-) by chemotrophic soil bacteria. The oxidation of ammonia is an energy-yielding reaction, and such energy released in this process is used by the bacteria (*Nitrosomonas*) to reduce carbon dioxide.

$$2NH_3 + 3O_2 \rightarrow 2NO_2^- + 2H^+ + 2H_2O$$

Nitrite is toxic to plants, but rarely accumulates in soil. *Nitrobacter* oxidizes nitrite to form nitrate ions, again with release of energy.

$$2NO_2^- + O_2 \rightarrow 2NO_3^-$$

Because of nitrification, nitrate is the form in which almost all nitrogen is absorbed by plants. Low soil temperatures and pH greatly reduce the rate of nitrification.

Nitrogen cycle Circulation of nitrogen atoms, brought about mainly by living organisms. Inorganic nitrogenous compounds (chiefly nitrates) are absorbed by autotrophic plants from soil or water and synthesized into organic compounds. These autotrophs die and decay or are eaten by animals, and the nitrogen, still in the form of organic compounds (e.g. proteins, nucleic acids), returns to the soil or water via excretion or by death and decay. Ammonifying and nitrifying bacteria then convert them to inorganic compounds (see **ammonification**, **nitrification**). Some nitrogen is lost to the atmosphere as nitrogen gas by **denitrification**. A great deal (about 140–700 $mg.m^{-2}.yr^{-1}$, more in fertile areas) is extracted from the atmosphere by N-fixing bacteria and blue-green algae (see **nitrogen fixation**). Lightning causes oxygen and nitrogen to react, producing oxides of nitrogen which react with water to form nitrate ions, adding on average approx. 35 $mg.m^{-2}.yr^{-1}$ of nitrogen to the soil.

Nitrogen fixation Incorporation of atmospheric nitrogen (N_2) to form nitrogenous organic compounds. The majority (about 10^8 tonnes globally per year) are produced by nitrogen-fixing prokaryotes. Of the various groups of nitrogen-fixers, the symbiotic bacteria are by far the most important in terms of the total nitrogen fixed; however, species of free-living bacteria and blue-green algae (**Cyanobacteria**) can also fix atmospheric nitrogen. The most common nitrogen fixing symbiotic bacterium is *Rhizobium*, which invades roots of leguminous plants (e.g. alfalfa, *Medicago sativa*; clovers, *Trifolium*; peas, *Pisum sativum*; and beans, *Phaseolus*). The beneficial effects upon soil fertility from growing leguminous plants has been recognized for centuries. The bacteria use carbohydrates supplied via the host phloem, providing in turn nitrog enous products to the host.

Blue-green algae are the only photosynthetic prokaryotes producing oxygen, which can readily inactivate the nitrogen-fixing enzyme **nitrogenase**; similar to nitrogenase of the anaerobic bacterium *Clostridium*; dissimilar to the nitrogenase of the aerobic bacterium *Azotobacter*, where the nitrogenase is stable in the presence of oxygen. Blue-green algae with respect to nitrogen fixation can be divided into two groups: (a) filamentous forms producing **heterocysts**, which are the site of nitrogen fixation; lacking photosystem II, they cannot evolve oxygen, yet by cyclic phosphorylation they form the necessary ATP for nitrogen fixation; and (b) non-filamentous blue-green algae that fix atmospheric nitrogen in the dark, when there is no production of nitrogenase-inhibiting oxygen by photosynthesis. Nitrogen-fixing blue-green algae are important in symbiotic relationships between liverworts, ferns and gymnosperms. The relationship between the blue-green alga *Anabaena azollae* and the small aquatic, free-floating fern *Azolla* is important because heavy growths are allowed to develop in rice paddies. As the fern dies after being shaded out by the rice, nitrogen is released for use by the rice plants themselves.

Other nitrogen-fixing symbioses occur between the fungus *Actinomyces* and alders, *Alnus*. See **nitrogen cycle**.

NK cell See **natural killer cell.**

Nociception The reception, conduction and central nervous processing of noxious signals, often producing the subjective sensation of pain. Receptors involved are termed *nociceptors*; the stimuli may arise externally (e.g. heat) or internally (e.g. inflammation). Neural structures involved constitute the nociceptive (nocifensive) system.

Nodal bract Modified leaf-like appendage emanating from a **node** on a stem, cone or fruiting apex.

Node (Bot.) Part of plant stem where one or more leaves arise.

Node of Ranvier Exposed region of axon of **nerve fibre**, where the **myelin sheath** is absent between Schwann cells. Only here does current flow through the membrane during passage of an **impulse** along the myelinated nerve, so that these axons have much faster impulse transmission than non-myelinated axons. The conduction is sometimes referred to as saltatory, since it leaps from one node to the next.

Nominalism Approach to **classification** which denies the reality of categories employed, holding instead that they are artificial constructs employed for convenience or for naturalistic reasons. Compare **essentialism.**

Non-disjunction Chromosomal **mutation** resulting in failure of either a) the two members of a bivalent to separate during the first meiotic anaphase, or b) the two sister chromatids of a chromosome to separate at second meiotic anaphase. Results in aneuploidy: resulting cells have either one too many or one too few chromosomes. Trisomies and monosomies may arise this way. In man, may cause **Down's syndrome.**

Noradrenaline (norepinephrine) Generally excitatory catecholamine **neurotransmitter** of sympathetic nervous system, produced at postganglionic termini; in the brain, from cell bodies located in the brainstem terminating in the cortex, limbic system, hypothalamus and cerebellum. Implicated in maintaining **arousal** in human brain, while dreaming, hunger, thirst, emotion and sexual behaviour may all involve its release. Like **adrenaline**, a tyrosine derivative; also secreted by the adrenal gland – but in smaller quantities.

Northern blot technique A technique following the same principles as the **southern blot technique** except that RNA is separated according to size in a denaturing gel prior to being blotted onto a solid support. The mRNA transcripts can then be detected by hybridization with a **DNA probe**, the intensity of the radioactive signal indicating mRNA abundance. See Fig. 88.

Nostrils See **nares**.

Notochord Rod of vacuolated tissue enclosed by firm sheath and lying along long axis of chordate body, between central nervous system and gut. Present at some stage in all chordates. In most vertebrates, occurs complete only in embryo (or larva) but remnants may persist between the vertebrae, which obliterate it. Found in larval and adult cephalochordates, larval urochordates, and perhaps adult hemichordates. In these latter forms it acts skeletally, antagonizing the myotomes. Usually regarded as mesodermal, and with it forming the *chorda-mesoderm*.

Notogea Australian **zoogeographical region**.

Nucellus Tissue surrounding the

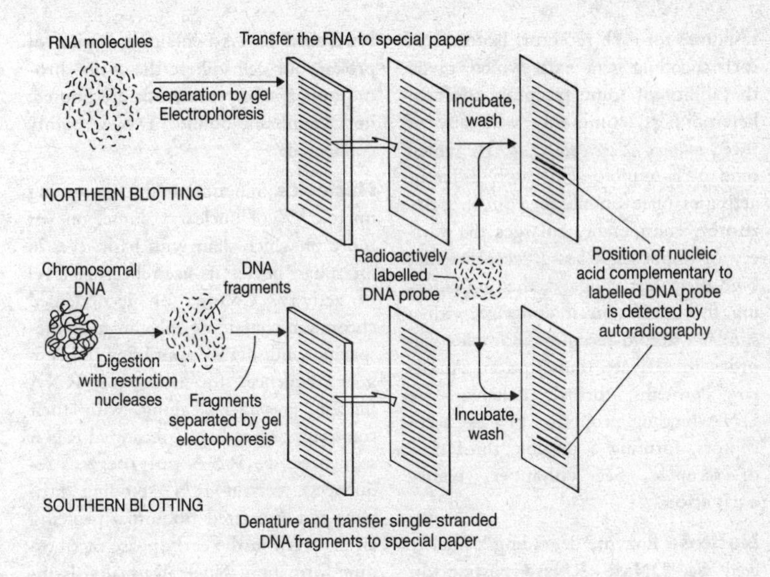

Fig. 88 *A comparison of the techniques of northern and southern blotting.*

megasporangium (ovule) in a seed of a gymnophyte or flowering plant; part of the sporophyte; surrounded by one or two integuments. Generally provides nutrition for megaspore.

Nuclear pore See **nucleus**.

Nuclear proteins (1) Proteins forming part of the nuclear envelope, pore complexes and nuclear lamina. (2) Proteins imported into the nucleus. They bear a short targeting sequence of amino acids (often basic: the nuclear localization signal) required to open the pore complex, ATP-dependently. Such signals, apparently permanently attached, may also be recognized by receptor proteins which carry imported proteins through the pore. These proteins include enzymes, nuclear **RNP**s, transcription factors (e.g. E2F, p53) and their modulators (e.g. retinoblastoma protein, Rb). See **tumour suppressor gene**.

Nuclear receptors DNA-binding proteins (of the *nuclear hormone receptor superfamily*) which, on binding an intracellular ligand, can bind a specific nuclear chromatin region and inhibit/enhance transcription of target genes (i.e. they are **transcription factors**). There are at least 35 members in the superfamily, in at least two classes: thyroid hormone/retinoic acid/vitamin D receptors (class 1) and steroid hormone receptors (class 2). Class 1 receptors can recognize a palindromic DNA sequence comprising an inverted repeat of the sequence AGGTCA, but each also recognizes the direct repeat of AGGTCA separated by either a 3-, 4- or 5-base pair spacer, enabling specific transcriptional responses for each receptor. Isoforms of each encoding gene exist which, given the ability of some receptors to form heterodimers (commonly with any of three *retinoid X receptors*, *RXR*, which bind 9-cis-retinoic acid) with different activities once bound to a given **promoter**, considerably enlarges the variety of responses. Class 2 receptors first bind their steroid ligand, then on entering the nucleus must dimerize with a similar steroid-receptor complex in order to activate transcription. Auxiliary proteins further enhance the DNA-binding properties of class 2 receptors, forming a possible third class of receptor. See **enhancer**, **transactivation**.

Nuclease Enzyme degrading a nucleic acid. See **DNase**, **RNase**, **restriction endonuclease**. Some occur within **lysosomes**.

Nucleic acid A polynucleotide. There are two naturally occurring forms, DNA and RNA, polymers formed by condensation of nucleotides by, respectively, DNA and RNA polymerase. Both forms occur in all living cells, but only one of them in a given virus. DNA (except in RNA viruses) carries the store of molecular information (the genetic material, a major component of **chromosomes**); RNA is involved in deciphering of this information into cell product during transcription and translation. See **gene**, **nucleoprotein**, **protein synthesis**.

Nucleic acid hybridization See **DNA hybridization**.

Nucleohistone See **chromatin**, **chromosome**.

Nucleoid DNA-containing region of prokaryotic cell, where the main chromosome and any **plasmids** are housed; not membrane-bound. DNA highly condensed.

Nucleolus Spherical body, occupying up to 25% of nuclear volume, one or more of which stain with basic dyes in interphase nuclei. Its size reflects its level of activity. Consists of decondensed chromatin containing the chromosome-specific and highly amplified tandem gene sequences for ribosomal RNA (*nuclear organizers*), along with their transcription products (ribosomal RNA sequences; see **RNA polymerases** sequences), certain RNA-binding proteins and associated ribosomal proteins, all involved in the early phase of ribosome formation. Nucleoli usually dwindle and disappear early in nuclear division, but reappear as RNA synthesis recommences in telophase. Much of the nucleolar RNA and protein seem to be carried by chromosomes while the nucleolus itself is disassembled. In the **Euglenophyta**, nucleoli (= endosomes) persist through the mitotic cycle. See **gene amplification**.

Nucleomorph Genome located between two membranes of the **plastid endoplasmic reticulum** in members of the **Cryptophyta** (see Fig. 2 in the **algae** entry). May be surrounded by ribosome-like particles. Appears to be the vestigial nucleus of an algal endosymbiont.

Nucleoprotein Complex of nucleic acid and associated proteins forming a **chromosome**. In eukaryotes the proteins include histones. See **chromatin**.

Nucleosides The base-ribose moieties

of **nucleotides**. Common naturally-occurring forms in RNA are adenosine, guanosine, cytidine and uridine, while *deoxy* forms (deoxyribose instead of ribose) of all but uridine occur in DNA, deoxythymidine replacing uridine.

Nucleoside triphosphate (nucleotide diphosphate) Molecule comprising a **nucleoside** to which three phosphate groups are bound, usually at the 5′ hydroxyl of the pentose. ATP, GTP, CTP and UTP are substrates in RNA synthesis, while the *deoxy* forms dATP, dGTP, dCTP and dTTP are substrates in DNA synthesis (see **DNA replication**).

Nucleosome Fundamental packing unit of a eukaryotic **chromosome**, comprising an octomeric **histone** core and 146 base pairs of DNA wound around it. Transcription of eukaryotic RNA occurs on DNA organized as nucleosomes.

Nucleotide A molecule comprising either a purine or pyrimidine base bonded to either a phosphorylated ribose or (in the case of a deoxynucleotide) deoxyribose moiety. A phosphorylated nucleoside. Nucleotides and deoxynucleotides comprise the monomers of RNA and DNA respectively, in which they are linked by phosphodiester bonds. In *cyclic nucleotides* such as cyclic **AMP**, the phosphate moiety forms a diester bond with the 3′ and 5′ hydroxyls of the ribose. Nucleotides are incorporated into nucleic acids by polymerases which use nucleoside or deoxynucleoside triphosphates as substrates. See **DNA replication**.

Nucleus (1) The organelle in eukaryotic cells making about 10% of the cell volume and containing the cell's **chromosomes**. Together with the **cytoplasm** of the cell it comprises the cell **protoplasm**. A eukaryotic cell may have no nucleus, or one or more; but each such cell ultimately derives from one which was nucleated.

Nuclei vary in size, even growing and diminishing within the same cell, tending to be largest in actively synthetic (e.g. secretory) cells. Each nucleus has an *envelope* of two membranes punctuated by *pore complexes* of 30–100 nm internal diameter (nuclear pores) where the inner and outer membranes are continuous. Pore complexes are involved in active transport of molecules in to and out from the *nucleoplasm* within the envelope, and appear to have an iris-like variable aperture; they are able simultaneously to transport protein into the nucleus and RNA out of it. The inner nuclear membrane has attachment sites for **intermediate filaments** (chiefly *lamins*) constituting the thin *nuclear lamina*, which support the **chromatin** fibres during interphase. Lamins are believed to be the target for kinases whose activity in the presence of cyclin initiates the **cell cycle** (see esp. **maturation promoting factor**). Other intermediate filaments surround the cytoplasmic surface of the outer membrane. Phosphorylation of the lamins at mitotic and meiotic prophase in higher eukaryotes (e.g. not diatoms, dinoflagellates or fungi) is associated with disintegration of the nuclear membrane, while their dephosphorylation at anaphase causes vesicles of nuclear membrane to assemble around chromosomes and gradually fuse to reform the nuclear envelope. The outer membrane of the envelope is usually continuous in places with the

endoplasmic reticulum. Nucleoplasm is regionally differentiated into domains (e.g. **nucleolus, coiled body**) and contains enzymes, **proteasomes, RNPS, high-mobility group proteins** and other proteins. Some of these are involved in transcription of ribosomal RNA from multiple copies of rRNA genes located, often tandemly as *nucleolar organizers*, on particular chromosomes. The nucleus cannot synthesize proteins, but can import them from the cytosol through pore complexes if they have appropriate peptide **sorting signals**.

If properly stained, chromosomes are visible in interphase nuclei as thin grains of chromatin. Interphase nuclei contain relatively decondensed chromatin. This condenses (shortens and thickens by coiling) during nuclear division to give chromosomes visible in light microscopy. Ciliates and some other protozoa contain two nuclear classes: a *macronucleus* concerned with the vegetative 'day-to-day' running of the cell, and one or more *micronuclei* involved in **conjugation**. The macronucleus, which contains much of the genome in multiple and often fragmented copies, divides amitotically (see **aberrant chromosome behaviour**), the micronucleus by mitosis. Mature eukaryotic cells in which the nucleus has disintegrated (e.g. mammalian red blood cells) may rely on 'long-lived' messenger RNA for protein synthesis. Nuclear transplantation has revealed much about cell differentiation, not least that the nucleus generally releases information to the cytoplasm (in the form of mRNA) only when 'instructed' by it to do so, as for example by **growth factors** or by steroid hormones complexed to **nuclear receptors**; and it is speculated that by permitting only mature mRNA to enter the cytoplasm, the nuclear envelope exerts a level of control over which proteins are translated that is unavailable to prokaryotes (see **RNA processing**).

(2) One of the several anatomically distinct aggregations of nerve cell bodies (ganglia) in the vertebrate **brain**. The *lateral geniculate nuclei* within the **thalamus**, for example, contain cell bodies of neurones of the two optic nerves. (3) The central part of an atom of a chemical element, containing at least one proton and usually at least one neutron.

Nude mice Developmentally abnormal mice, lacking thymus (*athymic*). Used in immunological work.

Null hypothesis Hypothesis constructed so as to predict from it a hypothetical set of experimental results (*expected results*) which would obtain under specified experimental conditions were the hypothesis true, and which may be compared with *observed results* derived from an actual experiment operating under those conditions. If there is a significant discrepancy between the two sets of results (see **chi-squared test**), then it is probable that the null hypothesis is incorrect. Testing explicit null hypotheses is reckoned to be less methodologically problematic than corroborating positive hypotheses, since they avoid the problems inherent in all inductive generalizations and are often more precisely stated. Because a null hypothesis is usually expressed largely in ignorance of the factors affecting the results of the experiment it is often stated in a way that suggests that varying the one under investigation makes no difference to those results. See **experiment**.

Nullisomy Abnormal chromosome complement in which both members of a chromosome pair are absent from an otherwise diploid nucleus. Type of **aneuploidy**; viable only in a polyploid (e.g. wheat, which is hexaploid).

Null mutation A **mutation** eliminating the activity of a wild-type **gene**. See **gene targeting**.

Numerical phyletics Erection of an inferred phylogeny of living organisms through numerical analysis of characteristics. See **phyletics**.

Numerical taxonomy Erection of classifications by numerical analysis of characteristics. See **classification**.

Nurse cells (Bot.) In some liverworts, sterile cells among the spores lacking any special wall thickening and frequently disintegrating before spores mature. (Zool.) For Sertoli cells, see **testis**.

Nut Dry indehiscent, single-seeded fruit, somewhat similar to an achene but the product of more than one carpel and usually larger, with a hard woody wall; e.g. hazel nut. A small nut is called a nutlet.

Nutation (circumnutation) Spiral course pursued by apex of plant organ during growth due to continuous change in position of its most rapidly growing region; most pronounced in stems, but also occurs in tendrils, roots, flower stalks, and sporangiophores of some fungi.

Nutrition Any process whereby an organism obtains from its environment the energy and atoms required for maintenance, growth, reproduction, etc. Either **autotrophic** or **heterotrophic**.

Nyctinasty Response of plants to periodic alternation of day and night; e.g. opening and closing of many flowers, 'sleep movements' of leaves. Related to changes in temperature and light intensity. See **photonasty**, **thermonasty.**

Nymph Larval stage of exopterygote insect. Resembles adult (e.g. in mouthparts, compound eyes), but is sexually immature. Wings absent or undeveloped.

Obligate Term indicating some type of restriction in an organism's way of life, from which it cannot depart and survive. Used particularly in the contexts of parasites, aerobes and anaerobes. Contrast **facultative**.

Occipital condyle Bony knob at back of skull, articulating with first vertebra (the atlas). Absent in most fish, whose skulls do not articulate with vertebral column. Single in reptiles and birds, double in amphibians and mammals.

Occiput (occipital region) (1) In vertebrates, an arbitrarily delimited region of the skull and head in the neighbourhood of the occipital condyle. (2) In insects, a plate of exoskeleton forming back of the head.

Oceanic Inhabiting the sea where it is deeper than 200 metres. Compare **neritic.**

Ocellus General term for several types of *simple eye*, as found in some coelenterates, flatworms, annelids, insects (the only eyes of larval endopterygotans), arachnids and other arthropods. Usually incapable of image-formation. See **eye**.

Ochre mutation A 'stop' mutation; UAA triplet of messenger RNA. The ochre codon forms the normal terminus of the coding regions of many codon. See **genetic codon**, **amber mutation.**

Octamer-binding proteins Proteins which bind a highly conserved 8-base pair sequence (octamer) found in every Ig promoter so far studied and within the IgH enhancer. Some at least of these proteins contain a **homeobox** motif similar to the homeodomain common to *Drosophila* regulatory proteins encoded by **segmentation genes**. Also present in adenovirus origin of replication.

Octopamine One of the monoamine family of **neurotransmitters**.

Oculomotor nerve Third **cranial nerve** of vertebrates. Supplies four of the extrinsic eye muscles and, by neurones of parasympathetic system, via the ciliary ganglion, intrinsic eye muscles of **accommodation** and pupil constriction. A ventral root.

Odonata Dragonflies and damselflies. Order of exopterygote insects with aquatic nymphs. Carnivorous as nymphs and adults, with biting mouthparts, hinged on an extensible 'mask' in the nymph. Large, compound eyes; two pairs of similar wings, folded over the back in damselflies (Zygoptera) and horizontally in dragonflies (Anisoptera). Some fossil (Carboniferous) forms had wingspans in excess of half a metre.

Odontoblasts Cells lying in pulp cavities of vertebrate teeth, sending processes into adjacent dentine, which they help form.

Odontoid process See **atlas.**

Oedema Swelling of tissue through increase of its **tissue fluid** volume. Can occur during **inflammation**, through blockage of vessels of **lymphatic system** returning tissue fluid to venous system (see **filarial worm**) and through low levels of blood proteins causing poor return of tissue fluid through **capillaries**, as in protein malnutrition (kwashiorkor).

Oesophagus Anterior part of gut, between pharynx and stomach. Usually concerned simply in peristaltic passage of food to stomach. May contain a **crop.**

Oestradiol, β-oestradiol Most potent **oestrogen** in female mammals.

Oestrogens Group of **steroids**, both naturally-produced ovarian hormones and artificial homologues of them. Synergists of **growth hormone**. High blood levels inhibit release of gonadotrophic releasing factors (GnRFs) from the **hypothalamus**. The most potent natural oestrogen in humans is β-oestradiol; two others are oestrone and oestriol. Produced by theca cells of developing **Graafian follicles** of the ovary and, after ovulation, by granulosa cells; adrenal cortices also produce small amounts.

Main roles of oestrogens in female mammals are: development of **mammary glands** (synergistically with **progesterone**), regulation of fat deposition, softening and smoothing of skin texture (in humans), regulation of uterine endometrium, again with progesterone (see **menstrual cycle**), promotion of flattening and widening of **pelvic girdle** and **growth** spurt at puberty (all *secondary sexual characteristics*). Effects on sexual behaviour depend on the species involved, but may cause 'maternal behaviour' if injected into males.

Oestrogens enter target cell nuclei in combination with specific cytoplasmic receptors, annealing to specific chromosome regions and initiating mRNA transcription (see **gene expression**). They are major components of the **contraceptive pill**, and some synthetic forms (e.g. stilboestrol) have been shown to be carcinogenic. See **nuclear receptors**.

Oestrous cycle Reproductive cycle of short duration (usually 5–60 days) occurring in sexually mature females of many non-human mammals, in the absence of pregnancy. Regulated by endocrine (often neuroendocrine) interactions involving the **hypothalamus**, pituitary and ovaries. The phases in the cycle include: *oestrus*, *metoestrus*, *dioestrus* and *pro-oestrus*. Compare human **menstrual cycle.**

Oestrus Follicular phase of the **oestrous cycle** during which sexual desire and attractions of the female may be heightened, leading to copulation (see **neuroendocrine coordination**); less so in some primates. In cats and dogs, periods of oestrus (when females are 'on heat') occur two or three times per year, separated by long periods of *anoestrus*.

Olecranon process Bony process on mammalian ulna, extending below elbow joint and employed in attachment of muscles straightening limb.

Olfaction Detection of odours; sense of smell. Appears to involve an **inositol 1,4,5 triphosphate** pathway. See **receptor**, **pheromones**.

Olfactory nerve First **cranial nerve** of vertebrates, running from olfactory area of cerebral cortex to olfactory bulb (see **olfactory organs**).

Olfactory organs Organs of smell. In vertebrates, consist of sensory epithelia in nasal cavities, whose cells respond to molecules dissolved in their moist mucous membranes. *Olfactory cells* are bipolar neurones whose dendrites synapse with branches of the olfactory nerve in a mass of grey matter termed an *olfactory bulb* within the braincase. The olfactory nerve travels to olfactory regions of cerebral cortex.

Oligo- Prefix indicating 'few'. Thus oligosaccharides, oligopeptides, oligonucleotides, referring to relatively short-chain polymers (i.e. each contains relatively few, perhaps up to twenty or so, condensed monomers).

Oligocene Geological epoch, sudivision of the **Tertiary** period lasting from 38–26 Myr BP, during which the Alps and Himalayas rose, South America separated from Antarctica, and volcanoes erupted in the Rocky Mountains. Browsing mammals and monkey-like primates existed. Many modern genera of plants evolved.

Oligochaeta Class of annelid worms which includes earthworms. See **Annelida.**

Oligonucleotides Laboratory-prepared oligonucleotides ('oligos'; short sequences of nucleic acid) are proving useful in clinical trials to treat viral infections, cancer, Alzheimer's disease and **malaria**. Oligos bind complementary nucleic acid sequences and form double-stranded complexes when designed to target mRNA, or triple-stranded complexes when targeting duplex DNA. Either way, oligos can very selectively close down expression of a specific gene. One antisense oligo under trial is designed to prevent expression of *gag* protein, crucial for HIV replication (see **HIV**). Another targets the mRNA encoding the enzyme dihydrofolate reductase-thymidylate synthase (DHFR-TS) specific to the malarial parasite *Plasmodium falciparum* (i.e. slightly different from the human form of the enzyme) required for its pyrimidine synthesis. Such treatment might sidestep the parasite's global resistance to the drug pyrimethamine, which also blocks this enzyme. Besides their probable clinical value, oligos will likely have great application as probes (e.g. see **DNA probe**) of important cellular processes.

Oligopeptide See **peptide.**

Oligosaccharide Carbohydrate formed by joining together monosaccharide units (4–20) in a chain by links termed *glycosidic bonds*, formed enzymatically. Are intermediate digestion products of **polysaccharides**, and some form side chains of **glycoproteins** and **glycolipids**. See **glycosylation**.

Oligotrophic (Of lakes and rivers) originally introduced to describe the phytoplankton assemblages characteristic of montane lakes, which were believed to be deficient in combined nitrogen and phosphorus. Nowadays used in a more general sense to describe lakes and rivers that are unproductive in terms of organic matter formed; nutrient status of the water and nutrient supply are very low. Compare **eutrophic**, **mesotrophic**, **dystrophic**.

Ommatidium See **eye**.

Omnivore Animal eating both plant and animal material in its diet. See **carnivore**, **herbivore**.

Onchoceriasis 'River blindness'. See **filarial worms.**

Onchosphere (hexacanth) Six-hooked embryo of tapeworms (**Cestoda**). Develops from egg, usually while still in proglottis. Will bore through gut wall of secondary host and is then carried in blood to host tissues, in which it lodges and develops into the **cysticercus**.

Oncogene A gene (or gene locus) in which mutation, loss of function or other reduction of expression induces neoplasia (transformation of a cell into a **cancer cell**). Oncogenic viral genes (v-*onc*) generally occur in retroviruses; a cellular oncogene (c-*onc*) is sometimes termed a proto-oncogene (proto-*onc*). They are generally referred to by three italicized letters indicating the retrovirus in which they were first identified, most v-*oncs* starting in a host cell genome and getting into the viral genome by a transduction-like event after which they can exert their effect in a host cell which becomes infected by the retrovirus. A v-*onc* sequence is normally very like its c-*onc* homologue but often lacks the latter's introns, suggesting its origin by reverse transcription from cellular mRNA. Most oncogenes (e.g. c-*abl*) encode **protein kinases**, involved in the regulation of the **cell cycle**, and may exert their effects in mitotic cells, meiotic cells or both; others direct the transition out of the cell cycle and into the differentiation pathway.

Most retroviral oncogenes originate as normal cellular (host) genes (proto-oncogenes or c-*oncs*) that are picked up by the retrovirus by a transduction-like process and subsequently undergo mutation (e.g. for c-*myc*, see **programmed cell death**). Some oncogenes may be expressed when **transposable elements** insert adjacent to them in the genome. See **suppressor mutation**, **oncoproteins**.

Oncogenic virus See **oncogene**, **virus**.

Oncoproteins Proteins, such as **tyrosine kinases** and **Ras proteins**, encoded by **oncogenes**.

One gene–one enzyme hypothesis See **gene**.

Ontogeny The whole course of an individual's development, and life history. Compare **phylogeny**. See **recapitulation**.

Onychophora Small group of animals (e.g. *Peripatus*) with annelid and arthropod affinities, sometimes regarded as a distinct phylum, but here placed as a class within the **Arthropoda**. Represent an early stage of *arthropodization*. Body segmented with soft, unjointed cuticle $1\mu m$ thick and permeable to water; body wall with longitudinal and circular smooth muscles; head not demarcated, bearing a pair of long, mobile *pre-antennae*, simple jaws and oral papillae. Limbs fleshy, unsegmented, with terminal claws; probably evolved from parapodial appendages. The heart has ostia. Coelom replaced in adult by haemocoele. Gaseous exchange by tracheae and spiracles. Excretory system by segmentally repeated *coxal glands*, similar to those of arachnids and crustaceans but with ciliated excretory ducts. The pair of ventral nerve cords lack segmental ganglia but have numerous connectives. Eyes simple. Forest-dwelling, generally nocturnal predators. About 70 species. Some aquatic Cambrian fossils, such as *Aysheaia* and *Opabinia*, show affinities. See **Tardigrada**.

Oocyte Cell undergoing **meiosis** during **oogenesis**. *Primary oocytes* undergo the first meiotic division, *secondary oocytes* undergo the second meiotic division. See **maturation of germ cells**.

Oogamy Form of sexual reproduction involving production of large non-motile gametes, which are fertilized by smaller motile gametes. Extreme form of **anisogamy**, occurring in all metazoans and some plants.

Oogenesis Production of ova, involving usually both meiosis and maturation. See **maturation of germ cells** and Fig. 78a.

Oogonium (1) (Bot.) Female sex organ of certain algae and fungi, containing one or several eggs (**oospheres**). (2) (Zool.) Cell in an animal ovary which undergoes repeated mitosis, giving rise to oocytes. Compare **spermatogonium**.

Oomycota (oomycetes) Water moulds and related organisms (Kingdom **Protista**) producing motile biflagellate cells possessing smooth and hairy flagella and including both saprotrophs and parasites. The cell wall consists largely of cellulose or cellulose-like polymers, thus markedly different from cell/hyphal walls of **fungi**. Range from unicells to highly branched, coenocytic filaments. Most reproduce asexually via zoospores and sexually via oogamy. The oogonium contains many eggs (**oospheres**), and the antheridium contains many male nuclei. The zygote is thick-walled (an **oospore**). Of the water moulds, the genus *Saprolegnia* includes both saprotrophic and parasitic species, causing disease of fish and fish eggs. Of the primarily terrestrial group, some genera are important plant pathogens, probably the best-known genus being *Phytophthora* ('plant-destroyer'), with species causing widespread destruction of many crops including pineapples, tomatoes, strawberries, onions, apples, tobacco and citrus fruits. The best-known is *P. infestans*, the cause of late blight of potatoes, which caused the great famine in Ireland in 1846–7.

Oosphere (Bot.) Large, naked, spherical non-motile macrogamete (egg); formed within an **oogonium**.

Oospore Thick-walled resting spore formed from a fertilized **oosphere**.

Opal mutation A 'stop' mutation. A UGA mRNA triplet. See **amber mutation**.

Open bundle Vascular bundle in which a vascular cambium develops.

Operator See **Jacob-Monod theory**.

Operculum (Bot.) Lid or cover; in certain fungi, part of cell wall; in mosses, a multicellular apical lid that opens the sporangium. (Zool.) (1) Cover of gill slits of holocephalan and osteichthyan fishes, and of larval amphibia. Contains skeletal support. (2) Horny or calcareous plate borne on back of foot in many gastropod molluscs, brought in over the body when animal withdraws into its shell. (3) A second **ear ossicle** of many urodele and anuran amphibians; flat plate fitting into oval window of inner ear. May replace the stapes.

Operon See **Jacob-Monod theory**.

Ophidia Snakes. Suborder of the Order **Squamata** (sometimes a separate order). Limbless reptiles, with exceptionally wide jaw gape due to mobility of bones. Eyelids immovable, nictitating membrane fused over cornea; no ear drums.

Ophiuroidea Brittlestars. Class of **Echinodermata**; star-shaped, with long, sinuous ambulatory arms radiating from clearly delineated central disc; mouth downwards; well-developed skeleton of articulating plates; tube feet without suckers; no pedicillariae; easily break up by autotomy; scavengers.

Ophthalmic Alternative adjective to optic.

Opiates See **endorphins**, **enkephalins**.

Opisthobranchia Subclass of gastropod molluscs, in which shell and mantle cavity are reduced or lost. Body bilateral and slug-like. Sea hares (e.g. *Aplysia*), other nudibranchs, and the actively swimming pteropods.

Opisthosoma One of the **tagmata** of chelicerate arthropods (Merostomata and Arachnids). Comprises the trunk segments, devoid of walking limbs; often (e.g. in scorpions) separable into an anterior *mesosoma* and posterior *metasoma*.

Opsins Proteins belonging to the superfamily of **G-protein**-coupled **receptors** which, bound to 11-*cis*-retinal, act as vertebrate visual pigments (see **rhodopsin**, **rod cell**, **cone**).

Opsonins Proteins coating foreign particles, such as surfaces of pathogenic microorganisms, rendering them more susceptible to ingestion by phagocytic leucocytes (*opsonization*). Many are **antibodies**.

Optic chiasma Structure formed beneath vertebrate forebrain by those nerve fibres of the right optic nerve crossing to left side of brain and those of left side crossing to the right side. In most vertebrates all the fibres cross; in mammals about 50% on each side do so. See **binocular vision**, **decussation**.

Optic nerve Second **cranial nerve** of vertebrates; really part of brain wall. See **retina**.

Optimization theory (optimality theory) Very generally the theory that, through natural selection, the behaviours of organisms are such that they tend to the most cost-effective use of time and available resources, given environmental circumstances. Despite charges of being vacuous and truistical, and although similar statements may be deducible from some formulations of Darwinian theory, their heuristic value has been great, prompting detailed and testing studies which have greatly enhanced understanding of adaptations involved in such diverse fields as foraging behaviour, diet choice, habitat selection and competition in animals, and reproductive behaviour more generally. See **game theory**.

Orbit Cavity or depression in vertebrate skull, housing eyeball.

Order Taxonomic category; inclusive of one or more families, but itself included by the class of which it is a member. See **classification**.

Ordovician Geological **period**, second of the **Palaeozoic** era, lasting approximately from 500–440 Myr BP. Period starts with the first major extinction event. Molluscs diversified while plants possibly invaded the land. The first fungi appeared. Climate was mild; seas were shallow and continents flat.

Organ Functional and anatomical unit of most multicellular organisms, consisting of at least two tissue types (often several) integrated in such a way as to perform one or more recognizable functions in the organism. Examples in plants are roots, stems and leaves; in animals, liver, kidney and skin. Organs may be integrated into functional **systems**. Sometimes it is debatable where the limits of organs are; thus the stomach is often regarded as an organ, but on occasions so is the whole alimentary canal.

Organ culture By partial immersion in nutrient fluid, the growth and maintenance *in vitro* of an organ after removal from the body. The organ must be small (usually embryonic) since nutrients must diffuse into it from outside. See **tissue culture**.

Organ of Corti See **cochlea**.

Organelle Structural and functional part of a cell, distinguished from the cytosol. Often membrane-bound (e.g. nucleus, mitochondria, chloroplasts, endoplasmic reticulum); sometimes not (e.g. ribosomes). The plasma membrane is itself an organelle. Those in cytoplasm are termed *cytoplasmic organelles*, in contrast to the nucleus. See Fig. 14 and Table 4.

Organizer Any part of an embryo which can induce another part to differentiate. Classic example is dorsal lip of the **blastopore**. See **evocation**, **induction**, **positional information**.

Organotrophic See **heterotrophic**.

Oriental Zoogeographical **region** comprising India and Indo-China south to **Wallace's Line**.

Origin of life Major hurdles to the origin of life-forms resembling even the simplest known cells are the production of a) a genetic system which is b) sufficiently discrete from other such systems to enable its collection of properties to be a unit in reproductive competition with them. These two requirements would seem to put a premium on simple genetic (information) systems engaging in, at least initially, autocatalysis, with subsequent expansion of the stored information to code for molecules which do not themselves store information but instead form part of the structure of the system.

Even the simplest living systems today are far too complicated to resemble at all closely the earliest forms of life. Three major features of present living systems (construction from complex, often polymeric, organic compounds; catalysis by proteins; storage of molecular information in the form of nucleic acid) are likely to be the end results of evolution by natural selection between systems which were originally far simpler and lacked all these features. The following account is highly conjectural, but raises some of the issues being debated.

For long it was thought that the early Earth's atmosphere consisted of reducing molecules (ammonia, methane, nitrogen and water vapour), and attempts to create likely precursor molecules of life (amino acids, bases) by passing electrical sparks through such gaseous mixtures were rewarding. However it is probable that four thousand million years ago the sun was weaker in terms of total thermal output than it is now, and more active in terms of ultraviolet output. Implications for the Earth's early atmosphere and the origin of life are that: a) without an atmospheric **greenhouse effect**, the surface temperature of the Earth would have been too low to avoid ocean freezing, and that b) incident ultraviolet radiation on the Earth's atmosphere would have resulted in photolysis of methane and ammonia, releasing free hydrogen out of the atmosphere, with probable retention of carbon dioxide and atomic and molecular nitrogen. The carbon dioxide could have provided the greenhouse effect: the early oceans were not frozen.

One of the precursors of adenine (a base present in the nucleic acid RNA) may well have been hydrogen cyanide (HCN). RNA has become a lively contender in the 'origin' debate on account of its catalytic ability (see **ribozyme**). It is now clear that, in principle at least, RNA is capable of self-catalysis (*autocatalysis*). HCN may have been formed from methane present as a trace gas in the early mantle. Hydrothermal vents (undersea hot springs) provide reducing compounds which, by permeating rock, could form complex molecules, cooled sufficiently by the water to reduce the likelihood of their dissociation. These, in association with the self-assembling crystalline organization provided by such clays as *kaolinites* and *illites*, could come to form polymers supported by the lattice of the crystal.

The first membranes might have arisen by self-assembly, forming spherical vesicles in much the same way that some phospholipids do when mixed with water. Such lipid vesicles might then have accumulated any organic molecules soluble in their membranes and if these polymerized inside they might not have been able to escape. Continued polymerization within might have put the vesicles under strain, promoting their enlargement by incorporation of further membrane components, as can happen with artificial **liposomes**. Rupture of the spheres followed by renewed growth of the rupture products would have achieved a simple form of reproduction.

In the past, RNA may have had wider catalytic ability than it has now (see **telomere**). If RNA were among early polymers, formed perhaps on clay crystals trapped within membrane spheres, and were it to catalyse amino acid polymerization, then protein synthesis might gain independence from clay crystals, RNA also supporting the protein. If some of these proteins were to associate with the membrane in such a way as to stabilize it and even promote entry of building materials, selection could have favoured those systems which in replicating retained catalytic production of the membrane proteins by RNA; that is, which achieved the reproducible link between genotype and phenotype. Something like a simple cell ('RNA life') would have been formed, and interestingly a motif that binds ATP in solution has been located on some cellular RNAs. DNA could eventually have been synthesized off the RNA (retroviruses do this today), the resulting duplex providing a more stable store of information.

Ornithine cycle (urea cycle) See **urea**.

Ornithiscia Extinct order of **Archosaurs** ('ruling reptiles'), lasting throughout Jurassic to late Cretaceous. 'Bird-hipped' and herbivorous dinosaurs. Some early forms were bipedal. Both the pubes and ischia of the **pelvic girdle** pointed downwards and backwards. Included stegosaurs (e.g. *Stegosaurus*), ankylosaurs and ceratopsids (e.g. *Triceratops*). Compare **Saurischia**.

Ornithology Study of birds.

Ornithophily Pollination by birds.

Orthogenesis Theory, prevalent in early decades of 20th century, that the evolutionary path of a lineage can acquire an 'impetus', or 'inexorable trend' carrying it in a direction independently of selective constraints imposed by the environment. It was once suggested that the genetic material itself might somehow be responsible. Now discredited.

Orthognathous 'Short-faced' mammals, lacking muzzles. Hominid evolution has been accompanied by the evolution of increasingly orthognathous faces from **prognathous** ones.

Orthoptera Large order of exopterygote insects, including locusts, crickets and grasshoppers. Medium or large insects, usually with two pairs of wings (sometimes absent), front pair narrow and hardened, hind pair membranous; hind legs usually large and modified for jumping. Stridulatory organs; usually involving wings being rubbed together (*alary*), or inner face of hind femur being rubbed against a hardened vein on forewing (*femoro-alary*). Paired auditory organs (*tympana*) either on anterior abdomen or tibias of front legs.

Orthoselection Directional selection. See **natural selection**.

Osculum Large opening through which water leaves the body of a sponge (**porifera**), having entered through *ostia*.

***oskar* genes** (*osk*.) A group of **maternal effect** genes in *Drosophila*, whose mRNA transcription products become localized in the posterior pole of the developing oocyte. Embryos from females mutant for any of these genes develop normal heads and thoraxes, but lack abdomens. The *osk* protein product activates transcription of the *nanos* gene, whose product in turn activates *knirps* transcription and represses *hunchback* translation (an example of **pleiotropy**). See ***bicoid* gene**, **eicosanoids**.

Osmium tetroxide Substance (OsO_4) which, in aqueous solution, blackens fat and is often used to demonstrate myelin sheaths of neurones. Used as a fixative in light and electron microscopy. Sometimes called osmic acid.

Osmophilic body Lipid-containing granules, appearing dark when treated with osmium fixation.

Osmoregulation Any mechanism in animals regulating a) the concentration of solutes within its cells or body fluids, and/or b) the total volume of water within its body.

Each major inhabited environment, freshwater, marine and terrestrial, poses its own osmotic problems for organisms. Cells approximate to sea water in **water potential**, but have lower water potentials than freshwater and far higher water potentials than air at normal humidities. Freshwater protozoans and sponges use **contractile vacuoles** to expel osmotic water from their cells; the **sodium pumps** of most animal cell membranes are important in limiting the cell's internal ion concentration. Protection from osmotic uptake of water and from desiccation is achieved by a body surface covered with an impermeable **cuticle** or **skin**, but there are usually soft parts (e.g. gut, gills) through which water enters. Adaptive behaviour may reduce osmotic dangers; thus limpets adhere closely to rocks at low tide; marine mussels and periwinkles retract into shells. *Euryhaline* animals tolerate wide fluctuations in water potentials of their surroundings, possibly through a limited ability to swell in hypotonic conditions and by a reduction in intracellular levels of some organic compounds, or inorganic ions such as sodium and chloride. Some engage in active ion uptake (e.g. of sodium) if the external medium becomes dilute enough to cause sodium loss by diffusion. Unless urine can be made hypotonic to body fluids, inorganic ions will be lost by this route, and a kidney which reabsorbs ions from the excretory fluid will reduce this loss. **Anadromous** animals, such as many cyclostomes and some fish, automatically take up water in freshwater but produce a copious hypotonic urine, active ion uptake by the gills replacing urinary loss. On return to marine conditions they lose water to the sea and correct this by drinking sea water: sodium, potassium and chloride are absorbed by the intestine and excreted across the gills (see **sodium pump**).

Restriction of water loss on land is associated with evolutionary change from excretion of ammonia (ammonotely) to urea (ureotely) and uric acid (uricotely). Amphibian metamorphosis includes a change from ammonotely to ureotely; land reptiles produce a urine hypotonic to body fluids, but because of uricotely (uric acid being insoluble requires only incidental water in its removal) the total volume produced is not great. Evolution of the **amniotic egg** would have been less likely without the uricotely of reptilian and bird embryos. Marine birds and reptiles do not regularly drink sea water, but they absorb much salt from their food and excrete it through *salt glands* above the orbits of the eyes.

Vertebrate **kidneys**, especially mammalian, can produce a urine which is variably copious and hypotonic to body fluids or sparse and hypertonic, depending upon the water balance of the body. The hypothalamus monitors blood concentration, and any increase above the homeostatic norm results in release of **antidiuretic hormone** by the posterior **pituitary**. This causes uptake of water by the collecting ducts and reduced volume of hypertonic urine. Moreover, the hypothalamic *thirst centre* now initiates drinking, restoring blood concentration. Sudden loss of blood volume (or pressure) causes *aldosterone* release from the **adrenal** cortex (see **angiotensins**). This increases potassium excretion and sodium retention by the kidney, tending to make extracellular (tissue) fluid more concentrated than body cell fluid, lowering its water potential and drawing water out from the cells to help restore blood volume. After a blood meal, the bug *Rhodnius* disposes of excess fluid by release of a *diuretic hormone* from its thoracic ganglia. This accelerates active secretion of sodium ions from the gut, chloride and water following them into the haemocoele, where they are dealt with by **Malpighian tubules**. Osmoregulation may therefore involve structural, physiological and behavioural adaptations.

Osmosis The net diffusion of water across a selectively permeable membrane (permeable in both directions to water, but varyingly permeable to solutes) from one solution into another of lower **water potential**. The *osmotic pressure* of a solution is the pressure which must be exerted upon it to prevent passage of distilled water into it across a semipermeable membrane (one impermeable to all solutes, but freely permeable to solvent), and is usually measured in pascals, Pa (1 Pa = 1 Newton/m^2). A solution's *osmotic potential* is always negative, the more so the greater the concentration of solute particles. The plasmalemma is selectively (i.e. not semi-) permeable and permits selective passage of solutes (see **cell membranes**), as does the *tonoplast* in plant cells.

The osmotic pressure of a solution depends upon the ratio between the number of solute and solvent particles present in a given volume: 1 mole of an undissociated (non-electrolytic) substance dissolved in 1 dm^3 of water at 0 °C has an osmotic pressure of 2.26 MPa (megapascals). But a solution of 0.01 mole of sodium chloride per dm^3 of water has almost twice the osmotic pressure of a solution of 0.01 mole of glucose per dm^3 of water. This is because sodium chloride in water dissociates almost completely into Na^+ and Cl^- ions (giving twice the particle number) whereas glucose does not dissociate in water. Solutions of equal osmotic pressure are *isotonic*; one with a lower osmotic pressure than another is *hypotonic* to it, the latter being *hypertonic* to the former.

When cells produce polymers from component monomers they may dramatically reduce the osmotic pressure of their cytoplasm; hence polymers (e.g. starch, glycogen) make good storage compounds because they are *osmotically inactive*. Glucose is far more *osmotically active* than is starch.

Osmosis is a physical process of great importance to all organisms, affecting relations with their environments as well as between their component cells. If an animal cell takes up water osmotically its plasma membrane will eventually rupture (*osmotic shock*); but the **sodium pump** reduces this problem under normal physiological conditions. Cells with cell walls are normally prevented from osmotic rupture. Those which lose water osmotically shrink and become *plasmolysed* (see **plasmolysis**). Water relations of cells and organisms are generally discussed in terms of water potential. See **turgor**, **diffusion**, **osmoregulation**.

Osmotic activity, osmotic pressure and osmotic shock See **osmosis**.

Ossicle A very small bone, or calcified nodule.

Ossification Formation of bone. May occur by replacement of **cartilage** (*endochondral ossification*) or by differentiation

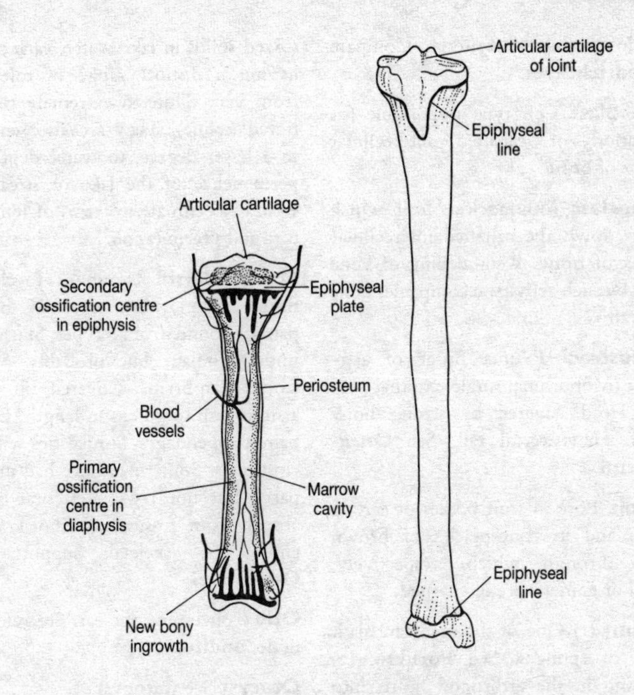

Fig. 89 *Final stages in ossification of a human tibia. The epiphyseal line represents the bony remains of the once actively dividing cartilaginous epiphyseal plate (growth plate).*

of non-skeletal mesenchyme (*intramembranous ossification*), forming **dermal bones**. In the former, cartilage first calcifies and hardens to form *ossification centres*. The cartilage cells die and those in the periochondrium develop into *osteoblasts* and lay down a ring of bone to form the *periosteum*, which enlarges as blood vessels invade and osteoblasts enter vacant lacunae to develop into *osteocytes*. The bone lengthens and thickens, finally becoming mineralized. Cartilage remains as *articular cartilage* forming **joints** at the bone ends, and in the *growth plate* in the bone **epiphysis**. See **diaphysis, growth**.

Osteichthyes Bony fishes. Largest vertebrate class. Some fossils come from the Upper Silurian but a considerable radiation had occurred by the Lower to Middle **Devonian**. Includes subclasses **Choanichthyes** (see **Dipnoi** and **Crossopterygii**) and **Actinopterygii** (which itself includes the **Teleostei**). Other fish groups (e.g. agnathans, placoderms, acanthodians) contain bone; but the osteichthyans are the only jawed fishes with bony vertebrae and gill arches and with paired fins, but lacking bony plates on head and body. **Acanthodii** are sometimes regarded as a subclass of the Osteichthyes, and do not really fall outside the above criteria. Compare **Chondrichthyes**.

Osteoblast Cell type responsible for formation of calcified intercellular matrix of **bone**.

Osteoclast Multinucleate cell which breaks down the calcified intercellular matrix of **bone**. Remodelling of bone shape by such activity accompanies bone **growth**.

Osteostraci Extinct order of **Agnatha** (Monorhina: single external nostril). Head covered by strong bony shield. Heterocercal tail. See **Ostracodermi**.

Ostiole Pore in fruit bodies of certain fungi, and in conceptacles of brown algae through which, respectively, spores or gametes are discharged.

Ostium (1) One of the many inhalant pores of sponges. See **Porifera**. (2) Opening in the arthropod heart, into which blood flows from the haemocoele.

Ostracoda Subclass of **Crustacea**. Most are a few mm long. Aquatic. **Carapace** bivalved, closable by adductor muscles, and completely covering body. **Filter feeding**. Trunk limbs few and small; food gathered by head appendages. About 2000 species. Shells of ostracods are calcified, and become preserved in lake sediments, particularly under calcareous or saline conditions (where the action by carbonic acid is minimal). Therefore, they are most abundant in marl-sediments; rare in the typical organic lake sediment. Useful **indicator** organisms and used in palaeolimnological studies because they are particularly sensitive to the concentration of dissolved solids in lake water, each species having a distinct range of tolerance, from very dilute to extremely concentrated (saline). They are also sensitive, to a lesser degree, to water depth and permanence of the lake or stream, as well as to climate in terms of temperature and precipitation.

Ostracodermi Group of fossil **Agnatha** (Diplorhina: external nostrils paired) found from Upper Silurian to mid-Devonian, but probably present from Cambrian. Covered in bony armour; up to 50 cm in length. Lacked paired appendages. Similarities with cyclostomes. Some may have had mouthparts, but not jaws. Vertebrae so far absent from fossils; notochord apparently the skeletal support. See **Osteostraci**.

Otic Concerning the ear. See reference under **auditory**.

Otocyst See **statocyst**.

Otolith Granule of calcium carbonate in the vertebrate **macula**. See also **statocyst**.

Outbreeding Effect of any mechanism which tends to ensure that gametes which fuse at fertilization produce zygotes with a higher degree of heterozygosity than they otherwise would. Darwin realized that many plants (e.g. orchids, primroses) are structurally adapted to cross-pollination, and subsequent work on the genetics of incompatibility mechanisms, preventing self-fertilization, has shown that selection must favour such plants under certain conditions. Mechanisms promoting outbreeding include having separate sexes (see **dioecious**), **dichogamy**, **heterostyly**,

incompatibility mechanisms and, commonly in animals, the social structure of the breeding group (e.g. incest taboos in humans). Advantage of outbreeding in sexual populations is that it serves to free genetic variability in the population to selection and thereby reduces evolutionary stagnation of the population (see **genetic variation**). Compare **inbreeding**.

Outcrossing (Bot.) Pollination between (normally genetically) different plants of same species.

Ovarian follicle See **Graafian follicle**.

Ovary (1) (Bot.) Hollow basal region of a carpel, containing one or more ovules. In a flower which possesses two or more united carpels the ovaries are united to form a single compound ovary. (2) (Zool.) Main reproductive organ in female animals, producing *ova*. In invertebrate development eventual **germ cells** often become distinct at early cleavage. In vertebrates a pair of ovaries develops from mesoderm in the roof of the abdominal coelom; they become invaded by cells from the endoderm, but it is debatable whether these or mesodermal cells developing *in situ* become the *primordial germ cells*; however, the ovary epithelium (*germinal epithelium*) invaginates and some of its cells begin to undergo *oogenesis* (see **maturation of germ cells**). Each *oogonial* cell becomes surrounded by other cells to form a *follicle*. The ovaries produce various sex hormones, for details of which in mammals see **menstrual cycle**. See **testis**.

Overturn Complete mixing of a body of water, from surface to bottom; the breakdown of **stratification**. Results from various external factors.

Oviduct (Zool.) Tube carrying ova away from ovary (or from the coelom into which ova are shed) to the exterior. See **Müllerian duct**.

Oviparity Laying of eggs in which the embryos have developed little, if at all. Many invertebrates are *oviparous*, as are the majority of vertebrates, including the prototherian mammals. See **ovoviviparity**, **viviparity**.

Ovipositor Organ formed from modified paired appendages at hind end of abdomens of female insects, through which eggs are laid (*oviposition*). Consists of several interlocking parts. Frequently long (e.g. in ichneumon wasps), and capable of piercing animals or plants, permitting eggs to be laid in otherwise inaccessible places. Stings of bees and wasps are modified ovipositors.

Ovotestis Organ of some hermaphrodite animals (e.g. the garden snail, *Helix*; sea bass, Serranidae) serving as both ovary and testis. In the citrus tree pest *Icerya purchasi* (cottony cushion scale) diploid ('female') individuals transform into functional self-fertilizing hermaphrodites, possessing an ovotestis in which the centre is haploid and testicular, the cortex diploid and ovarian. They are thus chromosomal **mosaics**. In order to achieve the haploid interior, one set of chromosomes is eliminated in some of the early ovotestis cells (see **aberrant chromosome behaviour** (3)).

Ovoviviparity Development of embryos within the mother, from which they may derive nutrition, but from which they are separated for most, or all, of development by persistent **egg membranes**. Examples include many insects, snails, fish, lizards and snakes. See **oviparity**, **viviparity**.

Ovulation Release of ovum or oocyte from mature follicle of vertebrate ovary. See **menstrual cycle** for hormonal details.

Ovule Structure found in seed plants which develops into a **seed** after fertilization of an egg cell within it (see **double fertilization**). In gymnophytes (gymnosperms), ovules are unprotected; in flowering plants (anthophytes), they are protected by the **megasporophyll**, which forms a closed structure (**carpel**) within which they are formed singly or in numbers. Each ovule is attached to carpel wall by stalk (funicle) which arises from its base (chalaza). A mature flowering plant ovule comprises a central mass of tissue (the **nucellus**) surrounded by one or two protective layers (integuments) from which the seed coat is ultimately formed. Integuments enclose the nucellus except at the apex, where a small passage (the micropyle) permits entry of water and oxygen during germination. Within the nucellus is a large oval cell, the **embryo sac**, developed from the megaspore and containing the egg cell.

Ovum Unfertilized, non-motile, egg cell. In many animals it is an **oocyte**. Product of the **ovary**.

Oxidative phosphorylation Process by which energy released during electron transfer in aerobic **respiration** is coupled to production of ATP. See **bacteriorhodopsin**, **NAD**, **mitochondrion**.

Oxidoreductases Major group of **enzymes**; catalyse **redox reactions**.

Oxygen debt Amount of oxygen needed to oxidize the **lactic acid** produced in the anaerobic work done by muscle during vigorous exercise, and to resynthesize *creatine phosphate* used (see **phosphagen**). Until then, oxygen intake remains above normal while lactic acid remains in the muscle and blood. May be oxidized to pyruvic acid, converted to carbohydrate by the liver in **gluconeogenesis**, or neutralized by blood bicarbonate **buffer** and excreted as *sodium lactate* by the kidneys.

Oxygen dissociation curves (oxygen equilibrium curves) See **haemoglobin**.

Oxygen quotient See Q_{O_2}.

Oxyhaemoglobin See **haemoglobin**.

Oxytocin Oligopeptide hormone secreted by *pars nervosa* of **pituitary** glands of birds and mammals. Involved in contraction of alveoli of mammary glands during expression of milk, and in promoting smooth muscle contraction of uterus during coitus, and during **parturition**.

Ozone A gas existing as a layer in the stratosphere, 20–40 km above Earth, formed when short wavelength ultraviolet (uv) light splits a molecule of oxygen gas (O_2) into its two atoms high in the stratosphere. These reactive atoms combine with oxygen molecules to form ozone molecules, O_3. Ozone gas absorbs uv of a longer wavelength and protects life on Earth from damaging effects of **ultraviolet radiation**. Ozone in the lower atmosphere (troposphere) appears to be generally increasing, largely through the rising levels of nitrogen oxides (NOx), hydrocarbons and hydrogen peroxide from which O_3 is produced through photochemical reactions, high light and temperature promoting production. It is an efficient greenhouse gas (see **greenhouse effect**), absorbing infrared radiation. Ozone damage to European herbaceous plants occurs when levels exceed 100 parts in 10^9 (2.5 times normal level), shifting sucrose distribution from roots to shoots and increasing drought-dependent stress, possibly through reacting with ethylene to produce damaging free radicals. See **acid rain**, **CFC**s.

P_1 Parental generation in breeding work. Their offspring constitute the F_1 generation.

Pacemaker (1) Group of modified cardiac muscle cells in *sinus venosus* of vertebrate **heart**. In mammals and birds, forms *sinuauricular* (*sinoatrial*) *node* and lies in right atrial wall near superior vena cava. Cells have wandering membrane potentials, with a built-in tendency to depolarize (the so-called pacemaker potential), and the cell with the fastest intrinsic rate of depolarization leads the rest to fire with it simultaneously. The electrical current generated spreads to adjacent *Purkinje fibres*, and from there to atrial muscle cells, which contract together. The intrinsic pacemaker discharge is increased by sympathetic nerves from the thorax and decreased by branches of the vagus (parasympathetic system). See **cardio-acceleratory** and **cardio-inhibitory centres**. (2) A *myogenic pacemaker* occurs in the smooth muscles of vertebrate gut; longitudinal and circular layers each have one.

Pachytene See **meiosis**.

Pacinian corpuscles Very rapid-adapting pressure receptors in subcutaneous adipose tissue of both hairy and hairless vertebrate skin. Also found in joint capsules, tendons, etc.

Paedogenesis Form of **heterochrony**, in which reproductive organs undergo accelerated development relative to rest of body, giving larval maturity. The alternative term **progenesis** has been advocated. See **neoteny**, **paedomorphosis**, **parthenogenesis**.

Paedomorphosis Evolutionary displacement of ancestral features to later stages of development in descendant organisms, either through **progenesis** or **neoteny**. See **heterochrony**.

Pair-rule genes A class of at least eight segmentation genes, particularly studied in *Drosophila*, mutants which bring about repetitive deletion of specific parts of alternating segments. The fact that they are expressed in seven or eight stripes during cellularization of the *Drosophila* **blastoderm** is a key event in the **pattern formation** process.

The earliest to be expressed in development (primary pair-rule genes) are those directly controlled by the prior expression and distribution of **gap genes**, and in *Drosophila* are *hairy* (*h*), *even-skipped* (*eve*) and *runt* (*runt*). Their **promoters** are recognized by gap gene products, and variations in the promoter 'strengths' for different gap gene products seem to enable pair-rule genes to respond to their varying (often very low) concentrations. Primary pair-rule gene products then act as regulators (whether positive or negative is not certain) of secondary pair-rule genes such as *fushi tazaru* (*ftz*) and *paired* (*prd*), producing in the end fourteen stripes (bands) of product. The result of this mutually interactive sequence of gene activation and repression is that blastoderm cells emerge with different combinations of gene product that can now serve as reference points for further pattern formation, particularly in defining the domains of the **segment-polarity genes** *engrailed* (*en*) and *wingless* (*wg*). The pair-rule segmentation mechanism may not exist in some **short-germ** insects. See ***bicoid* gene** for diagram of some of these relationships.

Palaearctic Zoogeographical region, consisting of Europe, north Africa, and Asia south to Himalayas and Red Sea.

Palaeoanthropology Combination of studies attempting to reconstruct social structures, beliefs and behaviour of prehistoric human (hominid) populations from their artefactual and fossil remains. See ***Homo***.

Palaeobotany (paleobotany) Study of fossil plants.

Palaeocene Geological epoch; earliest division of the **Tertiary** period, during which the early insectivorous mammals and primates appeared. Climate was mild to cool. Wide, shallow, continental seas largely disappeared.

Palaeoecology The study of relationships between past organisms and the environment in which they lived. Largely concerned with reconstruction of past ecosystems. Ecology and palaeoecology have similar aims, and invoke many of the same biological principles; however, there are differences since past ecosystems cannot be examined directly. Inferences have to be made from fossils. Thus, palaeoecology is limited to the study of those past organisms whose fossils preserve or of those that can be detected chemically. Further, the fossil or sedimentary record can be altered through diagenesis, transportation and redeposition.

Palaeogene Collective term for Palaeocene, Eocene and Oligocene epochs. See **geological periods**.

Palaeognathae Ratites. Superorder of the subclass **Neornithes**. Includes birds with a reduced breastbone, and which are therefore secondarily flightless. Examples are: ostrich, rheas, cassowaries, tinamous, kiwis, and the extinct moas and elephant birds. Their feathers lack barbs. Probably represent a **grade** rather than a **clade**. See **neognathae**.

Palaeolimnology (paleolimnology) A branch of **limnology** concerned with the interpretation of past limnology from changes that occurred in the ecosystem of the lake, and their probable causes. Palaeolimnologists are interested in understanding how the lake functioned; how this aquatic ecosystem responded to external changes in climate, catchment area changes, fire, volcanism, tectonic events and anthropogenic influences (e.g. clearing of forests and establishment of settlements; agriculture and industry). It is through the palaeolimnological approach that long-term trends can be ascertained; thus palaeolimnology provides an historic perspective and is today an important approach in studies of **acid rain**, **eutrophication**, and climate change.

Palaeolithic An archaeological term denoting the Old Stone Age of human prehistory. For over a century, Eurasian material has been classified into Lower, Middle and Upper Palaeolithic, and

African material into Early, Middle and Late Stone Age. The Lower Palaeolithic was characterized by the hand-axe and complementary pebble tools/chopping tools; the Middle Palaeolithic usually lacked these large artefacts but contained many tools classified on their probable functions as scrapers; the Upper Palaeolithic technology included long, thin parallel-sided blade-like objects, while artistic and decorative material also appeared. **Radiometric dating** methods have taken the Upper Palaeolithic back to > 40,000 yr BP; uranium-series dates for carbonates, electron-spin resonance dates for mammal teeth and thermoluminescence dates for sediments and burnt flint have taken the Middle Palaeolithic (in Europe, Mousterian) industries back to > 200,000 yr BP, while the Lower/Middle Palaeolithic division probably took place between 250,000–200,000 yr BP, the former traditionally extending back to European colonization at least 700,000 yr BP. It was long held that Middle Palaeolithic culture was associated with Neanderthals (see ***Homo***) while Upper Palaeolithic artefacts were the province of Cro-Magnon (modern humans). But Neanderthal remains from Kebara and modern crania and skeletons from Qafzeh (both in Israel) are each associated with Middle Palaeolithic tools. Again, outside Europe, there is no clear-cut artefactual boundary-line between Middle and Upper Palaeolithic, since blades and tools traditionally associated with the Upper Palaeolithic sometimes lie beneath strata with Mousterian culture. This lends support to those classifying tools less on appearance (typology) than on function.

Palaeontology Study of **fossils** and evolutionary relationships and ecologies of organisms which formed them.

Palaeozoic Earliest major geological era. See **geological periods**.

Palate Roof of the vertebrate mouth. In mammals and crocodiles the roof is not homologous with that of other vertebrates; a new (false) palate has developed beneath original palate, by bony shelves projecting inwards from bones of upper jaw. In mammals, bony part of false palate (*hard palate*) is continued backwards by a fold of mucous membrane and connective tissue, the *soft palate*.

Palatoquadrate (pterygoquadrate) Paired cartilage or cartilage bone forming primitive upper jaw (as in **Chondrichthyes** and embryo tetrapods). See **autostylic jaw suspension**, **hyostylic**.

Palea (superior palea, pale) Glume-like bract of grass spikelet on axis of individual flower which, with the **lemma**, it encloses.

Palindromic Reading the same forwards and backwards. Some DNA sequences are palindromic. See **inverted repeat sequence**, **insertion sequence**.

Palisade See **mesophyll**.

Pallium See **cerebral cortex**.

Palmelloid (Of algae) describing an algal colony comprising indefinite number of single, non-motile cells, embedded in mucilaginous matrix.

Palps Paired appendages of many invertebrates, on the head or around the mouth. In polychaete annelids, tactile (on head); in bivalve molluscs, ciliated flaps around mouth generating feeding currents; in crustaceans, distal parts of appendages carrying mandibles (locomotory or feeding); of insects, parts of first and second **maxillae**, sometimes olfactory.

Palynology See **pollen analysis**, **palaeoecology**.

Pancreas Compound gland of vertebrates, in mesentery adjacent to duodenum; endocrine and exocrine functions. Secretes *pancreatic amylase*, *lipase*, *trypsinogen* and *nucleases* from its **acinar cells** in an alkaline medium of sodium hydrogen carbonate, promoted by **cholecystokinin** and **secretin**, and by the vagus (cranial nerve X), when the gastric phase of digestion is complete. The two major pancreatic hormones are **insulin** and **glucagon**, secreted respectively from β and α cells of the Islets of Langerhans. See **digestion**.

Pancreatin Extract of pancreas containing digestive enzymes.

Pancreozymin See **cholecystokinin**.

Paneth cells Actively mitotic cells at the bottoms of the crypts of Lieberkühn of the ileum (see **intenstine**) which move slowly up the sides of the crypts, being finally sloughed off into the intestinal lumen, breaking up and releasing small quantities of digestive enzymes. While part of the crypt, they complete the final stages of digestion by enzymes in their microvilli. See **digestion**.

Pangaea Ancient land mass persisting over 200 Myr, until end of **Jurassic**. Began to break up in late Triassic, eventually forming **Laurasia** and **Gondwanaland**.

Pangenesis Theory adopted by Charles **Darwin** to provide for the genetic variation his theory of natural selection required. Basically Lamarckist, it supposed that every part of an organism produced 'gemmules' ('pangenes') which passed to the sex organs and, incorporated in the reproductive cells, were passed to the next generation. Modification of the body, as through use and disuse, would result in appropriately modified gemmules being passed to offspring. Severely criticized by Darwin's cousin, Francis Galton. Empedocles held a similar view, but Aristotle rejected it.

Panicle Type of **inflorescence**.

Panmixis Result of interbreeding between members of a species population, no important barriers to gene flow occurring within it. The whole population represents one **gene pool**. Panmictic animal populations tend to have high **vagility**.

Pantothenic acid **Vitamin** of the B-complex; precursor of **coenzyme A** (CoA), a large molecule comprising a nucleotide bound to the vitamin. See **Krebs cycle**.

Pantotheria Extinct order of primitive therian Jurassic mammals. Small and insectivorous. Have therian features: large alisphenoid on wall of braincase, and triangular molar teeth. Contemporaries of **Multituberculata**.

Papilla Projection from various animal tissues and organs. *Dermal papillae* project from dermis into epidermis of vertebrates, providing contact (and finger-print patterns); in feather and hair follicles, papillae provide blood vessels. *Tongue papillae* increase surface area for taste buds in mammals; they are cornified for rasping in cats, etc.

Pappus Ring of fine, sometimes feathery, hairs developing from calyx and crowning fruits of the Family Compositae (e.g. dandelion). Act as parachute in wind dispersal of fruit.

Parabiosis Surgical joining together of two animals so that their blood circulations are continuous. Each member of the pair is termed a *parabiont*. Often employed to monitor humoral influences in behaviour and development (e.g. in insect moulting).

Paracrine cell (1) Cell secreting a local chemical signal which is rapidly taken up at receptor site on another cell, but destroyed or inactivated within about 1 mm from release site. Includes several cells secreting **prostaglandins**, **mast cells** and nerves of the sympathetic system (see **varicosity**). (2) Any cell secreting a substance having an effect upon another cell. The substances secreted would be described as having a *paracrine action*.

Paradox of the plankton Phrase coined by G. E. Hutchinson as a result of observing that 10–50 phytoplankton species commonly co-exist in the same body of water at any one time. Ecological principle of competitive exclusion predicts that relatively homogeneous environments, such as the photic zone of lakes and oceans, should contain very few species which possess similar ecological requirements. Since all phytoplankton are essentially photoautotrophs, with similar needs, competition for resources, especially nutrients, should result in the elimination of all but one species (best able to use the limiting resources).

This 'paradox' can be explained in terms of each species having its own nutrient-utilization characteristics resulting in each different species being limited simultaneously by a different nutrient. Direct competition is avoided, and potentially as many species can co-exist as there are limiting nutrients. Grazing by zooplankton may also improve the chances for co-existence among planktonic algae. Other explanations stress that the photic zone is not homogeneous, since gradients in light intensity, nutrients and temperature occur providing spatial heterogeneity.

Parallel evolution Possession in common by two or more taxa of one or more characteristics, attributable to their having similar ecological requirements and a shared genotype inherited from a common ancestor: the common characteristics would be **homologous**. In **cladistics**, no distinction is made between parallel evolution and **convergence**.

Parallel venation Pattern of leaf venation, where principal veins are parallel or nearly so; characteristic of **Monocotyledonae**.

Paramylon (Of algae) storage polysaccharide composed solely of β-1,3 linked glucans in the **Euglenophyta**, **Xanthophyta**, and **Prymnesiophyta** (e.g. *Pavolva mesolychnon*). Occurs outside the chloroplast as water-soluble, single-membrane-bound inclusions of various shapes/sizes.

Parapatry Where the ranges of two populations of the same or different species overlap they are **sympatric**; but if the ranges are contiguous, i.e. if they abut for a considerable part of their length but do not overlap, the distributions show parapatry. Distributions of several organisms formerly regarded as single species have been shown to consist of several **sibling species** or **semispecies** with parapatric distributions, as in frogs of *Rana pipiens* species group.

Paraphyletic Term describing taxon or taxa originating from and including a single stem species (known or hypothetical) but excluding one or more smaller **clades** nested within it. E.g. if, as is commonly accepted, flowering plants arose from a gymnophyte ancestor, then gymnophytes are a *paraphyletic group* since the group does not include all descendants of a common ancestor. Such taxa are not permitted in **cladistics** but are much used by adherents of **evolutionary taxonomy.** See **clade**, **monophyletic**.

Paraphysis (Bot.) Sterile filament, numbers of which occur in mosses and certain algae, interspersed among the sex organs, and in the hymenia of certain fungi (**Ascomycota**, **Basidiomycota**).

Parapodium Paired metameric fleshy appendage projecting laterally from the body of many polychaete annelid worms (especially errant polychaetes). Usually comprises a more dorsal *notopodium* and a ventral *neuropodium*, each with bundles of chaetae and endowed with a supporting chitinous internal *aciculum* to which muscles moving the parapodium are attached.

Parasegment Unit in insect development which corresponds not to a morphological segment but to the posterior **compartment** of one segment and the anterior compartment of the next most posterior segment. See **homeobox**, **segment polarity genes**.

Parasexual cycle See **parasexuality**.

Parasexuality Fungal life cycle (*parasexual cycle*) discovered by G. Pontecorvo and J. A. Roper in 1952 which includes the following: occasional fusion of two haploid heterokaryotic nuclei in the mycelium to form a diploid heterozygous nucleus; mitotic division of this nucleus during which crossing-over (*mitotic crossing-over*) occurs, then restoration of haploidy to the nucleus by either mitotic **non-disjunction**, or some form of chromosome extrusion, removing a haploid set of chromosomes (see **aberrant chromosome behaviour**). The non-sexual spores produced differ genetically from the parent mycelium. It accounts for the variation of pathogenicity in certain fungal pathogens, and has enabled genetical studies to be made using members of the **Deuteromycota**.

Parasite One kind of symbiont. Organism living in (*endoparasite*) or on (*ectoparasite*) another organism, its *host*, obtaining nourishment at the latter's expense. Metabolically dependent upon their hosts, as are carnivores, herbivores, etc. Distinction between herbivorous caterpillar and ectoparasitic fluke is not clear-cut. Endoparasites generally display more, and more specialized, adaptations to parasitism than do ectoparasites, and often include both primary and secondary **hosts** in the life cycle. *Obligate parasites* cannot survive independently of their hosts; *facultative parasites* may do so. *Partial parasites* (e.g. mistletoe, *Viscum*) are plants which photosynthesize and also parasitize a host. Sometimes relationships between members of the same species are parasitic (e.g. males of some angler fishes live attached to the female and suck her blood). Placental reproduction shares features with parasitism, as do forms of viviparity in which young emerge causing death of the parent (e.g. in the midge *Miastor* and many aphids and water fleas). See **symbiosis**, **malarial**, **parasitoid**.

Parasitoid Insects, and some other animals, which introduce their eggs into another animal, in which they grow and develop in a slow and controlled manner using the host's resources without killing it. At maturation they emerge and usually do cause death of the host (unlike most **parasites**).

Parastichy See **phyllotaxy**.

Parasympathetic nervous system See **autonomic nervous system**.

Parathyroid glands Endocrine glands of tetrapod vertebrates, usually paired, lying near or within **thyroid** depending on species. Arise from embryonic gill pouches, and produce **parathyroid hormone**. Removal produces abnormal muscular convulsions within a few hours.

Parathyroid hormone (PTH, parathormone) Polypeptide hormone of parathyroid glands operating with vitamin D and **calcitonin** in control of blood calcium levels. Injection releases calcium from bone and raises blood Ca^{2+} level, inhibiting further parathyroid hormone release, apparently via direct negative feedback on parathyroid glands. Also reduces Ca^{2+} excretion by the kidneys. Deficiency produces muscle spasms.

Paratype Any specimen, other than the type specimen (**holotype**) or duplicates of this, cited with the original taxonomic description and naming of an organism.

Parazoa Subkingdom of the **Animalia**, containing the phylum **Porifera** (sponges).

Parenchyma (Bot.) Simple tissue, comprising parenchyma cells occurring in the primary plant body commonly as continuous masses in the cortex of stems, in roots, in the stem pith, in the leaf mesophyll, and in the flesh of fruits. Parenchyma cells also occur as vertical strands in the primary and secondary vascular tissues, as well as horizontal strands called rays in the secondary vascular systems. Cells are thin-walled, often almost as broad as they are long, and remain alive and capable of dividing at maturity. Some have a secondary wall in addition to the primary wall. Those with just primary walls, because of their ability to divide, play an important role in regeneration and wound healing; it is these cells from which adventitious structures (e.g. adventitious roots) arise. Other functions may include photosynthesis, secretion, storage, water movement and transport of food materials. See **transfer cell**. (Zool.) Loose tissue of irregularly shaped vacuolated cells with gelatinous matrix, and forming a large part of the bodies of some invertebrate groups, notably platyhelminths and nematodes.

Parietal (Bot.) Referring to peripheral position, as in chloroplasts of some algal cells located near the cell's periphery. (Zool.) *Parietal bones* lie one on each side of the vertebrate skull, behind and between the eye orbits. See also **pineal gland**.

Parietal eye See **pineal eye**.

Parietal placentation (Bot.) Attachment of ovules in longitudinal rows on carpel wall.

Parthenocarpy (Bot.) Development of fruits without prior fertilization. Occurs regularly in banana and pineapple (which are therefore seedless). Can be induced by certain auxins in unfertilized flowers, e.g. those of tomato.

Parthenogenesis Development of an unfertilized gamete (commonly an egg cell) into a new individual. One of a spectrum of forms of uniparental **sexual reproduction**.

A parthenogenetic egg cell (or nucleus) may become diploid either through nuclear fusion, or through a restitution division (see **restitution nucleus**). Sometimes cleavage products of a haploid egg may undergo fusion, producing a diploid embryo. Phenomenon includes non-gametic forms of **automixis** and in animals is a common cause of **male haploidy**. In animals, **thelytoky** (absence of males) enables rapid production of offspring without food competition from males. *Cyclical parthenogenesis* (as in some aphids and flukes) involves a combination of thelytoky and bisexual fertilization. In some aphids thelytoky prevails in summer, males only appearing in autumn or winter when fertilization occurs. In the midges *Miastor* and *Heteropeza*, larvae possess functional ovaries enabling progenetic reproduction by automixis, adults not appearing for generations; some larval flukes are progenetic. Some instances of thelytoky (automictic, or meiotic, thelytoky) involve meiotic egg-production, and two of the four meiotic products sometimes fuse to restore diploidy; in others (apomictic, or ameiotic, thelytoky), mitosis produces the egg cells. In some cases diploidy may be restored by **endomitosis** after meiosis.

In the protozoan *Paramecium*, fusion may occur of two of the micronuclei produced meiotically from the cell's parent micronucleus (*automixis*) but no new individual is produced.

Since development of unfertilized eggs can occur, on rare occasions, in many species (e.g. *Drosophila* and grasshoppers), and be induced artificially in many others (see **centriole**), it is still surprisingly rare, especially since it avoids the **cost of meiosis**. Thelytokous forms seem to be liable to early extinction compared with their bisexual relatives, probably through progressive homozygosity (see **genetic variation**).

Haploid egg cells have been recorded developing parthenogenetically in plants, but this seems to have had little evolutionary impact. Unreduced (diploid) gametophytes may arise either from an unreduced megaspore (*diplospory*) or from an ordinary unreduced somatic cell of the sporophyte (*apospory*). Both are genetically equivalent to apomictic (ameiotic) parthenogenesis in animals. In dandelions (*Taraxacum*) the megaspore mother cell undergoes meiosis, the first division producing a restitution nucleus, the second producing two cells, each unreduced, from one of which the 8-nucleate embryo sac is produced. See **gynogenesis**, **parthenocarpy**.

Parturition Expulsion of foetus from uterus at end of pregnancy (term) in therian mammals. In man and other primates it appears that the foetus determines pregnancy length by initiating the release of prostaglandins, principally PGF2a, from the endometrium (and later from the myometrium) of the uterus which initiates labour through its effects on smooth muscle contraction in the uterine wall. Oxytocin has also been implicated in the onset of labour. Successful transition from intrauterine to extrauterine life requires the previous maturation of lungs, the laying down of fat and carbohydrate reserves and the onset of maternal lactation (see **prolactin**). Foetal adrenal glands are involved in much of this regulation, becoming more active as term approaches.

Passage cells Cells of the **endodermis**, typically of older monocot roots, opposite protoxylem groups of stele, remaining unthickened and with casparian strips only after thickening of all other endodermis cells. Allow transfer of material between cortex and vascular cylinder.

Passerines Members of largest avian order, the Passeriformes. Perching birds, characterized by having large first toe directed back, the other three forward. Includes most of the common inland birds. See **Neognathae**.

Pasteur, Louis (1822–95). French chemist and microbiologist; professor of chemistry at the Sorbonne, but worked mostly at the Ecole Normale in Paris. Became director of the Institut Pasteur, Paris, in 1888. Championed the view that fermentation was a *vital* rather than a *simple* chemical process, as against the chemical theory of Liebig and Berzelius. Already aware, with others, that yeasts were associated with alcoholic fermentations, he demonstrated presence of microorganisms in other fermentations. In 1858 he demonstrated fermentation in the absence of organic nitrogen, destroying the chemical theory. Pasteur's many experiments supported the germ theory of fermentation, as against the theory of spontaneous generation, and its implications were appreciated by Joseph Lister in his work on antisepsis in the 1860s. Pasteur came to accept the role of microorganisms in disease, showing that attenuated forms of bacteria produced by serial culture could be used in inoculation to immunize the host. His vaccines against anthrax and rabies, like Jenner's earlier ones against smallpox, were instrumental in establishing the germ theory of disease. See **Pasteurization**, **Virchov**.

Pasteur effect Phenomenon whereby onset of aerobic respiration inhibits glucose consumption and lactate accumulation in all facultatively aerobic cells, conserving substrates. Depends upon the allosteric inhibition of glycolytic enzymes by high intracellular ATP to ADP ratio. See **ATP**, **glycolysis**.

Pasteurization Method of partial sterilization, after Louis **Pasteur**, who discovered that heating wine at a temperature well below its boiling point destroyed the bacteria causing spoilage without affecting its flavour. Widely used to kill some disease-causing bacteria in food, e.g. tubercle bacteria in milk in the 'holder' method by heating at 63–65°C for 30 mins; in the HTST method heating at 72°C for at least 15 secs followed by rapid cooling delaying its fermentation. Compare **Tyndallization**.

Patella Kneecap. Bone (sesamoid bone) over front of knee joint in tendon of extensor muscles straightening hind limb. Present in most mammals, some birds and reptiles.

Pathogen Disease-causing parasite, usually a microorganism.

Pathology Study of diseases or diseased tissue.

Patristic (Of similarity) due to common ancestry. See **cladistics**.

Pattern, pattern formation Describing phenomena whereby cells in different parts of embryo become locked into different developmental pathways, co-ordinated in such a way as to produce a viable multicellular system. Early animal embryos engage in a hierarchy of decisions involving progressive steps in *regional specification*, in which **positional information** is imparted to cells, which then respond to it. Cells of a particular histological type may have arrived at their condition via alternative, *non-equivalent* routes. Parts of an embryo acquire different **determined** states through regionalization processes. See **maternal effect**, **compartment**, **pair-rule genes**.

PCBs (polychlorinated biphenyls) Persistent and toxic pollutants, toxicity varying with the position and number of chlorine atoms bound to the biphenyl core. PCBs appear to interfere with plasma transport of vitamin A and thyroxine (whose structure is not dissimilar).

Peat Accumulated dead plant material which has remained incompletely decomposed owing, principally, to lack of oxygen. Occurs in moorland, bogs and fens, where land is more or less completely waterlogged; often forms a layer several metres deep. Of local value as fuel for burning.

Peck order See **dominance** (2).

Pectic compounds Acid polysaccharide carbohydrates, present in **cell walls** of unlignified plant tissues; comprise pectic acid and pectates, pectose (propectin) and pectin. Form gels under certain conditions. Principal components are galacturonic acid, galactose, arabinose and methanol. Form the basis of fruit jellies.

Pectoral fin See **fins**.

Pectoral girdle (shoulder girdle) Skeletal support of vertebrate trunk for attachment of fins or forelimbs; in fish, attaches to skull. Primitively, a curved bar of cartilage or bone on each side of body, fusing ventrally to form a hoop transverse to long axis, incomplete dorsally. Each bar bears a joint with fin or limb (see **glenoid cavity**). Components are: *scapulae* dorsal to the joint, *coracoids* ventrally. Additional dermal bones are the *cleithra* in fish and primitive tetrapods, and *clavicles*, usually on the ventral side. In mammals each clavicle joins a scapula at a process of the coracoid, the *acromion*. Scapulae do not articulate with the vertebral column or ribs (compare **pelvic girdle**). In tetrapods, coracoids and clavicles join mid-ventrally to the sternum. See Fig. 90.

Pedicel (Bot.) Stalk of individual flower of an inflorescence. (Zool.) Narrow tube-like 'waist' of many hymenopteran insects.

Pedipalps Second pair of head appendages of **Arachnida**. See **mouthparts**.

Peduncle Stalk of an **inflorescence**.

Peking man Early form of *Homo*

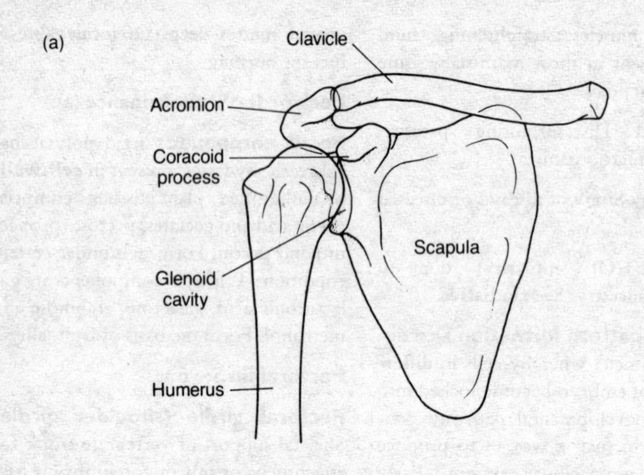

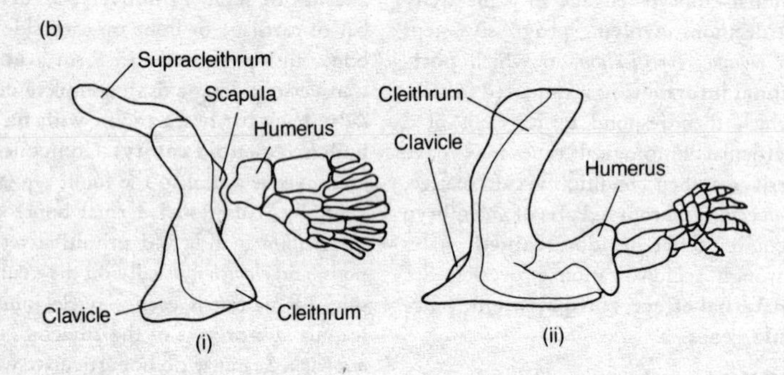

Fig. 90 (a) *Human shoulder region, viewed from the front.* (b) *Pectoral girdle and fin of* (*i*) *crossopterygian fish and* (*ii*) *early fossil amphibian. One side only shown in each case.*

erectus from China; formerly termed *Sinanthropus*. See ***Homo***.

Pelagic Inhabiting the mass of water of lake or sea, in contrast to the lake or sea bottom (see **benthos**). Pelagic animals and plants are divided into **plankton** and **nekton**.

Pelecypodia See **Bivalvia**.

P element A kind of **transposable element** found in the fruit fly *Drosophila* and responsible for **hybrid dysgenesis** in crosses between P-strain male and M-strain female flies. 0.5–1.4 kb in length, they are flanked by inverted repeats 31 base pairs long. They originate through deletions within larger *P factors*, a few copies of which occur in *P strains* of the fly. Appropriately injected into M-strain embryos they can be used as

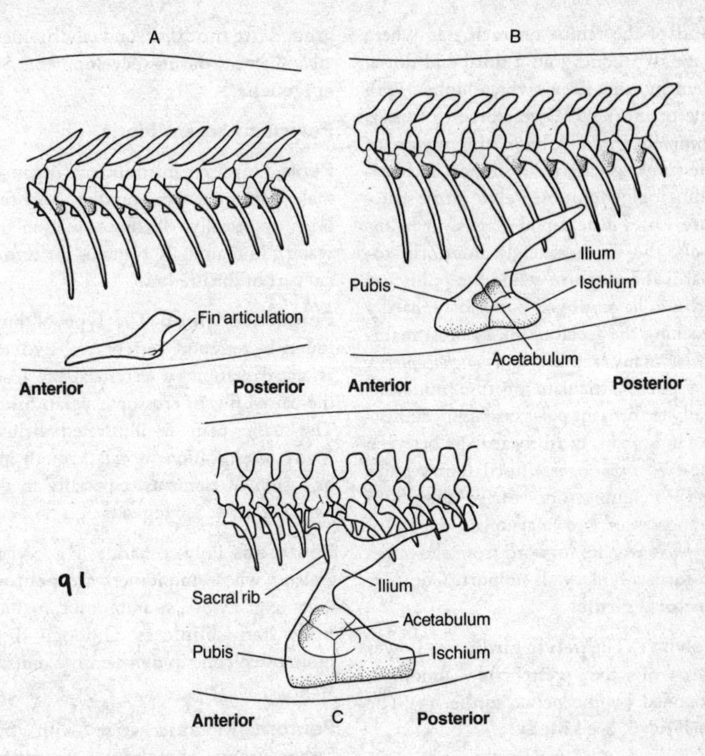

Fig. 91 *Pelvic girdles of (A) fish, (B) early tetrapod and (C) later tetrapod with attachment of girdle to enlarged sacral rib.*

gene vectors, the resulting fly's germ line tending to acquire the gene. See **copia**.

Pellicle (1) In the **Euglenophyta**, a proteinaceous surface layer, composed of overlapping strips, below the plasma membrane, which can be flexible, rigid, or in some instances ornamented. (2) In armoured Dinoflagellates, that portion of the cell covering surrounding the cell after the theca is shed in ecdysis. (3) A flexible proteinaceous surface layer helping to confer shape in some ciliate protozoans. Compare **periplast**.

Pelvic fin See **fins**, **pelvic girdle**.

Pelvic girdle (hip girdle) Skeletal support for attachment of vertebrate hindlimbs or pelvic fins. In fish (see Fig. 91), a pair of curved bars of bone or cartilage embedded in the abdominal muscles and connective tissue, fused to form a midventral plate, articulating with the fins but not with the vertebral column. In tetrapods, the ventral plate ossifies from two centres on each side: the *pubis* anteriorly, and *ischium* posteriorly. A large rounded socket (*acetabulum*) receives the head of the femur on each side where these two bones join a third and dorsal element, the *ilium*, which unites with one or more sacral vertebrae to form a complete girdle around this region of the trunk, giving rigid support to hind-limbs for locomotion. Pelvic girdle structure varies in tetrapod classes. In mammals, the ilium extends anteriorly towards the **sacrum**, while the pubis and ischium have moved posteriorly, hardly reaching the acetabulum. In most mammals, many reptiles and *Archaeopteryx* the pubes articulate or fuse mid-ventrally to form the *pubic symphysis*, consisting in humans of fibrocartilage between the two *coxal bones* (fused ilium, pubis and ischium on each side). In monotremes and marsupials a pair of *prepubes* reaches forward from the pubes to form a body wall support. Compare **pectoral girdle.**

Pelvis (1) The **pelvic girdle.** (2) Lower part of the vertebrate abdomen, bounded by the pelvic girdle. (3) The *renal pelvis*. See **kidney**.

Penetrance A dominant character determined by an **allele** is either always expressed in any individual where it occurs (*completely penetrant*), or is expressed in some individuals but not in others (*incompletely penetrant*). Once a character finds expression, it may be expressed to varying degrees in different individuals (*variable expressivity*). Possible factors affecting penetrance and expressivity include the genetic background (see **modifier**) and environmental influences during development. See **epigenesis**.

Penicillins See **antibiotic**.

Penis Unpaired intromittant organ of male mammals, some reptiles and a few birds (especially of those mating on water). In mammals, contains the terminal part of the urethra.

Pentadactyl limb The type of limb found in tetrapod vertebrates. Evolved as an adaptation to terrestrial life from the paired fins of crossopterygian fishes. The basic plan is illustrated below. Many modifications occur through loss or fusion of elements, especially in the terminal parts. See Fig. 92.

Pentosans Polysaccharide (e.g. xylan, araban) whose monomers are **pentose** units (e.g. xylose, L-arabinose). Include many **hemicelluloses**, although these more often contain non-pentose units as well.

Pentose Monosaccharide with five carbon atoms in molecule, e.g. ribose and deoxyribose (important constituents of **nucleic acids**), and ribulose. Found in various plant polysaccharide chains, e.g. pectin, gum arabic.

Pentose phosphate pathway (phosphogluconate pathway, hexose monophosphate shunt) An alternative route to **glycolysis** for glucose catabolism, involving initially conversion of glucose-phosphate to

Fig. 92 (*a*) Diagram of skeleton of pentadactyl limb of vertebrate, giving names of parts of fore-limb, and (*in brackets*) *of hind-limb. On left, common name of whole part; on right, names of bones.* (*b*) *Diagram indicating the adaptive radiation of vertebrate limb from a basic archetype.*

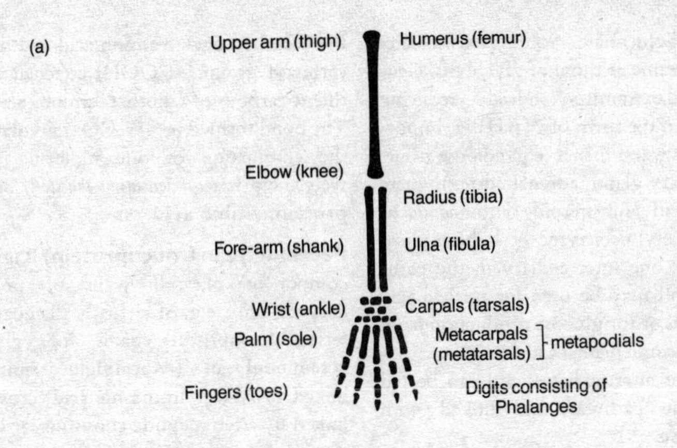
(a)
Upper arm (thigh)
Humerus (femur)
Elbow (knee)
Radius (tibia)
Fore-arm (shank)
Ulna (fibula)
Wrist (ankle)
Carpals (tarsals)
Palm (sole)
Metacarpals (metatarsals)
metapodials
Fingers (toes)
Digits consisting of Phalanges

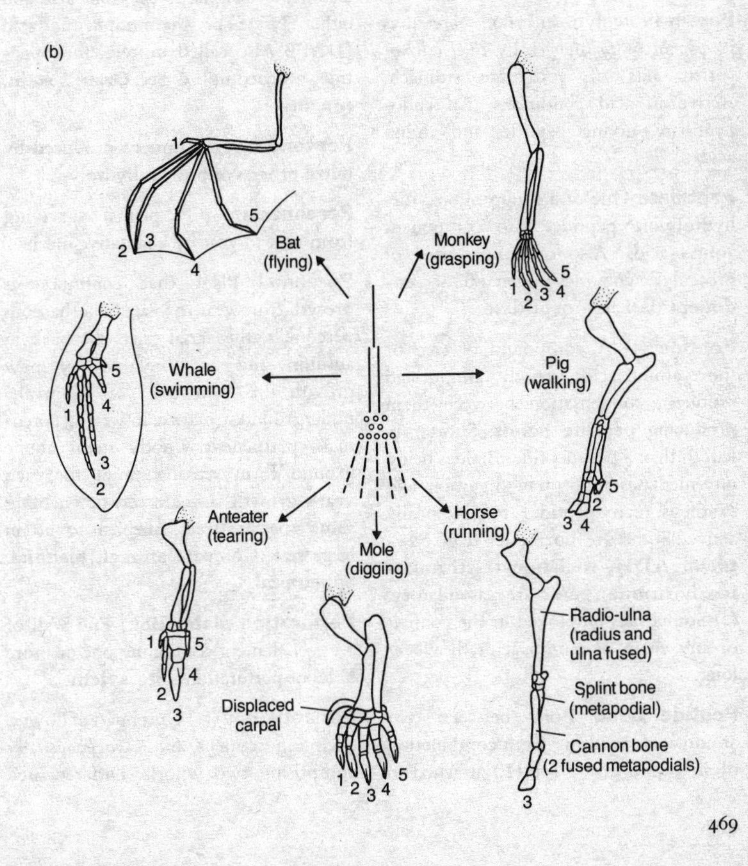
(b)
Bat (flying)
Monkey (grasping)
Whale (swimming)
Pig (walking)
Anteater (tearing)
Mole (digging)
Horse (running)
Displaced carpal
Radio-ulna (radius and ulna fused)
Splint bone (metapodial)
Cannon bone (2 fused metapodials)

phosphogluconate. Some intermediates are the same as those of glycolysis. Generates extramitochondrial reducing power in the form of NADPH, important in several tissues (e.g. adipose tissue, mammary gland, adrenal cortex) where fatty acid and steroid synthesis occur from acetyl coenzyme A. Ribose-phosphate is one intermediary in the pathway, and may be used for nucleic acid synthesis or for glucose production from CO_2 in some plants. Other monosaccharides are intermediates, and can be fed into the pathway and linked with glycolysis.

Pepsin Proteolytic enzyme, secreted as its precursor *pepsinogen* by *chief cells* of gastric pits of vertebrate stomach. Active in acid conditions. An endopeptidase, giving peptides and amino acids.

Peptidase One of a group of enzymes hydrolysing peptides into component amino acids. Also sometimes used of proteolytic enzymes in general. See **endopeptidase, exopeptidase**.

Peptide (1) A compound of two or more amino acids (strictly, amino acid residues), condensation between them producing **peptide bonds**. Shorter in length than a polypeptide. Besides being intermediates in protein digestion and synthesis, many peptides are biologically active. Some are hormones (e.g. **oxytocin**, **ADH, melanocyte-stimulating hormone**); some are vasodilatory. (2) Sometimes employed in the context of any chain of amino acids, however long.

Peptide bond Bond between two amino acids resulting from combination of an amino group ($-NH_2$) attached to the α-carbon of one amino acid and the carboxyl group (–COOH) attached to the α-carbon of another amino acid. The bond formed (–NH–CO–) involves the elimination of one molecule of water, and is a *condensation reaction*. See **protein**, **amino acid**.

Peptidoglycan (mucoprotein) Rigid component of cell walls of prokaryotes only, e.g. of virtually all eubacteria. Comprises chains of glycosaminoglycan (*N*-acetylglucosamine linked to *N*-acetylmuramic acid) cross-linked by a tetrapeptide consisting of L-alanine, D-alanine, D-glutamic acid and either lysine or diaminopilemic acid (DAP). Also called murein, mucopeptide, mucocomplex. See **Gram's stain**, **mucins**.

Peptone Large fragment produced by initial process of protein hydrolysis.

Perennation (Of plants) surviving from year to year by vegetative means.

Perennial Plant that continues its growth from year to year. In herbaceous perennials, the aerial parts die back in autumn and are replaced by new growth the following year from the underground structure. In woody perennials, permanent woody stems above ground form starting point for each year's growth, a characteristic enabling some species (trees, shrubs) to attain large size. Compare **annual**, **biennial**, **ephemeral**.

Perforation plate (Bot.) End wall of a vessel element containing one or more holes or perforations. See **xylem**.

Perianth (Bot.) (1) Outer part of flower, enclosing stamens and carpels; usually comprising two whorls. Differentiated in Dicotyledonae as an outer green **calyx** and an inner **corolla**, the latter usually conspicuous and often brightly coloured. In Monocotyledonae, usually no differentiation of calyx and corolla, both whorls looking alike. (2) In leafy liverworts, a tubular sheath surrounding the archegonia and, later, the developing sporophyte.

Periblem See **apical meristem**.

Pericardial cavity Space enclosing the heart. In vertebrates, is coelomic and bounded by a double-walled sac, the *pericardium*. In arthropods and molluscs, is haemocoelic, supplying blood to heart.

Pericarp (Bot.) Wall of an ovary after it has matured into a **fruit**. May be dry, membranous or hard (e.g. achene, nut), or fleshy (e.g. berry).

Perichaetium Distinct whorl of leaves surrounding sex organs in mosses.

Periclinal (Bot.) (Of planes of division of cells) running parallel to the surface of the plant. See **cell division** for diagram; compare **anticlinal**.

Pericycle (Bot.) Tissue of the vascular cylinder lying immediately within the **endodermis**, comprising parenchyma cells and sometimes fibres.

Periderm (Bot.) Cork cambium (phellogen) and its products; i.e. cork and secondary cortex (phelloderm).

Perigene development (Bot.) Development of **guard cells** in which the guard cell mother cell does not give rise to subsidiary cells.

Perigonium (Bot.) The antheridium, together with associated perigonial leaves or bracts (surrounding the antheridium), borne on a specialized branch (perigonial branch), in mosses and liverworts.

Perigynous See **receptacle**.

Perikaryon Alternative term for **cell body** of neurone.

Perilymph See **vestibular apparatus.**

Perineum Region between anus and urogenital openings of placental mammals.

Periosteum Sheath of connective tissue investing vertebrate bones, and to which tendons attach. Contains osteoblasts, and white and elastic fibres.

Peripatus See **Onychophora**.

Peripheral nervous system See **nervous system**.

Periplast A cell covering found in the algal division **Cryptophyta**, comprising cell membrane with an underlying layer of plates or membranes and an overlying layer of granular material; not as elaborate as a **pellicle**.

Perisperm Nutritive tissue surrounding the embryo in some seeds; derived from the **nucellus**. Compare **endosperm**.

Perissodactyla Order of eutherian mammals containing the odd-toed ungulates (e.g. horses, tapirs, rhinoceroses). Walk on hoofed toes, the weight-bearing axis of foot lying along the third toe, which is usually larger than the others (some of which may have disappeared): horses have very large third, but minute second and fourth, toes. Tapirs have four toes on front feet, three on hind feet. Rhinoceroses have three toes on each foot. See **Artiodactyla**.

Peristalsis Waves of contraction of smooth muscle, passing along tubular organs such as the intestines. Serve to move material from one end of the tube to the other.

Peristome (Bot.) Fringe of pointed appendages (teeth) around the opening of dehiscent moss capsule, concerned with spore liberation. (Zool.) Spirally twisted groove leading to cytostome in some ciliate protozoans.

Perithecium Rounded, or flask-shaped, fruiting structure (**ascocarp**) of some members of the **Ascomycota** and **lichens**; with an internal hymenium of asci and paraphyses, and opening via a pore (ostiole) through which ascospores are discharged.

Peritoneum Epithelium (serous membrane) lining posterior coelomic cavity (*abdominal*, *perivisceral* and *peritoneal cavities*) of vertebrates, around the gut. That covering the gut and other viscera is termed *visceral*; that lining the wall of the cavity is *parietal*. The **mesentery** carrying blood vessels to and from the viscera is also peritoneal.

Perivisceral cavity The vertebrate coelomic cavity lined by **peritoneum**.

Permanent teeth Second of the two successive sets of teeth of most mammals, replacing **deciduous teeth**. See **dentition**.

Permease **Transport protein**, or carrier molecule, assisting in transport across cell membranes without being permanently altered in the process.

Permian Last period of the **Palaeozoic** era, 270–225 Myr BP. In Britain, Permian and **Triassic** marine faunas are often united as New Red Sandstone (*Permo-Trias*). Among vertebrates, palaeoniscoid fish dominated and holosteans appeared. Reptiles diversified and included pelycosaurs and their replacements the **Therapsida**, but there were few representatives of the dominant **Mesozoic** reptiles. Conifers, cycads and ginkgos evolved, and the earliest forest types declined. Global climate was arid; extensive glaciation occurred in the Southern Hemisphere. At its close there was a considerable mass **extinction** of marine fauna; anoxia has been suggested as its cause.

Peroxisome **Microbody** containing **catalase**, especially in vertebrate liver and kidney cells, and those plant cells involved in **photorespiration**. Catalase uses hydrogen peroxide (itself produced by oxidative enzymes using molecular oxygen inside the organelle) to remove hydrogen atoms from substrates, so oxidizing (detoxifying) phenols, formic acid, formaldehyde and ethanol (oxidized to acetaldehyde). Biogenesis not well understood. All their component enzymes, including those involved in β-oxidation of fatty acids, are encoded by nuclear DNA and imported post-translationally, apparently by **protein targeting**. See **glyoxylate cycle**, **superoxides**.

Pest control Five major strategies of pest control are employed, each dependent for its effectiveness upon the ecological 'strategy' of the pest organism (see **pests**). They are: pesticide control; **biological control**; cultural control (where agricultural or other practices are used to change the pest's habitat);

	r-pests	*Intermediate pests*	*K-pests*
Pesticides	Early widescale applications based on forecasting	Selective pesticides	Precisely targeted applications based on monitoring
Biological control		Introduction or enhancement of natural enemies	
Cultural control	Timing, cultivation, sanitation and rotation	→ ←	Change in agronomic practice, destruction of alternative hosts
Resistance	General, polygenic resistance	→ ←	Specific, monogenic resistance
Genetic control			Sterile mating technique

Table 6. *Principal control methods appropriate for different pest strategies.*

breeding for pest resistance in cultivated organisms, and sterile mating control, where pest populations are variously sterilized to reduce their reproductive rates. These main approaches, and the types of pest against which they are employed, are indicated in Table 6.

Pests Species whose existence conflicts with human profit, convenience, or welfare. Some cause serious nuisance; their injuriousness is well established and their control is either a social or economic necessity. Pest status commonly arises from (a) entry of species into previously uncolonized regions; (b) some change in the properties of previously unproblematic species; (c) changes in human activities, bringing contact with species to which there was previous indifference; (d) increase in abundance of a species, with resulting nuisance value.

Pest status may be interpreted in terms of ecological strategies wrought by different selection pressures to which pest species are exposed; thus: *r-pests*, where ***r*-selection** influences are uppermost; *k-pests*, where **k-selection** influences dominate; and *intermediate pests*, lying somewhere between these two. See **pest control**, **weeds**.

Petal One of the parts forming corolla of flower; often brightly coloured and conspicuous. See **flower**.

Petiole Stalk of a **leaf**.

Peyer's patches Patches of secondary

lymphoid tissue in submucosa of amniote intestines.

pH A quantitative expression denoting the relative proton (hydrogen ion, H^+) concentration in a solution. The pH scale ranges from 0–14: the higher the pH value, the lower the acidity. A pH of 7 indicates a neutral solution. Defined as $-\log [H^+]$, where $[H^+]$ = concentration of protons, expressed as $g.dm^{-3}$. In consequence, a solution of pH 6 has ten times the H^+ concentration of a solution of pH 7. It is important that the pH of cells and body fluids is kept within acceptable values, one reason being the effect of pH change on the shapes of globular protein molecules. See **protein**, **buffer**.

Phaeophyta Brown algae (Kingdom **Protista**). Almost exclusively marine; only four freshwater genera (*Heribaudiella, Pleurocladia, Bodanella, Sphacelaria*); several marine forms found in brackish water and salt marshes. Derive their characteristic colour from large amounts of the carotenoid *fucoxanthin* in their chloroplasts, which also contain chlorophylls *a*, c_1 and c_2 as well as *β*-carotene and violaxanthin. There are two membranes of **chloroplast endoplasmic reticulum**, usually continuous with the outer membrane of the nuclear envelope in the Ectocarpales but apparently discontinuous in the Dictyotales, Laminariales and Fucales. Membrane-bound tubules are common in the region between the chloroplast endoplasmic reticulum and chloroplast envelope where the latter two are not closely appressed. Chloroplasts have three thylakoids per band. All orders have pyrenoids in the form of a star-like structure set off from the main body of the chloroplast, containing a granular substance not traversed by thylakoids. Surrounding the pyrenoid outside the chloroplast endoplasmic reticulum is a membrane-bound sac containing the main storage product, the polysaccharide *laminarin*. Mannitol is however the accumulation product (up to 25% dry weight) of some *Laminaria* in the autumn. Cell walls have cellulose as the main structural component; the amorphous component is made up of *alginic acid* and *fucoidin*. Calcification of the wall occurs in some species of *Padina*.

Morphologically, a very diverse group ranging from minute (< 1 mm long) filaments to very large (60–70 m long) and complex thalli, with a root-like holdfast, and stem-like stipe bearing branched or unbranched leaf-like blades or laminae, often with one or more air bladders and a relatively complex internal structure. There are no unicellular or colonial forms. Complex thalli show several **growth** forms (e.g. *diffuse*, *apical*, *trichothallic*, *promeristem*, *intercalary*, *meristoderm*).

Two types of reproductive structures occur: (a) unilocular, single-celled sporangia releasing haploid zoospores after meiosis (meiosporangium) which will germinate to form the gametophyte generation, and produce gametes; (b) plurilocular sporangium, each cell of which produces a single motile cell, functioning either as a gametangium (producing haploid gametes) if borne upon the gametophyte, or as a sporangium (producing diploid zoospores) if borne upon a sporophyte. Motile cells, always zoospores or gametes, have a long anterior hairy flagellum with tripartite hairs, and a shorter posteriorly directed smooth flagellum (except the Fucales where the posterior flagellum of the sperm is longer than the anterior flagellum). Flagella and eyespots are similar in structure to those of the xanthophytes and chrysophytes (40–80 lipid globules in a single layer between the outermost band of the thylakoids and chloroplast envelope). See **heterokont**.

Sexual reproduction is isogamous (Sphacelariales), anisogamous (Cutleriales), or oogamous (Desmarestiales, Laminariales, Fucales, Dictyotales). Both isomorphic and heteromorphic alternation of generations occur. In the Fucales, the gametophyte generation is reduced to the gametes (egg and sperm).

Most brown algae are epilithic, others are epiphytic and occur in the intertidal or subtidal zones, where they dominate in the Northern Hemisphere in colder waters. A species of the genus *Sargassum* is exceptional since it is pelagic, accumulating in large quantities in the Sargasso Sea.

Phaeophytes probably evolved from an organism in the Sarcochrysidales (**Chrysophyta**), which have motile cells similar to those found in the brown algae.

Phage Viruses infecting bacteria are termed **bacteriophages**; those infecting fungi are termed *mycophages*.

Phage conversion Phenomenon whereby new properties may be conferred upon host cells when infected by temperate phage. Each cell receiving the prophage also acquires the new property. See **bacteriophage**.

Phage restriction When phage are grown upon bacteria of one strain and then upon another, their titre may drop in the second host strain. They are then said to be *restricted* by the second host strain. Due to degradation of the phage DNA by host **restriction endonucleases**. See **modification**.

Phagocyte Cell which can ingest particles from its surroundings (phagocytosis), forming vacuole composed of the plasma membrane in which the material lies. The vacuoles may then fuse with **lysosomes** to form *heterophagosomes*. Receptor sites on the plasma membrane may be involved in vacuole formation, as with antibody markers on surfaces of phagocytic leucocytes. Many protozoans are phagocytic, as are those vertebrate white blood cells (*neutrophils*, *monocytes*) and **macrophages** which engulf bacteria and clumped antigens. See **leucocyte**, **immunity**.

Phagocytosis One form of **endocytosis** in which large particles/cell debris are taken up into large endocytic vesicles (*phagosomes*) which seems to occur by a 'membrane-zippering' process in which adhesion between engulfing cell and particle occurs progressively until the latter is completely surrounded by the cell's pseudopodia, in which actin and actin-binding proteins accumulate – sometimes with **clathrin**. The resulting endocytic vesicle (food vacuole) is normally converted to a heterophagosome as lysosomes fuse with it, degrading the contents to particles smaller than 1 Kd in size. The resulting concentration gradients of amino acid, sugars, etc., drive diffusion through specific carriers or channels into the cytoplasm. See also **potocytosis**.

Phagosome For autophagosomes and heterophagosomes, see **lysosome**.

Phalanges Bones of vertebrate digits (fingers and toes). Each finger has 1–5 phalanges (more in whales) articulating end-to-end in a row, the proximal of each row forming a joint with a *metacarpal bone*. See **pentadactyl limb**.

Phanerophytes Class of **Raunkiaer's life forms**.

Phanerozoic Geological division lasting from approx. 600 Myr BP to the present. Initiated by appearance of metazoan fossils. During it atmospheric oxygen level rose from about 0.1 of present levels to that of the present.

Pharynx (1) The vertebrate gut between mouth (buccal cavity) and oesophagus, into which opens the glottis in tetrapods and gill slits in fish. In man and other mammals is represented by throat and back of nose; partly divided by soft palate into upper (nasal) section and lower (oral, or throat) section. Contains sensory receptors setting off swallowing reflex. The **gas bladder** and **Eustachian tube**, where found, also open into it. See **gill pouch**. (2) Part of the gut into which gill slits open internally in urochordates and cephalochordates.

Phase contrast See **microscope**.

Phasmida Order of exopterygote insects containing stick insects and leaf insects.

Phellem See **cork**.

Phelloderm Tissue formed by the **cork cambium**; a cork skin. The cells of the phelloderm are living at maturity, lack **suberin**, and resemble **parenchyma** cells of the cortex, and can be distinguished from cortical cells by their inner position in the radial rows of other peridium cells.

Phellogen (cork cambium) Meristematic cells producing **cork**.

Phenetics Grouping of organisms into taxa on the basis of estimates of overall similarity, without any initial weighting of characters. Diagrams which result (*phenograms*) are devoid of necessary phylogenetic implications although may be interpreted phylogenetically. Phenetics is a branch of numerical taxonomy. See **classification**, **phylogenetics**.

Phenocopy Environmentally induced alteration in the phenotype of an organism (or cell), commonly resulting from abnormal developmental conditions. Mimics effect of a known mutation but is non-heritable. See **genetic assimilation**. Contrast **transdetermination**.

Phenogram See **phenetics**.

Phenology Study of periodicity phenomena in plants, such as timing of flowering in relation to climate.

Phenotype Total appearance of an organism, determined by interaction during development between its genetic constitution (**genotype**) and the environment. Different phenotypes may result from identical genotypes, but it is unlikely that two organisms could share all their phenotypic characters without having identical genotypes. Shared presence of a character in two organisms does not necessitate identical genotypes with respect to that character. See **dominance**, **penetrance**, **ecotype**, **phenotypic plasticity**.

Phenotypic plasticity Extent to which phenotype may be modified by expression of a particular genotype in different environments. Its often adaptive nature is illustrated by **heterophylly** in some members of *Ranunculus* (subgenus *Batrachium*) and by *Polygonum amphibium*. In animals, the phenomenon includes some cases of caste determination in insects and of **sex determination** in diverse groups. Any genotype probably has a characteristic degree of developmental plasticity (see **homeostasis**). In plants, dwarf, prostrate, thorny and succulent forms are often produced when **ramets** of a cloned genotype are grown in different conditions; genetic fixation (by selection) of an altered phenotype, preadapted to the conditions inducing it, may not be uncommon in wild populations although evidence is rather sparse. In a broad sense, animal **learning** could be included here; cell differentiation takes the range of the concept below the level of the individual. See **cline**, **deme**, **ecotype**.

Phenylketonuria Recessive human genetic disorder. The enzyme converting dietary phenylalanine to tyrosine is deficient, causing excretion of phenylpyruvate (or phenylalanine) in urine. Intellectual impairment common and epileptic attacks occur in about 25% of cases. Tendency to lighter hair and skin pigmentation than average. Can be detected soon after birth; a diet low in phenylalanine reduces symptoms.

Pheromones Chemical substances which, when released into an animal's surroundings, influence the behaviour or development of other individuals of same species. Include sexual attractants in many insect species. Worker and queen bees produce several different pheromones, each with its own effect; one deer produces pheromones from at least seven sites on its body, each with a different social function.

Phloem Principal food-conducting tissue in vascular plants. Mixed tissue, containing parenchyma and occasionally **fibres**, besides sieve elements, the main conducting cells (**sieve cells** in non-anthophytes, **sieve-tube members** in anthophytes), and their **companion cells**. Substances transported include sugars, amino acids, some mineral ions and growth substances. Phloem may be *primary* or *secondary* in origin, the former frequently being stretched and destroyed during elongation of the plant. Protoplasts of adjacent sieve elements connect via groups of narrow pores, the *sieve areas*, concentrated on the overlapping ends of these long, slender cells. Where the pores are large, the sieve area is termed a *sieve plate*. See **mass flow**, **vascular bundle**, **xylem**.

Phoronidea Small phylum of marine worm-like animals, unsegmented and coelomate, living in tubes (tubicolous) of chitin which they secrete. Resemble **Ectoprocta** in habit and appearance, but are only superficially colonial. Feed by ciliated **lophophores.** Planktonic larva resembles a trochophore. Vascular system contains haemoglobin.

Phosphagen Any of several of **high-energy phosphate** compounds which act as reservoirs of phosphate-bond energy in the cell, a **kinase** transferring their phosphate to ADP, forming ATP. Include *phosphocreatine* and *phosphoarginine*. Nerve and muscle, especially, contain phosphagens. Creatine and arginine become phosphorylated when ATP concentration in the cell is high,

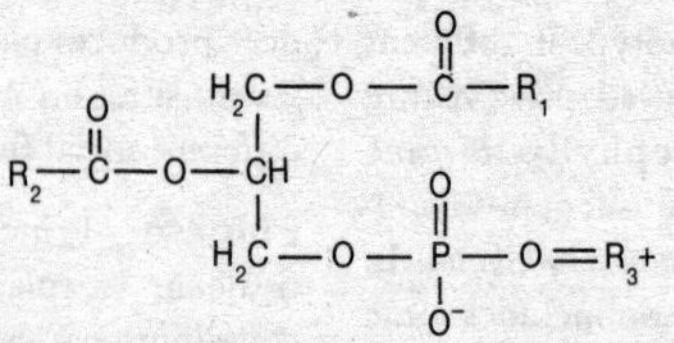

Fig. 93a *A generalized phospholipid.* R_1 *and* R_2 *represent fatty acid chains.* R_3 *represents one of several groupings.*

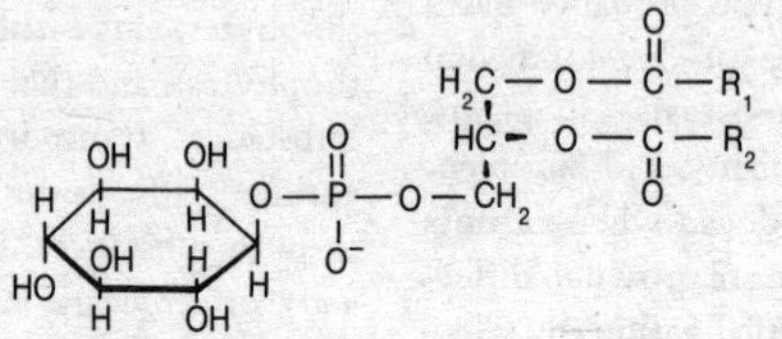

Fig. 93b *Phosphatidyl inositol.*

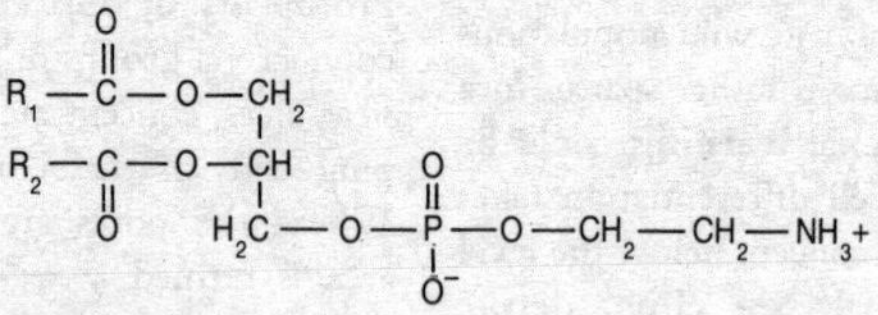

Fig. 93c *Phosphatidyl ethanolamine.*

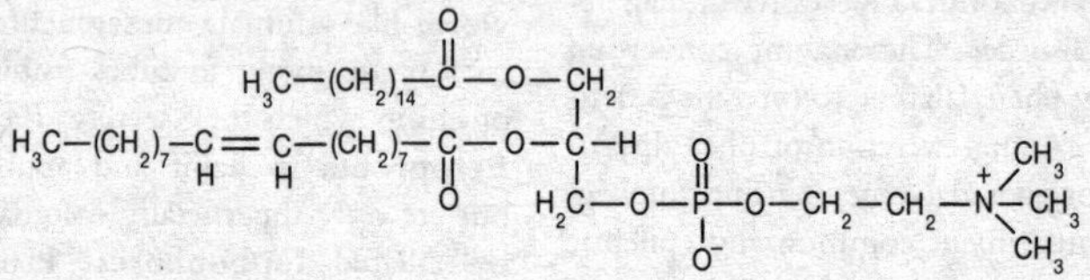

Fig. 93d *A phosphatidyl choline (= lecithin), with palmitic and oleic acid residues as the fatty acid chains.*

the reverse occurring when the cell's ATP to ADP ratio is low. Phosphoarginine is characteristic of invertebrates, phosphocreatine of vertebrates. Both occur in echinoids and hemichordates. See **creatine**.

Phosphatases Enzymes splitting phosphate from organic compounds. Compare **kinase**.

Phosphate-bond energy See **high-energy phosphate**.

Phosphatides See **phospholipids**.

Phosphatidylinositol See **phospholipids, phospholipid transfer proteins, inositol 1,4,5-triphosphate**.

Phosphocreatine See **phosphagen, creatine**.

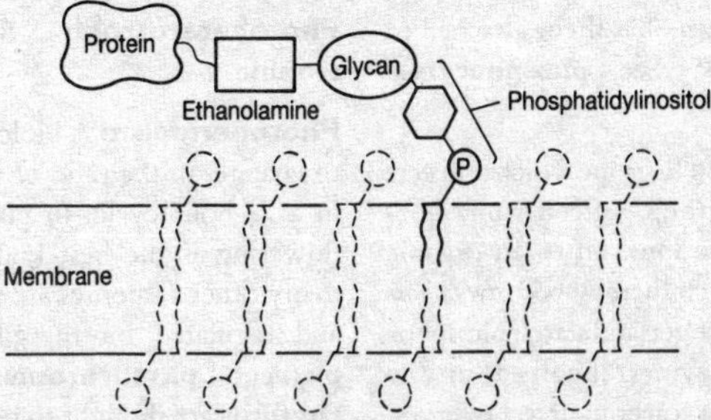

Fig. 93e *Figure indicating how a phospholipid molecule (here phosphatidylinositol) can form an attachment site for a membrane protein, in this case via a glycan and ethanolamine linker. Such attachments are common, notably in the glycoalyx.*

Phosphodiesterases Group of enzymes capable of hydrolysing cyclic AMP to (non-cyclic) 5′-AMP. Some also hydrolyse the phosphodiester backbone (sugar-phosphate backbone) of nucleic acids.

Phospholipase C (PLC) A **G-protein**-linked enzyme in the inner phospholipid layer of perhaps most eukaryotic plasmalemmas, releasing **inositol 1,4,5-triphosphate** and **diacylglycerol** (both **second messengers**) from phosphatidylinositol 4,5-bisphosphate located in the same layer. See **receptor**.

Phospholipids (phospholipins, phospholipoids, phosphatides) Those **lipids** bearing a polar phosphate end, commonly esterified to a positively charged alcohol group. Major components of cell membranes and responsible for many of their properties. There are two main groups: (a) the *phosphoglycerides*, derivatives of phosphatidic acid, include *lecithin* (phosphatidylcholine), *cephalin* (phosphatidylethanolamine) and phosphatidylinositol. Several important proteins are anchored in membranes by links between their C-terminal portions and phosphatidylinositol via a mannose-rich glycan and ethanolamine.

Lecithin, an important cell membrane component (and surfactant in vertebrate lungs), is a component of **bile**, helping to render cholesterol soluble; (b) the *sphingolipids*, containing the basic *sphingosine* instead of glycerol, forming components of plant and animal cell membranes, e.g. *sphingomyelins* (not to be confused with *myelin*) of liver and red blood cell membranes. See **cell membranes**, **cerebrosides**, and Fig. 93.

Phospholipid transfer proteins (PLTPs) Diffusible cytosolic eukaryotic proteins catalysing *in vitro* energy-dependent transfer of lipids between membrane bilayers. Some are very specific in their catalysis (e.g. PCTP, for phosphatidylcholine), some less so (e.g. PITP, for phosphatidylinositol), others much less so, carrying most PLs, glycolipids and cholesterol. Work *in vivo* suggests that in some yeasts the ratio of PITP/PCTP may affect **Golgi apparatus** secretory competence. See **cholesterol**.

Phosphoproteins Proteins with one or more attached phosphate groups. The milk protein *casein* has them attached to serine residues. See **phosphorylase kinase**.

Phosphorylase Enzyme which transfers a phosphate group, often from inorganic phosphate ions, on to an organic compound which thereby becomes *phosphorylated*. Glycogen and starch phosphorylases are enzymes involved in the mobilization of carbohydrate reserves, forming glucose-phosphate.

Phosphorylase kinase A **kinase** activating a **phosphorylase** by transfer of a phosphate group from ATP. Often AMP- or Ca^{2+}- dependent (see **calmodulin**). May exert its effect by allosteric change in the protein adjacent to the phosphorylase in a multiprotein complex. One of many regulatory enzymes in cells.

Phosphorylation Transfer of a phosphate group by a **phosphorylase** to an organic compound. Usually ATP-dependent, this produces compounds which are highly reactive in water with other organic molecules in the presence of appropriate enzymes, when their phosphate is transferred in turn and energy made available for work. The most important energy-transfer system in metabolism. See **kinase**, **high-energy phosphate**, **oxidative phosphorylation**, **photophosphorylation** and **ATP**.

Photic zone Upper portion of a lake, river or sea, sufficiently illuminated for photosynthesis to occur.

Photoautotroph See **autotrophic**.

Photoauxotroph Photosynthetic organism requiring an external vitamin source.

Photoheterotroph See **heterotrophic**.

Photoperiodism A biological response to changes in the ratio of light and dark in a 24-hour cycle. In plants, although flowering is the best known example, many other responses are photoperiodic and regulated by the photoreversible pigment **phytochrome**. **Circadian rhythms** are thought to be fundamental to photoperiodism.

With respect to flowering, plants may be grouped into three categories: (a) *short-day plants*, flowering in early spring or autumn (fall), requiring a dark period exceeding a critical length; (b) *long-day plants*, flowering mainly during summer, requiring a dark period less than a critical length; and (c) *day-neutral plants*, where flowering is unaffected by photoperiod. Different populations within a species are often precisely adjusted to the photoperiodic regimes where they live. Some plants require only a single exposure to the critical day–night cycle in order to flower. The stimulus is perceived by the leaves and transmitted (probably by some growth substance, or *florigen*) to growing points where flowering is initiated. Photoperiodic effects in animals include initiation of mating in aphids, fish, birds and mammals. See **proximate factor**.

Photophile Literally, light-loving; light-receptive phase of a **circadian rhythm**, lasting about 12 hours. Compare **skotophile**.

Photophosphorylation Coupling of phosphate with ADP to produce ATP, using light energy absorbed in **photosynthesis**. See **bacteriorhodopsin**.

Photoreactivation See **DNA repair mechanisms**.

Photoreceptor (1) (Bot.) Light-sensitive region of a cell receiving stimulus in phototaxis; usually a dense area in flagellar swelling. Compare **eyespot**. (2) In general, any light-sensitive **receptor organ** (e.g. **eye**), cell (e.g. **rod cell**) or molecule (e.g. **phytochrome**, **rhodopsin**).

Photorespiration Type of very active non-mitochondrial respiration occurring in conditions of high light intensity, reduced CO_2 levels and raised O_2 levels in temperate plants carrying out C_3 **photosynthesis**; usually absent (or low) in tropical C_4 plants. Involves oxidation of carbohydrates, takes place in **peroxisomes**, and yields neither ATP nor $NADH_2$; thus appears very wasteful. Main substrate, *glycolic acid*, is derived from oxygenation of ribulose bisphosphate (RuBP) through competitive inhibition of chloroplast ribulose bisphosphate carboxylase by molecular oxygen (see **enzyme**). Glycolic acid is oxidized by molecular oxygen in peroxisomes to yield hydrogen peroxide, which is destroyed by catalase. Up to 50% of photosynthetically fixed carbon may be reoxidized to CO_2 during photorespiration, lessening the efficiency of C_3 photosynthesis.

Photosynthesis The light-dependent synthesis of organic carbon from inorganic molecules occurring in **chloroplasts**, cells of blue-green algae, prochlorophytes, green sulphur bacteria, purple sulphur bacteria, and purple non-sulphur bacteria, in the presence of one or more types of light-trapping pigments (notably chlorophylls). Chloroplasts, blue-green algal and prochlorophycean cells contain chlorophyll *a* (occurs in all photosynthetic eukaryotes) as the major light-trapping pigment. Depending upon the group, other chlorophylls may or may not be present, e.g. chlorophyll *b* in the bryophytes, vascular plants, green and euglenoid algae, and prochlorophytes. When a molecule of chlorophyll *b* absorbs light, the excited molecule transfers its energy to a molecule of chlorophyll *a*, which then transforms it into chemical energy. Because chlorophyll *b* absorbs light of different wavelengths from chlorophyll *a*, it extends the range of light that can be used for photosynthesis. In other groups such as the **Chrysophyta**, **Bacillariophyta** and **Phaeophyta**, chlorophyll *c* serves the same function as chlorophyll *b*. Chlorophylls present in bacteria differ in several ways from chlorophyll *a* (e.g. chlorobium chlorophyll in the green sulphur bacteria; bacteriochlorophyll in purple sulphur and purple non-sulphur bacteria), but they have the same basic structure. Other accessory pigments include carotenoids and phycobilins. Photosynthesis is the route by which virtually all energy enters ecosystems. More than 150 billion metric tons of sugar are produced on a global scale per year.

It was Joseph Priestley (1771) who first reported that 'vegetation could restore air'; then F. F. Blackman (1905) demonstrated that photosynthesis was a two-step process, while B. Van Niel showed that purple sulphur bacteria reduced carbon to carbohydrates but released no oxygen. These bacteria require H_2S for their photosynthetic activity.

$$CO_2 + H_2S \xrightarrow{\text{light}} (CH_2O) + H_2O + 2S$$

Sulphur accumulated inside the bacterial

cells. Van Niel made the brilliant extrapolation to plants, proposing that water, not carbon dioxide, was split in photosynthesis in the 1930s. This was proven years later using ^{18}O isotopes:

$$CO_2 + 2H_2{}^{18}O \xrightarrow{\text{light}} (CH_2O) + H_2O + {}^{18}O_2$$

Thus, water acts as an electron donor in plants, algae, blue-green algae and prochlorophytes.

Photosynthesis occurs in two stages: (a) a light-trapping stage (*light phase*), in which light energy (photons) initiates photochemical reactions on pigment molecules, producing energy-rich compounds (ATP and NADPH) with the release of molecular oxygen (plants, algae, blue-green algae and prochlorophytes only); (b) a light-independent stage, in which enzymes located off the pigment molecules use products of the light phase and incorporate (fix) CO_2, using the atoms of its (inorganic) molecules to synthesize more organic molecules. (Since the carbon-fixing enzyme is not itself dependent upon light, this is often called the *dark phase*. It occurs in the light unless experimentally contrived so as not to.) Energy for this carbon-fixation comes from ATP and NADPH produced momentarily earlier in the light-phase.

Chloroplasts are thought to have evolved through **endosymbiosis** from prokaryotic cells resembling blue-green algae. Bacterial photosynthesis differs from others in not releasing molecular oxygen, for the pigment system (photosystem) needed to utilize hydrogen atoms in water is lacking, and these usually anaerobic bacteria use alternative hydrogen donors (see **autotrophic**).

In blue-green algae, prochlorophytes and chloroplasts, photons are absorbed on thylakoid membranes by chlorophylls and accessory pigments arranged in the form of **antenna complexes**. Energy is passed by resonance transfer between pigments until it reaches the reaction centre of one of two types of photosystem, each containing a specific form of chlorophyll *a*. Light energy enters photosystem II where it is trapped by the reaction centre, P_{680}, either directly or indirectly via one or more of the pigment molecules. When P_{680} is excited a pair of excited electrons (excitons) are donated to an organic receptor molecule, designated as 'Q' because of its ability to quench the loss of energy by fluorescence of excited P_{680} (probably a quinone). The electrons lost are replaced by a pair from a water molecule as protons and molecular oxygen are released into the thylakoid space. This light-dependent oxidative splitting of water molecules is termed *photolysis*. Manganese is an essential cofactor for the oxygen-evolving mechanism. In the other photosystem, photosystem I, chlorophyll a_{700} donates a pair of electrons to another organic receptor which, assisted by a **ferredoxin** molecule, reduces NADP to release NADPH into the chloroplast stroma. Electrons lost are replaced by those from P_{680} after they have passed along an **electron transport system** which includes cytochromes, iron-sulphur proteins, quinones, chlorophyll and the copper-containing protein plastocyanin. Energy released drives protons from the stroma across the thylakoid membrane into the thylakoid spaces, diffusing out through an ATPase in the membrane and releasing sufficient energy for synthesis of ATP from ADP and inorganic phosphate. The whole electron and proton flow sequence is called *non-cyclic photophosphorylation*. After the electrons have left P_{700} of photosystem I they may short-circuit back again via some of the electron transport molecules. This produces no NADPH, but proton-pumping and ATP formation do occur; this is termed *cyclic photophosphorylation*. See Fig. 32. The dark phase of photosynthesis differs between so-called C_3 and C_4 plants. In the former (including most temperate plants; about 85% of all plant species), CO_2 is fixed by the enzyme *ribulose bisphosphate carboxylase/oxidase* (Rubisco), acting as a carboxylase. The two substrates are CO_2 and ribulose 1,5-bisphosphate (RuBP), and the product is the 3-carbon (hence C_3) compound 3-phosphoglyceric acid (PGA). PGA is not energetic enough for further metabolism, but is converted using ATP and NADPH from the light phase to glyceraldehyde-3-phosphate. This can be used to recycle RuBP (the Calvin cycle – after its discoverer, the Nobel laureate Melvin Calvin) to synthesize starch, or sucrose. If Rubisco functions as an oxidase then RuBP serves as a substrate for **photorespiration**, and photosynthesis is less efficient.

In C_4 plants (e.g., many cereals and the rice grass *Spartina*), the first product of CO_2-fixation is 4-carbon oxaloacetate, produced by phosphoenolpyruvate carboxylase (PEP carboxylase), which uses phosphoenolpyruvate as its other substrate. In this Hatch–Slack pathway (named after the two Australians who played key roles in its elucidation), oxaloacetate is further reduced to malate or changed by addition of an amino group to aspartate. These reactions occur in the mesophyll cells of the leaf, and the malate (or aspartate) moves from them to bundle-sheath cells surrounding the vascular bundles (see **Krantz anatomy**). Here malate is decarboxylated to yield CO_2 and pyruvate, the CO_2 entering the Calvin cycle as substrate for RuBP-carboxylase while pyruvate reacts with ATP to form more PEP molecules. Such plants can generally photosynthesize at temperatures far higher than C_3 plants, and generally have a far lower CO_2 **compensation point**. Even though C_4 plants tolerate higher temperatures and more arid conditions than C_3 plants, at temperatures below 25°C they may not compete successfully, being more sensitive to cold than C_3 plants. See **bacteriophodopsin, crassulacean acid metabolism, carbon cycle**.

Photosystem See **photosynthesis** and Fig. 32b.

Phototaxis **Taxis** in which the stimulus is light.

Phototrophic See **autotrophic**.

Phototropism **Tropism** in which light is the stimulus, e.g. the bending of a stem of an indoor plant towards a window, brought about by increased cell elongation in the growth region of the shaded side. Has been shown to be caused by unequal distribution of the growth substance **auxin**, which migrates from the light side to the dark side of the shoot, particularly in light of wavelengths of 400–500 nm. Thus a pigment absorbing blue light mediates the effect; evidence suggests that it is a flavin pigment.

Phragmoplast (Bot.) Spindle-shaped system of microtubules, arising in all plant (bryophytes and vascular plants) and a few algal cells between two daughter nuclei at telophase, and within which the cell plate is formed. The microtubules in a phragmoplast are orientated perpendicular to the plane of cell division. See **phycoplast**.

Phycobiliprotein Water-soluble pigments located on (**Cyanobacteria**, **Rhodophyta**) or inside (**Cryptophyta**) thylakoids of the chloroplasts of the three aforementioned algal divisions. Coloured proteins (chromoproteins) in which the prosthetic group (non-protein part of the molecule or chromophore) is a tetrapyrrole (bile pigment) known as phycobilin. The prosthetic group is tightly bound by covalent linkages to its protein portion of the molecule (apoprotein). There are two apoproteins, α and β. The major chromophore of phycocyanin and allophycocyanin (blue pigments) is phycocyanobilin, and that of phycoerythrin is phycoerythrobilin. In addition in B- and R-phycoerythrin there is the chromophore phycourobilin. Each phycobiliprotein usually comprises a basic aggregate of three molecules of α-apoprotein and three molecules of β-apoprotein with the chromophore attached to the apoproteins. Phycobiliproteins have to be assembled into proper sequences and attached to the thylakoid membranes by means of linker polypeptides. These function as **accessory pigments**.

Phycobiont Term referring to the algal partner in a lichen.

Phycocolloid A polysaccharide colloid formed by an alga; e.g. **carrageenan** formed by various red algae.

Phycocyanin See **phycobiliproteins**.

Phycoerythrin See **phycobiliproteins**.

Phycology Study of **algae**.

Phycomycetes In older classifications, all lower fungi; included the distantly-related Mastigomycotina and **Zygomycotina**.

Phycoplast (Bot.) System of microtubules found in the largest class of green algae (**Chlorophyta**), the Chlorophyceae, which develops parallel to the plane of cell division between the daughter nuclei as they move towards one another and the non-persistent spindle collapses. Phycoplast functions in ensuring that the cleavage furrow passes between the daughter nuclei. See **phragmoplast**.

Phyllid Flattened leaf-like appendage in bryophytes.

Phylloclade See **cladode**.

Phyllode Flat, expanded petiole replacing blade of leaf in photosynthesis.

Phyllotaxy (phyllotaxis) (Bot.) Arrangement of leaves on the stem: whorled, opposite or spiral. In spiral phyllotaxy, a line connecting attachment points of successive leaves forms a spiral; individual leaves regularly positioned within this spiral. In the most simple, truly alternate, arrangement, leaves are 180° apart, and passage from one half leaf to that precisely above it involves one circuit of the stem and two leaves – a phyllotaxy of 1/2. Various forms of phyllotaxy occur, such as 1/3, 2/5, 3/8, 5/13, etc. (a Fibonacci series), each fraction representing an angle made by successive leaves with the stem (looking vertically downwards). At the end of each spiral, a leaf is directly above the one at the beginning. Looking down on a stem, these points of superimposition are identified as vertical rows of leaves known as *orthostichies*. At the growing point, though leaf primordia are spirally arranged, orthostichies do not occur; but looking down at the apex, the primordia are arranged in a series of descending curves, or *parastichies*, some clockwise. Parastichy in an apex may become orthostichy in a mature shoot by straightening during elongation.

Phylogenetics Approach to biological **classification** concerned with reconstructing **phylogeny** and recovering the history of speciation. This is possible when speciation is coupled with, and does not proceed faster than, character modification. *Phylogenetic trees* so produced should represent the hypothetical historical course of speciation and be open to rigorous testing. The system in ascendancy today is **cladistics**. Where appropriate, the techniques include comparative anatomy and embryology; DNA protein (and cytochrome) sequencing; **DNA hybridization**; immunodiffusion and immunoelectrophoresis. See **molecular clock**; compare **phenetics**.

Phylogeny Evolutionary history. Genealogical history of a group of organisms, in practice represented by its hypothesized ancestor-descendant relationships. Compare **ontogeny**. See **recapitulation**.

Phylotypic stage In an animal's development, the stage at which the precursor to the general body plan characteristic of the phylum to which it belongs is expressed; or, the stage at which all major body parts are represented in their final relative anatomical positions as undifferentiated cell groups (immediately after the principal morphogenetic tissue movements). Such stages have been proposed as: the tailbud stage (for vertebrates); the fully segmented germ band stage (for insects); the fully segmented ventrally closed stage (for leeches) and the completion of most embryonic cell divisions (for nematodes). See **zootype**.

Phylum Taxonomic category often restricted to the animal kingdom; includes one or more **classes** and is included within a **kingdom** in the taxonomic hierarchy. Corresponds to the category **division** in botany. See **phylotypic stage**.

Physiological saline See **Ringer's solution**.

Physiology Study of processes, many either directly or indirectly homeostatic, that occur within living organisms; in multicellular organisms, includes interactions between cells, tissues and organs and all forms of intercellular communication, both energetic and metabolic.

Phyto- Prefix indicating a botanical context.

Phytoalexins Non-specific compounds, generally phenolic, synthesized *de novo* or in greatly increased concentration by plants in response to infection by fungi, to which they are toxic. Believed to play a primary role in disease resistance.

Phytochrome Plant chromophore-containing protein pigment existing in two alternative, interconvertible, forms.

Debatable whether it is a **protein kinase** or merely intimately associated with kinase activity; but has ATP-binding site and mediates long-term alterations in gene transcription. P_r absorbs (is *receptive* to) red light (660 nm) and P_{fr} absorbs far-red light (730 nm), conversion from one to the other apparently modifying auto-phosphorylation sites. When a P_r molecule absorbs a photon it is converted in milliseconds to a molecule of P_{fr}, the reverse happening when P_{fr} absorbs a photon, they penetrate through the soil to the light, for plastid differentiation, anthocyanin development, stem elongation, leaf initiation and flowering. The P_r:P_{fr} ratio is instrumental in breaking **dormancy** in some seeds.

Phytogeography See **plant geography**.

Phytoplankton Algal plankton, comprising a diverse range of cell size and cell volume, including non-motile and motile unicells and colonies. Organisms complete all or the majority of their life cycle suspended in the open water. Phytoplankton can be arbitrarily classified according to size (e.g. microplankton, 60–500 μm diameter, nanoplankton, 5–60 μm diameter, ultraplankton, 0.5–5 μm diameter), and can be subdivided into several categories (*Euplankton* = Holoplankton, the permanent planktonic assemblage of organisms which completes its life cycle suspended in the water; *meroplankton*, includes those species which spend part of each season resting on the bottom sediments; *pseudoplankton* = tychoplankton, the assemblage of casual species derived from other habitats; quite common in lakes and rivers, especially after storms). Further the prefixes, *limno-*, *potamo-* and *heleo-* are used to refer to the plankton of lakes, rivers and ponds. There is also a flora, both in marine and freshwaters associated with flocs of organic detritus up to 1 mm or more in diameter. Algae, not planktonic taxa but benthic taxa associated with surfaces, can be found living in this unusual site and have been termed the *detritiplankton*.

Phytosociology See **plant sociology**.

Pia mater The innermost of the **meninges**.

Pileus Cap-like part of fruiting body of fungi of Basidiomycotina (e.g. mushrooms), bearing the hymenium on its undersurface.

Pili (sing. pilus) Proteinaceous filaments protruding the cell wall of mainly Gram-negative bacteria; shorter and straighter than flagella. The exact function is not known; however, they may serve as a bridge for transfer of DNA, or may pull conjugating cells together or may have an adhesive role, attaching to membrane surfaces. Pili (fimbriae) also occur protruding from the cell walls of some blue-green algae (**Cyanobobacteria**) but differ in several characteristics from those of bacteria. See **F-factor**.

Piliferous layer That part of the root epidermis bearing root hairs.

Pineal eye See **pineal gland**.

Pineal gland (epiphysis) Small mass of nerve tissue attached to roof of third ventricle of the vertebrate midbrain; loses all nervous connection with the brain but innervated by sympathetic nervous system. In amphibia, primitive reptiles (e.g. *Sphenodon*, the tuatara) and some snakes, pineal cells form a *parietal eye* (median eye) lying within a parietal foramen of the skull, with a lens-like upper epithelium and retina-like lower part. In higher vertebrates, the pineal organ has a glandular structure, and secretes *melatonin* which inhibits gonadotrophins and their effects. Melatonin production is inhibited by exposure of the animal to light. **Serotonin**, another pineal product, and melatonin demonstrate inverse circadian rhythms in their production. See Fig. 13.

Pinna (pl. pinnae) (Bot.) A primary division, or leaflet, of a compound leaf or frond. (Zool.) See **ear, outer**.

Pinnate Form of branching which occurs at uniform angles from different points along a central axis, all in one plane, as in a feather.

Pinnipedia Eutherian order (or suborder of the order **Carnivora**), containing specialized and aquatic mammals: seals (Phocidae), sealions (Otariidae) and walruses (Odobenidae). Limbs are broad flippers, with webbed feet; tail very short. In true seals, hind limbs are fused with tail.

Pinocytosis The bulk-phase component of **endocytosis**, involving ingestion of surrounding extracellular fluid by part of the eukaryotic plasma membrane by invagination to form an endocytic vesicle. See **coated vesicles, potocytosis** and Fig. 75.

Pistil Either each separate carpel (in *apocarpous gynoecia*), or two or more fused carpels (in *syncarpous gynoecia*). Each typically comprises an ovary, style and stigma.

Pistillate (Of flowers) naturally possessing one or more carpels but no functional stamens; also called carpellate. See **staminate**.

Pith Ground tissue located in the centre of stem or root, within the vascular cylinder. Usually comprises **parenchyma**. See **stele**.

Pithecanthropus See ***Homo***.

Pits Small, sharply defined depressions in wall of plant cell where the secondary wall is completely absent, permitting easier passage of material between adjacent cells. Usually occur over *primary pit fields* (where plasmodesmata are concentrated), but also in their absence. Often coincide with pits in the walls of adjacent cells, separated from them by a *pit membrane* comprising a middle lamella and a very thin layer of primary wall on either side, together forming a *pit pair*. Such *simple pits* connect living cells together and also occur in stone cells and some fibres. In *bordered pits*, characteristic of xylem vessels and tracheids, the pit cavity is partly enclosed by over-arching of the cell wall (the *pit border*) and the pit membrane may possess a central, thickened impermeable *torus*, closing the pit aperture if the pit membrane is displaced laterally.

Pituitary gland (hypophysis) Small but essential vertebrate gland, lying in a depression of sphenoid bone of skull and communicating with the hypothalamus (part of the diencephalon of the forebrain) by a stalk-like *infundibulum* (pituitary stalk). A composite gland, comprising an anterior lobe (*adenohypophysis*) deriving from pharyngeal ectoderm and secreting the bulk of pituitary hormones, and a posterior lobe (*neurohypophysis*) deriving from ectoderm of the hypothalamus and containing the termini of hypothalamic neurosecretory nerve axons.

The adenohypophysis releases its hormones into capillaries, drained by the hypophysial vein, under commands (*releasing factors*) from the hypothalamus which reach it via a pituitary portal system. Hormones secreted by the most anterior region (*pars distalis*) include **growth hormone**, **prolactin**, **thyroid-stimulating hormone** (**TSH**), **gonadotrophins** and **ACTH**. Where present, an intermediate region (*pars intermedia*) secretes **melanocyte-stimulating hormone**. The neurohypophysis, neurosecretory rather than endocrine in function, releases the hormones **oxytocin** and **antidiuretic hormone** (see **neurosecretory cells** for *neurophysins*).

The pituitary forms an integral link in many homeostatic feedback circuits in the body. See **hypothalamus** and Fig. 13.

Placenta That part of the ovary wall to which the ovules are attached and where they remain until mature. The arrangement of the placenta and ovules varies among different groups of flowering plants. May be *parietal* (ovules borne on the ovary walls or on extensions of it); *axile* (ovules borne upon a central column of tissue in a partitioned ovary with as many locules as carpels); *free central* (ovules borne upon a central column of tissue not connected by partitions with the ovary wall); *basal* (single ovule occurs at the very base of a unilocular ovary). Such differences are important in flowering plant classification. (Zool.) A temporary organ, consisting of both embryonic and maternal tissues, within the uterus of therian mammals and several other viviparous animals, which enables embryos to derive soluble metabolites by diffusion and/or active transport from the maternal blood supply. In placental mammals the organ is connected to the embryo by the *umbilical cord* and has an essential role in the immunological protection of the embryo. See Fig. 35.

In mammals (see Fig. 94), foetal components of the placenta derive initially from the **trophoblast**, connected with the embryonic bloodstream either through its contact with the **yolk sac** (*vitelline placentation*) or **allantois** (*allantoic placentation*). In the latter, the outer covering is termed the *allantochorion*, a term also used of placentae in which blood vessels, but little else, derive from the allantois (as in humans). In both vitelline and allantoic placentation, trophoblastic villi push out from the surface of the chorion and invade the uterine lining (*endometrium*; see **decidua**), greatly increasing the surface area (14 m^2 in humans) for exchange of solute molecules. In marsupials, vitelline placentation serves the entire, brief, intrauterine life; in eutherians it is soon replaced by allantoic placentation, and sometimes only certain restricted zones of the allantochorion (termed *cotyledons*) form villi and participate in exchanges. Functions of the mammalian placenta include: (a) allowing passage of small molecules from uterine capillaries, or the intervillous blood sinuses (depending upon the type of placentation), into the capillaries of the umbilical vein. These include O_2, salts, glucose, amino acids and small peptides (all by active transport), simple

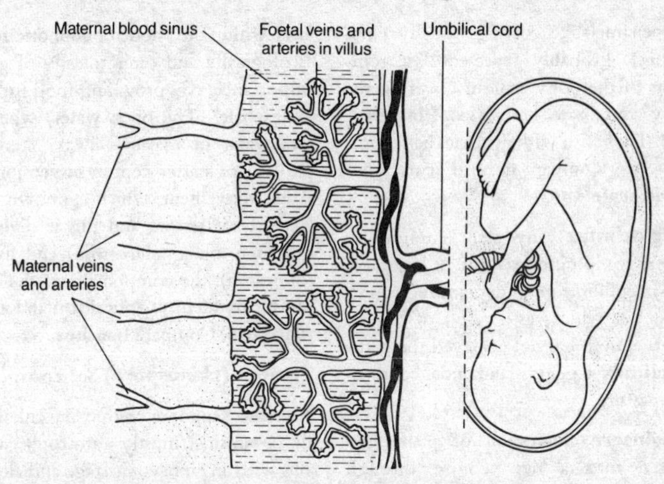

Fig. 94 *Diagram of human placenta and its relationship to the foetus. Left side of diagram is magnified compared to right side.*

fats, some antibodies (see **IgG**) and some vitamins (A, C, D, E and K), but in addition psychoactive (some dependency-including) drugs such as cocaine, **caffeine** and **nicotine**; (b) removal of embryonic excretory molecules, notably CO_2 and urea, by diffusion from the umbilical artery into the maternal circulation; (c) storage of glycogen, and its conversion to glucose if foetal glucose levels drop; (d) hormone production, substituting for maternal ovaries in production of the sex hormones **oestrogens** (largely oestriol) and **progesterone** (primates especially), and in humans producing **human chorionic gonadotrophin**, **human placental lactogen** and such 'pituitary' hormones as ACTH and TSH; (e) production of **endorphins**, and (f) prevention of red blood cell exchange between mother and foetus, avoiding **agglutination**. (The human placenta is exceptional in allowing a mutual exchange of leucocytes and blood platelets. See **parturition**.)

Placentalia 'Placental' mammals. See **Eutheria**.

Placentation (Bot.) Manner in which the ovules are attached in an ovary. See **placenta**. (Zool.) The type of arrangement of the **placenta** in mammals.

Placodermi Class of fossil bony fish, some large predators, mainly from the Devonian period. Bony shields on head and front part of body. **Autostylic jaw suspension**; paired fins; heterocercal tail. Possibly related more to Chondrichthyes than to Osteichthyes or Acanthodii.

Placoid scale (denticle) Tooth-like scale, characteristic of the **Chondrichthyes** and completely covering elasmobranch fish. Base bone-like, but made largely of dentine, with a pulp cavity and enamel-like coating (usually *vitrodentine*). Probably represent last remnants of the bony armour covering the early vertebrate body (see **Placodermi**). Those fish with denticles have similar teeth. Compare **ganoid scale**, **cosmoid scale**.

Plagioclimax Any plant community whose composition is more or less stable and in equilibrium under existing conditions but which, as a result of human intervention, has not achieved the natural **climax**; e.g. grassland under continuous pasture.

Plagiosere Succession of plants deflected into a new course through human intervention. Compare **prisere**.

Plagiotropism Orientation of a plant part by growth curvature in response to gravitational stimulus, such that its axis makes an angle other than a right-angle with the line of gravitational force. E.g. exhibited by branches of a main root which makes an acute angle with the vertical. Also used in a wider sense to mean the orientation of an organ so that its axis makes a constant angle with the vertical, and includes **diageotropism** as a special type. See **geotropism**.

Planarians Free-living members of the **Platyhelminthes**. Class Turbellaria.

Planation Dichotomies in two or more planes, flattening out into one plane, e.g. a leaf.

Plankton Organisms, algae and animals kept in suspension by water turbulence and dispersed more by such water movements than by their own activities. Most are small, non-motile or motile unicells or colonies, occurring particularly within the photic or euphotic zone. Ecologically and economically of great importance, e.g. providing food for fish and whales. The open water, whether freshwater or marine, always contains particulate matter kept in suspension by water movement, which is known collectively as seston, and this is divisible into non-living (abioseston) and living (bioseston); the components of the bioseston are the **phytoplankton** and **zooplankton**. Compare **benthos**.

Planont (planospore) See **zoospore**.

Plantae Kingdom comprising eukaryotic organisms (mostly autotrophs) usually with an embryonic stage and clearly defined cellulose-containing cell walls. Plantae include the **Bryophyta** (liverworts, mosses, hornworts), **Psilophyta** (psilopsids), **Lycophyta** (lycophytes), **Sphenophyta** (horsetails), **Pterophyta** (ferns), **Coniferophyta** (conifers), **Cycadophyta** (cycads), **Ginkgophyta** (*Ginkgo*), **Gnetophyta** (gnetophytes), and **Anthophyta** (flowering plants). All are adapted for life on land; their ancestors were specialized green algae (**Chlorophyta**). Reproduction in plants is primarily sexual, with cycles of alternating haploid and diploid generations, the former becoming reduced during evolution in the more advanced members. Structural differentiation into e.g. leaves, roots and support organs occurred during evolution of land plants.

Plant geography Study of plant distributions, past and present, and of their causes. Includes such fields as geology, palaeontology, plant genetics, evolution and taxonomy. See **zoogeography**.

Plantigrade Mode of walking involving the ventral surface of the whole foot, i.e. the metacarpus or metatarsus and digits. Humans are plantigrade. Compare **digitigrade**, **unguligrade**.

Plant sociology (phytosociology) Study of plant **communities** comprising the vegetation, their origins, compositions and structures.

Planula Solid free-swimming ciliated larva of most classes of Cnidaria and a few of the Ctenophora. Composed of an outer ectoderm and inner endoderm, the latter formed by migration of cells into the interior.

Plaque A clear area in a bacterial culture on nutrient agar caused by localized destruction of bacterial cells through **bacteriophage** activity. Similarly applied to a zone of lysis caused by a virus in animal tissue culture. Counting plaques provides a simple technique for estimating virus concentration in a fluid applied to the culture.

Plasma See **blood plasma**.

Plasma cell One kind of mature **B cell**, active in antibody secretion.

Plasmagel The gel-like state of the outer cytoplasm of many cells, as distinct from the more fluid inner cytoplasm (*plasmasol*) with which it is interconvertible. May be highly vacuolated in planktonic protozoans. Involved in several forms of **cell locomotion**, notably *amoeboid locomotion*.

Plasmalemma Outer membrane of a cell. See **cell membranes**.

Plasmid A piece of symbiotic DNA, mostly in bacteria but also in yeast, not forming part of the normal chromosomal DNA of a cell and capable of replicating independently of it. Plasmids carry a signal situated at their replication origin dictating how many copies (the copy number) are to be made, and this number can be artificially increased. It constitutes a form of *extrachromosomal DNA*, sometimes housing genes encoding enzymes (e.g. conferring drug resistance) of critical value to the host cell or organism. Sometimes confer no recognizable phenotype on the host cell (*cryptic plasmids*). Those that may be considered to do so include the bacterial sex factor (**F factor**) and some forms of **phage** (*temperate phages*, e.g. P1, in their non-temperate phase). A segment (T-DNA) of the tumour-inducing plasmid (T_i) can be transferred from the bacterium ***Agrobacterium*** to plant cells at a wound site. Some plasmids, if not all, can be transferred from one cell to another by conjugation (see **F factor**), **transduction** and **transformation**. Often act as vehicles for **transposable elements** such as **antibiotic resistance elements**. Plasmids which can pick up such transposable elements are termed *R plasmids*. Another type of plasmid, *Col plasmids*, are harboured by many bacterial cells and enable them to produce nutrient-harvesting proteins which effectively starve competing strains; e.g. the iron-scavenging colicin produced by one strain of *E. coli* tends to inhibit other strains of the species (see **transferrin**). The plasmid also renders cells which contain it immune to the colicins. The implications for public health of R plasmids, particularly their mobility between bacteria of different species and genera, are serious. Antibiotic resistance (selected for in non-pathogenic intestinal bacteria by too frequent administration of antibiotics) can be transferred to pathogenic bacteria (e.g. *Salmonella* and

Shigella). Such resistance has been recorded in sewers and polluted rivers and is a strong reason for restricting antibiotic administration to essential cases. See **cytoplasmic inheritance**.

Plasminogen activators Proteases secreted by endothelial cells, activating plasminogen to release the enzyme plasmin and enabling endothelial cells to digest the basal lamina of their capillary and invade new tissue (part of capillary increase, angiogenesis). Generally enables cells to dissolve their attachments and become mobile. Human urokinase plasminogen activator (of Ly-6 protein superfamily) is present on leucocytes and some non-lymphoid cells.

Plasmodesmata (sing. plasmodesma) Extremely fine cytoplasmic tubes (about 30–60 nm in diameter), which pass through the cell walls of living plant cells, and are lined by a plasma membrane of the two adjacent cells, connecting their cytoplasms (see **symplast**). They are traversed by a tubule of endoplasmic reticulum known as the desmotubule. Many plasmodesmata are formed during cell division as strands of tubular endoplasmic reticulum become trapped within the developing cell plate. They may occur throughout the cell wall, or may be aggregated in the primary pit fields or the membranes between pit-pairs.

Plasmodium (1) Generic name of malarial parasites. See **malaria**. (2) Vegetative stage of slime-fungi (**Myxomycota**). Multinucleate, amoeboid mass of protoplasm bounded by a plasma membrane but lacking definite shape or size. See **coenocyte, syncytium**.

Plasmogamy Union of two or more protoplasts, typically of gametes, unaccompanied by union of any nuclei they may have. Compare **karyogamy**.

Plasmolysis Shrinkage of a cell's protoplasm (in plants, away from cell wall) when placed in a hypertonic solution, as a result of osmosis. Compare **turgor**.

Plastid endoplasmic reticulum See **chloroplast endoplasmic reticulum**.

Plastids Small, variously shaped, organelles in the cytoplasm of plant and algal cells, one to many per cell. Each plastid is surrounded by an envelope of two membranes and contains a system of internal membranes, pigments and/or reserve food material, a more or less homogeneous ground substance (stroma) in which ribosomes of a prokaryotic type may be present. Arise either by division of existing plastids or from proplastids, and are classifiable into **chloroplasts** (sites of photosynthesis, and the chlorophyll and carotenoid pigments, as well as phycobiliproteins – in some algae), **chromoplasts**, non-pigmented **leucoplasts**, and **proplastids**. See **plastogene**.

Plastron (1) Flat horny scutes on underside of body of a turtle or tortoise, connected to the carapace above by a bony bridge. (2) An air store forming thin film over bodies of several aquatic insects (some beetles and heteropteran bugs), communicating with the spiracles and held by hydrofuge bristles or scales. Can act as a 'physical gill' if there is adequate dissolved oxygen, enabling the insect to remain under water.

Platelets (thrombocytes) Small cell-like blood-borne fragments (3–4 μm long) shed from megakaryocytes in vertebrate bone marrow. Discoid in shape, they contain secretory granules, express Class I MHC products and receptors, Fc receptors for IgG (see **antibody**) and IgE, and receptors for factor VIII (see **blood clotting**). Platelets adhere to injured endothelial cells and then release chemoattractants for leucocytes, platelet-derived **growth factor** (PDGF) and activators of **complement** as well as increasing permeability of capillaries (see **inflammation**). Platelets are destroyed by **reticulo-endothelial system** after about 10 days. See **eicosanoids**.

Platyhelminthes (flatworms) Invertebrate phylum, containing bilaterally symmetrical triploblastic acoelomates, dorso-ventrally flattened, with only one opening to the gut (when present), and without a blood system. Include planarians (Turbellaria), flukes (Trematoda) and tapeworms (Cestoda). They have bulky parenchymatous tissue derived from mesoderm, **flame cells** and complex hermaphroditic reproductive organs.

Platypus Duck-billed platypus (*Ornithorhynchus*). See **Prototheria**.

Platyrrhines Members of the anthropoid Infra-order Platyrrhini (Ceboidea, New World monkeys). Probably isolated from other primates since the Eocene. Characterized by broad nasal septum; 36 teeth (incl. 3 premolars in each jaw quadrant); often have prehensile tails. Include marmosets, tamarins, howler monkeys. See **Anthropoidea**, **catarrhine**.

Plecoptera Stoneflies. Small order of exopterygote insects with aquatic nymphs, long antennae, biting mouthparts and weak flight. Two pairs of wings, held flat over back at rest; hind pair usually larger. Adults tend to feed on lichens and unicellular algae. Good pollution indicators.

Plectenchyma Tissue composed of fungal hyphae; either *prosenchyma*, in which component hyphae are loosely woven and recognizable as such, or *pseudoparenchyma*, comprising closely-packed cells which can no longer be distinguished as hyphae and resemble parenchyma of higher plants.

Plectostele A protostele, split into many plate-like units.

Pleiomorphism (pleomorphism) Form of **polymorphism** in which there are several different forms in the **life cycle**; e.g., in the life cycles of rust fungi, where there is a succession of different spore forms.

Pleiotropy The ability of allelic substitutions at a gene locus, or a cell product, to affect or to be involved in the development of more than one aspect of phenotype. For any given allele some effects may be dominant, others recessive (see **dominance**). A gene locus may have pleiotropic effects if it encodes an enzyme or regulatory protein whose product is involved in several biochemical pathways, as commonly occurs in development (compare **polygenes**). Several **second messengers**, whose targets may not be restricted to a single molecule, have pleiotropic effects. **Phytochrome** and many **protein kinases** are pleiotropic.

Pleistocene Geological epoch; first of the Quaternary period (see **geological periods**). An upper value of duration would be 2.5 Myr–11,000 yr BP. Characterized by four major glacial advances (altogether more than two dozen glacial advances and retreats occurred) separated by interglacial periods lasting tens of thousands of years. The epoch saw rapid hominid evolution (see ***Homo***), the extinction of many large mammals and birds, the appearance of elephants, cattle and the modern horse. The climate fluctuated from cold to mild, and the final uplift of many mountain ranges occurred.

Plerome See **apical meristem**.

Plesiomorphic In **phylogenetics**, describing the original pre-existing member of a pair of homologous characters, the evolutionary novelty being the *apomorphic* member of the pair. See **cladistics**.

Plesiosauria Extinct suborder of marine euryapsid reptiles (Order Sauropterygia), prominent in the Mesozoic. Typical length 3 metres; long-necked, short-bodied, with limbs modified into powerful paddles. Derived from similar *nothosaurs*, they were not dinosaurs.

Pleura Serous membranes lining **pleural cavity** and covering the lung surfaces in mammals. Normally only a thin fluid layer separates the pleura from body wall.

Pleural cavity Coelomic space surrounding each lung in a mammal, separated from rest of perivisceral coelom by the diaphragm. Pleural cavities are separated from each other by the mediastinum and the pericardium.

Pleurocarpus Growth form in some mosses, where the gametophore is multibranched and creeping, with the sporophyte borne upon a very short lateral branch.

Pleuropneumonia-like organisms (PPLOs) See **mycoplasmas**.

Plexus See **nerve plexus**.

Pliocene The last epoch of the **Tertiary** period, during which the first **hominoids** appeared. Deserts formed. The climate became cooler, and widespread glaciation occurred in the Northern Hemisphere. Much uplift and mountain formation occurred; Panama joined North and South Americas.

Ploidy The number of haploid chromosome sets in a nucleus, or cell. A haploid nucleus has a ploidy of 1; a diploid nucleus 2, a triploid nucleus 3, etc. The ploidy of a eukaryotic cell influences its volume, approximately proportionately. See **aneuploid**, **polyploid**.

Plumule (Bot.) In a plant embryo, the apical meristem of the shoot terminates the epicotyl (the stem-like axis above the cotyledons). In some embryos the epicotyl bears one or more young leaves. The epicotyl, together with these young leaves, is called a plumule. (Zool.) Down feather of nestling birds, persisting in some adult birds between contour feathers. See **feather**.

Plurilocular Possessing many chambers, e.g. describing gametangia and mitosporangia in brown algae (**Phaeophyta**).

Pluripotency Property of a *pluripotent* cell which can give rise to many cell phenotypes of the organism to which it belongs when suitably challenged, but lacks complete **totipotency**. Pluripotent cell progeny eventually become **de-**

termined and show the same phenotype in different environments.

Pluteus Larva of echinoid (sea urchin) or ophiuroid (brittle star). Ciliated and planktonic.

Pneumatocyst Hollow region of stipes of some brown algae, containing gas helping to keep the alga afloat.

Pneumatophore (Bot.) Negatively geotropic specialized root branch, produced in large numbers by some vascular plants growing in water of tidal swamps, e.g. mangrove; grows into the air above the water and contains well-developed intercellular system of air spaces communicating with the atmosphere through pores on the aerial portion. (Zool.) Gas sac of a siphonophoran coelenterate.

Pogonophora Phylum of benthic and tubicolous worm-like animals; mostly very thin, but attaining large sizes near hydrothermal oceanic vents. Deuterostomes, the coelom in three parts (*oligomerous*), with a tentacular hydrostatic system comparable to the lophophore of phoronids. Lacking a gut, absorbing dissolved organic compounds across their tentacles.

Poikilothermy (ectothermy) Condition of any animal whose body temperature fluctuates considerably with that of its environment. Characteristic of invertebrates, and vertebrates other than birds and mammals among living forms. See **homeothermy**.

Polar body One of up to three small haploid nuclei, usually with a covering of cytoplasm, extruded from primary and secondary **oocytes** during **meiosis** in animals. The first polar body may undergo the second meiotic division, producing three polar bodies in all. They usually play no further genetic role, and get lost or broken down in early development.

Polarity (1) Having a difference (morphological, physiological, or both) between the two ends of an axis, e.g. shoot/root, head/tail. Characteristic of most organisms, unicellular and multicellular, and of many cells within organisms. (2) Non-random organization of the cytoplasm of an animal egg prior to fertilization, sometimes with consequences for future development of embryo. Determined largely by position of egg in ovary, its association with nutritive cells, and its attachment to ovary wall. May change at fertilization or cleavage. Sometimes marked by yolk and/or centriole distribution. Animal/vegetal axis may dictate future anterior–posterior **axis** of embryo. See **animal pole**, **maternal effect**.

Polar nucleus One of two nuclei that migrate to centre of megagametophyte of a flowering plant, fusing to form the central fusion nucleus prior to **double fertilization**.

Polar plasm (pole plasm) Cytoplasm at the posterior pole of the eggs of insects (e.g. *Drosophila*), the vegetal pole of eggs of nematodes (e.g. *Parascaris*), and fertilized frog eggs, which contains regionally specific molecules (proteins, mRNAs) usually as a result of localization by directional transport along **microtubules**; through specific attachment sites on mRNAs and/or by pole-specifying interactions of follicle cells with the oocyte. Involved in differentiation of **germ line** cells from somatic cells. See **maternal effect, positional information**.

Pollen analysis (palynology) Techniques for reconstructing past floras and climates using pollen grains (and other spores), especially those preserved in lake sediments and peat. Their resistance to decay and distinctive sculpturing enable quantitative and qualitative estimates of past species abundance, typically within a timescale of a few decades up to millennia.

Pollen grain Microspores of seed plants, containing a mature or immature microgametophyte. Outer **exine** coat sculptured. See **microsporangium**.

Pollen tube Tube formed on germination of a pollen grain, which transports the two 'male' nuclei to the egg. In flowering plants, each penetrates tissues of the stigma and style, entering the ovary and growing towards an ovule, where it passes through the micropyle, penetrating the nucellus and rupturing at the tip of the embryo sac to set free two nuclei. See **auxin**, **cutin**, **double fertilization**, **microgametophyte**.

Pollex 'Thumb' of pentadactyl forelimb. Often shorter than other digits. Compare **hallux**.

Pollination Transfer of pollen (by wind, water, insects, birds or other animals) from the anther where they were formed to a receptive stigma. Not to be confused with any later fertilization which may result.

Pollinium Coherent mass of pollen grains, as in orchids; generally transferred as a unit in pollination.

Polyadelphous (Of stamens) united by their filaments into several groups, as in St John's wort. Contrast **monadelphous**, **diadelphous**.

Polycarpic (iteroparous) (Of plants) reproducing more than once. Contrast **semelparous**. Compare **annual**, **perennial**.

Polychaeta Class of **Annelida**, including ragworms, tubeworms, fanworms and lugworms. Marine; free-swimming, errant, burrowing or tube-dwelling. Clearly segmented; typically a pair of **parapodia** per segment. Usually a well-marked head, often with jaws and eyes. Large coelom, usually divided by septa. Separate sexes; fertilization external; larva a trochophore. Compare **Oligochaeta**, **Hirudinea**.

Polyclave See **identification keys**.

Polyclone Well-defined anatomical region (e.g. insect wing **compartment**) consisting of several entire clones of cells.

Polyembryony (Bot.) The simultaneous occurrence of more than one embryo within a seed. Can occur when more than one egg cell is fertilized within a single ovule, as in most gymnosperms, where female gametophytes produce several archegonia each (generally only one embryo survives). Alternatively, as in orange seeds, one embryo is formed sexually and others by vegetative budding of the nucellus (which acts like a multiple suspensor); or again, the pro-embryo may branch early in its growth. (Zool.) Asexual formation of more than one embryo from each zygote produced; sometimes arises by budding at some early (or larval) stage of development. Common in some parasites. See **cercaria**, **hydatid cyst**, **monozygotic twins**, **progenesis**.

Polygenes Genes, at more than one locus, variations in which in a particular population have a combined effect upon a particular phenotypic character (said to be determined *polygenically*, or to be a *polygenic character*). Allelic substitutions at such loci tend to have little individual effect upon phenotypic differences between individuals, the normal situation in much of developmental biology, where expression of several different genes is required for production of a character. Such *polygenic inheritance* increases the likelihood that a character will exhibit continuous **variation** in the population. Human height is an example. Compare **pleiotropy**, **epistasis**.

Polygenic inheritance See **polygenes**.

Polymer A very large molecule comprising a chain of many similar or identical molecular subunits (monomers) joined together (polymerized). In biochemistry, this usually involves condensation reactions. Proteins, polysaccharides and nucleic acids are characteristic biopolymers.

Polymerase Enzyme joining monomers together to form a polymer; e.g. **DNA polymerase**, **RNA polymerase**, starch synthase.

Polymerase chain reaction (PCR) Very adaptable process for isolating and amplifying a specifically desired DNA sequence (see Fig. 95). Double-stranded DNA is first separated into its two complementary strands by heating to 72°C, while two short single-stranded DNA primers known to anneal upwards of the 3′-end of the DNA sequence to be amplified are added to either strand after the reaction mixture has cooled. A key to the process is that **DNA polymerase** requires such a priming sequence to extend the whole complementary strand in a 5′-wards direction, and *Taq* polymerase from the extreme thermophilic bacterium *Thermus aquaticus* is added (it can withstand the high temperatures involved in the later steps and consequently allows automation of the process). *Taq* extends the complementary strands beyond the position of the primer-binding site on the other template while the reaction is cool, and reheating separates the new DNA duplexes. On cooling, the primers are added again (four binding-sites are now available); *Taq* is added again and extension proceeds but is restricted to the desired target sequence on account of the limiting primers on the template strands. The whole process is repeated for many generations, producing over 10^6 double-stranded target molecules after thirty-two generations of the cycle. If the process is started with a large number of template molecules then fewer cycles are required (with less mutation risk) to produce sufficient amplified DNA for **vector**-cloning.

Polymorph (polymorphonuclear granulocyte) Any of several types of **leucocyte** with granular cytoplasms and lobed nuclei, produced in bone marrow; forming 60–70% of human blood leucocytes. Also found in tissue fluid and lymph (they can adhere to and penetrate capillaries). Predominantly phagocytic (esp. neutrophils), but include **eosinophils**, **basophils** and **mast cells**. **Platelets** also have granular cytoplasms, but are not polymorphs.

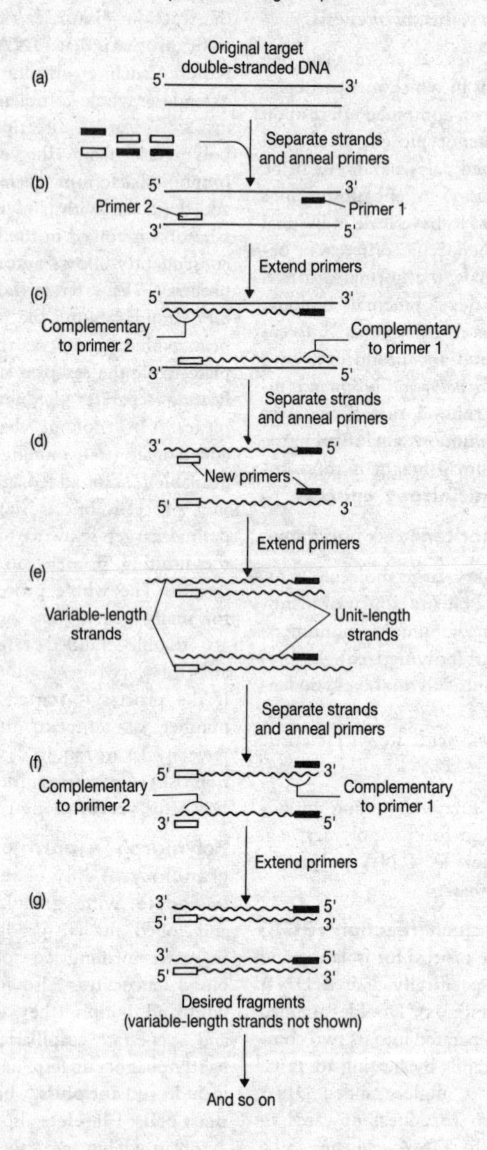
Amplification of target sequence
Original target double-stranded DNA
(a)
5'
3'
Separate strands and anneal primers
(b)
5'
3'
Primer 2
Primer 1
3'
5'
Extend primers
(c)
5'
3'
Complementary to primer 2
Complementary to primer 1
3'
5'
Separate strands and anneal primers
(d)
3'
5'
New primers
5'
3'
Extend primers
(e)
Variable-length strands
Unit-length strands
Separate strands and anneal primers
(f)
5'
3'
Complementary to primer 2
Complementary to primer 1
3'
5'
Extend primers
(g)
3'
5'
5'
3'
3'
5'
5'
3'
Desired fragments (variable-length strands not shown)
And so on

Neutrophils degranulate into their **phagosomes**; eosinophils, basophils and mast cells degranulate by exocytosis.

Polymorphism Major category of discontinuous **variation** within species. (1) *Genetic polymorphism*. Simultaneous occurrence in a species population of two or more discontinuous forms in such a ratio that the rarest of them could not be maintained solely by recurrent mutation. This excludes reproductively non-overlapping seasonal forms of a species (e.g. some **pleiomorphisms**), rare mutants, geographical races and all continuous variation. These polymorphisms generally involve clear genetic alternatives between individuals in the population, as controlled by **supergenes**, **inversions** or loci where allelic substitutions tend to bring about marked differences in phenotype (so-called 'major genes'). Examples include many forms of Batesian **mimicry**, human **blood groups**, the **MHC** system, some forms of **industrial melanism** and **heterostyly**, and many forms of **sex determination**. Enzyme and DNA polymorphisms, however, are often quite cryptic. The term *balanced polymorphism* describes polymorphism maintained by a balance of selective agencies whose effects tend to cancel each other out (e.g. see **frequency-dependent selection**).

Where a mutant gene increases in frequency in a population it may eventually replace its alternative allele(s). Such a situation is termed *transient polymorphism*, but may never go to complete fixation of the new mutant. The spread of melanic forms is an example. Relative fitnesses of different *morphs* can sometimes be calculated using the **Hardy-Weinberg theorem**. See **heterozygous advantage**, **neutralism**.

(2) *Non-genetic polymorphism*, where distinct forms of a species occur but without the definitional rigour of the cases above. Environmental influences play a large part in insect **caste** polymorphisms. Genetic factors (e.g. haplo-diploidy) may be important, but nutritional influences (e.g. in hymenopterans) and pheromones (e.g. in termites) are instrumental in developmental decisions. The term polymorphic is also applied to the different **zooids** in siphonophore and ectoproct colonies.

Polynucleotide Long-chain molecule formed from a large number of nucleotides (e.g. **nucleic acid**).

Polyp (hydranth) Sedentary form of the **Cnidaria**. Cylindrical stalk-like body attached to substratum at one end, mouth surrounded by tentacles at the other. Many bud asexually; some are sexual. Polyp-like stages in the scyphozoan life cycle (*scyphistomas*) produce ephyra larvae (which develop into sexual medusae) by strobilation. Most hydrozoans and anthozoans have (or are) polyps. See **polypide**.

Polypeptide Molecule consisting of a single chain of many amino acid residues linked by **peptide bonds**. See **protein**.

Fig. 95 **Polymerase chain reaction** *procedure*. (From *Recombinant DNA 2/E* by Watson, Gilman, Witkowski, and Zoller. Copyright © James D. Watson, Michael Gilman, Jan Witkowski and Mark Zoller, 1992. Reprinted with permission of W. H. Freeman and Company.)

Polypetalous (Of flowers) having petals free from one another; e.g. buttercup. Compare **gamopetalous**.

Polyphosphate granules Phosphate storage granules found in cells of blue-green algae (**Cyanobacteria**).

Polyphyletic Term for a group of taxa (species, genera, etc.) when, despite their being classified together as one taxonomic category, it is thought that not all have descended from a common ancestor which was also a member of the group. Such a taxon forms a **grade** rather than a **clade**. If the classification is to correspond with phylogeny, the group should be split into two or more distinct taxa. See **cladistics**, **homology**, **monophyletic**.

Polyphyodont See **dentition**.

Polypide Polyp-like individuals, forming a **colony** of ectoprocts or entoprocts.

Polyploidy Condition in which a nucleus (cell, individual, etc.) has three or more times the **haploid** number of chromosome sets characteristic of its species (or ancestral species). The haploid number is usually represented by the letter *n*, so a *triploid* is commonly represented as $3n$, a *tetraploid* as $4n$, a *pentaploid* as $5n$, and so on.

Tetraploidy can arise in otherwise normal diploid somatic tissue, both plant and animal, through failure of replicated chromosomes to separate in **mitosis**. This can be induced artificially by exposure to a compound such as colchicine, which inhibits formation of the spindle apparatus in mitosis through its disruptive effect upon microtubule formation; or by heat treatment. Such **autopolyploid** cells are much more likely to become part of the germ line in plants than in animals; but there may be some reduction in fertility due to unequal segregation of chromosomes at meiosis. However, some autotetraploid plants (e.g. cocksfoot grass, *Dactylis glomerata*; purple loosestrife, *Lythrum salicaria*) are fully fertile. **Allopolyploid** organisms are more likely to be fully fertile since each chromosome has only one fully homologous partner to pair with. Crossing and backcrossing between these and normal (usually diploid) individuals often produces a *polyploid complex*, resulting in gene **introgression**. Most instances of plant **apomixis** involve polyploid species, giving them an 'escape from sterility'. Polyploidy is comparatively rare in animals, probably because most animals are bisexual and obligate cross-fertilizers. A tetraploid ($4n$) animal could only hybridize with diploids and then only produce sterile triploid ($3n$) offspring, if any at all. Only a few amphibians and fish among vertebrates seem to be polyploid, possibly because of their method of **sex determination**. See also **hybridization**.

In plants, the range of variation is often narrower in allotetraploids than in a related diploid because each gene is duplicated. Polyploids are often self-pollinated, even when related diploids are mainly cross-pollinated, which reinforces the decrease in variability.

Polyploids are of interest because of the examples they may provide of 'instantaneous' **speciation**, and have been extremely important in the evolution of certain groups of plants. One of the most important polyploid (hexaploid) species is wheat, e.g. bread wheat, *Triti-*

cum aestivum, which has forty-two chromosomes. It was derived some 8000 years ago following the spontaneous hybridization of a cultivated wheat possessing twenty-eight chromosomes with a grass of the same group with fourteen chromosomes, followed by chromosome doubling.

Among plant polyploids are many of our most important crops, including, besides wheat, bananas, cotton, potatoes, sugar cane and tobacco. Many garden flowers are polyploids, e.g. chrysanthemums, day lilies. See **endomitosis**, **polyteny**.

Polyprotein Large protein from which different functional peptides may be cleaved enzymatically. Neuropeptides are cleavage products from a polyprotein synthesized in the neurone cell body. See **post-translational modification**.

Polysaccharide (glycan) Carbohydrate produced by condensation of many monosaccharide subunits to form polymers, forming long, often fibrous, molecules which are poorly soluble, osmotically inactive, and lacking a sweet taste. Extremely important structurally (e.g. **cellulose**, **chitin** and **glycosaminoglycans**) and as energy reserves (e.g. **starch**, **glycogen** and **inulin**).

Polysepalous (Of a flower) having sepals free from one another; e.g. buttercup. Compare **gamosepalous**.

Polysome (polyribosome) See **ribosome**.

Polysomy Condition in which one chromosome (rarely more) in a nucleus is represented more frequently than the remaining chromosomes. In a *trisomy* (see **Down's syndrome**) one chromosome is present three times in the nucleus as opposed to twice for other chromosomes. The nucleus would have a ploidy of $2n + 1$ (where n = haploid number). Polysomy can occur in nuclei which are otherwise of any ploidy. See **non-disjunction, polyploid**.

Polyspermy (Zool.) Entry of more than one sperm nucleus into an ovum at fertilization. Occurs normally only in very yolky eggs (e.g. shark, bird) and commonly in insects; only one nucleus fuses with egg nucleus, the rest taking no part in development. May upset normal development in other eggs. See **fertilization**, **pseudogamy**.

Polystelic (Bot.) Possessing more than one **stele**, comprising more than one (sometimes many) vascular bundles.

Polyteny Condition of some chromosomes, nuclei, cells, etc., in which many identical parallel copies of each chromosome are formed by repeated replication without separation, and lie side-by-side forming thick cable-like chromosomes. Best studied in salivary gland cells of the fruit-fly *Drosophila*, these and **lampbrush chromosomes** are often termed *giant chromosomes*, occurring only in cells which have lost the power to divide mitotically (permanent interphase). They tend to be longer than normal on account of being less coiled. Close pairing of sister strands may produce transverse 'banding', which enables both the detection of chromosome inversions and study of chromosome 'puffing' (see **ecdysone**). May occur in secretory tissues where increase in cell size rather than number has occurred. Compare **endomitosis, polyploidy**.

Polythetic (Of classification) a system where membership of a taxon depends upon possession of a large number of characters in common.

Polytopic Designating taxa occurring in two or more separate areas.

Polytypic Designating species which occur in a variety of geographical forms, or subspecies. See **infraspecific variation**.

Polyzoa (Bryozoa) See **Ectoprocta**, **Entoprocta**.

Pome 'False fruit', the greater part of which is developed from the receptacle and not the ovary; e.g. apple, pear. The edible fleshy part represents the receptacle, and the core represents the ovary.

Ponginae Subfamily of Hominidae (see **hominid**), including the Miocene fossil *Sivapithecus* (= *Ramapithecus*, 12.5–7 Myr BP) and the tribe Pongini (orang-utan). *Sivapithecus* is known from Indo-China to Turkey, seemingly from sub-tropical deposits, and was apparently an arboreal quadruped. Until fairly recently, it was thought that certain specimens (initially attributed to a distinct genus *Ramapithecus*, now placed in this subfamily) represented an early hominid line diverging from a common ape-human stock about 14–12 Myr BP. It is now thought that such a branching occurred far more recently (6–5 Myr BP?) and that the thick dental enamel of *Ramapithecus* (shared with **australopithecines**) is better considered a primitive feature of hominoids and hence of no taxonomic value in discriminating putative hominids. A large assemblage of ramapithecine fossils discovered in China in the 1980s indicates that in its facial characters it resembled the modern orang-utan with, *inter alia*, narrow interorbital distance, broad high zygomatic region and large size discrepancy between upper incisors. There is no clear evidence that the group were evolving towards bipedalism; the hallux (big toe) was opposable and capable of powerful gripping. The pongine clade of hominoids seems to have branched off earlier than the separation of humans and anthropoid apes. The molecular evidence (**DNA hybridization**, protein sequencing and immunological distancing) from living forms favours this view (see **hominid**); but controversy still surrounds the affiliations of **dryopithecines**.

Pons Floor of fourth ventricle of mammalian brain, lying in front of medulla oblongata and under cerebellum. Connects cerebrum to cerebellum and houses respiratory centres (see **ventilation**).

Population Group of conspecific individuals, commonly forming a breeding unit, sharing a particular habitat at a given time. See **-deme**, **infraspecific variation**.

Pore complex See **nucleus**.

Porifera Sole phylum of Subkingdom **Parazoa**. Sponges. Have an apparently loose diploblastic organization lacking both nerve and clear muscle cells, and comprising barely discrete tissues, there being no basement membranes. Basically vase-shaped, the outer 'epithelium' of flattened contractile cells (*pinacocytes*) with amoeboid locomotion; inner layer of cells consists of **choanocytes**. Internal skeleton may be of either calcareous (Class Calcarea) or siliceous (Class Demospongia) spicules, and frequently contains fibres of the protein spongin. Pinacocytes may develop into *porocytes* whose intracellular lumens form the surface pores (ostia) through which water enters propelled by flagella of the food-collecting choanocytes, and after entering the central cavity (spongocoel) leaves through one or more oscula at the top. Complexity of internal chambering increases from *ascon*, via *syconoid* and *sicon*, to *leucon* levels of organization. A gelatinous **mesogloea** between the two main layers contains amoeboid cells (see **amoebocytes**) whose varied functions include spicule secretion, gamete production and transport of sperm or food. Locomotion confined to flagellated larvae. See **multicellularity**.

Porphyrin Prosthetic group of several conjugated protein pigments; nitrogen atoms of tetrapyrrole nucleus often coordinated to metal ions (magnesium in chlorophyll, iron in haemoglobin and cytochromes).

Portal vein Blood vessel connecting two capillary beds, as in *pituitary portal system* and *hepatic portal system*.

Position effect Occurrence of phenotypic change resulting not from gene mutation as such but from a change in position of a piece of genetic material. Of two kinds. Stable position effects result from insertion of genes in the germ-line (e.g. inversion, translocation and crossing-over). *Position effect variegation* (PEV) produces mosaic phenotypes and can occur when a euchromatic gene finds itself close to a piece of **heterochromatin** (see **chromosomal imprinting**) or distal euchromatin. The heterochromatic condition may then 'spread' to adjacent euchromatin, inactivating nearby genes by repackaging their chromatin into heterochromatic domains. Other loci often act as modifiers of PEV, suppressing when in single dose and enhancing when in double dose. **Oncogenes** and **transposons** can have PEV-like effects, in the latter case especially when they insert in telomeric regions.

Positional information Embryological theory (perhaps currently out of favour) indicating those signals which enable cells in a developing metazoan to respond as though they appreciated their spatial positions in the embryo. Would normally involve specification of information as to where each cell lay on both antero-posterior and dorso-ventral axes. Such information would give cells their *positional values* and ultimately determine their fates. *Positional signals* are graded and can produce multiple outcomes, unlike *inductive signals*, which produce all-or-none responses and result in a single cell state. It is likely cells retain their positional values even when differentiated, making even cells of the same type (e.g. osteocytes) non-equivalent if they lie in different body regions. A cell might pick up and 'remember' different cues at different times prior to differentiation, such as chemical gradients (see **morphogen**) or those arising from key cell contacts. A cell's position on the various axes of the body might be given in sequence using a form of digital framework, evidence for which is emerging from, among others, studies of the genetics of *Drosophila* and mouse development (see **homeogene**, **zootype**). When the vertebrate body plan has been laid out in gastrulation, and the ***Hox* genes** have been regionally expressed, further migrations occur of neural crest, sclerotome, myotome and angioblast cells which in turn bring new inductions and organogenesis. Compare **prepattern**. See **axis**, ***bicoid* gene**, **gap genes**, **pair-rule genes**, **pattern**, **segment-polarity genes**.

Posterior (Bot.) Of lateral flowers, part nearest main axis. See **floral diagram**. (Zool.) Situated away from head region of an animal, or away from region foremost in locomotion (e.g. dorsal surface in bipedal animals such as humans).

Postorbital bar Bony enclosure of exterior surface of eye; in primate evolution, first appeared during Eocene and is present in all higher primates.

Post-transcriptional modification Changes occurring to some tRNA and rRNA transcripts prior to translation. May involve removal of nucleotides by hydrolysis of phosphodiester bonds. Compare **polyprotein**. See **RNA editing, RNA processing**.

Post-translational modification Changes occurring to protein structure after it has been produced on the ribosome. These may include (a) selective cleavage of the protein, sometimes removing inhibitory sequences (see **polyprotein, insulin**) or sorting signals (see **protein targeting**); (b) selective inactivation, possibly by ligand-binding; (c) covalent addition of further groups (e.g. glycosylation; see **Golgi apparatus**); (d) involvement with other polypeptides to produce functional quaternary structure (e.g. **haemoglobin**); (e) phosphorylation or dephosphorylation of selected tyrosine, serine or threonine residues by kinases and phosphorylases respectively, critical to progress through the **cell cycle** (e.g. see **maturation promoting factor**) and to the events of **signal transduction**.

Potentiation (1) Process whereby a given cellular response is achieved at a lower stimulus threshold. At **synapses**, a short burst of action potentials arriving at the presynaptic membrane can cause a greatly enhanced response to action potentials in the postsynaptic cell (lasting up to weeks in long-term potentiation), while earlier these had no effect. (2) See **synergism**.

Potocytosis Process enabling eukaryotic cells to take up and concentrate small molecules and ions prior to entry to the cytosol at special plasma membrane sites termed caveolae. These invaginated ~50 nm vesicles on the membrane's surface bear anchored proteins which form clusters and potocytosis begins when the caveolae pinch off and the anchored proteins internalize, the ions/molecules taken up being either (a) bound to a specific receptor within the caveola; (b) enzymatically converted within the caveola; (c) bound to carrier molecules ferrying them between cells and the caveola. Inside the small caveolae, they achieve a high concentration driving their diffusion out of the caveola into the cytosol.

PPLO See **mycoplasmas**.

P-protein (phloem-protein) Proteinaceous substance found in cells of flowering plant phloem, particularly in sieve-tube members, where it may block or hinder **translocation** through sieve pores.

Precambrian See **Archaean, Proterozoic**.

Precapillary sphincter Sphincter muscle around metarteriole where it gives rise to capillary bed. See **capillary**.

Precipitin Antibody combining with, and causing precipitation of, a soluble antigen. Such precipitations form the basis of *in vitro* procedures for separating and identifying antigens, often on gels (*precipitin reaction*).

Precocial Those young of mammals and birds which hatch or are born in a partially independent condition. Compare **altricial, nidicolous, nidifugous**.

Preformation Doctrine, generally accepted in Europe in 17th and 18th centuries, that all parts of the adult are already perfectly formed at the beginning of development. 'Ovists' believed these were present only in the ovum; 'spermatists' believed they were only in the sperm. Compare **epigenesis**.

Premaxilla Dermal bone forming front part of upper jaw in most vertebrates, bearing teeth (incisors in mammals). Forms most of upper beak in birds.

Premolar One of the crushing cheek teeth of mammals, anterior to the molars and posterior to the canines (or incisors). Unlike molars, has a predecessor in the milk teeth; usually has more than one root and a pattern of ridges on biting surface. See **dental formula**.

pre-mRNA, pre-rRNA, pre-tRNA See **RNA processing, ribosomes**.

Prepattern A theoretical model of animal **development** in which a prior spatial deployment of molecular gene regulators is 'read' by the cells of an embryo (rather than decoded, as with gradient models), the patterns of deployment dictating the gene activities and thereby the paths of differentiation of the cells. In the simplest form of the model, the 'plan' of the embryo would be laid out at the time of fertilization. More reductionist than the **positional information** model, the chief difficulties of the model arise in explaining regulative development (see **regulation**) and increases in spatial complexities of embryos. It also fails to account satisfactorily for the origin of prepattern itself. See **epigenesis**.

Preprotein Proteins which, normally destined for secretion or membrane-integration, are maintained in an unfolded conformation (see **chaperones**) and carry amino-terminal signal sequences which are subsequently removed to give the mature protein (see **protein targeting**).

Presumptive Term indicating that a particular cell or cell group in the early animal embryo will normally differentiate in a particular way. Thus, 'presumptive epidermis' indicates that a group of cells so described will normally become epidermal cells, even prior to their becoming **determined**.

Priapulida Phylum of burrowing marine worms of dubious affinity. Possibly pseudocoelomate. A distended proboscis, everted with great force, is used to anchor the worms while the body is drawn up over it.

Pribnow box Nucleotide sequence in prokaryote **promoter** regions, located about six bases upstream of a tran-

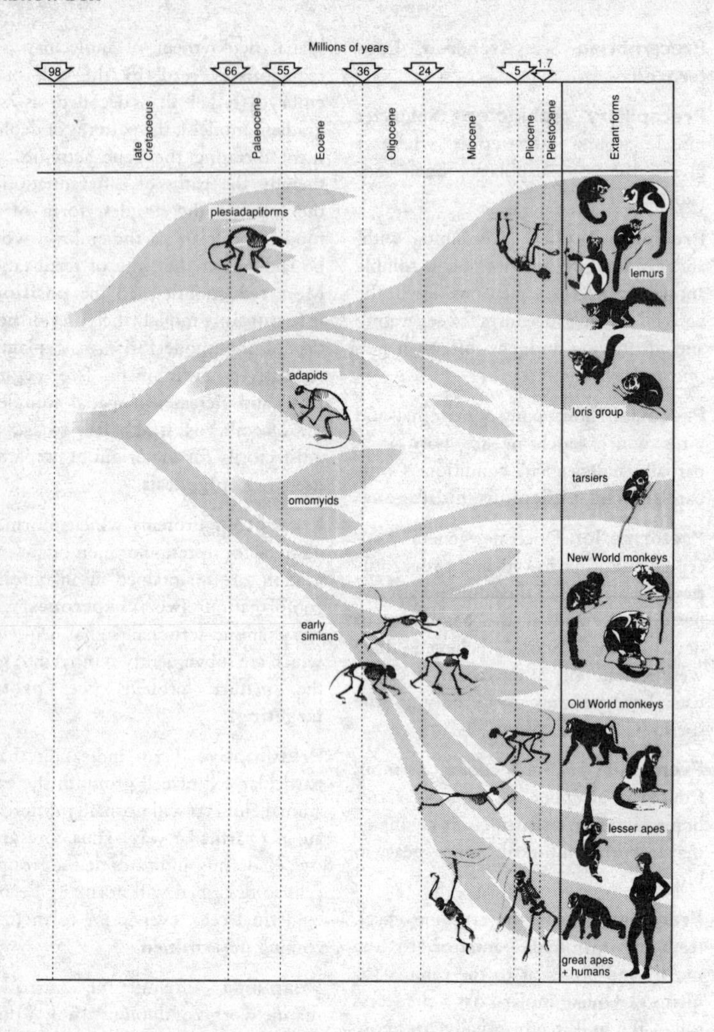

Fig. 96 *The main groups of living* **primates** *and their possible phylogenies. The fossil forms included are those with relatively complete skeletons, indicating the gaps which still exist in the fossil record. Illustration by Lucrezia Beerli–Bieler, reproduced with the kind permission of* Nature and Prof. R. D. Martin.

scribed region. Contains the sequence **TATAATG**. Also called a (−10) region, because of the invariant T residue at base 10 upstream from the start of the transcribed region. However, many promoters whose expression is controlled by **sigma factors** other than the normal vegetative one have different consensus sequences in different positions upstream of the actual transcription start point. See **TATA box**.

Primary endosperm nucleus Nucleus resulting from fusion of one generative nucleus with the two polar nuclei in a flowering plant embryo sac. Usually triploid, its mitotic products produce the endosperm. See **double fertilization**.

Primary immune response See **B cell**.

Primary pit field See **pits**.

Primary production The total organic material synthesized in a given time by autotrophs of an ecosystem. See **productivity**.

Primary structure See **protein**.

Primates Members of the mammalian Order Primates. Placentals, including today lemurs, lorises and tarsiers (Prosimii), and the **Anthropoidea** (simians). There is an on-going reinterpretation of primate origins based on new fossils and on calculations on the effectiveness with which fossils sample past primate species. Primate origins may be *c.* 85 Myr BP (Middle Cretaceous) instead of the more widely held *c.* 65 Myr BP (Lower Palaeocene). The superordinal relationships of primates remain unclear (see **Archonta**). Clavicle retained (lost in many mammalian orders), with shoulder joint permitting freedom of movement in all directions, and an elbow joint allowing rotation of forearm; five digits on all limbs, with generally enhanced mobility and usually opposability of thumb (pollex) and big toe (hallux); claws modified into flattened nails; reduced snout; binocular vision, great alertness, acuity and colour vision; brain enlarged relative to body, with expansion of cerebral cortex; two mammary glands only, and young usually born singly. Most of these features are adaptations to arboreal living (most primates are tree-dwellers). There have been independent moves towards terrestrialization, notably in the **hominoid** line. Among possible fossil antecedents of extant primates, plesiadapids now seem closer to colugos (Dermoptera); adapids have traditionally been linked to lemurs and lorises but some authors now suggest links with simians; omomyid relationships are highly controversial but have traditionally been linked to tarsiers – but some place tarsiers as the sister group of simians. See Fig. 96. See **australopithecine**, **hominid**, ***Homo***, **pongid**.

Primitive Referring to a characteristic present in an early ancestral plant or animal and which may be present, relatively unmodified, in living forms. A term indicating that a structure, taxon, etc. is regarded as similar in appearance to a stem line. Such a line may have given rise through adaptive radiation to diverse and often more specialized modifications of form. Has temporal rather than ecological implications.

Primitive groove See **primitive streak**.

Primitive streak Longitudinal thickening at gastrulation in early (disc-like) reptilian, avian and mammalian embryos caused by cell migration from posterior epiblast towards the centre. The streak narrows, thickens and extends anteriorly, marking the antero-posterior embryo axis. As cells converge on the streak, a depression (the *primitive groove*) forms within it and serves as a **blastopore** through which cells migrate into the blastocoele to form endodermal and mesodermal tissue. The anterior funnel-like tip of the streak (Hensen's node) serves as the amphibian dorsal blastopore lip, cells passing through it anteriorly to form head mesoderm and notochord. During late gastrulation, the streak regresses posteriorly as notochord and mesoderm are laid down progressively from anterior to posterior beneath the presumptive ectoderm, inducing **neural plate** formation and onset of **neurulation**. Posteriorly, Hensen's node forms the anal region (see **Deuterostomia**).

Primordial meristem See **promeristem**.

Primordium (Bot.) An immature cell or group of cells that will eventually give rise to a specialized structure. Thus, a leaf primordium will become a leaf.

Prion Minute protein fragment which may be able to direct its own replication in animal cells, without the presence of nucleic acids. Seemingly the cause of scrapie in sheep and Creutzfeldt–Jakob disease in humans. Possibly they activate a latent gene encoding the prion protein.

Prisere Primary sere. Complete natural succession of plants, from bare habitat to climax. Compare **plagiosere**.

Proband Propositus. Index case. Individual affected by a genetic disorder, whose presentation of it enables a pedigree for the disease to be drawn up.

Probasidium Cell in which karyogamy occurs in **Basidiomycota**. See **metabasidium**.

Probe DNA See **DNA probe**.

Proboscidea Elephants. Placental mammal order, probably derived during early Tertiary from the **Condylarthra**. Today characterized by trunk (formed from nose and upper lip), large size (largest terrestrial mammals), massive legs, greatly lengthened incisors (tusks) and huge grinding molars, only two pairs of which are in use at one time. Related to hyraxes.

Proboscis Suctorial feeding apparatus of some dipteran flies and most Lepidoptera. In the former most of the mouthparts contribute to its formation, and the labella contains food channels (pseudotracheae). In the latter it is formed largely from maxillae (mandibles being absent).

Procambium Tissue of narrow elongated cells in vascular plants, grouped into strands, differentiating just behind growing points of stems and roots and giving rise to vascular tissue. See **apical meristem**.

Procaryote See **prokaryote**.

Prochloron See **Prochlorophyta**.

Prochlorophyta Prokayotic organisms (with **Cyanobacteria**, in Kingdom **Monera**), but possessing both chlorophylls *a* and *b*, lacking phycobiliproteins, and releasing oxygen during photosynthesis, as in the eukaryotic algae and autotrophic plants. The carotenoid pigments differ from eukaryotic algae but resemble those of blue-green algae. Structurally, a prochlorophyte cell is similar to that of the blue-green algae, with muramic acid present in the cell wall (similar in structure to those of the Gram-negative bacteria and blue-green algae). The cells contain stacked thylakoid membranes, grouped in pairs or as stacks of many thylakoids. DNA microfibrils usually occur in the peripheral cytoplasm; ribosomes are of the 70S prokaryotic type, and carboxysomes (polyhedral bodies) are also present. Prochlorophytes occur as free-living planktonic filamentous algae in freshwater lakes, and as symbionts in colonial ascidians living in the common cloacal chamber and in the tunic of the colonial ascidians (sea squirts). *Prochloron* is an obligate symbiont, and the association appears also to be obligate for the host ascidian, with products from photosynthesis being utilized by the host animals. The prochlorophytes probably arose from the blue-green algae by acquisition of chlorophyll *b*, and loss of phycobiliproteins. Since prochlorophytes possess both chlorophylls *a* and *b*, as do chlorophytes and euglenoids, it has been suggested that prochlorophytes were involved in an endosymbiotic event that ultimately resulted in the evolution of chloroplasts in the green algae. However, there is no direct evidence, and there are no known advanced endosymbiotic associations involving prochlorophytes. Work on RNA polymerase and 16S rRNA suggests prochlorophytes are polyphyletic and that none is specifically related to chloroplasts. Intermediates leading to the evolution of green algal chloroplasts could have become extinct or they may have evolved independently of green algal chloroplasts. See **endosymbiosis**, **Cyanobacteria**, **Glaucophyta**.

Proconsul Genus of E. African Miocene primate (Suborder Hominoidea, Family Proconsulidae; 22–17 Myr BP); skull and dentition (e.g. thin molar enamel) resemble ancestral catarrhine, but postcranial features place it in the hominoid clade. At least two species, one small (*c.* 9 kg) and one larger (26–38 kg). Arboreal and quadrupedal, with a completely opposable thumb. Possibly ancestral to the later **dryopithecine** apes. A maxilla from earlier East African deposits (27–24 Myr BP) has been tentatively placed within the genus. It would constitute the earliest fossil hominoid yet discovered.

Proctodaeum Intucking of embryonic ectoderm meeting the endoderm of the posterior of the alimentary canal and forming anus or cloacal opening.

Procumbent (Of teeth) jutting forward, as in ape incisors.

Productivity (production) Rate at which solar energy and carbon dioxide are absorbed and utilized in photosynthesis. The *gross primary production* of an ecosystem is the total amount of organic matter produced by its autotrophs, and is usually recorded in kilojoules per hectare per year. *Net primary production* is gross primary production less that used by plants in respiration and represents food potentially available to consumers in the ecosystem. See **standing crop**.

Proembryo In seed plants, a group of cells formed by initial divisions of the zygote, which by further development differentiates into the suspensor and embryo proper.

Progenesis Form of **heterochrony** in animals, leading to **paedomorphosis** through precocious sexual maturity and a shortening of the developmental pathway, often resulting in a 'juvenilized' morphology. Believed to have been a source of important new taxa during evolution. Most successful paedomorphs seem to have been small progenetic larvae, e.g. six-legged myriapod larva (ancestral to insects) and the urochordate tadpole (ancestral to vertebrates). Several gall midges have progenetic larvae; wingless aphids and many parasites are progenetic. Progenesis tends to be favoured by ***r*-selection**. Compare **neoteny**.

Progesterone Steroid hormone secreted by **corpus luteum** of mammalian ovary, and made by **placenta** from maternal and foetal precursors. For timing of secretion, and roles in human female, see **menstrual cycle**. Broken down by **liver**. See **contraceptive pill**.

Progestogen Any substance with progesterone-like effects in the female mammal.

Proglottis (pl. proglottides) Segment-like unit of the tapeworm body which, when mature, leaves the gut of the primary host in the faeces. Normally each develops male reproductive organs first and later loses these as female parts develop. Proglottides are not true segments since they are budded off from the anterior of the worm (from the scolex). This does not amount to asexual reproduction (but see **hydatid cyst**). See **Cestoda**.

Prognathous 'Long-faced', the jaws protruding considerably in front of the brain case. In primates, an ape-like character; hominids tend towards **orthognathous** 'short' faces.

Programmed cell death While it has been known for some time that some animal cells (e.g. vertebrate neurons, adrenal cortical and developing cells) require specific signals from other cells to survive, it now seems that a general mechanism may exist whereby most animal cells are able to be killed by 'suicide programmes' activated or depressed by such signals, and that this is an efficient way for disposal of unwanted cells. Programmed (non-pathological) cell death involves nuclear and cytoplasmic shrinkage followed by phagocytosis by macrophages (*apoptosis*) with no leakage of cell contents and hence no inflammation; in some cases it also requires specific mRNA and protein synthesis. In the nematode *Caenorhabditis elegans*, activation of at least two particular (*ced*) genes must occur for cells to die, either in these cells themselves or in cells ancestrally close to them, and death fails to occur if these genes are mutated. A further *ced* gene plays a role in inhibiting cell death, but it is not certain whether its activation requires signals from other cells. One possible consequence of such a general mechanism for programmed cell death might be that ectopic cells would not survive, deprived of the trophic support from their proper neighbours. A further possibility is that after proliferation and somatic mutation in a germinal centre during an immune response, **B cells** may undergo apoptosis unless they have a membrane immunoglobulin which can bind the activating antigen, leading to their selection as producers of high-affinity antibody.

Deregulated expression of the potent oncogene c-*myc* (whose product c-Myc protein forms a heterodimer with Max protein) is often associated with increased incidence of cell death, usually by apoptosis, when cell growth conditions are sub-optimal (as for myeloid cells when the cytokine interleukin IL-3 is withdrawn). c-Myc protein is a (DNA-binding) transactivating transcription factor which can also bind retinoblastoma protein, RB (see **tumour suppressor genes**), is expressed at the G0–G1 transition, as well as throughout the **cell cycle** in response to mitogens, and disappears in their absence. c-Myc appears to induce apoptosis by bunding specific DNA sequences and thereby modulating apoptotic or anti-apoptotic genes. It seems that cell proliferation and cell death are inseparably c-Myc-driven events that are co-induced but separately regulated thereafter (see **cytostatic factors**), Fig. 28). Adenovirus E1A protein is also a potent apoptic factor and can replace c-Myc in this capacity (see **cancer cells**). Uncontrolled programmed cell death is involved in some neuropathological conditions, such as Alzheimer's disease. See **growth factors**.

Progymnospermophyta Progymnosperms. A group of fossil plants of the late **Palaeozoic** era, which had characteristics intermediate between those of the trimerophytes and gymnosperms. Morphologically some were similar to early seed ferns, while others were large trees with leafy branches. Reproduced by freely dispersed spores, and produced secondary xylem remarkably similar to that of gymnosperms. Unique in that they also produced secondary phloem. Progymnosperms and Palaeozoic ferns may have evolved from the more ancient trimerophytes, from which they differed mainly by having a more complex vascular system. Another suggestion is that the ferns evolved from progymnosperms. The most important evolutionary advance of progymnosperms over trimerophytes is the presence of a bifacial vascular cambium (producing both secondary xylem and phloem), characteristic of seed plants. Morphological evidence suggests that gymnosperms evolved from the progymnosperms upon evolution of the seed, but whether the seed evolved only once or several times is unknown; nor is it certain from which group of progymnosperms the gymnosperms evolved.

Prokaryote (procaryote) Prokaryotes (**bacteria**), blue-green algae (**Cyanobacteria**), **Prochlorophyta**, **mycoplasmas** and **Archaebacteria**) are typically either unicellular or filamentous, and small (up to 3 μm in diameter). Their DNA is not housed within a nuclear envelope (see **nucleoid**), and no prokaryotic cell is descended from such a nucleated cell. Within the plasma membrane, often folded and convoluted within cell interior, lies the rest of the cytoplasm, containing smaller ribosomes than those of **eukaryotes**, along with granular inclusions. They lack the tubulin, actin and histones diagnostic of eukaryotes and so have other methods of **cell division** and **cell locomotion**

(see **flagellum**). Their **genetic code** is remarkably similar to that of eukaryotes. Cell division commonly lags behind chromosome replication, so that prokaryotic cells commonly contain at least two chromosomes, each consisting of DNA and non-histone proteins, often attached for a time to the plasma membrane. Mitochondria and chloroplasts are absent, but they possess structures that function similarly (**mesosome**, **chromatophore**). Some authors recognize two superkingdoms, the Prokaryota and Eukaryota. The Monera would then form the one prokaryotic kingdom; but see **mesokaryote**.

Prolactin (lactogenic hormone, luteotropic hormone, ITH) A protein gonadotrophic hormone secreted by the vertebrate anterior pituitary gland. In mammals it promotes secretion of progesterone by the corpus luteum and is involved in **lactation**. In pigeons stimulates crop milk production and is necessary for maintenance of incubation (which itself stimulates its secretion) by both sexes.

Prolegs Stumpy unjointed appendages on ventral surface of abdomen of caterpillar.

Proliferation Growth by active cell division.

Promeristem Extreme tip of an **apical meristem** comprising actively dividing, but as yet undifferentiated, cells.

Promoter One type of ***cis*-acting control element**. About 100 base pairs long, it lies a short distance upstream of the coding part of the gene of which it forms a part, but downstream of the **operator**, and is the region to which an RNA polymerase molecule binds (often accompanied by one or more **transcription factors**) to initiate transcription. Promoters of genes which transcribe large amounts of product (strong promoters) are similar to one another (share similar **consensus sequences**) and have a **TATA box** about thirty base pairs upstream from where transcription begins. **Enhancers** regulate the promoter's rate of transcription. The retinoblastoma gene product (see **tumour suppressor gene**) is known to repress transcription of some cellular promoters by binding appropriate transcription factors. See **gene** for diagram.

Pronation Position of fore-limb, or rotation towards it, such that fore-foot (hand) is twisted through 90° relative to the elbow, the radius and ulna being crossed. In this way the fore-foot points forwards (the natural position for walking in many tetrapods). Humans and other primates can untwist the fore-arm (*supination*).

Pronephros See **kidney**.

Pro-oestrous See **oestrous cycle** and **menstrual cycle**.

Propagule (Bot.) A dispersive structure, such as a seed, fruit, gemma or spore, released from the parent organism.

Prophage Non-infectious phage DNA, integrated into a bacterial chromosome and multiplying with the dividing bacterium but not causing cell lysis except after excision from the chromosome, which can be *induced* by certain treatments of the cell. See **bacteriophage**.

Prophase First stage of **mitosis** and **meiosis**.

Proplastid Minute, self-reproducing, colourless or pale green, undifferentiated **plastids** occurring in meristematic cells of roots and shoots. They are the precursors of the more highly differentiated plastids (e.g. **chloroplasts**). Development of a proplastid may become arrested by a lack of light, then it may form one or more prolamellar bodies; semicrystalline bodies comprising tubular membranes. Such plastids are called etioplasts. Upon returning to the light the membranes of the prolamellar bodies develop into thylakoids.

Propositus See **proband**.

Proprioceptors (1) Receptors involved in detection of position and movement including **muscle spindles**, organs of balance in the vertebrate **vestibular apparatus**, and campaniform sensillae occurring in all parts of the cuticle of the insect body subject to stress (especially in joints, at wing and haltere bases). Do not usually show sensory **adaptation**. (2) In a wide sense, any receptor detecting changes within the body other than those caused by substances taken into the gut and respiratory tract; e.g. deep pain receptors and **baroreceptors**; **statocysts** in invertebrates.

Prop root Adventitious supportive root, arising from the stem above soil level.

Prosencephalon See **forebrain**.

Prosenchyma See **plectenchyma**.

Prosimian Of the primate suborder Prosimii. Primitive, nocturnal and mostly arboreal. **Orbits** not totally enclosed in bone at rear, and less frontally positioned than in anthropoids (hence poorer stereoscopic vision). Olfactory lobes larger than in anthropoids; snout longer, with a wet muzzle (rhinarium). Incisors and canines of lower jaw form a dental comb/scraper; diet more insectivorous than anthropoids, and brain size smaller relative to body. Neither frontal bone nor two halves of lower jaw fused. Tend to have 2–3 young and leave them in nests.

Prosoma One of the tagmata of **Chelicerata**, a combined head and thorax. Comprises one pre-oral segment and five post-oral segments, the latter bearing walking limbs. See **opisthosoma**.

Prostaglandins (PGs) Family of 20-carbon fatty acid derivatives continuously synthesized (in mammals at least) by most nucleated cells from precursor phospholipids of the plasma membrane. Rapidly degraded by enzymes on release; but if appropriately triggered a cell will increase its output of PG, raising local levels and influencing both itself and its neighbours. Like hormones, they may be released into the blood but, unlike them, are only effective over short distances. Promote contraction of smooth muscle, platelet aggregation, inflammation and secretion. Some bind membrane receptors and exert their effects by altering intracellular cyclic AMP levels. Of great clinical potential in alteration of blood pressure, in bronchodilation and constriction, inducing labour, reducing gastric secretion, etc. Also called local or tissue hormones. See **eicosanoids**, **parturition**.

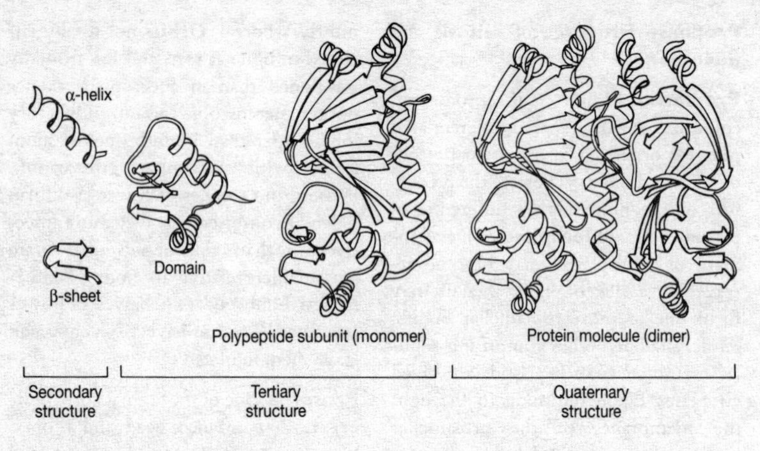

Fig. 97 *The 'ribbon model' method of representing protein structure. A domain is indicated, both isolated and as a part of a polypeptide subunit of the functional protein molecule.*

Prostate gland Gland of male mammalian reproductive system lying below the urinary bladder and around part of the urethra. Secretes alkaline fluid comprising up to a third of semen volume. Its effect on vaginal pH assists sperm motility. Rich source of **prostaglandins** in humans.

Prosthetic group Non-protein group which when firmly attached to a protein results in a functional complex (a conjugated protein). Many respiratory pigments (e.g. **haemoglobin**) are conjugated proteins, while many enzymes require prosthetic groups (some of them metal ions). The carbohydrates and lipids in glycoproteins and lipoproteins are prosthetic groups of their proteins. DNA is the prosthetic group of the histones in chromatin. Compare **coenzyme**.

Protamines Basic proteins of low molecular weight (about 5000), lacking prosthetic groups, containing many arginine and lysine residues, and found in association with DNA. See **chromosome**.

Protandry (1) In flowers, the situation in which anthers mature before carpels. See **dichogamy**. (2) In animals, the condition in sequential hermaphrodites in which sperm are produced prior to eggs (e.g. in many tapeworms).

Protease Enzyme that digests protein by hydrolysis of peptide bonds. See **proteolysis**.

Proteasomes An ATP-dependent **multienzyme complex** (700 Kd, 26S), variably composed of several subunits and located within both cytoplasm and nucleus, involved in selective protein degradation. Barrel-shaped, with a hole down its centre. **Ubiquitins** are subunits. Implicated in antigen-presentation through production of peptides presented on Class I MHC antigens, but some subunits also mapping in the MHC Class II locus.

Protein Polymer of very large or enormous molecular mass, composed of one or more polypeptide chains, and whose monomers are **amino acids**, joined together (in condensation reactions) by **peptide bonds**. In addition, some have covalent 'sulphur bonds' formed by oxidization between two cysteine radicals in the polypeptide. The potential variety of polypeptides is infinite: there are 20 common amino acids, so joining any two together would give 400 (= 20 × 20) possible *dipeptides*. Biological polypeptides are often several hundred amino acids long, so few of the possible polypeptides actually occur in organisms (see **protein synthesis**). Several proteins (e.g. **actin**, **tubulin**) form filaments of polypeptide molecules. The molecular mass which results (as in the protein coat of tobacco mosaic virus) may exceed 40 million daltons. See **biuret reaction**.

Despite considerable overlap between the following structural categories within proteins, they remain didactically useful. Each polypeptide has a *primary structure*: the number and sequence of its amino acids. There is an amino-terminal ($-NH_2$) end and a carboxy-terminal (–COOH) end to the molecule. During its production on a ribosome a polypeptide commonly assumes a corkscrew-like **alpha-helix** (α-helix) as its *secondary structure*, due to hydrogen bonding between the hydrogen atom attached to the nitrogen in one amino acid radical and the oxygen attached to a carbon atom three radicals along the chain. Other *intra*molecular hydrogen bonds may contribute to the secondary structure, as with antiparallel folding of the molecule back along itself in the same plane, hydrogen atoms of one side being linked to oxygen atoms of the side parallel to it. Such β-pleated sheets occur in many globular (spherical) protein molecules and commonly link several polypeptides of the same type (as in fibroin of silk, and some other fibrous proteins). Regions of α-helix and β-pleated sheets, along with less clearly organized (i.e. more random) stretches of the amino acid chain, may all contribute to the three-dimensional configuration (*tertiary structure*) of a polypeptide, the α-helical portions often being thrown into folds by electrostatic attraction and repulsion resulting from the charge distribution on the amino acid R-groups (see **amino acid**). Proteins often have hydrophilic and hydrophobic segments, and in such membrane-spanning species as **receptor** proteins the former tend to lie in the aqueous regions on either side of the membrane while the latter lie in the membrane's lipid interior. The tertiary structures of globular proteins in solution depend upon pH (since charges on R-groups depend upon pH) and upon sulphur (disulphide) bonds. **Enzymes** are, typically, globular proteins. Their functions depend upon their shapes and are affected by changes in pH and temperature (see **denaturation, isoelectric point**). Many important proteins (*conjugated proteins*) are formed by covalent union of a non-protein radical to the protein molecule, forming a hybrid molecule (e.g. **haemoglobin**, **cytochromes**). The *quaternary structure* of a protein is the shape adopted when two or more polypeptide chains associate (non-covalently) to produce the functional protein molecule. Both the non-enzyme haemoglobin and the enzyme lactate dehydrogenase (see **isoenzyme**) are proteins formed by quaternary association between four polypeptides of two different kinds. Indeed, functional enzyme molecules often consist of two or more different polypeptide subunits. For more on protein structure see **DNA-binding proteins**, **domain**, **receptor**.

Fibrous proteins often have major structural roles (e.g. in **cytoskeletons**, **connective tissue**, **striated muscle**, **chromosomes**). This is more obviously true in animals than in plants, although plant **cell walls** contain structural glycoproteins. All cells depend upon catalytic activities of enzymes, and have **transport proteins** in their membranes. Generally, **respiratory pigments** are conjugated proteins. Proteins also form major components of **antibodies** and other **glycoproteins**, and of **lipoproteins**. They commonly act as **buffers** (e.g. in blood plasma), and being colloids reduce the **water potentials** of cells and intercellular fluids. Although proteins are insoluble in lipid solvents, globular proteins (but not fibrous) dissolve in water and dilute salt solutions. Proteins are digested hydrolytically by **proteolytic enzymes** and mineral acids, and are usually separable by **chromatography** or **electrophoresis**. Only autotrophic organisms are capable of making the amino acid components of proteins from inorganic precursors. Two or more proteins may have very different amino acid sequences yet have similar structures and functions: we cannot yet predict a protein's structure from its amino acid sequence unless the sequence is clearly related to that of another protein of known structure. See **protein engineering**.

Protein engineering Techniques used in improving naturally occurring enzymes for human use. Isolation of the gene encoding a polypeptide by specific endonucleases is followed by site-specific **mutagenesis** (e.g. changing a particular glycine to an alanine or a lysine to an arginine may improve an enzyme's thermal stability). Likewise, increasing the number of disulphide (– SH) bridges (e.g. by one where the native molecule only has one) increases an enzyme's rigidity. Hope of progress in protein design depends upon increasing our understanding of how proteins fold to achieve their native state. Mimics of the enzymes chymotrypsin and trypsin have been designed in which short peptide sequences from their active sites are strung together between oligoglycine loops, the products hydrolysing alkyl ester and peptide bonds on a par with native enzymes. A different goal might be to replace the non-renewable resources so often used in manufacture by renewable engineered proteins providing fibrous and structural materials.

Protein family A molecular taxon containing proteins with > 50% amino acid sequence identity. Protein superfamilies contain proteins with 50% amino acid sequence identity.

Protein kinase Enzyme transferring a phosphate group from ATP to an intracellular protein (often also an enzyme), increasing or decreasing its activity. The **protein tyrosine kinase** family phosphorylate specific tyrosine residues on the target protein; the serine/threonine protein kinases phosphorylate specific serine or (less often) threonine residues respectively. These generally form parts of **receptor** protein kinases; but there are also non-receptor protein kinases (often associated with the inner surface of the plasma membrane) and these often contain SH2 and SH3 **domains**. See *cdc* **genes**, **cell cycle**, **incompatibility mechanisms**.

Protein synthesis Proteins are manufactured by cells on **ribosomes**, which involves joining together in the correct sequence possibly hundreds of amino acid molecules. Cells produce particular proteins either all the time (constitutively) or as and when required (see **gene expression**). Sequence of amino acids forming primary structure of a protein is encoded in the sequence of nucleotides of the genetic material, usually DNA (in some viruses it is RNA). When a piece of DNA becomes involved in protein synthesis, an **RNA polymerase** first breaks the hydrogen bonds holding the two DNA strands together, then uses one of the strands as a template on which to incorporate the nucleotides making a complementary RNA molecule, in an order dictated by **base pairing** rules. Once produced (the process is called *transcription*), this pre-RNA molecule is commonly modified (see **RNA processing**) to form a shortened *messenger* RNA (mRNA) molecule, always with the nucleotide triplet *start codon* AUG at one end (coding for the amino acid methionine). After transcription, the nuclear enzyme poly-A polymerase adds to most eukaryotic mRNAs a poly-A (poly-adenylyl) tail about 50–200 A-residues in length, improving the stability of the mRNAs and protecting them from ribonuclease degradation. This is bound (function?) in the cytosol by the RNA-binding protein, poly(A) tail-binding protein. This mRNA molecule passes into the cytoplasm (via the nuclear pore apparatus in nucleated cells) and triggers a ribosome to assemble upon it at the AUG codon nearest the 5′-end of the mRNA molecule.

Free amino acids are not assembled directly into protein but are first loosely bound to an *activating enzyme*, so-called because it (a) hydrolyses an ATP molecule, providing energy for (b) attachment of the activated amino acid to one of a small number of specific transfer RNA (tRNA) molecules, of which there is a pool in the cell. The result of this ATP-dependent catalysis is a pool of amino acyl- tRNAs from which protein synthesis proceeds.

Under the influence of **initiation factors**, each ribosome draws amino-acyl-tRNAs bound to **elongation factors** from its surroundings in an order determined by the nucleotide sequence of the mRNA molecule to which it attaches. AUG is the **codon** for the amino acid methionine, and the newly-assembling ribosome already has tRNA-methionine bound to it (see **P-site**). This tRNA base-pairs with the AUG codon of the mRNA by hydrogen bonds with a triplet of nucleotides exposed at one end of the molecule: its anticodon triplet. Then another tRNA molecule gets bound by the ribosome (see **A-site**), but only if its anticodon base-pairs with the codon next to AUG in the mRNA. It brings with it its own attached amino acid. This codon–anticodon hydrogen bonding provides the essential working principle of the **genetic code**.

The ribosome will continue to draw in appropriate tRNA molecules, joining each of their amino acids together to form a growing polypeptide chain. This chain elongation requires hydrolysis of two GTP molecules per hydrogen bond formed. Each of the tRNA molecules is released from the ribosome once it has donated its amino acid load.

As it draws in each amino acyl-tRNA in turn, the ribosome moves one codon further along the mRNA molecule, towards the 3′-end of the molecule. At an appropriate stop codon (termination codon) on the mRNA, translation ends and the ribosome releases the completed polypeptide. The poly-A tail mRNA is not translated and the N-terminal methionine is cleaved off. Downstream from the termination codon in eukaryotes, a variable nucleotide sequence is usually added containing the sequence AATAAA about 10–30 nucleotides upstream from the site of poly-A addition. Mutations here can cause transcription past the normal termination region (as in some **thalassaemias**).

This ribosomal phase of protein synthesis is called *translation*. Each mRNA molecule is simultaneously the site of attachment of many ribosomes (see **polysome**), and when each ribosome reaches the stop codon it releases another identical polypeptide molecule (see **protein targeting**). Both chloroplasts and mitochondria make some of their own proteins but, as with prokaryotic systems in general, their start codon (AUG) binds N-formylmethionine rather than methionine. Protein synthesis is energy-dependent, each amino acid incorporated into the polypeptide requiring hydrolysis of three **high-energy phosphate** bonds. Several **antibiotics** stop protein synthesis, during either transcription or translation. Much **post-translational modification** of proteins occurs in the **Golgi apparatus**.

Protein targeting During synthesis (co-translationally) or after synthesis (post-translationally), proteins may move from one cellular compartment to another, crossing the hydrophobic barrier of a membrane's phospholipid bilayer. Such behaviours are made possible by amino acid sequences (sorting signals) comprising small parts of the protein's structure and effectively targeting its destination in the cell. Signal sequences (signal peptides) are hydrophobic, occur at the amino-terminal end and direct the ribosome to the endoplasmic reticulum (ER). They also target the protein through the ER before they are cleaved off by small RNA **signal recognition** particles, e.g. 7SL-RNA. Some proteins cross the ER membrane completely; others (e.g. membrane **receptors**) do so only partly, remain integrated within it and end up in the plasmalemma or the membrane of another organelle. Because protein translocation beyond the ER is achieved via budded-off vesicles, the ER's is the only membrane a protein ever has to cross, even if destined for export from the cell (likewise some viruses). *Transit sequences* are amino-terminal and amphiphilic, targeting the protein into chloroplasts, mitochondria, etc. *Retention signals* keep a protein within a compartment despite the bulk flow of other proteins. *Signal (targeting) patches* are formed from different amino acid sequence regions which, through the protein's tertiary and quaternary structures, come to form part of the surface conformation of the molecule after folding in the ER (compare B cell and T cell epitopes). See Figs. 51 and 103, **sorting signals**, **glycosylation**.

Protein tyrosine kinase (PTK) Those **protein kinases** phosphorylating tyrosine residues in their target protein substrate (often an enzyme). They are enzymes providing a central switch mechanism in cellular **signal transduction** pathways, often involved in cell fate determination. The **receptor tyrosine kinases** (RTK) form intracellular components of cell-surface transmembrane receptors to growth factors and hormones, being activated when ligand binds the extracellular domain, causing autophosphorylation of the receptor's tyrosine residues, which are subsequently bound by SH2 domains of non-receptor tyrosine kinases and other proteins involved in signal transduction (see **domains**). Activation of a cell's PTKs can induce it to divide and migrate (e.g. fibroblasts treated with PDGF, see **growth factors**) or differentiate (neuronal precursors treated with NGF). Non-receptor protein tyrosine kinases (NRPTKs) include c-Src and c-Yes proteins, phosphorylated forms of which are actively involved in M-phase of the **cell cycle**. Over-expression or gain-of-function **mutations** in RTK genes can cause cancers (see **cancer cells**).

Proteoglycan Class of acidic **glycoproteins** found in varying amounts in extracellular matrices of animal tissues, notably connective tissues. Contain more carbohydrate than protein. One forms, with collagen, the rubbery material in cartilage preventing bone ends from grating together. As with **mucins**, a carbohydrate-free area serves for cross-links between the protein chains to produce aggregation. The fibrous polysaccharide **hyaluronic acid** serves as a chain along which many of these proteoglycan molecules align themselves. See **peptidoglycan**, **chitin**.

Proteolysis Protein hydrolysis, as achieved by mineral acids and – more widely and importantly – by proteases (proteolytic enzymes). Apart from its importance in nutrient digestion, proteolysis is essential to regulation of cellular protein concentration and distribution, in which **proteasomes** and **ubiquitins** are involved.

Proteolytic enzyme Any enzyme taking part in breakdown of proteins, ultimately to amino acids. Include **pepsin**, **trypsin**, **chymotrypsin**, **peptidases** and intracellular **cathepsins**. See **proteolysis**.

Proterozoic Geological division (eon) between the end of the **Archaean** eon (*c.* 2600 Myr BP) and onset of Phanerozoic (*c.* 600 Myr BP), from which it is separated by a discrete event: the radiation of eucoelomate animals. During it, atmospheric oxygen levels rose to about one tenth of present levels. Limestones became abundant for the first time, often containing **stromatolites**. Trace fossils permit recognition of perhaps three uppermost Proterozoic biozones, the best-known being the Ediacara fauna (soft-bodied) from Australia (590–560? Myr BP). Phytoplanktonic protists (acritarchs) and Cyanobacteria are quite well distributed from perhaps 630 Myr BP onwards. Chitinous sabelliditid worms and shelly fossils occur at or just below the Proterozoic/Cambrian boundary. There may have been two or more great global glacial periods in the late Proterozoic.

Prothallus Independent gametophyte stage of ferns and related plants. Small, green, parenchymatous thallus bearing antheridia and archegonia, showing little differentiation. Usually prostrate on the soil surface, attached by rhizoids. May be subterranean and mycotrophic.

Prothoracic gland See **ecdysone**.

Prothrombin See **blood clotting**.

Protista (Protoctista) Kingdom comprising eukaryotic and mesokaryotic unicellular or multicellular organisms. Nutritional modes are diverse, and include ingestion, photosynthesis and absorption. True sexuality is present in the majority of divisions (except euglenoids). Flagellated and non-flagellated forms occur. Includes all the eukaryotic and mesokaryotic algae, protozoa, slime moulds, and Oomycota and Chytridiomycota (a flagellated stage in their life cycle).

Protoctista See **Protista**.

Protochordata Subgroup of the **Chordata** comprising the **Hemichordata**, **Urochordata** and **Cephalochordata**: all three invertebrate chordate subphyla. Some would include pogonophorans and/or graptolites as well.

Protogyny (1) The condition in flowers (termed *protogynous*) whose carpels mature before their anthers, as in plantains. See **dichogamy**. (2) The condition in sequentially hermaphrodite animals in which first eggs are produced, then sperm. See **protandry**.

Protonema Branched, multicellular, filamentous or (less commonly) thalloid structure, produced on germination of a bryophyte spore, from which new plants develop as buds.

Protonephridium See **nephridium**.

Proto-oncogene (cellular oncogene) See **oncogene**.

Protoplasm Cell contents within and including the plasma membrane but usually taken to exclude large vacuoles, masses of secretion or ingested material. In most eukaryotic cells it includes, besides the **cytoplasm**, one or more nuclei. Prokaryotic cells lack nuclei. Cell walls, if present, are non-protoplasmic. Each protoplasmic unit constitutes a *protoplast*.

Protoplast (Bot.) Actively metabolizing part of a cell (its **protoplasm**), as distinct from cell wall. Equivalent to the whole cell in zoology. In **biotechnology**, the use of tissue culture has involved plant regeneration from protoplasts (the cell wall having been removed by enzymatic digestion). Protoplasts are also used in plant genetics; thus (a) protoplasts from different plants can be induced to fuse together to produce interspecific or, in some cases, intergeneric hybrid cells, while (b) the Ti **plasmid** from the crown gall bacterium *Agrobacterium tumefaciens*, or some other DNA injection technique, is used to introduce specific genes into protoplasts. Tissue culture is then used to regenerate plants from these individual protoplasts.

Protopodite See **biramous appendage**.

Protostele Simplest and most primitive type of **stele**, comprising a central core of xylem surrounded by a cylinder of phloem. Present in stems of some ferns and club mosses and almost universal in roots. In a *haplostele*, xylem forms a central rod; in an *actinostele*, xylem is ribbed and appears star-shaped in transverse section; in a *plectostele*, xylem is in several parallel, longitudinal strips embedded in the phloem.

Protostomia Those coelomate metazoans (sometimes termed an infragrade) in which the blastopore develops into the mouth of the adult, cleavage tends to be determinate, and the coelom tends to form by **schizocoely**. Includes annelids, arthropods, molluscs and, usually, those phyla with **lophophores**. Compare **Deuterostomia**.

Prototheria Mammalian subclass, of which only the monotremes (six species) survive. Includes extinct orders **Multituberculata**, Triconodonta and Docodonta, and the extant Order Monotremata. The latter comprise the duckbill, or platypus (*Ornithorhynchus*), of Australia and Tasmania, and the spiny anteaters (*Tachyglossus*, *Zaglossus*) of Australia, Tasmania and New Guinea. Fossil forms have been found only in Australia, and have not so far pre-dated the Pleistocene. All species have hair and mammary glands and are homoiothermic; but all have the reptilian features of egg-laying (ovipary), retention of separate coracoid and interclavicle bones in the **pectoral girdle**, and epipubic bones attached to the **pelvic girdle**. Brain size in relation to body size is lower than in placental mammals, resembling marsupials in this respect. A **cloaca** is present. Echidnas incubate in a pouch (marsupium); duckbills incubate in a nest. (See Fig. 77.)

Prototroph Any microorganism (esp. bacterium, fungus) expressing the normal (wild-type) phenotype with respect to its ability to synthesize its organic requirements when grown on nutritionally unsupplemented (i.e. *minimal*) medium. Contrast **auxotroph**.

Protoxylem The first elements to be differentiated from procambium; extensible. Described as *endarch* when internal to the later-formed metaxylem (as in roots), and as *mesarch* when surrounded by metaxylem (as in fern stems).

Protozoa Phylum, or subkingdom, of the **Protista**, comprising unicellular and colonial animals of varied form. Generally subdivided into four classes: **Sarcodina** (amoebae, radiolarians, foraminifera), **Mastigophora** (flagellates), **Ciliata** (ciliates and suctorians) and **Sporozoa** (e.g. coccidians, gregarines); but some of these at least represent grades rather than clades. The first two are sometimes united as the Sarcomastigophora. Reproduction commonly by binary or multiple **fission**, but sometimes by **conjugation**. Ubiquitous, inhabiting aquatic and damp terrestrial habitats. Several of them are serious pests of humans and their domestic animals (e.g. see **malaria**).

Protura Order of the **Apterygota**. Minute (0.5–2.5 mm long) whitish insects lacking antennae and eyes. Inhabitants of moist soils, leaf litter, etc. Stylet-like mandibles for piercing. Metamorphosis consists of addition of three abdominal segments and development of genitalia. No evidence that the adult moults.

Proventriculus Anterior part of the bird stomach, where digestive enzymes are secreted, posterior part being the **gizzard**. Used synonymously with gizzard in crustaceans and insects.

Provirus Viral genomes which are integrated into the host cell chromosome and most of the time remain there unexpressed (*latent*). Bacteria harbouring bacteriophages which do this are termed *lysogenic*. See **bacteriophage**.

Proximal Situated relatively near to a point of attachment or origin. Compare **distal**.

Proximate factor Explanations in biology are mostly either proximate or ultimate. Former are characteristically mechanistic and indicate how some outcome or change is intelligible in terms of antecedent causes. The latter are teleological, rendering phenomena intelligible in terms of probabilities of future states of affairs. Thus, change in photoperiod may be cited as the proximate factor causing change of winter coat colour in stoats, etc.; but better camouflage may be cited as an **ultimate factor** responsible for the change, bringing an increase in individual fitness.

Prymnesiophyta Haptophytes. Mostly marine, uninucleate flagellated algae (Kingdom **Protista**) possessing a **haptonema** between two smooth flagella. Most cells have two flagella of about the same length, except in the Pavlovales where one flagellum is longer than the other, and is usually covered with small scales. The chloroplasts are surrounded by two membranes of chloroplast endoplasmic reticulum, with the outer one continuous with the outer membrane of the nuclear envelope. Within the chloroplast, the thylakoids are aggregated into bands of three, and a **pyrenoid** is usually present in the centre of the chloroplast or as a bulge to one side. Eyespots are uncommon but where present comprise a group of lipid droplets inside the anterior end of the chloroplast (e.g. *Pavlova*). Chlorophylls a, c_1 and c_2 are present within the chloroplast along with fucoxanthin as the major carotenoid. The storage product is chrysolaminarin which is found toward the posterior of the cell in vesicles. The Golgi apparatus is located in the anterior of the cell, where sometimes a contractile vacuole is also found. In some species (e.g. *Chrysochromulina*), muciferous bodies occur under the plasmalemma that have the same structure as those found in the **Dinophyta**, **Chrysophyta** and **Raphidophyta**. Surface protrusions containing cytoplasm are also common (called pseudopodia). Cells can also extrude trailing filaments (filopodia). Many of these algae are phagocytic and consequently possess food vesicles; they feed upon bacteria and other small algae as well as detritus. Most members of the Prymnesiophyta have a cell covering of organic elliptical scales embedded in a mucilaginous substance, and in some a layer of calcified scales (coccolith) is outside the organic scales. The organic scales originate in the Golgi apparatus, and appear on the cell surface via exocytosis. Coccoliths, the calcified scales, were originally described as minute carbonate discs in Cretaceous deposits and thought to be of inorganic origin. They are basically organic scales that have $CaCO_3$ (in the form of calcite) deposited upon one surface in a species-specific pattern, and are characteristic of the Coccosphaerales. In the Cretaceous, coccolith-bearing algae dominated the calcareous nanoplankton.

Pseudoalleles Two mutations in the same cistron which give rise to different phenotypes when in the *cis* and *trans* conditions respectively. In the **cis-trans test** they fail to complement one another.

Pseudocoelom (pseudocoel) Fluid-filled cavity between body wall and gut with, however, an entirely different origin from true **coelom**. The pseudocoelom is a persistent blastocoel, lacking a definite mesoderm lining. In some cases, as in nematode worms, it may be filled with vacuolated mesenchyme cells. Pseudocoelomate invertebrates include **Aschelminthes**, **Endoprocta** and possibly priapulid worms. Their interrelationships are unclear, but it is likely that all pseudocoelomate animals have had progenetic origins (see **progenesis**), and that they are polyphyletic and represent a grade rather than a clade.

Pseudogamy Phenomenon where fertilization is required for development of sexually-produced offspring which derive all their genes from their maternal parent. In the grass *Poa*, and in *Potentilla*, apomictic plants produce perfectly functional pollen and fertilization precedes seed development, but only fertilization of the endosperm nucleus occurs, the egg cell nucleus remaining unfertilized. See **gynogenesis**, **parthenogenesis**.

Pseudogene A DNA sequence which, despite being largely homologous to a transcribed sequence elsewhere in the genome, is not transcribed. Some probably arise through gene duplication (*processed* pseudogenes); others, more numerous, must have originated by reverse transcription of mRNA and insertion into the gene, since they have lost their introns and are dispersed in location. The gene cluster for human haemoglobin contains two pseudogenes (designated by the symbol ψ in front of the gene symbol).

Pseudoparenchyma See **plectenchyma**.

Pseudopodium Temporary protrusion of some cells (in some sarcodine protozoans, e.g. *Amoeba*; **macrophages**) involved in amoeboid forms of **cell locomotion** and food capture. See **phagocyte**.

Pseudopregnancy State resembling pregnancy in female mammal, but in absence of embryos. Due to hormone secretion of **corpus luteum**, and occurs in species where copulation induces ovulation (e.g. rabbit, mouse), but when such copulation is sterile, or when normal oestrous cycle includes a pronounced luteal phase (e.g. bitch).

Pseudoscorpiones Arachnid order containing the pseudoscorpions: minute scorpion-like animals lacking tail and sting and common in soil where they are predatory, using pincer-like pedipalps.

Psilophyta Whisk ferns. Two living genera (*Psilotum* and *Tmesipteris*). *Psilotum* is tropical and subtropical in distribution; *Tmesipteris* is restricted to Australia, New Caledonia, New Zealand and other south Pacific islands. Sporophytes are very simple and may represent a reduction from a fern-like ancestor. *Psilotum* is unique among living vascular plants in lacking both roots and stems; the sporophyte comprises an underground rhizome system with many rhizoids, and a dichotomously branched aerial portion with small scale-like outgrowths. The main absorptive system is the mass of fungal hyphae that penetrates the cortex and interweaves among the rhizoids. The stele is **prostelic**; with a central, irregular mass of xylem surrounded by the phloem, which, in turn, is enclosed by several layers of parenchymatous cells that comprise the pericycle. The rhizomatous system can form minute, multicellular, ovoid gemmae, which arise at the tips of rhizoids, and are filled with starch, and readily break free. The chlorophyllous aerial portions are about 20–30 cm in length, and have a conspicuous cuticle; the epidermal cells lack chloroplasts and rhizoids, but stomata are present. Chloroplasts occur in a narrow band of outer cortical cells, and this forms the main photosynthetic tissue. Inside the cortex, and forming the main supportive tissue, is a layer of sclerenchyma cells. The endodermis forms the innermost layer of the cortex and surrounds the stele. The xylem is exarch, and comprises scalariform and pitted tracheids in the metaxylem; and annular and helical tracheids in the protoxylem. The phloem is between the endodermis and the xylem. Sporangia (or synangia) are borne upon the uppermost branches; spores germinate giving rise to bisexual gametophytes which resemble portions of the rhizome. The gametophyte usually lacks vascular tissue. Antheridia and archegonia are intermixed over the entire surface of the gametophyte. Sperm of *Psilotum* are multiflagellate, and require water to reach the egg. Sporophyte is initially attached to the gametophyte by a foot, which absorbs nutrients from the gametophyte. It eventually becomes detached from the foot. *Tmesipteris* is epiphytic, with larger, leaf-like appendages and a rhizomatous system similar to that of *Psilotum*, but it often has collenchymatous thickenings in the cortex.

P-site Binding site of ribosome for the tRNA corresponding to the **start codon** of mRNA (usually AUG, but sometimes GUG) and for the tRNA linked to the growing end of the polypeptide chain (hence P for peptidyl) during **protein synthesis**. The P-site receives this tRNA from the **A-site**. See **ribosome**.

Psocoptera Booklice (psocids). Order of exopterygote insects. Small, some wingless. Biting mouth parts. Feed on fragments of animal and vegetable matter and paste of book-bindings.

Psychoactive drugs Mood- and/or behaviour-modifying drugs. Lipid-soluble, they cross the **blood-brain barrier**. See **endoplasmic reticulum** (for detoxification), **synapse**.

Psychrophilic (Of microorganisms, e.g. *Flavobacterium*) with an optimum growth temperature, measured by generations per hour, in the 5–15°C range. Compare **mesophilic**, **thermophilic**.

Pteridophyta In older plant classifications, a division containing spore-bearing (as opposed to seed-bearing) tracheophytes: Psilophyta, Lycophyta, Sphenophyta and Pterophyta.

Pterodactyla See **Pterosauria**.

Pterophyta The ferns. Relatively abundant as fossils since the Carboniferous period; not known from the Devonian period. About 12,000 extant species, roughly two thirds in tropical regions, the other third in temperate regions. Display great diversity of form and habit, but are almost entirely terrestrial. The fern plant, the sporophyte, is the dominant generation; the gametophyte is the prothallus. Most temperate ferns comprise an underground siphonostelic rhizome which produces new leaves each year; roots are adventitious; leaves (fronds) are the megaphylls. Unique among seedless vascular plants in possessing megaphylls. All but a few ferns are homosporous, with sporangia variously placed and commonly borne in clusters called sori. Fern heterospory is restricted to two specialized groups of aquatic ferns. Spores of most homosporous ferns germinate to give rise to free-living gametophytes which develop into prothalli; these are monoecious, antheridia and archegonia developing on the ventral surface. Sperm are motile and require water to swim to the eggs. The sporophyte resulting from fertilization is initially dependent upon the gametophyte, but growth is rapid and it soon becomes independent. See **life cycle**.

Pterosauria Pterodactyls. Extinct order of **archosaurs**, originating in late Triassic and disappearing at end of Cretaceous. Winged reptiles, fourth finger of the fore-limb greatly elongated and supporting a membrane. Hind-limbs feeble. Long tail tipped by a membrane and acting as a rudder. Wing spread up to 16 metres.

Pterygota (Metabola) Insect subclass including all but the **Apterygota**, aome are secondarily wingless (e.g. fleas). Includes the **Endopterygota** and **Exopterygota**.

Ptyalin An **amylase** present in saliva of some mammals, including humans.

Pubic symphysis See **pelvic girdle**.

Pubis See **pelvic girdle**.

Puff (Balbiani ring) Swelling of giant chromosome (e.g. from dipteran salivary gland cell) normally regarded as representing regions being actively transcribed and consisting of many strands of decondensed DNA and its associated messenger RNA. Some recent work indicates some transcription of these chromosomes in regions lacking puffs, and some puff presence where transcription is lacking. See **polyteny**, **ecdysone**.

Pulmonary (Adj.) Relating to the **lung**. Vertebrate *pulmonary arteries* (where present, derived from sixth **aortic arch**) carry deoxygenated blood to the lungs from the right ventricle of the heart in crocodiles, birds and mammals (or from the single ventricle of lungfish and amphibians) to the lung capillaries; in tetrapods, *pulmonary veins* return oxygenated blood to the left atrium.

Pulmonata Order of **Gastropoda**. Lung develops from mantle cavity; e.g. snails, slugs.

Pulp cavity Internal cavity of vertebrate tooth or denticle, opening by a channel to the tissues in which it is embedded. Contains *tooth pulp* of connective tissue, nerves and blood vessels, with odontoblasts lining dentine wall of cavity. See **tooth**.

Pulse (Zool.) Intermittent wave of raised pressure passing rapidly (faster than rate of blood flow) from heart outwards along all arteries each time the ventricle discharges into the aorta. The increased pressure dilates the arteries, and this can be felt. Each pulse experiences resistance from the elastic walls of the arteries. See **artery**. (Bot.) Seed of leguminous plant (Fabaceae), rich in proteins; e.g. soya bean.

Pulvinus Joint-like thickening at base of leaf petiole (or petiolule of leaflet), playing important role in its movement.

Punctuated equilibrium See **evolution**, **cladistics**.

Pupa (chrysalis) Stage between larva and adult in life cycle of endopterygote insects, during which rearrangement of body parts (**metamorphosis**) occurs, involving development of **imaginal discs**. In some pupae (*exarate*) the appendages are free from the body; in many others (*obtect*) they are glued to the body by a larval secretion. Pupae of most culicine mosquitoes are active.

Pupil Opening in **iris** of vertebrate and cephalopod **eye**, permitting entry of light.

Pure line Succession of generations of organisms consistently homozygous for one or more characters under consideration. Initiated by crossing two appropriately homozygous individuals or, in plants, by selfing one such individual. Such organisms *breed true* for the characters, barring mutation. See **inbreeding**.

Purine One of several nitrogenous bases occurring in **nucleic acids**, nucleotides and their derivatives. Synthesis involves addition of glycine, and two transaminations from glutamne, to a ribose-phosphate precursor. By far the commonest are **adenine** and **guanine**; rarer *minor purines* including methylation products of these. See **pyrimidine**.

Purkinje fibres Modified **cardiac muscle** fibres forming the bundle of His conducting impulses from **pacemaker**, and the fine network of fibres piercing the myocardium of the ventricles. See also **heart cycle**. Not to be confused with *Purkinje cells*, which are large neurones of the cerebellar cortex.

Puromycins Antibiotics interrupting the translation phase of protein synthesis in both prokaryotes and eukaryotes by their addition to growing polypeptide chain, causing its premature release from the ribosome. Compare **actinomycin D**. See **antibiotic**.

Pusule A sac-like structure found in members of the **Dinophyta**, that opens via a pore into the flagellar canal. Like contractile vacuoles, has an osmoregulatory function; but is structurally more complex.

Putrefaction Type of largely anaerobic bacterial decomposition of proteinaceous substrates, with formation of foul-smelling amines rather than ammonia.

Pycnidium Flask-shaped structure in which conidia are formed in some members of the **Deuteromycota** and **Ascomycota**.

Pycnium Flask-like to variously shaped structure located subepidermally, wherein pycniospores or spermatia and special receptive hyphae develop. Produced by rust fungi (**Basidiomycota**, Order **Uredinales**). Pycniospores are exuded in a sugary liquid and are dispersed by insects. Those of one pycnium must be transferred to receptive hyphae of another to initiate the dikaryotic phase. See **Uredinales, aecium, uredium, telium**.

Pylorus Junction between vertebrate stomach and duodenum. Has a sphincter muscle within a fold of mucous membrane closing off the junction while food is digested in the stomach.

Pyramid of biomass Diagram, pyramidal in form, representing the dry masses in each **trophic level** of a community or food chain. The standing crops of populations are normally summated within each trophic level. Not all biomass has the same energy content, however. See **energy flow, pyramid of numbers, pyramid of energy**.

Pyramid of energy Diagram representing the energy contents within different **trophic levels** of a community or food chain. Generally pyramidal in form, it is difficult to represent the **decomposers**, and does not easily indicate the unavailability (to the next trophic level) of storage if standing crops are used. See **pyramid of biomass, pyramid of numbers**.

Pyramid of numbers Diagram, pyramidal (sometimes inversely) in form, representing the numbers of organisms in each **trophic level** of a community or food chain. All organisms are equated as identical units, a massive tree being equivalent to a flagellate cell. See **pyramid of biomass**.

Pyrenoid Proteinaceous region within chloroplasts of many algae, taking early products of photosynthesis and converting them to storage compounds. Occur in every eukaryotic algal group, and are considered a primitive characteristic. The pyrenoid is denser than the surrounding stroma and may or may not be traversed by thylakoids. In green algae (**Chlorophyta**), they are associated with starch synthesis and surrounded by starch deposits.

Pyridoxine One of the water-soluble vitamins in B6; precursor of *pyridoxal phosphate*, a **prosthetic group** associated with transaminase enzymes of mitochondria and/or the cytosol. Required by, among others, yeasts, bacteria, insects, birds and mammals.

Pyrimidine One of a group of nitrogenous bases occurring in **nucleic acids**, nucleotides their derivatives. Synthesis involves a step in which carbamoyl phosphate and aspartate form carbamoylaspartate prior to ring closure. The three commonest are cytosine, thymine and uracil. Methylated forms also occur; they are termed *minor pyrimidines*.

Pyrrophyta See **Dinophyta**.

Pyxidium Type of **capsule**.

Q_{10} (temperature coefficient) Increase in rate of a process (expressed as a multiple of initial rate) produced by raising the temperature by 10°C. Rate of most enzymatic reactions approximately doubles for each rise in temperature ($Q_{10} \simeq 2$), but this varies a little from one enzyme to another.

Qo_2. Oxygen quotient Rate of O_2 consumption, often of whole organisms or tissues. Often expressed in $\mu l.mg^{-1}.hr^{-1}$.

Quadrat (1) Delineated area of vegetation (standard size is one square metre) chosen at random for study of its composition. Quadrat frames of different dimensions are often 'thrown' on to vegetation to ascertain which frame size provides most information in least time.

Quadrate Cartilage-bone of posterior end of vertebrate upper jaw. Develops within the **palatoquadrate** and attaches to neurocranium, in most cases articulating with lower jaw. In mammals becomes the incus (see **ear ossicles**).

Qualitative inheritance See **variation**.

Quantasomes Granules occurring on inner surfaces of **thylakoids** of **chloroplasts**; thought to be basic structural units involved in light-dependent phase of photosynthesis rather than artefacts of electron microscope preparations.

Quantitative inheritance See **variation**.

Quantum In 1905, Albert Einstein proposed the particle theory of light in which light is considered to be composed of particles of energy (photons or quanta of light). The energy of a photon (quantum of light) is not the same for all kinds of light; instead it is inversely proportional to the wavelength.

Quaternary The **geological period** comprising both the Pleistocene and Holocene.

Quaternary structure See **protein**.

Quiescent centre Area at the tip of the root apical meristem where rate of cell division is lower than in surrounding tissue.

Quinine An important alkaloid extracted from the bark of *Cinchona* species (Family Rubiaceae) used in the treatment of malaria.

Quinoa *Chenopodium quinoa* (Family Chenopodiaceae) is an important grain crop in the Andes, and elsewhere in South America.

Quinone Important small hydrophobic components of the **electron transport chain**. Unlike **cytochromes** and **iron-sulphur proteins**, quinones (e.g. **coenzyme Q**, ubiquinone) carry the equivalent of a hydrogen atom. By alternating electron transfer between components that carry or do not carry a proton with the electron, protons can be moved across the membrane, setting up proton gradients. Most quinone molecules are not attached to proteins and diffuse rapidly in the plane of the membrane. See **chemiosmotic theory**.

Race See **infraspecific variation**, **ecotype**.

Raceme Kind of **inflorescence**.

Rachis (1) Main axis of an **inflorescence**. (2) The axis of a fern leaf (frond), from which the pinnae arise. (3) In compound leaves, the extension of the petiole corresponding to the midrib of an entire leaf.

Radial cleavage (bilateral cleavage) See **cleavage**.

Radial micellation (Bot.) The radial orientation of the cellulose microfibrils in the guard cell walls of stomata (see **stoma**), which allows the guard cells to lengthen while preventing them from expanding laterally. The common wall at the ends of the guard cells remains almost constant in length during opening and closing of the stomata. Therefore, increase in **turgor pressure** causes the outer (dorsal) walls to move relative to their common walls. As this occurs, the radial micellation transmits the movement to the wall bordering the stomatal opening, and the pore opens.

Radial section Longitudinal section cut parallel to the radius of a cylindrical body (e.g. a stem or root).

Radial symmetry Capable of bisection in two or more planes to produce halves that are approximately mirror images of each other. Characteristic of bodies of coelenterates and echinoderms; and of many flowers (see **actinomorphic**) and some algae, e.g. the **centric diatoms**, when viewed in valve view. Compare **bilateral symmetry**.

Radical (Bot.) Arising from the root or crown of the plant.

Radical Root of embryo seed plants.

Radioactive labelling See **labelling**.

Radioisotope (radioactive isotope) See **labelling**, **radiometric dating**.

Radiolaria Order of marine planktonic sarcodine protozoans, lacking shells but with a central protoplasm comprising chitinous capsule and siliceous spicules perforated by numerous pores through which spines project between vacuolated and jelly-like outer protoplasm. Between the spines radiate out branching pseudopodia exhibiting cytoplasmic streaming. Many house yellow symbiotic algal cells (zooxanthellae). *Radiolarian oozes* cover much of the ocean floor and are important in flint production.

Radiometric dating Methods employed to measure the amount of an isotope produced by radioactive decay, or the amount of the radioisotope itself. By assuming that the proportion of the radioactive isotope to the stable isotope is the same now as when the sediment was laid down, and that no subsequent addition or dilution has occurred, the proportion of the radioactive isotope or its product remaining today is a function of the time that it has had to decay at its constant known rate (the *half-life*).

Carbon-14 is the most commonly used radiometric dating technique in Quaternary palaeoecology, as it has a suitable decay rate (half-life = 5568 years), which allows dating to be made back to *c.* 40,000 yr BP. Accelerator mass spectrometry (AMS) has provided further analytical improvements in the ^{14}C method, since samples containing 1 mg or less of elemental carbon can be analysed. Previous methods have required nearly 1000 times that amount. Dating of recent materials using ^{14}C is difficult because of the relatively large errors in the measurements, and the large amounts of ^{14}C-deficient carbon that have been introduced into the atmosphere through burning of fossil fuels since the industrial revolution. So other isotopes have to be used, such as caesium-137 (^{137}Cs) or lead-210 (^{210}Pb). The former was introduced into the atmosphere as a result of nuclear testing in 1954. Since then, the amount of ^{137}Cs in undisturbed lake sediments reached maximum levels in 1963, since when there has been a decline. ^{210}Pb is more frequently used and can date sediments up to 150 years old. Radium decays in soils to radon-222, which escapes into the atmosphere, where it decays to ^{210}Pb. The latter enters a lake in precipitation, and eventually becomes incorporated into sediments. This is called unsupported ^{210}Pb. Supported ^{210}Pb is produced within the sediments, and is assessed by measuring its ^{226}Ra grandparent. The excess ^{210}Pb over the expected ^{210}Pb gives the amount of unsupported ^{210}Pb. The half-life of ^{210}Pb is 22.26 years. By assuming either a constant initial concentration in the sediments which have accumulated at a constant rate, or by assuming a constant rate of ^{210}Pb supply to the sediments, the age of the sediment can be determined. The second model provides more accurate dates. The thorium-228:thorium-232 ratio can be used to date sediments up to 10 Myr in age. In contrast, the method applicable to material from the early part of the **Pleistocene** is the potassium/argon (K/Ar) method, by means of which the earliest **hominid** remains from East Africa have been indirectly dated. The K/Ar method depends upon the decay of ^{40}K to the non-radioactive inert gas argon, and is particularly applicable to volcanic rocks in which the K/Ar clock was set to zero when they were formed. The half-life of ^{40}K is very long (1300 Myr), and in most cases the lower limit of the K/Ar dating is *c.* 250,000 yr BP. Uranium-series dating methods are applicable over the time range from a few thousand to *c.* 10^6 years, and can be applied to a wide range of materials. It is an open system using disequilibrium in the ^{238}U-^{234}U-^{230}Th (thorium) decay series.

Although not strictly radiometric, the following two techniques are now invaluable in dating. For material up to *c.* 1 Myr old, thermoluminescence (TL) and electron spin resonance (ESR) techniques are employed. TL depends upon the fact that heating and sunlight repair the damage to crystalline materials (e.g. flint) caused when natural radiation in them displaces electrons. The process of accumulating radiation damage then starts again, and the date of a burnt flint sample can be obtained when the material is heated to 500°C or more to release the energy of the electrons displaced since the first burning. ESR employs microwave radiation to measure the accumulated amount of radiation damage in crystalline materials (carbonates, tooth enamel), since it causes displaced electrons to emit a signal proportional to the damage.

Radius One of two long bones (the other being the ulna) in the tetrapod fore-limb. Articulates with the side of the fore-foot bearing the thumb. See **pentadactyl limb**.

Radula 'Tongue' of molluscs; a horny strip, continually renewed, with teeth on its surface for rasping food. Found in Amphineura, Gastropoda and Cephalopoda; absent in Bivalvia, which are microphagous. May be modified for boring in some species. Pattern of teeth helpful in identification.

Ramapithecus (***Sivapithecus***) See **Ponginae**.

Ramet An independent individual of a **clone**.

Raphe (1) In seeds formed from anatropous ovules, a longitudinal ridge marking position of the adherent **funicle**. (2) Elongated slit (or slit pair) through valve wall of some pennate diatoms, involved in movement of cell over the substratum.

Raphidophyta Chloromonads (Kingdom Protista). Biflagellated algae (some become palmelloid, e.g. *Vacuolaria*) possessing one anterior (commonly hairy) and one posterior (smooth) flagellum. Two membranes of chloroplast endoplasmic reticulum are present, and within the chloroplasts are chlorophylls *a* and *c*. Many cells possess mucocysts in the peripheral cytoplasm, which can burst out of the cell, releasing a fibrous material. Freshwater species are green in colour and common in the **epipelon**. Marine taxa are yellowish due to the carotenoid fucoxanthin, and are a common component of algal blooms in red tides in coastal waters. Sometimes included with the **Xanthophyta**, due to similar flagellation and pigmentation; but mucocysts are not found in xanthophytes.

Rarity A growing body of evidence suggests that locally rare (and geographically restricted) species tend to share some or all of the following characteristics compared with commoner taxa: (i) lower levels of self-incompatibility, (ii) tendency towards asexual reproduction, (iii) lower overall reproductive effort, and (iv) poorer dispersal abilities.

***Ras* genes** Oncogenes initially discovered through their ability to cause **rat** sarcomas (see **cancer cell**). Mutations in *ras* in mammalian cells tend to be oncogenic and are frequently associated with human cancers. Encode **Ras proteins**.

Ras proteins Oncoproteins encoded by ***ras* genes** which bind GTP and GDP and act as GTP-activated switches in **signal transduction** pathways in virtually all metazoan cells studied to date, and as key regulators in growth of eukaryotic cells (see **GAPs**). In one such signalling cascade, used to transduce both mitogenic and differentiation signals, both receptor and non-receptor **tyrosine kinases** lie upstream of Ras (see Fig. 99b in **receptor** entry). Often located on the cytoplasmic surface of the plasma membrane, Ras often requires an intermediary protein complex (including the GAP, Ras-activator or Sos) which binds both the activated receptor and Ras. Involved in transduction of mitogenic signals from cell periphery to the nucleus. See **domain** for SH2 and SH3.

Ratites See **Palaeognathae**.

Raunkiaer's life forms System of vegetational classification based on position of perennating buds in relation to soil level, indicating how plants survive the unfavourable season of their annual life cycle. The following are recognized: *phanerophytes*: woody plants whose buds are borne more than 25 cm above soil level (many trees and shrubs); *chamaerophytes*: woody or herbaceous plants whose buds are above soil level but less than 0.25 m above; *hemicryptophytes*: herbs with buds at soil level, protected by the soil itself or by dry dead portions of the plant; *geophytes*: herbs with buds below soil surface; *heliophytes*: herbs whose buds lie in mud; *hydrophytes*: herbs with buds in water; *therophytes*: herbs surviving the unfavourable season as seeds.

Ray Tissue initiated by **cambium** and extending radially in secondary xylem and phloem. Mainly parenchymatous, but may include tracheids in the xylem.

Ray flower See **disc flower**, **aster**.

Ray initials (Bot.) One of two types of meristematic cells of the vascular cambium; horizontally elongated or squarish cells that produce horizontally orientated ray cells, which form the vascular rays or radial system. The rays, composed largely of parenchyma cells, are variable in length. See **fusiform initial**.

Reaction centre (Bot.) The chlorophyll molecule of a photosystem capable of using energy in the photochemical reaction. See **photosynthesis**.

recA **Cistron** of the bacterium *E. coli* whose product, the RecA protein is involved in promoting general DNA recombination in the genome, and has ATPase activity in the presence of single-stranded DNA as well as catalysing DNA base-pairing and strand annealing (assimilation). RecA protein binds to single-stranded DNA and anneals it to any complementary sequence in a double-stranded (duplex) DNA in such a way as to replace one of the two DNA strands of the original duplex, forming a **heteroduplex**. Similar gene products are being found in eukaryotes.

Recapitulation See **biogenetic law**.

recB*, *recC **Cistrons** of the bacterium *E. coli* whose combined products make up the RecBC enzyme, which initiates DNA-unwinding at any free duplex end and has nuclease activities. Once bound to a free duplex end (not normally present in *E. coli*) RecBC proceeds to unwind the duplex, but wherever the enzyme encounters the DNA strand sequence 5′-GCTGGTGG-3′ (termed *Chi*), it cuts the strand, leaving it exposed to binding by RecA protein (see ***rec*A**). For this reason, recombination is promoted in *Chi* regions, of which there may be one thousand in the *E. coli* genome.

Recent See **Holocene**.

Receptacle (thalamus, torus) Apex of flower stalk, bearing flower parts (perianth, stamens, carpels). Its relation to the gynoecium determines whether carpels are *inferior* or *superior*. When carpels are at the apex of a conical receptacle and other flower parts are inserted in turn below, the gynoecium is

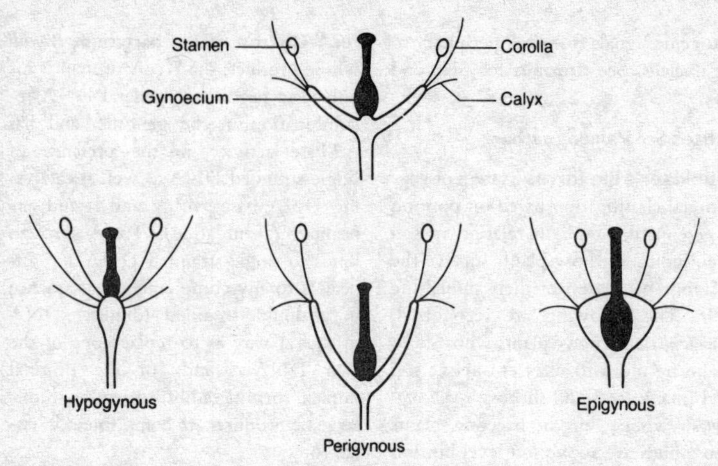

Fig. 98 *Diagrams of different types of floral* **receptacle**, *showing the position of the gynoecium (black) relative to other flower parts.*

superior and the flower *hypogynous* (e.g. buttercup, Fig. 98). When carpels are at the apex (centre) of a concave receptacle with other flower parts borne around its margin, the gynoecium is superior and the flower *perigynous* (e.g. rose, upper and lowest diagrams in Fig. 98). When receptacle completely encloses carpels and other flower parts arise from receptacle above, the gynoecium is inferior and the flower *epigynous* (e.g. apple, dandelion, Fig. 56). In this condition, carpel walls are intimately fused with the receptacle wall. (2) Describing the shortened axis of the **inflorescence** (capitulum) in Compositae. (3) In some brown algae (e.g. Fucales, Phaeophyta), the swollen thallus tip containing conceptacles.

Receptor (1) Sensory cell responding to some variable feature of an animal's internal or external environment by a shift in its membrane voltage. Sometimes (see Fig. 99), as in *primary receptors* (e.g. **muscle spindles**), only part of the cell is sensory and generates action potentials with a frequency related to the stimulus intensity; the rest of the cell is axonal, transmitting signals long distances. In *secondary receptors* the altered membrane voltage initiates action potentials in a synapsing neurone by bringing about voltage changes (*receptor*, or **generator potentials**) in the postsynaptic membrane of the neurone. The transduction event in all receptors is the production of a receptor potential on receipt of the signal (see **potentiation**, **signal transduction**). Receptors may be *interoceptors* or *exteroceptors* and classified by modality into *chemoreceptors*, *mechanoreceptors*, *photoreceptors*, etc. See **sense organ**.

(2) Membrane receptor molecules (receptor sites). Exquisitely adapted proteins. One classification distinguishes: (a) *channel-linked receptors*, associated with ion channels. Either ligand-gated (e.g. the **acetylcholine** receptor) or voltage-

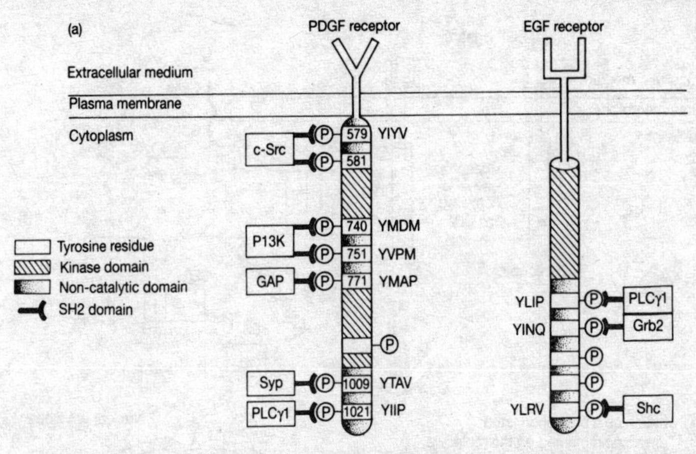

Fig. 99a *PGDF and EGF transmembrane receptor molecules with their kinase domains, non-catalytic domains and tyrosine residues (some numbered in PGDF, indicating the amino acid residue involved). Once phosphorylated, these tyrosine residues may bind specific SH2-bearing substrates, such as c-Src itself, GTPase-activating protein (GAP) of the GTPase Ras and phospholipase Cγ1 (PLCγ1). See* **receptor** *and specific references there.*

gated (e.g. the sodium channel). The action signalled by the stimulus is effected directly by the receptor molecule itself (unlike the *non-channel-linked receptors* below). One such class of receptor molecule (*seven transmembrane helix receptors*, or *heptahelicals*) span the membrane seven times with α-helices, and includes photoreceptors (e.g. **bacteriorhodopsin**, olfactory receptors and receptors for several neurotransmitters, all of them **G-protein**-linked and members of the rhodopsin superfamily of proteins; (b) *endocytic receptors*, such as the **LDL** and **transferrin** receptors, initiate clathrin-coated pit formation (see **coated vesicle** and Fig. 26) during receptor-mediated **endocytosis**. Neutrophils, macrophages and other mononuclear phagocytic white cells have cell-surface Fc receptors so that once **antibody** is bound to pathogen, the Fc end can bind the phagocyte; (c) *growth factor receptors*, such as those for insulin (see **diabetes**), epidermal **growth factor** and those used in intercellular signalling, e.g. CD4 and CD8 (see **accessory molecules**) span the membrane, linked covalently or by other direct coupling to an intracellular effector molecule, e.g. an enzyme such as a tyrosine kinase (when termed *catalytic receptors*; see **receptor tyrosine kinases**). Binding of ligand here initiates signalling cascades associated with the control of the **cell cycle**, gene expression and cell differentiation (see **signal transduction**). (3) For intracellular hormone receptors, see **nuclear receptors**.

Receptor potential See **generator potential.**

Receptor proteins See **receptor** (2) and reference there.

Receptor site See **receptor** (2).

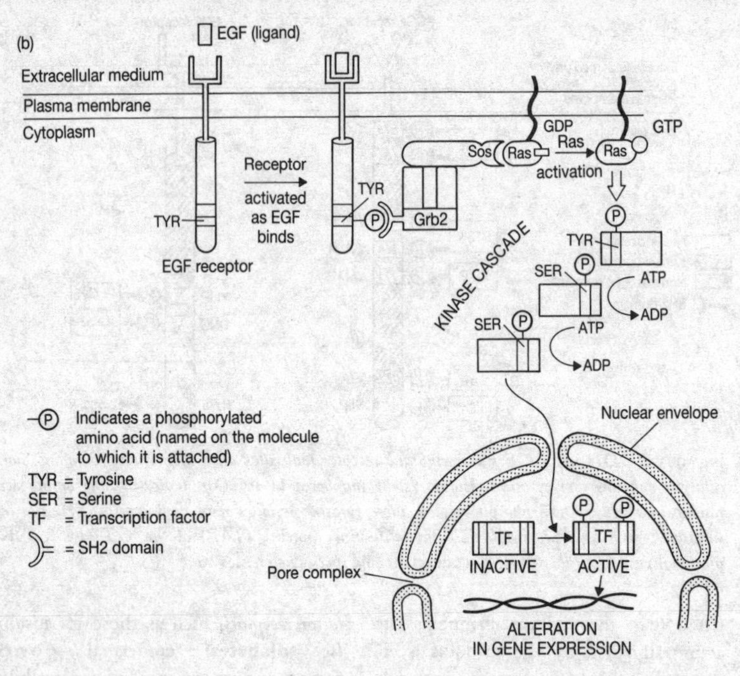

Fig. 99b *The involvement of EGF* **receptor** *in one of its signal transduction pathways involving the G-protein Ras. Ras-activation is required for proper functioning of many growth and differentiation factors, such as epidermal growth factor (EGF), shown here, insulin, PGDF, NGF, T cell receptor and many cytokines. Grb2 is also known as Sem5. Compare Fig. 104.*

Receptor tyrosine kinases (RTKs) Those **protein tyrosine kinases** having one **domain** which binds an extracellular ligand, one domain crossing the membrane bilayer (transmembrane) and one or more intracellular domains for phosphorylating tyrosine radicals of specific proteins. Each family of RTKs has probably evolved by **gene duplication**. See **signal transduction** and **receptor** (for Figs.).

Recessive (Of phenotypic characters) only expressed when the genes determining them are **homozygous** (complete recessivity). When heterozygous, the allele for the character is either 'silent', giving rise to no cell product, or its effect is masked by the presence of the other allele. Sometimes 'recessive' is used to describe alleles themselves, but since many gene loci are pleiotropic it would need to be made clear which aspect of phenotype was being described as recessive. Compare **dominance.** See **penetrance**.

Reciprocal altruism See **altruism**.

Reciprocal cross (1) Cross between two hermaphrodite individuals, in

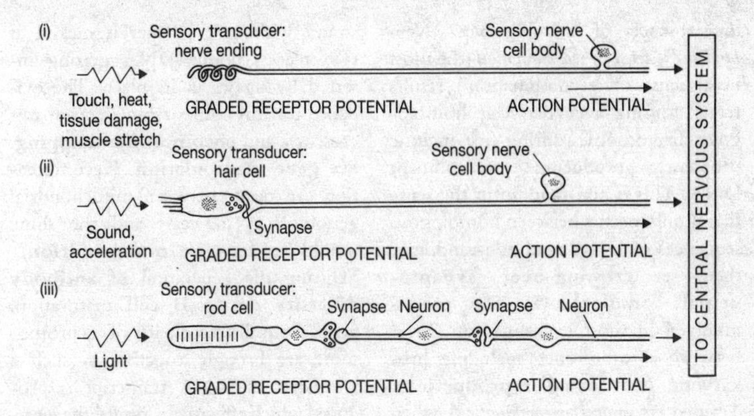

Fig. 99c *Different types of sensory cell. (i) Primary receptor; (ii) and (iii) secondary receptors. In all three cases, a graded receptor potential evoked in the sensory transducer is translated into an action potential firing frequency.*

which the male and female sources of the gametes used are reversed. (2) Crossing operation between stocks of two different genotypes, where each stock is used in turn as the source of male and female gametes. Employed when testing for **sex-linkage** (also sex-limited and sex-controlled inheritance), where one sex has a greater influence than the other in determining offspring phenotype. When reciprocal crosses give very different results (e.g. in F1 or F2) the character studied is likely to be sex-linked or under cytoplasmic control. See **cytoplasmic inheritance**.

Recombinant DNA DNA whose nucleotide sequence has undergone alteration as a result of incorporation of, or exchange with, another DNA strand. Such DNA occurs naturally as a result of **crossing-over** during **recombination**, and also during recombinant DNA techniques employed during **gene manipulation**.

Recombination Any process, other than point mutation, by which an organism produces cells with gene combinations different from any it inherited. Offspring resulting from such recombinant cells are *recombinant offspring*. A major source of **genetic variation**, its effectiveness is dependent upon mutation for initial gene differences, from which recombination events can generate further gene rearrangements.

(1) In meiosis (eukaryotes only), two kinds of recombination between chromosomes commonly occur. *Free combination* or *reassortment* (not always regarded as recombination) occurs when non-homologous chromosomes assort randomly to form the two haploid nuclei during anaphase of the first meiotic division. As a result, if N = the number of chromosome pairs in the parent cell, each chromosome pair being heterozygous at least at one locus, then the number of possible nuclear genotypes from the first division is 2^N, and this in

the absence of crossing-over. *Non-random* (*restricted*) *recombination* (the most usual sense of 'recombination') results from crossing-over between homologous chromosomes during first meiotic prophase, producing recombinant DNA. This is non-random in the sense that it only occurs between homologous sequences of DNA (and non-randomly then; see **crossing-over**, **synaptonemal complex**). (2) The process involved in most exchanges of DNA between chromosomes, including prokaryotic (see **sexual reproduction**). Termed *reciprocal recombination* when equivalent lengths of DNA are reciprocally exchanged between duplexes (i.e. between double helices), and *non-reciprocal recombination* (see **gene conversion**) when only one duplex retains its original length, as often happens in the immediate vicinity of crossovers (hence restricted, since crossovers themselves are of limited occurrence). In one model, homologous duplex DNA sequences first align themselves side-by-side; one strand of each duplex is then cut (nicked) by an enzyme and its broken ends joined up with their opposite partners by a DNA ligase (i.e. not merely rejoined again as before), to form two homologous duplexes whose nucleotide sequences have been altered. See ***recA***, ***recB*** and ***recC***.

In bacterial **transformation** and **transduction** (examples of *homologous* recombination) and some eukaryotic gene transfers, homologous DNA duplexes first align, the donor duplex undergoes **denaturation** (separates into its two strands), and one strand invades the host duplex, aligning with the host strand having the greater base-pairing conformity. It is then nicked while the host strand without a partner is nicked at two places, donor DNA getting inserted by ligases in its place. The evidence for this comes from electron microscopy and **chromosome mapping**. See **gene manipulation**. Recombination can occur between mitochondrial genomes, as in yeast and the slime mould *Physarum* (see **mitochondrion**).

During the generation of **antibody diversity** during **B cell** maturation, genes from different parts of a chromosome are brought together in such a way that an RNA transcript is produced which effectively 'omits' the intervening DNA sequences. Mitotic recombination between homologous chromosomes is very rare, but can be induced by **X-ray irradiation**. It can lead to production of two daughter cells homozygous for different alleles and for which the parent cell was heterozygous. See **twin spots**.

Recombination nodule See **synaptonemal complex**.

Recombination value Alternative for **cross-over value**.

Rectum Terminal part of intestine, opening via anus or cloaca and commonly storing faeces. In insects, often reabsorbs water, salts and amino acids from the 'urine' (see **Malpighian tubules**); some insect larvae have tracheal gills in the rectum, while larval dragonflies also eject water forcibly from the rectum for propulsion. Ectodermal in origin (see **proctodaeum**).

Red blood cell (red blood corpuscle, erythrocyte) Most abundant vertebrate blood cell; generated in bone marrow, usually from **reticulocytes**. Contains many molecules of **haemo-**

globin loading and unloading molecular oxygen (and carbon dioxide to a much lesser extent) and serving as a blood **buffer**. Mammalian erythrocytes are flattened, circular, biconcave discs (about 8 μm diameter in humans), lacking nuclei, mitochondria and most internal membranes. Tend to be larger and oval in shape in other vertebrates, retaining a nucleus. Damaged by passage through capillaries, they last about four months in humans (judged by radioactive tracers) before being destroyed by the liver's **reticuloendothelial system**. Their surface antigens specify **blood group**. Their membrane **sodium pumps** regulate cell volume, but hypotonic solutions cause osmotic swelling and rupture, leaving erythrocyte membranes as *ghosts*. The important enzyme **carbonic anhydrase** catalyses the reversible reaction:

$$H_2O + CO_2 \underset{\text{lungs}\longleftarrow}{\overset{\longrightarrow\text{tissues}}{\longleftrightarrow}} H^+ + HCO_3^-$$

enabling rapid exchanges of gases in the lungs and body tissues. Role of erythrocytes in CO_2 transport is primarily to generate HCO_3^- ions for carriage in the plasma and to reconvert them back to CO_2 molecules in the lungs, where exhalation occurs. However, about 23% of CO_2 carried in human blood is in the form of *erythrocytic carbaminohaemoglobin*, which breaks down in the lungs to release CO_2 again. See **haemopoiesis**.

Redia One of the larval types in endoparasitic Trematoda, developing asexually from the sporocyst and from other rediae (see **polyembryony**). Often parasitic in snails, developing into cercariae.

Redox reactions Oxidation-reduction reactions, in biology generally catalysed by enzymes. Involve transfer of electrons from an electron donor (reducing agent) to an electron acceptor (oxidizing agent). Sometimes hydrogen atoms are transferred, equivalent to electrons, so that dehydrogenation is equivalent to oxidation. Respiration involves many *redox pairs*, one member donating, the other accepting, electrons, determined by their relative *standard oxidation-reduction potentials*. See **electron transport system**.

Reducing sugar Sugar capable of acting as a reducing agent in solution, as indicated by a positive **Benedict's test** and ability to decolourize potassium permanganate solution. Depends upon presence of potentially free aldehyde or ketone group. Most monosaccharides are reducing sugars (and all are in weakly acid solution), as are most disaccharides except sucrose.

Reduction division First division of **meiosis**, the chromosome numbers of the daughter cells produced being half that of the parent cell.

Reductionism See **explanation**.

Reflex Innate (inherent) and often invariant neuromuscular animal response to an internal or external stimulus, usually with little delay involved. In its simplest form, mediated by a *reflex arc* involving input along a sensory neurone of a spinal nerve to the **central nervous system** (CNS) and output along a motor neurone of the same or a different spinal nerve to an effector (muscle, gland). Frequently in vertebrates an association (relay) neurone intervenes in the CNS, acting to transmit impulses via

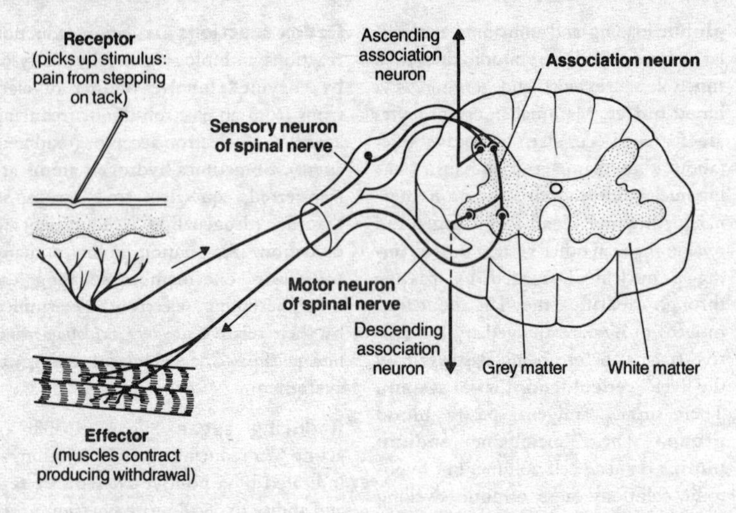

Fig. 100 *Diagram of section of spinal cord showing nervous pathways involved in a spinal* **reflex** *on one side. Ascending and descending association neurones permit involvement of higher and lower body regions.*

white fibres to the brain. This may enable perception and learning, as in **conditioned reflexes**. Vertebrate reflexes which involve only the spinal component of the CNS are termed *spinal reflexes*. Examples of innate reflexes in humans include blinking, sneezing, coughing, ventilation, regulation of heart rate and pupil diameter (the iris reflex), the thirst reflex (see **hypothalamus**) and complex postural reflexes involving the **cerebellum** during walking and running. See **nervous integration, nervous system**.

Refractory period Time taken for nerve and muscle membranes to recover their resting ionic imbalance after passage of an impulse. During this repolarizing period, voltage-gated Na^+ channels flip to a closed, inactive state (prior to becoming once again closed but not inactive) and cannot be reopened for a few milliseconds. See **impulse, muscle contraction**.

Refugium Locality (e.g. a tableland or mountain *nunatak*) which has escaped drastic alteration following climatic change (in this case glaciation), in contrast to the region as a whole. Refugia usually form centres for **relic** species populations or communities.

Regeneration Restoration by regrowth of parts of body which have been removed, as by injury or **autotomy**. Commoner in lower than higher animals, but extensive in some planarians (where new individuals can be regenerated from small body fragments), in the polychaete *Chaetopterus* (which can regenerate a whole animal from one segment) and in crustaceans (which can replace limbs) and echinoderms (which replace arms). Involves the re-establishment of local tissue differentiation. Some examples require long-range interactions and extensive cell movement (*morphallaxis*), while others involve short-range interactions and extensive growth (*epimorphosis*). Vertebrate embryonic limb regeneration involves production of a **blastema**. If a large part of an organ is removed (e.g. of liver, pancreas) the remaining organ commonly returns to normal size. This is *compensatory hypertrophy*. Removal of a mammalian kidney usually results in compensatory hypertrophy of the one remaining. Compare **regulation**. Very common in plants, occurring e.g. in higher plants by growth of dormant buds, formation of secondary meristems and production of adventitious buds and roots. Exploited on a large scale in plant propagation.

Regulation Ability of animal embryo to compensate for disturbance (e.g. involving removal, addition or rearrangement of cells) and to produce an apparently normal individual. Eggs may be *regulative*, meaning that removal of some early **cleavage** products has no effect on the eventual animal, other than decreasing its size. Embryonic regulation is a broad term, including *twinning* (production of two complete animals from bisected embryo), *fusion* (fusion of more than one embryo to give a single giant embryo), *defect regulation* (removal of part of the early embryo does not disturb **development**), and *inductive reprogramming* (see **induction**, where a grafted inducer causes surrounding parts to pursue a developmental route different from that expected from the **fate map**). Compare **mosaic development**.

Regulative development See **regulation**.

Regulator gene Genetic elements first identified in prokaryotes (see **Jacob-Monod theory**) by their control of the co-expression of a set of genes. On the basis of sequence comparisons and structural considerations of the proteins they encode it is now thought that, despite limited amino acid homology, some prokaryotic regulatory proteins do have homologues in eukaryotes. Several show remarkable evolutionary conservation at the structural level, from the bacterium *Escherichia coli* to the fruit fly *Drosophila* and to humans (see **homeobox, homeotic**, ***cdc*** **genes**). A *regulatory element* is any genetic element, or its product (commonly protein), involved in the control of **gene expression** and includes repressors, **enhancers** and **promoters**.

Regulatory element See **regulatory gene**.

Regulatory enzymes Enzymes specifically involved in switching metabolic pathways on or off. Include a) **allosteric** enzymes, where activity is modulated through the non-covalent binding of a specific molecule (e.g. see **ATP**, **ADP**, **glycolysis**) at a site other than the active site; b) *covalently-modulated enzymes*, which alternate between active and inactive forms as the result of other enzyme activity (e.g. see **kinase**). Regulatory enzymes are exquisitely responsive to alterations in the metabolic needs of the cell.

Reinforcement Process whereby

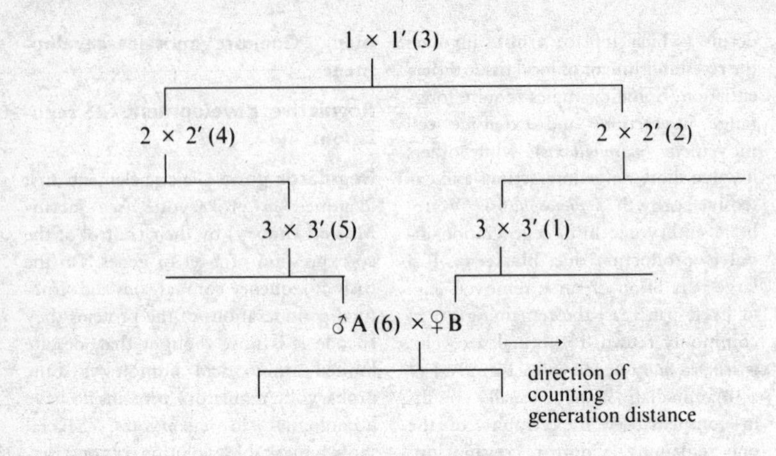

Fig. 101 *Diagram to show the calculation of* **relatedness** *between two first cousins.*

presentation of a stimulus (the *reinforcer*) to an animal just after it has performed some act alters the probability or intensity of future performances of that act when repeated either in isolation or in the presence of the stimulus. *Positive reinforcers* increase the probability and/or intensity of response; *negative reinforcers* decrease its probability and/or intensity. See **conditioning**.

Relatedness Incestuous and non-diploid matings excluded, the degree of relatedness between two conspecific individuals, male A and female B, is found by first locating their nearest common ancestor(s) and then counting the number of generations passed by moving from B to the common ancestor(s) and then on to A (the *generation distance*). This number is the power to which $\frac{1}{2}$ is first raised and then multiplied by the number of nearest common ancestors. For two second cousins, the calculation is as shown in Fig. 101. Generation distance is given by numbers in parentheses from B to A, i.e. 6. There are two nearest common ancestors of A and B, namely 1 and 1′, so degree of relatedness, $R = 2 \times (\frac{1}{2})^6 = \frac{1}{32}$, meaning that two second cousins (such as A and B) are expected on average to share 1 in 32 of their genes. See **Hamilton's rule**.

Releaser A standard external stimulus evoking a stereotyped response; commonly applied in the context of **instinct**. A *social releaser* emanates from a member of the same species as the respondent. Releasers include the *sign stimuli* (e.g. claw-waving in fiddler crabs, postures adopted by courting birds, etc.) which have been so valuable in **ethology**.

Releasing factor (RF, releasing hormone, RH) Substance stimulating release of hormone into blood. The *locus classicus* for their production is the vertebrate **hypothalamus**, several of whose neurosecretions initiate hormone release from the anterior pituitary.

Relic (1) Surviving organism, population or community characteristic of an earlier time. (2) A *relic* (*relect*) *distribution* of fauna or flora is one representing the localized remains of an originally much wider-ranging distribution such as those organisms, now confined to mountain tops, which were far more widespread in glacial times (see **refugium**).

Renal Pertaining to **kidney**.

Renin Enzyme produced by juxtaglomerular cells of kidney when blood pressure falls. See **angiotensins**.

Rennin Enzyme secreted by stomachs of young mammals; converts soluble caseinogen to the insoluble protein casein, which coagulates and therefore takes longer to leave the stomach. *Rennet* is impure rennin.

Repetitive DNA Apart from the repeated nucleotide sequences of **satellite DNA**, there exist in human and rodent genomes (at least) transposable sequences of about 300 nucleotides (*Alu* sequences) which are individually repeated and inserted throughout the chromosomes, making up about 5% of the genome mass (see **C-value**). They have changed far more slowly in mammalian evolution than have satellite DNAs and appear to have arisen from a 7SL RNA gene (see **signal recognition**). Much longer dispersed and transposable repetitive sequences occur in primates, notably the 'LINE' sequences, which comprise about 4% of the human genome mass and move by an RNA-requiring process which also involves a reverse transcriptase (which one sequence may encode). They are probably 'junk' (parasitic) sequences, but may have had significant effects on expressions of nearby genes (see **oncogene**). See **gene duplication**, **inverted repeat sequence**.

Replacing bone Synonym for cartilage bone. See **ossification**.

Replica plating Method employed to isolate mutant microorganisms sensitive to a component of, or deficiency in, a growing medium and which, by failing to grow on it, are difficult to detect. The diluted organism is plated on to a solidified medium suitable for growth and colony formation. An adsorbent material (e.g. filter paper) is then pressed lightly on to the medium surface to pick up cells from each colony, and then pressed on to fresh solidified medium with the additive or deficiency. The distribution of resulting colonies on this new medium can be compared with the original distribution, any gaps indicating sensitive strains. See **auxotroph**.

Replication origin See **DNA replication**.

Replicon Nucleic acid molecule containing a nucleotide sequence forming a replication origin, at which replication is initiated. Usually one per bacterial or viral genome, but often several per eukaryotic chromosome.

Repression (1) (Of enzyme) see **enzyme** feedback inhibition. (2) (Of gene) see **Jacob-Monod theory**.

Repressor molecule See **Jacob-Monod theory**, regulatory gene, **lambda repressor**.

Reptilia Reptiles. Vertebrate class including the first fully terrestrial tetrapods. Probably a **monophyletic** group, whose members lay **amniotic eggs**, have horny or scaly skin and metanephric kidneys. Stem reptiles (Cotylosaurs, Subclass Anapsida) evolved in the Lower Carboniferous, or even earlier, and it has been generally held that they had affinities with the fossil amphibian group, the anthracosaurs. They had solid skulls lacking fenestrations (modern turtles and tortoises are representatives). Synapsida had a single fenestration low on side of skull and included pelycosaurs. Therapsida (**mammal-like reptiles**) appeared from the mid-Permian onwards; some were large and herbivorous, others smaller and carnivorous. They were dominant in the Permian fauna and had varied cranial morphology. By the close of the Triassic, therapsids had become very mammal-like, as indicated by the *cynodonts* (dog-sized), which gave rise to mammals. The great Mesozoic reptilian radiation produced three subclasses: Euryapsida (marine reptiles such as ichthyosaurs and plesiosaurs), Archosauria (**archosaurs**), including dinosaurs and crocodiles, and Lepidosauria, including extinct crocodile-like forms as well as lizards and snakes (see **Squamata**). Modern forms are poikilothermic (ectothermic), but some extinct archosaurs were probably homoiothermic (endothermic). For *Sphenodon*, see **Rhynchocephalia**.

Resin duct Relatively large intercellular spaces lined by thin-walled parenchyma cells secreting resin into the duct. Wounding, pressure, frost and wind damage can stimulate their formation. Resin apparently protects the plant from attack by decay fungi and bark beetles. Found in conifers.

Resistance (Of pests, etc.) the spread within a pest population of genes (often latent in the population) through selection for their ability to destroy or otherwise reduce the effect of some artificial pesticide. See **LD_{50}**, **plasmid**.

Resistance genes Genes carried by plants such that any plant carrying one can recognize (presumably by its production of some kind of receptor), and be resistant to, pathogens carrying the complementary 'avirulence' gene. Resistance involves a hypersensitive reaction such that the plant can isolate the pathogen at the site of infection and prevent its spread through the plant. It appears that avirulence genes are involved in the production of a fully virulent infection. Cloning resistance genes is expected to be a valuable tool in crop husbandry. See **immunity**.

Resolution, resolving power See **microscope**.

Respiration (1) (*Internal*, *tissue* or *cellular* respiration.) Enzymatic release of energy from organic compounds (esp. carbohydrates and fats) which either requires oxygen (*aerobic respiration*) or does not (*anaerobic respiration*). Anaerobic respiration is sometimes used as a synonym of **fermentation**. Among eukaryotic cells only those with mitochondria, and among prokaryotes only those with mesosomes, can respire aerobically. All cells can respire anaerobically, using enzymes in their cytosol; but not all obtain sufficient energy release for their needs this way. A high proportion of the energy released during respiration is coupled to ATP synthesis (50% during anaerobic respiration in erythrocytes; 42% in aerobic respiration involving mitochondria). The major anaerobic pathway in cells is **glycolysis**. This is a metabolic 'funnel' into which compounds from a variety of original sources may be fed (e.g. from proteins, fats, polysaccharides). The energy-rich products are ATP (from ADP) and NADH (from NAD); pyruvate formed is either catabolized further in mitochondria (see **Krebs cycle**) or converted to lactate or alcohol. The actual ATP-generating steps of aerobic respiration occur during passage of electrons along the **electron transport system**. (2) (*External respiration.*) See **ventilation**. (3) See **photorespiration**.

Respiratory centres See **ventilation**.

Respiratory chain Alternative for **electron transport system**.

Respiratory movement See **ventilation**.

Respiratory organ Animal organ specialized for gaseous exchanges of oxygen and carbon dioxide. Include **lungs**, **lung books**, **gills**, **gill books** and the arthropod **tracheal system**. See **ventilation**.

Respiratory pigment (1) Substance found in animal blood or other tissue, involved in uptake, transport and unloading of oxygen (and/or carbon dioxide) in solution, always by weak and reversible bonds. Some visibly colour the blood (**haemoglobin**, **haemerythrin**, **chlorocruorin**); but **myoglobin** is not blood-borne, and does not noticeably colour the tissues where it is found. (2) Substances (e.g. **cytochromes**, **flavoproteins**) found in all aerobic cells (in insufficient concentrations to colour them), and involved in the **electron transport system** of aerobic respiration.

Respiratory quotient (RQ) Ratio of volume of carbon dioxide produced to oxygen consumed by an organism during aerobic respiration. The theoretical RQ value for oxidation of carbohydrates is 1; for fats 0.7; for protein 0.8. Cells can oxidize more than one of these at a time, making *in vivo* results problematic, although data on levels of nitrogen excreted can be helpful. Interconversion of carbohydrates to fats can also affect RQ values. With people on a normal diet an RQ of 0.82 is expected.

Response Change in organism (or its parts) produced by change in its environment. Usually adaptive. See **irritability**.

Resting nucleus (resting cell) See **interphase**.

Resting potential (membrane potential) Electrical potential across a cell membrane when not propagating an impulse. All plasma membranes exhibit such voltage gradients (inside negative with respect to the outside) generated by **transport proteins**, the more important in neurone and muscle membranes being the K^+ leak channel, which lets potassium ions flow back out of the cell along their electrochemical gradient until internal negativity of −75 mV (nerve axons) is achieved – the potential retarding loss of K^+ to the extent that its *equilibrium potential* results. Much less important is the Na^+/K^+ ATPase pump (**sodium pump**), which contributes slightly to the resting potential. The flow of ions needed to set up a resting potential is rapid and minute, leaving ion concentrations practically unaffected. Membrane potentials of 20–200 mV internally negative can be generated, depending on the species and cell type. See **impulse**.

Restitution nucleus (restitution meiosis, restitution division) Inclusion within the same nuclear membrane of all the chromatin after meiotic chromosome or chromatid separation has occurred, giving an unreduced egg which may develop by automictic **parthenogenesis** (e.g. in dandelions, *Taraxacum*). See **automixis**.

Restriction endonucleases (restriction enzymes) Class of nucleases originally extracted from the bacterium *E. coli* (where it digests phage DNA but leaves the cell DNA intact). Type I restriction enzymes bind to a recognition site of duplex DNA, travel along the molecule and cleave one strand only, about 75 nucleotides long, apparently randomly. Type II are more valuable in **gene manipulation** and cleave the duplex at specific target sites at or near the binding site. Their naming uses the first letter of the genus and the first two letters of the specific name of the organism (host) from which they were derived, with host strain identified by a full or subscript letter. Roman numerals then identify the particular host enzyme if there are more than one. Thus *Eco*RI is the first restriction enzyme from *E. coli* strain RY13, and *Hind*II comes from *Haemophilus influenzae* Rd. Target sites for these nucleases are very specific. See **modification, phage restriction, restriction mapping**.

Restriction mapping Method employed in **chromosome mapping**, usually of organelle and viral genomes. Complete nucleotide sequencing may result.

Rete mirabilis (pl. retia mirabilia) Network of arterioles running towards an organ and breaking up into capillaries adjacent to, but in the opposite direction from, a similar set of capillaries returning blood to the venous system. Exchanges between the two networks form a **countercurrent system**, passively increasing local value of some variable (e.g. metabolite concentration, temperature). They occur for example in teleost gills and **gas bladders** and in feet of birds which stand in cold water or on ice.

Reticular fibres Fine branching collagen-like protein fibres (*reticulin*) forming an extracellular network in many vertebrate connective tissues and holding tissues and organs together. Stains selectively with silver preparations. Abundant in **basement membranes** under many epithelia, sarcolemma around muscle and around fat cells. Forms framework for lymph corpuscles in **lymphoid tissue**.

Reticular formation (reticular activating system) Nerve cells scattered throughout the vertebrate **brainstem**, some forming nuclei, receiving impulses from the spinal cord, cerebellum and cerebral hemispheres and returning impulses to them. Stimulation of some of these cells produces **arousal** in unanaesthetized animals, the formation playing an important role in waking, attentiveness and, in higher vertebrates, consciousness. A *decerebrate* animal, whose brain is sectioned between basal ganglia and reticular formation, exhibits the effects of overactive stimulatory and underactive inhibitory mechanisms of the reticular formation, producing *decerebrate rigidity*.

Reticulate thickening Internal thickening of a wall of a xylem vessel or tracheid in the form of a network.

Reticulin See **reticular fibres**.

Reticulocyte Immature, non-nucleated, mammalian **red blood cell**. Develops from nucleated *myeloblasts* and is released into the blood during very active **haemopoiesis**. Cytoplasm basophilic, due to RNA.

Reticulo-endothelial system (RES) Tissue macrophages either circulating in blood, dispersed in connective tissue or attached to capillary endothelium. Of **myeloid** origin, these cells carry (F_c) receptors for IgG antibodies, complement and lymphokines. Attack foreign or tumour cells either by ingesting them or by lysis after adherence to them. Abundant in liver sinusoids (*Kupffer cells*), where e.g. they ingest worn erythrocytes, in kidney glomeruli (*mesangium cells*), alveoli of lung, attached to capillaries of brain (*microglial cells*), **spleen** and **lymph node** sinuses.

Reticulum (1) Second compartment of the **ruminant** stomach, receiving partially-digested material which has been in the rumen, rechewed, and swallowed a second time. Water is here pressed out before the food passes to the abomassum. (2) See **endoplasmic reticulum**.

Retina Photosensitive layer of vertebrate and cephalopod **eyes**, non-sensory in region of ciliary body and iris. Contains **rod cells** and **cones**, both of which synapse with intermediary neurones before impulses leave via the optic nerve. Also present are blood vessels and glial cells; in vertebrates light must traverse these tissues (the ganglion layer) before impinging upon photoreceptors, an arrangement termed an *inverted retina*. By contrast in cephalopod retinas, derived from external ectoderm, light impinges first upon photoreceptors. Vertebrate retinas arise as outpushings of the brain, photoreceptors having a pigmented layer behind them abutting the choroid. See **fovea**, **blind spot**.

Retinoblastoma A childhood cancer in which neural precursor cells in the retina differentiate into tumour cells. See **tumour suppressor gene**.

Retinoic acid (RA, vitamin A acid) Naturally occurring oxidation product of vitamin A, considered by some to be an endogenous vertebrate **morphogen**, in addition to playing a part in the growth and maintenance of epithelia and inhibiting growth of some malignant cells. In chick limb buds, where retinol can serve as a precursor, a gradient of RA levels provides the *positional information* specifying anterior–posterior digit pattern in the wing; Hensen's node (see **primitive streak**) and derivative midline structures can induce a second body axis and digit pattern duplications when grafted into chick wing bud, and is a source of retinoic acid. An RA receptor (RAR) protein, once bound to its ligand, then binds to specific DNA regions within the nucleus (see **nuclear receptors**).

Retinoids Diterpenoids derived from a monocyclic parent compound and containing five C–C double bonds and a functional group at the terminus of the acyclic portion. **Vitamin A** (retinol), retinene (retinal), **retinoic acid** and 3,4 didehydroretinoic acid (ddRA) are important animal examples; **bacteriorhodopsin** is a prokaryote retinoid.

Retinol (vitamin A) See **rhodopsin**, **vitamin A**.

Retroinhibition See **end-product inhibition**.

Retrotransposon (retroposon) See **transposable elements**.

Retrovirus See **virus**, **reverse transcriptase**.

Reverse transcriptase A **DNA polymerase** of retroviruses (initially Rous sarcoma virus), synthesizing complementary single-strained DNA (cDNA) on a single-stranded RNA template. Its inhibition could prevent retroviral infection. cDNA formed may be cloned and screened to detect differences in mRNAs between cells, mRNA acting as template for reverse transcriptase. See **gene manipulation**, **repetitive DNA**, **virus**.

RFLP (restriction fragment length polymorphism) In 1979, **DNA sequencing** by A. Jeffreys on the β- globin gene of human haemoglobin revealed an unexpected high sequence variation between individuals of approximately one base substitution every few hundred base pairs. These were effectively a new kind of **genetic marker** since restriction enzymes – by cleaving DNA at very specific base sequences (restriction sites) – cut DNAs from different people into fragments of different lengths. Such new cleavage sites turned out to be polymorphic and be revealed as RFLPs by the **southern blot technique**. **Genetic mapping** of mammalian (e.g. human) chromosomes has been transformed by this technique, since markers were previously far too isolated and rare for conventional **chromosome mapping** through linkage analysis. Mutations removing or adding a restriction site may be used as markers, too. Studies of genetic **relatedness** in both human and wild populations have also been greatly enhanced. However, novel restriction sites are not ideal markers since they are insufficiently polymorphic, i.e. they normally exist in only two alternative forms. Southern blotting is also difficult to automate. See **DNA fingerprinting**.

Rhabdovirus A bacilliform virus.

Rhesus system See **blood group**.

Rhipidistia Traditionally, an order or suborder of crossopterygians, appearing in early Devonian but extinct by end of Palaeozoic. Teeth with folded dentine and cranium divided into anterior and posterior parts. Many (osteolepids) were well covered in thick scales and adapted to life at the water's edge. Cladistic analysis of new fossils recommends placing coelacanths and traditional rhipidistians within the Sarcopterygii (see **Choanichthyes**), with tetrapods within the new more inclusive clade Rhipidistia.

Rhizine (Of lichens) a bundle of hyphae

attaching a lichen thallus to its substratum.

Rhizoid Single- or several-celled hair-like structure serving as a root. Present at bases of moss stems and on undersurfaces of liverworts and fern prothalli, as well as in some algae and fungi.

Rhizome More or less horizontal underground stem bearing buds in axils of reduced scale-like leaves. Serves in perennation and vegetative propagation; e.g. mint, couch grass, *Scirpus*.

Rhizomorph Root-like strand of hyphae produced by some fungi, transporting food materials from one part of the thallus to another, and increasing in length by apical growth. May be internally differentiated and complex.

Rhizoplane Root-surface component of the **rhizosphere**.

Rhizopoda Protozoan class, or subclass of the **Sarcodina**.

Rhizosphere Zone of soil immediately surrounding root, and modified by it. Characterized by enhanced microbial activity and by changes in the ratios of organisms compared with surrounding soil.

Rho A family of small **GTP**-ases, encoded by *rho* genes. See **G-protein**, **cytoskeleton**, **adhesion**. Contrast **rho factor**.

Rhodophyta The red algae (Kingdom **Protista**). The oldest division of the eukaryotic **algae**, probably evolving from a cyanome in the **Glaucophyta**. Primarily marine, occurring at all latitudes, but becoming more abundant in temperate and tropical regions. Of the more than 4,000 species, about 200 are found in flowing freshwater, attached to rocks or other substrata. Characterized by a complete absence of flagellated stages (including the male gametes, spermatia), by the presence of chlorophylls *a* and *d*, phycobiliproteins, floridean starch as a storage product (occurring as grains outside the chloroplast) and thylakoids occurring singly in the chloroplast. No chloroplast endoplasmic reticulum is present, and pit connections between adjacent cells in filamentous genera occur. Cell walls generally contain cellulose, which forms the fibrillar component (some have xylan), and amorphous mucilages (polysaccharides) e.g. **agar** and **carrageenan**. Chloroplasts are usually stellate, possessing a central **pyrenoid** in the morphologically simple red algae; commonly discoid in the others. Phycobilin pigments are localized in phycobilisomes on the surfaces of the thylakoids (similar to **Cyanobacteria**). Light absorbed by the phycobiliproteins is transferred primarily to chlorophyll *a* involved in photosystem II (O_2 production). Production of greater amounts of phycoerythrin under low light intensities allows the red algae to grow to greater depths than other algae because of their ability to utilize better the green and blue wavelengths, which penetrate furthest into the water.

Some red algae (all the Corallinaceae, Order Gigartinales, and some of the Nemaliales) deposit extracellular $CaCO_3$ in their cell walls. Calcite and aragonite are deposited by the Corallinaceae and Nemaliales respectively.

Sexual reproduction, if present, is oogamous and involves production of non-flagellated male gametes (spermatia), which are carried passively by water currents to the egg cell (carpogonium), which has an elongate trichogyne. When fertilized, the carpogonium produces gonimoblast filaments that form carposporangia and diploid carpospores at their tips. Germination of individual carpospores gives rise to the diploid tetrasporophyte, giving rise to haploid tetraspores borne in tetrasporangia. The more advanced members possess an auxiliary cell with which the fertilized carpogonium fuses to form a multi-nucleate fusion cell. Asexual spores are also produced; again non-motile.

Red algae are generally smaller than brown algae, but as diverse. The thallus in a few is unicellular, in others filamentous (often branched) or flattened and membranous. Some species are parasites, but most epiphytic or epilithic.

Rhodopsin (visual purple) Light-sensitive pigment of **rod cells**, whose molecules contain a protein component (*opsin*) whose prosthetic group, the chromophore 11-*cis* retinal (vitamin A, a component of **cone** pigments), undergoes isomerization to *trans* retinal in light, becoming free of opsin in the process. Opsin seems to undergo an allosteric change, closing Na^+ channels in the rod cell membrane previously kept open in the absence of light by **cyclic GMP**, terminating release of neurotransmitter. An intermediary (perhaps Ca^{2+}) probably links these events. Photons may initiate calcium release from the discs in the outer segment of the cell, while rhodopsin's shape change possibly starts a **cascade** of reactions reducing intracellular cyclic GMP levels. Conversion of *trans* retinal to *cis* retinal occurs in absence of light and involves reduction by reduced NAD, *retinol* being an intermediary in the regeneration. See **bacteriorhodopsin**, **signal transduction**.

rho factor (ρ) ATP-dependent hexamer of identical polypeptides, associating temporarily with some mRNAs during their termination of transcription (at a *termination signal*). Appears to help pull transcript away from the RNA polymerase. See **sigma factor**. Contrast **Rho**.

Rhombencephalon See **hindbrain**.

Rhombomeres Swellings dividing the vertebrate rhombencephalon (see **brain**) metamerically into smaller true **compartments** (polyclones respecting boundaries), demarcated by grooves. Regulatory genes are expressed in specific sets of rhombomeres (e.g. *Hox-2.9* is expressed only in rhombomere 4).

Rhynchocephalia Order of lepidosaurian reptiles, related to eosuchians. Now represented solely by the tuatara (*Sphenodon*), restricted to New Zealand (a **relic** from **Gondwanaland**), where it was in danger of extinction. Appeared in the Mesozoic, but never a dominant group. Lizard-like with many primitive features, e.g. lack of external ear opening, a pineal eye (see **pineal gland**) and lack of an intromittant organ. See **Reptilia**.

Rhyncota See **Hemiptera**.

Rhyniophyta Fossil plant division containing the earliest-known vascular plants understood in detail dated back to the late **Silurian** period, at least 420 Myr ago. They became extinct in the mid-**Devonian** period (*c.* 380 Myr ago). Earlier vascular plants go back at least another 15 Myr. Seedless, they comprised a simple dichotomously branched axis bearing terminal sporangia. Differentiation into stems, leaves and roots is not apparent, and they were homosporous. *Rhynia* is the best known example; leafless, but possessing a rhizome with water-absorbing rhizoids. The internal structure resembled extant vascular plants, having an epidermis surrounding the photosythetic cortex, while the centre of the axis contained the xylem. See **Trimerophyta**, **Zosterophyta**. *Rhynia* was probably a marsh plant; its aerial parts were stems about 20–50 cm long, 3–6 mm thick, and covered by a cuticle, bearing stomata and serving as the photosynthetic organs.

Rhytidome See **bark**.

Riboflavin (vitamin B_2) Precursor of FMN and FAD (prosthetic groups of **cytochromes**). A photoreceptive pigment in the zygomycote mould *Phycomyces*. Insufficient intake in humans causes inflammatory lesions in corners of mouth and blocking of sebaceous glands of nose and face. See **vitamin B complex**.

Ribonuclease proteins See **RNPs**.

Ribonuclease See **RN**ase.

Ribonucleic acid See **RNA**.

Ribose A pentose (5-carbon) monosaccharide; $C_5H_{10}O_5$. Component of RNA and its nucleotide triphosphate precursors, etc.

Ribosomes Non-membranous, but often membrane-bound, organelles of both prokaryotic and eukaryotic cells, of chloroplasts and mitochondria. Sites of **protein synthesis**, each is a complex composed of roughly equal ratios of *ribosomal RNA* (rRNA) and 40 or more different types of protein. Prokaryotic ribosomes are slightly smaller than eukaryotic, with a sedimentation rate (Svedberg number, S) of 70S as opposed to 80S. Each is composed of one large (50S or 60S) and one smaller (30S or 40S) sub-particle, forming ribosomes with diameters of approx. 29 nm in prokaryotes and 32 nm in eukaryotes respectively. Each sub-particle is formed from one to three species of rRNA plus associated proteins. A functional ribosome has one conformational groove to fit the growing polypeptide chain (see **A-site**) and another for the messenger RNA (mRNA) molecule. **Initiation factors** are required for complexing of the small ribosomal subunit with the initiator tRNA (see **P-site**) and mRNA prior to binding of the large ribosomal subunit. Several ribosomes on an mRNA strand form a *polysome*.

Ribosomes are not attached to membranes in prokaryotes, chloroplasts and mitochondria, but are commonly membrane-bound in eukaryotic cells (esp. actively secreting ones), forming rough **endoplasmic reticulum**, as well as being attached to outer surface of outer nuclear membrane. The origin of most eukaryotic rRNA is the **nucleolus** (see **RNA polymerases**), a single rRNA transcript being cleaved into one each of 18S, 5.8S and 28S rRNA subunits. The 18S rRNA molecules form the small ribosomal subunit; the 5.8S and 28S rRNAs are joined by a 5S rRNA subunit made in the nucleoplasm to form the larger ribosomal subunit. Ribosomal proteins associate with these rRNAs in the nucleolus and seem to be exported as pre-ribosomal particles (**RNPs**) through the

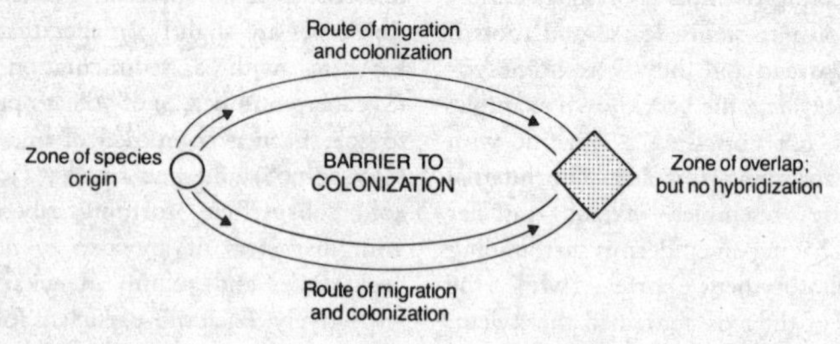

Fig. 102 *Ring species.*

nuclear pore complexes into the cytosol. Prokaryotic rRNA types are 5S, 16S and 23S and are cut from a 30S pre-rRNA transcripts. See **cell fractionation**.

Ribozyme Any catalytically active RNA molecule. Their discovery in 1981 has widened the extension of the term 'enzyme' beyond proteins. Several ribozymes are self-splicing introns (see **splicing**), causing speculation as to their possible roles as intermediates in the evolution of biological systems from prebiotic ones (see **origin of life**, **telomere**).

Rickettsiae Group of bacterium-like prokaryotes occurring naturally in tissues of arthropods (e.g. fleas, lice, ticks), which can transmit them to mammals where they can produce fatal diseases (e.g. typhus in humans). Some can be transmitted via food as well, notably those causing Q fever and a form of encephalitis, both in untreated goat's milk.

Ringer's solution Physiological saline; aqueous solution, containing sodium, potassium and calcium chlorides, employed in maintaining cells or organs alive *in vitro*. Always appropriately buffered, usually with bicarbonate or phosphate (see **buffer**).

Ring-porous wood Secondary **xylem** in which the vessels (pores) of the early wood, when viewed in cross-section, are distinctly larger than those of the late wood and form a well-defined zone or ring.

Ring species Species characterized by circular or looped geographical distributions (see Fig. 102), adjacent populations interbreeding on the two arms of the loop but not where arms overlap, i.e. those populations presumably furthest from the original area of spread. See **species**, **transitivity**.

Ritualized behaviour Behaviour which has become stereotyped in evolution through its role in communication. Many display and threat sequences are ritualized and some **displacement activities** have become so through selection as information-bearers.

RNA (ribonucleic acid) Nucleic acid class, differing from DNA in being usually either single-stranded or looped, in containing ribose not deoxyribose, and in that uracil replaces thymine. Synthesized by **RNA polymerases** from nucleoside triphosphates (ATP, GTP,

Type of RNA*	Function
mRNA	Transfers information from genes to protein-synthesizing machinery
tRNA	Carries activated amino acids for protein synthesis
rRNA	Protein synthesis
U1, U2, U4/6, U5 snRNAs	mRNA splicing
M1 RNA	Catalytic unit RNase P
Telomerase RNA	Template for telomere synthesis
Primer RNA	Initiation of DNA replication
7S RNA	Part of protein secretory complex
ATP	Carrier of energy-rich bonds
Coenzyme A	A key molecule in intermediate metabolism

* mRNA = messenger RNA; rRNA = ribosomal RNA; snRNA = small nuclear RNA; tRNA = transfer RNA.

Table 7 *Table indicating some functions of RNA and ribonucleotides*. (From *Recombinant DNA 2/E* by Watson, Gilman, Witkowski, and Zoller. Copyright © James D. Watson, Michael Gilman, Jan Witkowski and Mark Zoller, 1992. Reprinted with permission of W. H. Freeman and Company.)

CTP, UTP); with 5′- and 3′-ends to the molecule, as in DNA (see **DNA**).

The three RNA types, *messenger RNA* (mRNA), *transfer RNA* (tRNA) *and ribosomal RNA* (rRNA), are all involved in **protein synthesis**, in both prokaryotic and eukaryotic cells, but in eukaryotes especially, **RNA processing** accompanies mRNA production. In RNA viruses, RNA is sometimes double-stranded, serving as genetic material (see **reverse transcriptase**, **central dogma**); in some RNA viruses the RNA is transcribed into RNA and in others (*retroviruses*) it is reverse-transcribed into DNA. tRNA molecules fold back upon themselves by complementary base-pairing to form double-stranded 'stems' and single-stranded 'loops'. A loop at one end bears a specific nucleotide triplet (the *anticodon*) while the 3′-end of the molecule carries a tRNA-specific amino acid – both essential for protein synthesis to proceed by means of a **genetic code**. Ribosomal RNA subunits associate with protein molecules to form **ribosomes**. All tRNA and rRNA molecule types are encoded by DNA (see **gene**), and there are many more of these molecules per cell than there are of mRNA. Eukaryotic rRNAs are of four densities: 28S, 18S, 5.8S and 5S, while the three prokaryotic rRNAs are 23S, 16S and 6S. The genes for all these tend to be highly conserved and are often employed in phylogenetic studies. Some RNA molecules have catalytic activity (see **origin of life**, **ribozymes** and Table 7).

RNA binding proteins See **RNPs**, **protein synthesis**, **telomere**.

RNA capping Eukaryotic RNAs are processed in nuclei prior to release into the cytosol. This involves attachment of a *cap* of 7-methylguanosine triphosphate to their 5′-end. Ribosomes recognize this cap and commence translation at the AUG codon nearest to the cap, finishing at the first stop codon, ensuring that translation is usually monocistronic. See **codon**, **RNA processing**, **protein synthesis**.

RNA editing Process occurring in at least mitochondrial and nuclear compartments, which differs from **RNA processing** in that nucleotides are added, substituted or deleted between genomic DNA and mRNA stages of protein synthesis to ensure the correct reading frame is produced for translation. Sometimes involves interruption of an open reading frame by conversion of coding triplet into a stop codon.

RNA life See **origin of life**, **antibiotics**, **telomere**.

RNA polymerases Enzymes producing RNA from ribonucleoside triphosphates. Unlike DNA polymerases they do not require a polynucleotide primer. Three types occur in eukaryotic cells, *polymerase I* making large ribosomal RNAs, *polymerase II* transcribing structural genes (introns and exons), *polymerase III* making small RNAs such as tRNAs and rRNAs. See **RNA processing**, Fig. 108 (**TATA** box) and Fig. 113 (**transcription factors**).

RNA processing mRNA transcription within the nucleus produces RNAs of various sizes (*precursor RNA*, *heterogeneous RNA*) which are modified (processed) before passage to the cytosol for translation on ribosomes. The 5′-end of the molecule is first capped (see **RNA capping**) and then has a long poly-AMP sequence bound to the 3′- end, which may facilitate the rest of processing and passage to the cytosol. Major feature of nuclear processing is the excision from pre-mRNA of non-coding **intron** sequences by **spliceosomes**. This is achieved by cutting these sections out using a **phosphodiesterase**, and then splicing the transcript. This may rejoin one encoding region (exon) to another that is not its official nearest neighbour in the pre-mRNA. Alternatively, an exon may get cut out. This *alternative splicing*, especially common in higher eukaryotes, can produce two forms of the same protein at different times in development or in different cell types (see ***bithorax*** **complex**). Specific protein splicing factors, related to those in the spliceosome, control these splicing events. In B cells, the heavy chains of the cell surface IgM produced early in development differ at their 3′-ends from the secreted form produced later by the differentiated plasma cell – this is because an exon incorporated in the B cell mRNA is replaced in the plasma cell mRNA by two shorter exons encoding less hydrophobic amino acids, allowing the IgM to cross the plasma membrane. Alternative splicing plays a crucial part in **sex determination** in *Drosophila*. Ribosomal and transfer RNAs are derived from pre-rRNA and pre-tRNA, in both prokaryotes and eukaryotes (see RNPs).

Alternative splicing of RNA occurs when the primary mRNA transcript is spliced in different ways so that from a single gene different proteins may be produced at translation – often in tissue-specific ways. Involved *inter alia* in sex determination in *Drosophila* and in production of cell-surface as opposed to secreted forms of **antibody**. Contrast exon-shuffling (see **exons**).

RNase (ribonuclease) Any of several enzymes which hydrolyse RNA by breaking their phosphodiester bonds.

RNPs (ribonuclear proteins) Eukaryotic complexes of protein and RNA located especially within nuclei, but also in the cytoplasm; some are restricted to the nucleus. The proteins of heterogeneous RNPs (hnRNPs) are among the most abundant in nuclei and bind pre-mRNAs while these are processed (spliced) into mRNAs (see **spliceosome**). Several hnRNPs shuttle between nucleus and cytoplasm. Small nuclear RNPs (snRNPs) are found with hnRNPs on those pre-mRNAs transcribed specifically by RNA polymerase II in the nucleus and occur in high concentration in the *coiled body*, a domain within the nucleus (see **introns**). See **protein synthesis**, **telomere**.

Rod cell Highly light-sensitive secondary receptor of vertebrate **retina**, providing monochromatic vision even in very dim light. Its outer segment, closest to the choroid, contains parallel-stacked membranous discs with embedded molecules of *cis-retinal* (see **rhodopsin**), responding even to single photons. Attached to this by a cilium-containing restriction is the mitochondrion-rich *inner segment* which leads to the nuclear region (cell body) with its synaptic processes. Much integration of rod cell output is achieved by horizontal, bipolar and **amacrine cells** before ganglion cells lead off via the optic nerve. Each primate retina contains about 120 million rods, few of them in the fovea. See **scotopic vision**, **cone**.

Rodentia Most widespread and numerous of all mammalian orders. Placentals, they include rats, mice, capybaras, squirrels, beavers and porcupines. Herbivores or omnivores; one pair of large, continually growing, chisel-like incisors in both upper and lower jaws (see **dentition**, **Lagomorpha**). Large diastema; grinding molars.

Rolling circle See **gene amplification**.

Root The vascular plant organ that usually grows downwards into the soil, anchoring the plant and absorbing water and nutrient mineral salts. The first root originates in the embryo and is the primary root. In dicotyledons and gymnophytes, this root becomes a **taproot**, it grows vertically downwards, giving rise to lateral roots along the way. In monocotyledons, the primary root is normally short-lived, and the root system develops from **adventitious** roots, and their lateral branches give rise to a **fibrous** root system. Roots cannot be distinguished from stems on the basis of their position with respect to the soil; some plants have roots wholly above the ground, and others have underground stems (e.g. **rhizomes**). Externally, roots principally differ from stems in not bearing leaves or buds; they possess at their tips a protective layer of cells, the **root cap**. Internally, roots differ from stems in having their vascular tissue in a central region and with protoxylem exarch. Many roots are important storage organs, and some of these are economically important (e.g. potatoes, carrots, parsnips, sugar beet).

Root cap Cap of loosely arranged cells covering the apex of the growing point of a root; protects the apical meristem, and aids the root in penetrating the soil. As the root grows in length and the root cap is pushed forward, the cells of the periphery of the root cap are sloughed off, and these cells form a slimy covering around the root. Thus they lubricate its passage through the soil. As quickly as cells are sloughed off they are replaced by the apical meristem. This slimy substance is a highly hydrated polysaccharide (probably a pectin), that is secreted by the outer root cap cells. It accumulates in the Golgi apparatus vesicles, which fuse with the plasmalemma, and then release the slime into the wall; it then passes through the wall emerging on the outside as droplets. In addition to a protective role, the root cap plays an important role in controlling the response of the root to gravity.

Root hair (pilus) Tubular outgrowth of the root epidermal cell, possessing a thin, delicate wall in intimate contact with soil particles. Root hairs are produced in large numbers behind the region of active cell division at the root tip, forming the piliferous layer. They enormously increase the root's absorbing surface area, taking up ions and water. They are continually being replaced from new root tissue formed by the apical meristem. Water and dissolved ions move from the root hairs through the cortex, the endodermis and the pericycle into the xylem, and then upwards through the root and stem and finally into the leaves. This may occur apoplastically or symplastically. See **transpiration**, **transpiration stream**.

Root nodules Small swellings on roots of leguminous plants (e.g. pea, bean, clover) produced through infection by nitrogen-fixing bacteria. See **nitrogen fixation**.

Root pressure Pressure under which water passes from living root cells into the xylem; demonstrated by exudation of liquid from the cut end of a decapitated plant. May continue for long periods, arising through maintenance of water potential gradients by active transport of solutes. See **cavitation**, **guttation**, **translocation**.

Rotifera Abundant and widespread phylum (or class of **Aschelminthes**); microscopic unsegmented pseudocoelomates, all aquatic and many capable of producing resistant sexual eggs. Parthenogenesis common; one group (bdelloids) thelytokous. Feed usually by ciliated trochal disc; gut entire, commonly with a muscular pharynx and chitinous 'jaws'. Excretion involves flame cells. Nervous system very simple. Elastic cuticle covers most of body. Extraordinary in being of approx. protozoan size, but at the 'organ' level of organization.

Roundworms See **Nematoda**.

***r*-selection** Selection tending to operate in scattered and transient habitats, favouring organisms with ability to colonize rapidly, make use of short-lived resources, and complete their reproduction rapidly. Such organisms are selected for high r-values (r = intrinsic rate of increase); are clearly exploiters of their environments, have a tendency to **progenesis**, and are at opposite end of range of strategies from organisms subject to ***K*-selection**.

Ruderal (Of a plant) growing in disturbed soil, and so often in waste near human habitation.

Ruffini ending Slow-adapting pressure receptor in dermis and hairless vertebrate skins responding to both vertical pressure and skin stretching; often direction-specific.

Ruffled membrane See **lamellipodium**.

Rumen Diverticulum of **ruminant** oesophagus; the chamber of the 'stomach' in which storage and initial digestion of food occur.

Ruminant Artiodactyl (Suborder Ruminantia) characterized by very complex herbivorous stomach and feeding method. Grazed food is swallowed into the **rumen** and mixed with mucus, undergoing partial and anaerobic digestion by cellulase from a symbiotic bacterial flora. Some products are absorbed here; but 'pulp' can be returned to the mouth as 'cud' for chewing. After this (often done at leisure), the bolus is returned to the **reticulum** and omassum, where water is removed. The *abomassum* is the true stomach, in which further digestion occurs. In some ruminants, food can be seen regularly travelling up and down the oesophagus in the neck. Include deer, antelopes, giraffes, cattle, sheep and goats.

Runner A **stolon** that roots at its tip and forms a new plant which is eventually freed from connection with the parent through decay of the runner. Also used horticulturally of the daughter plant itself.

Rust fungus A member of Class Teliomycetes, Order **Uredinales**, of the **Basidiomycota**.

Saccharomyces Genus of **yeast** fungi (**Ascomycota**, Endomycetales), including organisms used in the production of bread and alcoholic drinks.

Sacculus See **vestibular apparatus**.

Sacral vertebrae Vertebrae of tetrapod lower back, articulating by rudimentary ribs with pelvic girdle. Just one occurs in most amphibia; two or more are fused in other vertebrae. See **sacrum**.

Sacrum Group of fused sacral vertebrae, ilia of the **pelvic girdle** being united to some or all of them. See **synsacrum**.

Sagittal crest Medial bony ridge on top of some primate skulls (notably *Australopithecus robustus*, gorilla, chimpanzee) anchoring the **temporalis muscle** on both sides.

St Anthony's fire See **ergot.**

Saliva Fluid secretion of salivary glands (labial glands in insects). In terrestrial vertebrates contains mucus and, in some of these and insects generally, contains amylases (e.g. *ptyalin* in mammals). *Invertases* occur in some insects (e.g. worker bees). Anticoagulants are present in salivas of blood-sucking leeches, insects and vampire bats, while those of the last two groups may contain pathogens.

Salivary gland chromosomes See **polyteny**.

Salmonella Genus of bacterium, causing typhoid and paratyphoid; spread in food and water.

Saltatory conduction See **node of Ranvier**.

Salt marsh The shores of protected portions of some estuaries are often occupied by a tidal salt marsh. These broad, flat areas extend from the water's edge to the uplands, at the extreme limit of tidal influence. Vegetation includes both flowering plants and algae.

SAM Abbreviation for S-adenosylmethioine, a compound produced in the biosysnthesis of **ethylene**.

Samara Simple, dry, one- or two-seeded indehiscent fruit, the pericarp bearing wing-like outgrowths (e.g. sycamore).

Saprophyte See **saprotroph**.

Saprotroph (saprophyte, saprobe) Organism obtaining organic matter in solution from dead and decaying organisms. Usually involves secretion of extracellular enzymes on to the material, followed by absorption of the digestion products. Common in bacteria and fungi engaged in decay; of enormous importance in recycling and making available carbon dioxide, nitrates, phosphates and other nutrients via the **decomposer** food chain.

Saprozoic (Of animal) taking up nutrients from the environment in solution;

e.g. gut parasites utilizing the host's digestion products, some free-living flagellate protozoans.

Sapwood Outer region of xylem of tree trunks, containing living cells (e.g. xylem parenchyma, medullary ray cells); functions in water conduction and food storage, as well as providing mechanical support. Usually distinguished from **heartwood** by its lighter colour.

Sarcodina Protozoan class characterized by **pseudopodia**. Some bear flagella in part of life cycle, supporting unification with flagellate protozoans in the single class Sarcomastigophora. Most are free-living, but some amoebae (e.g. *Entamoeba*) are parasitic. Many secrete shells. Includes subclasses Rhizopoda (amoebae, **Foraminifera**) and Actinopoda (**Radiolaria, Heliozoa**).

Sarcoma See **cancer cell**.

Sarcoplasmic reticulum Endoplasmic reticulum of muscle fibres; especially important in **striated muscle** in regulating calcium ion level in myofibril environment. Forms sac-like *terminal cisternae* around transverse tubule system, giving triad appearance in section. See **muscle contraction** and Fig. 85.

Sarcopterygii Lobe-finned fishes. Alternative term is **Choanichthyes**.

Satellite DNA Tandem repeats of a simple DNA sequence, which form minor 'peaks' (on account of its peculiar base composition and hence density) after DNA fragmentation and **density gradient centrifugation**. In general, any such large sequences are not transcribed and occur in **heterochromatin**, as e.g. associated with centromeres. In some mammals, a single satellite may comprise > 10% of the genome, perhaps even a whole chromosome arm (see **selfish DNA**). They seem to evolve rapidly and can shift position within the karyotype. The human myoglobin locus was shown by A. Jeffreys in 1985 to contain an unusual tandem repeat DNA sequence, based on a core repeat of just 10–20 base pairs, producing a 'minisatellite'. It turned out that individuals varied in the exact number of core repeats, and that this variation was polymorphic (hence VNTR, for variable number of tandem repeats). It has provided a useful **genetic marker** in **DNA fingerprinting**. Smaller repeats still, involving di-, tri- and tetranucleotides ('microsatellites') are also often highly polymorphic for repeat number, can be amplified by **PCR**, and seem to be more random in distribution through the genome than minisatellites, providing an almost ideal tool for **genetic mapping** and kinship studies. The human genome contains numerous $(CA)_n$ dinucleotide microsatellites, and considerable success has been achieved in combining maps from **RFLPs** and $(CA)_n$ repeats (see **human genome project**).

Saurischia Extinct order of lizard-hipped reptiles forming (with **Ornithischia**) the dinosaur archosaurs. Distinguished by pubes pointing downwards and forwards and ischia pointing downwards and backwards (see **pelvic girdle**). Included *theropods* (bipedal carnivores, e.g. *Tyrannosaurus*, *Allosaurus*) and herbivorous *sauropodomorphs* (e.g. *Apatosaurus* (= *Brontosaurus*), *Diplodocus*). The last two were the largest land animals yet discovered, a semi-aquatic life supporting their bulk.

Savanna Grasslands with scattered broad-leaved deciduous and evergreen trees, occurring singly or in small clumps. Sometimes trees dominate; others are dominated by shrubs. Savanna trees are generally deciduous, dropping their leaves during the dry season. Savannas are transitional regions between evergreen tropical rain forests and deserts, covering vast areas of East Africa, or in the regions between the prairies and the temperate deciduous forests, and between the **taiga** and the **prairies** in North America and Russia. They also cover much of eastern Mexico, Cuba, southernmost Florida, Australia, and parts of South America.

In savannas, both annual rainfall and temperatures vary widely seasonally. Also, light levels at ground level are high because of the sparse distribution of trees. Perennial grasses are common, along with bulbous plants. Trees of the savanna usually possess thick bark; seldom more than 15 m tall; nearly all deciduous, flowering when leafless.

The existence of savannas depends to a large degree upon periodic burning.

Scalariform thickening Internal thickening of cell wall of xylem vessel or tracheid, taking the form of more or less transverse bars, suggestive of the rungs of a ladder.

Scale (Bot.) In some algae, an external element covering the cell and sometimes the plasmalemma of the flagella. They may be inorganic or organic in composition, with intricate ornamentation. Inorganic scales are commonly siliceous as in the **Chrysophyta**, Class Synuracaeae, and possess species-specific ornamentation. Such scales can provide invaluable information concerning past environmental conditions. Since such species occur widely in oligotrophic, acidic lakes they are being used as indicators of lake acidification (see **acid rain**). Some organic scales have an inorganic covering of calcium carbonate deposited upon their surface (e.g. **Prymnesiophyta**, Order Coccosphaerales; see **coccolith**). In vascular plants, a plate-like outgrowth. (Zool.). As before.

Scale insects (mealy bugs) Hemipterans (Superfamily Coccoidea); females wingless and scale-like, gall-like or covered in waxy secretion. Males usually with one pair of wings and vestigial mouthparts. Important economic **pests**; cottony cushion scale (*Icerya purchasi*) damages citrus fruits, others coconuts. Control by coccinellid beetles (ladybirds) is usually successful.

Scanning electron microscopy See **microscope**.

Scape Leafless flowering stem arising from ground level (e.g. dandelion, daffodil).

Scaphopoda Small class of bilaterally symmetrical marine molluscs, (tusk shells, e.g. *Dentalium*) somewhat intermediate between bivalves and gastropods. The tubular shell opens at both ends, while the reduced foot is used in burrowing. Ctenidia absent; larva a trochosphere.

Scapula Dorsal component of vertebrate **pectoral girdle**; the shoulder-blade of mammals.

Schistosoma Blood fluke (see **Trematoda**) of Africa (esp. Egypt), S. America and China causing *bilharzia* (= *schistosomiasis*) in man. Smaller male lives attached to female within a groove of her body; eggs laid in abdominal blood vessels penetrate walls (causing most damage) and pass to bladder. Larval miracidia hatch in urine diluted on contact with freshwater and burrow into snail, in which rediae develop and from which cercariae emerge and penetrate skin of bathing/rice-planting people, debilitating over 200 million worldwide. Three common species infect humans. Education, support of irrigation engineers, molluscicides and medically prescribed drugs (e.g. antimony-containing) are all involved in prevention/cure.

Schizocarp Dry fruit formed from a syncarpous ovary that splits at maturity into its constituent carpels forming several partial fruits, which are usually single-seeded, resemble **achenes**, and are called *mericarps*; e.g. hollyhock, mallow, geranium.

Schizocoely Mode of formation of the **coelom** in most annelids, in arthropods, molluscs and higher chordates. The coelom arises *de novo* as a cavity in the embryonic mesoderm. Compare **enterocoely**.

Schizogenous (Bot.) (Of structure) formed by splitting due to separation of cells; e.g. oil-containing cavities in leaves of St John's wort. See **lysigenous**.

Schwann cell Specialized vertebrate **glial cell**, derived from **neural crest**, ensheathing axons of peripheral nervous system and responsible for formation there of the **myelin sheath**. See **nodes of Ranvier**.

Scion Twig or portion of twig of one plant grafted on to the stock of another.

Sclera (sclerotic) The white of the vertebrate **eye**. Coat of fibrous or cartilaginous tissue covering whole eyeball except for cornea.

Sclereid See **sclerenchyma**.

Sclerenchyma Tissue, composed of sclerenchyma cells, which may form in any or all parts of the primary and secondary plant bodies that provide mechanical support to plants. Cells often lack protoplasts at maturity, and possess thick cell walls, with often lignified secondary walls. Two major types of sclerenchyma cells are found: (1) *fibres*, which are typically very elongated cells with tapering ends, occurring singly or variously grouped in strands or bundles. Economically important fibres include hemp, jute and flax, which are derived from the stems of dicotyledons. From monocotyledons comes manila hemp extracted from leaves; and (2) *sclereids*, which occur singly or in groups in the **ground tissue**, and are of variable shape, often branched, and relatively short cells. They form the seed coats of seeds, shells of nuts, the stone or **endocarp** of stone fruits, and in pears. Compare **collenchyma**.

Sclerite Region of arthropod **cuticle** in which the exocuticle is fully differentiated and *sclerotized*. Between two such sclerites there is usually a region of flexible membranous (unsclerotized) cuticle where exocuticle is undeveloped, allowing for articulation at joints. The body wall of a typical arthropod segment is divisible into four sclerotized regions: dorsal tergum, ventral sternum and a lateral pleuron on each side. Sclerites of the tergum are *tergites*, those of the sternum are *sternites* and those of each pleuron are *pleurites*.

Scleroproteins Group of insoluble proteins forming major components of **connective tissues**. Include **collagen**, **keratin** and **elastin**.

Sclerosis (1) Hardening, as occurs in vertebrate tissues after injury. Involves deposition of **collagen**. (2) *Atherosclerosis* is the principal contributory process in the pathogenesis of gangrene and myocardial and cerebral *infarction* (death of a tissue area due to interruption of blood supply) as well as being a disease in its own right. Involves deposition of **cholesterol** and triglycerides in arterial walls (abnormal **LDL** receptors predispose individuals towards premature atherosclerosis). Normally a protective response to damage to endothelium and smooth muscle of arterial walls; involves inflammation and formation of fibro-fatty and fibrous lesions which, when excessive, become the disease – several **growth factors** (e.g. TGF β, PDGF) being involved. Endothelial cells, by producing **nitric oxide** (NO), tend to prevent platelet aggregation and promote vasodilation. Macrophages normally scavenge potentially toxic materials; but their uptake of too much oxidized LDL (oxLDL, produced by endothelial cells) can cause them to metamorphose into passive lipid-laden *foam cells*, which adhere to the artery's smooth muscle to form a fatty streak which may become artery-blocking plaque and/or produce cytokines and growth factors which attract monocytes and T cells and instruct the artery wall cells to multiply. This combination may be fatal.

Sclerotic See **sclera**.

Sclerotium (Of fungi) compact tissue-like mass of fungal hyphae, often possessing a thickened rind, varying in size from that of a pin's head to that of a man's head; capable of remaining dormant for long, perhaps unfavourable, periods, and commonly giving rise to fruiting bodies; e.g. **ergot**.

Sclerotization Tanning of arthropod **cuticle**. Involves cross-linking of protein by quinones (tyrosine derivatives). In some insects (e.g. *Calliphora* larvae), a quinone precursor is released onto the cuticle surface via pore canals and subsequently enzymatically oxidized by a phenol oxidase, tanning the protein cuticulin in the epicuticle and diffusing inwards to tan the outer procuticle to produce exocuticle. This is a major cause of cuticles becoming hard and brittle.

Sclerotome See **mesoderm**.

Scolex Part of tapeworm attached by suckers and hooks to gut wall of host; sometimes called the head. Proglottides are budded off behind it.

Scorpiones Order of **Arachnida** containing scorpions. Pedipalps form large pincers, chelicerae small ones. End part of abdomen a segmented flexible tail bearing sting. Viviparous; terrestrial. See **pseudoscorpiones**.

Scorpion flies See **Mecoptera**.

Scotophile See **skotophile**.

Scotopic vision Dark-adaptation of eye. Decrease in threshold of sensitivity of the **rod cells** with increasing length of time in the dark. Involves enzymatic resynthesis of **rhodopsin**, increase in sensitivity of cells proceeding faster than the resynthesis. See **eye**.

Scrotum Pouch of skin of perineal region of many male mammals, containing testes (at least during breeding season) and keeping them cooler (by 2 °C in humans) than body temperature, without which sperm formation is impaired. Female spotted hyaenas have a pseudoscrotum, important in formation of dominance hierarchy.

Scutellum Single cotyledon of a grass embryo, specialized for absorption of the endosperm.

Scyphistoma Polyp stage in life cycles of **Scyphozoa**, during which the ephyra larvae are budded off by a kind of strobilation.

Scyphozoa Class of **Cnidaria** containing jellyfish. Life cycle exhibits **alternation of generations**. Medusoid stage the rather complex adult jellyfish (gonads endodermal; enteron of four pouches, with complex connecting canals). Fertilized eggs develop into ciliated planula larvae which settle to become **scyphistomas**.

Sea anemones See **Actinozoa**.

Sea cucumbers See **Holothuroidea**.

Seals See **Pinnipedia**.

Search image Proposal by L. Tinbergen in 1960 that predators perform a highly selective 'sieving operation' on the visual stimuli reaching their retinas so improving their abilities to locate new (unfamiliar) cryptic prey when they are common, ignoring them when rare (see **apostatic selection**). The modern use of the phrase implies that predators individually undergo a perceptual change that enhances their detection of *familiar* prey items. Research today concerns analysing the possible role of alternative explanations of the data, such as that predators alter their search rate rather than acquire a search image in detecting such prey. The results will have consequences for traditional theories on the origin and maintenance of **crypsis** and **mimicry**.

Seaweeds Red, brown or green **algae** living in or by the sea.

Sebaceous gland Mammalian holocrine skin gland opening into hair follicle. Secretes oily lipid-containing *sebum* that helps to waterproof fur and epidermis. Epidermal in origin but projecting into dermis.

Secondary growth (secondary thickening) In plants, growth derived from secondary or lateral meristems, the vascular or cork cambia. Results in increase in diameter of gymnosperm and dicotyledonous stems and roots, providing additional conducting and supporting tissues for growing plants and, in most cases, makes up greater part of mature structure. Rare in monocotyledons.

Secondary meristem (Bot.) Region of active cell division that has arisen from permanent tissues, e.g. cork cambium (phellogen), wound cambium. See **primary meristem.**

Secondary metabolites A diverse array of chemically unrelated compounds such as alkaloids, quinones, essential oils (including terpinoids), glycosides (including carcinogen compounds and saponins), flavonoids, and raphides (needle-like crystals of calcium oxalate). The presence of such compounds can characterize families or groups of flowering plants. Such chemicals function in decreasing or restricting the palatability of the plants in which they occur, or in causing animals to avoid them. This ability to produce and then retain such compounds in their cells and tissues is an important evolutionary step, providing such plants with a biochemical defence from most herbivores. Some fungi produce secondary metabolites which are highly toxic and also carcinogenic, while the siphonaceous green alga *Halimeda* contains compounds that significantly reduce feeding by phytophagous fish. In microbiology, secondary metabolites are not produced during the organism's primary growth phase (not essential for growth/reproduction) but when the culture enters its stationary phase. They usually demand a large number of complex enzymic steps. Often of industrial interest: **antibiotics** are examples. See **coevolution**.

Secondary sexual characters Characters in which the two sexes of an animal species differ, excluding gonads, their ducts and associated glands. In humans include mammary glands, subcutaneous fat deposition, shape of pelvic girdle, voice pitch, mean body temperature, mass and extent of muscle development. See **oestrogens, testosterone**.

Secondary wall Innermost layer of plant **cell wall**, formed by some cells after elongation has finished; possesses highly organized microfibrillar structure.

Second messenger Organic molecules and sometimes metal ions, acting as intracellular signals, whose production or release usually amplifies a signal such as a hormone, received at the cell surface. Some hormones bind to the cell membrane and activate an enzyme there to generate the second messenger. Alternatively, the ligand may be a non-hormone which opens or closes a **gated channel** affecting membrane permeability to an ion. Calcium ion (Ca^{2+}) concentration is extremely important in control of many cell functions (see **calcium pump**, **calmodulin**). First organic molecule hailed as a second messenger was cyclic AMP (see **AMP**) but others have been discovered. See **cascade**, **diacylglycerol**, **inositol 1,4,5-triphosphate**.

Secretin Polypeptide hormone of intestinal mucosa, produced there in response to acid chyme from stomach. Inhibits gastric secretion, reduces gut mobility but stimulates secretion of alkaline pancreatic juice, bile production by liver and intestinal secretion. See **cholecystokinin, gastrin**.

Secretion (1) Production and release from a cell of material useful either to it or to the organism of which it is a part. Material is commonly packaged into *secretory vesicles* which bud off from the **Golgi apparatus** and fuse with the cell's apical plasmalemma (see **merocrine gland**); other methods are employed by **apocrine** and **holocrine glands**. Plant cell walls are secreted. (2) *Active secretion*. Process whereby a substance (often an ion) is actively pumped out of a cell against its concentration gradient, as in the ascending loop of Henle in the vertebrate kidney, and in fish gills. See **active transport**.

Secretory vesicles See **Golgi apparatus**.

Sedimentation coefficient See **Svedberg unit**.

Seed Product of fertilized ovule, comprising an embryo enclosed by protective seed coat(s) derived from the integument(s). Some seeds (castor oil, pine) are provided with food material in the form of **endosperm** tissue surrounding the embryo, while in other (*non-endospermic*) seeds food material is stored in the cotyledons (e.g. pea). The seed habit is the culmination of an evolutionary development involving, in sequence, heterospory, reduction of a free-living female gametophyte generation dependent on water for fertilization, and its retention within the tissues of the sporophyte by which it is protected and supplied with food. It occurred early in geological history, and in more than one group of plants. Of equal importance was the reduction of the male gametophyte to the pollen grain and its pollen tube, again avoiding the need for water in fertilization. Independence from water in sexual reproduction increased immensely the range of environments open for colonization. Biological advantages of the seed habit include: continued protection and nutrition of the embryonic plant during its development, provision of dispersal mechanisms, provision of food to tide over critical periods of growth of the embryo, and its establishment as an independent plant after the seed is shed. See **life cycle**.

Segmentation (1) Alternative term for **cleavage** of an egg. (2) In zoology, commonly synonym for *metameric segmentation* or *metamerism*, to indicate production of a body plan of repeating organizational units, variably distinguishable, along the antero–posterior body axis. Metamerism most marked in annelids and arthropods (see **tagmosis**) but vertebrates exhibit it in the segmental organization of nervous system and muscles, most clearly in the embryo (see **mesoderm**). Has often become reduced or wholly lost in evolution, notably in molluscs and echinoderms. Debate surrounds role of the **homeobox** in control of segmentation. See **homeobox**, **compartment**.

Segmentation genes Genes dividing the early *Drosophila* embryo into a series of repetitive segmental primordia along the anterior–posterior axis, mutations in which cause loss of parts of (even of entire) segments. **Gap genes, pair rule genes** and **segment polarity genes** form the three classes to date. For a diagram of one proposed cascade of developmental gene regulation involving these, see ***bicoid* gene**.

Segment polarity genes A class of segmentation genes, studied particularly in *Drosophila*, mutants in which cause not only a loss of the same part in each segment, but also a mirror-image duplication of the remaining part to be produced in its place. In *Drosophila* at least, domains of mRNA transcript expression of segment polarity genes are dependent largely upon prior distributions of **gap gene** expression domains. For instance, *engrailed* (*en*) and *wingless* (*wg*) express as stripes of transcript in the early embryo, *wg* stripes representing the posterior limits of **parasegments**, while *en* stripes represent anterior limits. It is likely that some at least of these genes encode transmembrane proteins involved in translating extracellular signals into changes in cell fate. Compare

pair-rule genes. See **positional information**.

Segregation (1) See **Mendel's Laws**. (2) Process whereby alleles usually present together in somatic cells of an organism separate into different cells during meiosis in the germ line. Such alleles are said to segregate.

Segregation distorter (Sd) See **aberrant chromosome behaviour** (4).

Seismonasty (Bot.) Response to non-directional shock stimulus; e.g. rapid folding of leaflets and drooping of leaves in *Mimosa pudica* when lightly struck or shaken.

Selachii Elasmobranch Order containing modern sharks. Members have five to seven gill openings on each side of head and an upper jaw not fused to skull. Group appeared in the Jurassic and is characterized by the *rostrum* (snout) that hangs over the mouth. Includes largest living fishes (over 20 metres in whale sharks, *Rhinedon*). Rays and skates (Batoidea) are sometimes also included. See **Elasmobranchii**.

Selectins A small family of **adhesion** molecules with a characteristic sequence of protein motifs comprising a lectin-related amino-terminal domain, an EGF-like repeat unit and two or more domains related to consensus sequences of several **complement** receptors. They seem to help regulate leucocyte binding to endothelium at sites of **inflammation**. See **integrins**.

Selection See **artificial selection**, **natural selection**.

Selection coefficient (s) See **coefficient of selection**, **fitness**.

Selection pressure See **natural selection**.

Self-assembly Usually reserved for biological structures which become assembled from components without help of enzymes or 'scaffolds' which do not form part of the functional structure. Some **viruses** (e.g. tobacco mosaic virus, TMV) assemble within the host cell from separate nucleic acid and capsomere protein molecules. Several other structures are either fully self-assembling (e.g. bacterial ribosome) or require surprisingly few enzyme steps or accessory proteins acting as 'jigs' in assembly. Some viruses, membranes, cilia, mitochondria and myofibrils are in this second category. It is thought that some steps in these more complex cases of partial self-assembly require appropriate timing, and are irreversible if disassembly is imposed upon them.

Self-fertilization Fusion of micro- and macrogametes from the same individual. See **inbreeding**.

Selfish DNA/genes If the **units of selection** are genes, or perhaps smaller-sized replicable hereditary material, a consequence appears to be that any mutation promoting their transmission into the next generation would automatically be selected for – at least in the short-to-medium term. Such promotion might involve simply the 'hijacking' (a kind of molecular parasitism) of normal genetic mechanics and machinery, which selection might be unable to prevent. It has been suggested that evidence for such 'self-promotion' of genes, even in the face of harmful phenotypic effects they may have on their bearers (uncoupling morphological and molecular evolution), is to be found in the **C-value** paradox, *mif*$^+$ plasmid and ω^+ intron (see **mitochondrion**), ***t* haplotypes, introns, chromosome diminution, gene conversion** and driving genes (see **aberrant chromosome behaviour** (3) and (4)). There is now talk of 'intranuclear warfare' between genetic elements, even between whole chromosomes.

Self-pollination Transfer of pollen from anther to stigma of same flower, or to stigma of another flower of same plant. See **inbreeding**.

Self-restriction See **T cell**.

Self-sterility (Of some **hermaphrodites**) inability to form viable offspring by self-fertilization. See **incompatibility**.

Semelparous Reproducing only once in its lifetime. Contrast **iteroparous**.

Semicell See **desmid**.

Semen Sperm-bearing fluid produced by the testes and accessory glands (e.g. **prostate gland, seminal vesicles** and Cowper's gland in mammals), particularly of animals with internal fertilization.

Semicircular canals Component of vertebrate **vestibular apparatus** detecting (directional) acceleration of the head.

Seminal vesicle (1) Organ of lower vertebrates and of some invertebrates (e.g. earthworms, insects) that stores sperm. (2) Diverticulum of the *vas deferens* (sperm duct) of male mammals, whose alkaline secretions lower the pH of the semen and counteract vaginal/uterine acidity.

Seminiferous tubules See **testis**.

Semispecies Populations in the process of acquiring mechanisms isolating one from the other reproductively. Regarded therefore as *incipient species* in the sense of the biological species concept (see **species**). Semispecies are generally expected to exhibit greater genetic differences than do geographical subspecies (see **intraspecific variation**).

Senescence See **ageing**.

Sense organ Group of sensory **receptors** and associated non-sensory tissues specialized for detection of one sensory modality (e.g. light in the **eyes**, sound in the **ears**, etc.).

Sensitization Process rendering an organism or cell more reactive to a specific **antigen** or antigenic determinant. See **immunity, allergic reaction**.

Sepal Component member of the calyx of dicotyledonous flowers; usually green and leaf-like. Cultures of pollinated flowers of several species on media containing only sucrose and mineral salts indicate that sepals have a role in nitrate assimilation and synthesize and export **auxin, gibberellins** and **cytokinins** to the developing fruit. See **flower**.

Septicidal (Bot.) Describing the dehiscence of multilocular capsules by longitudinal splitting along septa between the carpels, separating the carpels from one another, e.g. St John's wort. See **loculicidal**.

Septum Partition or wall. The structure concerned is said to be *septate*.

Sere Particular example of plant **succession**. Seres originating in water are referred to as **hydroseres**; those arising under dry conditions as **xeroseres**, of which those developing upon exposed rock surfaces are known as **lithoseres**.

Serial homology See **homology**.

Serial sections Series of successive microtome sections, from which three-dimensional structure can be built up.

Serine kinases See **protein kinases**.

Serine proteases (serpins) Family of homologous proteolytic enzymes (see **gene duplication**) including several involved in digestion (e.g. chymotrypsin, trypsin) and **blood clotting** (e.g. thrombin). Implicated in the genesis of **positional information** during *Drosophila* development. See **cascade**.

Serology The study of **antigen-antibody reactions** in vitro.

Serosa Outermost layer of most parts of vertebrate **gut**. A **serous membrane**.

Serotonin (5-hydroxytryptamine, 5-HT) L-tryptophan-derived **neurotransmitter** of vertebrate brain; esp. of *raphe nuclei* of **pons** and **medulla**, and of **pineal gland** (precursor of **melatonin**). General inhibitor of activity – apparently in opposition to **noradrenaline**; tricyclic antidepressant drugs probably potentiate neurotransmission of both by blocking their neuronal uptake. Hyperpolarizes post-synaptic membranes and activates *phosphofructokinase* in liver (see **glycolysis**), both mediated by cyclic AMP. See also **platelets, blood clotting**.

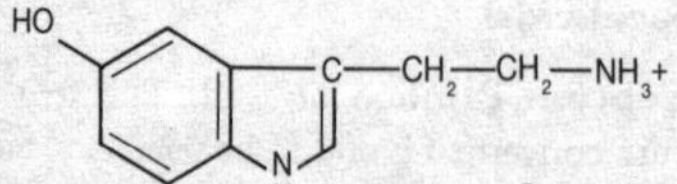

HO, C—CH_2—CH_2—NH_3^+, N, H

Serotonine

Serous membranes Mesothelial layers overlying deeper connective tissue and lining the vertebrate coelomic spaces (pericardial, perivisceral, pleural, peritoneal cavities).

Sertoli cells See **testis**, **maturation of germ cells**.

Serum See **blood serum**.

Sesamoid bone Bone (e.g. patella) developing within tendon of vertebrate, particularly where tendon operates over ridge of underlying bone.

Sessile (1) (Of animals) living fixed to the substratum, e.g. sponges, corals, barnacles, limpets, tunicates. (2) Lacking stalks, e.g. eyes of some crustaceans. Opposite of *pedunculate* in plants.

Seta (Bot.) Stalk of sporagonium in mosses and liverworts; in algae, a stiff hair, bristle or bristle-like process – an elongated hollow cell extension. (Zool.) Invertebrate epidermal bristle, consisting solely of cuticular material (**chaeta**), or of a hollow projection of cuticle enclosing epidermal cell or its part (e.g. in insects).

Sewall Wright effect See **genetic drift**.

Sex (1) Often used as synonym of **sexual reproduction**. (2) Used in several senses. *Germ cell sex* distinguishes individuals by their abilities to produce gametes of particular morphological types, namely, microgametes (sperm, generative nuclei, etc.), or macrogametes (eggs, egg cells, etc.). Males (with 'male' sex organs) produce the former; females (with 'female' sex organs) produce the latter. **Hermaphrodites** produce both, either simultaneously or sequentially. *Genetic sex* concerns an individual's genotype in so far as it bears on **sex determination**; *phenotypic sex* relates to anatomical appearances normally associated with one or other sex (e.g. **secondary sexual characters**), sometimes distinguished from *behavioural sex*, where the two sexes behave in distinctive ways. *Hormonal sex*, identified by the particular hormonal production from an individual's sex organs, is determined by sex organ appearance and physiology rather than by genotype (see **sex-reversal gene**). *Brain sex* refers to distinctive anatomical differences between the brains of the two sexes, itself sometimes causally related to behavioural sex.

The *origin and evolution of sex* are problematic, but the processes common to all forms of genetic recombination may have evolved from cellular **DNA repair mechanisms** or as an antidote to driving genes (see **aberrant chromosome behaviour** (4)). Some hold that since the only advantages of sexual reproduction over asexual seem to be long-term ones, some kind of **group selection** is necessary to account for prevalence of the former. Others argue that natural selection alone can account for eukaryotic sexual reproduction, despite the **cost of meiosis**. One view is that a parasitic gene (see **selfish DNA/genes**) might force a cell to fuse with another (plasmogamy), leave a copy of itself in this second cell, and then force the two cells to split apart (see **mitochondrion**). The evolution of nuclear fusion (karyogamy) might then have involved selection to avoid the deleterious consequences for a nucleus of bearing a recessive 'mating type' allele being condensed, envacuolated and eliminated from the cytoplasm – as happens in *Physarum* (see **mating type**). See **Müller's ratchet**, **recombination**.)

Separation of sexes (*dioecism*) has the effect of reducing the potential number of individuals between which fertilization may occur (see **incompatibility**).

Sex chromatin See **Barr body**.

Sex chromosome Chromosome having strong causal role in **sex determination**, usually present as a homologous pair in nuclei of one sex (the **homogametic sex**) but occurring either singly (or with a partial homologue) in those of the other sex (the **heterogametic sex**). In cases of partial homology one chromosome (Y) is often much smaller than the other (X), and both may resemble autosomes. In birds, where females are heterogametic, males are sometimes given the genotype ZZ and females ZW. Sometimes sex-determining loci are situated on such a short region of a single chromosome pair as to make sex chromosomes indistinguishable in appearance. In species with an XX/XY system (e.g. humans) sex chromosomes usually pair up at meiosis and form a bivalent; in mammals, there may be crossing-over and chiasma formation between homologous (often sub-terminal) regions. Organisms with both types of sex organ combined in one individual (e.g. monoecious plants, hermaphrodites) lack specialized sex chromosomes. See **autosome**, **hemizygous**, **sex linkage**.

Sex-controlled character See **sex-limited character**.

Sex determination Control of occurrence of, and differences between, sexes (see **sex**). Where male and female differentiation occurs within a single individual (e.g. a **hermaphrodite**), it is commonly restricted to the sex organs. See **gynandromorph**, **freemartin**.

Bisexual (dioecious) species often have genetic sex-determining mechanisms. These are occasionally cytoplasmic (see **cytoplasmic inheritance**), sometimes depend upon the individual's ploidy (see **male haploidy**), but most commonly involve a pair of **sex chromosomes**.

In some cases (e.g. many gonochorist species of fish, amphibians and reptiles) one or a few loci on an unspecialized chromosome pair may be responsible for sex determination through segregation of a pair or a few pairs of alleles (resembling some **incompatibility** mechanisms). In cases of sex organ differentiation within a single hermaphrodite individual (e.g. some moss and fern gametophytes) the mechanism involved resembles normal cytological **differentiation**. Sex-determining loci may not be restricted to sex chromosomes, even where these are differentiated (e.g. see **sex reversal gene**). But where sex is determined by a differentiated sex chromosome pair it may be absence of a second X or presence of a Y that results in sex differentiation. In humans it is the latter. Individuals with **Klinefelter's syndrome** (XXY) are male while those with **Turner's syndrome** (XO) are female. In mammals, the Y-chromosome encodes a testis-determining factor (see **testis** for effect of *Sry* gene) which causes embryos with a Y-chromosome to develop testes and become males, while those lacking a Y-chromosome develop ovaries and become females. But in the fruit fly *Drosophila* the X-chromosome is female-determining while the autosomes collectively are male-determining, an individual's sex resulting from the balance, or ratio, between the number of sets of autosomes and X-chromosomes. This was established by artificial production of **intersexes** which are triploid for their autosomes but diploid (XX) for their sex chromosomes. 'Superfemales' were diploid for their autosomes but triploid for sex chromosomes (XXX); 'supermales' were triploid for their autosomes but hemizygous for an X-chromosome (XO). So an XY/XX sex chromosome system may in itself tell us little about the sex-determining mechanism involved. In *Drosophila*, whatever measures the ratio of sex chromosomes to autosomes also controls the details of alternative splicing (see **RNA processing**) of two gene transcripts (*Sxl* and *tra*); for in males (the 'default' pattern) the pre-RNA of each contains an intron bearing a stop codon which makes for functionless gene product, whereas in females these introns are excised. The *tra* gene product is required for female development, which it does by regulating further RNA splicing events. The nematode *Caenorhabditis elegans* has a similar 'balance' sex-determining mechanism. The worm *Bonellia* (Echiuroidea) has environmental sex-determination, albeit with genetic involvement. If a larva develops independently it becomes female; if it is influenced by pheromones produced by the adult female's proboscis it develops into a male. In some reptiles (e.g. turtles, crocodilians) sex is strongly determined by the temperature at which the embryo develops.

Sexduction (F-duction) See **transduction, F Factor**.

Sex factor See **F Factor**.

Sex hormones See **oestrogens, testosterone**.

Sex-limited character (sex-controlled character) Character determined by a genetic element expressed differently between the sexes, commonly as a result of hormonal differences. *Pattern baldness* in humans is a character expressed in men who are either homozygous or heterozygous for the allele, but only in those women who are homozygous for it. See **sexual selection**.

Sex linkage A character is said to be sex-linked if it is determined by a genetic element occurring either (a) on an X-chromosome where the method of **sex determination** involves an XX/XO system, or (b) on any non-homologous region of a pair of **sex chromosomes** where the method involves an XX/XY system. Most readily detected by an appropriate **reciprocal cross**. In an XX/XY system, if the character in question is recessive it will appear in the heterogametic sex with the same frequency (say, n) as the chromosome bearing the element; hence it will appear in the homogametic sex with the square of that frequency (i.e. n^2). In humans, red-green colour blindness and haemophilia are such examples. *X-linked* characters cannot be transmitted from father to son since a father contributes a Y-chromosome to his male offspring. A mother may be a carrier, in which case half her male offspring would tend to be affected. Dominant sex-linked characters would be expressed in hemizygous, heterozygous and homozygous conditions.

Sex pili See **pili, F factor**.

Sex ratio Proportion of males to females in a species population, usually expressed as the number of males per 100 females. *Primary sex ratio* is assessed immediately after fertilization; *secondary sex ratio* is assessed at birth (or hatching); *tertiary sex ratio* is assessed at maturity. The sex ratio among the offspring of a particular female may be subject to fluctuations, as with queen bees and other female eusocial insects. This is the subject of much theorizing and investigation.

Sex-reversal gene *SRY* in humans, *Sry* in mice. Dominant autosomally determined character of mice causing the somatic gonadal tissues of females to develop as testis. This then secretes testosterone and brings about male phenotype despite the presence of female genotype. See **testis** for the role of testis-determining factor, *TDF*/*Tdy*, which is very likely the same gene.

Sex-role reversal Occurrence, notably in some bird species (e.g. some arctic wading birds), where females are more brightly coloured than males and display for mates, sometimes leaving them to incubate the offspring alone.

Sexual dimorphism Occurrence of populations where individuals differ in respect of two distinct sets of phenotypic characters, sex itself being one of them. Some cases may be attributable to **sexual selection**.

Sexual reproduction A misleading phrase, for the essential processes of sex (gene **recombination** and/or transfer) and reproduction (production of new individuals) are quite distinct, despite common association in **life cycles**. For example, during **conjugation** in protozoa such as *Paramecium*, sex and reproduction are only tenuously linked and are completely uncoupled during **automixis** in the *Paramecium aurelia* species complex. Again, nuclear fusion producing the second endosperm nucleus in **Anthophyta** is a sexual process (automixis), yet that nucleus contributes nothing genetically to the future offspring (reproduction).

Gene transfers here include only those between existing cells, excluding cell division by mitosis, binary fission and budding – none of which is sexual; but it includes **conjugation**, **fertilization**, **plasmid** transfer and viral infection (and even some forms of **gene manipulation**). Excision of plasmids and their incorporation within the genome of an unrelated recipient species raises questions about the validity of **species** definitions relying on the concept of a shared **gene pool**. In prokaryotes, lacking **meiosis** and **fertilization**, sexual processes commonly involve a specialized form of conjugation (e.g. see **F factor**). In eukaryotes, the distinctive features of sexual **life cycles** are meiosis and fertilization; but gametes are not invariably produced by meiosis, and fertilization is not always followed by mitosis. The genetic recombination normally associated with sexual processes tends to promote **genetic variation** among offspring (but see **inbreeding**). This is generally considered adaptive in unstable or patchy environments. However, meiosis has a potentially disruptive effect upon coadapted gene complexes. See **sex**.

Sexual selection Form of selection, generally contrasted with **natural selection** although also proposed by **Darwin**. Results from the exercise of mating preferences (by either sex, but most commonly by females) in favour of individuals expressing certain genetically determined characters. As a result, genes for these characters tend to spread through the population, being expressed as **sex-limited characters.** These could be defined as 'attractive' to the choosing sex.

This is a form of natural selection, a male's fitness being enhanced by his expression of 'attractive' characters, a female's fitness being improved by contributing genetically both to male offspring likely to have these characters and to female offspring likely to be genetic carriers of them. Sexual selection is often invoked to explain cases of extreme **sexual dimorphism** (e.g. where males are huge compared with females), and such dimorphisms do tend to evolve where males are polygynous. Sexual selection may enhance the directional component of selection, but its role in the spread of genes which have undramatic phenotypic effects is the subject of debate and experimentation. See **arms race**.

Shared homologue A character found in two or more taxa whose most recent common ancestor also had (or can be inferred to have had) it. See **symplesiomorphy**, **synapomorphy**.

SH2 and SH3 domains (*src* homology 2 and 3 domains) See **domain**, **receptor**.

Sharks See **Selachii**.

Shell No clear-cut definition; includes

many hardened animal secretions having protective and/or skeletal roles, often attached only to part of body surface. Structures termed shells are found in many animal groups, e.g. the **Sarcodina**, **Mollusca**, **Brachiopoda**, in **Crustacea** (a *carapace*) and in echinoid echinoderms (a *corona*).

Short-day plants See **photoperiodism.**

Short-germ Mode of **development** in some lower insect groups (e.g. Orthoptera) in which only the most anterior head structures are defined before or during gastrulation. Compare **long-germ**.

Shoulder girdle See **pectoral girdle**.

Shuttle vector See **vector**.

Siblings (sibs) Brothers and/or sisters; offspring of the same parents.

Sibling species Very closely related **species** differing only in minor respects, or appearing identical, but in fact reproductively isolated. Separation is often important where one of the species is vector to an economically important parasite (as in *Simulium* and *Anopheles* spp.), when DNA probes may be used to distinguish them. See **superspecies**.

Sickle-cell anaemia A hereditary, genetically determined disorder affecting many newborn African and American negroes and others of tropical climates where malaria is, or has recently been, endemic. Caused by homozygosity for allele Hb^S, producing a single amino acid substitution in the β-chain of the normal haemoglobin molecule determined by allele Hb^A. Individuals heterozygous for the allele (Hb^AHb^S) are resistant to the most serious form of **malaria** (subtertian malaria caused by *Plasmodium falciparum*; see **heterozygous advantage**), and allele Hb^S therefore attains a high frequency in malarial areas. However the homozygote Hb^SHb^S suffers from sickle-cell anaemia, so-called because red blood cells have a sickle shape even at high blood oxygen levels and block capillaries, becoming phagocytosed and causing anaemia and other symptoms. An example of **natural selection** in humans. See **thalassaemias**, **genetic load**.

Sieve area Area of sieve element wall containing clusters of pores through which protoplasts of adjacent sieve elements are connected. See **phloem**, **sieve plate**.

Sieve cell See **phloem**.

Sieve plate End-wall of **sieve tube member**, where pores are larger than in lateral sieve area; highly differentiated.

Sieve tube Series of **sieve tube members**, arranged end-to-end and interconnected by **sieve plates**. Functions in transport of food materials (e.g. sucrose, amino acids). Lacking nuclei at maturity. See **translocation**.

Sieve tube member Component of a **sieve tube**. Elongated, unlignified tube-like cells found primarily in flowering plants, but also in some brown algae (Laminariales); typically associated with a **companion cell**. See **phloem**.

Sigma factor (σ factor) Protein component of prokaryotic **RNA polymerases** binding loosely to the core enzyme and restricting mRNA transcription to one of the two DNA strands and

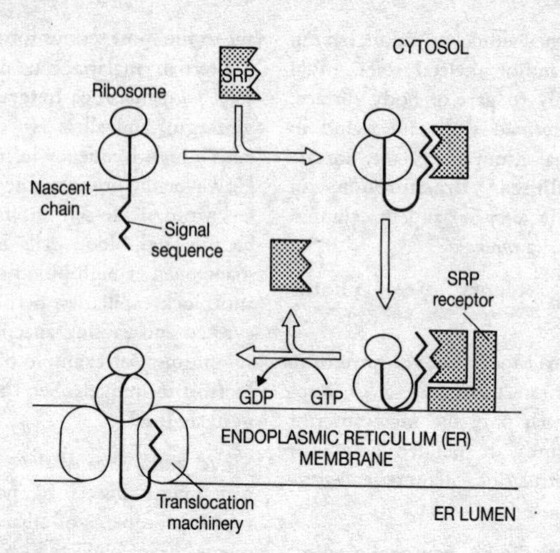

Fig. 103 *Scheme showing the dual roles of the eukaryotic* **signal recognition particle** *(SRP) in both chaperoning an unfolded nascent chain through the cytosol and targeting it to the SRP receptor in the ER membrane.*

appropriate promoter region (see **Jacob-Monod theory**). Core RNA polymerases (lacking sigma factor) tend to start transcription randomly, on either DNA strand. The σ factor tends to dissociate itself from the RNA polymerase after a few RNA nucleotides have been incorporated, when its place may be taken by elongation factors involved in chain elongation and termination. For eukaryotic sigma factor equivalents, see **TATA factor**, **transcription factors**. See also **rho factor**, **Pribnow box**.

Signal recognition. As a **signal region** of a **preprotein** comes away from the ribosome on which it was translated, it is bound by a specific elongated signal recognition particle (SRP) comprising a 7S RNA and six different polypeptides. Binding can slow down or stop the translation process. The SRP targets the nascent polypeptide–ribosome complex to the endoplasmic reticulum by its interaction with a receptor in the membrane. The SRP then leaves the polypeptide–ribosome complex as the signal regions contact the translocation mechanism in the ER membrane.

Signal regions (signal sequences, leader sequences) See **protein targeting**, **signal recognition**, **plasmids**.

Signal transduction The processes by which a cell responds to an external signal (be that molecular or a form of energy), all of which involve the 'leading across' (transduction) of information from outside the cell to bring about changes in the signals within. This commonly involves a cell's binding a signal molecule (ligand) at one of its cell-sur-

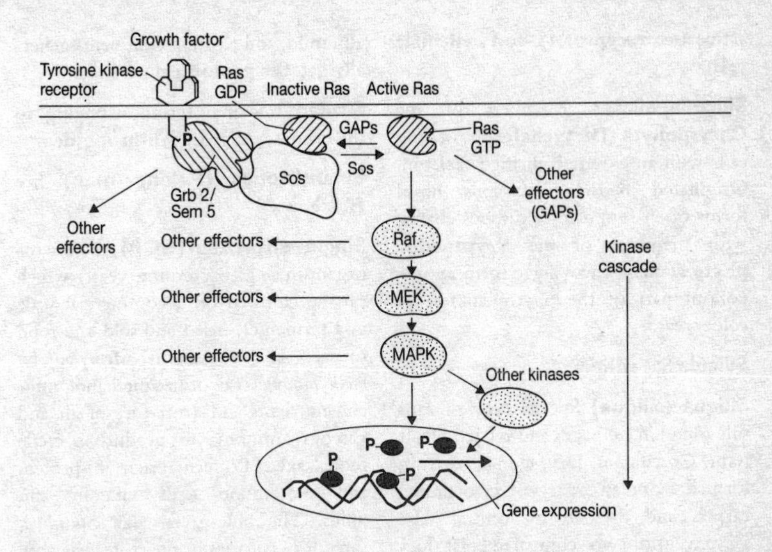

Fig. 104 *A* **signal transduction** *pathway involving a growth-factor-specific receptor tyrosine kinase whose autophosphorylation on binding its ligand enables it to recruit exchange factors, such as Sos, once an appropriate exchange control factor (here Grb2, also known as Sem5) has bound the tyrosine kinase. Such a complex can now activate Ras, turning it from inactive Ras-GDP to active Ras-GTP. Active Ras can then initiate a tyrosine kinase cascade, the ultimate signals entering the nucleus as phosphorylated transcription factors regulating gene expression. As indicated several of the proteins involved (many of them oncoproteins) are pleiotropic in their effects. Compare Fig. 99b.*

face receptor molecules (see **receptor** (2)) amplifying that signal at or close to the membrane and activating molecular pathways culminating in changes in, e.g., **gene expression**, **cell cycle** and cell fate. Alternatively, some signal molecules cross the cell membrane and bind intracellular **nuclear receptors** before activating gene expression. Signal molecules include traditional **hormones**, **pheromones** and **neurotransmitters**, but have expanded widely to include **growth factors, cytokines**, etc. Much research also involves bacterial (e.g. **bacteriorhodopsin**), protistan and plant signal transduction mechanisms. The alga *Chlamydomonas* has a **rhodopsin**-regulated signal transduction chain involving its 'eyespot', flagellum base and Ca^{2+}, controlling its behaviour to light, so it is of interest that much plant-cell signal transduction seems to be calcium-linked (see **calmodulin**). Although much recent research highlights the molecular signals arriving at cell surfaces (see **accessory molecules**, **major histocompatibility complex**, **incompatibility, interferons**), work on transduction mechanisms in photoreceptors and mechanoreceptors is also active (see **receptor**(1), **rod cell, hair cell**).

Silicoflagellates Members of the **Chrysophyta** (**Dictychales**) possessing cells with an external silicified skeleton. Originated in the Cretaceous. Fossil forms often found in calcareous chalks, with members of the **Prymnesiophyta**. Today, these algae form an important part of the phytoplankton of colder seas.

Silicula See **siliqua**.

Siliqua (silique) Special type of capsule found in cabbages and related plants (fam. Cruciferae). Dry, elongated fruit, formed from an ovary of two united carpels and divided by central false septum into two compartments (locules). Dehiscing by separation of carpel walls from below upwards, leaving the septum bounded by persistent placentas (replum), with seeds adhering to it. The *silicula* has a similar structure, but is short and broad; e.g. honesty, shepherd's purse.

Silk Fibroin (a β-keratin protein) produced by modified salivary glands of silkworms (silkmoth larvae) and by spiders, among other arthropods. Has small amino acid R-groups, while anti-parallel polypeptide molecules hydrogen-bond to form very stable β-pleated sheets.

Silurian Period of the **Palaeozoic** era (*c.* 440–400 Myr BP). The period commences with a major **extinction** event. The first fossil plants occur, and the first jawed fish appear. Other animal fossils include scales of ostracoderm fishes, corals, crinids, trilobites, brachiopods and graptolites. The climate was generally mild, and existing continents generally flat. See **geological periods**.

Simian Informal term; belonging to the primate suborder **Anthropoidea**.

Sinanthropus (Peking man) See ***Homo***.

Single cell protein (SCP) Any microorganism (e.g. bacterium, yeast) which can be cultured on a commercial scale in a fermenter, dried and sold as a food source. The bacterium *Methylophilus methylotrophus* can utilize methanol, mineral nutrients and a mixture of air and gaseous ammonia to produce a cattle food cake ('Pruteen') rich in protein and free amino acids, vitamins and lipids. The blue-green alga *Spirulina*, eaten for centuries in the tropics, is now commercially harvested in the warm water of cooling plants, with a huge yield of dry weight and protein. Eukaryotic SCP has less nucleic acid content than does prokaryotic and is therefore more suitable as a human food source, since excess nucleic acid causes health problems. See **bioreactor**.

Sinoatrial node See **pacemaker**.

Sinus (Bot.) Space or recess between two lobes of leaf or other expanded organ. (Zool.) A *blood sinus* is an expanded vein, particularly of selachian fish; sometimes also used of the **haemocoel**. *Nasal sinuses* are air-filled spaces within some facial bones of mammals lined by mucous membrane and communicating with the nasal cavity.

Sinusoid Type of **capillary**.

Sinus venosus Chamber of vertebrate **heart**, between veins and auricle(s).

Thin-walled; absent from adult birds and mammals. See **pacemaker**.

Siphonaceous (siphonous) Coenocytic; e.g. members of several orders of algae, which are filamentous, sac-like or tubular, without cross-walls.

Siphonaptera Order of endopterygote insects. Fleas. Secondarily wingless, with legless detritus-feeding larvae; exarate pupa enclosed within a cocoon. Mouthparts consist of long serrated mandibles, short palped maxillae and reduced palped labium. Laterally compressed, with legs adapted for running between body hair and for jumping. Include human flea *Pulex irritans* and rat flea (transmitter of bacterial plague) *Xenopsylla cheopis*.

Siphonophora Order of **Cnidaria** containing complex and polymorphic colonial animals. Colonies often form by budding from original medusoid individual.

Siphonostele (solenostele) Type of **stele** containing hollow cylinder of vascular tissue surrounding a pith. See **ectophloic** and **amphiphloic**.

Siphunculata (Anoplura) Sucking lice. Wingless exopterygotan insects, ectoparasitic on mammals. Eyes (and ocelli) reduced or absent; mouthparts modified for piercing and sucking. Thorax of fused segments. The human louse *Pediculus humanus* can carry *Rickettsia prowazeki*, distributed via the insect's faeces and causing epidemic typhoid fever.

Sirenia Order of placental mammals. Manatees, dugongs (seacows). Aquatic (coastal and river-dwelling) and herbivorous, with transversely expanded tail, front legs modified as flippers and vestigial hindlegs. One pair of mammary glands. Not closely related to cetaceans.

Sister group A species or higher monophyletic taxon believed to be the closest genealogical relative of a given taxon, excluding the species ancestral to both taxa. Two taxa forming sister groups are thus thought to share an ancestral species not shared by any other taxon.

Site-directed mutagenesis See **mutagen**.

Sivapithecus See **Ponginae**.

Skates See **Elasmobranchii**.

Skeletal muscle See **striated muscle**.

Skeleton Any component of an animal's anatomy serving for muscle attachment and/or tranmission of restoring forces extending a contracted muscle. See **endoskeleton**, **exoskeleton**.

Skin (cutis) Vertebrate organ covering most of body surface, comprising *epidermis* (ectoderm) above, produced by the **Malpighian layer**, with varying amounts of underlying connective tissue of *dermis* (mesoderm) and *subcutaneous fat* (adipose tissue). Often impervious to water, but permitting gaseous exchange in modern amphibia. Produces, variously, **scales**, **hair**, **feathers**, **nail** and sometimes bone (see **dermal bone**). Glands located here include mucus glands (fish, amphibia), and sebaceous and sweat glands (mammals). May be variously pigmented (see **chromatophore**) and contain several kinds of receptors in the dermis (e.g. pain, pressure, temperature). Attachment to underlying organs by loose connective tissue helps it move over them and return elastically. In amphibia the attachment is particularly loose. See **cornification**, **keratin**.

Skotophile Literally, dark-loving; dark-receptive phase of circadian rhythms, lasting about 12 hours. See **photophile**.

Sliding filament hypothesis Hypothesis that the apparently diverse activities of **muscle contraction**, **cyclosis**, **cytoplasmic streaming**, and various kinds of **cell location** all depend fundamentally upon energy-dependent sliding of microfilaments of **actin** over **myosin**. Similar mechanism, involving **microtubules** of tubulin, explains the beating of eukaryotic flagella (see **cilium**), and organelle and chromosome movement.

Slime fungi, slime moulds See **Myxomycota**.

Slow fibre (tonic fibre) See **muscle contraction**.

Small intestine See **ileum**, **jejunum**.

Smooth muscle (involuntary muscle) Contractile tissue (generally called *visceral muscle* in vertebrates) comprising numbers of elongated spindle-shaped cells (sometimes syncytial) lacking transverse striations or other obvious ultrastructure. Position of the one nucleus per cell varies widely. Responsible among much else for peristaltic movement of food along gut, of blood along some contractile vessels (esp. invertebrates) and in regulation of vertebrate blood pressure (see **vasomotor centre**). Often arranged in *muscle coats* as bundles of cells averaging 100 μm in diameter, frequently within vertebrate connective tissue but occurring widely in both invertebrates and vertebrates. Among the latter principally in visceral, vascular and other locations where hollow organs occur (e.g. uterus). Cell lengths vary from 30–450 μm; diameters from 2–6 μm. Cells in the muscle bundle are separated by a basement membrane of glycosaminoglycans, glycoproteins, collagen and elastic fibres about 60 nm wide; cell contacts include tight junctions and gap junctions. Proteins involved in its contraction are basically those of striated and cardiac muscle; but the myosin filaments are not as regularly arranged, while the actin filaments are more randomly distributed, lying along the longitudinal axis of the cell parallel to the myosin filaments; some of them terminate via accessory proteins at the cell membrane.

In vertebrates, almost always under control of **autonomic nervous system**, often with intrinsic **pacemakers**. Hormones may alter the threshold for contraction. The sympathetic nervous system is generally excitatory on vascular smooth muscle but inhibitory on gut muscle; the parasympathetic system may be either excitatory or inhibitory, depending on the organ. See **muscle contraction**.

snRNP See **RNPs**.

Society (Bot.) Minor climax community within a consociation, arising as a result of local variations in conditions and dominated by species other than the consociation dominant.

Sociobiology The systematic study of the biological basis of all social behaviour.

Sodium pump **Active transport** mechanism present in plasma mem-

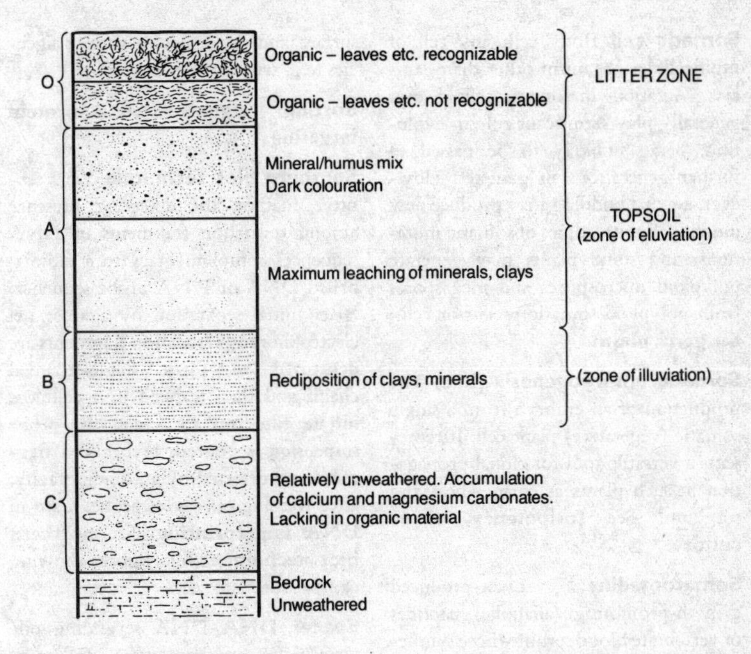

Fig. 105 *Generalized* **soil profile** *indicating the major horizons. Not all soils have all these and their relative thicknesses vary greatly.*

branes of most animal cells, consuming an estimated third of a cell's ATP production in pumping sodium ions (Na^+) out of the cell and potassium ions (K^+) into it in the ratio 3:2. It plays a minor role in establishment of a cell's **resting potential**, but helps regulate cell volume by casting out Na^+ which would tend to enter along its electrochemical gradient, adding to the negative osmotic potential of the cell and drawing water in. The cell's internal electrical negativity prevents chloride ions (Cl^-) from entering and having the same effect. The pump is blocked by external *ouabain*, and animal cells may therefore swell or burst if this or other inhibitors of ATP synthesis or hydrolysis are added. It is indirectly responsible for glucose and amino acid uptake by cells since it creates a sodium gradient necessary for Na^+-based symports (see **transport proteins**).

Soil profile Series of recognizably distinct layers or horizons visible in a vertical section through the soil down to the parent material. Study of soil profiles yields valuable information on the character of soils. See Fig. 105.

Solar plexus See **autonomic nervous system, nerve plexus**.

Solonostele See **siphonostele**.

Soma See **cell body, somatic cell**.

Somatic cell Body cell; any cell of multicellular organism other than gametes. Mutations in somatic cells do not generally play significant role in evolution, being unlikely to be passed to further generations in gametes. However, asexual budding may produce new individuals with copies of somatic mutations, and some plants may generate polyploid microspores and megaspores from polyploid somatic precursor cells. See **germ plasm**.

Somatic embryogenesis (Bot.) The production of an embryo from a single somatic (vegetative) plant cell. It represents a versatile tool for clonal propagation in such plants as corn, wheat and sorghum. See **totipotency, tissue culture**.

Somatomedins Liver-produced growth-promoting (anabolic) peptides of vertebrate blood serum whose synthesis and activity are dependent upon presence of **growth hormone** and whose activities they *mediate*. Cause instantaneous release of hypothalamic **somatostatins**. Enhance wound-healing.

Somatostatin Hypothalamic *release-inhibiting factor* (RIF) preventing release of growth hormone releasing factor (GHRF) and hence of **growth hormone**. Some pancreatic cells (delta cells) also release somatostatin, here inhibiting release of pancreatic hormones.

Somatotrophin See **growth hormone**.

Somite See **mesoderm**.

Soredia Organs of vegetative reproduction in **lichens**. Minute clusters of algal cells surrounded by fungal hyphae, formed in large numbers over lichen surface and dispersed by various agencies (e.g. wind, water, insects).

Sorting signals See **protein targeting**.

Southern blot technique Very sensitive method for detecting presence among restriction fragments of DNA sequences complementary to a radiolabelled DNA or RNA probe sequence. After initial separation by agarose gel electrophoresis, restriction fragments are denatured to form single-stranded chains and then trapped in a cellulose nitrate filter on to which the probe suspension is poured. Hybridized fragments are detected by autoradiography, after washing off excess probe. Used in **DNA fingerprinting**. See **northern blot technique** for figure showing comparison.

Spacer DNA DNA separating one gene from another, often not transcribed itself. Common where there is **gene amplification** (e.g. between ribosomal RNA genes).

Spadix Kind of **inflorescence**.

Spathe Bract enclosing inflorescence of some monocotyledons (e.g. *Arum*).

Spatial summation See **all-or-none response**.

Special creationism The view that each *kind* of organism (present and past) owes its existence to a unique creative act of God. Many influential scientists of the 17th and 18th centuries (Ray, Burnet, Whiston) believed science would vindicate a literal interpretation of the biblical creation story, as did many in the 19th (Buckland, Sedgwick, Agassiz). Darwin, whose views oppose special creationism, thought he had refuted this theory, holding that an intelligent Creator would have distributed organisms according to their physiological requirements rather than their taxonomy. Contrast **evolution**.

Specialized Having special adaptations to a particular ecological niche which often result in wide divergence from the presumed ancestral form. Such *specializations* evolve and may result in niche restriction.

Speciation The origin of **species**. If species are not real ('objective') entities, speciation cannot be regarded as a real process. If, however, species are lineages with temporal continuity, then speciation is whatever generates independence of lineages from one other.

The major models of speciation are the *allopatric* and *sympatric* models. In the former a parent species becomes physically separated into daughter populations by geography, restricting (or eliminating) gene flow between non-overlapping populations. In sympatric models, a parent species differentiates into lineages in the absence of any physical restriction on gene flow.

Some phenotypic characters bear directly on speciation. Shape of genital aperture, timing of breeding, degree of assortative mating, compatibility of pollen and stigma, degree of developmental homeostasis – all are aspects of phenotype subject to genetic variation. They are therefore responsive to disruptive or directional selection, and can alter cohesion of the population as an evolutionary unit or lineage. Debate surrounds whether speciation is adaptive, or merely a stochastic process which perhaps selection cannot prevent. Allopatric speciation, at least, seems to involve non-selective splitting of lineages. In time, the daughter populations may simply gain sufficient distinctive genetic and phenetic characteristics for taxonomists to recognize separate species. Sympatric speciation models more often invoke the adaptiveness of species formation. But some insist that in either model speciation is *de jure* incomplete until sufficient overlapping of daughter populations has occurred for the biological species concept to be applied. It is often said that after overlap, selection against hybrids either reinforces lineage independence or is too weak to prevent collapse of population identities; but much will hinge on the species concept, phenetic or biological, being employed.

Major sympatric models involve disruptive selection on a deme already polymorphic for an ecological requirement (e.g. food plant, oviposition site), and assortative mating in favour of individuals sharing that requirement. Its occurrence in the wild is slowly gaining acceptance.

Stasipatric speciation postulates that a widespread species may generate internal daughter species whose chromosomal rearrangements play a primary role in speciation (through reduced fecundity or viability of individuals heterozygous for the rearrangements). Daughter species might then extend their ranges at the expense of the parent species and might hybridize where ranges abut, although resulting offspring would be less fertile.

Polyploidy can be a form of 'instantaneous' speciation, since it may result in the complete reproductive isolation of an individual from other gene pools. Allopolyploids are more significant here than autopolyploids, permitting regular bivalent formation during meiosis as well as originating through combination of genomes from different species. They breed true for their hybrid character. Moreover, crosses of allotetraploids to their diploid progenitors give sterile triploids, preventing backcrossing and gene flow. They are widespread in plants, where vegetative propagation of sterile hybrids can enhance colonization prior to allopolyploidy, successful meiosis and improved fertility. See **polyploidy**.

Debate surrounds the role of speciation in phyletic (macro-)evolution. Apart from stasipatric and polyploid speciation, there are no clear accounts of the genetics of speciation.

Species Term used both of a formal taxonomic category ('the species') and of taxa exemplifying it (particular species). In the system of **binomial nomenclature**, taxa with species status are denoted by Latin binomials, each species being a member of a genus. The naturalist John Ray, writing of plants in 1672, did not entirely avoid circularity in stating that the true criterion of species as taxa is that they are 'never born from the seed of another species and reciprocally': the cross-sterility criterion. To a special creationist like Ray such cross-sterility was to be expected (compare the nominalism of **Buffon**). For Darwin, all degrees of sterility should be discoverable if sterility *barriers* (the quite different, evolutionary, concept of **isolating mechanisms**) take time to evolve. See **speciation**.

Darwin's writings sometimes reveal nominalism on the species question, to be expected on the view that species evolve gradually. At other times he drew a clear distinction between *taxa* and *categories*, and although doubtful about the possibility of defining the category 'species' (i.e. the taxonomic unit) he was in no doubt that taxa with specific status actually existed. He did not espouse **essentialism** with regard to species.

Zoologists find the criterion of reproductive isolation especially valuable in demarcating species in the wild. As such, the *biological species concept* includes as species groups of populations which are phenotypically similar and reproductively isolated from other such groups, but which are actually or potentially capable of interbreeding among themselves. Problems in its application arise (a) when reproductive isolation from other populations admits of degrees, being incomplete over part of a species range, (b) with **ring species**, (c) with obligately asexual species (*agamospecies*), (d) with animals which although sexual lack males (obligate **thelytoky**), and (e) with **anagenesis**. The biological species concept also fails to incorporate a historical dynamic into its account, thereby ignoring that species are genealogically unique. The majority of fungi, plants and marine invertebrates broadcast their gametes widely, making it impossible to establish reproductive isolation in the wild. So complexes of phenetic characters (the stock-in-trade of museum taxonomists) usually serve to identify such species. But **sibling species** pose problems, and classification rests here upon cytological techniques (e.g. use of **DNA probes**) rather than external morphology, although reproductive isolation is often the original clue to their existence.

Recently favour has developed for

ecological and evolutionary species concepts. The former fix a classification to independently existing environmental states or niches (difficult to isolate independently of the organisms which occupy them), equating speciation with niche change but leaving open the extent of change required; the latter stress the genealogical uniqueness of species (useful in asexual and thelytokous forms), but emphasize the cladistic (branching) nature of speciation at the expense of gradual anagenesis. Most species probably comprise two or more *subspecies* or *races* (see **infraspecific variation**, **semispecies**) and are said to be *polytypic*.

Inability to find a unified species concept is no disgrace, reflecting the variety of reproductive systems and the dynamic state of biological material. It is probably no accident that the species concept does not figure prominently in biological theory: species may be best regarded not as **natural kinds** (such as elements in chemistry) but as individuals, each historically unique and irreplaceable once extinct. If species are individuals then species names are proper names, so that properties of species would describe but not define them. Definitions of taxa would then be *necessary* in philosophical terms, even though there is no list of necessary properties for any biological taxon. See **transitivity**.

Species diversity index See **biodiversity**.

Species flock A species-rich, narrowly endemic and ecologically diverse group of organisms, usually considered to have radiated rapidly in geological terms from one, or more than one, very closely related stem species. They are sometimes quoted as *prima facie* candidates for sympatric speciation, but insufficient evidence is usually available to ascertain the mode of speciation which produced them. Examples include cichlid fish in African lakes, Galapagos finches, honeycreepers and fruitflies on Hawaii. Mitochondrial DNA (**mtDNA**) analysis on the fishes of Lakes Malawi and Victoria support monophyletic origins for these flocks against the alternative theory that they represent polyphyletic products of numerous invasions by already-differentiated taxa. Debate surrounds the involvement of **sexual selection** in speciation within such flocks. Sadly, introduction of the nile perch (*Lates niloticus*) into Lake Victoria has already led to the disappearance of about 200 endemic species.

Species group Informal taxonomic category used in preference to such formal categories as subgenus or infragenus.

Spectrins A family of cytoskeletal proteins (including spectrin, dystrophin and α-actinin) all of which seem capable of forming antiparallel homodimers or heterodimers. Several are involved in linking **actin** filaments together, and appear to be linked to one another by hinges enabling extension and retraction.

Sperm See **spermatozoid**, **spermatozoon**.

Spermateliosis See **maturation of germ cells**.

Spermatheca (seminal receptacle) Organ in some female or hermaphrodite animals which receives and stores sperm from the other mating individual.

Spermatid Haploid animal cell resulting from the meiotic division of a secondary spermatocyte. Undergoes extensive cytoplasm loss and condensation of its nucleus during spermiogenesis (see **maturation of germ cells**).

Spermatium Non-motile male sex cell in the **Rhodophyta** (red algae), and some fungi belonging to the **Ascomycota** and **Basidiomycota**.

Spermatocyte (Bot.) Cell which becomes converted into a spermatozoid (without intervention of cell division). See **spermatid**. (Zool.) Cell undergoing meiosis during sperm formation. *Primary spermatocytes* undergo first meiotic division; *secondary spermatocytes* undergo second meiotic division. See **maturation of germ cells**.

Spermatogenesis Formation of spermatozoa. See **maturation of germ cells**.

Spermatogonium Cell within testis, commonly lining seminiferous tubules, which either divides mitotically to produce further spermatogonia or else gives rise to primary spermatocytes. See **testis**, **maturation of germ cells**.

Spermatophore Small packet of sperm produced by some species of animals with internal fertilization, e.g. many crustaceans, snails, mites, scorpions, *Peripatus*, newts, etc. Those of cephalopod molluscs are very complex and pump seminal fluid into the female by syringe-like structures.

Spermatophyta In some classifications, the Division containing all seed-bearing vascular plants. See **Anthophyta**, **Gnetophyta**, **Ginkgophyta**, **Cycadophyta**, **Coniferophyta**.

Spermatozoid (antherozoid) (Bot.) Small, motile, flagellated microgamete.

Spermatozoon (Zool.) Microgamete (see Fig. 106), usually motile, produced by testes; commonly flagellated but amoeboid in nematodes and aschelminthes. Sperm of *Drosophila* are about 2 mm in length. Most use lipids as fuel; complex sperm store glycogen. See **acrosome**, **maturation of germ cells**.

Spermiogenesis See **maturation of germ cells**.

Spermogonium (spermagonium) Flask-shaped or flattened, hollow structure in which spermatia are formed.

Sphenodon See **Rhynchocephalia**.

Sphenophyta Horsetails. Sphenophytes extend back to the **Devonian** period, but reached their peak abundance and diversity later in the **Palaeozoic** era (*c.* 300 Myr ago). During the late Devonian and **Carboniferous** periods, they were represented by the calamites, a group of trees that reached 15 metres in height, with a trunk that could be more than 20 cm thick. Represented today by a single genus *Equisetum*, which comprises sixteen species widespread in moist or damp locations (e.g. by streams, or truly aquatic or along the edges of woodland). They are homosporous vascular plants possessing jointed stems with prominent nodes and elevated siliceous ribs. Small scale-like leaves, whorled at the nodes, have a simple structure, and are probably reduced **megaphylls**. When present, branches arise laterally at

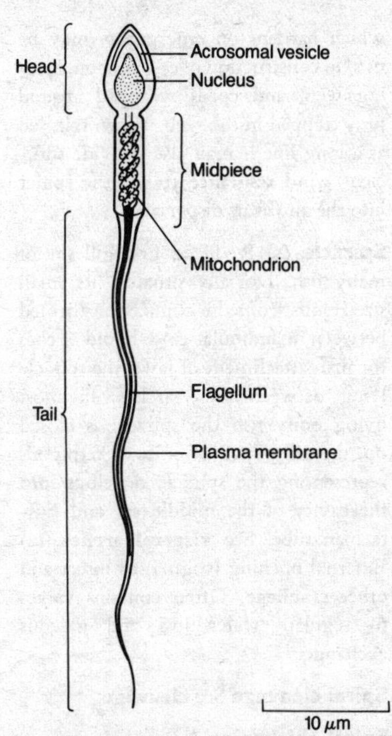

Fig. 106 *Diagrammatic illustration of a* **spermatozoon**.

the nodes, and alternate with the leaves. Sporangia are borne in a strobilus at the stem apex. Aerial stems arise from a branching, underground rhizome. Gametophytes are green, free-living, about the size of a pin-head. They become established mainly upon mud that has recently been flooded, and is rich in nutrients. They are bisexual or male. Sperm is multiflagellated, requiring water to swim to the egg. Horsetails have been used to scour pots and pans, and have acquired the name 'scouring rushes'.

Spheroplast Cell whose wall material has been largely but not entirely removed (protoplasts have theirs entirely removed). Often employed in **DNA cloning** (e.g. bacterium, yeast).

Spherosome Single, membrane-bound, spherical structure in plant cell cytoplasms. Many contain lipids and are apparently centres of lipid synthesis and accumulation.

Sphincter Ring of smooth muscle in wall of tubular organ, opening of hollow organ, etc., whose contractions and relaxations close and open the tube or aperture; e.g. pyloric and cardiac sphincters of stomach, anal sphincter.

Sphingolipids See **phospholipids**.

Sphingomyelin See **phospholipids**.

Sphingosine See **phospholipids**.

Spiders See **Araneae**.

Spike Indeterminate inflorescence in which main axis is elongated and flowers are sessile.

Spikelet Basic unit of grass inflorescences; small group of grass flowers.

Spinal column See **vertebral column**.

Spinal cord The part of the vertebrate **central nervous system** lying within the vertebral canal, protected by the vertebral column, and consisting of a hollow cylinder of mixed nervous tissue (derived from the **neural tube**) with walls of relatively uneven thickness. Paired and segmental spinal nerves leave it on each side between vertebrae. Contains both grey and white matter, former usually H-shaped in cross-section with the *cerebrospinal canal* running through the 'cross-bar', the latter surrounding the grey matter, the whole covered by the **meninges**. Carries sensory and motor information via ascending and descending tracts of white fibres; also provides for reflex arcs (intra- and intersegmental, ipsi- and contralateral). Continuous with the medulla oblongata of **brainstem**. See **reflex** for diagram (Fig. 100).

Spinal nerves Peripheral nerves arising from the vertebrate spinal cord between vertebrae; typically one on each side per segment. Each has a dorsal (afferent) root and a ventral (efferent) root, which typically fuse on emergence from the vertebral column. Compare **cranial nerves**.

Spindle (1) **microtubule** complex appearing during mitoses and meioses. Has two sources: regions around the **centrioles**, and the **kinetochores**. Microtubules from the former extend to the equator of cell where they apparently overlap and generate sliding forces which often push the poles of the cell apart; kinetochores somehow pull the centromeres of sister chromatids towards opposite poles. Microtubules (*polar*) developed from the centriolar region have their fast-growing (+) ends away from the pole; those from the kinetochore have their (+) ends attached to the kinetochore. While a cilium-like dynein has been implicated in the sliding of polar microtubules, the origin of the force pulling chromatids to the poles is not clear. (2) See **muscle spindle**.

Spindle attachment See **centromere**.

Spinnerets Maximum of three pairs of silk-spinning organs on the posterior opisthosomas of spiders (**Araneae**); most probably modified legs. The silk, which hardens on emergence, may be used in construction of egg cocoon, feeding web, and cords wrapped around prey trapped in the web. When released as a long line it may also provide sufficient wind resistance to lift the spider into the air for its dispersal.

Spiracle (1) Reduced first gill slit of many fish. Dorsally situated, its small size results from the connection formed between mandibular and hyoid arches for firm attachment of jaws, the spiracle lying between these arches. In most living bony fish the spiracle is closed up; the gill pouch of embryo tetrapods representing the spiracle develops into the cavity of the middle ear and Eustachian tube. See **visceral arches**. (2) External opening (stigma) of insect and other **tracheae**. Often contains valves to regulate water loss and gaseous exchange.

Spiral cleavage See **cleavage**.

Spiral thickening Internal thickening of wall of xylem vessel or tracheid, in form of a spiral. Occurs in cells of protoxylem and, while providing mechanical support, permits longitudinal stretching as neighbouring cells grow.

Spiral valve Spiral fold of mucous membrane projecting into the intestine of some fish, notably elasmobranchs, ganoids and Dipnoi. Probably serves to increase surface area for absorption.

Spirillum Long, coiled or spiral bacterium.

Spirochaetes Elongated, spirally twisted, unicellular bacteria with thin, delicate walls; up to 500 μm long (large for a bacterium); motility by a helical wave along the cell. Some are free-living, some parasitic and pathogenic (e.g. *Treponema pallidum*, causing syphilis).

Spleen Largest mass of **lymphoid tissue**, lying in mesentery of stomach or intestine of jawed vertebrates. Unlike **lymph nodes**, perfused by blood rather than lymph. Important lymphocyte and **plasma cell** reservoir, and component of the **reticulo-endothelial system**, its cells phagocytozing worn red blood cells and platelets. A store of red blood cells, sympathetic contraction of smooth muscle squeezing them into the circulation (often giving a 'stitch') in emergency. See **haemopoiesis**.

Spliceosomes Nuclear particles formed from a complex of five small nuclear RNAs (the snRNAs U1, U2, U4, U5 and U6) and many (50–100?) proteins (snRNPs); on a par with **ribosomes** in structural complexity. The snRNAs bind in sequence along with snRNPs at an intron on the pre-mRNA (of a Class II **gene**), fold the pre-mRNA and catalyse two transesterification reactions to yield mRNA and intron products. This is ATP-dependent. There are similarities between spliceosomes and certain self-splicing introns in organelles of plants and fungi. See **introns** (for possible origin of snRNAs), **RNPs**.

Splicing See **intron**, **RNA processing**.

Split gene Gene with at least one **intron** sequence embedded within it.

Sponge See **Porifera**.

Spongy mesophyll See **mesophyll**, **leaf**.

Spontaneous generation The view that life can arise from non-life, independently of any parent. In the sense of the **origin of life**, this may be regarded as a scientifically respectable view but it has been held in the past that many individual organisms arise abiogenically from e.g. fermenting broth, rotting meat, etc. (see **biogenesis**). In the 19th century, endorsement of spontaneous generation by Lamarck and Geoffroy St Hilaire resulted in its commonly being associated in France with any evolutionary theory. See **Lamarck**, **Pasteur**, **special creationism**.

Sporangiole (Of fungi) small sporangium containing only one or a few spores (sporangiospores), and lacking a columella.

Sporangiophore (Of fungi) hypha bearing one or more sporangia; sometimes morphologically distinct from vegetative hyphae.

Sporangium Organ within which are produced asexual spores; typically in fungi and plants.

Spore A single or several-celled reproductive structure (propagule) that detaches from the parent and gives rise, directly or indirectly, to a new individual; a general term. Spores are usually microscopic, of many different types, produced in various ways. Spores can be formed asexually or sexually. They may be motile (zoospore) or non-motile (aplanospore; conidium); thin- or thick-walled, they often serve for very rapid increases in the population, as when produced in enormous numbers and distributed far and wide by wind, water, animals, etc. Others are resting spores, enabling survival through unfavourable periods. Spores occur in all plant groups, algae, fungi, bacteria, Cyanobacteria, protozoans and other protists. The term sexual spore usually indicates a spore that can engage directly in fertilization; less commonly it indicates a spore produced by meiosis or fertilization. Asexual spores, therefore, either do not engage in fertilization, or are not produced by meiosis. See **microspore**, **megaspore**.

Spore mother cell (Bot.) Diploid cell giving rise by meiosis to four haploid spores or nuclei.

Sporic meiosis Pattern of **alternation of generations** where the spores are produced by meiosis and develop into multicellular gametophytes before the gametes are produced; occurs in many algae, all bryophytes and vascular plants.

Sporocyst (1) Cyst of some **Sporozoa**. (2) Stage in life cycles of many flukes (see **Trematoda**); lacks mouth and gut; can produce daughter sporocysts or rediae (see **polyembryony**).

Sporodochium (Of fungi) a cluster of conidiophores arising from a stroma or mass of hyphae.

Sporogonium Spore-producing structure of liverworts and mosses that develops after fertilization; the sporophyte generation of these plants.

Sporophore (Of fungi) general term for a structure producing and bearing spores; e.g. a sporangiophore, conidiophore (simple sporophore), or mushroom (complex sporophore).

Sporophyll Leaf, bearing sporangia. In some plants, indistinguishable from ordinary leaves except by presence of sporangia, e.g. in bracken fern; in others, much modified and superficially quite unlike ordinary leaves, e.g. stamens and carpels of flowering plants.

Sporophyte Spore-producing diploid phase in **life cycle** of a plant (see **alternation of generations**). Arises by union of sex-cells produced by haploid gametophytes.

Sporopollenin Tough substance of which the exine (outer wall) of spores and pollen grains is composed. One of the most resistant organic substances known, not affected by hot hydrofluoric acid or concentrated alkali. It consists of complex polymers with an empirical formula $[C_{90}H_{142}O_{36}]$. Formed by oxidative polymerization of carotenoids and their esters.

Sporozoa Class of parasitic protozoans, many intracellular, some with alternate hosts. Mature stages lack locomotor organelles, but young may be amoeboid or flagellated. Large numbers of young are produced, either naked or in spores, by multiple fission after syngamy, when transmission to another host may occur. Includes *Plasmodium* (see **malaria**).

Sport (rogue) Individual exhibiting the effect of a mutation, often an unusual or rare one.

Spot desmosome See **desmosomes**.

Springtail See **Collembola**.

Squamata Order of the **Reptilia** (Subclass Lepidosauria) containing lizards (Lacertilia), snakes (Ophidia), amphisbaenids (Amphisbaenia) and the tuatara (**Rhynchocephalia**). Males generally have unique paired copulatory organs. In *lizards*, mandibles are joined at a symphysis but, as in snakes, there is a freely movable quadrate bone to which lower jaw is hinged. *Snakes* lack eardrums and movable eyelids and their eyes are covered by transparent eyelids. The jaws are exceptionally mobile and only loosely attached to the skull; mandibles are only loosely attached anteriorly by elastic ligaments and laterally by skin and muscles, with great relative mobility. Snakes feeding on prey large enough to struggle are usually constrictors (e.g. boas, pythons, anacondas) or venomous (e.g. cobras, mambas, coral and sea snakes, pit vipers and true vipers). The group is apparently undergoing rapid speciation. *Amphisbaenids* are legless burrowing forms, mostly under 1.5 m in length, with a single tooth in midline of upper jaw fitting into a space in lower jaw (effective nippers). Head shape blunt or flattened. Eardrums absent and eyes rudimentary. The Squamata have bodies covered in horny epidermal scales.

Squamosal A **membrane bone** of skull, in mammals taking over from the quadrate the articulation of lower jaw (dentary). See **ear ossicles**.

Squamous cells Animal cells, commonly forming an epithelium, which are flattened and resemble paving stones in shape. Cells of the inside of the cheek form a squamous epithelium; those lining blood vessels are squamous endothelium. See **endothelium**, **epithelium**.

Squamulose Lichen growth form which is similar to the **foliose** form, but with numerous small, loosely-attached thallus lobes or *squamules*.

***Sry* genes** See **testis**.

Staining Treatment of biological material with chemicals (dyes, stains) that only colour specific organelles or parts of a structure, thus providing contrast, as between nucleus and cytoplasm, mitochondria and other organelles, or between cell wall and cytoplasm. For use in light **microscopy**, most stains are organic compounds (dyes) comprising a negative and positive ion. In acid stains, the colour arises from the organic anion; in basic stains it arises from the organic cation, while neutral stains are mixtures of both acid and basic stains. Stains are applied to a biological material by its immersion in a stain solution. *Vital staining* refers to staining of living tissue when no damage to the tissue occurs (e.g. neutral red stains vacuoles and granules in many cells; Janus green stains mitochondria specifically). *Non-vital staining* refers to the staining of dead tissue. Various modified techniques are required for staining different types of tissues for light microscopy; e.g. *counterstaining* (double staining), where two stains are used in sequence so as to stain different parts of the specimen. See **dehydration**, **fixation**.

Staining techniques may give quantitative data, as when the amount of stain taken up is proportional to the amount of stained component, and methods exist (microspectrophotometry) to measure through the microscope the amount of stain present at a given site within a cell. When the structure stained takes on a colour different from that usually produced by the stain (or dye), the staining is said to be metachromatic.

Stains in electron microscopy are not the coloured stains of light microscopy; rather they contain heavy metal atoms (e.g. lead, uranium) in forms that combine with chemical groups characteristic of specific structures in the cell. Presence of such atoms permits fewer electrons to pass through the specimen, so that an image is produced. 'Stained' (electron-dense) structures appear darker than their surroundings. *Negative staining* for electron microscopy is used to examine three-dimensional and surface aspects of cell structure. Specimens are not sectioned; rather they are placed directly on a thin plastic film, covered with a drop of solution containing heavy metal atoms and allowed to dry, leaving the specimen with a thin layer of electron-dense material.

Stamen Organ of **flower** which forms microspores (shed after development as pollen grains); a microsporophyll; comprises stalk or filament bearing anther at apex. Anther comprises two lobes united by a prolongation of the filament connective, and in each lobe there are two pollen sacs (microsporangia) producing pollen.

Staminate Flower which has stamens but no functional carpels, i.e. is male. See **pistillate**.

Staminode Sterile stamen; one that does not produce pollen.

Standard free energy See **thermodynamics**.

Standing crop Biomass per unit area (or per unit volume) at any one time. Not equivalent to biomass productivity, which is a *rate* measure. Standing crop values are often given in terms of energy content and commonly relate to populations or trophic levels.

Stapes Stirrup-shaped mammalian **ear ossicle**, representing columella auris of other tetrapods and the hyomandibular of fish. See **columella**.

Starch A complex insoluble polysaccharide carbohydrate of green plants, one of their principal energy ('food') reserve materials. Formed by polymerization (condensation) of several hundred glucose subunits ($C_6H_{10}O_5$) and easily broken down enzymatically into glucose monomers (see **amylase**). Comprises two main components: *amylose* and *amylopectin*. The former consists of straight chains of α[1,4]-linked glycosyl residues and stains blue-black with iodine/KI solution; the latter contains in addition some α[1,6] branches in its molecules and stains red with iodine/KI solution. Found in colourless plastids (**leucoplasts**) in storage tissue and in the stroma of **chloroplasts** in many plants. Formed into grains, laid down in a series of concentric layers. See **dextrin**, **statolith**.

Starch sheath Innermost layer of cells of cortex of young flowering plant stems, containing abundant and large starch grains; considered homologous with **endodermis**, it may sometimes lose starch and become thickened as an endodermis at later stage.

Starfish See **Asteroidea**.

Start codon See **protein synthesis**, **genetic code**.

Statoblasts Resistant internal buds with chitinous shells produced by some

Ectoprocta asexually and capable of withstanding unfavourable conditions, as during winter. They break open in spring to produce new colonies.

Statocyst (otocyst) Mechanoreceptor and/or position receptor evolved independently by several invertebrate groups (Cnidaria, Platyhelminthes, Crustacea) and vertebrates (see **macula**). Typically a fluid-filled vesicle containing granules of lime, sand, etc. (*statoliths*), which impinge upon specialized setae (*statolith hairs*) and stimulate sensory cells as the animal moves. Resulting nerve impulses initiate reflexes which often serve to right the animal after it has been turned upside down. See **statolith**.

Statocyte Plant cell containing one or more **statoliths**.

Statolith Gravity sensor in plant cells. Perception of gravity probably involves the sedimentation of amyloplasts (starch-containing plastids) within specialized cells of the shoot and root. Such cells are to be found in cells contiguous to or surrounding the vascular bundles all along the shoot. In roots, in contrast, they are localized in the root cap. How the gravity sensors, and their movement, is translated into hormonal gradients is not fully understood, although calcium may play a key role in roots through its control of **growth substance** transport.

Statospore (stomatocyst) Resting spore charcteristic of the **Chrysophyta**, produced asexually or sexually, and characterized by its siliceous wall, which may or may not be ornamented. Similar spores are produced by some members of the **Xanthophyta**.

Stele (vascular cylinder) (Bot.) Cylinder or core of vascular tissue in centre of roots and stems, comprising xylem, phloem, pericycle, and in some steles, pith and medullary rays; surrounded by endodermis. Structure of the stele differs in different groups of plants. See **dictyostele**, **protostele**, **siphonostele**, **vascular bundle**.

Stem Normally aerial part of axis of vascular plants, bearing leaves and buds at definite positions (nodes), and reproductive structures, e.g. flowers. Some are subterranean (e.g. rhizomes) but these, like all stems, are distinguished externally from roots by the occurrence of leaves (scale leaves on rhizomes) with buds in their axils, and internally by having vascular bundles arranged in a ring forming a hollow cylinder, or scattered throughout tissue of the stem, with the protoxylem most commonly endarch.

Stenohaline Unable to tolerate wide variations in environmental salinity.

Stenopodium See **biramous appendage**.

Stenothermous (-thermic) Unable to tolerate wide variations in environmental temperature. Compare **eurythermous**.

Stereocilium Specialized microvillus. See **hair cell**.

Sterigma (Of fungi) a minute stalk bearing a spore or chain of spores.

Sterile (1) Unable to produce viable gametes and/or sexual offspring, unlike normal individuals. (2) Free from micro-organisms. See **antiseptic**, **autoclave**, **disinfectant**.

Sternum (1) Breast bone. Tetrapod bone lying ventrally in mid-chest to which ventral ends of most ribs are attached. Attached anteriorly to **pectoral girdle**. (2) Cuticle on ventral side of each segment of an arthropod, often forming a thickened plate. See **tergum**.

Steroids Chemically similar but biologically diverse group of **lipids** originating from *squalene*. Include bile acids, vitamin D, adrenal cortex and gonadal hormones, active components of toad poisons and digitalis. Saturated hydrocarbons, with seventeen carbon atoms in a system of rings, three 6-membered and one 5-membered, condensed together. *Sterols* (e.g. **cholesterol** and ergosterol) form a large steroid subgroup, having a hydroxyl group at C_3 and an aliphatic chain of eight or more carbon atoms at C_{17}.

Sterols See **steroids**.

STH Somatotropic hormone. See **growth hormone**.

Stick insects See **Phasmida**.

Sticky ends See **DNA ligase**, **teleomere**.

Stigma (Bot.) (1) Terminal portion of the style; surface of the carpel which receives the pollen. (2) See **eyespot**. (Zool.) Rare alternative name for insect **spiracle**.

Stimulus Any change in the internal or external environment of an organism intense enough to evoke a response from it without providing the energy for that response. See **irritability**, **adaptation**, **habituation**.

Stipe Stalk. (1) Of fruit bodies of certain fungi, e.g. **Basidiomycota**. (2) Of the thallus of seaweeds (e.g. *Laminaria*), the organ connecting the holdfast system and the photosynthetic laminae or blades.

Stipule Small, usually leaf-like, appendage found one on either side of leaf stalk in many plants, protecting axillary bud; often photosynthetic.

Stock Part of plant, usually comprising the root system together with a larger or smaller part of the stem, on to which is grafted a part of another plant (the *scion*). See **graft**.

Stolon Stem growing horizontally along the ground, rooting at nodes (e.g. strawberry runner).

Stoma (pl. stomata) (Bot.) An opening or pore in the plant epidermis, regulated by specialized cells called **guard cells**; especially in the leaves, and through which gaseous (including water vapour) exchange occurs. Stomata are often associated with epidermal cells that differ in shape from the ordinary epidermal cells (termed subsidiary or accessory cells). Stomata may occur upon both sides of a leaf, but usually they are more numerous on the undersurface. They may occur only on the upper leaf surface of aquatic flowering plants, whose leaves float upon the water surface; immersed leaves usually lack stomata entirely. Leaves of **Xerophytes** tend to have a larger number of stomata compared to other plants, allowing for a higher rate of gaseous exchange during the infrequent periods of favourable water supply. However, these stomata are sunken (sometimes hairy) depressions on the lower leaf surface – adaptations to reduce water loss.

Dicotyledon leaves generally have a scattered arrangement of stomata, whereas the stomata in monocotyledons are arranged in rows parallel with the longitudinal axis of the leaf. See **radial micellation**, **transpiration**.

Stomach Enlargement of anterior region of **alimentary canal**. In vertebrates it follows oesophagus and usually has thick walls of **smooth muscle** to churn food and lining mucosal cells secreting mucus, pepsinogen and hydrochloric acid (see **gastric**). Cardiac and pyloric sphincters can close the ends during churning. See **gastrin**, **rennin**, **ruminant**.

Stomium Place in wall of fern sporangium where rupture occurs at maturity, releasing spores.

Stomodaeum Intucking of ectoderm meeting endoderm of anterior part of **alimentary canal**, forming mouth.

Stone Age Term denoting **Palaeolithic** and **Neolithic**.

Stone cell Type of sclereid. See **sclerenchyma**.

Stoneflies See **Plecoptera**.

Stop codon See **protein synthesis**, **genetic code**.

Stratum corneum Outer layer of epidermis of vertebrate skin. Cells undergo **cornification** (keratinization) and die, becoming worn off.

Strepsiptera (stylopids) Small endopterygote insects; females endoparasitic; males free-living, short-lived, with large metathorax, anterior wings haltere-like, hind wings large and fan-shaped. Females degenerate, apodous and larviform, enclosed in persistent larval cuticle. Several forms parasitize hymenopterans. Larvae emerge from host and probably find new hosts by waiting on flowers or through contact in nest. Probable affinities with the Coleoptera.

Streptococcus Genus of non-spore-producing, Gram-positive bacterium, forming long chains. Many are harmless colonizers of milk; some commensal in the vertebrate gut; but the *pyogenic* group are human pathogens, some producing haemolysins, destroying erythrocytes. The *viridans* streptococcal group lives usually non-pathogenically in the upper respiratory tract, but can cause serious infections (some chronic), and may cause bacterial arthritis.

Streptomycin **Antibiotic** inhibiting translation of mRNA on prokaryotic, but not eukaryotic, ribosomes and can be used to distinguish these translation sites. Streptomycin resistance is conferred upon prokaryotes by plasmid-borne transposon. See **antibiotic resistance element**, **chloramphenicol**.

Stretch receptor See **muscle spindle**, **proprioceptor**.

Striated muscle (skeletal/voluntary/striped muscle) Contractile tissue, consisting in vertebrates of large elongated muscle fibres formed by fusion of **myoblasts** to form syncytia. The cytoplasm of each fibre is highly organized, producing conspicuous striations at right angles to its long axis, and contains numerous longitudinal fibrils (*myofibrils*), each with alternating bands (A, anisotropic, I isotropic), H zones and Z discs caused by distributions of **actin** and **myosin** myofilaments and of

α-actinin (see Fig. 107). The cross-striations of a whole fibre result from similar bands lying side by side. Each fibre is bounded by a *sarcolemma* (plasmalemma and basement membrane), the plasmalemma of which is deeply invaginated into the fibre forming *transverse tubules* (T-system), generally between the Z discs and H zones in insects but over the Z discs in vertebrates. These bring membrane depolarizations right into the fibre, ensuring uniform contraction. Mitochondria abound between myofibrils. The endoplasmic reticulum is modified to form a confluent system of sacs (*sarcoplasmic reticulum*) controlling calcium ion concentration.

On stimulation, a striated muscle fibre contracts by shortening and thickening. Fibres are bound together by connective tissue to form muscle tissue, and bring about locomotion by moving the skeleton, to which they are attached in vertebrates by tendons. See **muscle contraction**, **cardiac muscle**, **smooth muscle**, **neuromuscular junction**.

string* gene** A gene in *Drosophila* whose product is homologous to yeast cdc2 protein (see ***cdc genes) and positively regulates **maturation promotion factor**. Translated from maternal mRNA stored in the oocyte for the first thirteen cell division cycles of the embryo, after which maternal mRNA is degraded and nuclear *string* mRNA is produced.

Strobilation Process of transverse fission which produces proglottides from behind the scolex of a tapeworm and ephyra larvae from the jellyfish scyphistoma (Scyphozoa). Regarded as a method of asexual reproduction in the latter, but less commonly in the former. The whole ribbon-like chain of tapeworm proglottides may be referred to as a strobila.

Strobilus (Bot.) Cone. Reproductive structure comprising several modified leaves (sporophylls), or ovule-bearing scales, grouped terminally on a stem.

Stroma (Bot.) (1) Tissue-like mass of fungal hyphae, in or from which fruit bodies are produced. (2) Colourless matrix of the **chloroplast**, in which grana are embedded. (Zool.) Intercellular material (matrix), or connective tissue component of an animal organ.

Stromatolites Macroscopic structures produced by certain blue-green algae (**Cyanobacteria**) where there is deposition of carbonates along with trapping and binding of sediments. Predominantly hemispherical in shape, they possess fine concentric laminations produced by growth responses to regular (often daily) environmental change. Fossil stromatolites occur from early Precambrian (more than 3000 Myr BP) to the Recent period.

Style Slender column of tissue arising from top of ovary and through which pollen tube grows.

Subarachnoid space Area between the arachnoid and the pia mater, filled with cerebrospinal fluid. See **meninges**.

Subcutaneous Immediately below dermis of vertebrate skin (i.e. the *hypodermis*). Such tissue is usually loose connective tissue, blood vessels and nerves and generally contains fat cells (see **adipose tissue**). In many tetrapods, also includes a sheet of striated muscle (*panniculus carnosus*) to move skin or scales.

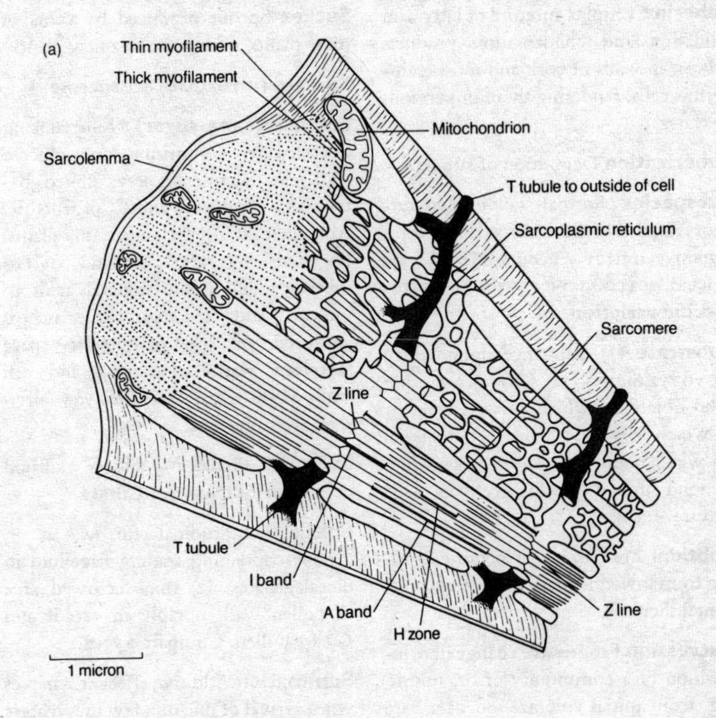

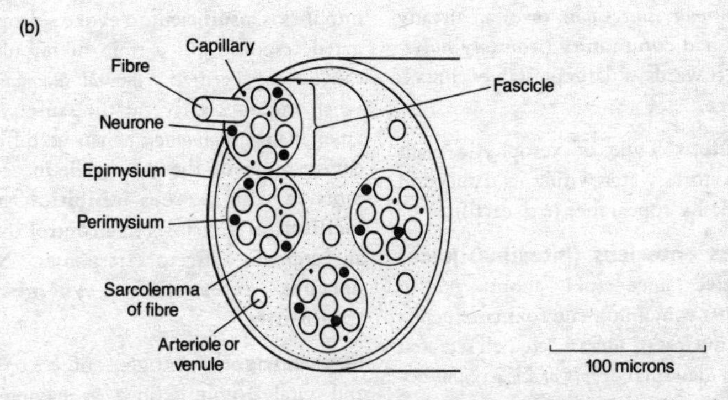

Fig. 107 (a) *Diagram of the ultrastructure of a vertebrate* **striated muscle** *fibre showing a complete sarcomere and two adjacent parts of sarcomeres. Several such fibres together, with appropriate connective tissue, form a muscle fascicle several of which in turn comprise a striated muscle as shown in* (b). (b) *Four striated muscle fascicles and surrounding connective tissue forming a small muscle.*

Suberin Complex mixture of fatty acid oxidation and condensation products present in walls of cork and most endodermis cells, rendering them impervious to water.

Suberization Deposition of **suberin**.

Subspecies Formal taxonomic category used to denote the various forms (types), usually geographically restricted, of a polytypic species. See **infraspecific variation**.

Substrate (1) Substance upon which an **enzyme** acts. (2) Ground or other solid object on which animals walk or to which they are attached. (3) Material on which a microorganism is growing, or solid surface to which cells in tissue culture attach.

Subtidal Zone in sea or ocean extending from low-tide mark to edge of continental shelf.

Succession Progressive change in composition of a community of organisms, e.g. from initial colonization of a bare area (*primary succession*), or of an already established community (*secondary succession*), towards a largely stable climax. See **sere**.

Succulent Type of xerophytic plant which stores water within its tissues and has a fleshy appearance (e.g. cacti).

Succus entericus (intestinal juice) Digestive juice (pH about 7.6 in humans) containing enterokinase, peptidases, nucleases, sucrase, etc., all secreted by the glandular *crypts of Lieberkuhn* between intestinal villi. Completes hydrolysis of food molecules begun higher in the gut. About 2–3 litres per day secreted in humans. See **digestion**.

Sucker Sprout produced by roots of some plants, giving rise to a new plant.

Sucrase (invertase) See **sucrose**.

Sucrose (cane sugar) Non-reducing disaccharide, comprising one glucose and one fructose moiety, linked between C_1 of glucose and C_2 of fructose. Abundant transport sugar in plants. Digested by the enzyme *sucrase* (*invertase*) and dilute mineral acids to glucose and fructose: invertase is anchored in the periplasmic space (between plasm membrane and cell wall) of the yeast. *Succharomyces cereviseae*. See **invertase**.

Suctoria Predatory ciliates, ciliated only in larval stage. See **Ciliata**.

Sulcus Longitudinal furrow, as in groove containing trailing flagellum in dinoflagellates; (2) thin furrowed area of pollen wall, notably in cycads and *Gindgo* pollen. Compare **gyrus**.

Summation Additive effect at synapses when arrival of one or a few presynaptic impulses is insufficient to evoke a propagated response but a train of impulses can do so. Termed *temporal summation* when impulses arrive at the same synapse, *spatial summation* when at different synapses of the same cell. In conjunction with nervous **inhibition** and **facilitation** it enables fine control over an animal's effector responses. See **nervous integration, synergism, transmitter**.

Superantigens Antigens of bacterial and viral origin defined as having a potent stimulatory effect on **T cells**. Unlike normal antigens, they are not processed and associate with MHC Class II molecules outside the peptide binding groove and recognition is not MHC-restricted. See Figure 110.

Supergene Group of gene loci with mutually reinforcing effects upon phenotype that have come (through selection) to lie on the same chromosome, increasingly tightly linked so as to be inherited as a block (e.g. supergenes for the heterostyle or homostyle system in primroses (*Primula*) and shell colour and banding pattern in the snail (*Cepaea*). See **polymorphism**.

Superior ovary See **receptacle**.

Supernumerary chromosome (accessory chromosome) Chromosome additional to normal karyotype of the species (B chromosome in botany). Either not homologous, or only partially homologous, with members of normal karyotype. In some populations of a species most of the individuals carry such chromosomes; in others, frequency is low. Most are heterochromatic. Their presence does not seem to affect markedly the individual's appearance. Geographical distribution often non-random.

Superspecies Informal taxonomic category usually applied to allopatric arrays of species where evidence suggests common ancestry and where the species are sufficiently similar. Such species complexes tend to be discoverable with difficulty due to cryptic distinguishing characteristics between the species (usually sibling species), such as *Anopheles gambiae* and *Stimulium damnosum* complexes. See **DNA probe**.

Supination See **pronation**.

Suppressor mutation (1) (*Intragenic.*) A mutation (nucleotide addition or deletion) at a site in a chromosome sufficiently close to a prior nucleotide deletion or addition to restore the reading frame and thus suppress the effect of the original mutation. (2) (*Intergenic.*) A mutation at one locus which prevents the expression of a mutation at another locus. (3) (Uncommonly) a mutation preventing local or complete crossing-over in meiotic cells.

Some *suppressor genes* seem to be responsible for preventing oncogenic phenotype in normal cells; their loss may result in **oncogene** expression. See **aberrant chromosome behaviour** (1), **mutation, hypostasis**.

Suprarenal gland See **adrenal gland**.

Survival value Characters and genes are said to have positive survival value if they increase in individual's **fitness**, and (less commonly) negative survival value if they decrease it. Neutral survival value is attributable to characters and genes with no effect on fitness. See **natural selection**.

Suture Area of fusion of two adjacent structures. (1) (In flowering plants) the line of fusion of edges of carpel is known as a ventral suture. Mid-rib of carpel is known as the dorsal suture, not implying any fusion of parts to form it but to distinguish it from the ventral (true) suture. (2) Junction between the irregular interlocking edges of adjacent skull bones, or between plates of hardened cuticle of exoskeleton. Suture-lines occur on shells of ammonites, marking edge of a septum with side-wall of shell. (3) (Surgical) to sew a wound together.

SV40 See **virus**.

Svedberg unit The unit (S), related to sedimentation coefficient, used to indicate the time taken for large molecules, small organelles, etc., to reach an equilibrium level when ultracentrifuged in water at 20°C. The unit (S) is defined as a sedimentation coefficient of 1×10^{-13} seconds. Sedimentation coefficient is given by the equation:

$$s = \frac{dx/dt}{\omega^2 x}$$

where x = the distance of the sedimenting boundary from the centre of rotation (cm), t = time (s) and ω = angular velocity (radian s^{-1}). The S-value of an object is termed its *Svedberg number*. See **ribosomes**.

Swarm spore See **zoospore**.

Sweat Aqueous secretion of mammalian **sweat glands** containing solutes in lower concentrations than in blood plasma, to which it is hypotonic. In humans, contains 0.1–0.4 % sodium chloride, sodium lactate and urea; about one litre is lost per day in temperate climates (up to 12 litres in hot dry conditions when salt and water supply permit). The urea and lactate may be regarded as excretory, but may also reduce colonization by skin microflora. Its production, along with vasodilation of skin capillaries, removes body heat as latent heat of vapourization of water. See **homeothermy**.

Sweat glands Epidermal glands, projecting into mammalian dermis, releasing **sweat** when stimulated by sympathetic nervous system. (1) *Apocrine sweat glands* are simple, branched tubular glands of (mainly) armpits, pubic region and areolae of breasts. Secretion more viscous than from eccrine sweat glands. (2) *Eccrine sweat glands* are simple, coiled tubular glands. Widely distributed; rows of them open via pores between the papillary ridges (responsible for fingerprints) on hands and feet of many mammals. Absent from cats and dogs except between toes. See **mammary glands**.

Swim bladder See **gas bladder**.

Symbiont A symbiotic organism. See **symbiosis**, **mycobiont**, **phycobiont**.

Symbiosis The living together in permanent or prolonged close association of members (*symbionts*) of usually two different species, with beneficial or deleterious consequences for at least one of the parties. Included here are: *commensalism*, where one party (the commensal) gains some benefit (often surplus food) while the other (the host) suffers no serious disadvantage; *inquilinism*, where one party shares the nest or home of the other, without significant disadvantage to the 'owner'; *mutualism*, where members of two different species benefit and neither suffers (symbiosis in a restricted sense) and *parasitism*, where one party gains considerably at the other's expense (see **parasite**). Some include intraspecific relationships within symbiosis.

Symmetrodonta Order of extinct mammals (Infraclass Trituberculata). Small Mesozoic forms; insectivorous. Possibly ancestral to **Pantotheria**.

Sympathetic nervous system See **autonomic nervous system**.

Sympatric Of populations of two or more species, whose geographical ranges or distributions coincide or overlap.

From an ecological and genetic viewpoint, often more valuable to distinguish between sympatry as defined above and *effective* sympatry. Where adults are settled in different geographical regions but gametes, developmental stages, etc., are widely dispersed, there may be effective sympatry despite adult distribution. In lice, several species may inhabit the same host but settle in well-defined and species-specific anatomical regions. They are sympatric, yet **allotopic**. See **allopatric, syntopic, speciation**.

Sympetalous See **gamopetalous**.

Symphysis Type of joint allowing only slight movement, in which surfaces of two articulating bones (both covered by layer of smooth cartilage) are closely tied by collagen fibres. Symphyses occur between centra of vertebral column. For *pubic symphysis*, see **pelvic girdle**.

Symplast Interconnected protoplasts and their plasmodesmata which effectively result in the cells of different plant organs forming a continuum (e.g. from root hairs to stele). See **translocation**.

Symplesiomorphy Any character which is a **shared homologue** of two or more taxa, but which is thought to have occurred as an evolutionary novelty in an ancestor earlier than their earliest common ancestor. Compare **synapomorphy**. See **plesiomorphic, cladistics**.

Sympodium Composite axis produced, and increasing in length, by successive development of lateral buds just behind the apex. Compare **monopodium**.

Symport See **transport proteins**.

Synangium Compound structure formed by lateral union of sporangia; present in some ferns.

Synapomorphy Term used in phylogenetics (see **cladistics**) to denote what are otherwise termed *homologous* characters in many writings on evolution. Any character which is a **shared homologue** of two or more taxa and is thought to have originated in their closest common ancestor and not in an earlier one. Synapomorphy has been used in a more inclusive sense than has homology, characters involved including, e.g., geographical ranges. See **homology, symplesiomorphy**.

Synapse Region of functional contact between one neurone and another, or between it and its effector. Commonly a gap (*synaptic cleft*) of at least 15 nm occurs between the apposed plasma membranes, the direction of impulses defining *presynaptic* and *postsynaptic* membranes. An individual neurone commonly receives 1000–10,000 synaptic contacts with 1000 or more other neurones, most synapses being between axon terminals of the stimulating neurone and dendrites of the receiving neurone, but axon–axon and dendrite–dendrite synapses occur. Most synapses are chemical (humoral), involving release of neurotransmitter via *synaptic vesicles* in response to Ca^{2+} entry at the presynaptic membrane: the more Ca^{2+} entry, the more transmitter release (see **acetylcholine, impulse**); others are electrical, impulses passing without delay from one neurone to another via gap junctions (see **intercellular junction**). Chemical synapses, although encountering greater resistance due to diffusion time of transmitter, do permit **nervous integration**. Most psychoactive drugs exert their effects at synapses, many binding to specific receptors. $GABA_A$ receptors mediate rapid **inhibition** and are ligand-gated chloride channels acted on by GABA, by benzodiazapine tranquillizers (e.g. Valium, Librium) and by barbiturates – each binding a different site on the same receptor protein and apparently operating by allowing reduced amounts of GABA to open the channel (see **potentiation**). See **summation**, **neuromuscular junction**.

Synapsis Pairing-up of chromosomes. See **meiosis**.

Synaptic vesicle Vesicle produced by the Golgi apparatus of a nerve axon, in which neurotransmitter is stored prior to release into synaptic cleft. Bud off endocytotically again and re-fuse with the Golgi apparatus after release of transmitter. See **coated vesicles**.

Synaptonemal complex Ladder-like (zip-like) protein complex holding chromosomes together during synapsis of **meiosis**, chromosome loops pointing away from it. Within it lie *recombination nodules*, apparently determining where **chiasma** formation occurs. The complex dissolves at diplotene.

Syncarpous (Of the gynoecium of a flowering plant) with united carpels (e.g. tulip). See **flower**.

Synchronous culture Culture (of microorganisms or tissue cells) in which, through suitable treatment, all cells are at any one time at approximately the same stage of development, or of the **cell cycle**. In mammalian tissue culture, cells in mitosis (M phase) round up and can be separated from others by gentle agitation to start a synchronous culture. They will rapidly enter G I.

Syncytium Animal tissue formed by fusion of cells, commonly during embryogenesis, to form multinucleate masses of protoplasm. May form a sheet (as in mammalian **trophoblast**) or cylinder (**striated muscle**), or network of more or less discrete cells linked by intercellular bridges (as in mammalian spermatogenesis; see **maturation of germ cells**). Smooth muscle is occasionally syncytial. See **acellular**, **symplast**, **domain**.

Synecology Ecology of communities as opposed to individual species (autecology).

Synergidae (synergids) Two short-lived cells lying close to the egg in mature **embryo sac** of flowering plant ovule.

Synergism Interaction of two or more agencies (e.g. hormones, drugs) each influencing a process in the same direction. Their combined effects are either greater than their separate effects added (*potentiation*) or roughly the sum of their separate effects (*summation*). Compare **antagonism**.

Syngamy See **fertilization**.

Syngeneic Strains of organisms produced by repeated inbreeding, each pair of autosomes being identical.

Syngenesious (Of stamens) united by their anthers; e.g. Compositae.

Syngraft See **isograft**.

Synovial membrane Connective tissue membrane forming a bag (*synovial sac*) enclosing a freely movable joint, e.g. elbow joint, being attached to the bones on either side of the joint. Bag is filled with viscous fluid (*synovial fluid*) containing glycoprotein, lubricating the smooth cartilage surfaces making contact between the two bones.

Syntopic Two or more organisms are syntopic if they share the same habitats (or microhabitats) within the same geographical range. They are *microsympatric*. This may be the expression of different phenotypes within a single species. See **sympatric**.

Syntype Each of several specimens used to describe a new species when a single type specimen was not chosen. See **holotype**.

Syrinx Sound-producing organ of birds, containing typically a resonating chamber with elastic vibrating membranes of connective tissue (**vocal cords**); situated at point where trachea splits into bronchi. Compare very different **larynx** of mammals, which in birds lacks vocal cords.

System (organ system) Integrated group of **organs**, performing one or more unified functions, e.g. **nervous system**, **vascular system**, **endocrine system**.

Systematics Term often used as synonym of **taxonomy**, but sometimes used more widely to include also identification, practice of classification, nomenclature. See **biosystematics**, **classification**.

Systemic Generally distributed throughout an organism.

Systemic arch Fourth **aortic arch** of tetrapod embryo, becoming in adult main blood supply for the body other than the head. In Amphibia both left and right arches persist in the adult; in birds only the right; in mammals only the left (the aorta).

Systole (1) Phase of **heart cycle** when heart muscle contracts. (2) Phase of contraction of contractile vacuole.

Tachycardia Abnormally rapid heart rate. See **beta-blocker**.

Tactile (Adj.) Referring to the sense of touch. Sense organs include many surface pressure receptors.

Tadpole Term given to aquatic larval stages of tunicates (**Urochordata**) and of most modern amphibians. Metamorphoses into the adult.

Tagma (pl. tagmata) Functional and anatomical regionalizations of the arthropod body associated with division of labour; thus appendages with similar functions tend to be grouped on adjacent segments. Tagmata include *head*, *thorax* and *abdomen*, but each class within the phylum has a characteristic pattern of *tagmosis*. Insects have distinct head, thorax and abdomen; crustaceans generally lack a clear head–thorax distinction; arachnids have **prosoma** and **opisthosoma**, etc. Tagmosis in post-head region is closely related to mode of locomotion. See **Arthropoda**.

Tagmosis See **tagma**.

Taiga The northern coniferous forest or boreal forest that extends over vast areas of Russia, Scandinavia and North America. Taiga is the Russian word for vegetation found in this **biome**. This forest is primarily evergreen, except in large areas of northeastern Siberia, where larches (*Larix*), which are deciduous conifers, are dominant. Common trees of the taiga include *Picea* (spruce), *Abies* (fir), *Populus* (poplar) as well as *Larix*; shrubs include *Salix* (willows), *Betula* (birch) and *Ledum* (Labrador tea). The members of all these genera of trees and shrubs are ectomycorrhizal. Herbaceous vegetation is dominated by mosses and lichens. Flanked by montane forests, savannas or grassland to the south, grading unevenly into tundra to the north. The northern limits are determined by the severity of the arctic climate. Characterized by severe winters, with most precipitation falling during the summer. Evaporation rates are low, consequently lakes, bogs and marshes are common. More than 75% of the northern taiga is underlain by permafrost (permanent ice), usually within less than a metre of the surface, thus even though precipitation is low, the ground is usually moist, since the water cannot percolate through the soil. Fire occurrence is common and results in generally warmer, more productive areas for 10–20 years. Soils are generally highly acidic and low in nutrients.

Tangential section Longitudinal section cut at right angles to radius of cylindrical structure (e.g. root or stem).

Tannins Group of complex astringent substances occurring widely in plants, dissolved in vacuoles. Particularly common in tree bark, unripe fruits, leaves and galls; also found in animals (e.g. insect cuticles). Include phenols and hydroxy acids or glycosides. Some deter herbivores. Employed in the tanning process, which is usually enzymatic (see **cuticle**). Commercial leather production involves cross-linking collagen molecules in skins by treatment with benzoquinone, and a similar process occurs naturally.

Tapetum (tapetum lucidum) (Zool.) Reflecting layer of vertebrate choroid (retina in some teleosts), especially in nocturnal forms and deepwater fish. Generally contains guanine crystals reflecting light back through the retina, increasing its stimulatory effect but reducing visual acuity. The pigment melanin is commonly expanded over the crystals in bright light. See **chromatophore**. (Bot.) In vascular plants, a layer of cells, rich in reserve food, surrounding a group of spore mother cells; e.g. in fern sporangium, pollen sacs of the anther. Gradually disintegrate and liberate their contents, which are absorbed by developing spores.

Tapeworm See **Cestoda**.

Tap root Primary root of a plant, formed in direct continuation with root tip or radicle and forming prominent main root, directed vertically downward, bearing smaller lateral roots; e.g. dandelion. Sometimes swollen with food reserves; e.g. carrot, parsnip. See **fibrous root**.

Tardigrada Order of minute aquatic arthropods of uncertain status, but unique among these in having a terminal mouth. The four pairs of stumpy appendages are simple and lobopod (resembling **Onychophora**), ending in claws. Clear circulatory and respiratory systems lacking. Most pierce and suck plant juices. Resistant to desiccation, living among mosses, etc.

Tarsal bones (tarsals) Bones of proximal part of tetrapod hind-foot (roughly the ankle). Primitively 10–12 bones in a compact group, reduced during evolution by fusion and/or loss. There are seven in humans, one (calcaneum) forming the heel. They articulate proximally with tibia and fibula, distally with metatarsals. See **pentadactyl limb**.

Tarsus (1) Region of tetrapod hind-foot containing **tarsal bones** (approx. the ankle). Compare **carpus**. (2) A segment (fifth from the base) of an insect leg.

Taste bud Vertebrate taste receptor, consisting of group of sensory cells, usually located on a papilla of the **tongue**, but in aquatic forms often widely scattered (especially in barbels around fish mouths). Buds are often (e.g. in humans) localized in regions sensitive to salt, sweet, bitter and acid (sour) tastes; other taste modalities are detected in fish.

TATA box Short nucleotide consensus sequence in eukaryote **promoter** sequences bound by RNA polymerases II & III, about 25–30 base pairs upstream of transcribed sequence (hence -25 to -30). An AT-rich region, commonly including the interrupted sequence TATAT ... AAT ... A. Recognized and bound by TATA-binding protein, TBP (see **TATA factor**). See **Pribnow box**, **transcription factors**, Fig. 113.

TATA factor (TATA-binding protein, TBP, transcription factor IID) Eukaryotic protein with key role in **promoter** recognition by RNA polymerase II, where its binding to the **TATA box** is the first step in assembly of the multiprotein transcrip-

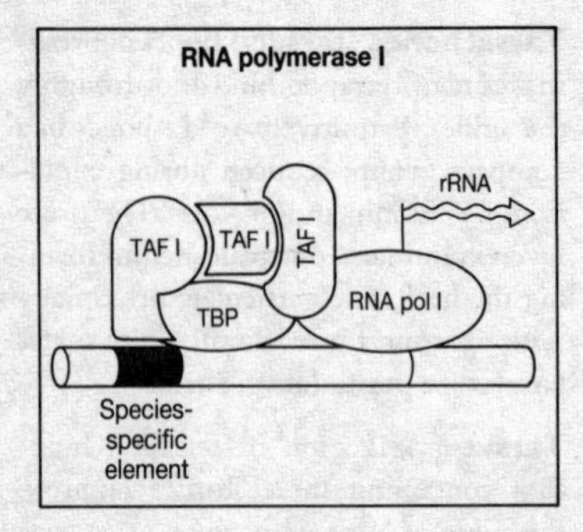

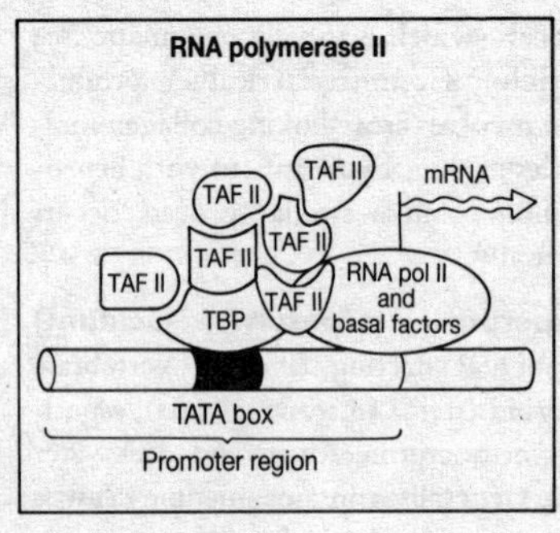

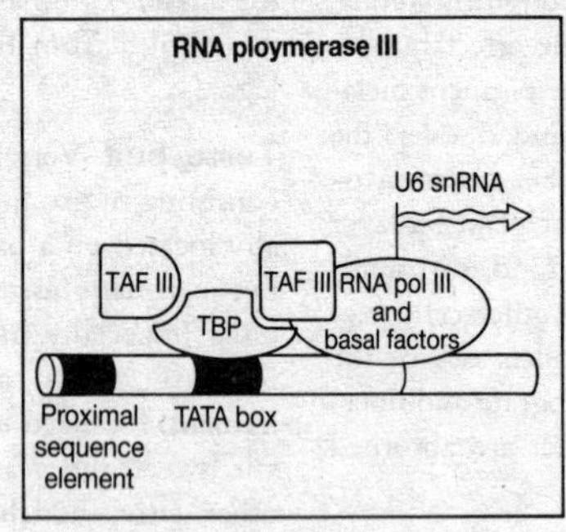

Fig. 108 *Distinct multiprotein complexes with a common subunit, the TATA-binding protein (TBP), participate in specific promoter recognition by the three eukaryotic RNA polymerases. For clarity, only a single type of minimal promoter is shown for each RNA polymerase.*

tion initiation complex (see Fig. 108). Also required for specific transcription by RNA polymerases I and III, even when promoter regions lack a TATA box. Its C-terminal 180 amino acids are thought to have some sequence homology to prokaryotic **sigma factors**. See **transcription factors**.

Taxis Directional locomotory response, as opposed to growth, to external stimulus, such as temperature gradient (thermotaxis), light (phototaxis), aerial or dissolved substances (**chemotaxis**). Movement into increasing stimulus intensity is a *positive* response; movement into decreasing stimulus intensity is a *negative* response. Compare **tropism**.

Taxon The organisms comprising a particular taxonomic entity, e.g. a particular class, family or genus. Members of a *particular* species form a taxon, but the **taxonomic category** species does not.

Taxonomic category A category, formal or informal, used in **classification**. Formal categories include Kingdom, Phylum (Division), Class, Order,

Family, Genus and Species. Informal categories include superspecies and race. Instances of these (e.g. Phylum Arthropoda) are taxa, not taxonomic categories. See **infraspecific variation**.

Taxonomy Theory and practice of **classification**. Classical taxonomy is concerned with morphology (including cytological, biochemical, behavioural), and may involve weighting phenetic characters in some scheme of relative taxonomic importance or value. *Numerical taxonomy* dispenses with such weighting and involves computerized analysis of data obtained from observing whether or not organisms being compared have or do not have any of the 'unit characters' involved in the comparison. Unit characters have an 'all-or-none' nature, being either present or absent. The data tend to arrange themselves into sets or phenons, which can then be organized into a **dendrogram**. Relationships between organisms which can be thus evaluated are termed *phenetic* if determined by overall morphological similarity between organisms, or *cladistic* if they depend upon community of descent. In *experimental taxonomy*, breeding work and field experiments may be used to clarify the taxa to which organisms belong.

T cell (T lymphocyte) Lymphocytes which travel from the bone marrow via the blood and enter the thymus, after which they enter the circulation again and settle in spleen and lymph nodes. Unlike **B cells**, T cells do not recognize native antigen conformation directly, but only in association with self-antigens of the **major histocompatibility complex** (see also **antigen-presenting cells**). There are two major subsets of T cell: those expressing the **accessory molecule** CD4 and those expressing CD8. The former are mainly *T helper cells* (T_H); the latter are generally effector cells recognizing and destroying infected cells (i.e. cytotoxic (T_C) cells). T_H cells recognize a specific MHC-antigen complex on the surface of a B cell and then induce its maturation and proliferation into specific antibody-secreting (plasma) cells (see **HIV**). While in the thymus a T cell 'learns' during *T cell maturation* both to treat its body's Class I molecules as 'self-antigens' and to recognize as foreign a specific **epitope** of a 'non-self antigen' when this is bound to MHC molecules. T cells do not produce antibody, but antibody production by B cells often requires T cell help. **T cell receptors** (see Fig. 110) bind antigen on the surfaces of other cells only after it has been degraded or otherwise processed by that cell, and only after it has become physically associated with molecules of the MHC. *MHC restriction* refers to the process during T cell maturation in the thymus when an individual's T cells come to recognize, and be activated by, antigen only when presented in physical association with that individual's particular MHC molecules, often in conjunction with certain **accessory molecules**. *Self-restriction* occurs when T cells preferentially recognize foreign antigens when bound to MHC molecules encountered during their own development in the thymus. *Cytotoxic T-cells* (T_C) recognize tumour or virus-infected cells by their surface antigens in combination with their MHC markers, and will kill them (see **cytolysins**). Other T cells (*macrophage-activating cells*) produce **lymphokines** which promote macrophage activity.

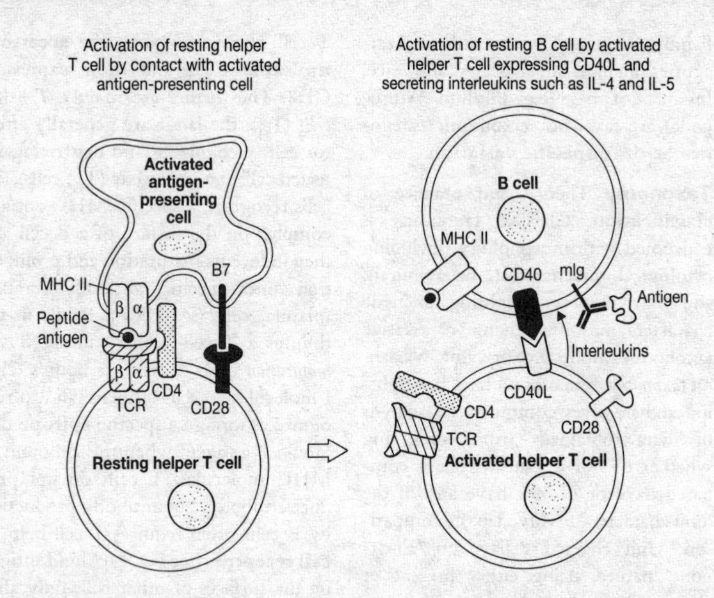

Fig. 109 *Cell contact-dependent signals in the initiation of an immune response. In the first step the resting helper* **T cell** *contacts an antigen-presenting cell, such as a macrophage, presenting peptide antigens in the context of MHC Class II molecules. A second contact-dependent signal is provided by the B7 molecule, expressed by antigen-presenting cells that have themselves previously been activated and transmitted to the T cell by CD28. In the second step the activated T cell, which now expresses CD40L, provides an initial activating signal to a resting B cell via CD40. Binding of antigen to mIg on the B cell is not essential for activation but enhances the response. The interleukins, such as IL-4 and IL-5, secreted by the activated T cell enhance B cell proliferation, differentiation to a phenotype with a higher rate of antibody secretion and immunoglobulin class switching to IgG and IgE.*

Suppressor T cells (T_S) specifically suppress the immune response, probably through affecting antigen-presenting cells and/or through more direct interactions with T_H or B cells. See Fig. 71.

T cell receptor (TCR) T cells need to recognize a wide variety of antigens, doing so through the cooperation of a membrane receptor (TCR) and specific **accessory molecules.** The TCR comprises one α and one β polypeptide (TCR-2) or one γ and one δ polypeptide (TCR-1). The genes encoding the receptor resemble those for antibodies and comprise variable and constant regions (see **antibody diversity**, Fig. 4) and, as in that system, production of the TCR repertoire involves both germ line diversity and gene rearrangements. The β and δ chains are encoded by V, D and J segments while α and γ chains

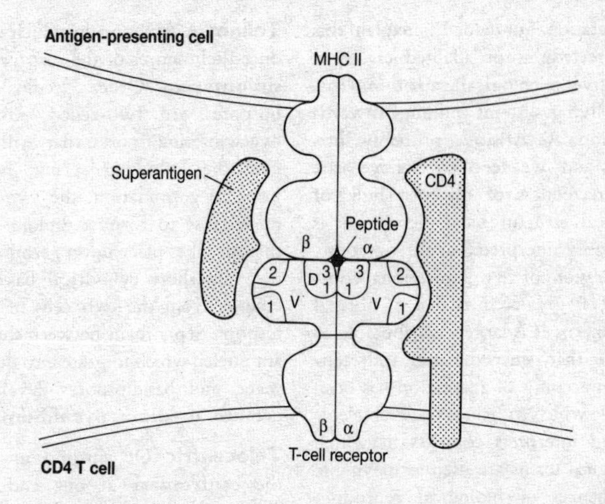

Fig. 110 **T cell receptor** *interaction with an MHC Class II molecule presenting either a peptide in the antigen-binding groove or a superantigen. 1, 2 and 3 indicate the three hypervariable regions of the T cell receptor α and β chains. The receptor interacts with a Ras protein on the inner surface of the T cell membrane to mediate further T cell response. See* **receptor** *figures.*

are encoded by V and J segments only. Antigen recognition by the T cell receptors involves their binding that antigen (often a peptide fragment) when presented on another cell's surface stably bound to a protein encoded by the organism's **major histocompatibility complex**. Antigen-binding by the TCR activates a **protein tyrosine kinase** and generation of phosphatidyl-derived second messengers (see **inositol 1,4,5-triphosphate**).

T cell restriction Alternative for MHC-restriction of **T cell**.

***TDF* (testis-determining factor)** Called *TDF* in humans, *Tdy* in mice. See **testis**.

T-DNA A DNA segment of a Ti plasmid which can be transferred by recombination into DNA of the host organism (usually a plant). See ***Agrobacterium***, **plasmid**.

Tectum (optic tectum) Dorsal region of vertebrate midbrain (see **brain**). A correlating region integrating vestibular, tactile and visual information and initiating or modifying reflex motor responses. Except in mammals, it is the primary visual centre. See Fig. 13.

Teeth See **dentition, deciduous teeth, permanent teeth**.

Teleology Biological discourse becomes teleological when reference is made to function, purpose and design, as when structure is related to function. In recent history, vitalists, organismic biologists and others have stressed that biological systems demand a functional interpretation, but tended to explain this goal-directedness or adaptedness by a retroactive (teleological) causation peculiar to living systems 'pulling' towards perfection. As Aristotle probably intimated, and we tend to believe, the goal-directedness of the vast bulk of biological structure and behaviour is more easily interpreted as the outcome, or expression, of an organizational complexity far greater than that of normal inert objects. It is largely attributable to **Darwin** that, since the mid 19th century, an account of adaptation has been available which trivializes classical teleology and interprets contexts in which teleological terms are used as inevitable consequences of biological reproduction, accompanied as that is by selective transmission of heritable variation from one generation to the next as organisms compete for finite environmental resources. The concept of teleology is difficult to explicate without circularity. See **function**, **proximate factor**, **ultimate factor**, **adaptation**.

Teleostei Higher bony fishes (about 20,000 living species). A fish taxon of uncertain category (e.g. suborder, series) but within Subclass **Actinopterygii** of Class **Osteichthyes** (bony fishes). Probably arose from a holostean stock in the Mesozoic, but became abundant in the Cretaceous. Vertebral axis turns upwards in the tail, which is superficially symmetrical; paired fins are small; scales are usually rounded, without ganoid covering, thin and bony. Among living teleosts soft-rayed fish (e.g. salmon, herring, carp) are regarded as relatively primitive, spiny-rayed fish, which include the **Acanthopterygii** (e.g. perches), as more progressive. See **fins**.

Telium A structure in which are produced teliospores of the rust fungi (**Basidiomycota**, Order Uredinales). Teliospores are two-celled, dikaryotic, overwintering spores that will germinate the following spring; however, prior to germination, the two haploid nuclei fuse to form a diploid nucleus; meiosis takes place upon germination in the two short cylindrical basidia that emerge from the two cells of the teliospore. Septa form between the resultant nuclei, which migrate into the sterigmata, and basidiospores develop. See **aecium**, **uredium**, **pycnidium**.

Telocentric Of chromosomes, with the **centromere** at one end (terminal).

Telolecithal Of eggs (e.g. amphibian) with yolk more concentrated at one end and a marked **polarity** along the axis of yolk distribution.

Telome One of the distal branches of a dichotomized plant axis; a morphological unit in a primitive vascular plant. Telomes are regarded as the basic units from which the diverse types of leaves and sporophylls of vascular plants have evolved.

Telomere Originally, term used by cytogeneticists for an end of a eukaryotic chromosome. Such ends lacked any tendency to fuse together spontaneously (i.e. they were 'non-sticky'). Telomeres appeared never to become incorporated within the chromosome (e.g. by terminal inversion). It is now known that they are examples of simple tandem DNA repeat sequences (regular or irregular), replicated by a specific *telomerase* enzyme (in a 5′ to 3′ end direction), which is a specialized reverse transcriptase providing the template for new telomeres. Telomerase is an enzyme composed of both RNA and protein, and as such may be an evolutionary link between systems using **ribozymes** only and those using purely protein enzymes (see **origin of life**). The ability to replicate without loss the 5′-ends of each DNA strand in the chromosome is a requirement, given the normal machinery of semiconservative DNA replication – which works in a 5′→3′ direction – and the need of cellular **DNA polymerases** for an RNA primer. It is eukaryotic chromosome ends that are described as telomeres, in contrast to the various termini of linear viral, non-nuclear plasmid and mtDNA genomes. Telomeric DNA is regionally organized into non-nucleosomal chromatin (the *telosome*) which can interact with the envelope of the **nucleus** and probably facilitates chromosome pairing during early **meiosis**. Telomeres are composed of **heterochromatin** and may be involved in transcriptional regulation and nuclear architecture. Human telomeres have been cloned in **yeast artificial chromosomes**. See **prion**.

Telophase Terminal stage of **mitosis** or **meiosis**, when nuclei return to interphase.

Telson Hindmost segment of arthropod abdomen, forming sting of scorpions and part of tail fan of lobsters and their allies. In insects present only in the embryo.

Temperate Of **bacteriophage** that can insert its genome into that of its host so that it is replicated with it. See also **episome**.

Temperature coefficient See Q_{10}.

Temperature-sensitive mutant Mutant form expressed only under certain temperature conditions, the wild-type phenotype being expressed at others. Normal function occurs within the permissive temperature range (commonly low), malfunction within the restrictive temperature range (commonly high). Many *cdc* mutations (see ***cdc* genes**) are temperature-sensitive. Commonly due to heat-labile gene product. See **chaperones**.

Template In nucleic acids, the strand used by a polymerase to build a new and complementary polynucleotide strand.

Temporalis muscle Muscle originating on temporal bone of mammalian skull, inserting on the dentary; elevates and retracts mandible and assists in its side-to-side movements. Compare **masseter muscle**. See **sagittal crest**.

Tendon Cord or band of relatively inextensible vertebrate connective tissue attaching muscle tissue to another structure, often bone. Consists almost entirely of closely packed collagen fibres with rows of fibroblasts between. With high tensile strength and coefficient of elasticity.

Tendril Stem, or part of leaf, modified as a slender branched or unbranched thread-like structure; used by many climbing plants for attachment to a support, either by twining around it (e.g. pea, grape vine) or by sticking to it by an adhesive disc at the tip (e.g. virginia creeper). See **haptotropism**.

Teratogen Any factor or agent causing malformation in embryos (e.g. **X-ray irradiation**, certain chemicals).

Teratoma Growth of cell mass from an unfertilized egg within mammalian ovary (or of germ cells in the male's testis) which becomes disorganized and uncontrolled. Many differentiated cell types may be represented, all mixed up. Teratomas may become malignant (*teratocarcinomas*). See **parthenogenesis**.

Tergite A **sclerite** composing an arthropod **tergum**.

Tergum Thickened plate of cuticle on dorsal side of arthropod segment.

Termination See **protein synthesis**.

Termite See **Isoptera**.

Terpene Unsaturated **lipid** consisting of multiples of *isoprene* (a 5-carbon hydrocarbon), sesquiterpenes having three such units. Includes vitamins A, E and K, carotenoids and many odorous substances in plants. *Squalene* (see **cholesterol**) is another. Rubber and gutta-percha are polyterpenes.

Territory Area or volume of habitat occupied and defended by an animal, or group of animals of same species. Territories commonly arise during or prior to breeding, with an important spacing role in many vertebrate and arthropod populations. Compare **home range**.

Tertiary Geological period, lasting from about 70–3 Myr BP. With Quaternary it comprises the Cenozoic era. Often called the Age of Mammals on account of the radiation which occurred during it. Birds dominated the air throughout, almost all major modern groups being present early in the period.

Testa Seed coat. Protective covering of embryo of seed plants, formed from the integument(s); usually hard and dry.

Test cross Cross between an individual of unknown genotype (or one heterozygous at a number of loci) and another homozygous for recessive characters at loci of interest. Offspring ratios are then used to deduce the genotype of the unknown, and/or the linkage relationships of its heterozygous loci. Not necessarily a **backcross**. See **chromosome mapping**.

Testicular feminization Syndrome of some mammals, including humans. Due to presence of the *Tfm* allele at an X-linked locus, individuals chromosomally male (XY) are phenotypically female since male genital (Wolffian) ducts fail to develop, lacking cell membrane receptors capable of binding androgens (testosterone and its derivatives). Uterus absent or rudimentary; gonads usually abdominal or inguinal testes. For role of *Sry* gene, see **testis**.

Testis Sperm-producing organ of male animal; in vertebrates producing also 'male' sex hormones (androgens) and derived in part from *genital ridges* of coelomic epithelium in dorsal abdomen adjacent to the mesonephros, forming the testis cortex. There is a 35 Kb sex-determining region (*SRY* in humans, *Sry* in mice) on the Y-chromosome expressed briefly in the Sertoli cell precursors, which interact with other gonadal somatic cells to cause testis production and block ovary development (see **Müllerian inhibiting substance**). Germ cells then develop as egg or sperm upon signals from somatic cell neighbours. Subsequent male sexual differentiation is a consequence of the hormonal products of the testis. It is very likely that the testis-specific *Sry* gene product is a **transcription factor**. Amoeboid *primordial germ cells* invade the ridges from the endoderm of the yolk sac and as the ridges hollow out, primitive sex cords develop from mesenchyme to form *seminiferous tubules* (see Fig. 111), to which primordial germ cells attach. Tubules are separated by septa and discharge sperm into the **epididymis** via ducts (*vasa efferentia*). Each testis is attached to the abdominal wall by a ligament (the *gubernaculum*) and under testosterone influence is drawn into the scrotum through a canal (the inguinal canal) as body elongation occurs. Endocrine cells (*interstitial cells*, or *Leydig cells*) between the tubules secrete androgens, most potent being **testosterone**. See **maturation of germ cells, Wolffian duct, ovary**.

Testosterone The main vertebrate androgen ('male' sex hormone). Anabolic steroid produced largely by the Leydig cells of the testes from cholesterol or acetyl coenzyme A, and in smaller quantities by the adrenal cortex. Responsible for maintaining testes and for male growth spurt at puberty by stimulation of longitudinal bone growth and deposition of calcium. Closes epiphyses a few years after puberty, arresting growth. Promotes protein synthesis (hence muscle development), sexual behaviour and spermateliosis (see **maturation of germ cells**). In humans promotes hair growth in pubic, axillary, facial and chest regions and enlargement of laryngeal cartilage and voice deepening.

Tetanus (1) Disease caused by toxin from anaerobic spore-forming bacterium *Clostridium tetani*. Increasing muscle spasms make opening of jaw difficult (hence lock-jaw). Progressive convulsions may be fatal, often by asphyxia or exhaustion. (2) Sustained muscle contraction resulting from nervous stimulation at a rate too great to allow muscle relaxation. Results from overabundance of calcium ions in sarcoplasm. See **muscle contraction**.

Tetrad Group of four haploid cells or nuclei produced by meiosis, while they are adjacent.

Tetraploid Of nuclei, cells, individuals, having four times ($4n$) the haploid number of chromosomes. See **polyploid**.

Tetrapod Four-limbed vertebrate (see Fig. 112). Includes amphibians, reptiles, birds and mammals. Secondary loss of one or both pairs of limbs, or modification into wings, flippers, etc., has occurred in some taxa. See **pentadactyl limb, Rhipidistia**.

Tetraspores In some red algae, the four spores formed through meiosis in a tetrasporangium. Borne usually on a free-living and diploid sporophyte.

Tetrodotoxin (TTX) Highly specific puffer fish toxin, blocking voltage-gated Na^+ channels in striated muscle membrane and causing paralysis. Some marine dinoflagellates make a toxin (saxitonin) with the same effect. Valuable tools for studying muscle membranes and counting channel density.

Thalamus (Bot.) Receptacle of flower. (Zool.) Part of telencephalon of vertebrate forebrain, forming roof and/or lateral walls of third ventricle and composed largely of grey matter. Organized into nuclei relaying sensory information from spinal cord, brainstem and cerebellum to cerebral cortex. See

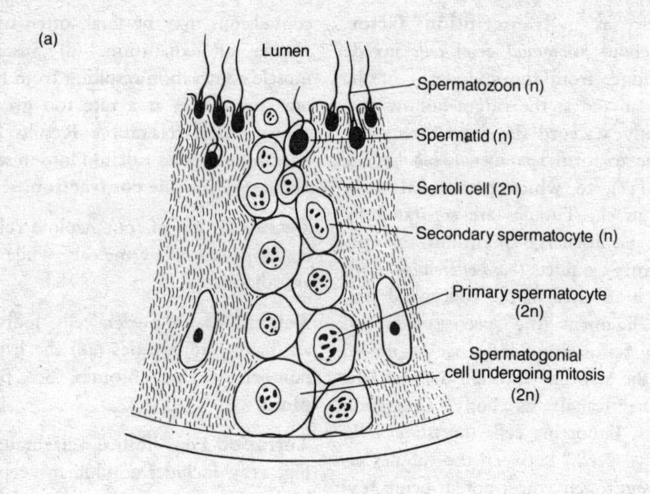

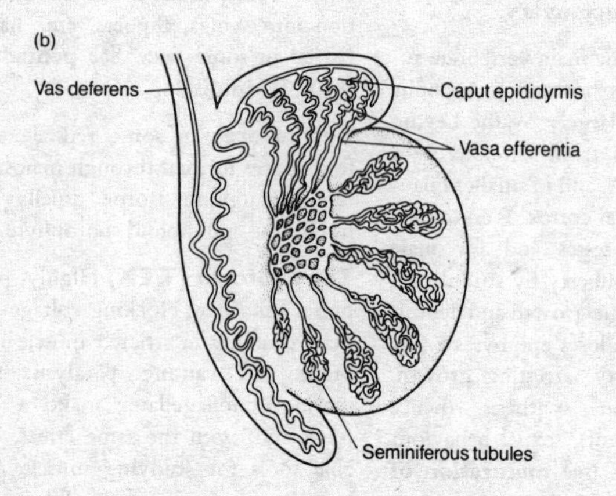

Fig. 111 *(a) Section through a vertebrate seminiferous tubule indicating the various cell stages involved in the production of spermatozoa. (b) Section through mammalian testis.*

hypothalamus and Fig. 13.

Thalassaemias Inherited disorder whereby the two globins (α and β) involved in haemoglobin production are synthesized in unequal amounts. Normally presents as a heterozygous disease in the Mediterranean region and Asia. In parts of Italy, over 10% of the population are carriers of the α-globin gene. In β^+-thalassaemia, the β-globin is in low amounts, sometimes because too little

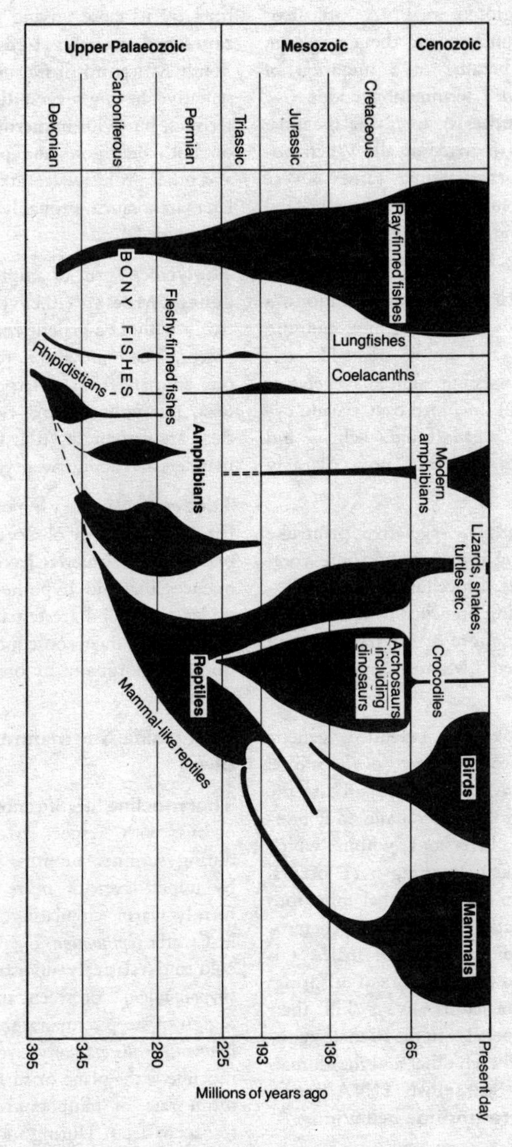

Fig. 112 *Diagram indicating approximate tetrapod phylogenies. The width of a group's entry indicates its approximate species abundance.*

of the β-globin mRNA precursor leaves the nucleus for the cytoplasm, sometimes because of a mutation of codon 39 to a termination codon (see **protein synthesis**). In β°-thalassaemia, no β-globin is made at all. When homozygous, thalassaemia causes severe haemolytic anaemia and is usually fatal in childhood. See **RFLP**, **genetic counselling**, **sickle-cell anaemia**.

Thallophyta In older classifications, a Division of plant kingdom housing prokaryotes and simple plant-like eukaryotes possessing simple vegetative bodies (thalli). Included bacteria and cyanobacteria, algae, fungi, lichens and slime fungi. Term now largely abandoned.

Thallus Simple, vegetative plant-like body, lacking differentiation into root, stem and leaf. Unicellular or multicellular, comprising branched or unbranched filaments; or more or less flattened and ribbon-shaped. Monostromatic when one cell thick.

t haplotype An extended genetic entity on chromosome 17 of the mouse (*Mus musculus*) which, compared to the normal homologue, contains four non-overlapping inversions within which only infrequent crossing-over occurs. Most contain recessive lethal mutations acting prenatally. But in the heterozygous condition ($+/t$) in males, t is transferred to 99% or more of offspring: t-bearing spermatids always 'kill' their meiotic partners (so in the t/t male genotype they kill each other and the animals are sterile). See **selfish DNA/genes**, **aberrant chromosome behaviour**.

Thecodontia 'Stem' order of **archosaurs**, arising in early Triassic and extinct by its close, giving rise to **dinosaurs** and probably birds (see **Aves**). South African fossil *Euparkeria* is representative, having several **diaspid** characteristics, but with numerous small teeth on both pterygoid and palatine bones and other primitive features absent from later archosaurs. Probably at least partially bipedal.

Thelytoky Form of animal **parthenogenesis**. Males are either very rare, effectively without a genetic role, or entirely absent. There are about 1000 thelytokous animals from a large number of taxa, but only approx. twenty-five of these are vertebrates (four fish, two salamanders and about twenty lizards).

Theory Explanatory hypothesis, usually firmly founded in observation and experiment. They tend to have more consequences than do hypotheses, being of wider scope, and are tested by examining whether their consequences (predictions) are borne out by observation and experiment.

Therapsida See **mammal-like reptiles**.

Thermocline Stratification of lakes and oceans with respect to temperature during summer months, characterized by upper layer of more or less uniformly warm, circulating, fairly turbulent water (*epilimnion*) overlying deeper, cold and relatively undisturbed region (*hypolimnion*). Between the two is a region of steep temperature drop (*metalimnion*, or *discontinuity layer*). The thermocline is the plane or surface of maximum rate of temperature drop with respect to depth. During autumn, epilimnetic temperatures decrease, water density rises and mixing (*overturn*) of water and nutrients results in a more even temperature distribution.

Thermodynamics Classical thermodynamics deals with *closed systems* – those which do not interact with their surroundings in terms of energy or matter. By contrast, living systems (in so far as they are alive) are *open systems*, interacting in both ways.

The first law of thermodynamics states that in any process the total energy of a system and of its surroundings remains constant, even though energy may be transformed from one form to another (transduced). The second law states that during any process the combined *entropy* (S) of a system and its surroundings (its disorder, randomness) tends to increase until equilibrium is attained, at which point no work can be done. The tendency for entropy in the universe to be maximized could be thought of as the 'driving force' of all chemical processes. Living cells exist in states of thermodynamic non-equilibrium, and in different steady states, in which rates of energy/matter input from the surroundings equal their output. Biology is concerned primarily with reactions taking place at constant temperature (*isothermally*) but heat changes accompany even isothermal reactions. If under these conditions heat is lost to the surroundings the system is said to lose *enthalpy* (H) and the reaction is said to be exothermic. Absorption of heat characterizes endothermic reactions. The form of energy capable of doing work in a system under constant temperature and pressure is its *free energy* (G), whose value is the key to predicting the direction of a chemical reaction. These relationships may be summarized:

$$\Delta G = \Delta H - T\Delta S$$

where ΔG = the free energy change of the system, ΔH = its change in enthalpy, T = absolute temperature and ΔS = its change in entropy. Reactions are *exothermic* if ΔH is negative, *endothermic* if ΔH is positive; they are *exergonic* (and may do work) if ΔG is negative, and *endergonic* (not doing work) if ΔG is positive. Spontaneous reactions are characterized by a loss of free energy (exergonic). Endergonic reactions cannot proceed spontaneously.

The free energy change of any reaction at constant temperature and pressure involves a fixed constant for that reaction known as the *standard free energy change* of the reaction (ΔG_0), giving the loss of free energy when the reaction is allowed to go to equilibrium starting with certain standard conditions – in particular, when all reactants and products are present at 1.0 molar concentrations. It is a measure of the difference between the sum of free energies of products and of reactants and is related to the equilibrium constant K (since $\Delta G_o = - RT\ln K$). Free energy of a reaction varies with concentrations of reactants and products, affecting its probability of occurrence. A chemical reaction only occurs if its ΔG is negative in sign, and the maximum amount of work it then does equals this decrease in free energy.

Thermoluminescence (TL) See **radiometric dating**.

Thermonasty (Bot.) Response to a general, non-directional, temperature stimulus; e.g. opening of crocus and tulip flowers with temperature increase.

Thermophilic (Of microorganisms, e.g. *Thermus*) with optimum growth temperature (measured by generations per hour) in the 55–65°C range. Compare **mesophilic**, **psychrophilic**.

Therophytes Class of **Raunkiaer's life forms**.

Thiamine (vitamin B) Vitamin precursor of coenzyme *thiamine pyrophosphate* (TPP), serving enzymes transferring aldehyde groups during decarboxylation of α-keto acids (such as pyruvate in mitochondria), and formation of α-ketols. Deficiency causes beri-beri in man, and polyneuritis in birds. See **vitamin B complex**.

Thigmotropism See **haptotropism**.

Thoracic duct Main mammalian lymph vessel, receiving lymph from trunk (including lacteals) and hindlimbs and running up the thorax close to vertebral column, discharging into left subclavian vein (humans), or another major anterior vein. Often paired in fish, reptiles and birds.

Thorax (1) In terrestrial vertebrates, part of body cavity containing heart and lungs (i.e. the chest); in mammals, clearly separated from abdominal cavity by diaphragm. (2) In arthropods, body region between head and abdomen; often not clearly separable, or tagmatized. In adult insects comprises three segments, each typically with a pair of walking legs and one or two of them commonly bearing a pair of wings. See **tagma**.

Thread cell See **cnidoblast**.

Threadworms See **Nematoda**.

Threonine kinases See **protein kinases**.

Threshold Critical intensity of a stimulus below which there is no response by a tissue (e.g. nerve, muscle).

Thrombin Proteolytic enzyme derived from prothrombin during sequence of reactions involved in **blood clotting**, converting fibrinogen to fibrin.

Thrombocytes See **platelets**.

Thromboplastin (thrombokinase) Phospholipid and protein, liberated from **platelets** and damaged tissues on wounding, initiating the **cascade** of reactions resulting in **blood clotting**.

Thylakoid Flattened membranous vesicle containing **chlorophyll** pigments; site of photochemical reactions in **photosynthesis**. Vary in form and arrangement between different groups of organisms.

Thymidine Nucleoside comprising the base **thymine** linked to ribose by a glycosidic bond.

Thymine Pyrimidine base of the **nucleotide** thymidine monophosphate, a monomer of DNA but not RNA. Also a component of thymidine di- and triphosphates.

Thymus gland Bilobed vertebrate organ containing primary **lymphoid tissue**, usually situated in the pharyngeal or tracheal region. Originates from gill pouches (in mammals, the third pouches) or gill clefts. In mammals lies in thorax (mediastinum), the gill pouch cells mixing with lymphocytes. Attains maximum size at puberty, slowly diminishing afterwards. Responsible for maturation of **T cells**, and possibly with an endocrine function.

Thyroglobulin (TGB) Transport

protein for thyroid hormones produced by thyroid gland. Stored there and carried in blood plasma.

Thyroid gland Vertebrate endocrine gland, either single (e.g. in mammals) or paired (e.g. most amphibians and birds), in neck region. Derived from endoderm of pharyngeal floor; probably homologous with **endostyle** of cephalochordates, etc. Comprises sac-like *thyroid follicles* in which **thyroid hormones** (thyroxine and triiodothyronine) are stored, colloidally complexed with **thyroglobulin**, surrounded by parafollicular cells (secreting **calcitonin**). Inadequate dietary iodine causes enlargement of the thyroid (goitre) due to raised TSH secretion (see **thyroid stimulating hormone**).

Thyroid hormones In addition to the hormone **calcitonin**, most vertebrate thyroids produce small iodine-containing hormones as derivatives of the amino acid tyrosine. These seem to act like steroid hormones, passing through cell membranes and binding to **nuclear receptors**, the complex then binding specific chromosomal DNA sites to bring about **gene expression**.

Thyroxine (T_4) is produced in greater quantities than *triiodothyronine* (T_3), but is about a quarter as potent. About one third of circulating T_4 is converted to T_3 in lungs and liver. Both are homeostatically controlled, and if circulating hormone levels or blood temperature are too low, the hypothalamus releases thyrotropin releasing factor (TRF) into the anterior pituitary portal system, stimulating release of TSH (**thyroid stimulating hormone**), which, in turn, releases bound and unbound hormones from the thyroid. These increase **basal metabolic rate** (especially in the liver) and oxygen consumption, uncoupling electron transport from ATP synthesis in liver mitochondria. Respiration then releases heat into the blood (raising its temperature) rather than generating ATP.

They have general catabolic effects upon fat and carbohydrate, promoting gluconeogenesis; but they stimulate protein synthesis. Are important in tissue growth and development, and in metamorphosis of amphibian tadpoles. Deficiency in growing vertebrates causes reduced nerve and organ development; presence promotes neural activity, heart rate, blood pressure. Dietary intake of iodine (in humans 100–400 μg day^{-1}) is required for their normal production.

Thyroid stimulating hormone (TSH, thyrotropin) Protein product of anterior pituitary gland, stimulating thyroid production and release of thyroxine (T_4) and triiodothyronine (T_3). Secretion influenced by TSH-releasing factor (TRF) from the hypothalamus. Negative feedback occurs between levels of (a) (T_4) and (T_3) and (b) TSH and TRF in blood. Fall in T_3 and T_4 raises outputs of TRF and TSH; rise of the last two shuts off TRF and TSH release.

Thyroxine (T_4) See **thyroid hormones**.

Thysanoptera Thrips. Order of minute exopterygote insects, most with piercing mouthparts for sucking plant juices (and occasionally aphids). Have prepupal and pupal stages despite being exopterygote. Of economic importance as transmitters of disease, and ecologically in studies of population regulation.

Thysanura Silverfish. See **Apterygota**.

Tibia (1) Shinbone; the more anterior of the two long bones below knee of tetrapod hindlimb, the other being the fibula. See **pentadactyl limb**. (2) Fourth segment from base of an insect's leg.

Ticks See **Acari**.

Tight junction See **intercellular junction**.

Tiller (Of grasses) a side shoot arising at ground level.

Tissue Association of cells of multicellular organism, with a common embryological origin or pathway and similar structure and function. Often, cells of a tissue are contiguous at cell walls (plants) or cell membranes (animals) but occasionally the tissue may be fluid (e.g. blood). Cells may be all of one type (a *simple tissue*, e.g. squamous epithelium, plant parenchyma) or of more than one type (a *mixed tissue*, e.g. connective tissue, xylem, phloem). Tissues aggregate to form organs. See **acellular**.

Tissue culture (explantation) A technique for maintaining fragments of animal or plant tissue or individual cells from the organism. A sterile bathing medium (commonly **Ringer's solution** in the case of animal tissues or cells or a culture medium for plant tissues or cells) surrounds the cells or tissues, while appropriate temperature, pH and nutrient levels are maintained and waste products removed. With respect to plants, tissue culture is one of the most important procedures in **biotechnology**, and this is feasible only because of an understanding of how hormones function, and regulate development. The greatest impact of tissue culture is in the area of plant multiplication (clonal propagation). There are three different methods of tissue culture propagation, each taking advantage of the control of development that can be accomplished with plant hormones, and include: (a) **somatic embryogenesis**: (b) shoot multiplication, where growing shoot tips are exposed in culture to a high enough concentration of **cytokinin**, not only to sustain shoot growth, but also to stimulate lateral bud development. Commonly used for multiplying ferns, orchids and woody plants; and (c) callus and/or protoplasts, where cytokinin is used to promote formation of growing shoots in unorganized masses of parenchymatous tissue, which is known as callus; once shoots form they can be treated with **auxin** to initiate root formation. Protoplasts can also be used for plant regeneration, after cell walls have been removed by enzymatic digestion, in a technique known as protoplast fusion; protoplasts from differing plants are allowed to fuse together to produce a hybrid cell. The first plant developed using this technique was a hybrid of tobacco. Intergeneric hybrid plants have also been developed using protoplast fusion (e.g. between the potato and tomato).

Tissue fluid (interstitial fluid) Fluid derived from blood plasma by filtration through capillaries in the tissues. Differs from blood chiefly in containing no suspended blood cells and in having lower protein levels. Bathes tissue cells in appropriate salinity and pH and acts as route for reciprocal diffusion of dissolved metabolites between them and blood. Most of the water filtered from capillaries is reabsorbed by them osmotically; that which remains (along with some solutes) enters the **lymphatic system**, when it is termed *lymph*. See **oedema**.

Titre Usually, a relative measure of the amount of antibody in a fluid, usually serum. The greatest dilution capable of producing a particular detectable antibody–antigen reaction is its titre. This assay technique can be used to compare titres in different samples. See **precipitin**.

TMV Tobacco mosaic virus. See **virus**.

TNF See **tumour necrosis factor**.

Toadstool Common name for fruiting bodies of fungi (other than mushrooms) belonging to the Agaricaceae (**Basidiomycota**). Often wrongly assumed that all are poisonous, although some are.

Tocopherol Vitamin E. See **vitamins**.

Tolerance See **immune tolerance**.

Tone (tonus) See **muscle contraction**.

Tonoplast (Bot.) The membrane that surrounds the vacuole in plant cells. It plays an important role in the active transport of certain ions into the vacuole, and their retention there.

Tonsils Masses of non-encapsulated **lymphoid tissue** in mouth or pharynx of tetrapods, lying close underneath mucous membrane. Humans have a pair of palatine tonsils at the junction of mouth and pharynx, and a single pair of pharyngeal tonsils (adenoids) at the back of the nose. Sites of lymphocyte production. See **Peyer's patches**.

Tonus (tone) See **muscle contraction**.

Tooth See **dentition**.

Tornaria Ciliated planktonic larval stage of the **Hemichordata**. Its development by radial cleavage and **enterocoely** suggests links with echinoderms.

Torus (1) Receptacle of a **flower**; (2) central thickened portion of pit membrane in bordered **pits** of conifers and some other gymnophytes.

Totipotency (Bot.) The inherent capability of a single cell to provide the genetic programme required to direct the development of an entire individual. See **somatic embryogenesis**.

Toxin Any poisonous substance produced by an organism; commonly injurious to potential competitors/predators. May be of microbial, plant or animal origin; some are among the most potent poisons known. Antibodies raised to toxins are *antitoxins* and can be used to neutralize diphtheria and tetanus toxins. See **Cyanobacteria**, **endotoxin**, **exotoxin**, **toxoid**.

Toxoid Toxin modified so as to retain its antigenicity (its epitopic domain) but to lose its toxicity. Used in **vaccines** against human diphtheria and tetanus.

Trabeculae Slender bars of tissue lying across cavities. (Zool.) They form the mesh-like interiors of spongy **bone**, commonly growing along stress lines. (Bot.) Slender rod-shaped supporting structures found in various plant organs: (1) radially elongated endodermal cells, each becoming a row of cells by division, that support the stele in stems of the lycopod *Selaginella*, linking stele to cortex; (2) membranes, comprising rows of cells, traversing interior of sporangia in the lycopod *Isoetes*; (3) rod-like outgrowths of cell wall lying across lumen of tracheids.

Trace element Element required in minute amounts for an organism's normal healthy growth. A *micronutrient*. Often a component or activator of an **enzyme**. *Essential* plant trace elements (without which death eventually ensues) include zinc, boron, manganese, copper and molybdenum. Fluorine is essential for hardening tooth enamel. Iodine is a component of thyroxine and triiodothyronine. For cobalt, see **cyanocobalamin**. Compare **vitamin**.

Tracer Term sometimes used for an isotopic label. See **labelling**.

Trachea (1) Vertebrate windpipe. A single tube, covered and kept open by incomplete rings of cartilage; with smooth muscle in its wall, and a ciliated lining. Leads from glottis down neck, branching into two bronchi. (2) In most insects and some other arthropods (e.g. woodlice), hollow tubes of epidermis and cuticle conducting air from spiracles directly to tissues. The finest branches (*tracheoles*, diameter about 0.1 μm) are intracellular. Both tracheae and tracheoles may be permeable but tracheoles are more important in gaseous exchanges. See **ventilation**, **ecdysis**.

Tracheary element General term for the water-conducting cell in vascular plants. Comprise **tracheids** and **vessel** members.

Tracheid Non-living **xylem** element, characteristic of vascular plants other than flowering plants. Formed from a single cell, it is elongated with tapering ends and thick, lignified and pitted walls: an empty firm-walled tube running parallel to long axis of organ in which it lies, overlapping and in communication with adjacent tracheids by means of pits. Functions in water conduction and mechanical support. Primitive compared with a **vessel**.

Tracheole See **trachea**.

Tracheophyta A division in older classifications which regard possession of vascular tissue as of greater taxonomic significance than the seed habit. Would include all vascular plants, thus comprising the majority of the world's terrestrial vegetation.

Training See **lipoprotein**, **muscle contraction**.

Trait A particular phenotypic character, as opposed to character mode. Thus eye colour in *Drosophila* is a character mode, whereas red eye colour is a character trait. Traits can be *autosomal* or *sex-linked*, and determined either by a single locus or polygenically. Sometimes, in heritable disorders, *trait symptoms* are contrasted with *disease symptoms*, in which case the former are usually milder and occur in heterozygotes while the latter are more severe and occur in homozygotes (see **sickle-cell anaemia**).

***trans*-acting control elements** Regulatory genetic elements (e.g. **repressor** genes and genes for other **promoter**-binding proteins) whose products are not limited in their binding to the DNA molecule of which they form a part. See **transcription factors**. Contrast ***cis*-acting control elements**.

Transactivation Interaction of steroid receptors with ***trans*-acting control elements** on DNA to stimulate transcription of hormonally responsive genes. See **nuclear receptors**.

Transaminase Enzyme carrying out **transamination**.

Transamination Transfer by a *transaminase* of an amino group from an amino acid to a keto acid (e.g. pyruvate, ketoglutarate), producing respectively a keto acid and an amino acid. See **pyridoxine**.

Trans-configuration (Of mutations) see ***cis*-trans test**.

Transcriptase An **RNA polymerase**. Compare **reverse transcriptase**.

Transcription Production of an RNA molecule off a DNA template by an RNA polymerase. For regulation of transcription, see **transcription factors**, references there, and Fig. 206, in the **chromosome** entry. See **protein synthesis**.

Transcription factors (TFs) Regulatory proteins determining the efficiency with which RNA polymerases bind to DNA promoter regions during transcription. *Transcription activators* are involved in positive, and *transcription repressors* in negative control of gene expression. The DNA-binding ability of these proteins often comprises the common structural motif of an α-helix, a short turn and a second α-helix (the 'helix-turn-helix'). The first helix interacts with the DNA backbone while the second lies in a major groove of the DNA molecule. Many eukaryotic transcription factors contain in addition tandemly repeated amino acid sequences each producing a metal-binding (often Zn^{2+}) finger-like protrusion in the functional molecule. These finger sequences (*multifinger loops*) often include identical or near-identical (*consensus*) subsequences which are thought to be conserved and homologous. Some may have oncogenic properties if their amino acid sequence is altered by mutation in the encoding gene. Several eukaryotic transcription activators have been shown to have one nucleic acid-binding region and a second 'activating region' binding additional transcription factors or perhaps RNA polymerase itself. See Fig. 108 (**TATA box**) for the various complexes of TATA-binding protein, transcription activating factors (TAFs) and RNA polymerase required for transcription of different gene classes. See **DNA-binding proteins**, **nuclear receptors**, **homeobox**, **promoter**, **enhancer**, **sigma factors** and Fig. 113).

Transdetermination Change brought about in groups of cells (e.g. insect imaginal disc cells) which when subsequently cultured differentiate into a structure normally produced by cells from a different lineage within the organism (e.g. a different disc). The effect resembles mutation in a stem cell, but occurs to a group of cells which shifts from one heritable state to another as a result of environmental influences. See **homeotic mutation**.

Transduction (1) Process in which usually a **bacteriophage** picks up DNA from one bacterial cell and carries it to another, when the DNA fragment may become incorporated into the bacterial host's genome. Two basic types: (a) *generalized transduction*, where the phage DNA-packaging mechanism picks up 'by mistake' any phage-sized fragment of chromosomal DNA, which can be integrated by homologous

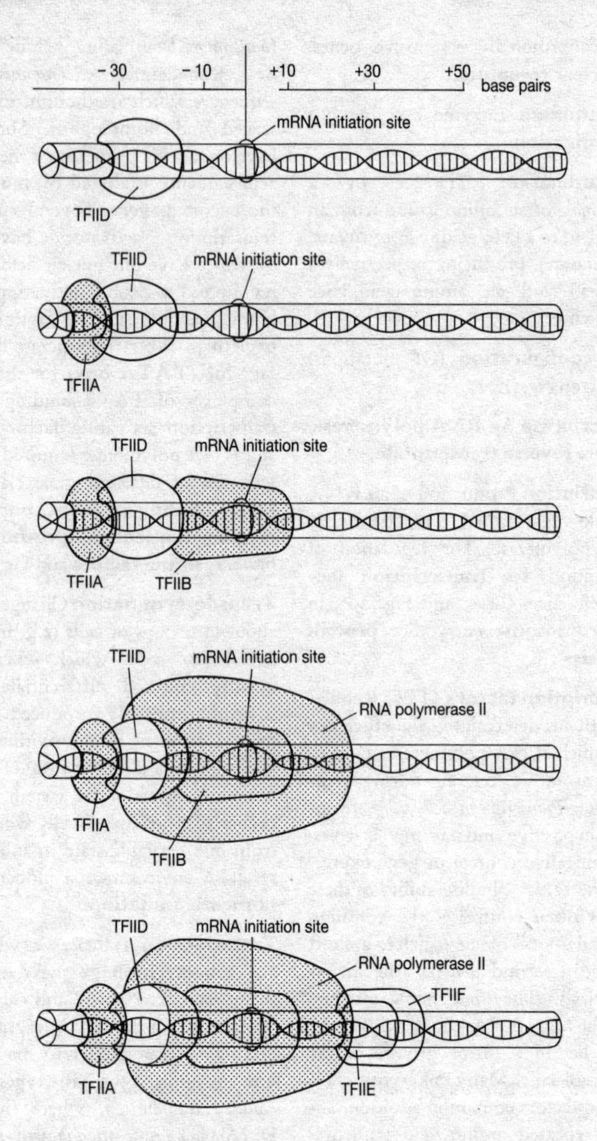

Fig. 113 *Development of active eukaryotic transcription initiation complex for Class II genes. TFIID = TATA factor or TATA-binding protein (TBP). Some notations have all TFs replaced by TAF (= transcription activating factor).*

recombination into the recipient genome after injection into the cell by the phage apparatus; (b) *specialized transduction*, in which, on induction, integrated phage DNA (e.g. λ) genome is imprecisely excised from the chromosome, carrying adjacent chromosomal DNA with it; since the phage is generally integrated at a specific site in the chromosome, only a few bacterial genes can be transduced this way. A similar process, not usually regarded as transduction, occurs in *F-mediated sexduction* (see **F factor**) in which an F factor (a plasmid, not a phage) carries bacterial DNA from one cell to another. All three types are employed in **chromosome mapping**. See **human gene therapy**. (2) (Of energy) the conversion of energy from one form to another. In this sense, chloroplasts and muscle fibres are two among many biological transducers. (3) For amplification of signals by a cell (e.g. all sensory cells), see **signal transduction**.

Transect Line or belt of vegetation selected for charting plants; designed to study changes in composition of vegetation across a particular area. See **quadrant**.

Transfection In prokaryotes, reserved for the uptake by transformation of naked phage DNA to produce a bacteriophage infection. See **transformation** for use in eukaryotes.

Transferase Enzyme (e.g. acyl transferases, transaminases) transferring a non-hydrogen functional group from one molecule to another. Coenzymes are often involved.

Transfer cells Specialized parenchyma cells possessing ingrowths of the cell wall and producing unusually high surface-to-volume protoplast ratios. Believed to play an important role in transfer of solutes between cells over short distances. Exceedingly common; probably serving a similar role throughout the plant. Occur in association with **xylem** and **phloem** of small veins in cotyledons and leaves of many herbaceous dicotyledons; in xylem and phloem of leaf traces, at nodes in both dicots and monocots; in various tissues of reproductive structures, e.g. embryo sacs, endosperm; in various glandular structures, e.g. nectaries and glands of carnivorous plants, and in absorbing cells of haustoria of dodder (*Cuscuta*).

Transferrin Iron-binding β-globulin protein transferring iron from reticulo-endothelial cells in bone marrow to immature red blood cells, where it binds and passes iron (Fe^{2+}) into the cell. Cells short of iron produce transferrin receptor molecules for import of iron-bound transferrin to endosomes, whose low pH frees the iron while transferrin bound to its receptors returns to the cell membrane. See **ferritin**.

Transfer RNA See **RNA**, **genetic code**, **protein synthesis**.

Transformation (1) Change in certain bacteria (occasionally other cells) which, when grown in presence of killed cells, culture filtrates or extracts from related strains, take up foreign DNA and acquire characters encoded by it. Transformed cells retaining this DNA may pass it to their offspring. When cultured mammalian or other animal cells take up foreign DNA the term *transfection* is often used. Discovery implicated DNA as a genetic material. (2) Event producing **cancer cell** properties in an otherwise normal animal cell. May involve viral infection. See **carcinogen**, **transgenic**.

Transformation series See **evolutionary transformation series**.

Transgenic Describing an organism whose normal genome has been altered by introduction of a gene by a manipulative technique (microinjection of DNA into egg; use of plasmid or virus-based DNA vector), often involving introduction of DNA from a different species. The foreign DNA may then integrate into the host genome. Such organisms are potentially valuable in plant and animal husbandry. The technique is yielding results in the genetics of animal development. See **transduction**.

Transition Replacement of one purine nucleotide by another, or of one pyrimidine nucleotide by another, during nucleic acid synthesis. Compare **transversion**.

Transitivity A logical relation of importance in the context of **species** definitions. The relational property of populations '*interbreeds with . . .*' is transitive if, when population A interbreeds with population B and population A interbreeds with population C, then population B interbreeds with population C. **Ring species** indicate that '*interbreeds with . . .*' is not transitive at the population level and so cannot do the job which the biological species concept demands of it.

Translation Ribosomal phase of **protein synthesis**.

Translocation (1) (Bot.) Long-distance-transport of materials (e.g. water, mineral ions, photosynthates) within a plant. Absorption of mineral ions from the soil solution occurs through the root epidermis to the endodermis symplastically. The ions travel from the epidermal cells to the first layer of cortical cells through plasmodesmata in the epidermal-cortical cell walls. Radial movement of the ions continues in the cortical **symplast** from cell to cell (protoplast to protoplast) via plasmodesmata, through the endodermis and into parenchyma cells of the vascular cylinder by diffusion. Most inorganic (mineral) ions are taken up by active transport (support for this hypothesis comes from the fact that mineral uptake is an energy-requiring process). In fact, ion transport from the soil to the xylem requires two carrier-mediated membrane events: first at the plasmalemma of the epidermal cells; second, secretion into xylem vessels at the plasmalemma of the vascular parenchyma cells. Active transport of ions across the plasmalemma may result in electrochemical gradients being established. When this occurs, a voltage difference (the transmembrane potential) develops across the membrane. The hydrogen ion (H^+) is one of the most important cations involved; with pumping of H^+ ions to the outside, a negative potential develops on the inside of the cell, which will attract positively charged ions (e.g. Na^+, K^+). Once mineral ions are in the xylem vessels, they are rapidly distributed upward and throughout the plant in the **transpiration stream**. Movement of organic molecules (e.g. sugars, amino acids) and some inorganic ions occurs upwards and downwards in the phloem from leaves to storage organs to regions of active growth (or storage), the pattern changing with the season or the stage of development. Rates of movement in the phloem (60–120 cm h^{-1}) greatly exceed normal sucrose diffusion rates in water. Assimilate movement is said to follow a source-to-sink pattern. The *pressure-flow hypothesis* (see Fig. 114) offers the simplest and most widely accepted explanation for long-distance assimilate transport in the phloem. According to this hypothesis, assimilates move from a source to a sink along an osmotic turgor pressure gradient. In the leaves (the source), sugars produced by photosynthesis are actively secreted into sieve tubes. This active process is termed phloem loading and creates a more negative water (and osmotic) potential in the sieve tube, causing water entering the leaf in the transpiration stream to move into the sieve tube by osmosis. Sugars are transported passively (by mass flow) to a sink, where they are unloaded from the sieve tube. The removal of sugar results in a less negative water (and osmotic) potential in the sieve tube at the sink, so water moves out of the sieve tube. Thus the sieve tubes appear to play a passive role in the movement of sugars; however, active transport is involved in the loading and possibly unloading of sugars and other compounds into and out of the sieve tubes at the sources and sinks. Evidence suggests the phloem loading at the source is provided by a proton pump (energized by ATP and mediated by ATPase at the plasmalemma), involving a sucrose-proton co-transport (symport) system. Metabolic energy required for this is expended by the companion cells or parenchyma cells bordering the sieve tubes. Loading occurs across the plasmalemma of the companion cells; but some sieve tubes are capable of loading themselves, the site of active transport being their own plasmalemmas. See **transpiration**. (2) Form of chromosomal **mutation** in which an excised chromosome piece either rejoins the end of the same chromosome or is transferred to another.

Transmitter See **neurotransmitter**.

Transpiration Loss of water vapour from plant surfaces. More than 20% of the water taken up by roots is given off to the air as water vapour, most of that transpired by plants coming from the leaves. Differs from simple evaporation in taking place from living tissue and in being influenced by the plant's physiology. Most occurs through **stomata**, and to a much lesser extent through the leaf cuticle. Transpiration rates are affected by levels of CO_2, light, temperature, air currents, humidity and availability of soil water. Most of these affect stomatal behaviour, whose closing and opening are controlled by changes in the turgor pressure of guard cells (see **radial micellation**), in turn correlated with the potassium ion (K^+) levels inside them. So long as stomata are open (during exchange of gases between leaf and atmosphere), loss of water vapour to the atmosphere must occur and, although for a healthy plant this is inevitable, it is harmful when excessive and causes wilting and even death. See **transcription stream**.

Transpiration stream The **translocation** of water and dissolved inorganic ions from roots to leaves via xylem, caused by **transpiration**. Water enters plant through root hairs, much of it moving along a **water potential** gradient to the leaves via the **xylem**. This

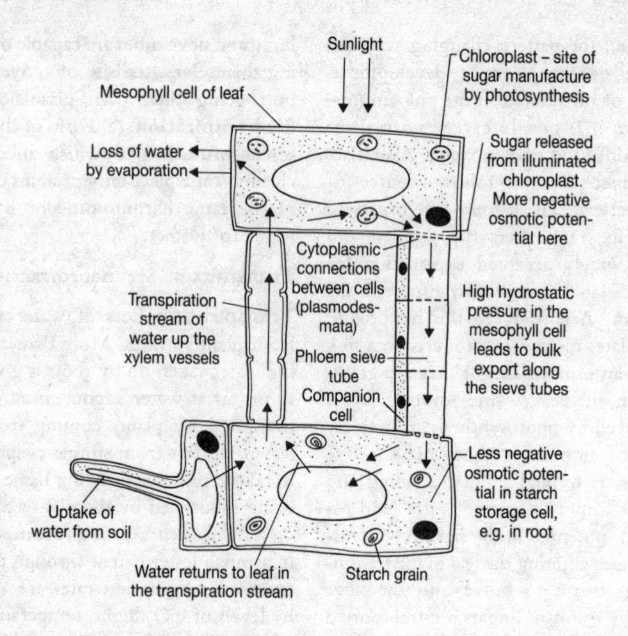

Fig. 114 *Illustration of the pressure-flow (mass flow) hypothesis of phloem translocation.*

gradient is via both the **symplast** and **apoplast** (see **root hair**). The theory of water movement is known as the cohesion–tension theory or cohesion–adhesion–tension theory. Water within xylem vessels is under considerable tension because polar water molecules cling together in continuous columns pulled by evaporation above. Because of the cohesiveness of water, this tension is transmitted all the way down the stem to the roots, so that water is withdrawn from the roots, pulled up the xylem, and distributed to those cells losing water to the atmosphere. The actual loss makes the water potential of the roots negative, thus increasing their ability to extract water from the soil. Adhesion of water molecules to the walls of the xylem, and to the walls of the leaf and root cells is just as important for the rise of water as cohesion and tension. Cell walls along which the water moves are very effective water-attracting surfaces, and take advantage of water's adhesiveness. Thus this amplifies the expression of cohesiveness. Evidence for this theory of water movement is that (1) water possesses sufficient tensile strength to withstand such tensions when contained within narrow diameter conduits, (2) water in the xylem is under tension, (3) flow begins in the uppermost branches of trees, and (4) the trunk shrinks slightly at the beginning of water movement. If the water column breaks, **cavitation** results. See **translocation**.

Transplantation (1) Artificial transfer of part of an organism to a new position either in the same or a different organism. Practically synonymous with *grafting*; but no close union with tissues in new position is implied. See **graft**. (2) For nuclear transplantation, see **nucleus**.

Transport proteins (carrier proteins) Proteins intrinsic to some cell membrane (not just plasmalemma), mediating either passive diffusion, **facilitated diffusion** or **active transport** of a solute across it. All appear either to traverse the membrane or be part of a structure which does. Similar to enzymes in having a specific binding site for transported solute, which may be saturated similarly to an active site. Reversible allosteric changes in the protein are probably responsible for carriage, and for actively transported molecules these conformational changes are in part linked to ATP hydrolysis. Transport proteins are either *uniports* (taking one solute type one way), *symports* (transport of one solute depends upon simultaneous carriage of another in the same direction) or *antiports* (carriage of one solute into the cell depends upon simultaneous transport of another out of the cell, e.g. **sodium pump**). See **bacteriorhodopsin**, **gated channels**, **impulse**, **ionophore**.

Transport vesicles Membrane-bound vesicles, many originating either from plasmalemma or Golgi apparatus, involved in endocytosis or exocytosis.

Transposable elements Prokaryotic **insertion sequences** and **antibiotic resistance elements**, and several eukaryotic DNA sequences (e.g. the **copia** elements among over thirty elements in *Drosophila* making $\simeq$ 30% of its total genome, Ty elements in yeast, IAP and VL30 in rodents, *Alu* sequences in humans, Ac, Spm and other elements from maize, and many others), all being incapable of replication independently of the host cell genome, and inherited only when physically integrated into that genome or that of a **plasmid** or **bacteriophage**. They mediate a wide range of genetic effects in their host, notably inactivation or change of expression of genes, and sometimes chromosomal rearrangements (mutations) such as inversions, translocations and deletions. Their insertion into, or adjacent to, another gene often blocks its expression, but insertion into a gene promoter region may increase that gene's expression. If recombination occurs between copies of a transposable element at different sites in a chromosome, then *non-homologous crossing-over* sometimes occurs, possibly a source of reciprocal translocations, deletions, duplications or **gene amplification**. Some eukaryote transposable elements (*retrotransposons* or *retroposons*), such as the Tyl element in yeast, transpose by a reverse transcription of their RNA to yield a complementary DNA chain which inserts after conversion to a double helix. Their genomes resemble those of retroviruses in encoding a reverse transcriptase and in being flanked by two identical terminal repeat sequences, which may each contain a **promoter**. It is now thought that **crossing-over** is often associated with previously unsuspected transposable elements. One class of elements, e.g. *Drosophila* copia, yeast Ty (sometimes called *Class I elements* in eukaryotes), have structural affinities with retroviruses and seem to transpose via RNA intermediates, being flanked by direct (i.e. not inverted) repeat DNA sequences. Another class, e.g. *Drosophila* **P element**, maize Ac and Spm (sometimes called *Class II elements* in eukaryotes), seem to transpose directly from DNA to DNA and are flanked by either long or short **inverted repeat sequences**. Copying and insertion of these elements into a genome may involve processes similar to mRNA **splicing**. See **repetitive DNA, transposition, transposon**.

Transpositions Phenomena in which genetic elements insert and excise themselves from chromosomal and plasmid DNA. Those that can replicate independently of the host genome or plasmid are termed episomal (e.g. prophages and temperate viruses); those without their own replication origin are termed **transposable elements**.

Transposon A genetic element, varying from 750 base pairs to 40 kilobase (kb) pairs in length, having at least the genes necessary for its own transposition (movement from a site in one genome to another site in the same or in a different genome). *Simple* transposons (**insertion sequences**) carry this information alone; *complex* transposons house additional genes, such as **antibiotic resistance elements**. Many if not all encode the enzyme *transposase* which facilitates their insertion, although its level may be kept low in the host cell by repression from a (*resolvase*) gene also carried on the transposon. In simple transposition, the transposon is moved to the new DNA site but leaves a lethal gap in the old DNA; in replicative transposition the transposon is replicated so that one copy is left *in situ* while the other is moved to a new DNA site. Replicative transposition often requires the enzyme resolvase. See **transposable elements**.

Transverse division Cell division in the plane of the long axis of a plant (or plant organ). See **cell division** for diagram.

Transverse process Lateral projection, one on each side, of neural arch of tetrapod vertebra, with which the head of the rib articulates.

Transverse section Section cut perpendicular to longitudinal axis of an organism.

Transversion The insertion into a growing polynucleotide of a purine in the place of a pyrimidine, or vice versa. Compare **transition**.

Trematoda Flukes. Class of parasitic **Platyhelminthes**. Ectoparasites (Monogenea) or endoparasites (Digenea) of vertebrates. Mouth and gut are retained, adults having a thick epidermal cuticle. All have a sucker around mouth. Adult *digeneans* live in liver, gut, lung or blood vessel of primary host and may cause serious disease (see **Schistosoma**). Ciliated larvae (miracidia) pass out of the host with egesta and parasitize snails (secondary host) in which they reproduce asexually (see **polyembryony**). Sporocysts then give rise to **rediae** and these to **cercariae**. *Monogeneans* normally have a large multiple central sucker for attachment, *Polystomun* alone among them being endoparasitic (frog bladder).

Triacylglycerol Alternative term for **triglyceride**.

Triassic (Trias) Geological period, lasting from about 225–180 Myr BP. Permian and Triassic are sometimes united as the Permo-Trias. Gymnosperms and ferns dominated the forests; the first dinosaurs and mammals appeared. Continents were mountainous and were united in a single landmass, with large areas being arid.

Tribe Taxonomic category between subfamily and genus, used to indicate a group of closely related genera. Tribe names normally end-eae in botany and -ini in zoology.

Tricarboxylic acid cycle (TCA cycle) See **Krebs cycle**.

Trichogyne A receptive, sometimes elongate projection extending from the female sex organ that receives the male gamete prior to fertilization in some fungi (e.g. **Ascomycota**, **Basidiomycota**), red algae (**Rhodophyta**), green algae (**Chlorophyta**) and **lichens**.

Trichome In blue-green algae (**Cyanobacteria**), a filament comprising a uniserate or multiserate chain of cells. See also **hair**.

Trichoptera Caddis flies. Order of moth-like, weakly-flying endopterygote insects. Two pairs of membranous wings covered in bristles (hairs) rather than scales. Mandibles reduced or absent in adult. Larvae aquatic, often encased, with biting mandibles and important as herbivores, carnivores and as prey items. Some are indicators of unpolluted water.

Triconodonta Extinct mammalian order (Lower Cretaceous–Upper Triassic), apparently lacking descendants. Up to four incisors in each half-jaw; molars typically with three sharp conical cusps in a row. Cat-sized or smaller; possibly derived from therapsids. Brain small.

Tricuspid valve Valve between atrium and ventricle on right side of bird or mammalian heart. Consists of three membranous flaps preventing backflow of blood into the atrium on ventricular contraction. Compare **mitral valve**.

Trigeminal nerve Fifth vertebrate **cranial nerve**.

Triglyceride (**triacylglycerol**, **neutral fat**) Ester formed by condensation of three fatty acid molecules to the trihydric alcohol glycerol. See **fat**, **glycerol**.

Trilobita Extinct class of marine arthropods, abundant from Cambrian–Silurian, surviving until Permian. The conservative body, oval and depressed, comprised the following tagmata: shield-like *cephalon* with one pair of antennae and paired sessile compound eyes, a *trunk* of freely movable segments terminating in the united segments of the *pygidium*. Along length of body, longitudinal grooves separated large pleural plates (covering the biramous limbs) from a higher middle region on each trunk segment. Average length 5 cm, but some reached 0.6 m. Probably benthic, feeding on mud and suspended matter. See **biramous appendage** (for diagram), **Arthropoda**, **Merostomata**.

Trimerophyta Primitive fossil vascular plants whose main axis formed lateral branch systems that dichotomized several times; lacked leaves, and some of the smaller branches terminated in elongate sporangia, while others were vegetative. The cortex was wide, with cells having thick walls; in the centre was the vascular strand. Homosporous, the trimerophytes probably evolved directly from the **Rhyniophyta** and seem to represent ancestors of ferns, progymnosperms and perhaps horsetails. They appeared in the early **Devonian** period *c.* 395 Myr BP but became extinct at the end of the mid-Devonian, *c.* 20 Myr later.

Triploblastic Animals with a body organization derived from three **germ layers** (ectoderm, endoderm, mesoderm). Includes all metazoans except coelenterates, which are diploblastic.

Triploid Nuclei, cells or organisms with three times the haploid number of chromosomes. A form of **polyploidy**. Triploid individuals are usually sterile due to failure of chromosomes to pair up during meiosis. See **endosperm**.

Trisomy Nucleus, cell or organism where one chromosome pair is represented by three chromosomes, giving a chromosome complement of $2n + 1$. May result from **non-disjunction**. See **Down's syndrome**.

Trochanter (1) Any knob for muscle attachment on the vertebrate femur. Three occur on mammalian femurs, largest in humans being conspicuous at hip joint. (2) Segment lying second from base of an insect leg.

Trochlear nerve Fourth vertebrate **cranial nerve**.

Trochophore (trochosphere) Ciliated larval stage of polychaetes, molluscs and rotifers. Usually planktonic, developing by spiral cleavage. The main ciliary band (*prototroch*) encircles the body in front of mouth (preorally). Commonly a second ring (*telotroch*) surrounds the anus. Coelom arises by schizocoely, not enterocoely. Two protonephridia may be present.

Trophic cascade hypothesis. The controversial hypothesis that food webs are more strongly influenced 'from above' (e.g. by predation) than 'from below' (e.g. by nutrient supply). Nutrient input, it is agreed, sets the potential productivity of lakes; but deviations from this potential are held to be due to food web effects. Nutrient and food web effects are complementary but act on different timescales. Species variations at the top of the food chain (top predator) bring about differences in selective predation, the effects of which cascade through the zooplankton and phytoplankton to influence the ecosystem dynamics.

Trophic level Theoretical term in ecology. One of a succession of steps in the transfer of matter and energy through a **community**, as may be brought about by such events as grazing, predation, parasitism, decomposition, etc. (see **food chain**). For theoretical and heuristic purposes, organisms are often treated as occupying the same trophic level when the matter and energy they contain have passed through the same number of steps (i.e. organisms) since their fixation in photosynthesis. Primary producers, herbivores, primary, secondary and tertiary carnivores and decomposers all commonly figure as trophic levels in the analysis of ecosystems. Different developmental stages and/or sexes within a species may occur in more than one trophic level, and so some consider trophic levels to be convenient abstractions lacking even approximate material counterparts. The number of trophic levels in a community is thought to be limited by inefficiency in **energy flow** from one trophic level to the next; however, food chains are no longer in tropical communities, where energy input is high, than they are in Arctic communities, where energy input is low.

Trophoblast Epithelium surrounding the mammalian **blastocyst**, forming outer layer of chorion and becoming part of the embryonic component of the **placenta** or of the **extra-embryonic membranes**. In humans, produces **human chorionic gonadotrophin** and **human placental lactogen**.

Tropical rain forest Forests found in three major regions of the world, covering only 7% of the land surface but containing more than half the total number of living species. Largest is in the Amazon basin of South America, with extensions into coastal Brazil, Central America, eastern Mexico and some islands in the West Indies; the second is in the Zaire basin of Africa, with an extension along the Liberian coast; the third extends from Sri Lanka and eastern India to Thailand, the Philippines, the large islands of Malaysia and a narrow strip along the north-east coast of Queensland in Australia. Neither water nor low temperatures are limiting factors for photosynthesis, and rainfall averages from 200–400 cm. yr^{-1}. It is the richest biome in terms of both plant and animal diversity, it being not unusual for 1 km^2 in Central or South America to contain several hundred bird species, and many thousands of butterfly, beetle and other insect species. A total of 700 tree species were identified in ten one-hectare forest plots in Borneo. The trees are evergreen, characterized by large leathery leaves. There is a poorly developed herbaceous layer on the forest floor because of low light levels; but there are many vines and epiphytes at higher levels. Many of the animals inhabit tree tops. Little accumulation of organic debris occurs because decomposers rapidly break it down, released nutrients being quickly absorbed by mycorrhizal roots or leached from the soil by rain (very little in undamaged forest). Tropical rain forest now forms about half the forested area of the planet, but is in process of being destroyed by human activities. Many of the soils of tropical rain forests are conditioned by high and constant temperatures and abundant rainfall and are relatively infertile (*latosols* – red clays largely leached of their nutrients). When the forest is cleared, this leaching process accelerates and the soils either erode or form thick, impenetrable crusts on which cultivation is impossible. It is estimated that by the end of this century most of the tropical rain forests (with huge untapped genetic resources) will have disappeared to produce fields which will be completely useless to agriculture within a very few years and then turn to desert, with considerable effect on global climate. See **biodiversity**.

Tropism Response to stimulus (e.g. gravity, light) in plants and sedentary animals by growth curvature; direction of curvature determined by direction of origin of the stimulus. See **nastic movement**.

Tropomyosin See **muscle contraction**.

Troponin See **muscle contraction**.

Truffle Fungal genus (**Ascomycota**, Order Tuberales) that produces a subterranean fruiting body; prized as a gastronomic delicacy.

Trumpet hyphae Drawn-out sieve cells, wider at the cross-walls than in the middle, in some brown algae (Order Laminariales). Active transport of photosynthates (mostly mannitol) through sieve cells occurs in this Order.

Trypanosomes Flagellated protozoans containing a kinetoplast. Parasites in blood of vertebrates and their blood-sucking invertebrates (e.g. insects, leeches) which act as vectors. The genus *Leishmania* parasitizes vertebrate lymphoid/macrophage cells, causing visceral leishmaniasis (kala-azar) in humans; *Trypanosoma* is a vertebrate blood parasite, different species causing sleeping sickness and Chagas' disease in humans and nagana in domesticated cattle. Notorious for **antigenic variation**, making search for vaccines against them almost pointless.

Trypsin Protease enzyme secreted in its inactive form (trypsinogen) by vertebrate pancreas and converted to active form by **enterokinase**. Converts proteins to peptides at optimum pH 7–8.

Trypsinogen Inactive precursor of **trypsin**.

Tsetse fly Genus (*Glossina*) of dipteran fly, sucking vertebrate blood and acting as vector for trypanosomes (spreading sleeping sickness).

TSH See **thyroid stimulating hormone**.

Tuatara See **Rhynchocephalia**.

Tube-feet (podia) Soft, hollow, extensile and retractile echinoderm appendages, responding to pressure changes within **water vascular system**. In some groups (starfish, sea urchins) they end in suckers and are locomotory; in other groups (brittle stars, crinoids) suckers are absent. Those around mouth may be used for feeding (sea cucumbers, brittle stars). Ciliated in crinoids, forming part of food-collecting mechanism.

Tuber Swollen end of **rhizome**, bearing buds in axils of scale-like rudimentary leaves (stem tuber), e.g. potato; or swollen root (root tuber), e.g. dahlia. Tubers contain stored food materials and are organs of vegetative propagation.

Tubulidentata Aardvarks. Order of fossorial termite-eating African mammals with long snouts and tongues; once regarded as edentates but now considered related to the extinct **Condylarthra**. One family (Orycteropidae) and species (*Orycteropus afer*). Large digitigrade mammal, with large ears and strong, digging claws. Skin nearly naked. No incisors or canines in the permanent dentition; cheek teeth rootless, growing throughout life. Arrangement of dentine around tubular pulp cavities gives Order its name.

Tubulin A characteristic eukaryotic protein. Filamentous (F) tubulin is formed by GTP-dependent polymerization of globular (G) protein monomers, each of these being a heterodimer ($\alpha\beta$) of two tubulin subunits whose genes are in turn multiply allelic. All α subunits appear to be able to copolymerize with all β subunits. Thirteen parallel tubulin filaments make up each of the composite cylindrical **microtubules** which play a major role in the eukaryotic cytoskeleton.

Tumour (neoplasm) Swelling caused by uncontrolled growth of cells; may be malignant (see **cancer cells**) or benign.

Tumour necrosis factors (TNFs) Misleading term, deriving from the nineteenth-century observation that visible tumours sometimes regressed dramatically if the patient suffered a simultaneous acute infection. This results from an effect of TNFs on the endothelial linings of blood vessels within certain tumours. But TNFs also have powerful systemic effects, causing 'systemic' or 'endotoxic' shock in high doses. One TNF group comprises **cytolysins** produced by cytotoxic **T cells** and macrophages; others, from **myeloid** cells, cause many tumour cells to be engulfed by macrophages and activate blood polymorphs and macrophages – critical in the elimination of intracellular pathogens (e.g. *Listeria monocytogenes*, mycobacteria and *Leishmania*) from macrophages. TNFs have direct effects on the hypothalamus and pituitary, and one or both TNF receptors (members of the nerve **growth factor** receptor superfamily) are located on most human cell types. TNF-α, produced by **mast cells**, is a multifunctional **cytokine** with effects on **inflammation**, sepsis, lipid and protein metabolism, **haemopoiesis**, angiogenesis, host resistance to parasites and malignancy. The TNF protein product p53 (see **cancer cells**) is probably a transcription factor, for it binds specific DNA sequences and is required for transcription of cyclin-dependent kinase (cdk) inhibitors (see **CKIs**).

Tumour suppressor gene (TSG) A gene normally restraining a cell from excessive proliferation, by involvement of its product in the transduction of an anti-mitotic signal. Mutants in, or loss of, these genes seem to provoke cancerous behaviour only if the system controlling cell division is already disturbed. One such gene, implicated in many human tumours (see **cancer cells**) is *P53*, whose protein product p53 is probably a **transcription factor** involved in **cell cycle** expression of cyclin-dependent kinase inhibitor genes (see **tumour necrosis factor**). The retinoblastoma gene *Rb* is an example, and its product (RB, or p105) plays a key role in cell proliferation by binding cellular transcription factor domains (notably on E2F) and modulating their behaviour positively or negatively (see **promoter**). The unphosphorylated form of RB, which binds E2F, may suppress entry to G1 of the cell cycle (see Fig. 15) unless it is first bound by transforming proteins encoded by certain viral **oncogenes**. Cells with poorly phosphorylated RB tend to be differentiated or non-dividing. Other TSG products are secreted, or are membrane-associated **signal transduction** proteins. See **suppressor mutation**, **adhesion**.

Tundra Treeless region to the farthest limits of plant growth. A vast biome, occupying about one fifth of the planet's surface, and best developed in the northern hemisphere, being found mostly north of the Arctic circle. The Arctic tundra essentially comprises a huge band across Eurasia and North America, with alpine tundra, more closely related to adjacent mountain forests, extending southward in the mountains. In general, permafrost (permanent ice) lies within a metre of the surface and ground conditions in the tundra are usually moist since water is unable to drain through the soil due to the ice. The soils are acid to neutral, low in nutrients and with available nitrogen in very short supply. Evaporation is low due to the high moisture content of the air at low temperatures. In contrast, some tundra regions are so dry they constitute true polar deserts. For plant survival, the mean temperature must be above freezing for at least one month per year; the growing season may be only two months per year. Plants include low shrubs, e.g. birch (*Betula*), willow (*Salix*), blueberry (*Vaccinium*), and Labrador tea (*Ledum*), perennial herbs, mosses and lichens. Few annuals are found. Many of the plants, including several grasses and sedges, are evergreen, which enables them to initiate photosynthesis immediately conditions of light, temperature and moisture allow. Vegetative reproduction is characteristic of many perennials, and much of the biomass of tundra plants is underground, consisting not only of roots but also of rhizomes.

Tunica corpus Organization of shoot of most flowering plants and a few gymnophytes, comprising one or more peripheral layers of cells (*tunica layers*) and an interior (*corpus*). Tunica layers undergo surface growth and corpus undergoes volume growth.

Tunicata See **Urochordata**.

Tupaiidae The family of modern tree shrews (Order Scandentia). See **Archonta**.

Turbellaria Class of **Platyhelminthes**. Mostly aquatic and free-living. Epidermis ciliated, especially ventrolaterally, for locomotion. Gut a simple or branched enteron, with ventral mouth and protrusible pharynx.

Turbidity Level of cloudiness of a suspension, commonly of cells. Often used as indicator of density of cells in suspension.

Turgor State of plant cell in which cell wall is rigid, stretched by an increase in volume of vacuole and protoplasm during absorption of water. The cell is described as *turgid*. An essential feature of mechanical support of the plant tissues, loss of turgor (when water loss exceeds absorption) being followed by wilting. Aids expansion of young cells in growth. Compare **plasmolysis**. See **water potential**.

Turgor pressure The hydrostatic pressure that develops within a plant cell through osmosis and/or imbibition. Plant cells tend to concentrate relatively strong solutions of salts within their vacuoles along with sugars, amino and other organic acids. Consequently, plant cells continually absorb water by osmosis and this creates their inner hydrostatic pressure which, in turn, keeps the cell rigid (turgid). This is a major role of the vacuole and its membrane (tonoplast). When turgid, the turgor pressure is equal and opposite to the pressure of the cell wall (wall pressure). See **water potential**.

Turion Detached winter bud by means of which many aquatic plants survive the winter; e.g. *Myriophyllum*.

Turner's syndrome Human genetic disorder caused by absence of a sex chromosome, leaving one X chromosome per cell (XO). **Non-disjunction** is often the cause. Individuals usually have webbing of the neck, narrowing of the aorta, and reduced height. Gonads undifferentiated secreting no hormones; so no menstrual cycle or pubic hair occur. Abnormal XO conditions also occur in mice, horses and rhesus monkeys. See **sex determination**.

Turnover number See **molecular activity**.

Twin spots A form of tissue mosaicism in which two tissues of different phenotype lie adjacent to one another against a tissue background of a third (commonly wild-type) phenotype. Caused by mitotic crossing-over between homologous chromosomes in a stem cell heterozygous for recessive alleles at two linked loci (e.g. + sn/y +) which normally complement to give wild-type tissue (background) but in this case give patches of singed (sn) and yellow (y) tissue phenotypes resulting from recessive homozygosity after the cross-over event and subsequent mitosis of mutant daughter cells.

Twitch fibres See **muscle contraction**.

Tylose (tylosis) Balloon-like enlargement of membrane of a **pit** in wall between a xylem parenchyma cell and a vessel or tracheid, protruding into the cavity of the latter and blocking it; wall may remain thin or become thickened and lignified. Tyloses occur in wood of various plants, often abundantly in heartwood of trees, and may be induced to form by wounding. Often occur in the xylem below developing abscission layer before leaf-fall.

Tympanic cavity See **ear, middle**.

Tympanic membrane (tympanum) See **ear drum**.

Tyndallization Fractional sterilization devised by the physicist John Tyndall (1820–93). Method involved lengthy immersion of the material in boiling brine (over 100°C), killing vegetative microorganisms, followed by a period in which unkilled spores were allowed to germinate, followed by further immersion in boiling brine. This could be repeated two or three times, gradually eliminating all bacterial spores. Compare **pasteurization**.

Type specimen See **holotype**.

Tyrosine kinase See **protein tyrosine kinase**.

Ubiquinone See **coenzyme Q**.

Ubiquitins A family of highly conserved proteins involved in the degradation of most cellular proteins, initially by binding to them covalently, which targets them for rapid hydrolysis by a (26S) proteolytic complex which contains a 20S degradative core termed a **proteasome**. They are bound, and often modulated, by ubiquitin-conjugating enzymes such as ubiquitin kinases, of which Cdc34 is required for **DNA replication** and Ubc9 is needed for M-phase of the **cell cycle**, in which they play a crucial role.

Ulna Bone of tetrapod fore-limb, lateral to radius. Articulates distally with carpus and proximally with humerus, above which it may protrude as the olecranon process ('funny bone'). Serves in attachment of main extensor muscle of fore-limb. May fuse with radius or, particularly in mammals, be lost altogether. See **pentadactyl limb**.

Ultimate factor Factor considered to be a cause of some biological structure or phenomenon, whose presence it serves to explain in terms of future states of affairs rather than (mechanistic) antecedent conditions. **Adaptation** and **ontogeny** require consideration of future conditions. Compare **proximate factor**. See **teleology**.

***Ultrabithorax* gene (*Ubx* gene)** A **homeotic** gene of the bithorax complex (BX-C) of *Drosophila* expressed in cells forming the adult haltere (posterior thoracic segment), whose loss of function causes the haltere to be replaced with a wing (homologous structure of middle thoracic segment). See **homeogene**.

Ultracentrifuge High-speed centrifuge, devised in 1925 by Svedberg, capable of sedimenting (and separating) particles as small as protein or nucleic acid molecules. See **cell fractionation**, **Svedberg unit**.

Ultramicrotome See **microtome**.

Ultrastructure Fine structure of a cell or tissue, as obtained by instruments with higher resolving power than the light **microscope**.

Ultraviolet irradiation Ultraviolet light (uv) has wavelenghts in the band between visible light and X-rays, i.e ~ 13-397 nm. Solar radiation is high in these wavelengths (see **ozone**), produced by emission from excited atoms and ions and so generated by discharge tubes. It is capable of ionizing atoms and molecules, of spliting molecules into free radicals (some biologically harmful) and of initiating polymerizations. In human skin, converts ergosterol into **vitamin D**. Because nucleic acids absorb energy in the uv range of the electromagnetic spectrum, DNA absorption peaking at ~ 260 nm, this can result in production of the thymidine dimers, chain breakage and cytosine hydration (see Fig. 115), with consequent impairment of DNA replication unless repaired (see **DNA repair mechanisms**, **melanin**, **cancer cells**).

Umbilical cord Connection between ventral surface of embryo of placental mammal and placenta. Consists mainly of allantoic mesoderm and blood vessels (umbilical artery and veins), covered by amniotic epithelium. Surrounded by amniotic fluid, it usually breaks or is bitten through at birth. See **extraembryonic membranes**.

Undulipodium Alternative name for eukaryotic **flagellum**.

Ungulate Non-formal term for a hoofed mammal; usually herding and adapted for grazing. Run on firm, open ground on tips of digits (*unguligrade* locomotion). Includes **Artiodactyla** and **Perissodactyla**.

Unguligrade See **ungulate**. Compare **digitigrade**, **plantigrade**.

Uniaxial (Of algae) having an axis composed of a single filament.

Unicellular (Of organisms or their parts) consisting of one cell only. Compare **acellular**, **multicellularity**.

Uniparous Producing one offspring at birth.

Uniport See **transport proteins**.

Uniramous appendage Appendage of myriapods and insects, consisting of a single series of segments (podomeres) as opposed to two series as in a **biramous appendage**.

Unisexual (Bot.) (1) (Of flowering plants and flowers) having stamens and carpels in separate flowers. Can be either monoecious or dioecious. (2) (Of algae) having only one type of gametangium formed on a plant. (Zool.) (Of an individual animal) not producing sperm or eggs but both. Compare **hermaphrodite**.

Unit membrane See **cell membranes**.

Unit of selection Whatever it is that **natural selection** ultimately distinguishes between and whose frequencies thereby not only change but may as a result be causally responsible for changing frequencies of character in the population. These causal chains would reflect our best theories at the time. Controversy surrounds this issue. Strong initial contenders for such units are individual organisms themselves. However, since selection discriminates between these on the grounds of phenotype, and since total phenotype is not what sexual reproduction reproduces, phenotypic characters themselves might better qualify as units of selection, individual organisms being composites of character traits. Strictly, however, it is usually not traits but their material causes (physical encodings, **genes**) which pass from parent to offspring in reproduction. Yet if selection acts at the level of classical Mendelian genes, many of which are pleiotropic, then as dominance relations between genetically determined characters are subject to background genetic modifiers, genetic backgrounds as much as individual alleles (alternative genes at a locus) may qualify as units of selection. However, alleles themselves may not be durable enough to be the units, since crossing-over, when it occurs, normally does so within genes. Instead, the

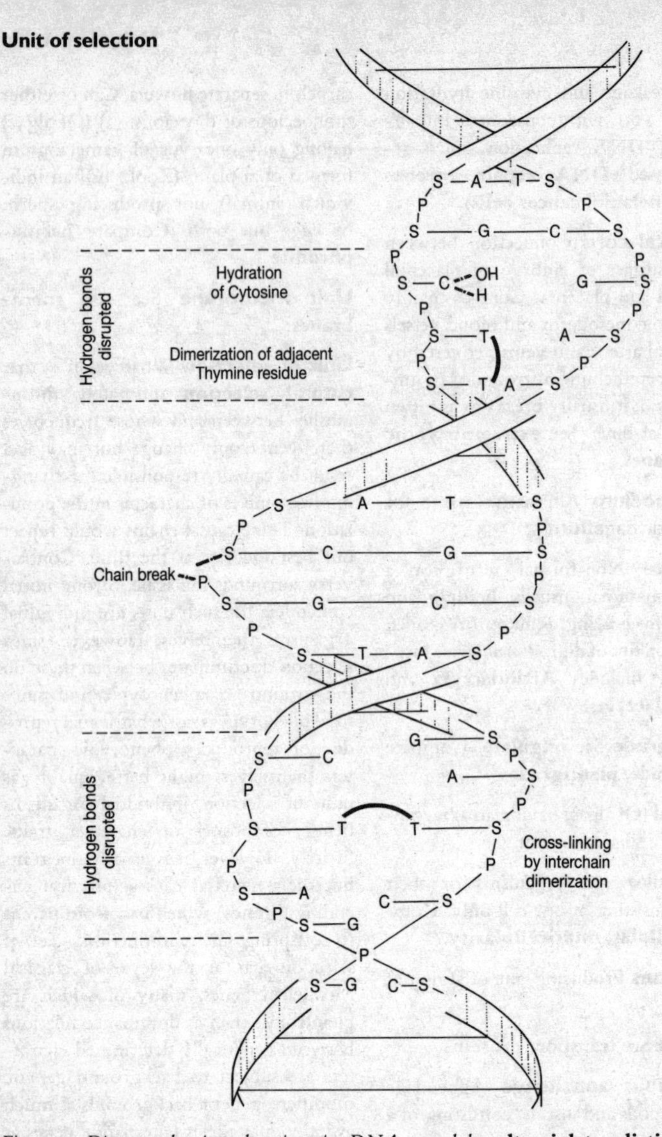

Fig. 115 *Diagram showing alterations in DNA caused by* **ultraviolet radiation**. *The formation of thymidine dimers is the change most likely to do damage to living cells and viruses. S-P-S-P is the sugar-phosphate backbone; A, T, G and C are the four bases; hydrogen bonds are shown on pale horizontal bars. Thymidine dimers are shown as curved bars.* (Adapted from *Ultraviolet Radiation and Nucleic Acid* by R. A. Deering. Copyright © Scientific American, Inc., 1962. All rights reserved.)

genetic unit favoured by selection may simply be any length of DNA which persists intact through a greater than average number of meioses. But a mutant molecular 'trick' (e.g. **gene conversion**) producing such better-than-average persistence would tend to be favoured automatically by selection, thereby disrupting the normal control selection exerts on genotype via expression of mutants in the phenotype: for such a mutant might have no phenotypic effect at all. In conclusion, a plurality of entities may qualify as units of selection, producing a range between neo-classical Mendelian genes and 'selfish' DNA. See **fitness**, **altruism**, **selfish DNA/genes**.

Univoltine (Of animals, typically insects) completing one generation cycle in a year and then diapausing through winter, the adults dying.

Uracil Pyrimidine base occurring in the nucleotide uridine monophosphate (UMP), whose radical is an integral component of RNA but not of DNA. The UMP radical, derived from UTP, base pairs with AMP and is incorporated into RNA by **RNA polymerase**.

Urea Main nitrogenous excretory product, $CO(NH_2)_2$, of ureotelic animals, produced by liver cells from deaminated excess amino acids via the *urea cycle*, part of which occurs in mitochondria. Degradation product of both pyrimidines and, via uric acid, purines (especially fish, amphibians). Elasmobranch fish have a high urea content in body fluid, enabling water to enter osmotically through soft body surfaces. See Fig. 116.

Uredinales Rust fungi. **Basidiomycota**, Class teliomycetes, comprising higher plant parasites. Like the smut fungi (**Ustilaginales**), and in contrast to other members of this division, they do not form basidiocarps. They do form dikaryotic hyphae, and basidia which are septate. Spores are produced in masses. The life cycles of rust can be complex; species can be heteroecious (require two different hosts to complete a life cycle) or autoecious (require only one host.) Once thought to be all obligate parasites; however, several species have been maintained in artificial culture. Many rusts are of great economic importance, causing extensive damage to crops throughout the world, e.g. *Puccinia graminis*, the cause of black stem rust of wheat and other cereals (this is but one of about 7000 species of rusts).

Uredinium Rust-coloured, linear streaks on the leaves and stems of wheat, containing dikaryotic urediniospores in rust fungi (**Basidiomycota**, Order Uredinales). These spores are capable of reinfection. See **aecium**, **pycnidium**, **telium**.

Ureotelic (Of animals) whose main nitrogenous waste is **urea**, such as elasmobranchs, adult amphibians, turtles and mammals. Compare **ammonotelic**, **uricotelic**.

Ureter Duct conveying urea away from kidney. In vertebrates, usually restricted to duct of amniotes leading from metanephric kidney to urinary bladder. Develops as outgrowth of **Wolffian duct**. See **cloaca**.

Urethra Duct leading from urinary **bladder** of mammals to exterior; joined by vas deferens in males, conducting both semen and urine. See **penis**.

Uric acid Major purine breakdown

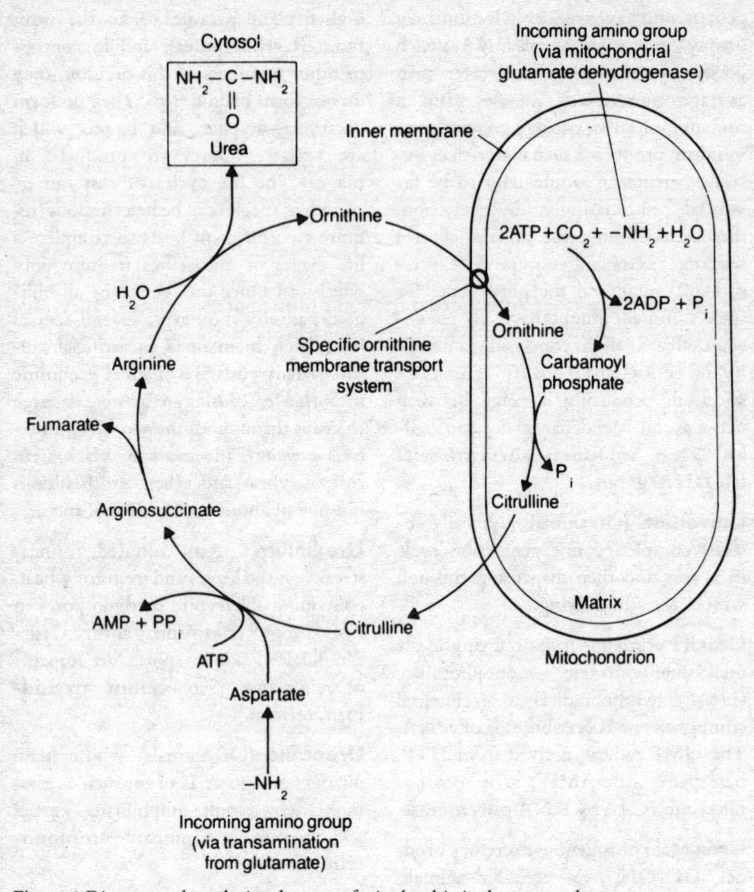

Fig. 116 *Diagram to show the involvement of mitochondria in the* **urea** *cycle.*

product of adenine and guanine; also the form in which nitrogen from excess amino acids is excreted in *uricotelic* animals. Almost insoluble in water, it is an adaptive nitrogenous waste in environments where water is at a premium.

Uricotelic (Of animals) whose main nitrogenous waste is **uric acid**; e.g. snakes, lizards, a few mammals (some desert rodents), embryonic and adult birds, terrestrial gastropods, and insects (see **Malpighian tubules**).

Urinary bladder See **bladder**.

Uriniferous tubule (kidney tubule) Convoluted tube forming bulk of each *nephron* (excluding Malpighian corpuscle) of vertebrate **kidneys**.

Urkaryote Term reserved for proto-eukaryotes, organisms (presumably single-celled) from which eukaryotes evolved after endosymbiotic incorporation of whichever prokaryotes gave rise to mitochondria, chloroplasts and perhaps some other membranous structures now included within the category of organelles. See **endosymbiosis**.

Urochordata (Tunicata) Subphylum of the **Chordata** (see **Protochordata**). Sea squirts, etc. Adults sedentary (some free-swimming forms, the Larvaceae, are probably progenetic), feeding microphagously by a ciliary-mucus method employing an **endostyle**, with pharyngeal gill slits, reduced nervous system, no notochord. Larvae active and tadpole-like, with well-developed nervous system and notochord. See **progenesis**.

Urodela Order of **Amphibia**. Newts and salamanders; with elongated bodies, tail and short limbs. Mostly confined to moist terrestrial habitats and/or ponds and lakes. Tadpoles resemble adults more than do those of frogs and toads. *Ambystoma*, a salamander, is an example of **neoteny**.

Urostyle Rod-shaped bone (representing fused caudal vertebrae) in anuran vertebral column; important in rigidifying posterior part of the animal during jumping. Flanked by two ilia, with which it articulates.

Ustilaginales The smut fungi. **Basidiomycota**, Class Teliomycetes. Like the rust fungi, they do not form complex basidiocarps, but do have dikaryotic hyphae and basidia, which are septate, and form spores in masses. They cause much economic loss of crop plants.

Uterus (womb) Muscular expansion of **Müllerian duct** of female mammals (except monotremes) in which embryo develops after implantation. Usually paired, each connecting to a Fallopian tube; single in humans, through fusion of lower part. Connected by vagina to exterior, controlled by sex hormones (see **menstrual cycle**), oxytocin and prostaglandins; enlarging at sexual maturity or in breeding season. See **cervix**. Glandular lining membrane (endometrium, decidua) nourishes early embryo. Smooth muscle in its wall greatly increases during pregnancy; its contraction ultimately expels embryos and their placentae (**parturition**).

Utriculus See **vestibular apparatus**.

Vaccine Material producing in an animal an immune reaction and an acquired **immunity** to a natural microorganism (*immunization*; compare **inoculation**). Almroth Wright (1861–1947) first showed that dead, rather than living (typhus), bacteria could produce immunity. BCG vaccine (against tuberculosis) contains live, attenuated (weakened) tubercle bacilli, while vaccination against smallpox has eliminated the disease. The ideal vaccine is one in which a single administration at birth provides lifetime protection from multiple diseases. Efforts at producing vaccines for the Third World centre upon genetically engineering live attenuated vaccine vectors (*multi-vaccine vectors*) that immunize against many antigens. Viral vaccines have the advantage of expressing naturally folded, glycosylated, assembled and secreted antigens in eukaryotic cells as well as stimulating cytotoxic **T cells** and antibody production. Immunity to poliomyelitis, yellow fever, measles and mumps has been achieved with 'live' attenuated virus; but antibodies to the numerous strains of **HIV** and influenza virus cross-react minimally or not at all (see **Pasteur**). Bacterial vaccines immunize for longer periods and can provide local immunity, e.g. in the gut. In mice, engineered BCG vaccine containing *Mycobacterium tuberculosis* with a plasmid expression **vector** for β-galactosidase, tetanus toxins and HIV proteins among others, has been shown to elicit antibody and cell-mediated responses; however, such recombinant vaccines may not provide immunity from pathogens showing **antigenic variation**. Some work suggests that the technique of *genetic immunization* may offer hope: gold microprojectiles coated in plasmids containing a gene encoding the antigenic protein (linked to a suitable **promoter**) are injected into the recipient animal to elicit antibody. See **toxoid**.

Condition	*Deaths per year (million)*	*Estimated episodes or incidence per year (million)*
Respiratory disease	10	15
Diarrhoea	4.3	28
Tuberculosis	3	10
Measles	2	67
Malaria	1.5	150
Hepatitis B	0.8	3.7
Meningitis	0.35	1
Schistosomiasis	0.33	10
AIDS	0.1	0.75
Worm infections	< 0.06	4,900

NOTE: These data were published in 1989 and the AIDS death rate is only an estimate.

Table 8 *Diseases of the Third World that potentially could be prevented by vaccines.*

Vacuolar membrance (Bot.) See **tonoplast**.

Vacuole Membrane-bound regions within the cytoplasm. In plant cells the membrane is termed the **tonoplast**. The immature plant cell typically has numerous small vacuoles, which increase in size and fuse into a single vacuole as the cell enlarges. A vacuole may take up as much as 90% of a plant cell's volume, with the cytoplasm forming a thin peripheral layer closely pressed against the cell wall. A direct consequence is the development of turgor pressure and maintenance of tissue rigidity. Within the vacuole the cell sap comprises water, with other components depending upon the species of plant, as well as its physiological state. Salts and sugars are usually present, along with, in some species, dissolved proteins. Ions (e.g. Ca^{2+}) accumulate in the cell sap in concentrations far in excess of those in the surrounding cytoplasm, even to the extent that crystals form, e.g. calcium oxalate crystals are particularly common (see **inositol 1,4,5-triphosphate**). The cell sap is slightly acidic, and very acidic in citrus fruits. Vacuoles are important storage sites for various metabolites (e.g. malic acid in CAM plants); they also remove toxic secondary products, and are often the sites of pigment deposition (e.g. anthocyanins), as well as being involved with the breakdown of macromolecules, and recycling of their components with a cell. Entire organelles (ribosomes, mitochondria, plastids) may enter a vacuole where they are broken down (see **lysosome**). Vacuoles are absent from the cells of bacteria and **Cyanobacteria**. See **contractile vacuole**, **pusule**, **food vacuole**, **water potential**.

Vagility Distance in a straight line between either (a) an individual's birth place and the place where it dies, or (b) its site of conception and the place where it gives rise to a new zygote. Populations or demes of low mean vagility have more probability of recruiting rare homozygous mutants.

Vagina Duct of female mammal connecting uteri with the exterior via short vestibule. Usually single and median due to fusion of lower part of **Müllerian duct** in the embryo. Receives the male's penis during copulation. Lined with stratified non-glandular epithelium, which may undergo cyclical changes during oestrous cycle under the influence of sex hormones.

Vagus nerve Large, mixed (sensory and motor), vertebrate **cranial nerve** (CN X) innervating much of the gut, ventilatory system (including fish gill muscles) and heart. Each emerges within skull from the medulla, running back on each side as an abdominal ramus, the main parasympathetic route. **Accessory nerve** is basically a motor element of the vagus, emerging further back along the spinal cord. Vagi tend to terminate in plexi, such as in the sympathetic ganglia of the solar plexus. See **autonomic nervous system**.

Valve (Bot.) (1) One of several parts into which a **capsule** separates after dehiscence by longitudinal splitting. (2) One of two halves of diatom cell wall (frustule). (Zool.) (1) (Of heart) flap or pocket of vertebrate heart wall at entry and exit of each chamber, preventing back-flow of blood (e.g. **mitral valve, tricuspid valve**). (2) (Of veins, lymphatics) pocket-like flap in birds and mammals, preventing back-flow.

Vanilla Genus of orchid, the seed pods of which are the natural source of the popular flavouring of the same name. Extracted from the dried fermented seed pods. Originally used by the Aztecs in what is now Mexico. Cultivated now in Madagascar and other islands in the western Indian Ocean. Synthetically produced vanilla flavouring (*vanillin*) is now more widely used.

Variation Phenotypic and/or genotypic differences between individuals of a population. *Continuous* (*qualitative*) characters are those exemplified throughout the phenotypic range, tending to be determined by **polygenes**. *Discontinuous* (*quantitative*) characters are those unrepresented in all parts of the phenotypic range (a lack of intermediates), tending to be *polymorphic*, determined by genetic 'switch-mechanisms' (see **polymorphism**). Controversy over the relative influence of heredity and environment in producing phenotypic differences fuels the Nature-Nurture debate. Problems arise in obtaining acceptable control populations to test hypotheses. By starting with genetically uniform material (e.g. by cloning or repeated inbreeding) it is often possible to compare phenotypes produced under different environmental regimes and to estimate **heritability**. See **genetic variation**, **phenotypic plasticity**, **infraspecific variation**.

Varicosity (Of some motor nerve axons) a swelling, releasing the neurotransmitter noradrenaline via synaptic vesicles – not at specialized active zones on the membrane (as at synapses) but seemingly randomly over its surface. An axon may have several varicosities where it passes close to smooth muscle, the transmitter acting locally on **G-protein**-linked **receptors**.

Variegation Irregular variation in colour of plant organs (e.g. leaves, flowers) through suppression of normal pigment development in certain areas. Commonly due to **transposable elements**, or to chloroplast- (esp. cpDNA-)based mosaicism, or maybe a disease symptom (e.g. mosaic diseases due to viral infection).

Variety Taxonomic category involving groups of plants or animals of less than species-rank. Some botanists view plant varieties as equivalent to subspecies, while others consider them divisions of subspecies. See **infraspecific variation**.

Vasa recta Capillaries taking blood into the **kidney** medulla from the efferent arterioles of juxtaglomerular nephrons (those whose loops of Henle extend into the inner zone of the medulla – about 15% of nephrons in humans). Blood flow is slow, totalling *c.* 10% of kidney blood volume at any time. Originally isotonic with cortical fluid, blood in descending vasa recta becomes increasingly concentrated as water leaves by osmosis and small solutes enter down concentration gradients. At a hairpin bend the vasa recta turns upwards and flows towards the cortex, where water is reabsorbed into the blood from the peritubular fluid and solutes leave by diffusion. This produces a passive **countercurrent system** of exchange which preserves the medullary solute gradient (higher in the inner medulla than the outer). Water tends to leave the medulla via the vasa recta because of the colloid osmotic pressure due to their contained proteins.

Vascular (Bot.) Adjective pertaining to any plant tissue or region comprising or giving rise to conducting tissue (e.g. **xylem**, **phloem**, **vascular cambium**).

Vascular bundle Longitudinal strand of conducting (vascular) tissue comprising essentially **xylem** and **phloem**; unit of stelar structure in stems of gymnophytes and flowering plants and occurring in appendages of leaves (e.g. veins in leaves). In gymnophyte stems, arranged in a ring surrounding the pith; in monocotyledons, scattered throughout stem tissue. May be (a) *collateral*, with the phloem on same radius as xylem and external to it (typical condition in flowering plants and gymnophytes); (b) *bicollateral*, with two phloem groups, external and internal to xylem, on same radius (uncommon, occurring in some dicotyledons, e.g. marrow); (c) *concentric*, with one tissue surrounding the other. The latter are *amphicribral* when phloem surrounds the xylem, as in some ferns, and *amphivasal* when xylem surrounds phloem, as in rhizomes of certain monocotyledons. Vascular bundles are further described as *open* when cambium is present, as in dicotyledons, and *closed* when cambium is absent, as in monocotyledons.

Vascular cambium (Bot.) One of two lateral meristems in vascular plants which produces secondary tissues that comprise the secondary plant body. Cells of the vascular cambium are highly vacuolated and occur in two forms: (1) *fusiform initials*, vertically elongate cells; (2) *ray initials*, squarish cells. The vascular cambium arises from procambial cells that remain undifferentiated between the primary xylem and primary phloem, as well as from **parenchyma** of the interfasicular regions. It produces secondary xylem and secondary phloem through periclinal divisions of the cambial initials and their derivatives (i.e. the cell plate that forms between the dividing cambial cell's initials is parallel to the stem or root surface). Cambial initials divided off towards the outside and inside of the root or stem become phloem or xylem cells, respectively. Thus, a long continuous radial row of cells is formed, extending from the cambial initial outwards to the phloem and inwards to the xylem. Xylem and phloem cells produced by fusiform initials have their long axes orientated vertically and form the axial system of the secondary vascular tissues, while those cells produced by the ray initials form the vascular rays or radial system. That portion of the cambium arising within the **vascular bundles** is the fasicular cambium, and that arising in the interfasicular region is the interfasicular cambium. In woody stems, the formation of secondary xylem and secondary phloem results in the formation of a cylinder of secondary vascular tissues, with rays extending radially throughout this cylinder. More secondary xylem than secondary phloem is formed. See **cork cambium**.

Vascular cylinder See **stele**.

Vascular plant Any plant containing a **vascular system**. See **Tracheophyta**.

Vascular system (Bot.) See **stele**. (Zool.) System of fluid-filled vessels or spaces in animals where transport and/or hydrostatics is involved; e.g. **blood system**, **lymphatic system**, **water vascular system**.

Vas deferens (pl. vasa deferentia) One of a pair of muscular tubes (one on each side) conveying sperm from testis to exterior. In male amniotes the vasa lead from the epididymis to the cloaca or urethra. See **Wolffian duct**, **vas efferens**.

Vas efferens (pl. vasa efferentia) Tube (many on each side), developing from a mesonephric tubule and conveying sperm from the testis to the epididymis in male vertebrates. See **kidney**, **testis**, **Wolffian duct**.

Vasoconstriction Narrowing of blood vessel diameter (commonly arteriole) through smooth muscle contraction. Brought about in skin and abdomen by adrenaline, but more generally by vasoconstrictor nerves of sympathetic nervous system. **Antidiuretic hormone** and **angiotensins** may also be involved. See **vasomotor centre**, **endothelin**.

Vasodilation Increase in blood vessel diameter (commonly arteriole), often through reduction in its sympathetic nervous stimulation, by cholinergic nervous stimulation, or (in heart and skeletal tissue) by adrenaline. **Histamine** and **kinins** are vasodilators associated with inflammatory response. See **capillary**, **vasomotor centre**.

Vasomotor centre Group of neurones in vertebrate medulla (see **brainstem**) whose reflex outputs control muscle tone of arteriole walls, producing normal **vasoconstriction** (vasomotor tone), but increasing this tone when required by raised sympathetic output. Inputs include those from **baroreceptors** influencing cardiac centres.

Vasopressin See **antidiuretic hormone**.

Vector (1) Organism (animal, fungus) housing parasites and transmitting them from one host to another, commonly acting as a host itself. Insects, ticks and mites often act as vectors, many transmitting parasites to man, crops and domestic animals. See **malaria**, **filarial worms**. (2) A vehicle, e.g. a drug-resistance **plasmid** or modified **bacteriophage**, which has its own replication origin and is capable of use for insertion and cloning of other pieces of DNA. *Cloning vectors* are those in which the foreign DNA is cloned prior to further study, e.g. for construction of **gene libraries**. These were first constructed by cloning DNA fragments into a λ phage vector; but only fragments of about 15 kb can be packaged into this phage head. *Cosmid vectors* are cloning vectors carrying the cos sites (which leave GC-rich 'cohesive ends' after specific endonuclease treatment) from λ phage, plus a plasmid origin of replication and a drug-resistance gene. These enable cloning of fragments up to about 45 kb in length. For cloning of fragments over 100 kb in length, **yeast artificial chromosomes** are employed. *Expression vectors* are those in which the foreign DNA is not just cloned, but also expressed

Fig. 117 *Expression* **vector** *technique employed to ensure that a previously cloned cDNA sample is expressed within an* E. coli *cell so that the protein encoded by the original mRNA (which acted as a template for cDNA production) is produced in large amounts. An* E. coli *promoter is used for the expression of foreign protein in the cell. IPTG inactivates the repressor once large enough numbers of* E. coli *have been grown.* (From *Recombinant DNA 2/E* by Watson, Gilman, Witkowski, and Zoller. Copyright © James D. Watson, Michael Gilman, Jan Witkowski and Mark Zoller, 1992. Reprinted with permission of W. H. Freeman and Company.)

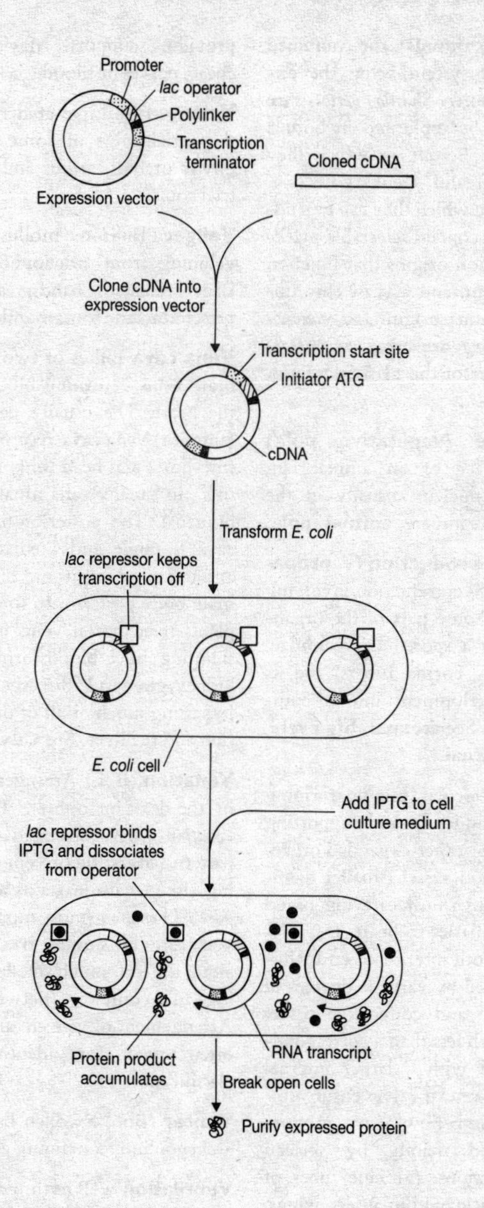
Promoter
lac operator
Polylinker
Transcription terminator
Expression vector
Cloned cDNA
Clone cDNA into expression vector
Transcription start site
Initiator ATG
cDNA
Transform E. coli
lac repressor keeps transcription off
E. coli cell
Add IPTG to cell culture medium
lac repressor binds IPTG and dissociates from operator
Protein product accumulates
RNA transcript
Break open cells
Purify expressed protein

(see **expression signals**), the combined vector-and-host system being the **expression system**. *Shuttle vectors* are those that can be replicated in both a bacterium (e.g. *E. coli*, in which they are propagated) and a eukaryote (e.g. *Saccharomyces*) in which they can be studied. They must contain selectable markers and replication origins that function in each organism, and a yeast chromosomal replication origin (an *autonomously replicating sequence*, or ARS) often forms part of the plasmid vector. See **PCR**.

Vegetal pole (vegetative pole) Point on surface of an animal egg furthest from nucleus; usually in the yolkiest end. Compare **animal pole**.

Vegetative reproduction (v. propagation) (Bot.) Reproduction involving detachment of some part of the organism other than a spore (e.g. gemmae, rhizomes, bulbs, corms, tubers) and its subsequent development into a complete organism. See **asexual**, **life cycle**. (Zool.) See **asexual**.

Vein (Bot.) Vascular bundle forming part of the conducting and supporting tissue of a leaf or other expanded organ. (Zool.) (1) Blood vessel (smaller diameter than venous sinus) carrying blood back from capillaries to heart. In vertebrates, has smooth interior of **endothelium** surrounded by variable degrees of smooth muscle and connective tissue, but always much less than a corresponding artery and with a larger internal diameter. Non-return valves ensure unidirectional blood flow, low pressure being provided mainly by skeletal muscle contractions. (2) Fine tubes of toughened cuticle within insect wings, providing support. May contain tracheae, nerves or blood.

Velamen Multiple epidermis covering the aerial roots of some tropical epiphytic orchids, aroids and some terrestrial roots.

Veliger Planktonic molluscan larva, developing from trochophore and with larger ciliated bands; adult organs present include foot, mantle and shell.

Vena cava Either of two major vertebrate veins, returning blood directly to the heart. The usually paired superior (anterior) vena cava returns blood from fore-limbs and head (only the right persists in many mammals, including humans). The posterior (inferior) vena cava is single and median in lungfish and tetrapods, returning blood from the main body posterior to fore-limbs. Supplants renal portal veins in phylogeny, shunting much blood formerly entering kidney glomeruli directly to the heart, bypassing much or all of the renal filtration and the liver. See **Cuvierian duct**.

Venation (Bot.) Arrangement of veins in the leaf mesophyll. Termed *netted venation* where arranged in branching pattern (largest dicot vein often extending along main leaf axis as midrib); or *parallel venation* (most monocots) where veins are of similar size and parallel along the leaf, much smaller veins forming a complex network. (Zool.) Arrangement of veins in an insect wing; often used for identification and classification.

Venter (Bot.) Swollen basal region of archegonium, containing egg cell.

Ventilation Rhythmic breathing movements of an animal's muscles and skeleton, increasing gaseous exchange across respiratory surfaces. Often termed *external respiration*. In large active insects the abdomen commonly vibrates and undergoes reversible 'telescoping' as the sternum and tergum of each segment move, causing compression and expansion of the tracheal system, pumping air in and out of the spiracles. Thoracic pumping may also be involved during flight. In fish, gill ventilation is achieved using a pressure pump in front and a suction pump behind. Amphibians at rest on land employ throat movements to renew air in the mouth, where gaseous exchange occurs; when active, a buccal force-pump ventilates the lungs, possibly assisted by flank movements. Bird ventilation is very complex (see **air sacs**). Only mammals have a **diaphragm** (stimulated by phrenic nerves from the cervical spinal cord) which, with **intercostals**, ventilates the lungs by changing thorax volume during inspiration and expiration under the control of the medulla (see **brainstem**). Here an *inspiratory centre* receives inputs from stretch receptors in the lungs (via the vagus), from **carotid bodies** and from an inhibitory *expiratory centre* in the pons. Both centres are directly sensitive to gaseous tensions in the blood.

Ventral Generally the surface resting on, or facing, the substratum. (Bot.) See **adaxial**. (Zool.) In chordates, body surface furthest from nerve cord.

Ventral aorta See **aorta**, **ventral**.

Ventricle (1) Chamber of vertebrate heart; either single or paired (tendency to separate into two in phylogeny). Thicker-walled than **atrium**. See **heart**, **heart cycle**. (2) Chamber of molluscan heart responsible for blood flow to tissues. (3) One of the cavities of vertebrate brain filled with **cerebrospinal fluid**. A pair in the cerebrum, another in the rest of the forebrain, and one in the medulla.

Venule Small vertebrate blood vessel, intermediate in structure and position between capillary and vein. Most have non-return valves. Highly permeable, comprising layer of endothelium with coat of collagen fibres. Larger venules also have smooth muscle in their walls.

Vernalization (Bot.) Promotion of flowering by application of cold treatment; hormones, cold and day length interact to modify the response. First studied in cereals, some of which if sown in spring will not flower in the same year but continue to grow vegetatively. Such plants (winter varieties) need to be sown in autumn of the year preceding that in which they are to flower, and contrast with spring varieties which, planted in the spring, flower in the same year. By vernalization, winter cereals can be sown and brought to flower in one season. Seed is moistened sufficiently to allow germination to begin, but not enough to encourage rapid growth, and when tips of radicles are just emerging the seed is exposed to a temperature just above 0°C for a few weeks. Seed thus vernalized acquires properties of seed of winter varieties; sown in the spring, it produces a crop in summer of same year. Even after vernalization, plant must be subjected to suitable photoperiod for flowering to occur. In some plants, gibberellin treatment can substitute for cold exposure, while vernalization effects can be reversed in some plants by exposure to high temperature in anaerobic conditions.

Vernation Way in which leaves are arranged in relation to one another in the bud.

Vertebra See **vertebral column**.

Vertebral column (spinal column) Segmentally arranged chain of bones or cartilages near vertebrate dorsal surface, surrounding and protecting the spinal cord. In most vertebrates, vertebrae form a hollow rod attached anteriorly to skull, largely replacing the notochord. Each vertebra straddles the junction of two embryonic somites, a central mass (*centrum*) replacing the notochord, and an arch above (*neural arch*) enclosing the spinal cord. A similar arch below (*haemal arch*), or ventral projections, often encloses major axial blood vessels (aorta, posterior cardinal veins/vena cava). Projections from vertebrae are for muscle attachment. Vertebrae join at their centra (see **symphysis**), often also by projections of neural arch; only restricted movement is possible between any two. Fish vertebrae are very similar from skull to tail; but in tetrapods there are usually these regional differences: atlas and axis; *cervical vertebrae*, with very short ribs; *thoracic vertebrae*, bearing main ribs; *lumbar vertebrae*, without ribs; *sacral vertebrae*, attached by rudimentary ribs to pelvic girdle; and tail (*caudal*) vertebrae. See **bone**, **cartilage**.

Vertebrata (Craniata) Major subphylum of **chordata**. Cyclostomes, fish, amphibians, reptiles, birds and mammals. Differ from non-vertebrate chordates in having a skull, vertebral column and well-developed brain. All but agnathans have paired jaws. Classes include **Agnatha**, **Placodermi**, **Acanthodii**, **Chondrichthyes**, **Osteichthyes**, **Amphibia**, **Reptilia**, **Aves** and **Mammalia**.

Vessel (trachea) Non-living element of xylem comprising tube-like series of cells arranged end-to-end, running parallel with long axis of the organ in which it lies and in communication with adjacent elements by means of numerous pits in side-walls. Components of a vessel, *vessel members*, are cylindrical, sometimes broader than long, with large perforations in end walls, and function in **translocation** of water and mineral salts, and in mechanical support. They have evolved from tracheids, principal features of this evolution being elimination of multiperforate end walls to form a single large perforation, reduction or disappearance of taper of end walls and a change from a long, narrow element to one relatively short and wide. Generally thought to be more efficient conductors of water than tracheids because water can flow relatively unimpeded from one vessel member to another through the perforations. With few exceptions, confined to flowering plants.

Vestibular apparatus (v. organ) Complex part of the *membranous labyrinth* forming part of inner ear of most vertebrates generating sensory input for reflexes of balance and general posture. Lying on either side within otic region of the skull, each comprises a closed system of cavities forming sacs and canals lined by epithelium and filled with fluid *endolymph* resembling tissue fluid. Major sac-like parts are the *utriculus* (utricle) and the *sacculus* (saccule), the latter normally lower, each containing a **macula** with **hair cells** and otoliths. These respond to head tilt and accleratory movements; but turning movements are detected by the *semicircular canals*, three of which are generally present on each side. Two canals lie in vertical planes at right angles, the third horizontal. Each has a swollen *ampulla* containing a sensory *crista* with hair cells embedded in a jelly-like *cupula* which swings to and fro under influence of endolymph in the canals, pulling the hair cells whose combined outputs along the **auditory nerve** register turning movements in all spatial planes. The **cochlea** is the other component of the membranous labyrinth.

Vestibulocochlear nerve (auditory nerve, acoustic nerve) Eighth vertebrate cranial nerve, innervating the inner ear. Essentially a dorsal root of facial nerve, relaying impulses from hair cells of the cochlea (via cochlear branch) and from vestibular apparatus (via vestibular branch).

Vestigial organ An organ whose size and structure have diminished over evolutionary time through reduced selection pressure. Only traces may remain, comparative anatomy providing evidence of phylogeny. Remains of pelvic girdles in snakes and whales indicate tetrapod ancestries. See **appendix** (not vestigial in humans), **evolution**.

Vibrissae Stiff hairs or feathers, usually projecting from face. Tactile.

Villus (1) Finger-like projection from the lining of the small intestines of many vertebrates. Their great number (up to 4–5 million, each up to 1 mm long in humans) gives the mucosa a velvet-like appearance. Each is covered by epithelium whose brush borders increase the surface area for absorption and digestion 25-fold (see **microvillus**, **digestion**). Each contains blood vessels and a lacteal, extending and retracting by means of smooth muscle. (2) For chorionic (trophoblastic) villi, see **placenta**.

Virchow, Rudolph German medical microscopist (1821–1902) who did most to dispel the notion, current from the time of Hippocrates, that disease resulted from imbalance in body 'humours'. Replaced it by a cell-based theory, arguing that cells are derived only from other cells (*omnis cellula a cellula*), broadening and deepening **cell theory**.

Virion See **virus**.

Viroid Smallest known agents of infectious disease; smaller than the smallest viral genome, and lacking capsids (protein coats). Known only from plants, they comprise small, single-stranded molecules of RNA that replicate autonomously in susceptible cells. The RNA may be a closed circle or a hairpin-shaped structure open at one end, and have 300–400 nucleotides. Viroids are found almost exclusively in the nucleus of infected cells, replicating themselves by forming a complementary strand that acts as a template. Presumably host enzymes catalyse the process. It is suggested that viroids may cause their symptoms by interferences with gene regulation in the infected host cells. They have been identified as causes responsible for some very important plant diseases.

Virus One of a group of minute infectious agents (20–300 nm long and/or wide), unable to multiply except inside living cell of a host, of which they are obligate parasites and outside of which they are inert. Pass through filters which trap bacteria, and resolvable only by electron microscopy. Not normally regarded as 'living' since none has any enzyme activity away from its host. When inside a cell, may pack together in a crystalline condition through symmetry and regularity of their capsids. They are biological systems, since each contains molecular information in the form of nucleic acid (but unlike cells, never both DNA and RNA), transcribed and replicated within the host cell. The fully formed virus particle (*virion*) contains its nucleic acid or nucleoprotein core either within a naked coat of protein (*capsid*) consisting of protein subunits (*capsomeres*) or within a capsid enveloped by one or more host cell membranes acquired during exit from the cell and often modified, as by addition of specific glycoproteins. Capsid symmetry may be *helical* (capsomeres forming a rod-shaped helix around the nucleic acid core, e.g. tobacco mosaic virus, TMV); or *icosahedral* (*spherical*, capsomeres forming a 20-sided structure), capsomere number ranging from 12 (φX174) to 252 (adenovirus). Some viruses (*phages*) infect fungi (*mycophages*) or bacteria (**bacteriophages**).

Viruses may be classified in terms of the type of nucleic acid core they contain. In RNA viruses this may be either single-stranded (e.g. picornaviruses, causing polio and the common cold) or double-stranded (e.g. reovirus, causing diarrhoea); likewise DNA viruses (e.g. single-stranded in φX174 phage and parvovirus; double-stranded in adenovirus, herpesvirus and poxvirus). RNA viruses are of three main types: + *strand RNA viruses*, whose genome serves as mRNA in the host cell and serves as template for a minus (−) strand RNA intermediate; − *strand RNA viruses*, which cannot serve directly as mRNA but rather as templates for mRNA synthesis via a virion transcriptase; and *retroviruses*, which are + strand and can serve as mRNAs but on infection immediately act as templates for double-stranded DNA synthesis (which immediately integrates into the host chromosome) via a contained or encoded **reverse transcriptase**. Human T-lymphotropic viruses (HTLVs) are single-stranded retroviruses and can cause acquired immune deficiency syndrome (AIDS). Any human *immunodeficiency virus* (see **HIV**) forms part of the *lentivirus* subgroup of RNA retroviruses. See Fig. 118.

While bacteriophages inject their genomes into the host cell, animal viruses are engulfed by endocytosis or, if encapsulated by a membrane, fuse with the host plasmalemma to release the nucleoprotein core into the cell. HIV and poliovirus are among many with **virus receptors** on host cell membranes. Once inside, the genome is commonly initially transcribed by host enzymes; but virally encoded enzymes usually then take over. Host cell synthesis is generally shut down, the viral genome replicated, and capsomeres synthesized prior to assembly into mature virions (see **self-assembly**). The virus commonly encodes a late-produced enzyme, rupturing the host plasma membrane (the *lytic phase*) to release the infective progeny; but a viral genome may become integrated into the host chromosome and be replicated along with it (a *provirus*). Many eukaryotic genomes have proviral components. Sometimes this results in *neo plastic transformation* of the cell (see

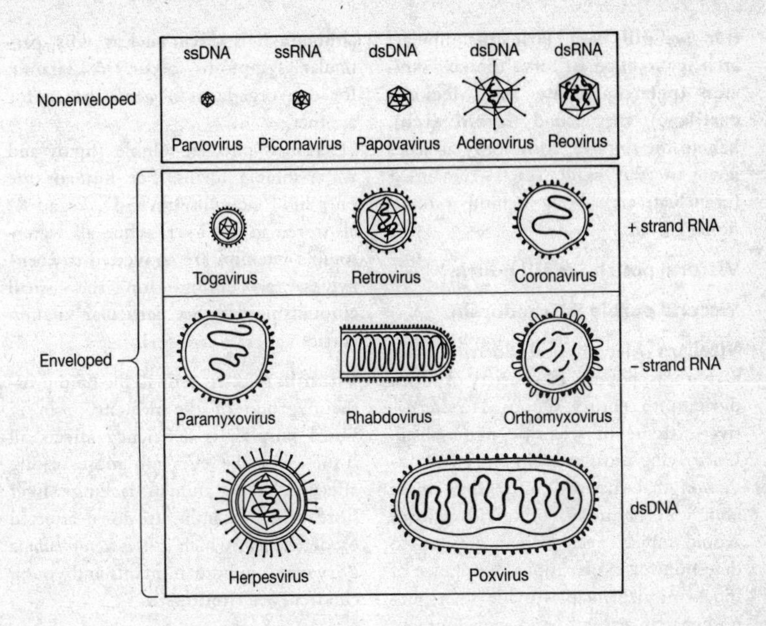

Fig. 118 *Illustration of the important families of animal viruses. ss = single-stranded; ds = double-stranded. See* **virus** *for clarification.*

oncogene, cancer cell) through synthesis of proteins normally produced only during viral multiplication. DNA tumour viruses include *adenoviruses* and *papovaviruses*; RNA tumour viruses are capsulated and include some retroviruses (e.g. Rous sarcoma virus). The DNA of viruses invading eukaryotic cells often contains **enhancers** activated by host cell proteins. Viruses may be used as genetic tools in **transduction**. They may have evolved from **plasmids** that came to encode capsid proteins. See **interferons**.

Virus receptors Cell **adhesion** receptors subverted as virus receptors, notably the CD4 **accessory molecule** in the case of HIV, and ICAM-1 in the case of rhinoviruses causing the common cold.

Viscera The large organs housed within an animal's coelomic cavities (i.e. within the thoracic and abdominal cavities of humans). The adjective *visceral* is often contrasted with *somatic*, where the latter in this context relates to structures lying external to the coelomic cavities.

Visceral arch (1) One of a series of partitions in fish and tetrapod embryos on each side of the pharynx. In fish, lying between mouth and spiracle and between this and adjacent gill slits; in tetrapod embryos, between corresponding gill pouches; (2) skeletal bars (cartilage or bone) lying in these partitions in fish (see **gill bar**). First (mandibular) arch is modified to form the jaw skeleton (**palatoquadrate** and **Meckel's cartilage**); the second (**hyoid arch**), behind the spiracle, commonly attaches jaws to the skull, each remaining (branchial) arch lying behind a functional gill slit.

Visceral pouch See **gill pouch**.

Visceral purple See **rhodopsin**.

Vitalism Metaphysical doctrine with early roots, popular in a variety of forms during 19th century. Opposed to alternative extreme of scientific materialism. Underlying most vitalisms was the conviction that life was more than mere complex chemistry, otherwise science would subject even human activity to deterministic explanations. Ignorance of biochemical principles made such phenomena as growth and development mystifying, employing terminology beyond physics or chemistry. Such causal agencies as 'entelechies' and 'vital forces' were invoked for really baffling phenomena but added nothing to understanding. Vitalists, often religiously inspired, eschewed Darwin's mechanistic philosophy.

Vital staining See **staining**.

Vitamin Organic substance not normally synthesized by an organism, which it must obtain from its environment in minute amounts (a micronutrient). **Essential amino acids** are required in larger amounts, and are not vitamins. *Provitamins*, closely related precursors of some vitamins, may occur in the diet. Absence of a vitamin from the diet for sufficient time gives symptoms of a resulting *deficiency disease*, although it is often unclear why particular symptoms occur. A vitamin for one organism may not be so for another.

Classified into fat-soluble (lipid) and water-soluble forms. For humans the only lipid vitamins are A, D, E and K, all stored in the liver, while all water-soluble vitamins are converted to coenzymes, accounting for the small amounts needed. See particular vitamin entries.

Vitamin A 11-*cis*-retinal, the lipid prosthetic group of the protein opsin in visual purple. Its deficiency affects all tissues, but the eyes are most readily affected. Young animals, lacking a liver store of the vitamin, are most affected by deficiency, which causes *xeropthalmia* ('dry eyes') in human infants and young children. See **rhodopsin**.

Vitamin B complex Several water-soluble vitamins (currently 12). B_1(aneurine, or thiamine) prevents beri-beri. Group includes **biotin**, **cyanobalamin** (B_{12}), **folic acid**, **nicotinic acid**, **pantothenic acid**, **pyridoxine** (B_6) and **riboflavin** (B_2).

Vitamin C See **ascorbic acid**.

Vitamin D Small group of fat-soluble (lipid) vitamins of humans and some other animals, deficiency causing rickets. Some dietary in origin, but also synthesized in the skin under ultraviolet light. One form (D_2, *ergocalciferol*) derives from provitamin ergosterol; another (D_3, *cholecalciferol*) derives from provitamin dehydrocholesterol. Both act as hormones (see **nuclear receptors**). People receiving insufficient sunlight can supplement their vitamin D levels by eating liver, particularly fish liver oils. The main circulating form of vitamin D in animals, *25-dihydroxycholecalciferol*, is very active and derived by modification of cholecalciferol in the liver. Like other forms of the vitamin it promotes uptake of calcium ions in the ileum and is required for calcification of bones and teeth. See **calcitonin**.

Vitamin E (tocopherol) Group of fat-soluble vitamins obtained principally from plant material (seed oils, wheat germ oil) but also found in dairy produce. Deficiency produces infertility in male and female rats, and probably in most vertebrates, kidney degeneration and other general wasting symptoms. They seem to prevent oxidation of highly unsaturated fatty acids (and their polymerization in cell membranes) in presence of molecular oxygen.

Vitamin K Fat-soluble vitamins (K_1 and K_2), required for liver synthesis of a substance (proconvertin) required for prothrombin production, and so for conversion of fibrinogen to fibrin during **blood clotting**. Produced by many plants (K_1), and by microorganisms (K_2), including those in the gut.

Vitelline membrane See **egg membranes**.

Vitreous humour Jelly-like semi-transparent substance filling cavity of the vertebrate eye behind the lens.

Viviparity (vivipary) (Zool.) Reproduction in animals whose embryos develop within the female parent and derive nourishment by close contact with her tissues. Involves **placenta** in placental mammals; occurs also in some reptiles, amphibians, elasmobranchs, waterfleas (cladocerans), aphids and tsetse flies. See ***r*-selection**. (Bot.) Having seeds germinating within the fruit, e.g. mangrove; or producing shoots (e.g. bulbils) for vegetative reproduction, instead of inflorescences; e.g. some grasses.

VNTR (variable number of tandem repeats) See **satellite DNA**.

Vocal cords See **larynx**, **syrinx**.

Volkmann's canals See **Haversian system**.

Voltage clamp Technique whereby the potential difference across a membrane is kept steady so that movements of ions across it can be measured. Employed a great deal in nerve and muscle research.

Voluntary muscle Alternative term for **striated muscle**.

Volva Cup-like fragment of the universal veil at base of stipe in some members of the **Basidiomycota** (Order Agaricales).

Wallace, Alfred Russel British naturalist and traveller (1823–1913). According to T. H. Huxley, his short essay on the mechanism of evolution, received by Charles **Darwin** from the Moluccas in June 1858, 'seems to have set Darwin going in earnest' on writing his *The Origin of Species* . . . Their joint paper of 1858, read at the Linnean Society, was the first publicized account of the theory of natural selection. At first, Wallace held that human evolution could be explained by this theory, but later departed from Darwin on this, believing a guiding spiritual force necessary to account for the human soul. Wallace considered **sexual selection** to be less important in evolution than did Darwin, holding that, unlike Darwin, it had no role in the evolution of human intellect. His contributions to zoogeography were of great importance (see **Wallace's Line**), and his book *Darwinism* (1889) attributes a wider role for natural selection in evolution than did Darwin's works.

Wallace's Line Boundary drawn across the Malay Archipelago, separating Arctogea and Notogea (formerly between Australian and Oriental **zoogeographical regions**). Not a precise demarcation line, lying along the edge of the south-east Asian continental shelf, between Bali and Lombok, Borneo and Celebes, the Philippines and the Sangai and Talaud Islands. Named by T. H. Huxley after the zoogeographical work of **Wallace** (see **Darwin**).

Wall pressure (Bot.) Pressure exerted against the protoplast by plant cell wall; opposed to turgor pressure. See **water potential**.

Warm-blooded See **homoiothermy**.

Warning colouration See **aposematic, mimicry**.

Water potential Term indicating the net tendency of any system to donate water to its surroundings. The water potential, ψ_{ω}, of a plant cell is the algebraic sum of its wall pressure (pressure potential), ψ_{p}, and of its osmotic (solute) potential, ψ_{o};

$$\psi_{\omega} = \psi_{p} + \psi_{o}.$$

Since the water potential of pure water at atmospheric pressure is zero pressure units by definition, any addition of solute to pure water reduces its water potential and makes its value negative. Water movement in nature is therefore always from a system with higher (less negative) water potential to one with lower (more negative) water potential. The units normally employed are *megapascals* (MPa; 1 Pa = 1 Newton/m^2), or *millibars* (1 mbar = 10^2 Pa). Standard atmospheric pressure, equal to the average pressure of air at sea level, is 1.01 bar.

The term has wide applicability. The water potential of any cell is its net tendency to donate water to its surroundings, one with a higher water potential than an adjacent cell tending to donate water to it. Plant cells in equilibrium with pure water have zero water potential and will be fully *turgid* (see **turgor pressure**). The volume of water entering or leaving a plant cell during equilibrium attainment is negligible and does not alter the cell's solute potential.

When applied to non-cellular systems such as solutions separated by a selectively permeable membrane, water potentials dictate the net direction in which water moves by osmosis, since net movement of water is always from higher to lower water potential values. Osmotic effects can therefore be subsumed under the wider category of water potential effects, for membranes may or may not be involved in phenomena explained in terms of water potential. It is therefore valuable in predicting water movement between cells, between cells and soil solutions and between cells and the atmosphere. It reduces to one the number of units employed in analysing the **transpiration stream** in plants. Compare **osmosis**.

Water vascular system System of water-filled canals forming tubular component of echinoderm coelom, usually with direct connection (the *madreporite*) with the exterior. A circumoesophageal water ring internal to the skeleton usually gives off radial water canals which pass through pores in the skeleton wherever *tube-feet* protrude. Each tube-foot is expanded by a muscular ampulla internal to the skeleton and is cooordinated with others in feeding and locomotion. See **tube-feet**.

Waxes **Fatty acid** esters of **alcohols** with high molecular weights. Occur as protective coatings of arthropod **cuticle**, skin, fur, feathers, leaves and fruits, often reducing water loss. Also found in beeswax and lanolin.

W-chromosome An alternative notation for the Y-chromosome in organisms where the heterogametic sex is female. See **sex chromosome**.

Weeds Plants growing where they are not wanted; compete for space, light, water and nutrients with garden or crop plants. Many adapted to exploit disturbed land, often being the first to colonize waste ground (when they may be beneficial by preventing soil erosion). Tend to produce vast numbers of seeds/fruits per plant, these being easily and widely disseminated, often remaining dormant until rapid growth and further seed production occur. Annuals, biennials and perennials, they often provide habitats for insect and fungal pests. Leguminous weeds can add nitrogen to the soil through N-fixing bacteria, and all increase soil fibre content on ploughing in. See **pest control**, **artificial selection**.

Weismann, August Theoretical biologist (1834–1914); one of the first to appreciate that chromosomes are the physical bearers of hereditary determinants and to forge a comprehensive theory of heredity. With Wilhelm Roux, he advocated the theory of **mosaic development**. Best remembered for his theory of the continuity of **germ plasm**. Prolific coiner of terms and entities.

Whalebone See **baleen**.

Whales See **Cetacea**.

White blood cell See **leucocyte**.

White matter Myelinated nerves forming through-conduction pathways of vertebrate **brain** and **spinal cord**, with associated glia and blood vessels. Compare **grey matter**, to which it normally lies superficially.

Whorl Three or more leaves, flowers, sporangia or other plant parts at one point on an axis or node.

Wild type Originally denoting the type (genotype or phenotype) most commonly encountered in wild populations of a species, but commonly applied to laboratory stock from which mutants are derived. As in natural populations, the laboratory wild type is subject to change. In genetics, term is usually employed in contexts of alleles and particular phenotypic characters. Such alleles are normally represented by the symbol + rather than by a letter, a genotype homozygous for a wild type allele being given by either ++ or +/+. Wild type characters are by no means always genetically dominant (see **dominance**).

Wilting Condition of plants in which cells lose turgidity and leaves, young stems, etc. droop. Results from excess of transpiratory water loss over absorption. Excessive wilting may be stressful, even irreversible.

Wishbone (furcula) Fused clavicles and interclavicles, diagnostic of birds. Present in ***Archaeopteryx***.

Wobble hypothesis Hypothesis, due to F. H. C. Crick, that the nucleotide at the 5′-end of a tRNA anticodon plays a less important role than the other two in determining which tRNA molecule will align by base-pairing with a codon triplet in mRNA. Helps to explain pattern of degeneracy observed in the **genetic code**, which is restricted to the 3′-ends of codons. See **hypoxanthine**, **inosinic acid**.

Wolffian duct (archinephric duct) Vertebrate kidney duct; one on each side. Develops in all vertebrates (initially in both sexes) from region of pronephros, becoming the mesonephric duct. In male amniotes, conveys both urine and semen to the exterior (opening into the cloaca). In male amniotes it becomes the vas deferens (vasa efferentia and epididymis representing mesonephric tubules which now lead into it from the testis). In all adult amniotes an outgrowth from lower end of the Wolffian duct develops into the **ureter**, invading the metanephric kidney. Hormonal influences from the gonads determine whether the Wolffian duct or the **Müllerian duct** will persist and differentiate in the adult.

Womb See **uterus**.

Wood See **xylem**.

Woodlice See **Isopoda**.

Worker See **caste**.

Xanthophyll **Carotenoid** pigment; each an oxygenated derivative of carotenes.

Xanthophyta Yellow-green algae (Kingdom **Protista**). Primarily freshwater, eukaryotic algae comprising one class, Xanthophyceae (Tribophyceae). Flagellated, coccoid, filamentous and siphonaceous forms. Motile cells characteristically possess an anteriorly pointing hairy flagellum, and a posteriorly pointing smooth flagellum. Chloroplasts contain chlorophyll *a*, (and in one case, *Vaucheria*, chlorophyll *c*), the major carotenoids (diadinoxanthin, heteroxanthin and vaucheriaxanthin ester), lack fucoxanthin and are coloured yellowish-green. The 'eyespot' of motile cells is always within the chloroplast, and the latter are surrounded by two members of **chloroplast endoplasmic reticulum**, the outer membrane of which is typically continuous with the outer membrane of the nucleus. Mannitol and glucose accumulate during photosynthesis in the plastids, and the major storage product is probably a β-1,3 linked glucan similar to paramylon. Where cell walls are found, they typically comprise two overlapping halves. Asexual reproduction occurs frequently via fragmentation, as well as via zoospores and aplanospores. There are few reliable reports of sexual reproduction, and it has only been established for three genera (*Botrydium*, isogamous, involving flagellated gametes; *Tribonema*, isogamous or anisogamous, involving flagellated gametes; and *Vaucheria*, oogamous).

X chromosome See **sex chromosome**.

Xenograft Graft between individuals of different species (*xenogeneic* individuals). See **allograft**.

Xeromorphy (Bot.) Possession of morphological characters associated with **xerophytes**.

Xerophyte Plant of arid habitat, able to endure conditions of prolonged drought (e.g. in **deserts**), due either to a capacity during the brief rainy season for internal water storage, used when none can be obtained from the soil, plus low daytime transpiration rates associated with stomatal closure (e.g. in such succulents as cacti); or ability to recover from partial desiccation (e.g. such desert shrubs as the creosote bush). As long as water is freely available, it seems that non-succulent xerophytes transpire as much as, or more freely than, mesophytes but unlike the latter they can endure periods of permanent wilting during which transpiration is reduced to minimal levels. Features associated with this reduction, in different xerophytes, include dieback of leaves covering and protecting perennating buds at soil surface, shedding of leaves when water supply is exhausted, heavily cutinized or waxy leaves, closure or plugging of the stomata (which are sunken or protected), orientation or folding of leaves to reduce insolation, and a microphyllous habit reducing the possibility of drought necrosis of the mesophyll. See **hydrophyte**, **mesophyte**.

Xerosere **Sere** commencing on a dry site.

Xiphosura See **Merostomata**.

X-ray diffraction (X-ray crystallography) Technique where a narrow beam of X-rays is fired through a crystalline source (e.g. of DNA or protein), the arrangement of atoms within the crystal being interpreted from the X-ray diffraction pattern produced (usually by computer reconstruction). Loose quasi-crystalline structures such as cell walls and cell membranes are sometimes analysed this way.

X-ray irradiation A form of ionizing electromagnetic radiation with beneficial effects in microbiological strilization and food preservation (although here some toxicity may result) but also a potentially hazardous mutagen in that it breaks chromosomes and causes developmental abnormalities (teratologies). There is no evidence for congenital abnormalities caused by diagnostic (clinical) X-ray irradiation treatment. Mammalian DNA damage caused by mild X-ray irradiation may block passage through two **cell cycle** 'checkpoints', namely the restriction point (START) and the entry to mitosis at G2-M. Moderate X-ray irradiation in S phase causes failure of new DNA replication origins to develop which would otherwise allow DNA repair before S phase continues. Mild uv and X-ray irradiation elevate the level of protein p53 (see **cancer cell**) by increasing its half-life, inducing G1 arrest of the cell cycle.

Xylem Wood. Mixed vascular tissue, conducting water and mineral salts taken in by roots throughout the plant, which it provides with mechanical support. Of two kinds: *primary*, formed by differentiation from procambium and comprising protoxylem and metaxylem, and *secondary*, additional xylem produced by activity of the cambium. Characterized by presence of tracheids and/or vessels, fibres and parenchyma. In mature woody plants, makes up bulk of vascular tissue itself and of entire structure of stems and roots. See **transpiration stream**.

YAC See **yeast artificial chromosome**.

Y-chromosome See **sex chromosome**.

Yeast artificial chromosome (YAC) A cloning **vector** used for cloning huge fragments of DNA within yeast cells. Each contains, in addition to selectable markers and a cloning site, an *autonomously replicating sequence* (ARS, derived from yeast chromosomal DNA), a yeast centromere (for stability) and a yeast **telomere** at each end (enabling replication as linear chromosome-like molecules). YACs behave like normal yeast chromosomes and replicate each time the cell goes through mitosis. The DNA for cloning is produced by digestion with rare cutting endonucleases, enabling production of 'YAC libraries' (see **gene library**). It is possible to identify overlapping sequences, even those making up entire human chromosomes. Human telomeres have been cloned within YACs, and YACs of >1 Mbp are now common. YACs may offer scope for human gene therapy, providing an unaffected allele to replace the affected one by homologous recombination: the large homologous regions would be advantageous. Construction of mammalian artificial chromosomes (MACs) is under investigation. See **genome mapping**.

Yeasts Widely distributed unicellular members of the **Ascomycota**, which reproduce asexually by fission or by budding, rather than by spore formation. Sexual reproduction occurs when either two cells or two ascospore units unite and form a zygote. The zygote may also produce diploid buds or it may function as an **ascus**, undergoing meiosis and producing four haploid nuclei; a mitotic division may then take place, and then within the zygote wall, which is now the functional ascus, cell walls are formed around each of the nuclei so that either four or eight ascospores are formed. Yeasts are of great economic importance because of their ability to ferment carbohydrates, breaking down glucose to produce ethanol (ethyl alcohol) and carbon dioxide in the process. The brewing, wine-making and baking industries depend upon this capacity, with brewers and vintners utilizing yeasts to produce alcohol, and bakers utilizing yeasts to produce carbon dioxide. Most of the yeasts important in the production of wine, beer, cider and sake are strains of just one species, *Saccharomyces cerevisiae*, although other species also play a role. Some species of yeasts are important human pathogens, causing such diseases as thrush and cryptococcosis. Others, especially *S. cerevisiae*, are important laboratory organisms for genetic research, and some are used as hosts in **DNA cloning** techniques of **biotechnology**. Others are used commercially as sources of proteins and vitamins. See **mating type**.

Yolk Store of food material, mainly protein and fat, in eggs of most animals. See **polarity**, **centrolecithal**.

Yolk sac Vertebrate **extraembryonic membrane**, conspicuous in elasmobranchs, teleosts, reptiles and birds. Contains yolk and hangs from ventral surface of the embryo. Has outer layer of ectoderm, inner layer (usually absorptive) of endoderm, with mesoderm containing coelom and blood vessels between. A gut diverticulum; yolk usually communicating with the intestine. In mammals it is normally devoid of yolk, forming part of the **chorion**; in marsupials, it forms an integral component of the **placenta**. As yolk is absorbed, the yolk sac is withdrawn, eventually merging into the embryo.

Y-organ Pair of epithelial endocrine glands in the heads of malacrostracan crustaceans, anterior to the brain. Appear to secrete an **ecdysone**-like hormone involved in moulting.

Zeaton A **cytokinin.**

Z-form helix A left-handed form of double helical DNA containing the alternating same-strand sequence dG-dC. It zigs and zags (hence the name); although found in crystals of synthetically prepared DNA, its presence in naturally occurring DNA is uncertain. See **B-form helix.**

Zinc finger see **DNA-binding proteins**.

Zinjanthropus Genus of hominid containing the species *Z. boisei*, now renamed *Australopithecus boisei.* See **australopithecine**.

Z lines See **striated muscle**.

Zona pellucida Glycoprotein membrane around mammalian ovum, disappearing before implantation. Secreted by cells of the **Graafian follicle**.

Zonula adhaerens One kind of **desmosome**.

Zoochlorella Symbiotic green algae assignable to the **Chlorophyta** (e.g. *Chlorella, Oocystis*); both freshwater and marine. See **Zooxanthellae**.

Zoogeographical regions (faunal provinces) Subdivisions of world surface into regions identified by differences in their dominant animals. They are: (a) Arctogea (Europe, Asia, Africa, Indochina and North America). It includes the Palaearctic (Asia, Europe and North Africa), Oriental (Indochina) and Nearctic (Greenland and North America south to Mexico) regions; (b) Neogea (Central and South America), equivalent to the Neogropical region; (c) Notogea (Australasia), normally taken to include Australia, Tasmania, New Zealand, eastern Indonesia and Polynesia. The island of Madagascar is often considered sufficiently distinct faunally to be regarded as a minor region: the mMalagasy region. Botanical equivalents are known as *floral regions.* See **Wallace's Line**, **Zoogeography**.

Zoogeography Study of, and attempt to interpret, global distributions of animal taxa. Takes into account **continental drift** and other major geological processes (e.g. orogenies, ice ages and other factors affecting sea level), as well as aspects of ecology and evolution (e.g. adaptive radiation, competition, speciation, geographical isolation, dispersal and vagility). It is largely due, for example, to isolation since the Cretaceous of the fauna of Notogea (see **zoogeographical regions**) on the Australian plate that its present fauna is so distinctive. Likewise, separation of Malagasy region from the rest of Africa prior to higher primate radiation results in a relict lemur population persisting there today, Island faunas, as Darwin and Wallace knew, have high levels of endemism, particularly if there is considerable separation from mainland faunas. See **plant geography**, **Gondwanaland**, **Laurasia**.

Zooid Member of a colony of animals (chiefly ectoprocts and entoprocts) in which individuals are physically united by living material. See **colony**.

Zoology Branch of biology dealing specifically with animals. There is considerable overlap, however, with botany.

Zooplankton Animal members of plankton.

Zoosporangium (Bot.) Sporangium in which zoospores are formed.

Zoospore (swarm spore) Naked spore produced within a sporangium (zoosporangium); motile, with one, two or many flagella; present in certain algae and oomycotan protists.

Zootype A proposed defining character (synapomorphy) of Kingdom Animalia. Animals would, on this suggestion, be those eukaryotes whose genomes contain the **positional information** system of *Hox* gene clusters, colinearly expressed along the chromosome and in their positions within the embryo during development (see **homeobox**). The stage at which the zootype is most clearly displayed appears to be the **phylotypic stage** for that animal phylum.

Zooxanthellae Algae varying in colour from golden, yellowish, brownish to reddish, living symbiotically in a variety of aquatic animals; includes algae assignable to the **Bacillariophyta, Chrysophyta, Dinophyta, Cryptophyta;** especially important in coelenterates of coral reefs. See **Zoochlorella**.

Zoraptera Small order of very small exopterygote insects (metamorphosis slight) of Subclass Orthopteroidea. Occur under bark, in humus, etc. Many species are dimorphic, apterous or winged, although wings can be shed.

Zosterophyta Extinct, seedless, vascular plants found in strata from the early to late **Devonian** period, from *c.* 408–370 Myr ago. Zosterophytes were leafless, dichotomously branched and possibly aquatic plants, whose aerial stems possessed a cuticle. Stomata were present but confined to the uppermost branches. Named because of their general resemblance to modern seagrasses (e.g. *Zostera* – a marine flowering plant which resembles grasses). The sporangia were globose or kidney-bean-shaped, and were borne laterally upon short stalks. Plants were **homosporous** and probably ancestors of the lycophytes.

Zwitterion Ion which has both positively and negatively charged regions. All **amino acids** are zwitterions, although their charge distributions are much affected by pH.

Zygomatic arch Bony arches formed by the temporal processes of the zygomatic bones (malars, 'cheekbones', with part of the outer wall and floor of the orbits) and the zygomatic processes of the temporal bones; for attachment of **masseter muscles**. The **temporalis muscles** pass under it.

Zygomorphic Bilateral symmetry of flowers.

Zygomycota Formerly Zygomycotina (Zygomycetes). Kingdom **Fungi**. Sexual reproduction characteristic, involving two gametangia which fuse via **conjugation**, with a zygosporangium (zygote) resulting. Inside the zygosporangium, the gametes, which are just nuclei, fuse to form one or more diploid nuclei. Asexual reproduction occurs through production of non-motile sporangiospores borne within a sporangium. Most species possess coenocytic (multinucleate) hyphae, within which the cytoplasm can often be seen streaming rapidly. Include a large number of saprophytes and parasites. Commercially important in production of organic acids, pigments, fermented foods, alcohols and modified steroids. Members of one class (Trichomycetes) are commonly commensals of arthropd guts. Some are common mould fungi (e.g. *Mucor, Rhizopus*) causing spoilage of stored grain, bread and vegetables.

One of the most important groups of zygomycetes is *Glomus*, and related genera, which occur in mycorrhizal associations.

Zygospore Thick-walled resting spore; product of conjugation in the **Zygomycota** and some green algae. Also used for product of fertilization in the isogamous *Chlamydomonas* and its relatives.

Zygote Cellular product of gametic union. Usually diploid.

Zygotene Stage in the first prophase of **meiosis**.

Zygotic meiosis Meiosis occurring during maturation or germination of a zygote.

Zymogen Any inactive enzyme precursor (e.g. pepsinogen, trypsinogen or prothrombin). Activation to the functional enzyme generally involves excision of a portion of polypeptide. The several zymogens in secretory vesicles of pancreatic **acinar cells** are termed *zymogen granules*, although some active enzyme is contained there too. Zymogen granules fuse with the cell apex under influence of acetylcholine or cholecystokinin.

READ MORE IN PENGUIN

In every corner of the world, on every subject under the sun, Penguin represents quality and variety – the very best in publishing today.

For complete information about books available from Penguin – including Puffins, Penguin Classics and Arkana – and how to order them, write to us at the appropriate address below. Please note that for copyright reasons the selection of books varies from country to country.

In the United Kingdom: Please write to *Dept. EP, Penguin Books Ltd, Bath Road, Harmondsworth, West Drayton, Middlesex UB7 ODA*

In the United States: Please write to *Consumer Sales, Penguin USA, P.O. Box 999, Dept. 17109, Bergenfield, New Jersey 07621-0120.* VISA and MasterCard holders call 1-800-253-6476 to order Penguin titles

In Canada: Please write to *Penguin Books Canada Ltd, 10 Alcorn Avenue, Suite 300, Toronto, Ontario M4V 3B2*

In Australia: Please write to *Penguin Books Australia Ltd, P.O. Box 257, Ringwood, Victoria 3134*

In New Zealand: Please write to *Penguin Books (NZ) Ltd, Private Bag 102902, North Shore Mail Centre, Auckland 10*

In India: Please write to *Penguin Books India Pvt Ltd, 706 Eros Apartments, 56 Nehru Place, New Delhi 110 019*

In the Netherlands: Please write to *Penguin Books Netherlands bv, Postbus 3507, NL-1001 AH Amsterdam*

In Germany: Please write to *Penguin Books Deutschland GmbH, Metzlerstrasse 26, 60594 Frankfurt am Main*

In Spain: Please write to *Penguin Books S. A., Bravo Murillo 19, 1° B, 28015 Madrid*

In Italy: Please write to *Penguin Italia s.r.l., Via Felice Casati 20, I–20124 Milano*

In France: Please write to *Penguin France S. A., 17 rue Lejeune, F–31000 Toulouse*

In Japan: Please write to *Penguin Books Japan, Ishikiribashi Building, 2–5–4, Suido, Bunkyo-ku, Tokyo 112*

In Greece: Please write to *Penguin Hellas Ltd, Dimocritou 3, GR–106 71 Athens*

In South Africa: Please write to *Longman Penguin Southern Africa (Pty) Ltd, Private Bag X08, Bertsham 2013*

READ MORE IN PENGUIN

REFERENCE

The Penguin Dictionary of Literary Terms and Literary Theory
J. A. Cuddon

'Scholarly, succinct, comprehensive and entertaining, this is an important book, an indispensable work of reference. It draws on the literature of many languages and quotes aptly and freshly from our own' – *The Times Educational Supplement*

The Penguin Spelling Dictionary

What are the plurals of *octopus* and *rhinoceros*? What is the difference between *stationery* and *stationary*? And how about *annex* and *annexe*, *agape* and *Agape*? This comprehensive new book, the fullest spelling dictionary now available, provides the answers.

The Roget's Thesaurus of English Words and Phrases
Betty Kirkpatrick (ed.)

This new edition of Roget's classic work, now brought up to date for the nineties, will increase anyone's command of the English language. Fully cross-referenced, it includes synonyms of every kind (formal or colloquial, idiomatic and figurative) for almost 900 headings. It is a must for writers and utterly fascinating for any English speaker.

The Penguin Dictionary of English Idioms
Daphne M. Gulland and David G. Hinds-Howell

The English language is full of pitfalls for the foreign student – but the most common problem lies in understanding and using the vast array of idioms. *The Penguin Dictionary of English Idioms* is uniquely designed to stimulate understanding and familiarity by explaining the meanings and origins of idioms and giving examples of typical usage.

The Penguin Wordmaster Dictionary
Martin H. Manser and Nigel D. Turton

This dictionary puts the pleasure back into word-seeking. Every time you look at a page you get a bonus – a panel telling you everything about a particular word or expression. It is, therefore, a dictionary to be read as well as used for its concise and up-to-date definitions.

READ MORE IN PENGUIN

REFERENCE

Medicines: A Guide for Everybody Peter Parish

Now in its seventh edition and completely revised and updated, this bestselling guide is written in ordinary language for the ordinary reader yet will prove indispensable to anyone involved in health care – nurses, pharmacists, opticians, social workers and doctors.

Media Law Geoffrey Robertson, QC, and Andrew Nichol

Crisp and authoritative surveys explain the up-to-date position on defamation, obscenity, official secrecy, copyright and confidentiality, contempt of court, the protection of privacy and much more.

The Slang Thesaurus

Do you make the public bar sound like a gentleman's club? The miraculous *Slang Thesaurus* will liven up your language in no time. You won't Adam and Eve it! A mine of funny, witty, acid and vulgar synonyms for the words you use every day.

The Penguin Dictionary of Troublesome Words Bill Bryson

Why should you avoid discussing the *weather conditions*? Can a married woman be celibate? Why is it eccentric to talk about the aroma of a cowshed? A straightforward guide to the pitfalls and hotly disputed issues in standard written English.

The Penguin Dictionary of Musical Performers Arthur Jacobs

In this invaluable companion volume to *The Penguin Dictionary of Music* Arthur Jacobs has brought together the names of over 2,500 performers. Music is written by composers, yet it is the interpreters who bring it to life; in this comprehensive book they are at last given their due.

The Penguin Dictionary of Physical Geography John Whittow

'Dr Whittow and Penguin Reference Books have put serious students of the subject in their debt, by combining the terminology of the traditional geomorphology with that of the quantitative revolution and defining both in one large and comprehensive dictionary of physical geography ... clear and succinct' – *The Times Educational Supplement*

READ MORE IN PENGUIN 

SCIENCE AND MATHEMATICS

The Character of Physical Law Richard P. Feynman

'Richard Feynman had both genius and highly unconventional style . . . His contributions touched almost every corner of the subject, and have had a deep and abiding influence over the way that physicists think' – Paul Davies

Fearful Symmetry Ian Stewart and Martin Golubitsky

'Symmetry-breaking is an important unifying idea in modern abstract mathematics, so it's a tribute to the authors' expository gifts that much of the book reads like a captivating travelogue' – *The New York Times Book Review*

Bully for Brontosaurus Stephen Jay Gould

'He fossicks through history, here and there picking up a bone, an imprint, a fossil dropping, and, from these, tries to reconstruct the past afresh in all its messy ambiguity. It's the droppings that provide the freshness: he's as likely to quote from Mark Twain or Joe DiMaggio as from Lamarck or Lavoisier' – *Guardian*

Are We Alone? Paul Davies

Since ancient times people have been fascinated by the idea of extraterrestrial life; today we are searching systematically for it. Paul Davies's striking new book examines the assumptions that go into this search and draws out the startling implications for science, religion and our world view, should we discover that we are not alone.

The Making of the Atomic Bomb Richard Rhodes

'Rhodes handles his rich trove of material with the skill of a master novelist . . . his portraits of the leading figures are three-dimensional and penetrating . . . the sheer momentum of the narrative is breathtaking . . . a book to read and to read again' – *Guardian*

READ MORE IN PENGUIN

SCIENCE AND MATHEMATICS

Bright Air, Brilliant Fire Gerald Edelman

'A brilliant and captivating new vision of the mind' – Oliver Sacks. 'Every page of Edelman's huge wok of a book crackles with delicious ideas, mostly from the *nouvelle cuisine* of neuroscience, but spiced with a good deal of intellectual history, with side dishes on everything from schizophrenia to embryology' – *The Times*

Games of Life Karl Sigmund
Explorations in Ecology, Evolution and Behaviour

'A beautifully written and, considering its relative brevity, amazingly comprehensive survey of past and current thinking in "mathematical" evolution . . . Just as games are supposed to be fun, so too is *Games of Life*' – *The Times Higher Education Supplement*

Gödel, Escher, Bach: An Eternal Golden Braid
Douglas F. Hofstadter

'Every few decades an unknown author brings out a book of such depth, clarity, range, wit, beauty and originality that it is recognized at once as a major literary event' – Martin Gardner. 'Leaves you feeling you have had a first-class workout in the best mental gymnasium in town' – *New Statesman*

The Doctrine of DNA R. C. Lewontin

'He is the most brilliant scientist I know and his work embodies, as this book displays so well, the very best in genetics, combined with a powerful political and moral vision of how science, properly interpreted and used to empower all the people, might truly help us to be free' – Stephen Jay Gould

Artificial Life Steven Levy

'Can an engineered creation be alive? This centuries-old question is the starting point for Steven Levy's lucid book . . . *Artificial Life* is not only exhilarating reading but an all-too-rare case of a scientific popularization that breaks important new ground' – *The New York Times Book Review*

READ MORE IN PENGUIN

DICTIONARIES

Abbreviations
Archaeology
Architecture
Art and Artists
Astronomy
Biology
Botany
Building
Business
Challenging Words
Chemistry
Civil Engineering
Classical Mythology
Computers
Curious and Interesting Numbers
Curious and Interesting Words
Design and Designers
Economics
Electronics
English and European History
English Idioms
Foreign Terms and Phrases
French
Geography
Historical Slang
Human Geography
Information Technology
Literary Terms and Literary Theory
Mathematics
Modern History 1789–1945
Modern Quotations
Music
Musical Performers
Physical Geography
Physics
Politics
Proverbs
Psychology
Quotations
Religions
Rhyming Dictionary
Saints
Science
Sociology
Spanish
Surnames
Telecommunications
Troublesome Words
Twentieth-Century History